RFID HANDBOOK

Applications, Technology, Security, and Privacy

RFID HANDBOOK

Applications, Technology, Security, and Privacy

EDITED BY

SYED AHSON
MOHAMMAD ILYAS

CRC Press
Taylor & Francis Group
Boca Raton London New York

CRC Press is an imprint of the
Taylor & Francis Group, an **informa** business

CRC Press
Taylor & Francis Group
6000 Broken Sound Parkway NW, Suite 300
Boca Raton, FL 33487-2742

Printed in the United States of America on acid-free paper
10 9 8 7 6 5 4 3

International Standard Book Number-13: 978-1-4200-5499-6 (Hardcover)

Library of Congress Cataloging-in-Publication Data

RFID handbook : applications, technology, security, and privacy / edited by Syed Ahson and Mohammad Ilyas.
p. cm.
Includes bibliographical references and index.
ISBN-13: 978-1-4200-5499-6
ISBN-10: 1-4200-5499-6
1. Radio frequency identification systems--Handbooks, manuals, etc. I. Ahson, Syed. II. Ilyas, Mohammad, 1953- III. Title.

TK6570.I34R485 2008
658.7'87--dc22 2007036582

Visit the Taylor & Francis Web site at
http://www.taylorandfrancis.com

and the CRC Press Web site at
http://www.crcpress.com

Contents

Section III Applications

Preface

Realization in the business community of the benefits of widespread adoption coupled with advances in manufacturing techniques and efficient data-handling methodologies is fostering explosive growth of radio frequency identification systems. RFID-enabled applications have grown at a tremendous rate with system deployments in a number of industries such as pharmaceuticals, health care, transportation, retail, defense, and logistics. An important aspect of RFID technology is its utilization in a wide spectrum of applications. RFID technology can help a wide range of organizations and individuals realize substantial productivity gains and efficiencies. Existing system components integrate the benefits provided by RFID while maintaining system modularity and efficiency. Radio frequency tags allow objects to become self-describing, communicating their identity to a close at hand RF reader. RFID is replacing bar-code–based identification mechanisms, as communication between a reader and a tag is not limited by the requirement of "line-of-sight" reading and each tag has a unique ID. RFID technology enables the optimization of multiple business processes through the improvement, the automation or even the elimination of existing processes, and the emergence of new processes called intelligent processes or smart processes, which are automatically triggering actions or events.

The major areas that have driven the commercial deployment of RFID technology are logistics, supply chain management, library item tracking, medical implants, road tolling (e.g., E-Z Pass), building access control, aviation security, and homeland security applications. These systems are used for a wide range of applications that track, monitor, report, and manage items as they move between different physical locations. From inventory management to theft detection, RFID has been applied in many areas such as in the automotive industry and logistics, as well as in warehouses and retail stores. Most cars are equipped with a remote control to open and lock a door. Money cards are used for public transportation payments. Although there is no RFID association in their names, both a car remote control and money cards are RFID applications. RFID technology has become more and more widely used in real-world applications without people realizing it. Potential has also been seen for the application of RFID in capital asset management applications such as keeping track of maintenance tools in the aircraft maintenance sector. RFID has the ability of giving a unique identity to each tagged object. Hence, there is a vision to extend this technology to item-level tagging (other than pallet and case-level tagging).

Radio frequency identification is revolutionizing supply chain management, replacing bar codes as the main object tracking system. RFID technologies better manage supply chain operations by tracking the movement of products or assets through a system. Over the last four years, major organizations in the U.S. (e.g., Wal-Mart and U.S. Department of Defense) and in Europe (e.g., Metro AG and Tesco) have mandated the use of RFID. Since then, the motivations for RFID adoption have moved from mandatory compliance to voluntary undertakings as companies are increasingly exploring the true potential of the technology, especially in the context of supply chains. RFID enabled automated receiving optimizes the handoff of products between supplier and client. Products maybe received at a manufacturing facility, a distribution center's warehouse or a retail store without manually scanning or verifying the merchandise. Although current state-of-the-art receiving systems are highly optimized by using bar coding and wireless communications to a central computer, the process is error-prone and time-consuming because of human intervention.

RFID presents security and privacy risks that must be carefully mitigated through management, operational, and technical controls in order to realize the numerous benefits the technology has to offer. Each RFID system has different components and customizations so that it can support a particular business process for an organization; as a result, the security risks for RFID systems and the controls available to address them are highly varied. The RFID handbook provides an overview of RFID technology, the associated security and privacy risks, and recommended practices that will enable organizations to realize productivity improvements while safeguarding sensitive information and protecting the privacy of individuals. While RFID security is a rapidly evolving field with a number of promising innovations expected in the coming years, these guidelines focus on controls that are commercially available today. Implementing the recommendations presented in this handbook should help organizations improve the security of their RFID systems.

The *RFID Handbook* provides technical information about all aspects of RFID. The areas covered in the handbook range from basic concepts to research grade material, including future directions. This handbook captures the current state of RFID technology and serves as a source of comprehensive reference material on this subject. It comprises four sections: Introduction, Technology, Applications, and Security and Privacy. It has a total of 36 chapters authored by 83 experts from around the world.

I Introduction

Chapter 1 (Physics and Geometry of RFID) introduces RFID tags and presents an overview of backscatter communication, antenna directivity, and gain.
Chapter 2 (EPCglobal Network) outlines the components of the EPCglobal Network.
Chapter 3 (Design Automation for RFID Tags and Systems) describes design automation techniques for the creation of customized RFID tags.
Chapter 4 (Far-Field Tag Antenna Design Methodology) considers RFID label antennas for near-field operation in the HF region and far-field operation in the UHF frequency range.
Chapter 5 (Contemporary RFID Reader Architecture) studies several aspects of a software-based RFID reader, including experimentation, implementation, and theoretical work.
Chapter 6 (Progress in RFID Education) addresses the educational issues in highlighting the advantages of RFID and smart labels. An overview of business drivers, trends and processes for RFID is presented.

II Technology

Chapter 7 (RFID Reader Synchronization) identifies synchronization of RFID readers as a mechanism to assist in RFID reader deployment in dense reader environments.
Chapter 8 (Adaptive Tag Anticollision Protocols for RFID Passive Tags) introduces adaptive splitting protocols, which use tree search for tag identification.

Chapter 9 (Comparative Performance Analysis of Anticollision Protocols in RFID Networks) presents a detailed description of tree-based and probabilistic anticollision protocols.
Chapter 10 (Maximizing Read Accuracy by Optimally Locating RFID Interrogators) discusses the significance of proper RFID interrogator location and examines the relationship between RFID reader antenna placement and the ability of a tag to be read accurately.
Chapter 11 (Minimum Energy/Power Considerations) describes several techniques to save power in existing RFID systems or extend capabilities in a power efficient manner.
Chapter 12 (Electromagnetic Coupling in RFID) outlines the fundamental principles governing electromagnetic coupling between an interrogator and its labels in an RFID system.
Chapter 13 (RFID Tags for Metallic Object Identification) considers the behavior of electromagnetic waves near metallic surfaces and the effects of metallic surfaces on RFID tag antennas.
Chapter 14 (WISP: A Passively Powered UHF RFID Tag with Sensing and Computation) describes WISP (Wireless Identification and Sensing Platform), a wireless, battery-free platform for sensing and computation that is powered and read by a standards-compliant ultrahigh frequency RFID reader.

III Applications

Chapter 15 (From Automatic Identification and Data Capture to "Smart Business Process": Preparing for a Pilot Integrating RFID) examines the logic underlying the rules configured in an RFID middleware to support smart business processes in a retail supply chain.
Chapter 16 (Technological Requirements and Derived Benefits from RFID Enabled Receiving in a Supply Chain) focuses on the implementation of RFID enabled automated receiving, which has been identified as one of the quickest profitable Supply Chain Management (SCM) RFID applications.
Chapter 17 (A Prototype on RFID and Sensor Networks for Elder Health Care) describes a project that integrates both sensor network and RFID technologies.
Chapter 18 (Triage with RFID Tags for Massive Incidents) describes a triage system using RFID in which triage tags are used to classify and transport the injured as well as obtain and publish the state and the scale of the casualty incident.
Chapter 19 (RFID Tagging and the Design of "Place") evaluates how RFID tagging can transform the way we look at place and the negotiation and design of place.
Chapter 20 (Photosensing Wireless Tags for Precise Location and Geometry Queries) presents a radio frequency identity and geometry transponder that can also communicate geometry, intertag location history or context-sensitive user annotation.
Chapter 21 (RFID and NFC on Mobile Phones) introduces RFID and NFC in relation to mobile phones and provides details of the associated standards, software development tools, and environments.
Chapter 22 (Applying RFID Techniques for the Next-Generation Automotive Services) presents ideas for new RFID applications in the automotive industry.
Chapter 23 (Application of RFID Technologies for Communication Robots) introduces applications of RFID technologies for communication robots through two field trials involving communication robots and active-type RFID tags.
Chapter 24 (Browsing the World with RFID Tags: Design and Implementation of an RFID-Based Distributed Environmental Memory) describes the design and implementation of a distributed environmental memory for browsing the world.

Chapter 25 (RFID-Enabled Privacy-Preserving Video Surveillance: A Case Study) describes the design of a privacy preserving video surveillance system that monitors subjects in an instrumented space only when they are involved in an access violation.

IV Security and Privacy

Chapter 26 (Is RFID Technology Secure and Private?) presents case studies where information stored on RFID tags has been compromised.
Chapter 27 (Privacy and Personal Information Protection in RFID Systems) presents personal information protection in RFID systems. Privacy protection and personal information protection are contrasted and two properties—anonymity and unlinkability—for personal information protection are introduced.
Chapter 28 (Multilateral Approaches for Reliable Mobile RFID Service Systems) presents an analysis of security and privacy threats, and multilateral security approaches to promoting a globally mobile RFID service.
Chapter 29 (ONS Security) discusses privacy and security risks introduced by the current Object Name Service (ONS) design and investigates possible countermeasures.
Chapter 30 (Practical Steps for Securing RFID Systems) discusses security controls that can mitigate the business risks associated with RFID systems.
Chapter 31 (Lightweight Cryptography for Low Cost RFID: A New Direction in Cryptography) expounds upon a new direction in cryptography needed to address the security and privacy needs of networked low cost RFID systems.
Chapter 32 (Low Overhead RFID Security) surveys various approaches to RFID security and presents two examples of low overhead authentication algorithms.
Chapter 33 (Layers of Security for Active RFID Tags) surveys common attacks to RFID tags, existing security techniques, and security requirements of RFID standards.
Chapter 34 (Cryptographic Approaches to RFID Security and Privacy) surveys studies pertaining to the security and privacy for RFID tags from the context of cryptography.
Chapter 35 (RFID Authentication: Reconciling Anonymity and Availability) explores the issues attending to the provision of two security services, namely privacy and availability, in the context of RFID applications.
Chapter 36 (Security and Privacy of RFID for Biomedical Applications: A Survey) discusses an innovative set of biomedical RFID applications and the relevance of current security solutions to these emerging disciplines.

The targeted audience for the *RFID Handbook* includes professionals who are designers and/or planners for RFID systems, researchers (faculty members and graduate students), and those who would like to learn about this field.

The handbook contains the following specific salient features:

- To serve as a single comprehensive source of information and as reference material on RFID technology
- To deal with an important and timely topic of emerging technology of today, tomorrow, and beyond
- To present accurate, up-to-date information on a broad range of topics related to RFID technology
- To present the material authored by the experts in the field
- To present the information in an organized and well-structured manner

Although the handbook is not precisely a textbook, it can certainly be used as a textbook for graduate courses and research-oriented courses that deal with RFID. Any comments from the readers will be highly appreciated.

Many people have contributed to this handbook in their unique ways. The first and the foremost group that deserves immense gratitude is the group of highly talented and skilled researchers who have contributed 36 chapters to this handbook. All of them have been extremely cooperative and professional. It has also been a pleasure to work with Ms. Nora Konopka, Ms. Jessica Vakili, and Mr. Richard Tressider of CRC Press and Ms. Chitra Subramaniam of SPi, and we are extremely gratified for their support and professionalism. Our families have extended their unconditional love and strong support throughout this project and they all deserve very special thanks.

Syed Ahson
Mohammad Ilyas

Editors

Syed Ahson is a senior staff software engineer with Motorola, Inc., in Plantation, Florida. He has extensive experience with wireless data protocols (TCP/IP, UDP, HTTP, VoIP, SIP, H.323), wireless data applications (Internet browsing, multimedia messaging, wireless e-mail, firmware over-the-air update), and cellular telephony protocols (GSM, CDMA, 3G, UMTS, HSDPA). He has contributed significantly in leading roles toward the creation of several advanced and exciting cellular phones at Motorola. Prior to joining Motorola, he was a senior software design engineer with NetSpeak Corporation (now part of Net2Phone), a pioneer in VoIP telephony software.

Ahson is a coeditor of the *WiMAX Handbook* (CRC Press, 2007) and the *Handbook of Wireless Local Area Networks*: *Applications, Technology, Security, and Standards* (CRC Press, 2005). He has authored "Smartphones" (International Engineering Consortium, April 2006), a research report that reflects on smartphone markets and technologies. He has published several research articles in peer-reviewed journals and teaches computer engineering courses as an adjunct faculty at Florida Atlantic University, Boca Raton, Florida, where he introduced a course on smartphone technology and applications. Ahson received his BSc in electrical engineering from Aligarh Muslim University, Aligarh, India, in 1995 and completed his MS in computer engineering in July 1998 at Florida Atlantic University.

Mohammad Ilyas received his BSc in electrical engineering from the University of Engineering and Technology, Lahore, Pakistan, in 1976. From March 1977 to September 1978, he worked for the Water and Power Development Authority, Pakistan. In 1978, he was awarded a scholarship for his graduate studies and he completed his MS in electrical and electronic engineering in June 1980 at Shiraz University, Shiraz, Iran. In September 1980, he joined the doctoral program at Queen's University in Kingston, Ontario, Canada, and completed his PhD in 1983. His doctoral research focused on switching and flow control techniques in computer communication networks. Since September 1983, he has been with the College of Engineering and Computer Science at Florida Atlantic University, Boca Raton, Florida, where he is currently the associate dean for research and industry relations. From 1994 to 2000, he was chair of the Department of Computer Science and Engineering. From July 2004 to September 2005, he served as interim associate vice president for research and graduate studies. During the 1993–1994 academic year, he took a sabbatical leave to work with the Department of Computer Engineering, King Saud University, Riyadh, Saudi Arabia.

Dr. Ilyas has conducted successful research in various areas including traffic management and congestion control in broadband/high-speed communication networks, traffic characterization, wireless communication networks, performance modeling, and simulation. He has published 1 book, 8 handbooks, and over 150 research articles. He has supervised 11 PhD dissertations and more than 37 MS theses to completion. He has been a consultant to several national and international organizations. Dr. Ilyas is an active participant in several IEEE technical committees and activities. He is a senior member of IEEE and a member of ASEE.

Contributors

Jeroen van Baar Mitsubishi Electric Research Labs, Cambridge, Massachusetts

Kensuke Baba Department of Informatics, Kyushu University, Fukuoka, Japan

Paul Beardsley Mitsubishi Electric Research Labs, Cambridge, Massachusetts

Ygal Bendavid ePoly Centre of Expertise in Electronic Commerce, École Polytechnique de Montréal, Montreal, Quebec, Canada

Surinder Mohan Bhaskar Department of Information Technology, Ministry of Communications and Information Technology New Delhi, India

Harold Boeck Department of Marketing, Université de Sherbrooke, Sherbrooke, Quebec, Canada; ePoly, École Polytechnique de Montréal, Montreal, Quebec, Canada

Mike Burmester Department of Computer Science, Florida State University, Tallahassee, Florida

James T. Cain Department of Electrical and Computer Engineering, University of Pittsburgh, Pittsburgh, Pennsylvania

Jihoon Choi Department of Computer Science and Engineering, Korea University, Seoul, Republic of Korea

Thomas Clouser Department of Computer Science, Kent State University, Kent, Ohio

Peter H. Cole Auto-ID Laboratory, School of Electrical and Electronic Engineering, North Terrace, University of Adelaide, South Australia, Australia

Paul Coulton Department of Communication Systems, Infolab21, Lancaster University, Lancaster, United Kingdom

Gerold Joseph Dhanabalan Department of Electrical and Computer Engineering, University of Pittsburgh, Pittsburgh, Pennsylvania

Paul Dietz Mitsubishi Electric Research Labs, Cambridge, Massachusetts

Shlomi Dolev Department of Computer Science, Ben-Gurion University, Beer-Sheva, Israel

Swapna Dontharaju Department of Electrical and Computer Engineering, University of Pittsburgh, Pittsburgh, Pennsylvania

Reuben Edwards Department of Communication Systems, Infolab21, Lancaster University, Lancaster, United Kingdom

Bernie Eydt Booz Allen Hamilton Inc., McLean, Virginia

Benjamin Fabian Institute of Information Systems, Humboldt University, Berlin, Germany

Brian J. Garner School of Engineering & IT, Deakin University, Geelong Campus, Victoria, Australia

Raja Ghosal Auto-ID Laboratory, School of Electrical and Electronic Engineering, North Terrace, University of Adelaide, South Australia, Australia

Alfio R. Grasso Auto-ID Laboratory, School of Electrical and Electronic Engineering, North Terrace, University of Adelaide, South Australia, Australia

Oliver Günther Institute of Information Systems, Humboldt University, Berlin, Germany

Norihiro Hagita ATR-IRC, Sourakugun, Kyoto, Japan

Peter Harliman Compiler and Advanced Computer Systems Laboratory, School of Electrical Engineering, Korea University, Seoul, Republic of Korea

Peter J. Hawrylak Department of Electrical and Computer Engineering, University of Pittsburgh, Pittsburgh, Pennsylvania

Loc Ho Department of Computer Science, San Jose State University, San Jose, California

Raymond R. Hoare Concurrent EDA, LLC, Pittsburgh, Pennsylvania

Sozo Inoue Library, Kyushu University, Fukuoka, Japan

Hiroshi Ishiguro ATR-IRC, Sourakugun, Kyoto, Japan, Osaka University, Japan

Behnam Jamali Auto-ID Laboratory, School of Electrical and Electronic Engineering, North Terrace, University of Adelaide, South Australia, Australia

Kyong Jin Jo Compiler and Advanced Computer Systems Laboratory, School of Electrical Engineering, Korea University, Seoul, Republic of Korea

Alex K. Jones Department of Electrical and Computer Engineering, University of Pittsburgh, Pittsburgh, Pennsylvania

Takayuki Kanda ATR-IRC, Sourakugun, Kyoto, Japan

A. Karygiannis U.S. Department of Commerce, National Institute of Standards and Technology, Gaithersburg, Maryland

Seon Wook Kim Compiler and Advanced Computer Systems Laboratory, School of Electrical Engineering, Korea University, Seoul, Republic of Korea

Shingo Kinoshita NTT Laboratories, Nippon Telegraph and Telephone Corporation, Musashino-shi, Tokyo, Japan

Marina Kopeetsky Department of Software Engineering, Sami-Shamoon College of Engineering, Beer-Sheva, Israel

Donghwan Lee Department of Computer Science and Engineering, Korea University, Seoul, Republic of Korea

Joon Goo Lee Compiler and Advanced Computer Systems Laboratory, School of Electrical Engineering, Korea University, Seoul, Republic of Korea

Wonjun Lee Department of Computer Science and Engineering, Korea University, Seoul, Republic of Korea

Élisabeth Lefebvre ePoly Centre of Expertise in Electronic Commerce, École Polytechnique de Montréal, Montreal, Quebec, Canada

Louis-A. Lefebvre ePoly Centre of Expertise in Electronic Commerce, École Polytechnique de Montréal, Montreal, Quebec, Canada

Kin Seong Leong Auto-ID Laboratory, School of Electrical and Electronic Engineering, North Terrace, University of Adelaide, South Australia, Australia

Marco Mamei Dipartimento di Scienze e Metodi dell'Ingegneria, Universita'di Modena e Reggio Emilia, Reggio Emilia, Italy

Anijo Punnen Mathew Design Research & Informatics Lab (DRIL), College of Architecture, Art, and Design, Mississippi State University, Mississippi

Leonid Mats Department of Electrical and Computer Engineering, University of Pittsburgh, Pittsburgh, Pennsylvania

Breno de Medeiros Department of Computer Science, Florida State University, Tallahassee, Florida

Sharad Mehrotra Bren School of Information and Computer Sciences, University of California, Irvine, California

Marlin H. Mickle Department of Electrical and Computer Engineering, University of Pittsburgh, Pittsburgh, Pennsylvania

Melody Moh Department of Computer Science, San Jose State University, San Jose, California

Teng-Sheng Moh Department of Computer Science, San Jose State University, San Jose, California

Jihoon Myung Department of Computer Science and Engineering, Korea University, Seoul, Republic of Korea

Mikhail Nesterenko Department of Computer Science, Kent State University, Kent, Ohio

Mun Leng Ng Auto-ID Laboratory, School of Electrical and Electronic Engineering, North Terrace, University of Adelaide, South Australia, Australia

Yasunobu Nohara Department of Computer Science, Kyushu University, Fukuoka, Japan

Bryan A. Norman Department of Industrial Engineering, University of Pittsburgh, Pittsburgh, Pennsylvania

Miyako Ohkubo Information-Technology Promotion Agency, Bunkyo-ku, Tokyo, Japan

Namje Park Information Security Research Division, Electronics and Telecommunications Research Institute, Daejeon, South Korea

Ted S. Phillips Booz Allen Hamilton Inc., McLean, Virginia

Jayant Rajgopal Department of Industrial Engineering, University of Pittsburgh, Pittsburgh, Pennsylvania

Damith C. Ranasinghe Auto-ID Lab, Institute of Manufacturing, Cambridge University Engineering Department, Cambridge University, United Kingdom

Omer Rashid Department of Communication Systems, Infolab21, Lancaster University, Lancaster, United Kingdom

Ramesh Raskar Mitsubishi Electric Research Labs, Cambridge, Massachusetts

Alberto Rosi Dipartimento di Scienze e Metodi dell'Ingegneria, Universita'di Modena e Reggio Emilia, Reggio Emilia, Italy

Alanson P. Sample Department of Electrical Engineering, University of Washington, Seattle, Washington

Timothy K. Shih Department of Computer Science and Information Engineering, Tamkang University, Taiwan

Masahiro Shiomi ATR-IRC, Sourakugun, Kyoto, Japan

Joshua R. Smith Intel Research Seattle, Seattle, Washington

Akihito Sonoda DNP LSI Japan Co. Ltd., Tokyo, Japan

Sarah Spiekermann Institute of Information Systems, Humboldt University, Berlin, Germany

Ellen Stuart Department of Computer Science, San Jose State University, San Jose, California

Koutarou Suzuki NTT Laboratories, Nippon Telegraph and Telephone Corporation, Musashino-shi, Tokyo, Japan

Shenchih Tung Department of Electrical Engineering, University of Pittsburgh, Pittsburgh, Pennsylvania

Nalini Venkatasubramanian Bren School of Information and Computer Sciences, University of California, Irvine, California

Zachary Walker Department of Computer Science, San Jose State University, San Jose, California

Samuel Fosso Wamba ePoly Centre of Expertise in Electronic Commerce, École Polytechnique de Montréal, Montreal, Quebec, Canada

Lin Wang Department of Industrial Engineering, University of Pittsburgh, Pittsburgh, Pennsylvania

Jehan Wickramasuriya Pervasive Platforms & Architectures Lab, Applications Research, Motorola Labs, Schaumburg, Illinois

Dongho Won School of Information and Communication Engineering, Sungkyunkwan University, Suwon, South Korea

Hiroto Yasuura Department of Computer Science, Kyushu University, Fukuoka, Japan

Daniel J. Yeager Intel Research Seattle, Department of Electrical Engineering, University of Washington, Seattle, Washington

Franco Zambonelli Dipartimento di Scienze e Metodi dell'Ingegneria, Universita'di Modena e Reggio Emilia, Reggio Emilia, Italy

Section I

Introduction

1

Physics and Geometry of RFID

Marlin H. Mickle, Leonid Mats, and Peter J. Hawrylak

CONTENTS

1.1 Overview

Radio frequency identification (RFID) is rooted in discoveries made by Faraday during the mid-nineteenth century and discoveries made between 1900 and 1940 in radio and radar technologies. Faraday discovered the concept of mutual induction, which forms the basis for powering passive tags operating in the near field. Technological advances enabling the development of far-field tags occurred during the first half of the twentieth century. This chapter focuses on passive tags operating in the far field.

Far-field tags must harvest energy to operate and backscatter the interrogation signal transmitted by the reader to communicate with the reader. Two discoveries form the basis for far-field passive RFID tags. First, the development of crystal set radios provides the basis for a tag to power itself. Crystal set radios used energy contained in the radio frequency (RF) signal to move a diaphragm in the headset enabling those without electricity (most people outside a major city during the early twentieth century) to listen to radio broadcasts. Second, discoveries in the field of radar during the Second World War form the basis for the backscatter communication employed by passive RFID tags. All objects reflect radio waves, and the tag can change the characteristics of the radio waves it reflects by changing the matching at the connection between the chip and the antenna making up the tag. These two discoveries form the basis of far-field passive RFID tags.

There are many types of RFID tags in existence, and although this chapter focuses on passive RFID tags, a brief introduction of the other types of RFID tags is presented in this section. EPCglobal designates four classes of RFID tags. Class 1 and Class 2 tags are purely passive tags with Class 2 being a Class 1 tag with additional memory or supporting additional protocol commands. Semipassive tags, designated as Class 3 and Class 4, are active tags.

Semipassive tags have an onboard power source and may have onboard sensors. The onboard power source serves two purposes: (1) it provides continuous power for the sensors and (2) it allows the intelligence contained in the chip to function without harvesting energy. When monitoring an asset, it is critical to take sensor readings at the required intervals to obtain a complete history of the asset with respect to a given phenomenon. The onboard power supply ensures that the semipassive tag can take these readings even in the absence of a reader to power the tag. Another use of semipassive tags is to increase the read range or to read the tag in an unfriendly environment. Because the chip in a semipassive tag is powered by the battery, the semipassive tag is not required to harvest energy for operation from the reader signal. Hence, a semipassive tag does not need to harvest energy to power the circuitry for the backscatter communication to produce a stronger signal that is easier for the reader to detect, resulting in increased range or the capability to be read in an unfriendly environment. Semipassive tags are currently investigated for use in the cold chain, where items (such as frozen foods or drugs) must be kept below a given temperature.

Active tags have an onboard power supply, active receiver, distinct active transmitter, and may talk with each other and form a network. Active tags are very similar to nodes in a sensor network. ZigBee and IEEE 802.11 networks can be considered as containing active tags using a very broad definition of active tags. One standard defining a network of active tags, ISO 18000–7, does not permit tag-to-tag communication, but only communications between a tag and a reader. Because of the active transmitter, active tags can transmit a signal to a reader several hundred meters away. Similarly, the active receiver enables the active tag to receive a very weak signal from devices up to several hundred meters away. The active communications hardware enables active tags to be used in places with large amounts of metal, which is typically very unfriendly to passive and semipassive tags. As more memory can be incorporated into an active tag, it is often used to store information about a shipment of goods in a shipping container. Active tags may incorporate sensors. With active tags, battery life is critical as an active tag cannot harvest energy from the reader signal as a passive or semipassive tag can.

Passive tags are the cheapest of the four but have the least capability. Passive tags are commonly used to track items functioning as a wireless barcode. Semipassive tags are more expensive than passive tags, but cheaper than active tags, with capabilities falling in the middle as well. Because of their cost, semipassive tags are ideal in cases where a large number of assets must be monitored (with sensors), or in situations where the tag cannot be reused. Active tags are the most expensive, but offer the greatest capabilities. Currently, the military and highway drivers (EZ-Pass) are the primary users of active tags. The remainder of the chapter focuses on far-field passive RFID tags.

1.2 Backscatter Communication

All objects reflect radio waves, where RF energy and these reflected waves are the basis for pulsed radar systems. Pulsed radar allows distance and direction to be determined. Two basic types of antennas exist, an omnidirectional and a directional antenna. An omnidirectional antenna emits RF energy in all directions, whereas a directional antenna emits RF energy in a specific direction.

Backscatter takes advantage of the reflection of radio waves. A passive tag contains an antenna, which is used for two purposes: (1) to harvest energy from the reader signal, command, and carrier wave (CW) and (2) to communicate with the reader. The amount of energy that the tag receives depends on many factors, but the distance between the reader and tag, the reader transmitter power, and the efficiency of the RFID tag antennas are the keys. The following section elaborates on *all* the factors governing the received power.

The impedance matching between the antenna and the tag circuitry determines the amount of energy that can be transferred between the antenna and tag circuitry. When matched, the maximum amount of energy is transferred between the antenna and the tag circuitry and this occurs when the imaginary parts of the complex tag circuitry impedance and antenna impedance completely cancel each other resulting only in a real resistance. The target input impedance value is not 50 Ω as in a typical RF system, which provides a degree of freedom and flexibility in an antenna design and matching networks. When the matching is not optimal, a parasitic imaginary part is present in the impedance quantity. This can be either capacitive or inductive, and results in less power being transferred from the antenna to the tag circuitry. Hence, this reduces the read range of the tag.

Matching and deliberate mismatching are used for backscatter communication. The tag can alter the matching by adding or removing an impedance, typically a capacitor, by means of a switch. When the capacitance is included in the circuit, the matching is not optimal and the tag reflects an amount of energy *B*. When the capacitance is not in the circuit, the matching is optimal and the tag reflects an amount of energy *A*. The energy amounts *A* and *B* are not equal. Using this difference, the tag can modulate data onto the reflected radio waves and communicate with the reader. Amplitude shift keying (ASK) or phase shift keying (PSK) are possible using this difference in backscatter.

1.2.1 General Overview of the Physics of RFID

The reading of an RFID tag is more than recognizing the backscatter by a simple RF receiver. In order for a passive tag to operate, the reader/interrogator must supply the operating energy to the tag through the transmission of an RF continuous wave (CW). The magnitude of the transmission of this energy is determined by the Friis equation as shown in Figure 1.1. There are three key elements involving physical aspects of antennas in energy harvesting: (1) antenna gains, (2) reflection coefficients, and (3) polarization.

Consider first the antenna gains, G_T and G_R, for the transmitter and receiver, respectively. These gains are dependent on the simple relative physical orientations of the two antennas for energy supply and energy harvesting. The polar coordinates of the two antennas are shown as functions of θ and ϕ for both tag and reader. The maximum energy transfer normally occurs when the center lines of the planes associated with each antenna are coincident. As the angles change from this most favorable orientation, the respective gains will begin to decrease. Thus, in order to evaluate the amount of power that is really available at the tag, it is necessary to know the relative physical (structural) orientations of the reader and the tag antennas.

Reflection coefficients

Antenna gains

$$P_R = P_T \frac{G_T(\theta_T,\phi_T)\, G_R(\theta_R,\phi_R)\lambda^2}{(4\pi r)^2} (1-|\Gamma_T|^2)(1-|\Gamma_R|^2)|\hat{p}_T \bullet \hat{p}_R|^2$$

Polarization

FIGURE 1.1
The Friis equation showing key elements.

In addition to the antenna gains which deal with the physical (structural) antenna orientations, there is a question of the polarization of the two antennas again as indicated in Figure 1.1. Polarization is concerned with the relative orientations involving the electric fields generated by the conducting elements of the interrogator and tag antennas. The polarization is a somewhat complicated facet and will be considered in a separate section to follow.

The term from the Friis equation that most people are familiar with is the distance, r, between the interrogator and the tag. This is important both for the powering of the tag and the ability of the receiver in the interrogator to see the changes in radar cross section of the tag antenna. In the case of powering, the P_R must be sufficient to operate the tag where the interrogator is supplying P_T. For the interrogator to read the backscatter from the tag, the change in the tag reflection coefficient, Γ_T, alters the power received by the interrogator, P_R, which is used to decode the data/number from the tag.

There are differences between tags, which operate close to the interrogator based on the wavelength of the frequency being used. The closer is termed the near field and the more distant operation is the far field which is the object of this presentation. This discussion is concerned with the far field. In particular, in the case of far-field ultrahigh frequency (UHF), there is always an effort to reduce the power required by the chip and thus the tag. Each time the matching impedance within the tag is changed to provide the modulation of the backscattered RF, the reflection coefficient is changed. Thus, power harvested by the tag is changed (reduced) by this change in reflection coefficient indicated in Figure 1.1. The reflection coefficient is again somewhat more complicated and will be discussed in a later section dedicated to that subject.

The power transmitted by the antenna, P_T, in Figure 1.1, is governed by the Federal Communications Commission (FCC). This power is spread over a three-dimensional area as determined by the antenna radiation pattern as shown in Figure 1.2. The frequency used, 915 MHz in EPCglobal Gen2, has a wavelength, λ, which is calculated from the relationship:

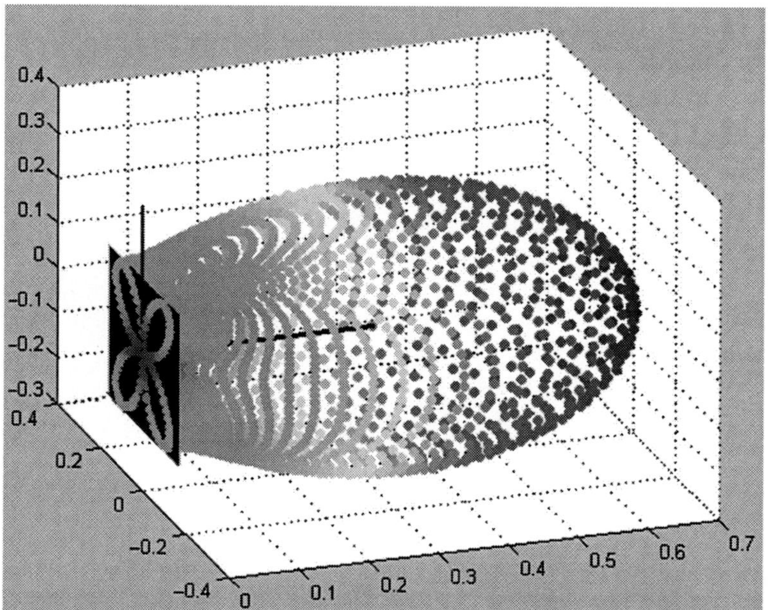

FIGURE 1.2
A radiation pattern.

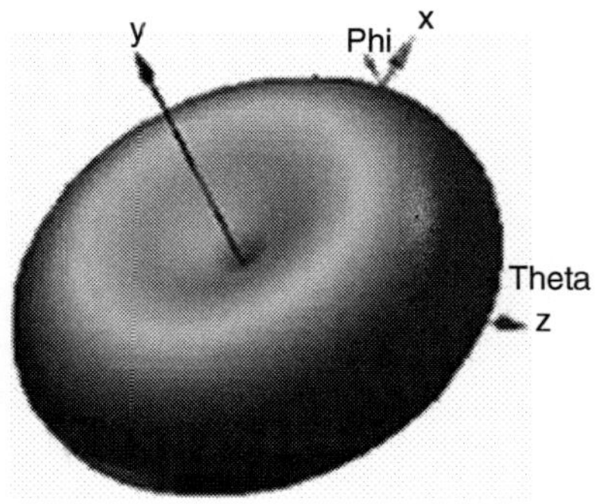

FIGURE 1.3
Radiation pattern of an ideal dipole.

$$c = f \times \lambda, \tag{1.1}$$

where c is the speed of light. Thus, λ of the Friis equation is 0.3278+ m. The gain of the transmitting antenna, $G_R(\theta_T,\phi_T)$, means that the energy provided to the antenna is focused to some extent thus causing the energy density to be greater in some area than would have been if it had been radiated in all directions equally (an isotropic antenna). The angular orientations θ_T, ϕ_T make it clear that the tag orientation will receive power that will be varied over all the angles of orientation with respect to transmitting energy (power). In the pattern of Figure 1.2, an RFID tag placed at the orientation (0,0,0.7) would receive the most energy from the transmitter. This location is part of the terminology of the most favorable location which is used by most vendors as the distance at which the tag can be read.

The next term of the Friis equation considered is the gain of the tag antenna, $G_T(\theta_T,\phi_T)$, where a pattern is likewise to be produced. Many RFID tags today are some form of a dipole. Figure 1.3 is an illustration of an ideal dipole. The dipole can be thought of as two collinear wires end to end with a small space between them. The two wires would be oriented along the Y axis of Figure 1.3. Figure 1.4 illustrates an interrogator patch antenna and the tag antenna simultaneously in the most favorable relative orientation. It is assumed that the Y and Z plane of the tag antenna illustration is parallel with the plane of the interrogator antenna. It is further assumed that the X axis of the tag antenna is perpendicular to the plane of the interrogator antenna at the (0,0) origin of that plane.

Thus, in Figure 1.4, the most favorable orientation of the interrogator and tag antenna implies that $G_R(\theta_T,\phi_T)$ and $G_T(\theta_T,\phi_T)$ are at their maximum values. That is the best you can do for the antennas specified. Thus, the distance, r, is the relative positional difference along the X axis as discussed earlier.

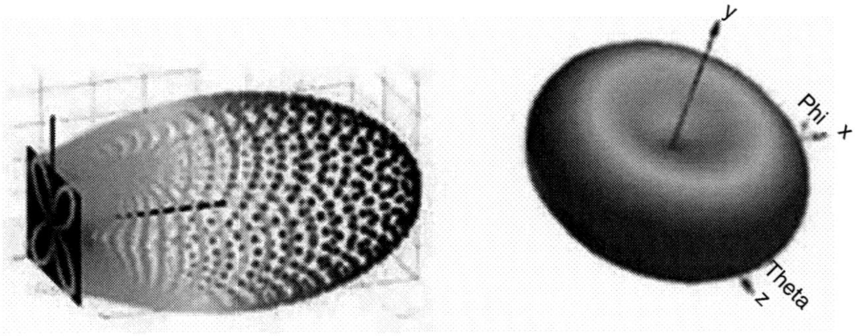

FIGURE 1.4
Interrogator radiation pattern (*left*), an ideal dipole (*right*).

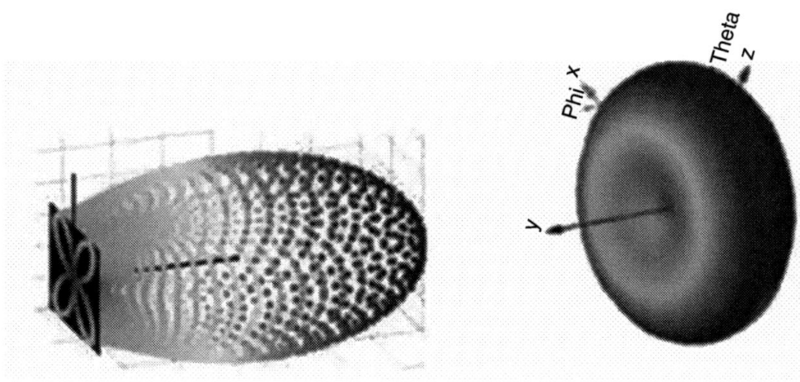

FIGURE 1.5
Most unfavorable orientation for the tag.

From the energy pattern of Figure 1.5, it can be understood that the energy along the *Y* axis is essentially 0, thus backscatter from the tag has very little if any practical energy in that orientation (Figure 1.6).

By this time, it should be clear to the reader that there are many scenarios in which the interrogator/tag combination will not work. Figure 1.7 provides a most favorable orientation for the two dipole antennas.

1.2.2 Polarization

Polarization is the term used to describe the motion of the tip of the electric field vector when electromagnetic (EM) energy is transmitted from an antenna, in this particular case, radio frequency waves for RFID. Consider the two sinusoidal waveforms in Figure 1.8, where the two waves are 90° out of phase. Each of these waveforms can be represented as a

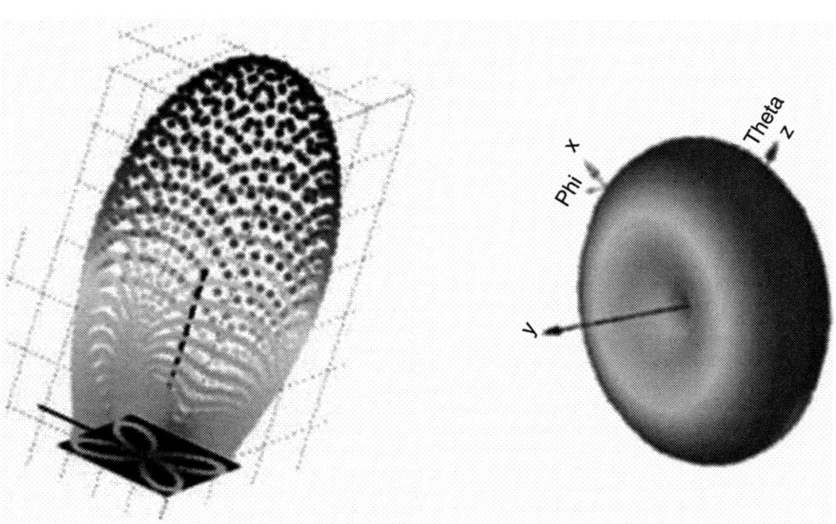

FIGURE 1.6
Another highly unfavorable orientation.

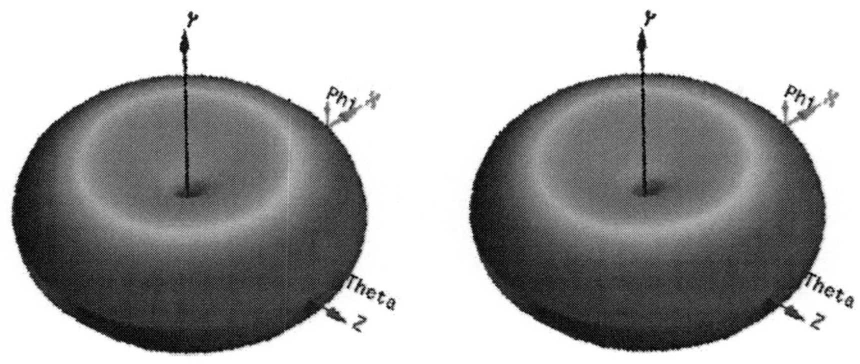

FIGURE 1.7
Most favorable polarization—reader (*left*) and tag (*right*).

vector with magnitude, |E| and |H|, and angles, ∠Θ and ∠Φ, respectively, where the two vectors will always be 90° out of phase.

Consider two dipole antennas A and B, aligned on a center line as shown in Figure 1.9, where we wish to view the E field between the two antennas from point a looking to point b.

The field pattern generated by an ideal dipole is shown in Figure 1.10 (*left*). The field patterns of the two aligned dipoles are shown in Figure 1.10 (*right*). While the full explanation of how the pattern is calculated is not necessary here, there is an intuitive understanding that can be derived by the color scheme where red indicates the strongest signal level, that is, the largest magnitude of |E|.

As shown in Figure 1.10, the *Y* axis is the vertical axis of Figure 1.9. Now consider the view from point a to point b (Figure 1.9 looking *left* to *right*) looking along the line connecting the two points as shown in Figure 1.11.

The magnitude of the E field, |E|, will appear as a vertical line. The antenna A is said to be *linearly polarized*. Based on the orientations shown in Figure 1.9, the two antennas are aligned such that the E field transmitted by A will impinge on antenna B in the same orientation as viewing the vector in Figure 1.11. Under certain conditions, it is possible to cause the E field of a particular type of antenna to rotate in angle such that the tip of the vector of Figure 1.11 will trace a circle as shown in Figure 1.12. In this case, the antenna is said to be *circularly polarized*. Any mathematical discussion of how this is done is beyond

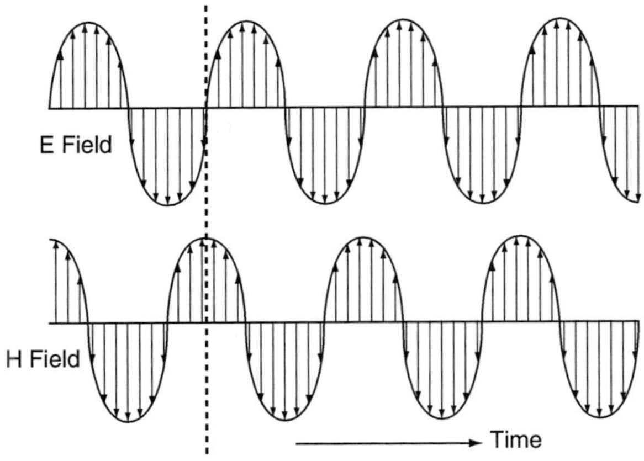

FIGURE 1.8
The E and H fields showing phase and time.

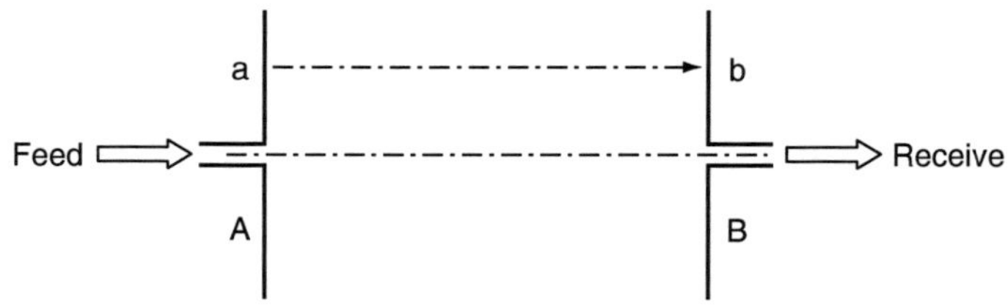

FIGURE 1.9
Two dipole antennas aligned vertically and horizontally as shown.

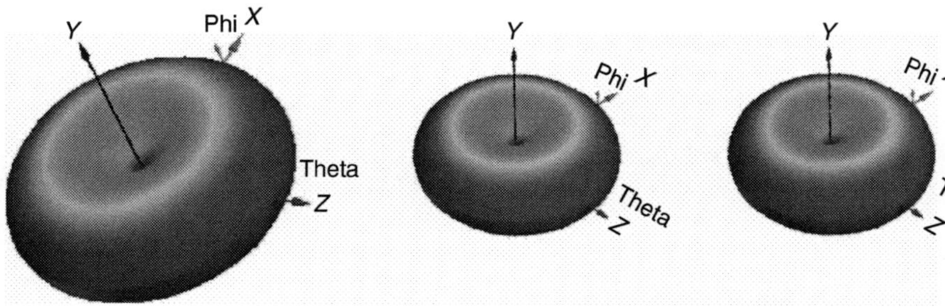

FIGURE 1.10
Radiation patterns ideal dipole (*left*) and aligned dipoles (*right*).

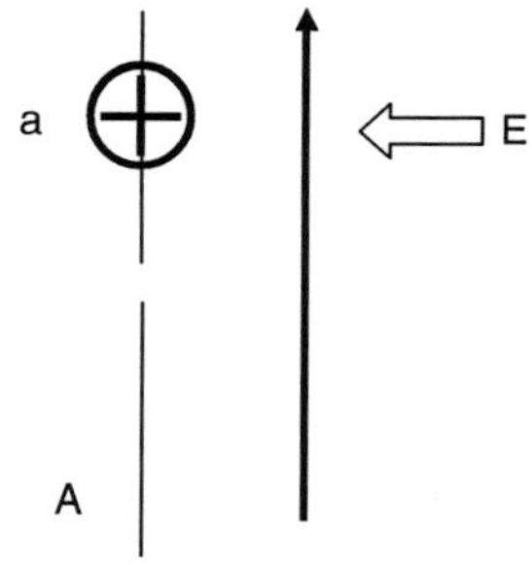

FIGURE 1.11
The E field in time and angle.

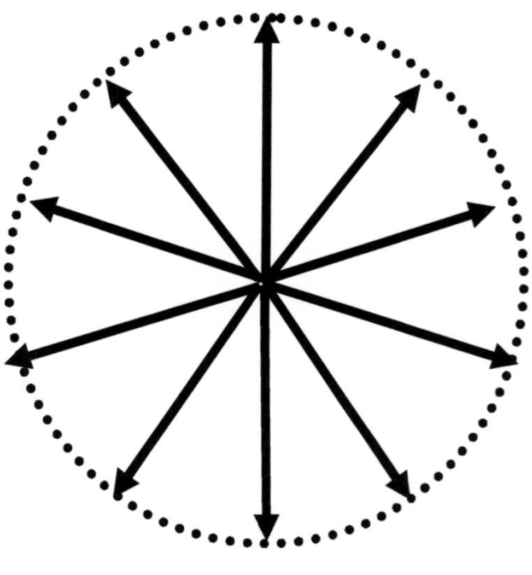

FIGURE 1.12
The E field with a circularly polarized antenna.

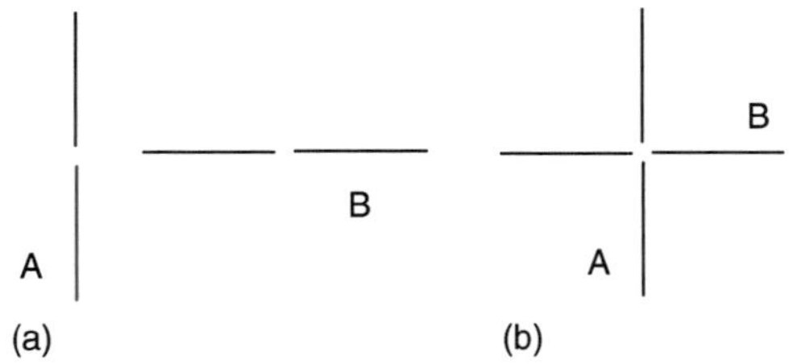

FIGURE 1.13
Two orthogonal dipoles.

the scope of this chapter. However, in general, the means by which this can be accomplished is to use a flat planar antenna, typically circular, with two feed attachments where the two attachments are electrically 90° out of phase. The feed points can be coupled to the antenna surface through the air with appropriate spatial distance and orientation.

The reason why the two antenna polarizations are important in RFID is because of the various possible orientations of the item that is tagged relative to the interrogator antenna. For example, consider the view from antenna A to antenna B under the condition where antenna B has been rotated by 90° in space as shown in Figure 1.13.

From Figure 1.13, it is obvious that very little of the E field vector as pictured in Figures 1.8 (E field) and 1.11 (E vector) from antenna A will impinge on antenna B. Thus, the amount of energy delivered from antenna A to antenna B will be minimal. In general, this means the RFID tag will likely not have received sufficient energy from the continuous wave (CW) of the interrogator antenna to power the RFID chip on the tag (Figure 1.14).

In the traditional use of radio, the station transmitting antennas are typically vertical, and most radio receivers are going to be placed on a flat surface parallel to the earth in which case the manufacturer will include the antenna with the proper polarization with respect to vertically polarized transmitting antennas, which are mounted vertically to be in the same polarization as the transmitting antennas. However, when RFID tags are placed on an item, it is most difficult to fix the orientation of the item throughout the item life cycle at all possible locations where the tag may be read. Thus, the relative polarizations will be at the best as in Figure 1.9, and at the worst as in Figure 1.13, with any other orientations in between these two extremes. Thus, except in very special circumstances, one does not want both the RF interrogator transmitter and the RFID tag to be dipole antennas due to the possibility of the orthogonal or other unfavorable orientations as illustrated in Figure 1.13, when transmitter or receiver may not be permanently aligned.

While radio transmitting antennas are typically mounted vertically for vertical polarization, television antennas are horizontally polarized. Before the days of cable and satellite TV, rooftops were adorned with antennas formed with multiple horizontal elements resulting in what is termed a Yagi antenna. The plane formed by the parallel elements

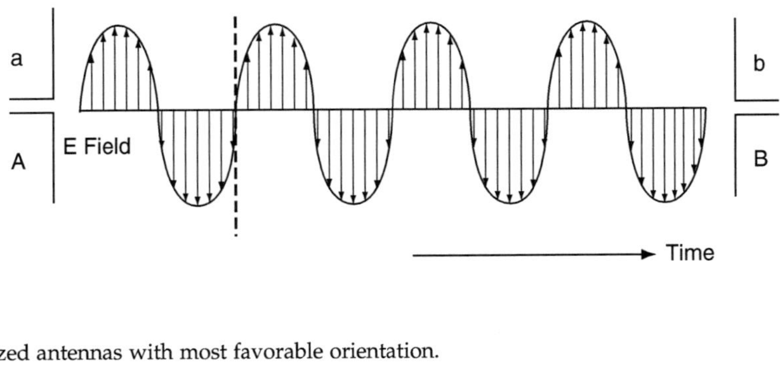

FIGURE 1.14
Linearly polarized antennas with most favorable orientation.

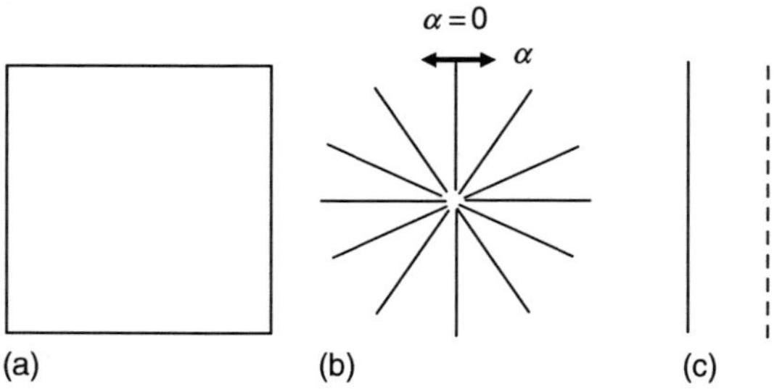

FIGURE 1.15
Planes of the interrogator antenna (a) and rotated tag antenna (b) and (c).

was (is) parallel to the earth, that is, horizontally polarized. Thus although radio and television do not share the same frequency bands, there is still little or no interference due to the respective polarizations.

The orientation issue in RFID is normally addressed by providing a circularly polarized antenna on the interrogator and a dipole antenna on the tag. Thus, the dipole will receive energy more favorably through virtually any possible orientation where the plane of the dipole if rotated in such a manner that the plane in which it is rotated will be parallel to the plane of the circularly polarized transmitting antenna. Due to the phasing and dual feed configuration requirements, the tag antenna cannot practically be circularly polarized and is typically some form of a dipole antenna.

The use of a circularly polarized antenna for the interrogator with the tag antenna orientations maintained as parallel planes insures suitable tag performance in many circumstances. However, the relative angles of the planes on Figure 1.15c vary, when the tag plane is rotated through the angles indicated as α-variations in efficiency will exist.

Consider the circularly polarized and dipole antennas of Figure 1.16. The pattern of the dipole on the right allows it to be rotated about the Y axis while maintaining a favorable orientation with the circularly polarized antenna on the left. In addition, the dipole can be rotated about the X axis of the dipole diagram.

Hence, in summary, a circularly polarized antenna reduces the number of unfavorable reader/tag antenna orientations allowing an increased success of reading the tag. A circularly polarized antenna does have a cost in terms of the distance at which the tag can be read. Typically, circularly polarized antennas do not have as great a range as linearly polarized antennas. One use of a circularly polarized antenna is when reader/tag orientations vary. In practice, linearly polarized antennas are used when maximum distance is required and reader/tag orientation is fixed. This is often the case in a manufacturing plant. Further, portals, which employ multiple antennas, typically four, can position antennas to cover all three, or at least two of the three coordinate axes. Here, the extra

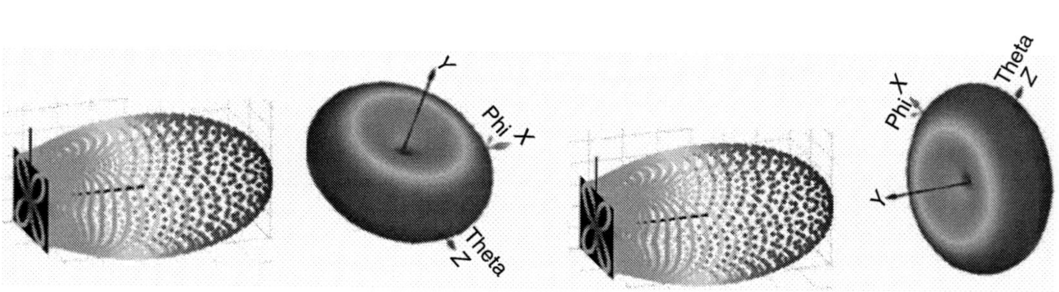

FIGURE 1.16
Patch/dipole orientations favorable (*left*) and unfavorable (*right*).

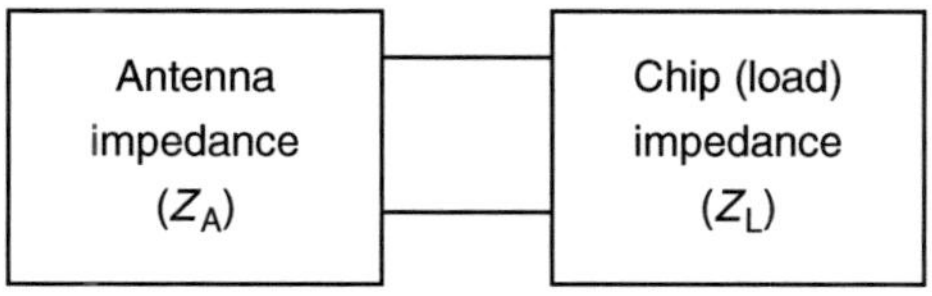

FIGURE 1.17
Antenna and chip (load) connection.

range, or power capabilities, of the linearly polarized antenna can help in reading a tag through material that is RF-unfriendly.

1.2.3 Reflection

Reflection is a property of the matching of the impedance of the antenna to the impedance of the load (chip) on the RFID tag. Consider the connection of the two components as shown in Figure 1.17.

The reflection coefficient is defined as:

$$\Gamma = (Z_L - Z_A)/(Z_L + Z_A). \tag{1.2}$$

If the load impedance is shorted (0), $\Gamma_{short} = -1$. If the load impedance is open (∞), $\Gamma_{open} = -1$, and if the two impedances are matched, that is, complex conjugates, $\Gamma_{match} = 0$. Thus, all other variations of matching the antenna and load impedances result in a Γ somewhere between +1 and −1, which is the value used in the Friis equation of Figure 1.1. There is a Γ_T for the transmitting device and a Γ_R for the receiving device.

1.3 Antenna Directivity and Gain

Two critical antenna characteristics are the directivity and the gain. The definitions for antenna directivity and antenna gain are essentially identical except for the power terms used in the definitions.

The directivity of the antenna [$D(\theta,\phi)$] is defined as the ratio of the antenna radiated power density at a distant point to the total antenna radiated power density isotropically.

Figure 1.18 demonstrates the coordinate system on which the directivity and the gain are based. The directivity is typically expressed in decibels (dB) above a reference, where $D(\theta,\Phi)[\text{dB}] = 10 \log_{10} D(\theta,\phi)$. The directivity of an isotropic radiator is equal to one. An isotropic antenna with the spherical pattern is the typical reference antenna, which is specified in the dBi units (Figure 1.19).

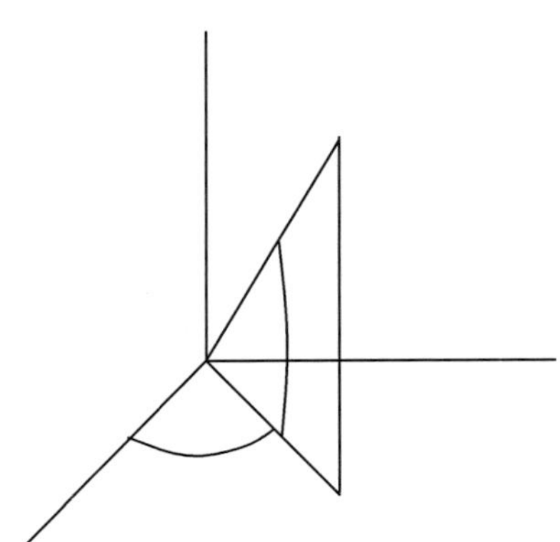

FIGURE 1.18
Coordinate system.

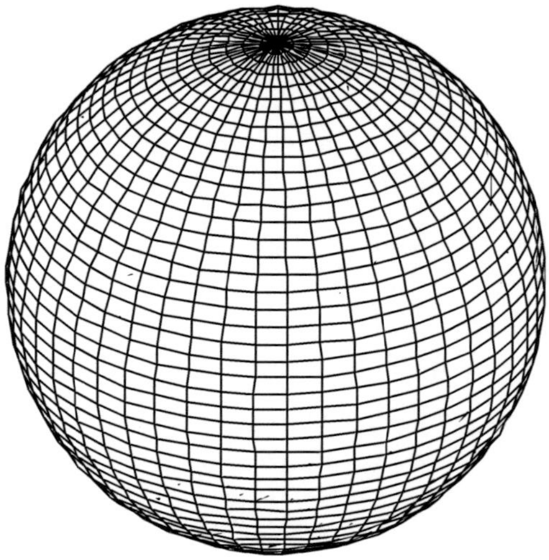

FIGURE 1.19
The isotropic radiation pattern.

In real world applications, an antenna is driven by a source (power generator), where the total radiated power of the antenna is not the total power available from the generator. The loss factors which affect antenna's efficiency can be attributed to the effects of the mismatch (reflection) losses at the connection of the antenna and the source. Therefore, the total power delivered to the antenna terminal is equal to the ohmic (P_{ohmic}) losses plus the radiated power (P_{rad}) by the antenna, $P_{\text{in}} = P_{\text{rad}} + P_{\text{ohmic}}$.

The gain $[G(\theta,\phi)]$ is defined as the ratio of the antenna radiated power density at a distant point to the total antenna input power (P_{in}) radiated isotropically. Thus, the antenna gain represents the actual efficiency of the antenna, where the gain and directivity are related with the efficiency factor that is applied to the directivity value to calculate the gain of the antenna. The graphical representation of the antenna gain is accomplished with the radiation patterns, which are typically drawn for the vertical and horizontal polarizations or three-dimensional polar plots.

The antennas can be classified into two principal groups: directional and omnidirectional. The directional antennas have a high gain, but a narrow field view (i.e., patch). The omnidirectional antennas have a low gain, but a wide field view (i.e., dipole).

In a typical RFID system, the tags are designed to use either a dipole, omnidirectional type antenna or a combination of two-dipole antennas to eliminate the problem of nulls in the radiation pattern of a single dipole (Figure 1.18). The radiation pattern of the dipole antenna allows the tag to intercept most of the incident signal regardless of the tag orientation with respect to the reader's antenna, where a dipole antenna typically has a gain of about 2.2 dBi. In addition, the RFID readers are designed to operate with the patch (directional) type antennas. Thus, the power of radiated signal is focused toward a passive tag. The antenna manufacturers normally specify the antenna gain for the circularly polarized patch antennas to be on the order of 6 dBi.

1.4 Summary

The successful operation of a passive RFID system is dependent on numerous factors most of which are interrelated and incorporated in the Friis equation of Figure 1.1. The typical hands on manipulation of tags in the presence of a reader involves various factors illustrated in this chapter. Such manipulation simultaneously varies multiple factors that can only be separated by an in-depth analysis of the physical factors of the tag and reader design. In addition, the multiple orientations and distances increase the dimensionality of the problem.

A successful read or nonread is an amalgamation of these many factors. One means of separating these factors is a well-documented testing procedure, providing a geometric positional log in conjunction with a real-time spectrum analyzer that can monitor the air medium and document the reader to tag energy profile and the tag to reader backscatter.

This chapter presents an overview of many interrelated factors that relate to the successful operation of a passive RFID system.

1.5 Future Directions

Primarily driven by the cost-insensitive products, the Class 3, battery-assistant smart passive technology offers an excellent solution for applications requiring more functionality than passive systems that are used today. The Class 3 is a proposed EPCglobal standard, which targets the battery-assistance passive tags with additional functionality on the chip (i.e., memory and sensors), longer operating distance, and moderate cost. The cold-chain visibility is the ideal candidate for the Class 3 type devices where temperature sensors can be used to insure that the temperature was adequate through the cold supply chain.

Class 3 devices require advances in lowering cost, increasing battery life, and lowering power sensors. Lowering cost is critical because the cost of using the tag must be recouped and each participant is only willing to pay so much. Lowering the cost will open doors for semipassive technology in areas where it was previously too expensive. Battery life is critical to powering the sensors to take the required sensor readings. The battery must last at least as long as time that the asset must be monitored to prevent incomplete or inaccurate results. For some assets, their lifetime is too long for the semipassive tag to monitor them over their entire lifetime. This is also an issue for active tags. Finally, lower power sensors are one way to reduce energy consumption and extend battery life. Advances in sensor technology will enable lower power sensors to be incorporated into semipassive tags.

2

EPCglobal Network

Alfio R. Grasso

CONTENTS

2.1 EPCglobal History

EPCglobal Inc. is a joint venture between GS1 US (http://www.gs1us.org/) and GS1 (http://www.gs1.org/).

EPCglobal was formed to commercialize and set up a user-driven standards process based on intellectual property developed by the Auto-ID Center at the Massachusetts Institute of Technology (MIT). The Auto-ID Center at MIT (http://autoid.mit.edu/cs/) was established in 1999 to develop the "Internet of Things." Led by Prof. Sanjay Sarma, Dr. Daniel Engels, David Brock, and Kevin Ashton, the center quickly expanded with the addition of other RFID laboratories around the world:

- Institute for Manufacturing, University of Cambridge, UK (http://www.autoidlabs.org.uk/), headed by Prof. Duncan McFarlane
- RFID Laboratory, University of Adelaide, Australia (http://autoidlab.eleceng.adelaide.edu.au/), headed by Prof. Peter Cole
- Keio University Shonan-Fujisawa Campus Murai Laboratory, Keio University, Japan (http://www.kri.sfc.keio.ac.jp/en/lab/AutoID.html) headed by Prof. Jun Murai
- Fudan Auto-ID Center, China (http://www.autoidlab.fudan.edu.cn/), headed by Prof. Hao Min
- St. Gallen University and Swiss Federal Institute of Technology, ETH Zurich, Switzerland (http://www.autoidlabs.org/the-labs/stgallen/), headed by Prof. Elgar Fleisch

In October 2003, the Auto-ID Center ceased to operate and the Intellectual Property developed by the six laboratories was licensed to the newly formed joint venture, EPCglobal Inc. In addition, the six Auto-ID Centers were renamed Auto-ID Labs. In April 2005, a seventh Auto-ID Lab was admitted to the Federated Auto-ID Labs. That laboratory was from the Information and Communications University (ICU) (http://www.autoid.or.kr/), headed by Prof. Sang-Gug Lee.

During the Auto-ID Center years of operation (1999–2003), more than 110 Auto-ID Center research papers were published. Those papers available from the Auto-ID Labs Web site (http://www.autoidlabs.org/publications/page.html) formed the basis of today's EPCglobal Network. Before its hand over to EPCglobal, the Auto-ID Center also published the following specifications:

- 900 MHz Class 0 Radio Frequency (RF) Identification Tag Specification
- 13.56 MHz ISM Band Class 1 Radio Frequency (RF) Identification Tag Interface Specification
- 860–930 MHz Class 1 Radio Frequency (RF) Identification Tag Radio Frequency & Logical Communication Interface Specification

These specifications are available from http://www.epcglobalinc.org/standards/specs/

Today, EPCglobal has established a network of organizations, both end users and solution providers. As of April 2007, EPCglobal had over 1300 member organizations, in which over 4000 individuals contribute to $\sim$100 work groups. Some of these groups require the participant to sign an opt-in agreement (see Section 2.2.1). The work groups are divided into a number of classes:

- Discussion Groups (open to everyone)
- Industry Action Groups (EPCglobal Subscribers some with optional opt-in)
- Joint Requirement Groups (opt-in required)

- Cross Industry Adoption & Implementation Groups (EPCglobal Subscribers only, but no opt-in)
- Technical Action Groups (opt-in required)

Some of these work groups gather business requirements and document use cases, others develop specifications that address those requirements and use cases, formulate conformance and performance test procedures, and certify solution provider's solutions. EPCglobal maintains a list of certified solution providers (hardware, software, and test centers) at http://www.epcglobalinc.org/certification/

Apart from the specifications (http://www.epcglobalinc.org/standards/), EPCglobal also publishes information documents, mainly from the Cross Industry Adoption and Implementation Groups, in the form of an RFID Cookbook (http://www.epcglobalinc.org/what/cookbook/), Key Learnings (only available to EPCglobal Subscribers), specific information for European or Asian adoption and other information such as EPCglobal Recommended Occupational Use Best Practices for Complying with Limits on Human Exposure to Electromagnetic Fields (http://www.epcglobalinc.org/public/bestpractice/).

2.2 EPCglobal Network

The EPCglobal Architecture Framework specification (EPCglobal, 2005a) is a collection of interrelated standards for hardware, software, and data interfaces, together with core services that have a common goal of enhancing the supply chain through the use of Electronic Product Codes (EPCs).

2.2.1 EPCglobal Standards Development Process

EPCglobal has also established a Standards Development Process (EPCglobal, 2007a). This process starts with end user requirements and develops a specification, which is prototyped, tested, and evaluated before the standard is ratified by the EPCglobal Board of Governors. EPCglobal works with retailers, manufacturers, and hardware, software, and integration solution providers to create and share intellectual property that will benefit the entire subscriber base. Similarly, EPCglobal is committed to open use of the EPCglobal Network. The EPCglobal Intellectual Property (IP) Policy (http://www.epcglobalinc.org/what/ip_policy/) ensures that all companies subscribing to the organization have open, neutral access to EPCglobal Network technology and standards. The agreement guarantees that the technology remains nonproprietary for the benefit of industry as a whole. Companies with a desire to participate in any Action Group or Working Groups are required to sign the appropriate EPCglobal Intellectual Property (IP) Policy. At key points in the process, calls for Intellectual Property are made and if IP is noted, every attempt is made to ensure that the specification developed is free from IP claims. Every participant in the technical work groups must execute an opt-in agreement, in which it is agreed that IP will be divulged at those key points. The objective is to have EPCglobal specifications that are free from IP claims.

The EPCglobal Network consists of standards that specify both hardware and software elements. The hardware specification details the air interface protocol, communication between the reader and tags, and is known as UHF Class-1 Generation-2 V1.0.9 (or C1G2 V1.0.9) (EPCglobal, 2005c). At the time of writing, other hardware specifications were in development, namely the HF Air Interface Protocol for Item Management, and the

UHF Air Interface Protocol for Item Management UHF C1G2 V1.1.2. Both of these specifications build on the command set of C1G2 V1.0.9 with HF in particular defining a signaling layer at 13.56 MHz, but using almost the same command set. The advantage of using the command set is that other parts of the EPCglobal Network may be able to be protocol agnostic.

The rest of the standards that comprise the EPCglobal Network are software elements namely:

- Tag Data Standards
- Tag Data Translation
- Reader Protocol
- Low-Level Reader Protocol
- Reader Management
- Application Layer Events
- ECPIS EPC Information Services
- Security Certificate Profile
- Object Naming Service
- Drug Pedigree Standard
- EPCIS Discovery
- Subscriber Authentication

The rest of this chapter is devoted to providing a short description of each element in the EPCglobal Network and a status of the specification as of April 2007.

2.3 Class 1 Generation 2

The Class 1 Generation 2 Specification (EPCglobal, 2005c) details the communications between the RFID tag and the RFID reader. It contains specifications of an air interface. Communication between readers and tags is based on "Reader Talks First" principles, whereby the reader issues commands (along with communication parameters) to a population of tags that are in the reader's RF field. The objective of the protocol is to singulate a tag in a multiple tag environment to read its identity or other information stored in the tag's memory.

The frequency of transmissions of the RFID reader operate anywhere from 860 to 960 MHz according to local regulations, whereas in general RFID tags should be able to operate anywhere in that frequency band, responding to locally configured readers. There are three frequency bands of interest:

- Region 1 (EU and Africa), typically 865–868 MHz, 2 W ERP Frequency Agile, LBT
- Region 2 (FCC—United States and Canada) 902–928 MHz, 4 W EIRP Spread Spectrum Frequency Hopping
- Region 3 (Asia). Most follow Europe, some follow FCC

Usually, the available spectrum in a local region is divided into channels or sub-bands. In the United States, channels are 500 kHz wide, whereas in Europe sub-bands are 200 kHz wide. In the United States, the 50 available channels, with operation at 4 W EIRP, are used according to a random selection process known as Spread Spectrum Frequency Hopping, in which operation in one channel is limited to 400 ms after which the reader must select another channel to transmit. In Europe, only 10 channels are available for operation at 2 W ERP, and are used according to a Frequency Agile operation. In addition, readers must follow a Listen before Talk operation, in which a reader must first listen for transmissions in the intended sub-band of operation, and only transmit if no transmissions are detected above a very low threshold. However, work within an ETSI Technical Group has begun to produce a Technical Report (TR 102 436) that recommends a code of practice that will enable RFID readers to be synchronized. EPCglobal Inc. also maintains a database of worldwide UHF regulations (EPCglobal, 2007b).

Tags must understand three different modulation schemes; one of these would be selected by the reader and sent to the tags that are in the reader's RF field as parameters of the command that establishes communication with the tags. The reader selects the most appropriate modulation scheme, based on its own operating parameters, such as regulatory environment, received noise information, and preferences established by the system engineer. The three different modulation schemes are

- Double Sideband Amplitude Shift Keying DSB-ASK
- Single Sideband Amplitude Shift Keying SSB-ASK
- Phase Reversal Amplitude Shift Keying PR-ASK

Data sent to the tag by the reader is encoded using a coding scheme known as pulse interval encoding (PIE). This coding scheme defines the duration of a zero-data bit pulse, also known as T_{ari}, as part of the parameters sent when communication is established with tags by the reader. The duration of a one-data bit is at least 50% longer. This enables the reader to specify variable data rates of 40, 80, 160, 320, and 640 kbits (kilobits per second).

The specification also caters for different modes of operation, Single Reader Use when isolated readers are deployed in an environment, or Dense Reader Use, in which the transmit spectrum masks are adjusted to allow for dense operation.

The commands that establish communication with tags also define the communication parameters for the tag's response. Again, these choices are made by the reader, based on local regulations, required application and noise performance of previous communication techniques. Tags may respond to reader commands using one of two tag backscatter modulation schemes, either ASK or PSK modulation, with the format being selected by the tag vendor, and readers are capable of demodulating either modulation type. Tags may be commanded by the reader to encode the backscattered data as either FM0 base band or Miller modulation of a subcarrier at the data rate (either 2,4, or 8 cycles of subcarrier per bit). A variety of data rates can be selected by the reader from 40 to 640 kbps.

Additionally, the specification details the reader commands and tag responses to those commands. Most of those commands are concerned with reading or writing data to the tag. Tags are identified (singulated) in a population of multiple tags, by the anticollision protocol, which is called the Q protocol. This protocol is an ALOHA-based reply protocol, in which tags receive anticollision parameters from the command by the reader and then randomly self-select a period to respond. This reply period is called a slot, one of many slots in a round. The round size (Q value) is provided by the reader and slots are numbered from 0 to $2^Q - 1$. Tags provide a response to anticollision commands, by virtue of a handle, known as RN16. This is a 16 bit random value selected by the tag for this communication session. This handle can also be used by the tag, under reader control, to cover-code (encrypt) the response data from the tag.

The anticollision commands include selection commands, in which tags are chosen to be in a round, based on parameters in the command, which dictate which bits are used for the selection process. Complex selection criteria can be developed by a succession of select commands.

There are four memory banks defined:

- Reserved memory (contains the kill and access passwords)
- EPC memory (contains CRC-16, Protocol-Control (PC) bits, and a code (such as an EPC)) that identifies the object to which the tag is or will be attached
- TID memory (contains an 8 bit ISO/IEC 15963 allocation class identifier, 11100010_2 for EPCglobal) and sufficient identifying information to uniquely identify the custom commands and/or optional features that a tag supports
- User memory

Other commands implement access and kill passwords, which allow readers access to memory, for reading and locking. Other passwords are used to kill the tag at the point of sale, if requested by the consumer. Tag memory can be selected for access and locking, such that only readers that issue the correct access password will be able to read the tag data. Some memory can be locked for either read or write access.

Tags implement a number of internal states: Ready, Arbitrate, Reply, Acknowledge, Open, Secured, and Killed. Tags also have four sessions, which can be used by a succession of interrogators to inventory tags, and a number of flags, some of which have persistent states (keep the state for at least a minimum period of time, after RF power is lost to the tag), and each session flag has two values, known as A and B.

2.4 Tag Data Standards

At the time of writing this chapter, there were two Tag Data Standards ratified by EPCglobal. The first is EPCglobal Tag Data Standards Version 1.3 (EPCglobal, 2006a) and the second is an older version EPC Generation 1 Tag Data Standards Version 1.1 Rev 1.27 (EPCglobal, 2005b). The main difference between the two standards is that V1.3 is aimed for use in UHF Class 1 Generation 2 Tags, whereas V1.1 was aimed for use in UHF Class 1 Generation 1 Tags. V1.3 maintains compatibility with V1.1 at the identity level, continuing to support the GS1 system* and DoD† identity types. Other differences between V1.1 and V1.3 are

- The deprecation of 64 bit encodings. EPCglobal has determined that 64 bit headers allocated in V1.1 will no longer be used after July 1, 2009 (EPCglobal SAG TDTS WG October 2006)
- The elimination of tiered header rules
- The encoding of EPC to fit the structure of Gen 2 Tags

* Previously known as the EAN.UCC system.

† DoD stands for the U.S. Department of Defense.

- The addition of the extension component to the Serialized Global Location Number (SGLN)
- Addition of SGTIN-198, SGLN-195, GRAI-170, GIAI-202 (defined later in this chapter) and corresponding changes in Uniform Resource Identifier (URI) expression for alphanumeric serial number encoding

The rest of this section is devoted to describe V1.3, since it is the most recent version.

The document defines the standards for encoding data onto RFID tags which conform to C1G2 and define completely that portion of EPC tag data that is standardized, including how that data is encoded on the EPC tag and how it is encoded for use in the information systems layers of the EPC Systems Network. The EPC encoding includes a Header field followed by one or more Value Fields. The Header field defines the overall length and format of the Value Fields. The Value Fields can contain a unique EPC code and if required a Filter Value. The specification defines four categories of URI:

- URIs for pure identities, which contain only the unique information that identifies a specific physical object, location, or organization, and are independent of tag encodings
- URIs that represent specific tag encodings, which are used in software applications where the encoding scheme is relevant
- URIs that represent patterns or sets of EPCs are used when instructing software how to filter tag data
- URIs that represent raw tag information are generally used only for error reporting purposes

In V1.3, the specific coding schemes include:

- 196 bit General Identifier (GID)
- Serialized version of the GS1 Global Trade Item Number (GTIN)
- GS1 Serial Shipping Container Code (SSCC)
- GS1 Global Location Number (GLN)
- GS1 Global Returnable Asset Identifier (GRAI)
- GS1 Global Individual Asset Identifier (GIAI)
- The DoD Construct

The specification details levels of identity:

- *Pure identity level*: Pure identity is the identity associated with a specific physical or logical entity and is independent of the identity carrier (RF tag, bar code, or database field), and takes the form of a URI (a character string). For the EPC General Identifier, the pure identity URI representation is urn:epc:id:gid:GeneralManagerNumber.ObjectClass.SerialNumber
- *Encoding identity level*: The encoding identity is achieved by encoding a pure identity together with additional information such as filter value into a specific syntax, which again is independent of the identity carrier.
- *Physical realization level*: The encoding of the identity into a physical implementation such as an RF tag.

2.4.1 General Identifier

The General Identifier (GID-96) is a generic scheme which can be used when the other schemes may not be applicable. It consists of three fields:

- General Manager Number which identifies an organizational entity that is responsible for maintaining the numbers in subsequent fields and ensuring uniqueness among the fields under its domain
- The Object Class used by that entity to identify a class or "type" of object, and which must be unique within each General Manager Number domain
- The Serial Number code, or serial number, is unique within each object class

For the rest of the encoding schemes, there are rules for converting an EAN.UCC (GS1) scheme to a serialized EPC code.

2.4.2 Serialized Global Trade Item Number

The Serialized Global Trade Item Number (SGTIN) consists of the following fields:

- The Company Prefix, assigned by GS1 to a managing entity
- The Item Reference, assigned by the managing entity to a particular object class
- The Serial Number, assigned by the managing entity to an individual object

The serial number is not part of the GTIN code, but is formally a part of the SGTIN.

2.4.3 Serial Shipping Container Code

The SSCC as defined by the General EAN.UCC Specifications is already intended for assignment to individual objects and therefore does not require any additional fields to serve as an EPC pure identity. The SSCC consists of the following fields:

- The Company Prefix, assigned by GS1 to a managing entity
- The Serial Reference which is derived from the SSCC by concatenating the Extension Digit of the SSCC and the Serial Reference digits

2.4.4 Serialized Global Location Number

The SGLN consists of the following fields:

- The Company Prefix, assigned by GS1 to a managing entity.
- The Location Reference, assigned uniquely by the managing entity to an aggregate or specific physical location.
- The GLN Extension, assigned by the managing entity to an individual unique location. The use of the GLN Extension is intended for internal purposes by the managing entity.

2.4.5 Global Returnable Asset Identifier

The GRAI consists of the following fields:

- The Company Prefix, assigned by GS1 to a managing entity
- The Asset Type, assigned by the managing entity to a particular class of asset
- The Serial Number, assigned by the managing entity to an individual object

2.4.6 Global Individual Asset Identifier

The GIAI consists of the following fields:

- The Company Prefix, assigned by GS1 to a managing entity
- The Individual Asset Reference, assigned uniquely by the managing entity to a specific asset

2.4.7 DoD Identity Type

The DoD Construct identifier is defined by the United States Department of Defense. This tag data construct may be used to encode 96 bit Class 1 tags for shipping goods to the United States Department of Defense by a supplier who has already been assigned a CAGE 477 (Commercial and Government Entity) code. At the time of this writing, the details of what information to encode into these fields is explained in a document titled *United States Department of Defense Supplier's Passive RFID Information Guide* that can be obtained at the United States Department of Defense Web site (http://www.dodrfid.org/supplierguide.htm).

2.4.8 Detailed Description of Tag Data Standards

The specification provides a number of tables defining each of the fields in making up an EPC code, such as all of the headers (256 available) that are currently defined, detailed structure of C1G2 EPC Memory Bank, such as CRC16, Protocol Control (PC) bits (16 in total), Numbering System Identifier (NSI), and reserved bits.

Then for each type, fine details of how each of the fields are further split, for example the SGTIN-96 consists of 8 bits for the header, 3 bits for the filter value, 3 bits for the partition value, 20–40 bits for the company prefix, 24–4 bits for the item reference, and 38 bits for the serial number.

The filter value is used to provide for a fast filtering and preselection of basic logistic types, where 000 defines all types, 001 is reserved for Retail Consumer Trade Items, 010 for Standard Trade Item Groupings, 011 for Single Shipping Trade Items, and the other four are reserved. Different identity types have different encodings for the filter value.

The partition value is to provide combinations of company prefix and item reference; for example, some organizations may need a large number of reference items, whereas others may need fewer than 16. For example, the SGTIN 7 combinations are defined. Different identity types have different encodings for the partition value.

2.5 Tag Data Translation

The Tag Data Standards earlier describe how to translate between three representations of the EPC, namely the binary format (stored in the tags EPC Memory Bank) and two formats of URI, one for tag-encoding and another for pure identity. URIs are intended for communicating and storing EPCs in information systems, databases and applications, to insulate those systems from knowledge about the physical nature of the tag. The tag-encoding URI provides a one-to-one mapping with the binary number recorded in the physical tag and as such indicates the bit length of the tag and may also include an additional filter field. The tag-encoding URI is therefore intended for low-level applications which need to write EPCs to tags or physically sort items based on packaging level. The pure-identity URI format isolates the application software from details of the bit length of the tags or any fast filtering values. The Tag Data Translation Specification (EPCglobal, 2006b) provides a machine-readable version of the EPC Tag Data Standards specification. The machine-readable version can be readily used for validating EPC formats as well as translating between different levels of representation in a consistent way. This specification describes how to interpret the machine-readable version. It contains details of the structure and elements of the machine-readable markup files and provides guidance on how it might be used in automatic translation or validation software, whether standalone or embedded in other systems.

Tag Data Translation capabilities may be implemented at any level of the EPC Network stack. Tag Data Translation converts between different levels of representation of the EPC and may make use of external tables. It is envisaged that Tag Data Translation software will be able to keep itself up-to-date by periodically checking for and downloading TDT markup files, although a continuous network connection should not be required for performing translations or validations, since the TDT markup files and any auxiliary tables can be cached between periodic checks; in this way, a generic translation mechanism can be extensible to further coding schemes or variations for longer tag lengths, which may be introduced in the future.

2.6 Reader Protocol

The Reader Protocol Specification (EPCglobal, 2006c) defines the protocol by which tag readers interact with EPCglobal compliant software applications (host). The terms "tag reader" or "reader" include RFID tag readers, supporting any combination of RF protocols, fixed and handheld, and so on. Also included are readers of other kinds of tags such as bar codes. Tag readers may also have the ability to write data into tags. The goal of the Reader Protocol is to insulate the host from knowing the details of how the reader and tags interact. The Reader Protocol is specified in three distinct layers:

- Reader layer (specifies the content and abstract syntax of messages exchanged between the reader and host)
- Messaging layer (specifies how messages defined in the reader layer are formatted, framed, transformed, and carried on a specific network transport)
- Transport layer (corresponds to the networking facilities provided by the operating system or equivalent)

Each messaging layer and transport layer pair is called a messaging/transport binding (MTB). Different MTBs provide different kinds of transport, for example, TCP/IP versus

Bluetooth versus serial line. Different MTBs may also provide different means for establishing connections (e.g., whether the reader contacts the host or the host contacts the reader), initialization messages required to establish synchronization, and means for provisioning of configuration information.

The specification also defines two message channels: a control channel (carries requests issued by the host to the reader, and responses from the reader back to the host) and a notification channel (carries messages issued asynchronously from the reader to the host).

The protocol specification can conceptually be divided into subsystems, the read subsystem (which consists of Sources [read tags], ReadTriggers [initiate read operations], and TagSelectors [defines and maintains a list of filter patterns], and acquires data from one or more tags from a single source called a read cycle), event subsystem (provides event smoothing, for example, when a tag first enters or leaves the RF field), the output subsystem (which determines what data to communicate with the host), and the communication subsystem (which consists of a specific MTB). The specification contains a list of commands (or functions) along with parameters (data field) and objects (variables). Also defined are a number of MTBs.

2.7 Low-Level Reader Protocol

The Low-Level Reader Protocol (LLRP) (EPCglobal, 2007e) defines a protocol to provide control of RFID air protocol operation timing and access to air protocol command parameters, and is the second generation of the Reader Protocol defined earlier. The LLRP features provide the following:

- A means to command an RFID reader to inventory tags, read tags, write to tags, and execute other protocol-dependent access commands (such as "kill" and "lock" in C1G2)
- A means for robust status reporting and error handling during tag access operations
- A means for conveying tag passwords necessary to effect commands that may require them, such as the "kill" command in C1G2
- A means to control the forward and reverse RF link operation to manage RF power levels and spectrum utilization, and assess RF interference, among RFID readers in a system
- A means to control aspects of Tag Protocol operation, including protocol parameters and singulation parameters
- A means to facilitate the addition of support for new air protocols
- A means for the retrieval of reader device capabilities
- A means for vendors of reader devices to define vendor-specific extensions to the protocol in a manner that is noninterfering among vendors, and which, to the extent possible, is vendor-administered

While LLRP is component of the second-generation Reader Protocol, the other component will be the High-Level Reader Protocol (HLRP). Reader devices may implement HLRP, LLRP, or both. LLRP is air-protocol-aware (V1.0 is C1G2 protocol aware), whereas HLRP will be air-protocol-unaware, and will incorporate only those commands and facilities that are generic across air protocols.

LLRP is specifically concerned with providing the formats and procedures of communications between a client and a reader. The LLRP protocol data units are called messages. LLRP operation consists of the following phases of execution: capability discovery; device configuration; an optional inventory and access operations setup; inventory cycles; RF survey operations and reports returned to the Client.

The specification defines a number of objects and commands, along with their parameters (data fields).

2.8 Reader Management

The Reader Management Specification (EPCglobal, 2006d) defines a protocol used by management software to monitor the operating status and health of EPCglobal compliant tag readers, and defines the EPCglobal Simple Network Management Protocol (SNMP) RFID Management Information Bases (MIB), and specifies the set of SNMP MIBII (a subtree that contains basic objects) groups required to comply with this EPCglobal Reader Management Specification over SNMP. The Reader Management Protocol specifies the interaction between a device capable of interfacing with tags (reader) and management software (host). The specification defines two separate but related management protocol specifications; the first specifies the EPCglobal SNMP MIB for monitoring the health of a reader, whereas the second specifies the EPCglobal Reader Management Protocol for monitoring the health of a reader. Just like the Reader Protocol described earlier, the reader management specifies three distinct layers:

- Reader layer
- Messaging layer
- Transport layer

However, unlike the Reader Protocol, three message channels are used. The first is the control channel; the second is the notification channel; and the third (and new channel) is one or more alarm channels (which carries alarms issued asynchronously from the reader to the host). The specification defines a number of objects and commands along with their parameters (data fields).

2.9 Application Layer Events

The Application Layer Events Specification (EPCglobal, 2005d) provides an interface through which clients may obtain filtered, consolidated EPC data from a variety of sources. In most EPC processing systems, there is a level of processing that reduces the volume of data that comes directly from EPC data sources such as RFID readers into coarser "events" of interest to applications. The processing done at this layer typically involves:

- Receiving EPCs from one or more data sources such as readers
- Accumulating data over intervals of time, filtering to eliminate duplicate EPCs and EPCs that are not of interest, and counting and grouping EPCs to reduce the volume of data
- Reporting in various forms

The ALE interface has a number of features:

- Provides a means for clients to specify what EPC data they are interested in
- Provides a standardized format for reporting accumulated, filtered EPC data that is largely independent of where the EPC data originated or how it was processed
- Abstracts the sources of EPC data into a higher level notion of "logical reader," often synonymous with "location," hiding from clients the details of exactly what physical devices were used to gather EPC data relevant to a particular logical location

The specification includes a formal processing model, an application programming interface (API) described abstractly via a Unified Modeling Language (UML), and bindings of the API to a Web Services Interoperability (WS-i) compliant Simple Object Access Protocol (SOAP) with associated bindings of the key data types to XML schema.

The client of the ALE will be responsible for interpreting and acting on the meaning of the report (i.e., the "business logic"), and may be a traditional "enterprise application," or it may be new software designed expressly to carry out an EPC-enabled business process but which operates at a higher level than the "middleware" that implements the ALE interface. The ALE interface revolves around client requests and the corresponding reports that are produced. Requests can either be (1) immediate, in which information is reported on a one-time basis at the time of the request or (2) recurring, in which information is reported repeatedly whenever an event is detected or at a specified time interval. The results reported in response to a request can be directed back to the requesting client or to a "third party" specified by the requestor.

A read cycle is the smallest unit of interaction with a reader. The result of a read cycle is a set of EPCs. The output of a read cycle is the input to the ALE. From the ALE perspective, a read cycle is a single event containing a set of EPCs, with nothing more implied. An event cycle is one or more read cycles, from one or more readers that are to be treated as a unit from a client's perspective. It is the smallest unit of interaction between the ALE interface and a client. Clients in the Application Business Logic specify the boundaries of event cycles to the ALE as part of a request for a report. A report is data about an event cycle communicated from the ALE implementation to a client. A client can specify how event cycle boundaries may be extended for a specified duration, or report periodically; may be triggered by external events; be delimited when no new EPCs are detected by any reader specified for that event cycle for a specified interval of time.

The specification defines a number of commands, along with their parameters (data fields), to define event cycle specifications (ECSpec) and corresponding reports.

2.10 ECPIS EPC Information Services

The EPC Information Services (EPCIS) (EPCglobal, 2007d) specifies a means to enable disparate applications to leverage EPC data via EPC-related data sharing, both within and across enterprises. Ultimately, this sharing is aimed at enabling participants in the EPC-global Network to gain a shared view of the disposition of EPC-bearing objects within a relevant business context. Version 1.0 is intended to provide a basic capability to meet the requirements of a basic set of use cases.

The EPCIS specification is organized into several layers:

- Abstract Data Model Layer specifies the generic structure of EPCIS data, specifying the general requirements for creating data definitions within the Data Definition Layer.
- Data Definition Layer specifies what data is exchanged through EPCIS, what its abstract structure is, and what it means.
- Service Layer defines service interfaces through which EPCIS clients interact. In the present specification, two service layer modules are defined, Capture and Query. The Core Capture Operations Module defines a service interface (the EPCIS Capture Interface) through which EPCIS Capturing Applications use to deliver Core Event Types to interested parties. The Core Query Operations Module defines two service interfaces (the EPCIS Query Control Interface and the EPCIS Query Callback Interface) that EPCIS Accessing Applications use to obtain data previously captured.
- Bindings specify concrete realizations of the Data Definition Layer and the Service Layer. A total of nine bindings are specified for the three modules defined in the Data Definition and Service Layers.

Generically, EPCIS deals in two kinds of data: event data and master data. Event data arises in the course of carrying out business processes, and is captured through the EPCIS Capture Interface and made available for query through the EPCIS Query Interfaces. Master data is additional data that provides the necessary context for interpreting the event data. It is available for query through the EPCIS Query Control Interface, but the means by which master data enters the system is not specified in the EPCIS 1.0 specification.

Vocabularies are used extensively within EPCIS to model conceptual and physical entities that exist in the real world. There are two kinds of vocabularies: Standard Vocabulary (a vocabulary whose elements, definition, and meaning are agreed to in advance by trading partners) and a User Vocabulary (a vocabulary whose elements, definition, and meaning are under the control of a single organization).

The specification defines a number of commands, along with their parameters (data fields), to define EPCIS events and then a query command to extract events from an EPCIS Events Repository.

2.11 Security Certificate Profile

The Security Certificate Profile (EPCglobal, 2006e) documents a security certificate that can be used in the authentication of entities (subscribers, services, physical devices) operating within the EPCglobal Network. The EPCglobal Architecture allows the use of a variety of authentication technologies, but it is expected that X.509 authentication framework will be widely employed. This specification defines a profile of X.509 certificate, based on two Internet standards, defined in the IETF's PKIX Working Group, RFC3280—Internet X.509 Public Key Infrastructure Certificate and Certificate Revocation List (CRL) Profile and RFC 3279—Algorithms and Identifiers for the Internet X.509 Public Key Infrastructure Certificate and CRL Profile. The profile supports the RSA asymmetric algorithm, with key lengths of 1024 (before 2010), 2048 (before 2030), and 3072 (after 2030).

2.12 Object Naming Service

The Object Naming Service (EPCglobal, 2005e) specifies how the Domain Name System (DNS) is used to locate authoritative metadata and services associated with a given EPC. It provides a service that returns a list of network addresses that may contain pertinent data about the EPC. ONS is also authoritative in that the entity that has change control over the information about the EPC is the same entity that assigned the EPC to the item to begin with. This is different to the Discovery Service (which is not universally authoritative) as it returns locations that have some data related to an EPC, and may contain pointers to entities other than the entity that originally assigned the EPC code.

In order to use DNS to find information about an item, the EPC is converted to its pure identity URI form as defined by the EPCglobal Tag Data Standard (EPCglobal, 2006a). Since ONS contains pointers to services, it uses the Naming Authority Pointer (or NAPTR) DNS record type. The ONS specification consists of

- A procedure that an ONS Client must follow to present a query to ONS, DNS NAPTR query
- A set of rules that ONS Publishers must follow to represent pointers to services for EPCs as DNS NAPTR records
- A procedure that an ONS Client must follow to interpret the results of an ONS query

2.13 Drug Pedigree Standard

The Drug Pedigree Standard (EPCglobal, 2007c) specifies an architecture for the exchange of electronic pedigree documents for use by pharmaceutical supply chain participants, and is targeted for use in complying with document-based pedigree laws. A pedigree is a certified record that contains information about each distribution of a prescription drug, such as the sale of an item by a pharmaceutical manufacturer, any acquisitions and sales by wholesalers or repackagers, and final sale to a pharmacy or other entity administering or dispensing the drug. The pedigree contains product information, transaction information, distributor information, recipient information, and signatures. The digital signature process conforms to the X.509 certificate profile defined in the EPCglobal Certificate Profile (EPCglobal, 2006e).

Two XML Schemas define a standard electronic pedigree format:

- A standard electronic envelope format that can be used by supply chain partners to package multiple pedigree documents for exchange
- A standard electronic pedigree response format that can be used by supply chain partners to respond to each pedigree transmitted

A high level, simplified pedigree process involves the creation of the pedigree, adding information to pedigree, certifying (digitally signing) the pedigree, sending pedigrees for products in shipment to customer, receiving the pedigrees, electronically authenticating each pedigree (including manually authenticating transactions that were not electronic), verifying products received against authenticated pedigrees, certifying the pedigree for receipt and authentication.

2.14 EPCIS Discovery

At the time of writing, the role and function of the EPCIS Discovery had not been defined. It is a placeholder for an architecture that provides a means to locate all EPCIS services that may have information about a specific EPC.

2.15 Subscriber Authentication

At the time of writing, the role and function of the Subscriber Authentication had not been defined. Functionality envisaged is the need to authenticate the identity of an EPCglobal Subscriber, by providing credentials so that one EPCglobal Subscriber may authenticate itself to other EPCglobal Subscribers, without prior arrangement.

2.16 Conclusions

The EPCglobal Network is a collection of specifications that are developed by thousands of persons in a cooperative and global way. The network is expanding with many new specifications expected to be ratified in 2007. It is anticipated that some of the specifications will be revised in the near future to accommodate the newer specifications, changes to business processes, new industries, new applications, and new technologies. The common thread to the EPCglobal Network is that the identifier, the EPC tag, acts as a simple licence plate or key to a database, and multiple databases spread geographically can be accessed by applications to retrieve or discover the most up-to-date information about the object that is identified by the EPC code.

References

EPCglobal Inc. 2005a. The EPCglobal Architecture Framework Version, July 1, 2005. Downloaded from http://www.epcglobalinc.org/standards/Final-epcglobal-arch-20050701.pdf

EPCglobal Inc. 2005b. EPC Generation 1 Tag Data Standards, Version 1.1, Revision 1.27. Downloaded from http://www.epcglobalinc.org/standards/EPCglobal_Tag_Data_Standard_TDS_Version_1.1_Revision_1.27.pdf

EPCglobal Inc. 2005c. EPC Radio-Frequency Identity Protocols Class-1 Generation-2 UHF RFID Protocol for Communications at 860 MHz–960 MHz, Version 1.0.9, January 31, 2005. Downloaded from http://www.epcglobalinc.org/standards/Class_1_Generation_2_UHF_Air_Interface_Protocol_Standard_Version_1.0.9.pdf

EPCglobal Inc. 2005d. The Application Level Events (ALE) Specification, Version 1.0, September 15, 2005. Downloaded from http://www.epcglobalinc.org/standards/Application_Level_Event_ALE_Standard_Version_1.0.pdf

EPCglobal Inc. 2005e. Object Naming Service (ONS), Version 1.0, October 4, 2005. Downloaded from http://www.epcglobalinc.org/standards/Object_Naming_Service_ONS_Standard_Version_1.0.pdf

EPCglobal Inc. 2006a. EPCglobal Tag Data Standards, Version 1.3, March 8, 2006. Downloaded from http://www.epcglobalinc.org/standards/EPCglobal_Tag_Data_Standard_TDS_Version_1.3.pdf

EPCglobal Inc. 2006b. EPCglobal Tag Data Translation (TDT), Version 1.0, January 21, 2006. Downloaded from http://www.epcglobalinc.org/standards/EPCglobal_Tag_Data_Translation_TDT_Standard_1.0.pdf

EPCglobal Inc. 2006c. Reader Protocol Standard, Version 1.1, June 21, 2006. Downloaded from http://www.epcglobalinc.org/standards/Reader_Protocol_Standard.pdf

EPCglobal Inc. 2006d. Reader Management, Version 1.0, December 05, 2006. Downloaded from http://www.epcglobalinc.org/standards/RM_Ratified_Standard_Dec_5_2006.pdf

EPCglobal Inc. 2006e. EPCglobal Certificate Profile Version 1.0, March 8, 2006. Downloaded from http://www.epcglobalinc.org/standards/EPCglobal_Certificate_Profile.pdf

EPCglobal Inc. 2007a. Standards Development Process, Version 1.3, Approved by EPCglobal Board on February 27, 2007. Downloaded from http://www.epcglobalinc.org/standards/sdp/EPCglobal_SDP_10002.3_Feb_27_2007.pdf

EPCglobal Inc. 2007b. Regulatory status for using RFID in the UHF spectrum, February 25, 2007. Downloaded from http://www.epcglobalinc.org/tech/freq_reg/RFID_at_UHF_Regulations_20070225.pdf

EPCglobal Inc. 2007c. Pedigree Standard, Version 1.0, January 5, 2007. Downloaded from http://www.epcglobalinc.org/standards/Ratified_Drug_Pedigree_Standard_Jan_5_2007.pdf

EPCglobal Inc. 2007d. EPC Information Services 1.0, April 12, 2007. Downloaded from http://www.epcglobalinc.org/standards/EPCglobal_EPCIS_Ratified_Standard_12April_2007_V1.0.pdf

EPCglobal Inc. 2007e. Low Level Reader Protocol 1.0, April 12, 2007. Downloaded from http://www.epcglobalinc.org/standards/EPCglobal_LLRP_Ratified_Standard_20April_20042007_V1.0.pdf

3

Design Automation for RFID Tags and Systems

Swapna Dontharaju, Shenchih Tung, Raymond R. Hoare, James T. Cain, Marlin H. Mickle, and Alex K. Jones

CONTENTS

3.1 Introduction

Radio frequency identification (RFID) systems are expanding rapidly with their applications in a wide range of areas. RFID systems consist of radio frequency (RF) tags and RF readers or interrogators. These systems are used for a wide range of applications that track, monitor, report, and manage items as they move between different physical locations. The tags consist of integrated circuits and an RF antenna. A wide range of extensions such as memory, sensors, encryption, and access control can be added to the tag. The interrogators query the tags for information stored on them, which can include items like identification numbers, user written data, or sensory data.

The major areas that drive the commercial deployment of RFID technology are logistics, supply chain management, library item tracking, medical implants, road tolling (e.g., E-Z

Pass), building access control, aviation security, and homeland security applications. Each of these RFID systems has customized requirements that currently are defined ad hoc. In addition, multiple, often competing, standards exist (ISO/IEC JTC1, ANSI, EPC, etc.) for RFID hardware, software, and data management. Thus, most of the RFID systems are deployed for closed loop applications using either proprietary protocols or nonintersecting standards with nonreusable tags and readers. As a result, in most applications, RFID tag and reader hardware and software must be specifically designed for each particular application, and must be physically modified or redesigned every time the specification for the current application is adjusted, as new applications are introduced, and the standards are modified or new standards are developed. This keeps the overall design time long and the system costs high.

Figure 3.1 presents a comparison of different RFID tag design methodologies. The current state of the art tag development shown in Figure 3.1a requires lengthy design, fabrication, and testing cycles, which can take months with intellectual property (IP) reuse to years if developing new IP. A customizable RFID tag, as shown in Figure 3.1b, can handle variations in standards and requirements as they are developed with a significantly shorter time to market than current flows. Such a tag could be mass produced and tailored to a particular RFID use after fabrication. With the use of automation to program the device, the design time could be reduced to hours or days.

This chapter presents an extensible RFID tag system, as shown in Figure 3.2. The tag can be easily customized to work with different existing or newly developed standards and even concurrently with proprietary commands tailored to the desired application. The tag

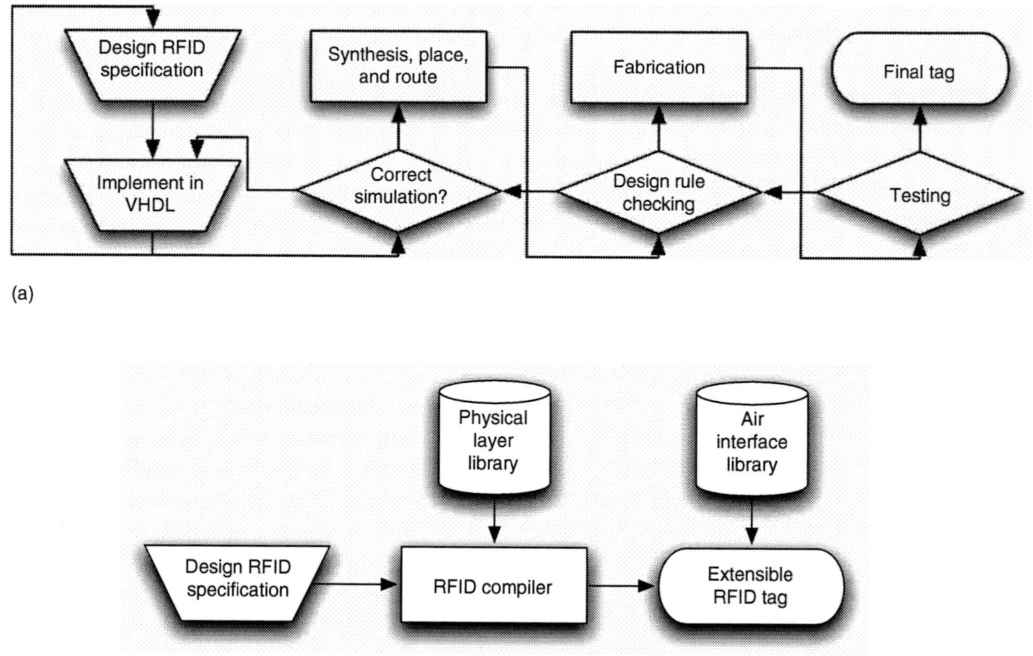

FIGURE 3.1
Comparison of RFID tag design philosophies. (a) Current RFID tag design flow. All tag components integrated manually. Estimated time: months or years. (b) Automated RFID tag design flow. Prepackaged extensible silicon device. Estimated time: hours or days. (From Jones, A.K., Hoare, R., Dontharaju, S., Tung, S., Sprang, R., Fazekas, J., Cain, J.T., and Mickle, M.H., *J. Microprocess. Microsys.*, 31, 116, 2007. With permission.)

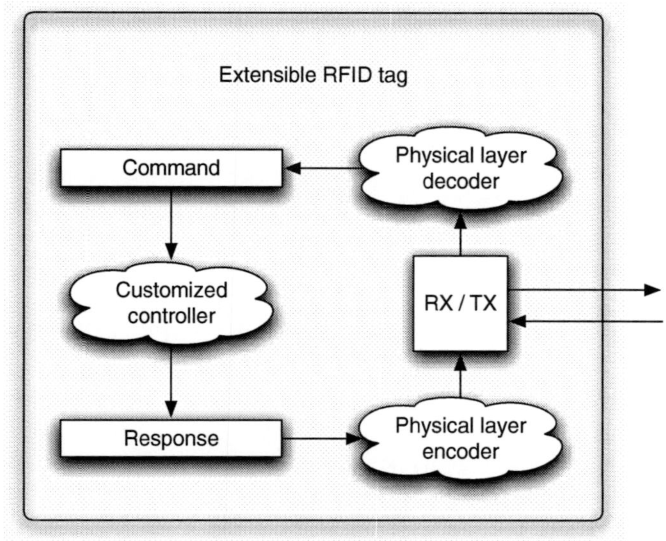

FIGURE 3.2
Extensible RFID tag. Balloon objects are automatically generated with RFID compiler.

consists of automatically generated controller and physical layer hardware that works with existing air interface blocks. To generate the controller and physical layer blocks, a design automation methodology has been developed.

For the physical layer, the methodology allows the specification of waveform features of the encoding, which are specified in a textual format. These features are then used as inputs to parameterized hardware blocks and combined with automatically generated customized hardware blocks. The result is automatically generated encoding and decoding hardware blocks.

For the controller, the methodology allows RFID primitives or the commands employed by the RFID system to be specified using RFID macros, an assembly-like format. These RFID macros are processed to generate a template file to specify the behavior for each primitive or macro. All behavior is specified using ANSI-C allowing the user to create arbitrarily complex behaviors. Finally, the RFID compiler generates the final controller used for managing the tag. This technique allows the seamless coexistence of several RFID standards such as the American National Standards Institute (ANSI) standard 256–2001 [1] and the International Standards Organization (ISO) standard 18000 Part 7 [2].

The chapter is organized as follows: Section 3.2 contains some background on research and related work pertaining to RFID. The physical layer design automation flow is presented in Section 3.3 with examples of encodings used in active, passive, and near-field communication standards. Section 3.4 describes the controller design automation flow with examples demonstrated for various RFID standards such as ISO 18000 Part 7 and Part 6C. Finally, conclusions are given in Section 3.5.

3.2 Background and Related Work

There has been an explosion of interest in RFID in the recent years. Research and development in RFID has made possible its deployment in numerous applications including supply chain management [3], automatic toll collection [4], retail stores [5], location sensing [6], libraries [7,8], healthcare [9], airports [10,11], animal tracking [12], and building access control.

Some of the open issues in the RFID domain are the existence of multiple standards, handling of stored data, tag orientation, reader collision [13], range, cost, and security concerns [14]. For active tags, maximizing battery life is an important concern. Recent research has been focused on finding solutions for some of these issues. To overcome privacy concerns, research groups have proposed novel solutions such as blocker tags [15] and authentication using novel encryption techniques [16]. A survey of the state of RFID security can be found in Chapter 33. Recent advancements in tag hardware contribute to improved range and power consumption [17], improved tag antenna design [18,19], packaging, and tag orientation. For the performance characterization of the RFID systems under active interference, a test protocol has been presented by Porter et al. [20] and its effectiveness is verified. For RFID readers, a solution is developed for the problem of Tx/Rx isolation at the physical layer [21].

The existence of multiple RFID standards calls for comparative studies that can help the end user in choosing the suitable approach for implementing the communication between interrogator and tag. There has been some previous work discussing the features of the ISO 18000 Part 6C (previously Gen 2) protocol and their implications for asset management [22]. The features discussed are tag memory, security techniques, operational modes, communication methods, global applicability, and other improvements with respect to the EPC class 0 and class 1 generation 1 protocols. The necessity for a uniform organization of user memory and the related omissions in the EPC protocols are addressed by Harmon [23].

Another issue is that proprietary hardware and software are used in specific application domains. These devices must be physically modified or redesigned for adjustments in the specification, introduction of new applications, and modifications to relevant standards. A customizable RFID tag, as shown in this chapter, can handle variations in standards and requirements as they are developed with a significantly shorter time to market than current ad hoc techniques. This RFID design automation flow is demonstrated toward the rapid prototyping of a tag with a representative set of commands from the ISO 18000 Part 7 and Part 6C protocols, among others.

In addition, the ability to rapidly prototype a custom RFID tag provides an economically viable means for individual companies to provide specialized features to differentiate their products from the competition in that particular application. The tools described here give the company the ability to obtain this differentiation without hiring large teams of skilled personnel.

3.3 RFID Physical Layer Design Automation

One of the main components of an RFID system communication is the physical layer protocol employed to encode bits of information. The physical layer features for the bit encoding mechanism vary across various RFID standards. For example, the ISO 18000 Part 7 active tag standard specifies Manchester encoding [24] to transmit encoded data RFID interrogators and tags [2], whereas the ISO 18000 Part 6C standard defines different physical layer features of transactions among readers and tags. Pulse-Interval Encoding (PIE) [25] is used to encode data transmitted from readers to tags, and either FM0 [26] or Miller encoding [27] is used to encode the backscattered data from tags back to readers [25]. Additionally, many other possible physical layer encodings can be considered for RFID communications.

This section describes how the physical layer decoder and encoder blocks can be automatically generated from a high-level specification of the protocol. This design flow

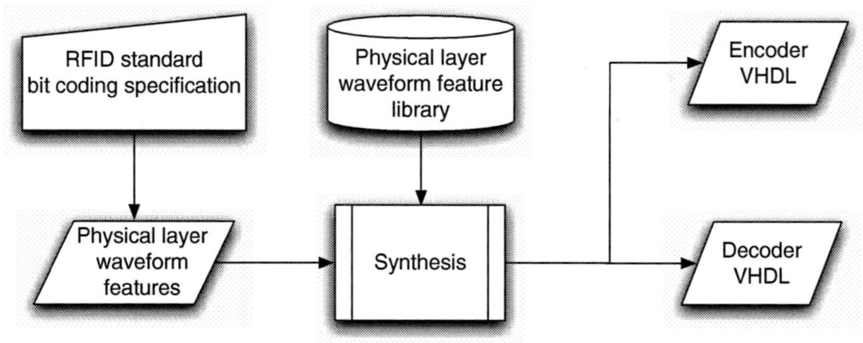

FIGURE 3.3
Generation flow for an RFID data encoder and decoder.

is described in Figure 3.3. The user describes the waveform features of the encoding scheme such as edge transitions, level detection, pulse width detection, etc., from a physical layer specification. The user can then combine one or more wave features to represent bits or groups of bits. The physical layer synthesis tool then automatically generates hardware blocks for encoding and decoding the signal in VHDL. These VHDL descriptions are created from the combination of predefined parameterized hardware libraries and automatically generated hardware blocks for detecting and generating the waveform features in the encoding.

3.3.1 Specification of Waveform

The user describes the features of the encoding scheme using a textual representation. This representation is created from a physical layer specification such as an RFID standard. It may include edge transitions, level detection, pulse width detection, etc. After this file has been created, it becomes the input into the physical layer synthesis tool shown in Figure 3.3. The textual file contains three major segments: (1) declaration of the waveform to encode data values, (2) declaration of the preamble waveform, and (3) transmission characteristics for serial to parallel conversion.

Manchester encoding [24] is a fixed-bit-window encoding scheme specified in ISO 18000 Part 7 for transactions among active RFID readers and tags. The waveforms of encoded bit 0 and bit 1 are illustrated in Figure 3.4a.

The waveform for encoding a bit as either 0 or 1 is described in Figure 3.5. `Sig` represents the nonreturn to zero (NRZ) value of the signal. The keyword `after` describes the delay from the beginning of a bit window. The length of the bit window is specified by a period `T`. Changes in the signal are represented by an `&` with a nonzero `after` parameter. Finally, `A` specifies how accurately each measurement must be as a percentage of `T`. In the example from Figure 3.5, a 0 is represented by a 50% duty cycle clock with a falling edge in the middle of the bit window. The edge must be within 12.5% of the total period, which means 6.75% of the period before or after the expected transition. In this case, the transition occurs at 18 ± 2.43 μs. A 1 is similar except with a rising edge.

Differential Manchester encoding shown in Figure 3.4a is a modification of Manchester encoding used most prominently in token ring networks [28]. At first glance, Manchester and differential Manchester encodings are indistinguishable as they both require a transition in the middle of a bit window for synchronization, and may or may not have a transition at the edge of a window. However, differential Manchester determines its

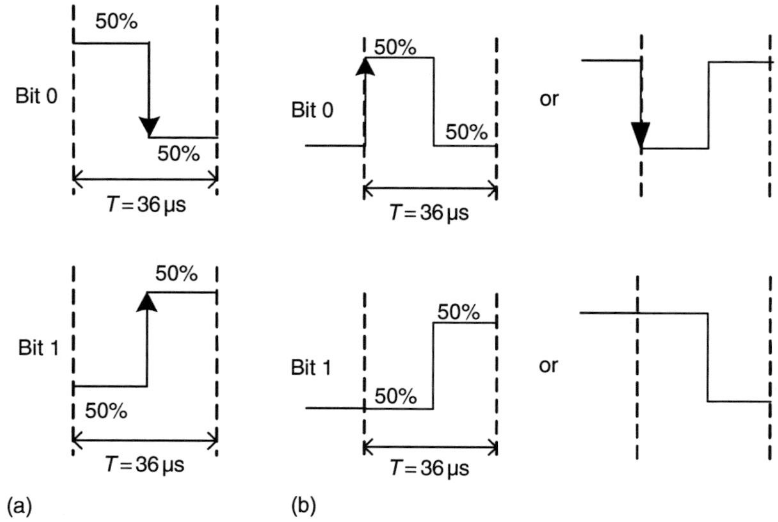

FIGURE 3.4
Continuous waveform for bits 0 and 1 of NRZ encodings. (a) Manchester encoding. (b) Differential Manchester encoding.

value by examining the level between windows. As shown in Figure 3.4a, if there is a transition between windows this encodes a 0 and if there is no transition this encodes a 1.

To support this case, we add an `Lprev` condition to our waveform representation as shown in Figure 3.6. Depending on the previous level, our description describes a level change for a 0 encoded bit and no level change for a 1 encoded bit. The `if` statement determines which waveform to consider based on `Lprev`.

PIE and FM0 encodings are two different encodings used in the ISO 18000 Part 6C standard. These encodings are shown in Figure 3.7. PIE is used for the data transmission and FM0 is used for backscattering the response, through absorption or reflection of the transmitted RF energy.

The physical layer characteristics of PIE are shown in Figure 3.7a. Encodings for both 1 and 0 are based on an active high pulse followed by a fixed width space called `PW`. The length of the pulse determines whether a 1 or 0 is encoded. Thus, the period `T` is different for each value. Unlike previous encodings, there is a large amount of flexibility in the pulse lengths to make a valid PIE-encoded value.

For PIE encoding, in the textual representation shown in Figure 3.8 we introduce the `error` keyword, which allows a transition to take place within a range of times. For example, `7.5 us error 1.6 us` means that the transition could occur anywhere from 5.9 to 9.1 μs. This is different from the `A` which corresponds to jitter associated with the RF transmission. For the PIE encoding described in the ISO 18000 Part 6C standard, there are three possible periods for encoding the values, each described in a separate statement in

```
'0': Sig='1' after 0 us & '0' after 18 us;
T=36 us; A=12.5%;

'1': Sig='1' after 0 us & '1' after 18 us;
T=36 us; A=12.5%;
```

FIGURE 3.5
Textual description of Manchester encoding.

```
'0': if Lprev='0' then Sig='1' after 0 us & '0' after 18 us;
     if Lprev='1' then Sig='0' after 0 us & '1' after 18 us;
T=36 us; A=12.5%;

'1': if Lprev='0' then Sig='0' after 0 us & '1' after 18 us;
     if Lprev='1' then Sig='1' after 0 us & '0' after 18 us;
T=36 us; A=12.5%;
```

FIGURE 3.6
Textual description of differential Manchester encoding.

Figure 3.8. In some cases, the period itself may fall within a range that is described by the `error` keyword.

The physical layer encodings for FM0 are shown in Figure 3.7b. Unlike PIE, FM0 is a fixed period encoding. The data rate of FM0 can be one of several discrete values as specified by the ISO 18000 Part 6C standard, including 160, 256, 320, and 640 kbps. The corresponding bit window period are 6.2, 3.9, 3.1, and 1.5 μs with the error tolerance of 7%, 10%, 10%, and 15%, respectively. To encode a 0, there must be a transition at the middle of a bit window. To encode a 1 there is no transition within a bit window; however, between two adjacent bits, there must be a transition at the edge of the bit window.

The textual representation of FM0 is shown in Figure 3.9. The description for FM0 is similar to how differential Manchester is described except that each value has four different representations corresponding to each of the four data rates. In addition, the encoding for 1 is simpler, as it does not have a transition within the bit window and as such has no `&` in the waveform description.

Modified Miller encoding [29] is an encoding scheme that is often employed in near field communication, or communication of 10 cm or less [30]. Modified Miller encoding has a low pulse at the beginning of the bit window to encode a 0 or the low pulse is delayed by half a period to encode a 1. However, if a 0 is preceded by a 1 the 0 is encoded with no low pulse at all. This is shown in Figure 3.10.

To represent this in our text, we introduce the new field `Vprev` which corresponds the previously encoded value. In the example from Figure 3.11, the modified Miller encoding is

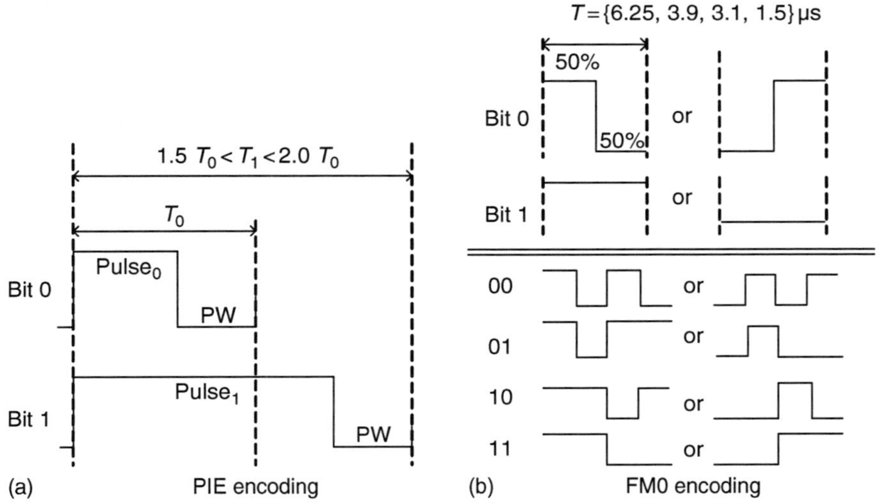

FIGURE 3.7
Continuous waveform for bits 0 and 1 of PIE and FM0 encodings.

```
'0': Sig='1' after 0 us & '0' after 3.55 us error 0.65 us;
T=6.25 us; A=1%;
    Sig='1' after 0 us & '0' after 7.5 us error 1.6 us;
T=12.5 us; A=1%;
    Sig='1' after 0 us & '0' after 15.1 us error 3.3 us;
T=25 us; A=1%;

'1': Sig='1' after 0 us & '0' after 8.3 us error 2.2 us;
T=10.95 us error 1.55 us; A=1%;
    Sig='1' after 0 us & '0' after 19.95 us error 7.75 us;
T=21.85 us error 3.15 us; A=1%;
    Sig='1' after 0 us & '0' after 33.8 us error 9.5 us;
T=43.75 error 6.25 us; A=1%;
```

FIGURE 3.8
Textual description of PIE.

shown for a period of 9.4 μs. To encode a 0 we either see a pulse at the beginning of the window if the current bit was preceded by a 0 or no pulse if preceded by a 1. The encoding 1 does not specify a `Vprev` value.

The preamble in a transmission alerts the system that a transmission packet is beginning. It typically includes a sequence of several pulses that are different from the encoded values in the encoding. The preamble must be matched exactly before any data transmission can

```
'0': if Lprev='0' then Sig='1' after 0 us & '0' after 3.1 us;
     if Lprev='1' then Sig='0' after 0 us & '1' after 3.1 us;
     T=6.2 us; A=7%;

     if Lprev='0' then Sig='1' after 0 us & '0' after 2 us;
     if Lprev='1' then Sig='0' after 0 us & '1' after 2 us;
     T=3.9 us; A=10%;

     if Lprev='0' then Sig='1' after 0 ns & '0' after 1.5 us;
     if Lprev='1' then Sig='0' after 0 ns & '1' after 1.5 us;
     T=3.1 us; A=10%;

     if Lprev='0' then Sig='1' after 0 ns & '0' after 0.7 us;
     if Lprev='1' then Sig='0' after 0 ns & '1' after 0.7 us;
     T=1.5 us; A=15%;

'1': if Lprev='0' then Sig='1' after 0 us;
     if Lprev='1' then Sig='0' after 0 us;
     T=6.2 us; A=7%;

     if Lprev='0' then Sig='1' after 0 us;
     if Lprev='1' then Sig='0' after 0 us;
     T=3.9 us; A=10%;

     if Lprev='0' then Sig='1' after 0 us;
     if Lprev='1' then Sig='0' after 0 us;
     T=3.1 us; A=10%;

     if Lprev='0' then Sig='1' after 0 us;
     if Lprev='1' then Sig='0' after 0 us;
     T=1.5 us; A=15%;
```

FIGURE 3.9
Textual description of FM0 encoding.

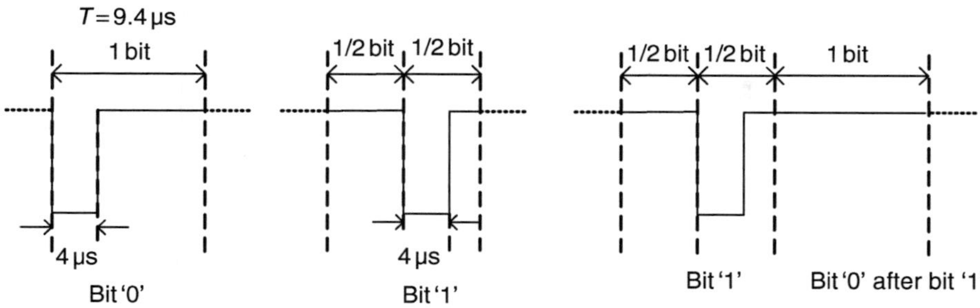

FIGURE 3.10
Continuous waveform for bits '0' and '1' of modified Miller encoding.

commence. Figure 3.12 shows an example preamble using a textual preamble description starting with a 15 μs pulse followed by 5 μs pulses separated by 10 μs. A typical preamble could continue for several more pulses.

RFID standards may also specify transmission protocols between readers and tags. Thus, corresponding transmission characteristics must be declared. For example, the transmission protocol defined in ISO 18000 Part 7 specifies that an RFID reader transmits the least significant bit (LSb) first within a byte and sends the most significant byte (MSB) first within a packet. Each byte is followed by a stop bit within a packet. The transmission order determines the sequence of the serial-to-parallel process for receiving, and the parallel-to-serial sequence for responding.

The textual description of each protocol ends with transmission characteristics. For example, the characteristics for ISO 18000 Part 7 are shown in Figure 3.13. This declares that in the byte the LSb is first, in the packet the MSB is first, each byte contains 9 bits and one of these bits is a stop bit. Thus, a complete physical layer description file contains a waveform description, preamble description, and a transmission characteristics description.

3.3.2 Waveform Features Library

The physical layer waveform feature library in Figure 3.3 is a collection of basic hardware-based components corresponding to various waveform features. For example, this set contains a hardware-based edge detector, sampling counter, sampling registers, serial-to-parallel hardware blocks, first-in-first-out (FIFO) blocks, and other basic blocks as predefined components in the library. These components are programmable and designed to fit different user-defined parameters specified in the textual description from Section 3.3.1. A hardware-based edge detector, for instance, can be used for detecting only a rising edge, only a falling edge or either edge. Similarly, a sampling counter can vary based on different sampling rates corresponding to the different data toggling rates and different duty cycles of bit data in the description.

```
'0': if Vprev='0' then Sig='0' after 0 us & '1' after 4 us;
     if Vprev='1' then Sig='1' after 0 us;
T=9.4 us; A=5%;

'1': Sig='1' after 0 us & '0' after 4.7 us & '1' after 4 us;
T=9.4 us; A=5%
```

FIGURE 3.11
Textual description of modified Miller encoding.

```
preamble: pre='1' after 0 us & '0' after 15 us & '1' after 10 us
          & '0' after 5 us & '1' after 10 us & '0' after 5 us. . . . ;
```

FIGURE 3.12
Preamble textual representation example.

Waveform variables are parameters that are converted from the waveform description file. These parameters describe the physical layer characteristics of an encoding mechanism in a way that the hardware blocks used to detect the waveform can understand. Thus, the synthesis process translates the textual description into the parameters for the feature library.

A bit window period is a simple example of a waveform characteristic. A bit window can contain a transition at a particular time during the period. Similarly, the direction of this transition is another characteristic. From Figure 3.4a, the Manchester encoding requires a 50% duty cycle waveform with a period of 36 μs where the direction determines the value. A 1 has identical features with a 0 except that it is composed of a rising edge at the middle of a bit window as opposed to a falling edge. Because the waveform is a continuous wave, a transition may also occur at the edge of two adjacent bits depending on the values of these windows.

The bit rate for a waveform with a period of 36 μs is 27.7 kbps. Due to the 50% duty cycle, a transition occurs at the middle of a bit window, and logic levels may change between bit windows. Therefore, the data signal toggling rate is 55.4 kbps. The sampling rate (f_s) must be least two times faster than the toggling frequency (f_t) of the target signal: $f_s = 2 \times f_t$. Since the data signal toggling rate is 55.4 kbps (f_t) the minimal sampling rate is 110.8 kbps or four times oversampling. However, the `A` parameter of the description tells us the typical fluctuation that might occur in the signal, which was specified as 12.5%. Thus, a minimum of eight times oversampling should be used for edge detection and synchronization.

An overview of the general waveform detection circuit is shown in Figure 3.14. This circuit contains a preamble detection circuit, an edge detection circuit used for synchronization with the incoming waveform, a timer circuit for signaling the controlling state machine to change states, a sampling register file for converting levels and edges into decoded bit values, and a serial-to-parallel converter for building bytes from the incoming bits. The system clock speed is also variable based on the required sampling by the circuit.

The data transmission begins after a valid preamble is detected by the start signal. The edge detection circuit shown in Figure 3.15 is used to keep in synchronization with the incoming waveform. Every system cycle in which the value changes signals the timer that an edge has occurred.

The timer circuit shown in Figure 3.16 controls the sampling times to check for levels or value changes in the incoming waveform. The timer circuit contains a counter that counts system cycles until its next sampling window. Thus, the hardware library contains parameters for the number of time points n to sample, the number of cycles to count between each sample $N_0, N_1, \ldots, N_{n-1}$ and the bit width of the counter which is $log_2\ \mathtt{max}(N)$.

The timer signals become inputs into an automatically generated finite state machine (FSM) controller circuit. The FSM uses the timer signals and the sampling registers block to determine the actual encoded values. The sampling registers circuit, shown in Figure 3.17,

```
transmission: bitOrder=least; byteOrder=most;
              byteSize=9; stopBits=1;
```

FIGURE 3.13
Serial transmission characteristics.

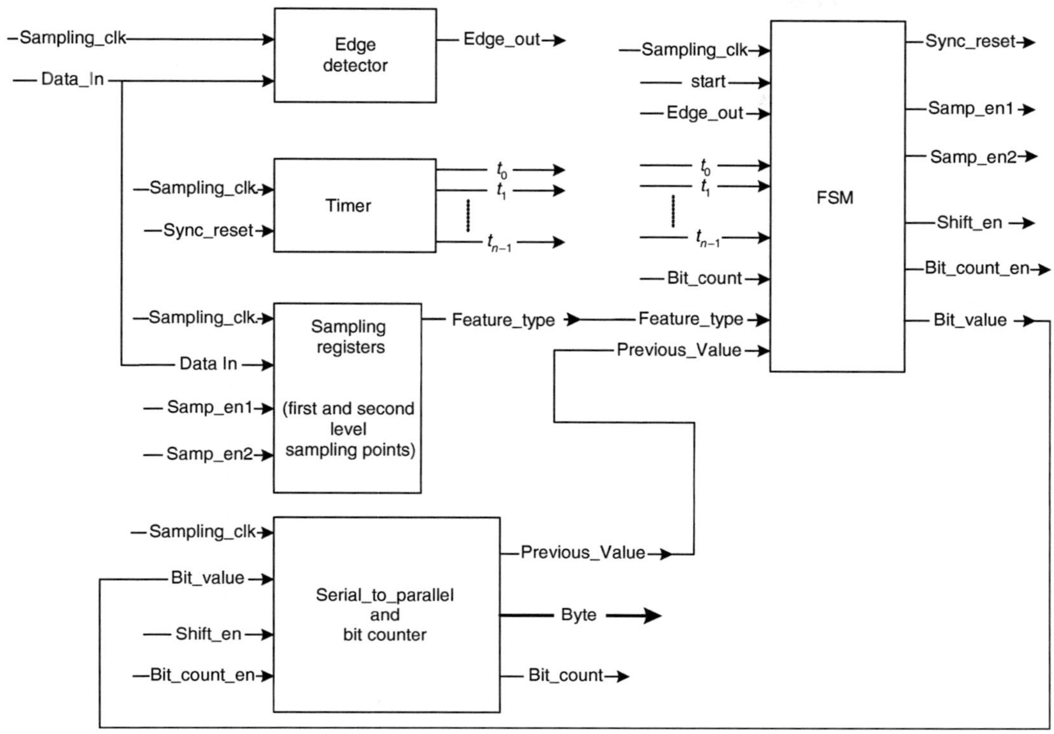

FIGURE 3.14
General waveform detection circuit.

samples two data values when signaled by the controller and reports back whether there has been a rising edge (0 followed by a 1), a falling edge (1 followed by a 0), or a constant level 1 or 0. Based on these values, the controller traverses states to match one of the bit conditions specified in the textual description. It can also base this on one or more previous values seen in the serial stream with the `previous_value` signal.

The controller uses the synchronization signal `Edge_out` to tell the timer when to reset with the `Sync_reset` signal. For cases where the bit window is not fixed, a bit may be determined before all timers have expired as with PIE encoded values of 0. The controller can reset the timer early in these cases to begin looking for the next bit.

Finally, once a bit is determined it is fed into the serial-to-parallel circuit. This circuit, shown in Figure 3.18, has parameters such as the number of bits per byte and the direction of shifting. For MSb first the circuit shifts bits into the register from the left and for LSb first the circuit shifts bits from the right. When buffering the whole packet a similar shifting technique is used for most and least significant bytes (MSB and LSB).

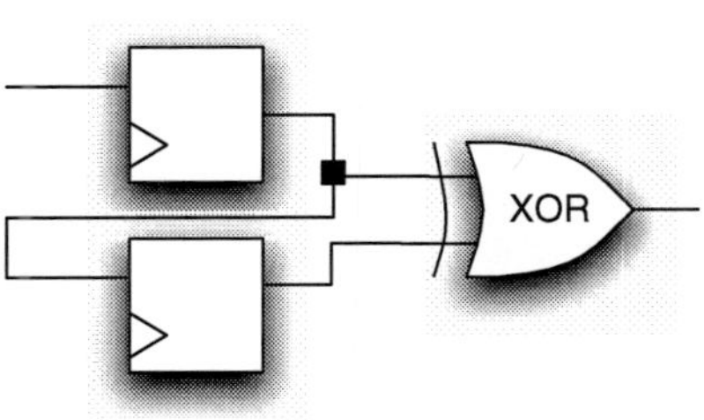

FIGURE 3.15
Edge detection circuit.

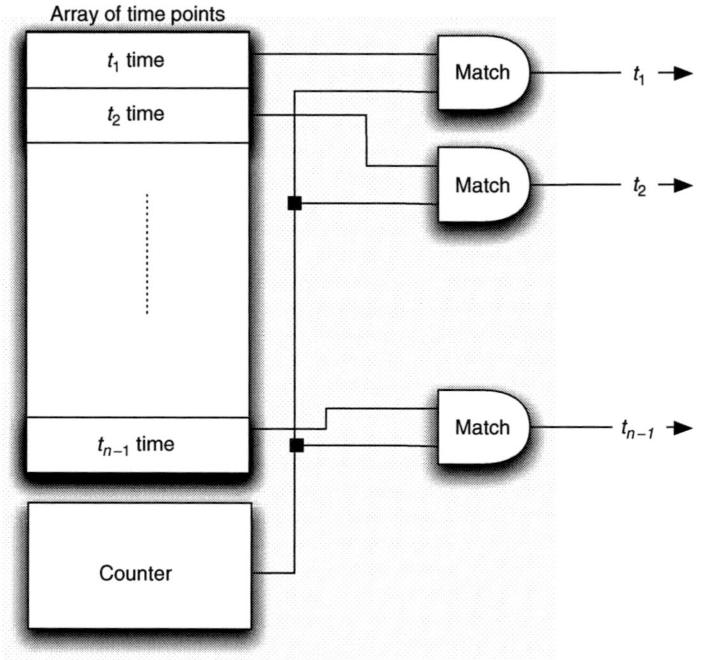

FIGURE 3.16
Timer circuit schematic.

3.3.3 Physical Layer Synthesis and VHDL Generation

The synthesis process from the textual description is primarily based on discovering sampling points based on the waveform properties. For example, consider the modified Miller encoding in Figure 3.10. The basic waveforms for 0 and 1 with an inverse pulse indicate sampling both during the pulse and outside the pulse. The synthesis process first selects sampling points in the center of a level region, thus at 2 and 6.7 μs for a 0 and at 2.7, 6.7, and 9.05 μs for a 1. First, the 9.05 μs is discarded, because the signal is high for all three descriptions. 6.7 μs is retained directly, because it matches both the 1 and 0 directly. The 2 and 2.7 μs values are determined to represent the same sampling window as they are within stable regions for both the 1 and 0 waveforms. As a result, any value within the range of 2–2.7 may be selected for a sampling point.

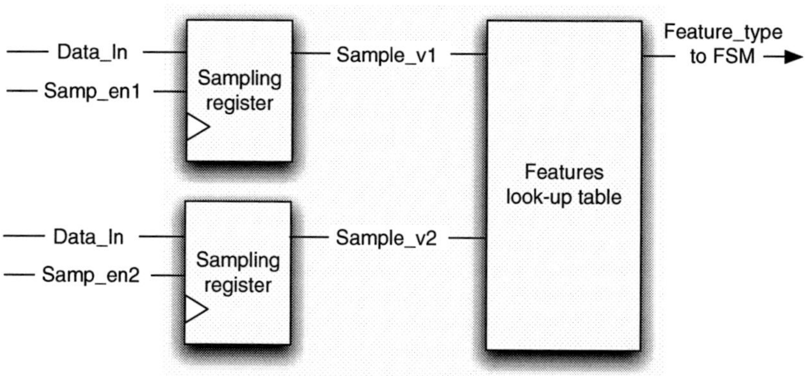

FIGURE 3.17
Sampling registers circuit schematic.

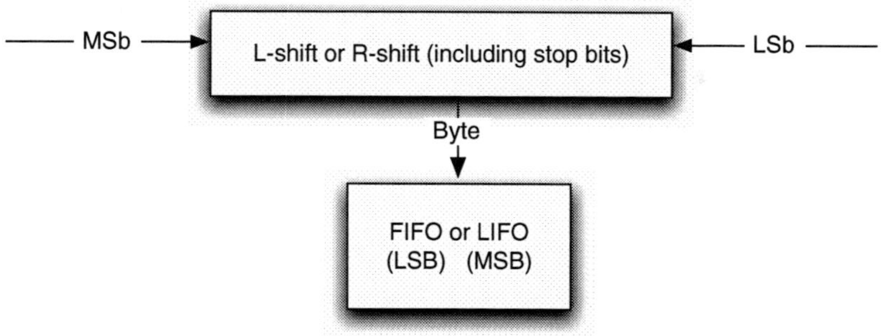

FIGURE 3.18
Serial-to-parallel circuit schematic.

The controller FSM is generated to detect different sequences of features for each encoded bit in a similar manner as a numeric sequence detector. For example, if there is a rising edge between the samples, it is a 0, a falling edge indicates a 1, and a constant level high is a 0. Both versions of 0 can be checked against the previous value.

For a PIE encoding (Figure 3.7a) there are three sampling points to consider. This is demonstrated for the first description for each 0 and 1 from Figure 3.8. According to the description, a falling edge occurs between 2.9 and 4.2 μs making this an invalid detection region. Thus for a 0 the sampling points are 1.45 and 5.225 μs. Because the period is not fixed for a 1 and the invalid detection region and period completion time overlap, the system can move into active sampling mode. Because a falling edge has not occurred by 4.2 μs, one must occur between 6.1 and 10.5 μs. Therefore a timer indicates when 6.1 μs have elapsed, and then the FSM looks for an edge before a timer indicates 10.5 μs. Finally, on seeing a second edge, the FSM begins a new bit window.

The process for encoding these values into the appropriate encoding is a much simpler subset of the detection process. The process requires a parallel-to-serial block complementary to the serial-to-parallel block. Each bit waveform is generated with a very simple controller FSM that uses timers to traverse states and each state outputs one particular level. This is a fairly straightforward conversion process from the textual representation.

The VHDL code generation is the final phase of the synthesis flow. The structure of the block is shown in Figure 3.14. Several of the libraries included are parameterized with VHDL `generic` constructs for specified parameters including the timer and serial-to-parallel blocks. The FSM controller is entirely generated by the synthesis engine using a `generic` for the number of timer signals to include.

3.3.4 Results

Hardware was generated for the five encodings, Manchester, differential Manchester, PIE, FM0, and modified Miller. The results for the decoders are shown in Table 3.1. Encoders

TABLE 3.1
Decoder Hardware Block Statistics for Five Different Encodings

Encoding	Clock (MHz)	Area (μm^2)	Power (μW)
Manchester	0.5	3780	2.9
Differential Manchester	0.5	3520	2.7
Pulse interval (PIE)	6	9264	84.8
FM0	6	6052	44.6
Modified Miller	2	3780	11.9

have been omitted because their hardware is trivial compared with the decoders for each type of encoding.

The Manchester and differential Manchester decoders have similar architectures. The size and power consumption are within 10% of each other requiring approximately 3500 μm^2 in area and 3 μW in power. These decoders also require relatively low sampling rates of 500 kHz. The modified Miller encoding requires a similar area as Manchester encoding but requires about 12 μW of power likely because of the four times higher sampling frequency. FM0 and PIE encoding requires significantly higher area (2–3 times more) than the other encodings. Due to the increased sampling rate and complexity these encodings require significantly higher power consumptions of 45 and 85 μW.

3.4 RFID Controller Design Automation

The RFID communication system consists of a transponder or tag and an interrogator or reader. The format for exchanges between the interrogator and the transponder is a set of commands or primitives that requests that the transponder perform a set of actions. The specifications of these commands vary from one standard to another. The flow of the RFID compiler is specified in Figure 3.19. This particular compiler can accept virtually any set of commands as input and target a microprocessor or hardware device to provide the RFID tag controller functionality.

Figure 3.20 shows an example of the ISO 18000 Part 6C protocol for inventory and access of a single RFID tag. In step 1, the interrogator (reader) issues a query. In step 2, the tag responds with a randomly generated 16 bit number. The reader acknowledges by returning a random 16 bit number in step 3. This selects only one tag with which to communicate. This random number is designed to avoid contention between multiple tags and to ensure that the reader is communicating with only a single tag. In step 4, only the tag that issued the matching random number responds with its PC/EPC, essentially its identifier. In step 5, the reader issues a transaction request with the same random number. The tag responds with a transaction handle in step 6. In step 7, the actual transaction is issued with the handle as a parameter. Finally, the tag responds to the transaction in step 8.

3.4.1 Specification of Macros

The communication transactions between the RFID reader and the tag can be broken down into a series of RFID primitives. To automate the generation of the tag controller for the prototype, these primitives are implemented as simple, assembly-like instructions called RFID macros. For example, the RFID macros required for executing the *Write* command of the ISO 18000 Part 6C standard are shown here. The format of the respective fields

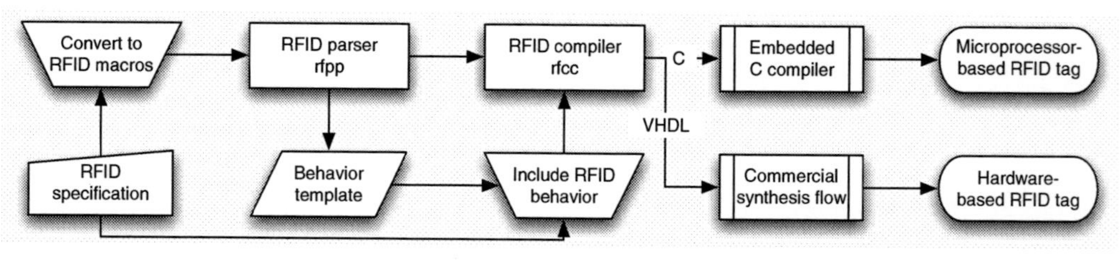

FIGURE 3.19
Specification methodology and compilation flow. (From Jones, A.K., Hoare, R., Dontharaju, S., Tung, S., Sprang, R., Fazekas, J., Cain, J.T., and Mickle, M.H., *Proceedings of FCCM*, 165, 2006. With permission.)

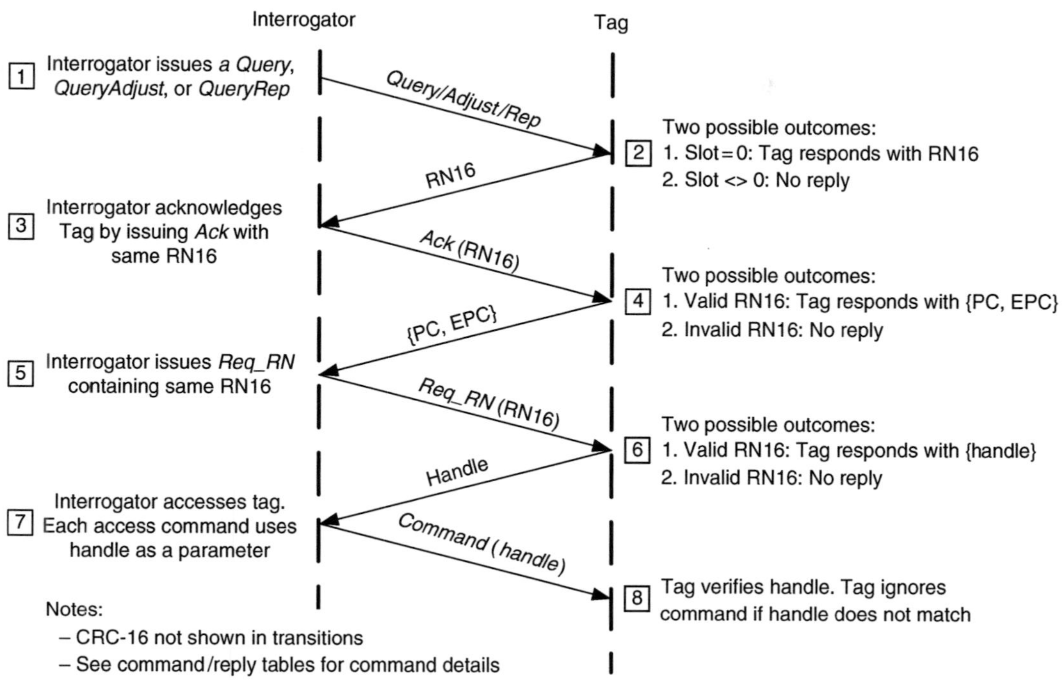

FIGURE 3.20
Example transaction for ISO 18000 Part 6C.

of each necessary primitive and its corresponding response are both illustrated in Figure 3.21.

The command code of each RFID primitive is a unique field or opcode that serves as the identifier. Each of the RFID primitives also contains a subset of fields with varying lengths providing positions for data present in a command, as can be inferred from Figure 3.21. Similarly, the tag response to each RFID primitive has fields of varying lengths.

Each RFID macro description contains a relatively short character string corresponding to the specific name of the primitive, a number indicating how many bits are used to represent the opcode of this particular primitive as well as the distinct number corresponding to the value of the opcode. Additionally, a set of operands that correspond to the primitive is included. On the next line, a set of operands is included that corresponds to the standard response.

Figure 3.22 shows an example RFID macros file containing the basic primitives of the ISO 18000 Part 6C standard for initiating a transaction as well as the *Write* primitive. The macros file has a `declarations` section and a `main` section. The `declarations` section allows the user to predeclare the lengths of all of the corresponding fields that occur in each of the primitives and responses. In the section identified as `main`, the primitives and their specific responses are defined in terms of their fields.

The specific fields can be easily described, as illustrated in Figure 3.22. This provides the user with the ability to adopt any level of granularity to manipulate the primitives and their corresponding responses. For example, in the macros illustrated in the figure, the string denoting the *Write* command is write. The decimal value of the command code for this specific command is 195, stored using 8 bit.

Query command

QCmd	DR	M	TRext	Sel	Session	Target	Q	CRC-5
4 bit	1 bit	2 bit	1 bit	2 bit	2 bit	1 bit	4 bit	5 bit

Response

RN16
16 bit

Ack command

AC md	RN
2 bit	16 bit

Response

PC	EPC	CRC-16
16 bit	96 bit	16 bit

Req_RN command

Cmd	RN	CRC-16
8 bit	16 bit	16 bit

Response

RN	CRC-16
16 bit	16 bit

Write command

Cmd	Mem Bank	Word Ptr	Data	RN	CRC-16
8 bit	2 bit	8 bit	16 bit	16 bit	16 bit

Response

Header	RN	CRC-16
1 bit	16 bit	16 bit

FIGURE 3.21
Selected primitives and response formats from ISO 18000 Part 6C.

3.4.2 RFID Controller Behavior

Communication from the RFID interrogator (reader) is accomplished by transmitting the primitive to the RFID tag using a standard air interface. The tag responds by changing the current state and transmitting a designated response message to the interrogator. The user has the capability of specifying the tag behavior in ANSI-C.

To simplify the user interaction, the RFID parser generates a template for the response behavior indicating where the user must specify such custom behavior. Any C language constructs (conditionals, loops, etc.) can be added (or left unchanged) by the user to check the values of the fields of the incoming primitive and to specify the values of the fields of the response. The template generated for the *Query* command is shown in Figure 3.23. A file containing similar templates for all the macros that were included in the macros specification file will be generated for the user.

The details involving size and field position in the command of the interrogator and the corresponding response packet are handled by the compiler. Therefore, complexities encountered in unpacking the command and subsequently packing the response can be abstracted from the user, as shown in Figure 3.24. However, the ability of the user to manipulate each of the individual fields in the response is intact. Therefore, the response customization along with the corresponding state changes can increase in complexity with ease.

```
declarations
DR(1)
M(2)
TRext(1)
Sel(2)
Session(2)
Target(1)
Q(4)
CRC-5(5)
RN16(16)
RN(16)
PC(16)
EPC(16)
CRC-16(16)
MemBank(2)
WordPtr(8)
Data(16)
Header(1)

main
query
(4,4)      DR        M          TRext   Sel   Session   Target   Q   CRC-5
                RN16

ack(2,1)        RN
                PC         EPC        CRC-16

req_rn(8,193)   RN         CRC-16
                RN         CRC-16

write(8,195)    MemBank    WordPtr    Data    RN    CRC-16
                RN         CRC-16
```

FIGURE 3.22
Macros specification.

3.4.3 Compiler-Generated RFID Tag Program

The code generation is the final compiler phase determined by the tag behavior and the input macros specification. Code generation can be in the form of ANSI-C for general-purpose microprocessor controllers or VHDL in the case of hardware controllers. The decode instructions generated by the compiler identify the received RFID primitive. For each incoming command, routines are generated by the compiler that unpack the command generating the fields that it is expected to contain. The corresponding behavior is attached to each field, and the corresponding routines for packing the response are then generated.

By virtue of the fact that C is a significant and universally known language compared with hardware description languages, the primitive behaviors are specified in C in the case of a target described with VHDL. Therefore, the C code must be converted to a readily synthesizable hardware code.

```
RN16 =
inventoryFlag =
current_state =
. . .
```

FIGURE 3.23
Template generated for *Query* command.

```
if (current_state ! =KILLED) {
  if (((current_state = = ACKNOWLEDGED) || (current_state = = OPEN)
     || (current_state = = SECURED)) && ((sel_var = = sel)
     && (target_var = = target))) {
    if (session = = last_session) {
      if (inventoryFlag = = 'A')
        inventoryFlag='B';
      else if (inventoryFlag = = 'B')
        inventoryFlag='A' ;
    } else {
      slot_counter=rand ((1 << Q) - 1) ;
      if (slot_counter ! = 0)
        current_state=ARBITRATE;
      else
        current_state=REPLY;
      if (current_state = = REPLY)
        RN16=slot_counter;
    }
  }
}
```

FIGURE 3.24
Tag behavior for *Query* command.

During the hardware conversion, the C code is converted into a control and data flow graph (CDFG). Compilers commonly use CDFGs to perform optimizations and transformations. Typically, behavioral synthesis tools will also use CDFGs as an internal representation [31]. In many cases, the control dependencies present in a CDFG create cycle boundaries during high-level synthesis.

In contrast, in the RFID compiler the CDFG is transformed into an entirely combinational representation by the SuperCISC compiler. The result is a super data flow graph (SDFG). The SuperCISC compiler [32,33] takes advantage of well-known compiler transformations including loop unrolling, function inlining, and *hardware predication* to convert each control dependency into a data dependency creating a combinational representation. The SDFG for the *Query* command is shown in Figure 3.25. The need for many potentially high-power consuming sequential constructs such as registers and clock trees are removed by this technique. Thus, the resulting SDFG-based hardware implementations are extremely power efficient [34].

The RFID compiler contains both power and area optimization routines. The power optimizations are described in detail by Jones et al. [34]. The area optimizations attempt to discover the maximum precision used by signals in the design and propagate that information through the design to reduce the size of storage elements and synthesized functional units. The automatically generated design is expected to be less optimal than a hand design, but provides a reasonable estimate for a system designer to compare different protocols and different implementation targets.

3.4.4 Results

The RFID design automation flow has been used to implement RFID primitives from a variety of different standards such as ISO 18000 Part 7, ANSI NCITS 256, ISO 18000 Part 6C, and ISO 18185 Part 1. The most critical metrics for success with the resulting implementations are area and power of the resulting tag controller. Performance is rarely an issue because of the limited transmission speeds of the RF protocols.

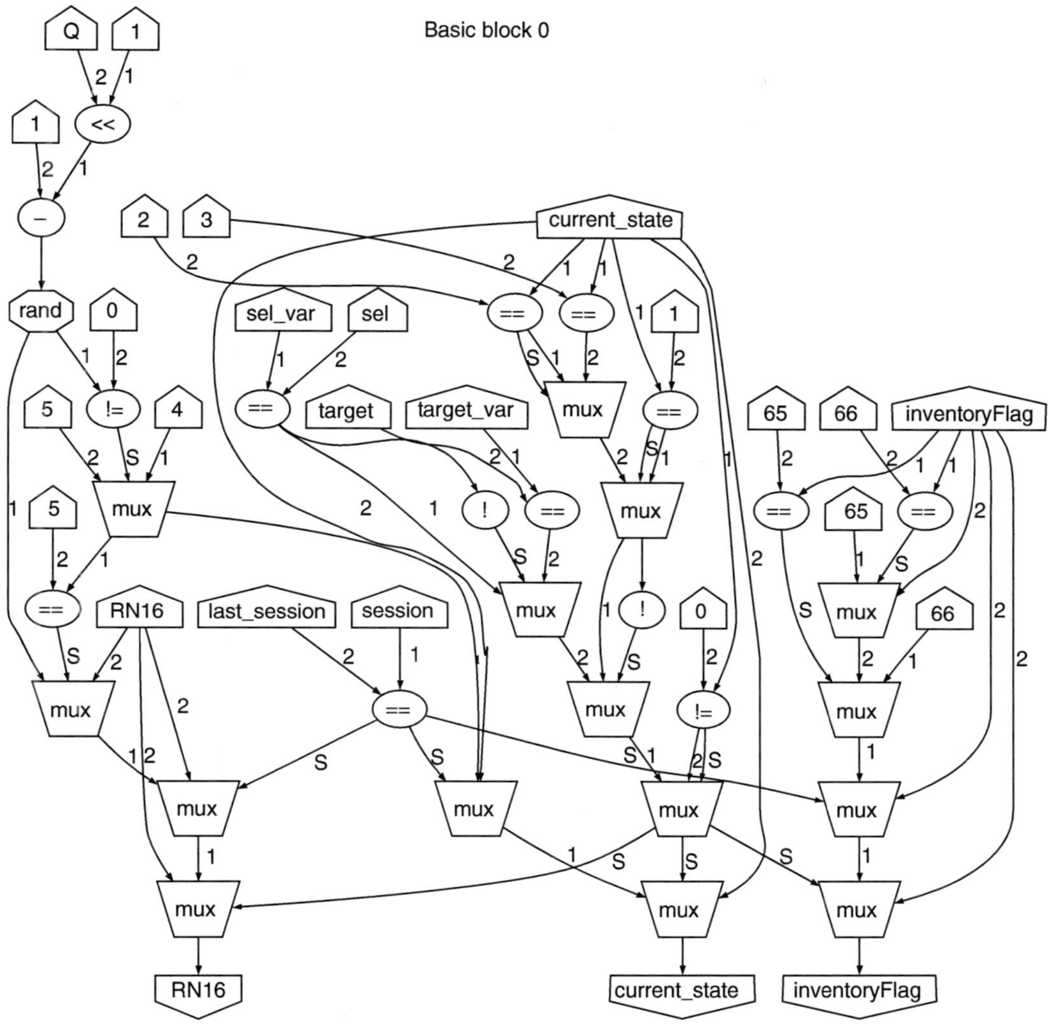

FIGURE 3.25
SDFG for the *Query* command.

For RFID tags, area is the primary concern as the size of the circuit results in a direct per part cost impact. In many cases, the memory components of the controllers dominate the area impact. In some cases this is dominated by the firmware for a microprocessor-based solution and in others by the memory used for storage in memory-enabled RFID tags.

Power optimization is important in RFID systems as the power supplied to the tags is fixed and battery drain needs to be limited. Because active systems are designed for extremely low-cost large-scale applications, frequent replacement of batteries is not feasible. Although the transceiver power dominates the power consumption in active tags, the controller power consumption is also a significant consideration.

3.4.4.1 RFID Compiler Prototype Targets

The prototype microprocessor-based system is composed of an Altera Apex FPGA prototyping board used for logic buffering the packets from the air interface, 16 bit EISC microprocessor development board from AD Chips [35], and a custom development board created at the University of Pittsburgh for the active air interface.

The prototype FPGA-based system uses a Spartan 3 FPGA development board from an Opal Kelly for the controller logic and any buffering logic that is required for the tag. For this prototype, the air interface is an off-the-shelf ultra high frequency (UHF) transceiver connected to the FPGA board through a custom board created at the University of Pittsburgh. This board does the analog-to-digital conversion.

3.4.4.2 ISO 18000 Part 7 RFID Standard

The ISO 18000 Part 7 standard [2] is an international standard that defines the air interface for RFID devices used in item management applications. The standard defines the forward and return link parameters for an active RFID air interface at 433 MHz and the communications protocol used.

Figure 3.26 shows the interrogator-to-tag command format for the *Collection* command. The command contains a `command code` to signal the tag what type of command is issued. Additionally, the command contains a `CRC` to ensure the command packet is properly formed. The remainder of the packet contains particular fields appropriate to the command. Similarly, the tag response includes the `command code`, `CRC`, and other data fields. The tag response also includes a `tag status` field, which consists of nested fields such as `acknowledge`, `tag type`, `battery`, etc.

Figure 3.27 shows an example RFID macros file containing the *Collection* primitive. The corresponding SDFG representation is shown in Figure 3.28. Thus, RFID compiler has been used to implement all the commands from ISO 18000 Part 7 for microprocessor-based as well as custom hardware-based tags.

3.4.4.2.1 Microprocessor-Based Tag

The RFID compiler was used to generate three programs: A with 24 primitives, B with 12 primitives, and C with 4 primitives. Experiments were conducted by executing one primitive of Program A, one primitive of Program B, and one primitive of Program C [36–38].

The sim-panalyzer [39] and XTREM [40] tools were used to estimate the power dissipation of the microprocessor-based tag for the ARM-based cores. Sim-panalyzer is a cycle-accurate, architecture-level power simulator built on top of the SimpleScalar simulator. XTREM is a SimpleScalar-based power and performance simulator tailored for Intel XScale microarchitecture. SimpleScalar's sim-profile was used to obtain ARM instruction and

Collection command

Prefix	Type	Owner Id	Interrogator Id	Command code	Size	Reserved	CRC
8 bit	8 bit	24 bit	16 bit	8 bit	16 bit	8 bit	16 bit

Response

Tag status	Message length	Interrogator Id	Tag Id	Command code	CRC
16 bit	1 bit	2 bit	1bit	2 bit	2 bit

Tag status

Modefield	Reserved	Acknowledge	Reserved	Tag type	Reserved	User Id	Battery
4 bit	3 bit	1 bit	2 bit	3 bit	1 bit	1 bit	1 bit

FIGURE 3.26
Collection command and response format from ISO 18000 Part 7.

```
declarations
prefix(8)
type(8)
ownerid(24)
interid(16)
tagid(32)
comcode(8)
siz(16)
res(8)
crc(16)
tagstatus(16)[
   modefield(4)
   reserved1(3)
   acknowledge(4)
   reserved2(2)
   tagtype(3)
   reserved3(1)
   userid(1)
   battery(1)
   ]
mesglen(8)

main
icol(16) prefix    type    ownerid  interid  comcode  siz  res  crc
         tagstatus mesglen interid  tagid    ownerid  crc
```

FIGURE 3.27
Macros specification for *Collection* from ISO 18000 Part 7.

instruction class profiles for tag software. Because an instruction set simulator was not available for the EISC, the application was run on the development board and the execution time was measured by setting a pin output from low to high on each iteration. Thus, the duration was measured using an oscilloscope. The energy consumed by the EISC was

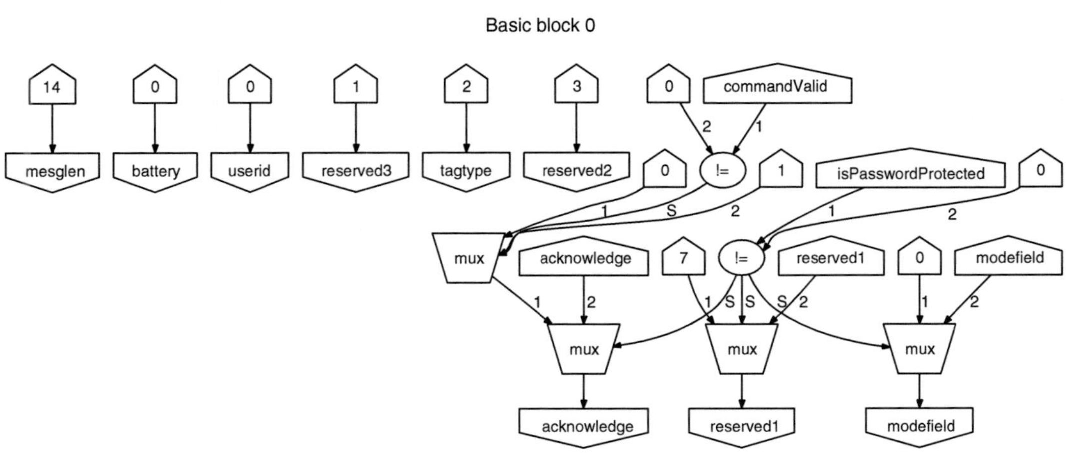

FIGURE 3.28
Super data flow graph for the example ISO 18000–7 command. (From Jones, A.K., Hoare, R., Dontharaju, S., Tung, S., Sprang, R., Fazekas, J., Cain, J.T., and Mickle, M.H., *Proceedings of the 43rd Design Automation Conference (DAC)*, 131, ACM, 2006. With permission.)

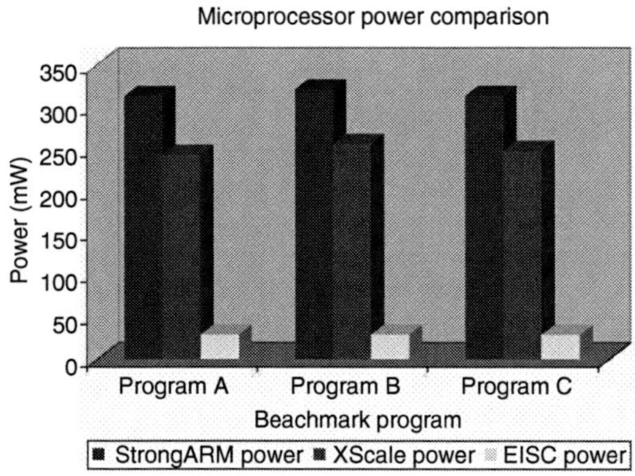

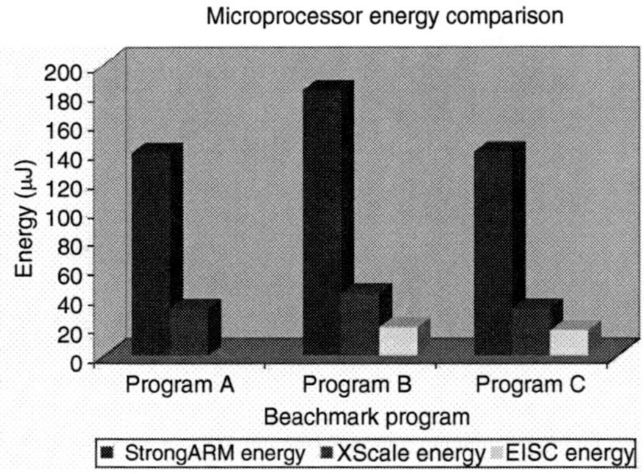

FIGURE 3.29
Power and energy comparison of tags with different microprocessor cores. (a) Power. (b) Energy.

estimated* based on a static power estimate from ADC [35], which should be within about 10% accuracy of an instruction level power estimation approach [41].

Figure 3.29 shows the power consumption (Figure 3.29a) and energy consumption (Figure 3.29b) of the tag programs on the StrongArm, XScale, and EISC processors. Both ARM-based processors operate in the 250–400 mW range, whereas the XScale uses significantly less energy. The EISC processor uses an order of magnitude less power, but operates much slower. However, the energy consumed is still less than half of XScale. The energy of executing Program A on the EISC could not be calculated because of the program size exceeding the program memory of the EISC board. This is probably reflective of a real area constraint in an RFID tag, as the chip size would be increased because of memory required for the firmware. Thus, a program size optimization might be necessary for actual implementation.

It can be seen that the power consumption of XScale is less than that of StrongArm though they both implement the ARM Instruction Set Architecture. This is because the XScale family of microprocessors uses deep pipelines and microarchitectural optimizations for high performance [42]. Further, the reduced power consumption and greater clock speed of XScale result in its far lower energy consumption.

3.4.4.2.2 *Custom Hardware-Based Tag*

Using the RFID compiler, the number of primitives for the custom hardware-based compiler were scaled up to 40 and implemented in 3 hardware targets, a Xilinx Coolrunner II CPLD, an Actel Fusion FPGA, and custom cell-based ASIC hardware at 0.16 μm. Initial results appear in Jones et al. [37]. It should be noted that to fit all 40 primitives, the largest Coolrunner II device available (XC2C512) had to be used, but all the primitives comfortably fit into the smallest Fusion device available (AFS090).

The area and power results for the implementations are presented in Tables 3.2 and 3.3, respectively. As previously mentioned, the area required by the custom hardware for ASIC implementation directly impacts the cost. As shown in Table 3.2, the ASIC controller area is quite low, less than 4300 cells for a 40 primitive controller, which is smaller than the 10s of kilobytes of memory required in software-based designs. The three implementations provide levels of power consumption, as shown in Table 3.3. The CPLD power hovers

* Energy calculation is static power consumption multiplied by measured execution time.

TABLE 3.2

Area for Implementing the Primitive Logic on a Coolrunner II XC2C512, an Actel Fusion AFS090, and 0.16 μm ASIC

Prims	2	4	6	8	10	12	15	20	24	30	35	40
Xilinx Coolrunner II XC2C512												
MCs	332	335	338	340	350	366	426	447	447	447	449	447
% Used	66	66	67	67	69	72	84	88	88	88	88	88
PTs	444	477	514	506	477	552	772	953	993	1106	1181	1213
% Used	25	27	29	29	27	31	44	54	56	62	66	68
Regs	262	267	271	271	283	307	422	443	443	443	443	443
% Used	52	53	53	53	56	60	83	87	87	87	87	87
FBIs	360	379	408	391	338	396	611	767	801	870	900	914
% Used	29	30	32	31	27	31	48	60	63	68	71	72
Actel Fusion AFS090												
VTs	256	258	265	292	300	317	329	371	380	411	434	442
% Used	11.1	11.2	11.5	12.7	13.0	13.8	14.3	16.1	16.5	17.8	18.8	19.2
0.16 μm ASIC												
Cells	3809	3836	3841	3835	3859	3961	3990	4140	4170	4235	4264	4298
Area	1.092	1.097	1.098	1.097	1.105	1.116	1.121	1.158	1.161	1.174	1.182	1.187

Note: Macrocells (MCs), product terms (PTs), registers (Regs), function block inputs (FBIs), VersaTiles (VTs). ASIC area is 100 μm^2.

around 1 mW when active, whereas the FPGA implementation stays between 6 and 9 mW when active. When idle, the quiescent power of the CPLD is 0.05 mW and the FPGA has standby/sleep modes dropping the power to 0.03 mW [43]. The direct ASIC implementation drops the power consumed to 0.065 mW when operating and to 0.0004 mW when idle.

3.4.4.3 ISO 18000 Part 6C RFID Standard

ISO 18000 Part 6C standard [25] is a recent amendment to ISO 18000 Part 6 that describes the RFID air interface for devices operating at 915 MHz and the associated communication protocols used. Part 6C extends the existing Part 6 standard, which previously contained type A and B devices with a type C modeled after the EPCGlobal Class 1 Generation 2 specification.

An interrogator manages tag populations using three basic operations, select, inventory, and access. The *Select* command is applied successively to pick a particular tag population based on user-specific criteria, enabling union-, intersection-, and negation-based tag

TABLE 3.3

Power in MilliWatts for Implementing the Primitive Logic on a Coolrunner II XC2C512, an Actel Fusion AFS090, and 0.16 μm ASIC

Prims	2	4	6	8	10	12	15	20	24	30	35	40
Xilinx Coolrunner II XC2C512: Quiescent Power 0.05 mW												
Total	1.06	1.06	1.06	1.07	1.06	1.06	1.24	1.24	1.24	1.24	1.24	1.24
Actel Fusion AFS090: Quiescent Power 3 mW												
Total	6.76	6.82	6.87	7.40	7.41	7.47	7.47	8.23	8.27	8.35	8.36	8.48
0.16 μm ASIC: Quiescent Power <0.4 μW												
Total	0.063	0.063	0.063	0.063	0.063	0.063	0.064	0.064	0.064	0.064	0.064	0.065

partitioning. An interrogator begins an inventory round by transmitting a *Query* command in one of four sessions. One or more tags may reply to this. The interrogator then detects a single tag reply and requests more information from the tag. The access operation allows the issuing of commands that read from or write to a tag once the tag is uniquely identified. ISO 18000 Part 6C tags implement features such as accessing and killing passwords, checking the electronic product code (EPC), CRC checking, manipulating the slot counter, pseudorandom number generation, etc. The states and keys of the target device are used to facilitate tag singulation, collision arbitration, security encoding, etc.

The communication primitives of ISO 18000 Part 6C standard are significantly different and more complex than the ISO 18000 Part 7 standard. The complexity of the Part 6C standard makes the design of these tags extremely time consuming and challenging for reducing power consumption and silicon area. Dontharaju et al. [44] examines various features of the ISO Part 6C standard and compares it with the ISO 18000 Part 7 standard for active tags for the purpose of evaluating generic interrogator/tag protocol complexity.

An example transaction of this protocol is shown in Figure 3.30. The output shown in this figure is generated from a special piece of equipment housed in the University of Pittsburgh RFID Center of Excellence called a real-time spectrum analyzer (RTSA). The example tag uses passive RFID technology requiring the RF energy generated by the reader to be used to power the tag, and to be used for a backscatter-based response. Backscattering uses the reflection of RF energy provided by an external source (in this case the reader) to transmit information. The backscattering device (in this case the tag) either absorbs or reflects the energy of the incoming RF to generate low and high values in the backscattered response.

Section 3.4.1 describes some of the commands such as *Query*, *Ack*, *Req_rn*, and *Write*. The corresponding macros representation is shown in Figure 3.22. Figures 3.23 through 3.25 show the synthesis process of the *Query* primitive.

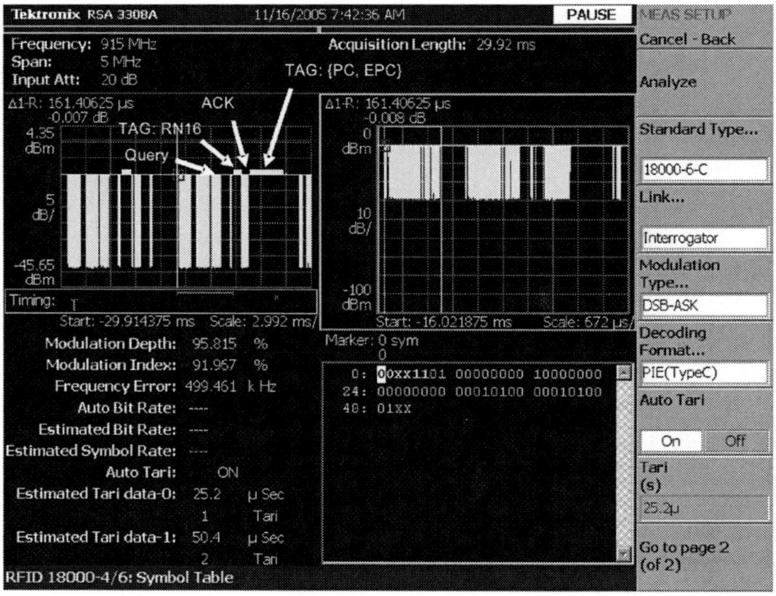

FIGURE 3.30
Real-time spectrum analyzer output of an example ISO 18000 Part 6C transaction.

TABLE 3.4

Area for Implementing the Gen-2 Primitive Logic on a 0.16 μm ASIC

Prims	1	2	3	4	5
Manual	1.1642	1.1933	1.2288	1.2313	1.3212
Automated	1.1326	1.2159	1.2842	1.2942	1.4606
% Increase with automation	−2.71	1.89	4.51	5.11	10.55

Note: ASIC area is 100 μm^2.

3.4.4.3.1 *Custom Hardware-Based Tag*

Using the RFID compiler, up to five inventory commands of the ISO 18000 Part 6C standard were implemented with 0.16 μm custom cell-based ASIC hardware and a Spartan 3 FPGA. To evaluate the effectiveness of the automated approach in providing a rapid prototype and accurate estimate of resource requirements of the RFID system, the areas of the automated tag designs generated by the RFID Compiler have been compared with the areas of our own manual tag designs for the above targets. The tools used for estimating area are design compiler and precision synthesis.

Table 3.4 shows the total area of ISO 18000 Part 6C tag designs for a 0.16 μm ASIC. Table 3.5 shows the resource utilization of Gen-2 tag designs for a Spartan 3 FPGA. The FPGA resource utilization is almost the same for all the designs and actually decreases slightly with the automated approach. For the ASIC implementation of five primitives, there is a nominal increase in area of up to 10.55%. We note that there is a trend that as primitives are added the area increase percentage rises. This is in part due to how Design Compiler does resource sharing. We noticed that this did not occur with other synthesis tools for FPGAs.

TABLE 3.5

Resource Utilization for Implementing the Gen-2 Primitive Logic on a Spartan 3 FPGA

Prims	1	2	3	4	5
IOs					
Manual	419	419	419	419	419
Automated	419	419	419	419	419
% Increase with automation	0	0	0	0	0
Global Buffers					
Manual	2	2	2	2	2
Automated	2	2	2	2	2
% Increase with automation	0	0	0	0	0
Function Generators					
Manual	720	757	787	789	817
Automated	713	742	814	814	825
% Increase with automation	−0.97	−1.98	3.43	3.17	0.98
CLB Slices					
Manual	569	572	580	580	588
Automated	559	563	571	571	571
% Increase with automation	−1.76	−1.57	−1.55	−1.55	−2.89
Dffs or Latches					
Manual	1138	1143	1159	1159	1176
Automated	1118	1125	1141	1141	1141
% Increase with automation	−1.76	−1.57	−1.55	−1.55	−2.98

TABLE 3.6

Power and Energy Results for Implementing *Query*, *Collection*, and 10 ISO Part 7 Primitives (Inclusive of *Collection*) as a 0.16 μm ASIC

Primitives	Dynamic Power (mW)	Area (100 μm^2)
Query	0.06752	1.1642
Collection	0.06308	1.0944
10 primitives	0.06495	1.1232

Source: Dontharaju, S., Tung, S., Jones, A.K., Mats, L., Panuski, J., Cain, J.T., and Mickle, M.H., *IEEE Communications Magazine*, 1, 4, 2007. With permission.

Note: ASIC area is in 100 μm^2. Dynamic power is in milliWatts. Quiescent power <0.4 μW.

Design Compiler does allow resource sharing through use of specialized controls, which provide an opportunity to reduce this overhead.

3.4.4.3.2 Comparison with ISO 18000 Part 7

To understand and compare the complexity of different standards such as the ISO 18000 Part 7 and Part 6C, the RFID compiler has been used to implement commands from both of them. A representative command, *Query*, was selected from ISO 18000 Part 6C and has been implemented in hardware and synthesized for the ASIC target. Similarly, the *Collection* command, which realizes similar functionality from ISO 18000 Part 7 standard, was implemented in hardware and targeted to the same ASIC process. Table 3.6 shows the power and area results for implementing these two commands. The *Query* command is much larger and higher power consuming than the *Collection* command. We also compared the *Query* command and the *Collection* command with nine additional primitives from the ISO 18000 Part 7 standard. As shown in Table 3.6, the *Query* command is still larger and higher power consuming than these 10 primitives from ISO 18000 Part 7.

3.4.4.4 ISO 18185 RFID Standard

The ISO 18185 Part 1 standard [45] is an international standard that provides a system for the unique identification and presentation of information about freight container electronic seals. It is used in conjunction with the other parts of ISO 18185 such as Part 4 that specifies data protection and Part 7 that specifies the physical layer protocol.

The electronic seal mandatory data includes `seal id`, `seal status`, `battery status`, details on the sealing and opening times, and protocol information. The `seal id` is a combination of the `tag manufacturer id` and the `tag id` and is used to uniquely identify the seal. It is permanently programmed into the seal during manufacturing. The `seal status` indicates the open, closed, or sealed state of the seal.

Figure 3.31 shows the interrogator to tag command/response formats for the *Sleep All But* and *Get Seal Model* commands. The command contains fields such as a `protocol` to identify the data link layer packet structures, an *opcode* `code` to identify the command, and `options` to indicate whether it is a point-to-point or a broadcast command and whether the command duration fields are present. The command duration fields are specified by the interrogator in point-to-point commands so that the tag may switch to *sleep* mode after waiting for the described duration. The *Sleep All But* command is a broadcast command. In response to this command, the specified seal remains awake while all the other seals return to *sleep* mode. This command does not require a response back to the interrogator. In the

Sleep All But command

Protocol	Options	Interrogator	Code	Length	Manufacturer	Tag	CRC
8 bit	8 bit	16 bit	8 bit	8 bit	16 bit	32 bit	16 bit

Get Seal Model command

Protocol	Options	Manufacturer	Tag	Interrogator	Code	Min time	Max time	Length	CRC
8 bit	8 bit	16 bit	32 bit	16 bit	8 bit	16 bit	16 bit	8 bit	16 bit

Get Seal Model response

Protocol	Status	Length	Interrogator	Manufacturer	Tag	Code	Model	CRC
8 bit	16 bit	8 bit	16 bit	16 bit	32 bit	8 bit	16 bit	

Status

Modefield	State	Reserved	Acknowledge	Reserved	Seal type	Reserved	Reserved	Battery
4 bit	2 bit	1 bit	1 bit	2 bit	3 bit	1 bit	1 bit	1 bit

FIGURE 3.31
Example command/response formats from ISO 18185 Part 1.

case of *Get Seal Model* command, the tag response includes the *opcode* `code`, the nested `seal` status, and other data fields.

Figure 3.32 shows an example RFID macros file containing the *Sleep All But, Get Seal Version, Get Seal Model,* and *Collection* primitives. The RFID compiler has been used to implement the commands from ISO 18185 Part 1 standard for a custom hardware-based tags target.

3.4.4.4.1 *Custom Hardware-Based Tag*

Using the RFID compiler, up to 10 commands of the ISO 18185 Part 1 standard were implemented with 0.16 μm custom cell-based ASIC hardware. The tool used for synthesis and area estimates is the Design Compiler. Power was estimated using PrimePower.

Table 3.7 shows the total area and the power consumption of ISO 18185 Part 1 seal designs for a 0.16 μm ASIC. The area of this seal design is much smaller than the area of ISO 18000 Part 6C tag design. The 18185 design is smaller than even the relatively simple ISO 18000 Part 7 tag design. The power consumption is also correspondingly lower. This could be in part because the commands in ISO 18185 Part 1 do not incorporate security mechanisms such as passwords.* For example, ISO 18000 Part 7 provides a password style security mechanism by the *set password, set password protect,* and *unlock* commands. There is also an additional layer of privacy introduced by the user id field. The logic required to implement the checking of these fields increases the total area and the power consumption of the tag designs.

3.5 Conclusions

This chapter presents an extensible RFID tag with associated design automation flow. The RFID compiler automatically generates RFID tag software or hardware for both microprocessor- and FPGA-based extensible tags. The compiler takes as input simple descriptions called RFID macros of the RFID primitives described in the standard and behavior for each primitive written in C. The system is extensible, in that it allows for addition (or removal) of a set of custom RFID primitives that may be a subset or superset of the original standards. The physical layer blocks are automatically generated through the description

* The security features of various RFID standards are described in Chapter 33.

```
declarations
protcl(8)
options(8)
manuf(16)
tagid(32)
interid(16)
opcode(8)
mindur(16)
maxdur(16)
arglen(8)
crc(16)
status(16)[
  modefield(4)
  state(2)
  reserved(1)
  acknowledge(1)
  reserved2(2)
  sealtype(3)
  reserved3(1)
  reserved4(1)
  battery(1)
]
paclen(8)
wsize(16)
criteria(8)
nodata(0)
version(16)
model(16)

main
sleepbt (22) protcl options interid opcode  arglen  manuf  tagid    crc
             nodata
sealver (12) protcl options manuf   tagid   interid opcode mindur   maxdur  arglen crc
             protcl status  paclen  interid manuf   tagid  opcode   version crc
sealmd (14)  protcl options manuf   tagid   interid opcode mindur   maxdur  arglen crc
             protcl status  paclen  interid manuf   tagid  opcode   model   crc
collect (16) protcl options interid opcode  arglen  wsize  criteria crc
             nodata
```

FIGURE 3.32
Macros specification for *Sleep All But* command from ISO 18185 Part 1.

of waveform features of the encoding. We have presented several prototype systems including a software-based system using an EISC microprocessor and a custom hardware-based system using a Spartan 3 FPGA. Case studies that implement widely used standards such as ISO 18000 Part 7 and ISO 18000 Part 6C using this automation technique are presented.

TABLE 3.7
Area and Dynamic Power for Implementing the ISO 18185 Part 1 Primitive Logic on a 0.16 μm ASIC

Prims	1	2	3	4	5	6	7	8	9	10
Area	0.7171	0.7284	0.7398	0.7678	0.7779	0.8125	0.8349	0.8509	0.8664	0.8815
Power	0.0338	0.0371	0.0379	0.0381	0.0382	0.0382	0.0383	0.0384	0.0385	0.0386

Note: ASIC area is 100 μm^2. Dynamic power is in milliWatts. Quiescent power <0.4 μW.

The RFID compiler and design automation flows allow comparison of different configurations of the tag and the impact on area and power for microprocessor or ASIC tag implementations. The design automation flow also allows for evaluating interrogator/tag complexity with respect to different protocols and encodings.

References

1. American National Standards Institute, "ANSI NCITS 236:2001." Standard Specification, 2002.
2. International Standards Organization, "ISO/IEC FDIS 18000-7:2004(E)." Standard Specification, 2004.
3. X. Gao, Z. Xiang, H. Wang, J. Shen, J. Huang, and S. Song, "An approach to security and privacy of RFID system for supply chain," *Proceedings of 2004 IEEE International Conference on E-Commerce Technology for Dynamic E-Business*, pp. 164–168, September 2004.
4. P. Blythe, "RFID for road tolling, road-use pricing and vehicle access control," *IEE Colloquium on RFID Technology*, October 1999.
5. G. Roussos, "Enabling RFID in retail," *Computer*, 39, 25–30, March 2006.
6. L.M. Ni, Y. Liu, Y.C. Lau, and A.P. Patil, "LANDMARC: indoor location sensing using active RFID," *Wireless Networks*, 10, pp. 701–710, November 2004.
7. D. Molnar and D. Wagner, "Privacy: Privacy and security in library RFID: issues, practices, and architectures," *Proceedings of 11th ACM conference on Computer and Communications Security*, 2004.
8. TI, "Texas Instruments' RFID Technology streamlines management of Vatican Library's treasured collections," 2004. www.ti.com/tiris/docs/news/news_releases/2004/rel07-07-04.shtml
9. C. Li, L. Liu, S. Chen, C.C. Wu, C. Huang, and X. Chen, "Mobile healthcare service system using RFID," *IEEE International Conference on Networking, Sensing and Control*, 2, 2004.
10. A. Cerino and W.P. Walsh, "Research and application of radio frequency identification (RFID) technology to enhance aviation security," *Proceedings of IEEE National Aerospace and Electronics Conference*, 2000.
11. Y.F. Wong, P.W.K. Wu, D.M.H. Wong, D.K. Chan, L.C. Fung, and S.W. Leung, "RFI: Assessment on human safety of RFID system at Hong Kong International Airport," *Proceedings of 17th International Zurich Symposium on Electromagnetic Compatibility*, pp. 108–111, February 2006.
12. A. Alu, C. Sapia, A. Toscano, and L. Vegni, "Radio frequency animal identification: electromagnetic analysis and experimental evaluation of the transponder-gate system," *International Journal of Radio Frequency Identification Technology and Application*, 1(1), 90–106, 2006.
13. D. Engels and S. Sarma, "The reader collision problem," November 2001. White paper MITAUTOID-WH-007, Auto-ID Center.
14. R. Want, "The magic of RFID," October 2004, ACM Queue.
15. A. Juels, R. Rivest, and M. Szydlo, "The blocker tag: Selective blocking of RFID tags for consumer privacy," *Proceedings of 10th ACM Conference on Computer and Communications Security*, pp. 103–111, 2003.
16. M. Feldhofer, S. Dominikus, and J. Wolkerstorfer, "Strong authentication for RFID systems using the AES algorithm," *Proceedings of 6th International Workshop Cryptographic Hardware and Embedded Systems (CHES 2004)*, pp. 357–370, 2004.
17. A.K. Jones, S. Dontharaju, S. Tung, P.J. Hawrylak, L. Mats, R. Hoare, J.T. Cain, and M.H. Mickle, "Passive active radio frequency identification tags (PART)," *International Journal of Radio Frequency Identification Technology and Application (IJRFITA)*, 1(1), 52–73, 2006.
18. S.A. Delichatsios, D.W. Engels, L. Ukkonen, and L. Sydanheimo, "Albano multidimensional UHF passive RFID tag antenna designs," *International Journal of Radio Frequency Identification Technology and Application*, 1(1), 24–40, 2006.
19. L. Ukkonen, M. Schaffrath, J. Kataja, L. Sydanheimo, and M. Kivikoski, "Evolutionary RFID tag antenna design for paper industry applications," *International Journal of Radio Frequency Identification Technology and Application*, 1(1), 107–122, 2006.
20. J.D. Porter, R.E. Billo, and M.H. Mickle, "Effect of active interference on the performance of radio frequency identification systems," *International Journal of Radio Frequency Identification Technology and Application*, 1(1), 4–23, 2006.

21. K. Penttila, L. Sydanheimo, and M. Kivikoski, "Implementation of Tx/Rx isolation in an RFID reader," *International Journal of Radio Frequency Identification Technology and Application*, 1(1), 74–89, 2006.
22. G. Barber and E. Tsibertzopoulos, "An analysis of using EPCglobal class-1 generation-2 RFID technology for wireless asset management," *Proceedings of IEEE Military Communications Conference (MILCOM 2005)*, pp. 245–251, October 2005.
23. C.K. Harmon, "The necessity for a uniform organisation of user memory in RFID," *International Journal of Radio Frequency Identification Technology and Application*, 1(1), 41–51, 2006.
24. IEEE Computer Society, "IEEE Standard 802.3–2005," December 2005.
25. International Standards Organization, "ISO/IEC FDIS 18000–6:2004/Amd 1:2006(E)," Standard Specification, 2006.
26. American National Standards Institute, "ANSI/TIA/EIA-422-B," Standard Specification, May 1994.
27. ATIS Committee T1A1, "ATIS Telecom Glossary 2000," Technical Report T1.523–2001, Alliance for Telecommunications Industry Solutions (ATIS), 2001.
28. IEEE Computer Society, "IEEE Standard 802.5–1998E," May 1998.
29. D.M. Levis, B.W. Thomson, P.I.P. Boulton, and E.S. Lee, "Transforming bit-serial communication circuits into fast parallel VLSI implementations," *IEEE Journal of Solid-State Circuits*, 23, 549–557, April 1988.
30. E. Haselsteiner and K. Breitfuß, "Security in near field communication (NFC): Strengths and weaknesses," *Proceedings of the Workshop on RFID Security*, 2006.
31. X. Tang, T. Jiang, A. Jones, and P. Banerjee, "Compiler optimizations in the PACT HDL behavioral synthesis tool for ASICs and FPGAs," *IEEE International SoC Conference (IEEE-SOC)*, September 2003.
32. A.K. Jones, R. Hoare, D. Kusic, J. Fazekas, and J. Foster, "An FPGA-based VLIW processor with custom hardware execution," *ACM International Symposium on Field-Programmable Gate Arrays (FPGA)*, pp. 107–117, 2005.
33. R. Hoare, A.K. Jones, D. Kusic, J. Fazekas, J. Foster, S. Tung, and M. McCloud, "Rapid VLIW processor customization for signal processing applications using combinational hardware functions," *EURASIP Journal on Applied Signal Processing*, 2006, Article ID 46472, 2006.
34. A.K. Jones, R. Hoare, D. Kusic, G. Mehta, J. Fazekas, and J. Foster, "Reducing power while increasing performance with SuperCISC," *ACM Transactions on Embedded Computing Systems (TECS)*, 5, 1–29, August 2006.
35. Y. Cha, "EISC core," Presentation to University of Pittsburgh, February 2005.
36. A.K. Jones, R.R. Hoare, S.R. Dontharaju, S. Tung, R. Sprang, J. Fazekas, J.T. Cain, and M.H. Mickle, "A field programmable RFID tag and associated design flow," *Proceedings of FCCM*, pp. 165–174, 2006.
37. A.K. Jones, R. Hoare, S. Dontharaju, S. Tung, R. Sprang, J. Fazekas, J.T. Cain, and M.H. Mickle, "An automated, FPGA-based reconfigurable, low-power RFID tag," *Proceedings of the 43rd Design Automation Conference (DAC)*, pp. 131–136, ACM, July 2006.
38. A.K. Jones, R. Hoare, S. Dontharaju, S. Tung, R. Sprang, J. Fazekas, J.T. Cain, and M.H. Mickle, "An automated, FPGA-based reconfigurable, low-power RFID tag," *Journal of Microprocessors and Microsystems*, 31, 116–134, March 2007.
39. Sim-panalyzer, "SimpleScalar-ARM power modeling project," http://www.eecs.umich.edu/panalyzer
40. C. Gilberto, M. Martonosi, J. Peng, R. Ju, and G. Lueh, "XTREM: A power simulator for the Intel XScale core," *Proceedings of ACM LCTES*, 2004.
41. J. Russell and M. Jacome, "Software power estimation and optimization for high performance, 32-bit embedded processors," *Proceedings of ICCD*, 1998.
42. Intel, *Intel PXA27x Processor Family Developers Manual*, 2004. ftp://download.intel.com
43. Actel, *Fusion Family of Mixed-Signal Flash FPGAs: DC and Power Characteristics*, 0.5v ed., 2006.
44. S. Dontharaju, S. Tung, A.K. Jones, L. Mats, J. Panuski, J.T. Cain, and M.H. Mickle, "The unwinding of a protocol," *IEEE Communications Magazine*, 1, 4–10, April 2007.
45. International Standards Organization, "ISO/IEC FDIS 18185-1:2006," Standard Specification, 2006.

4

Far-Field Tag Antenna Design Methodology

Damith C. Ranasinghe and Peter H. Cole

CONTENTS

4.1 Introduction

Antennas used in the HF region operate at 13.56 MHz whose frequency has an electromagnetic wavelength of around 22 m giving a near-field far-field boundary of around 3.5 m. Thus, given reading distance requirements of <3 m and using the regulated radiation power at the HF ISM band, reader antennas are almost always near-field creation structures that aim to create large energy density fields with the minimum amount of radiation. However, at UHF frequencies the scenario is different. At UHF frequencies, the near-field far-field boundary is around 50 mm. Thus the region of operation in the UHF spectrum is almost always in the far field, and therefore reader antenna designs are far-field creation structures that aim to operate at the highest possible efficiency.

This chapter considers RFID label antennas for both near-field operation in the HF region and far-field operation in the UHF frequency range. To aid in the development of UHF tag antennas, the chapter contains material on the formulation of antenna equivalent circuits and presents an RFID label antenna design methodology, illustrated in the far field with the design of long-range, bow-tie antennas for tagging cases and pallets.

4.2 RFID Label Antennas

There are numerous label antenna designs, each with their own set of characteristics. Antenna designs are influenced by a range of issues, such as the region of label operation (near or far), the coupling field (electric field or magnetic field), the regulatory constraints, and the environment in which they operate. For example, an environment with many metal structures can affect time varying electromagnetic fields, and thus affect the performance of an antenna. Designing antennas to suit metallic structures requires special consideration and is presented in Chapter 13 of this book. A number of examples of antennas to suit their environment are outlined in Cole (2003), Pope et al. (1997), and Ranasinghe et al. (2004). A vital aspect of the design process is to allow maximum coupling between the reader antennas and label antennas for the coupling field used. Before considering the subject matter of label antenna design, the following sections discuss a number of different label antenna structures and the merits of their designs.

4.2.1 Magnetic Field-Sensitive Antennas

A common example of a magnetic field-sensitive HF label suitable for operation in the HF ISM band (13.56 MHz) is shown in Figure 4.1. The label is 42 mm in width and 47 mm in length. The label is designed to have a sufficient number of turns to provide the resonating inductance for the microcircuit input capacitance, as well as a flux collecting area in the interior, which is as large as practicable and consistent with the size requirement for the label.

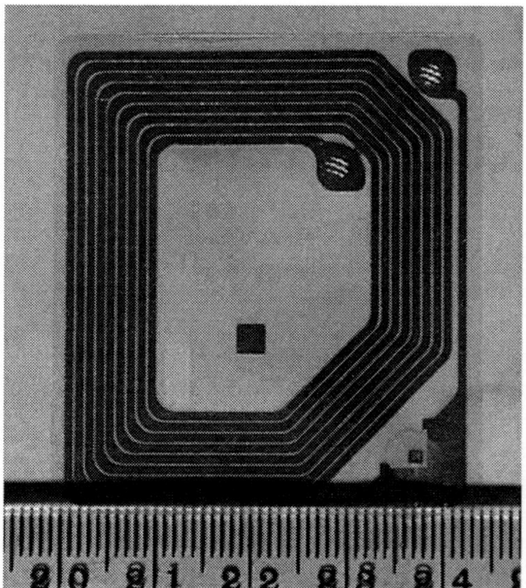

FIGURE 4.1
A magnetic field-sensitive antenna.

FIGURE 4.2
A large-loop antenna for an HF label. (From Cole, P.H., Jamali, B., and Ranasinghe, D., Coupling relations in RFID systems, 2003 Auto-ID Center White Paper Series, © 2003 by Auto-ID Center.)

Advantages of working in the near field at HF rather than at LF are that the number of turns required to resonate the microcircuit capacitance is small enough for low-resolution lithography to be used in antenna construction, and that no additional external resonating capacitance is required.

When a larger space is available for the tag label, a larger coil area should be used. As shown in Figure 4.2, fewer turns are then needed to obtain the required tuning inductance. A figure of merit for near-field antennas is readily obtainable from the coupling volume theory (Eshraghian et al., 1982; Cole et al., 2003; Ranasinghe, 2007) outlined in Chapter 12 of this book. It can be observed that the figure of merit (the coupling volume) for a planar coil operating in the near field varies as the third power of size, since the inductance of a coil is dependent on the equivalent coil diameter. Thus, the antenna of Figure 4.2 is about 18 times more sensitive than that of Figure 4.1. Unfortunately, this increased sensitivity does not translate to a corresponding increase in near-field range, as small coil interrogator antennas have an inverse sixth power decrease in energy density per unit volume as distance from the interrogator increases.

Clearly, both of the designs illustrated earlier are unsuitable for being placed flat against metal, as the boundary conditions will not allow a normal component of magnetic flux density at the metal surface. For this situation, the label antenna employing a solenoid with a magnetic core design shown in Figure 4.3 can be employed.

Eshraghian et al. (1982), Ranasinghe (2007), and Cole et al. (2003) show that without the magnetic core the coupling volume of a long solenoid is just the physical volume, but when a magnetic core is inserted, the coupling volume increases by a factor equal to the effective permeability of the magnetic core.

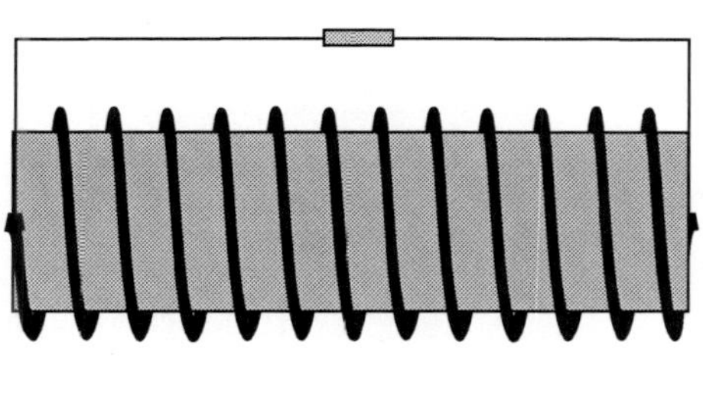

FIGURE 4.3
An antenna for HF operation against metal. (From Cole, P.H., Jamali, B., and Ranasinghe, D., Coupling relations in RFID systems, 2003 Auto-ID Center White Paper Series, © 2003 by Auto-ID Center.)

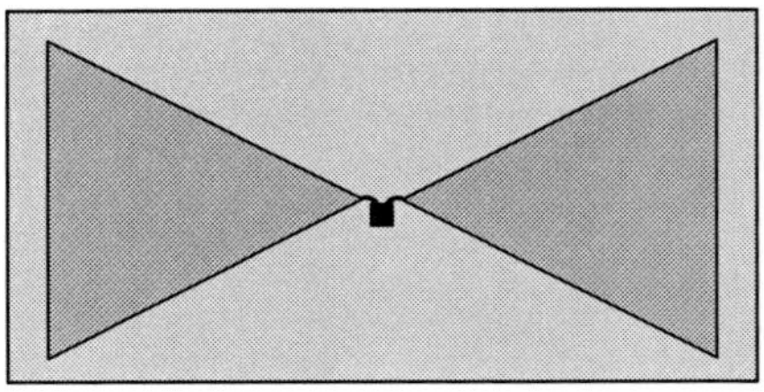

FIGURE 4.4
Electric field-sensitive label. (From Cole, P.H., Jamali, B., and Ranasinghe, D., Coupling relations in RFID systems, 2003 Auto-ID Center White Paper Series, © 2003 by Auto-ID Center.)

4.2.2 Electric Field-Sensitive Antennas

Two varieties of electric field-sensitive antennas are shown in Figures 4.4 and 4.5. Figure 4.4 shows a small bow-tie antenna that is intended to be sensitive to electric fields in the horizontal direction.

Figure 4.5 shows an electric field-sensitive antenna that is suitable for placement against a horizontal metal plate.

Analysis of the structure in Figure 4.4 is provided later in this chapter. The figure of merit for these antennas, when placed in the energy storage electric field is a coupling volume (Ranasinghe, 2007); for Figure 4.5 it is equal to the physical volume of the structure and for Figure 4.4 it is derived from the label dimensions, even though the antenna itself has no physical volume.

Both of the antennas will also have an effective electric flux collecting area but this area should not be confused with the effective area concept of a radiating antenna or of a far-field antenna. The effective area for a near-field electric field-sensitive antenna describes the extent to which the antenna can extract current from the displacement current density of the driving electric field.

4.2.3 Electromagnetic Field Antennas

An antenna can be considered as an electromagnetic field antenna on a couple of different bases. Firstly, if the antenna is capable of responding to both electric and magnetic fields, we would consider it to be an electromagnetic field antenna. It is almost invariably true that unless the antenna is very small, it does have this property. Proper analysis requires that it should be analyzed using the full set of Maxwell's equations rather than the subset or simplified versions that pertain to electrostatic or magnetostatic problems. A good example of this phenomenon is provided by the electromagnetic field-sensitive antenna shown in Figure 4.6, in which there is no obvious effort to couple to either electric or magnetic field alone.

Such electromagnetic antennas are generally useful for operation in the far field because far-field interrogation systems have shorter wavelengths, and antennas of acceptable size can no longer be considered to be electrically very small but are merely small.

Despite the earlier distinction, there are electromagnetic label reading environments in the UHF region in which, through reflections, either the electric or magnetic field is emphasized at the expense of the other. For such situations it is normally useful to take into account the nature of the driving fields in antenna design, and shape the design so that it is recognizably attuned to one or other of those fields.

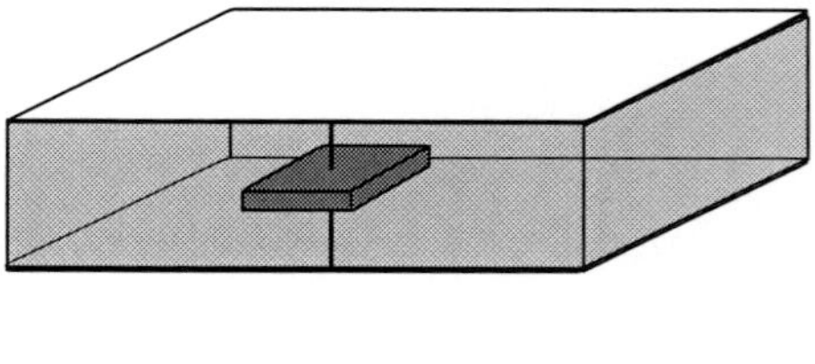

FIGURE 4.5
A parallel plate electric field-sensitive label. (From Cole, P.H., Jamali, B., and Ranasinghe, D., Coupling relations in RFID systems, 2003 Auto-ID Center White Paper Series, © 2003 by Auto-ID Center.)

FIGURE 4.6
An electromagnetic antenna.

4.3 Tag Antenna Design Considerations

Use of RFID in the identification of objects in various supply chains around the world has created research avenues into consumer product packaging (CPP) to find novel ways of integrating RFID labels into packaging and developing labels to suit packaging and the contents of packaged goods. The earlier sections have described the results of previous label antenna design developments to illustrate the multifaceted world of tag antenna design.

In addition to the antenna designs presented previously, there is an accumulating index of publications on RFID label antennas, for both active and passive tags, such as the slot antenna designs (Chen and Hsu, 2004), inverted F-antenna designs (Hirvonen et al., 2004), and the folded dipole antenna design (Xianming and Ning, 2004), to consider a few. Most of these publications only cover aspects of antenna analysis and practical aspects such as the suitability of the antenna for a specific application (Leong et al., 2005). However, what is not covered is a methodology for designing a tag antenna and a clear view of tag antenna design criteria.

The discussions in the following sections of this chapter are aimed at bridging that knowledge gap. Finally, Section 4.5 illustrates a successful antenna design for a passive RFID label that can be placed on corrugated cardboard boxes containing dry goods. The antenna designs presented are of a "credit card size," considered generally to be a suitable size for labeling a majority of cases used in supply chain applications.

4.3.1 Nature of Antennas Suitable for RFID

This section will consider, in general, types of antennas suitable for RFID applications. The evaluation will be based on both practical aspects and performance aspects. Considering practicable antennas for RFID applications restricts us to mostly planar structures that can be attached to items, cases, and pallets. In addition, it is important to consider the RFID chip input impedance at the threshold of operation to realize a conjugate antenna impedance to achieve maximum power transfer to the RFID label IC. Unlike other applications, it is not cost effective to design elaborate matching circuits based on lumped circuit elements to match the antenna impedance to the chip input impedance because it is both expensive and structurally unsuitable for labels used in a majority of supply chain applications.

Since passive RFID tags operate in a power-constrained environment created by electromagnetic compatibility regulations and the power required to operate the tags is obtained from the incident electromagnetic waves, maximum power transfer is of vital importance. Therefore, before postulating a theoretical "best" antenna, it is important to consider the load to which an RFID label antenna must provide power. The resulting load impedance presented to a tag antenna is considered later.

A UHF RFID label IC with an antenna terminal and a rectifying circuit that is intended to produce a rectified voltage used for powering the label circuits can be modeled as

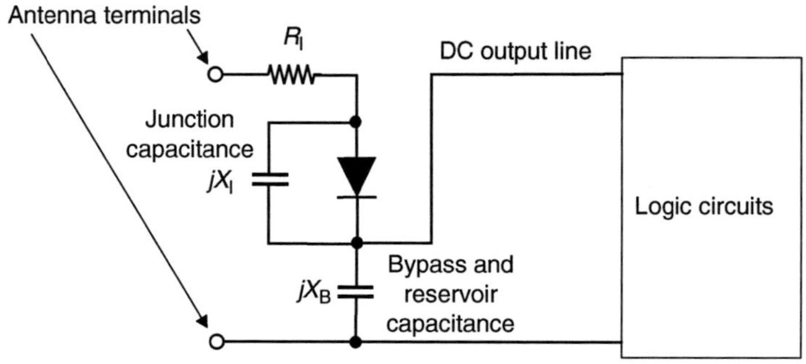

FIGURE 4.7
A simplified RFID label IC schematic. (From Ranasinghe, D.C., Leong, K.S., Ng, M.L., and Cole, P.H., *IEEE 2005 International Workshop on Antenna Technology: Small Antennas and Novel Metamaterials*, New York, USA, 2006, © 2005 by IEEE. With permission.)

indicated in Figure 4.7. Here, X_l represents the reactance of the diode capacitance, X_B is the reactance of the reservoir capacitor that also serves as an RF bypass, and R_l represents the loss in bringing reactive power into and out of the diode junction capacitance. It is clear from Figure 4.7 that the input impedance of an RFID chip is largely dictated by the junction capacitance of the rectification diode. The rectifiers on modern UHF RFID ICs are fabricated using Schottky diodes with a junction capacitance value in the range of a few picofarads or less. Due to the sensitivity of the junction capacitance to the biasing voltage, the input impedance of an RFID chip is a complex function of both the operating frequency and the input power to the chip from the antenna. Thus in general, the chip impedance, Z_c is measured at the threshold of operation so that the antenna impedance is a conjugate match at the lowest power level at which the chip will operate successfully. This ensures that the chip receives the most amount of power possible when the tag is furthest from the powering RF wave.

As illustrated in Figure 4.7, the input impedance of an RFID chip at the threshold of operation (minimum input sensitivity) is capacitive. The input impedance of an RFID IC can be modeled as indicated in Figure 4.8 as a series equivalent circuit or a parallel equivalent circuit. Depending on the fabrication technology and the IC design, the typical impedance of RFID ICs will vary. Some of the typical values expected are listed, using the series equivalent circuit in Figure 4.8b, $Z_c = R_{series} + (1/j\omega C_{series})$, as follows.

- $17-149j\ \Omega$ at 915 MHz (EPC Class I Gen I IC offered from Alien Technology, $R = 1300\ \Omega$, $C = 1.5$ pF)
- $36-117j\ \Omega$ at 866.5 MHz (EPC Class I Gen 2 from Impinj (registered trademark of Impinj Inc., Seattle, Washington) (Impinj, 2005)
- $33-112j\ \Omega$ at 915 MHz (EPC Class I Gen 2 from IMPINJ) (Impinj, 2005)

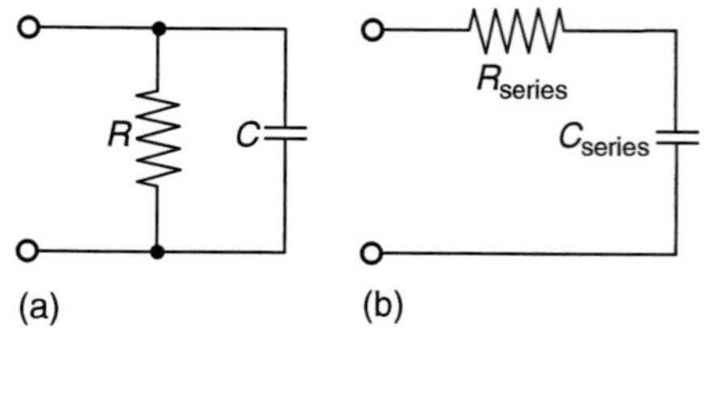

FIGURE 4.8
(a) A parallel equivalent circuit of an RFID IC input impedance where (b) is a series equivalent circuit of the chip input impedance.

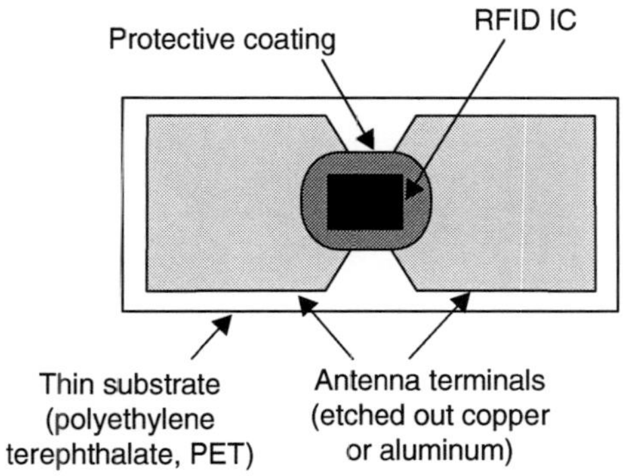

FIGURE 4.9
A direct chip attachment of an RFID IC.

The final chip impedance seen by the antenna is also affected by the technique used to attach the RFID IC to the tag. Generally, there are two different types of attachments employed. When the IC is in a flip-chip package (which is the industry standard technique for low-cost packaging) as shown in Figure 4.9, the RFID ICs can be directly attached to the antenna. The RFID ICs may also be obtained as a "strap" where the IC is connected to two mounting pads with a thin superstrate as shown in Figure 4.10.

Typically a resistance R of around 1300 Ω in parallel with a 1.1 pF capacitor C—which is that quoted for an Alien Class I Gen I RFID IC fabricated with CMOS technology and at the threshold of operation of the IC (Alien Technologies, 2005)—resulting in a series equivalent circuit impedance of 18.95–155.8j Ω can be expected from an RFID IC strap. Generally, it is good practice before designing an antenna to measure the input impedance of the chip at various operating frequencies using a network analyzer. Such a measurement method is outlined in detail in Eunni (2004).

Maximum power transfer requirements dictate that the antenna impedance should be a conjugate match to ensure the greatest possible performance from the RFID label (measuring the performance of an RFID label is discussed in Section 4.3.5). Hence considering the requirements of practicability and maximum power transfer, the "best" antenna for an RFID IC is a planar inductive antenna with a reactance that is able to tune out the capacitance of the label IC and also provide an adequate match for the real impedance of the label IC. The following section will consider modeling the input impedance of a typical label antenna by formulating a three-parameter circuit model.

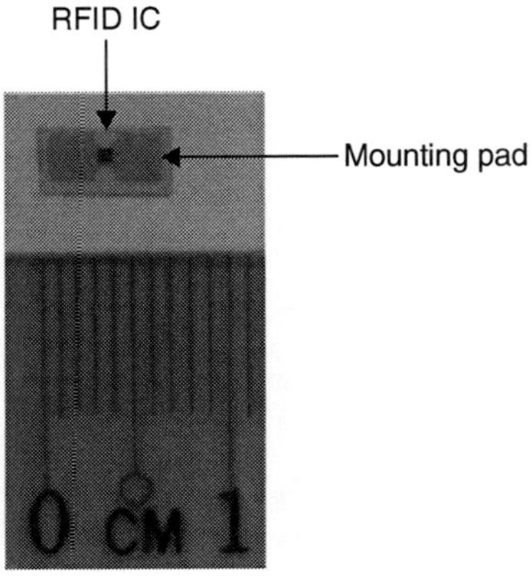

FIGURE 4.10
An RFID strap.

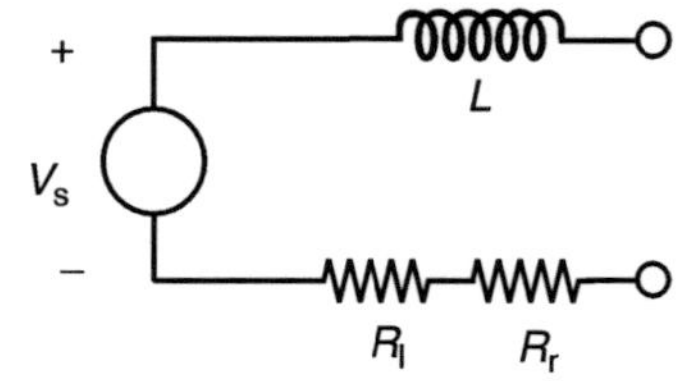

FIGURE 4.11
An equivalent circuit for a small magnetic field-sensitive antenna.

4.3.2 Label Antenna Equivalent Circuits

Equivalent circuits for small magnetic field-sensitive antennas and electric field-sensitive antennas are shown in Figures 4.11 and 4.12. The range of validity of these equivalent circuits is where the reactance properties of the antenna may be described by a single parameter, L or C. When the antenna is larger, the reactance properties can be described by an appropriate mixture of L and C.

In Figure 4.11, the source voltage is the voltage induced in the flux-collecting area of the coil by magnetic fields other than those fields which resulted from current flowing within the coil itself. Those induced voltages are represented by the voltage drop in the inductor L. The parameters R_l and R_r are loss and radiation resistances, respectively. Figure 4.1 is an example of a magnetic field-sensitive label antenna.

In Figure 4.12, the source voltage is the voltage developed across the self-capacitance of the antenna when it is open-circuited, as a result of the current injected into it, when it is short-circuited, by the displacement current density of the electric field in which the antenna is immersed. The parameters R_l and R_r are loss and radiation resistances, respectively. Figure 4.4 is an example of an electric field-sensitive label antenna.

Calculating the parameters of the magnetic field-sensitive antenna is straightforward, the relevant formulae being obvious and outlined in many elementary electromagnetic text books (Stutzman and Thiele, 1988; Cheng, 1989; Balanis, 1996; Kraus and Marhefka, 2002). For the electric field antennas, determination of the relevant parameters is sometimes not so simple, as electrostatic field solutions for the relevant geometries are not readily available. Therefore, empirical results or numerical modeling are more commonly employed for useful shapes.

Thus far, the nature of RFID ICs and the nature of the impedance of label antennas have been considered. Clearly, matching a label's input impedance to a label antenna's impedance is vital. While an antenna impedance may be adjustable by design variations (as illustrated in Section 4.5), the input impedance of an RFID IC cannot be altered without using external circuit components. This is an undesirable result. The following section investigates the practicability of matching to an RFID IC's input impedance at UHF frequencies allocated for RFID around the world.

4.3.3 Matching to an RFID Chip Impedance

Before embarking on the design of tag antennas, it is useful to consider the impedance to which a tag antenna must be matched. A useful tool for evaluating the practicability of

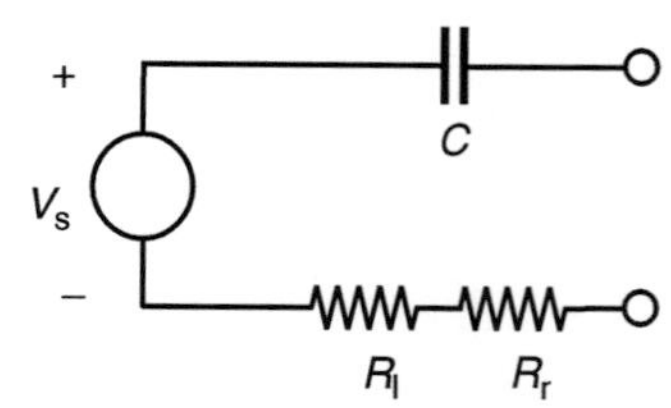

FIGURE 4.12
An equivalent circuit for a small electric field-sensitive antenna.

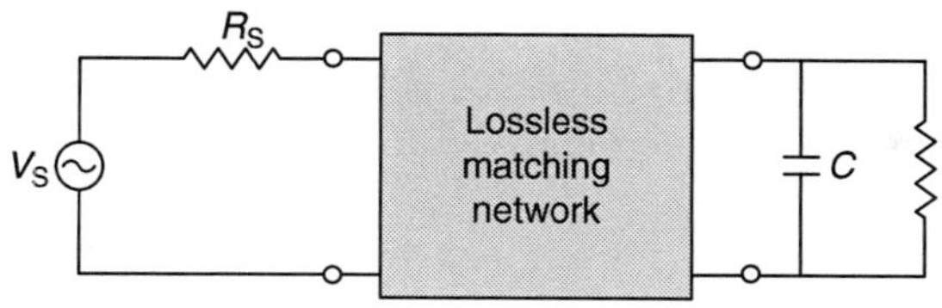

FIGURE 4.13
A circuit with a lossless matching network and a parallel *RC* load. (From Ranasinghe, D.C., Leong, K.S., Ng, M.L., and Cole, P.H., *IEEE 2005 International Workshop on Antenna Technology: Small Antennas and Novel Metamaterials*, New York, USA, 2006, © 2005 by IEEE. With permission.)

such a match can be derived from the Bode and Fano theorem (Fano, 1950). The theorem can be used to investigate the existence of any theoretical limitations to matching to an RFID IC's chip impedance (Ranasinghe et al., 2006). Figure 4.13 shows a circuit with a real source impedance, a lossless matching network, and an input impedance of an RFID IC represented by a parallel *RC* load. According to Bode and Fano, the fundamental limitation on impedance matching takes the form given in the following equation (Fano, 1950):

$$\int_0^{\infty} \ln\frac{1}{|\Gamma|}\,d\omega \leq \frac{\pi}{RC}. \tag{4.1}$$

In Equation 4.1, Γ is the reflection coefficient of the load and its assumed lossless matching network with respect to the source impedance R_S, and R and C represent the resistance and capacitance, respectively, in the parallel *RC* load (Figure 4.8).

Equation 4.1 places a maximum limit on the integral to π/RC. In order to completely use the given limit of π/RC for a desired angular frequency bandwidth ($\Delta\omega$), $|\Gamma|$ should be unity along the entire band except for the bandwidth, $\Delta\omega$ under consideration, thus implying a complete mismatch outside $\Delta\omega$. Considering a minimum achievable mismatch over $\Delta\omega$ and thus a minimum bound on the reflection coefficient of $|\Gamma|_{\Delta\omega}$ over the bandwidth, $\Delta\omega$ (refer to Figure 4.14) yields Equation 4.2 which reveals that for a given *RC* load there is a compromise between the maximum matching bandwidth and the maximum power transfer to the load:

$$|\Gamma|_{\Delta\omega} \geq e^{-\frac{1}{2\Delta f RC}}. \tag{4.2}$$

If matching is to be performed to satisfy a certain acceptable $|\Gamma|_{\Delta\omega}$ (and hence, amount of power transfer), the bandwidth may have to be reduced. On the other hand, if matching is to be performed over a certain given bandwidth, the amount of power transfer to the load may have to be compromised.

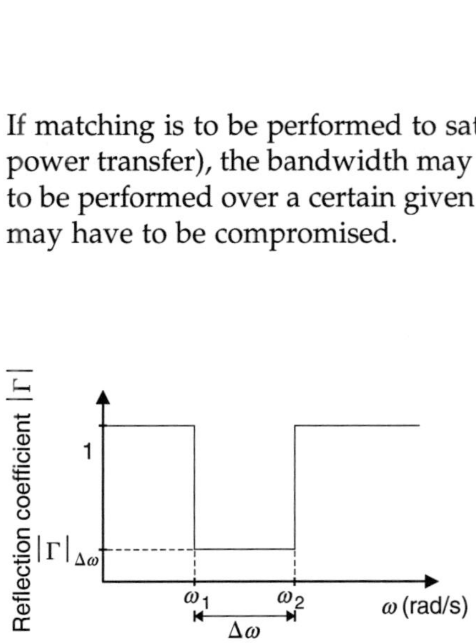

FIGURE 4.14
Theoretically possible minimum reflection coefficient for the best usage of π/RC over a bandwidth from ω_1 to ω_2. (From Ranasinghe, D.C., Leong, K.S., Ng, M.L., and Cole, P.H., *IEEE 2005 International Workshop on Antenna Technology: Small Antennas and Novel Metamaterials*, New York, USA, 2006, © 2005 by IEEE. With permission.)

TABLE 4.1

Regulated Frequency Allocations in the UHF Band for RFID Use in a Number of Major Geographic Regions around the World

Region	Frequency Range (MHz)	Bandwidth (MHz)
Europe	865–868	3
United States	902–928	26
Japan	952–954	2
Australia	918–926	8

Using the bandwidths outlined in Table 4.1, calculations of reflection coefficient limit established in Equation 4.2 are performed for the four regions: the United States, Europe, Japan, Australia, and all of these regions simultaneously. The results are outlined in Table 4.2. In the calculations, a chip resistance R of 1.3 kΩ and a chip capacitance C of 1.1 pF is assumed.

All the values for $|\Gamma|_{\Delta\omega}$ therein are small. This implies that the allocated bandwidths for RFID usage pose no theoretical limitations toward achieving a good impedance match to the input impedance of the RFID IC.

However, recent advances in the fabrication of Schottky diodes and low-power CMOS processes have yielded RFID chips with input chip impedance values where the resistance R is about 2500 Ω with a parallel capacitance C of around 500 fF. In this light, the cases presented earlier are reevaluated and the results are presented in Table 4.3.

From the values in Table 4.3, it can be observed that in practice, if the tag chip has an $R = 2500\ \Omega$ and a $C = 500$ fF, the theoretically achievable minimum reflection coefficient becomes even smaller and still presents no practical limit to the maximum power transfer to an RFID chip across the UHF RFID bands in the regions considered earlier.

4.3.4 Environmental Constraints

Understanding the effects of various environmental factors on a tag antenna is important so that suitable antennas can be developed to overcome any difficulties.

Liquids and metals play an important role in the performance with respect to the manner in which they affect electromagnetic waves. High dielectric and lossy materials such as liquids absorb or attenuate UHF RF energy and detune tag antennas reducing radiation efficiency, while metals can either absorb or reflect RF energy depending on the amount and shape of the metal. The unwanted result of a tag's reduced performance when attached to materials with high dielectric constants or metallic objects needs to be taken into consideration during the tag design process.

TABLE 4.2

Theoretical Bounds on Reflection Coefficients ($R = 1300\ \Omega$, $C = 1.1$ pF)

Region	Theoretical Bound on $|\Gamma|_{\Delta\omega}$
United States	1.44×10^{-6}
Europe	2.42×10^{-51}
Japan	1.18×10^{-76}
Australia	1.04×10^{-19}
All regions	0.032

TABLE 4.3

Theoretical Bounds on Reflection Coefficients ($R = 2500\ \Omega$, $C = 500$ fF)

Region	Theoretical Bound on $\|\Gamma\|_{\Delta\omega}$
United States	2.08×10^{-7}
Europe	1.24×10^{-58}
Japan	1.38×10^{-87}
Australia	1.93×10^{-22}
All regions	0.0014

Even a carton of photocopy paper may prove problematic for RFID labeling because the liquid that affects RFID performance does not have to be an actual liquid. Paper typically has high moisture content, and it does absorb RF energy. Wooden pallets made with green or indeed anything but oven-dried wood present the same challenge because of the moisture content. Fresh fruits and vegetables and frozen items will also pose liquid-related problems for RFID. It is important to evaluate whether the items have the potential to hold or attract moisture when considering the design of RFID tag antennas for tagging the items.

Metal is perhaps more of a challenge because it may either reflect or absorb electromagnetic waves. However, the behavior of electromagnetic waves next to a metal surface is predictable as opposed to the effects mentioned previously. In certain situations, the presence of metal can actually improve the performance of an RFID tag. Irregular metal, on the other hand, will either absorb the signal or reflect it in random directions. Metallized foil bags and even antistatic bags can act as metal. Some materials have metallic contents or coatings that need to be considered. Rice, for example, has been stated as having a high mineral iron content that affects RFID performance (Clarke et al., 2005).

The choice of void fill can affect RFID. Bubble wrap and loose Styrofoam void fill have very little effect on RFID, whereas dense foam will absorb some RF energy. Crushed or formed paper, corrugate cardboard will have little effect unless it is very densely packed (and then may pose the potential of the liquid problem).

While it may seem obvious that certain products will have an adverse effect on RFID, it is possible to design antennas that take advantage of the nature of the surrounding material properties. There are RFID tag designs that can be placed directly on flat metal surfaces. These tags employ a relatively thin layer of dielectric insulation between the tag and the metal surface. This effectively turns the metal surface into part of the antenna (a finite ground plane) and can significantly improve performance by using the metal to reflect the RF signal back to the interrogator that would otherwise radiate into the item. An example of an antenna design for tagging metallic objects can be found in Ranasinghe et al. (2004) which resulted from the investigation into the tagging of drill strings employed in oil rigs.

4.3.5 Performance Measure

While addressing the topic of RFID label antenna design, it is important to consider the practical performances of such antennas when attached to an RFID label IC. The accepted metric for such performance comparisons involves taking a read range measurement. The read range of a tag is the maximum distance between a reader antenna and the tag before the reader fails to decode the tag responses while the tag antenna is favorably oriented to the reader antenna propagation field. These read range measurements may be taken in an anechoic chamber, or may be performed in a more practical environment where the tag is to be deployed.

In the theoretical estimation of read range for systems operating in the UHF spectrum, two scenarios can be analyzed. These are

- Tag power constrained analysis
- Reader sensitivity constrained analysis

In the tag power constrained analysis, it is assumed that the system is adequately designed so that the sensitivity of the interrogator's RF receiver is not a limiting factor (this might be the case in the event of a reader using a bistatic antenna configuration). In such a scenario, the theoretical read range of a tag with a lossless antenna may be calculated from the Friis equation given as

$$\frac{P_r}{P_t} = g_r g_t \left(\frac{\lambda}{4\pi r}\right)^2, \tag{4.3}$$

where
P_r is the available source power from the tag antenna
P_t is the transmitted power
λ is the associated wavelength of the frequency used
r is the distance between the transmit and the receive antennas
g_r is the gain of the transmit antenna
g_t is the gain of the receiving antenna

Thus the read range of a tag may be evaluated as given in Equation 4.4 where the radiated power of the reader and the reader antenna gain are replaced by EIRP and the available source power required at the tag antenna is $P_{r(tag)}$:

$$r \leq \frac{\lambda}{4\pi}\sqrt{\frac{\mathrm{EIRP}_{\mathrm{reader}} g_{\mathrm{tag}}}{P_{r(tag)}}}. \tag{4.4}$$

Equation 4.4 is useful only if an expression can be obtained for $P_{r(tag)}$. This requires careful consideration in case of passive tag technology. Thus if the minimum amount of power required to operate a tag is known to be P_{IC}, and it can be assumed that the tag is receiving that power, and the efficiency of the rectifier structure is η and k_m is the power transfer factor from the antenna to the tag circuit in the presence of modulation, that is, the ratio of the power reaching the tag circuit in the presence of modulation at the greatest mismatch to the available source power from the tag antenna, then Equation 4.5 gives the minimum power required by a tag at its threshold of operation. Then it is possible to use $P_{r(tag)}$ from Equation 4.5 in Equation 4.4 to obtain a maximum possible read range when having enough power to energize the tag is the constraint:

$$P_{r(tag)} = \frac{P_{IC}}{\eta k_m}. \tag{4.5}$$

If however it is the sensitivity of the interrogator's RF receiver that is limiting, it is necessary to find the distance at which the received signal at the interrogator's receiver just meets the SNR of the receiver. Hence, the minimum received power $P_{\text{r(reader)}}$ at which the SNR of the receiver is satisfied is given in the following equation:

$$P_{\text{r(reader)}} = (S/N)_{\text{min}} kTB(\text{NF}), \tag{4.6}$$

where

NF is the noise factor of the receiver
B is the bandwidth of the receiver
k is Boltzman's constant
T is the absolute reference temperature used in the definition of the receiver noise factor NF
$(S/N)_{\text{min}}$ is the minimum signal-to-noise ratio needed to decode a tag reply successfully

Thus the minimum read range can be calculated by considering the one-way signal strength for a transmission from a tag to an interrogator with the required interrogator received power given by Equation 4.6. The result is the read range given in Equation 4.7, wherein the power $P_{\text{t(tag)}}$ is the power scattered back from the tag and where it is assumed that the tag is just sufficiently energized:

$$r = \sqrt{\frac{P_{\text{t(tag)}} g_{\text{reader}} g_{\text{tag}}}{P_{\text{r(reader)}}}} \left(\frac{\lambda}{4\pi}\right). \tag{4.7}$$

To calculate $P_{\text{t(tag)}}$, we need the ratio of the effective modulated power radiated by the tag antenna to the available source power from the tag antenna. We call this ratio k_{b}. Thus,

$$P_{\text{t(tag)}} = k_{\text{b}} P_{\text{r(tag)}}. \tag{4.8}$$

The value of k_{b} depends on how good we are in designing the modulator, and exactly how we define the effective modulated power, and that depends in turn on the form of modulation employed. In an inefficient design (in which most of the available source power is going to power the tag, so not much is backscattered), its value could be small. It could also be small if most of the power is backscattered but not in a way that is time varying or is a good expression of the type of modulation desired. Alternatively its value could be up to about 1, in the case where not much of the available source power goes into powering the tag and most is backscattered, and we are successfully using, for example, binary phase shift keying modulation.

In practice, the read ranges always need to be verified using practical measurements as practical RF propagation losses have not been taken into account in the Friis equation given in Equation 4.3 and it is not easy to model the propagation loss in various environments without extensive experimental data.

One simple method for evaluating tag performance in terms of tag read range in an ideal propagation context is to use an anechoic chamber. The reader antenna can be placed at one end of the anechoic chamber, whereas the tag is placed along the axis of maximum radiation from the reader antenna. The tag should be correctly oriented on a polystyrene stand so that there is maximum coupling between the tag antenna and the reader antenna. The reader output can be monitored while the tag is moved away from the reader antenna to obtain the read range. However, it may be more practical to conduct the read range measurement in an environment suitable to that in which the tag is to be deployed as it would give a more useful indication of the performance of the tag with respect to the application.

The following section considers identifying necessary tag antenna design requirements for a given application. Such identification should be undertaken before embarking on the antenna design process. Identifying requirements will help the translation of application requirements to design requirements for an RF engineer allocated to the task of designing a tag antenna.

4.4 Label Antenna Design Methodology

4.4.1 Design Requirements

Table 4.4 provides a necessary set of considerations that should be deliberated upon to identify design requirements before embarking on the tag antenna design process.

4.4.2 Design Methodology

Generally, RFID label size requirements are restricted by costs and practical aspects that depend on the application and the frequency of operation. Most practical tag antennas for UHF operation are physically small and hence tend to have a gain of around 1.76 dBi. These antennas also have moderately small bandwidths of operation. Generally, designing an antenna that is tunable in manufacture is highly desirable (Rao et al., 2005). This implies

TABLE 4.4

An Outline for Evaluating Antenna Design Requirements

Operational frequency band	Operational frequency of the tag will depend on the country or countries in which the tag is deployed. Table 4.1 outlines a list of such frequency bands in the UHF spectrum
Tag dimensions	Tag size requirements will depend on the application. For instance, the tag may need to be printed on a label for sticking on a cardboard box, embedded in plastic casing or placed within the confined space of a bottle cap
Label cost constraints	Keeping the cost of a tag to a minimum will limit the choice of RFID ICs that can be used, as well as the type of material that can be used for constructing the antenna. Generally, tag antennas are constructed using copper, aluminum, or silver ink, while the material used for the substrate may be anything from paper, polyester to FR4 dielectrics
Read range requirements	Consider the read range required by the particular application. Generally less efficient, smaller antennas may be used for smaller read range requirements. The read range is also affected by the electromagnetic compatibility regulations which control the EIRP or ERP of the reader antenna. Another important consideration is the tag orientation during an interrogation. This requires understanding the radiation pattern of the tag antenna; certain applications may require an almost omnidirectional directivity pattern, whereas others may only require tags to radiate in a particular direction
Objects to be tagged	It is important to consider the nature of the item on which the tag antenna is placed as the tag antenna can be designed to be suitable to the surface on which the tag is placed (such as metal or cardboard) or tuned for optimal performance based on the contents of the tagged item (such as liquids)
Operational conditions	The antenna design, material used, and the antenna package need to take into account the environmental condition to which a tag may be exposed. For instance, tags may be subjected to a range of temperature, pressure, or mechanical stresses depending on the application

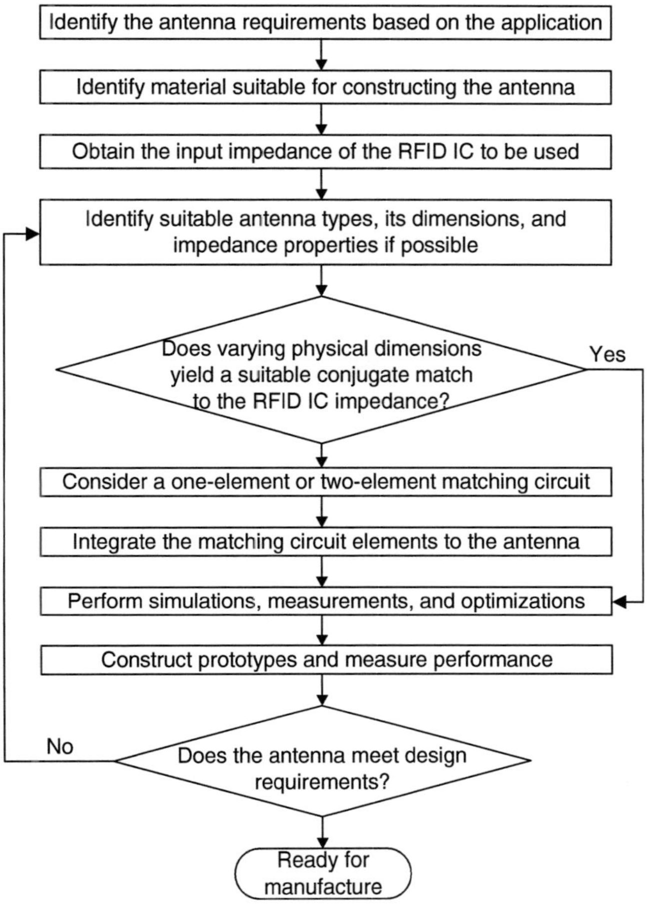

FIGURE 4.15
A label antenna design methodology illustrated using a simple flow chart to outline the process steps.

that the antenna impedance can be varied easily in a deterministic or a predictable manner by tuning a certain physical dimension or dimensions of an antenna. This will allow the same antenna design to achieve optimal performance when coupled to a range of different RFID ICs, or when the antenna is to be placed on different packaging materials or used in different frequency bands (as may be the case in porting an antenna design suitable for operation from the United States to Japan).

A systematic method for designing an RFID label antenna is shown in the form of a flow chart in Figure 4.15. This methodology is an expansion of the approach to tag antenna design found in Rao et al. (2005).

Once the antenna requirements are established by extracting them from the required application scenario, it is possible to look at the types of materials suitable for constructing the antenna in terms of cost constraints and operational conditions of the application. It is then important to determine the input impedance of the RFID IC in a selected package at the threshold of operation; this might be obtained from the manufacturer or may be obtained by direct measurement using a network analyzer.

Selecting a suitable antenna type is now possible; there are numerous designs available such as simple loops, dipoles, meander lines, spirals, and patch antennas. However, if an antenna parametric study reveals that it is not possible to obtain the required input impedance with a standard design within the design constraints, a designer is required to think more imaginatively.

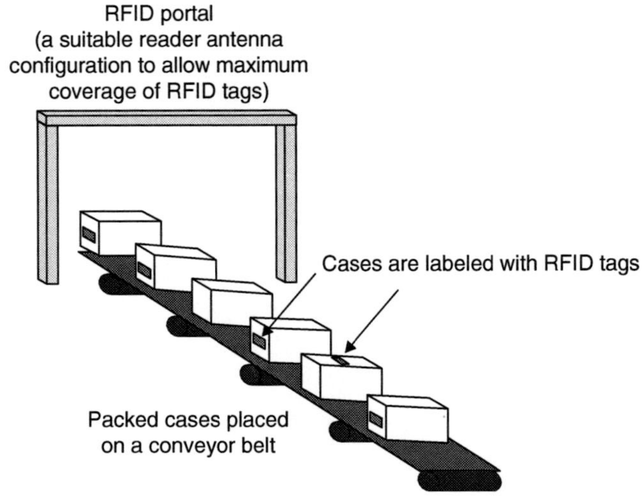

FIGURE 4.16
An illustration depicting the use of RFID labels in a supply chain application for tracking cases. Here the RFID portal is constructed by using an array of reader antennas at various orientations to ensure maximum coupling to tags placed at, possibly, different orientations.

A simple approach is to consider a single element or two elements matching network that will transform the antenna impedance to form a conjugate match to the chip impedance. Such a matching network should then be physically implemented as part of the antenna because the use of separate lumped circuit elements is an expensive, lossy, and less area efficient method.

The resulting RFID tags inevitably tend to be too complex for analytical investigation and various numerical electromagnetic analysis software based on method of moments for planar two-dimensional structures (MoM), finite element method (FEM), or finite difference time domain (FDTD) method for more complicated three-dimensional structures may be used.

Before using design tools, it is important to develop a simulation strategy and evaluate the performance of the tools by comparing simulated results with those from analytical and measurement results. Then, new antenna designs can be modeled and simulated to obtain desired antenna gain, input impedance, and to understand the relationship between tag antenna dimensions and antenna input impedance, which is critical for delivering maximum power to the tag.

Once an optimal antenna design has been developed, prototypes of the antenna can be built and their performance can be evaluated by taking read range measurements under controlled conditions (such as an anechoic chamber) or in the practical environment in which the tags are to be deployed. In the event the tag design is unsatisfactory, the whole design process needs to be iterated to obtain an antenna of adequate performance.

The following sections illustrate, in detail, the design of two tag antennas suitable for tagging cases in a supply chain application, such as those required in fast-moving consumer goods (FMCG) applications, depicted in Figure 4.16.

4.5 Illustrating a Novel Antenna Design

The antenna design methodology is best illustrated with an example. Considering the design of an RFID label for labeling cases constructed from corrugated cardboard boxes at a distribution center, a number of antenna requirements can be found.

4.5.1 Antenna Requirements, Material, and RFID IC Impedance

Assuming the tags have to operate under the electromagnetic compatibility constraints enforced by the FCC (2005), the following requirements can be outlined:

- A convenient size for an antenna for labeling most cases is approximately a credit card size label (86 mm × 54 mm).
- The frequency range of operation required is 902–928 MHz.
- While using a reader radiating 4 W EIRP to meet general application requirements, tags should have read range of not <2 m.
- Since tag orientation can be fixed on boxes, there are no constraints on the directivity of the antenna.
- Losses in the antenna should be kept small.

Given the earlier requirements, copper was chosen as the material of choice for the antenna due to its superior conductivity. Considering the skin depth of copper, copper sheets of thickness 32 μm should be used. The substrate considered needs to be flexible, thin, and have a low dielectric loss. Polyesters with low dielectric constants such as polyethylene terephthalate (PET) of thickness 50 μm are considered for the application.

The RFID straps used will be those from Alien Technologies. The straps have a Class I Generation 1 chip where the input impedance of a strap is typically $18.95 - 155.8j\ \Omega$ based on the parallel input impedance values of $R = 1300\ \Omega$ and $C = 1.1$ pF (refer to Figure 4.8a) at 915 MHz.

4.5.2 Antenna Type

It is possible to extend the analysis and use the results of Brown and Woodward (1952) to analyze bow-tie antennas. Then, the Ansfot HFSS simulation tool (Ansoft, 2005) can be used to fine-tune the antenna impedance properties to form a match to the label IC's input impedance.

A bow-tie antenna may be thought of as a construction of a monopole wedge above ground where the perfect ground plane is removed and the image under the ground plane is replaced by a physical structure as illustrated in Figure 4.17.

A three-parameter model for the bow tie can be developed using a detailed study undertaken of results first published by Woodward for a monopole wedge above ground. The empirical model for an equivalent circuit for a bow-tie antenna is shown in Figure 4.18. The model, which is derived from our experimental results that confirm Woodward's results, has an associated reactance $X(\omega)$ which is that of a capacitor C_B, whose value is that of the self-capacitance of the bow tie, and an inductor L_B, placed in the series circuit shown. The radiation resistance of the bow tie is represented by R_{Br}.

Calculating the self-capacitance as depicted by the field lines of Figure 4.19 of the bow-tie antenna by seeking analytical solutions to Laplace's equation presents a difficult problem. Nevertheless a numerical solution is tractable and has been performed using the method of moments to provide a suitable numerical approximation to the self-capacitance of a bow-tie antenna.

The radiation resistance outlined in the model parameters has significance in two ways. It allows the amount of radiated power to be calculated for a transmitting antenna, and also provides for a label antenna a means of calculating, using the reciprocity theorem, the effective electric flux collecting area of the antenna as depicted in Figure 4.20.

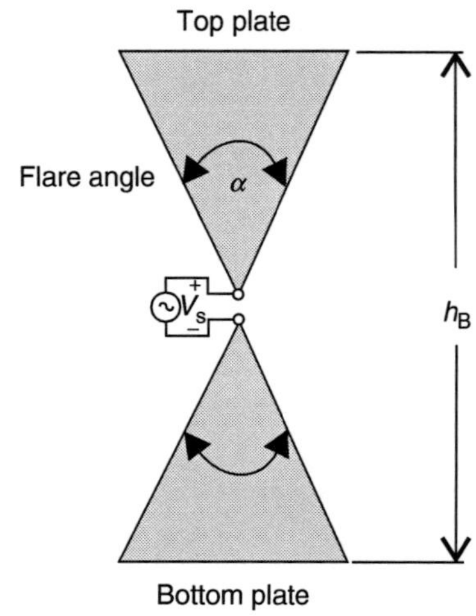

FIGURE 4.17
A bow-tie antenna with the height h_B and flare angle α.

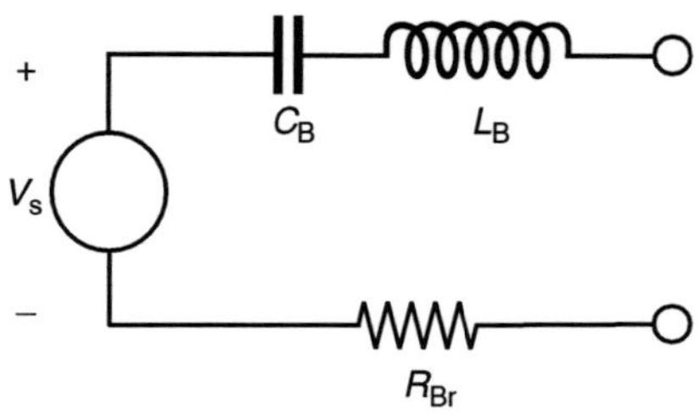

FIGURE 4.18
A three-parameter equivalent circuit model for a bow-tie antenna.

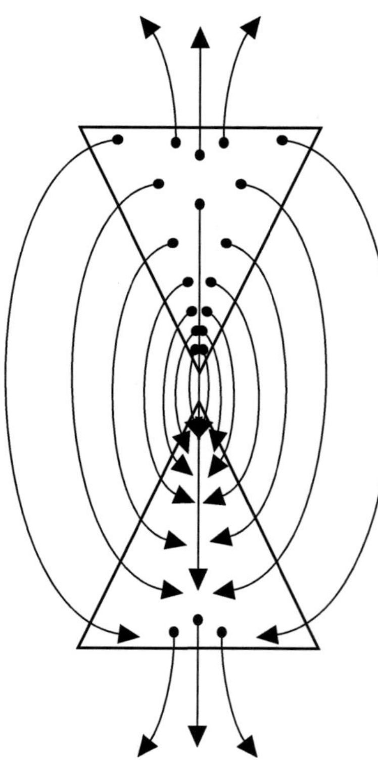

FIGURE 4.19
Field configuration around a bow-tie antenna used for the calculation of its self-capacitance.

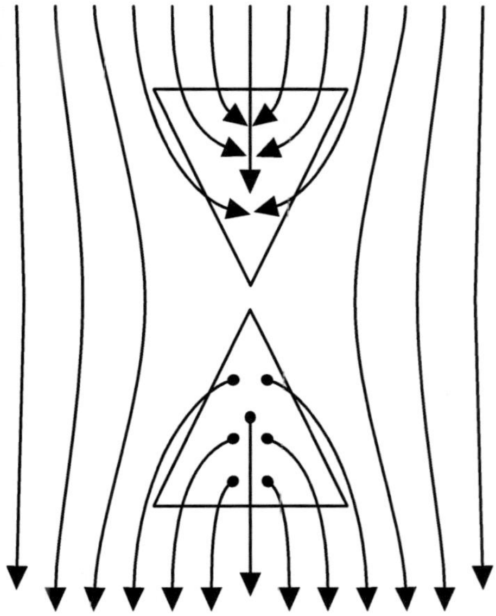

FIGURE 4.20
Field configuration for calculating the effective area of a bow-tie antenna.

The model parameters outlined in Figure 4.18 will vary for different flare angles and heights of the bow-tie antenna, as is the case for a wedge above a ground plane antenna. However, within the range of validity of the equivalent circuit, which depends on the dimensions of the structure in relation to a wavelength, the radiation resistance, the capacitance, and the inductance can be expected to scale up with increasing height for a specific flare angle. The radiation resistance, the capacitance, and the inductance variation for a bow-tie antenna can be summarized by the general expressions provided in Table 4.5, where the height h_B refers to the height of the bow-tie antennas as depicted in Figure 4.17.

In Table 4.5 the constants K_{BC} and K_{BL} are dimensionless quantities, whereas K_{BR} is measured in ohm. Table 4.5 is a summary of the parameters obtained for a bow-tie antenna using the relationship between a bow-tie antenna and a wedge above a ground plane antenna, using image theory based on the results from Brown and Woodward (1952), and using our own experiments.

However, the derived expressions are suitable only for electrically small antennas obeying the strict limit given by the following equation:

$$h_B \ll \frac{\lambda}{3}, \tag{4.9}$$

where h_B is the height of the antenna as indicated in Figure 4.17.

TABLE 4.5
Expressions for Evaluating Bow-Tie Antenna Circuit Model Parameters

Description	Expression
Capacitance (C_B) in farads	$K_{BC}\varepsilon_0 h_B$
Inductance (L_B) in henrys	$K_{BL}\mu_0 h_B$
Radiation resistance (R_{Br}) in ohms	$K_{BR}\,(\beta h_B)^2$

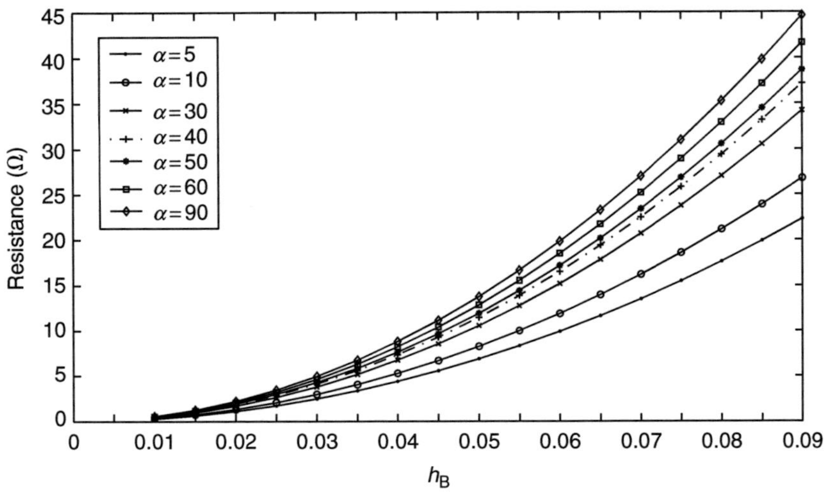

FIGURE 4.21
R_{Br} of bow-tie antennas of various flare angles evaluated using the expressions in Table 4.5.

4.5.3 Bow-Tie Antenna Design

While it is possible to find a bow-tie antenna with an adequate radiation resistance to match to an RFID chip impedance of 18.95 Ω (refer to Figure 4.21), it is not possible to find a bow tie with a flare angle and a height that will provide a conjugate match to the RFID chip's reactance as an examination of Figure 4.22 reveals that all bow-tie antennas of <90 mm in height appear to be capacitive. Hence any resulting bow-tie antenna, while possibly having the correct matching real impedance, will not be inductive to form a conjugate match to the chip impedance (Table 4.5).

The following sections will consider a different design of a bow-tie antenna. The design illustrates the design process for an RFID tag antenna and shows how a useful antenna can

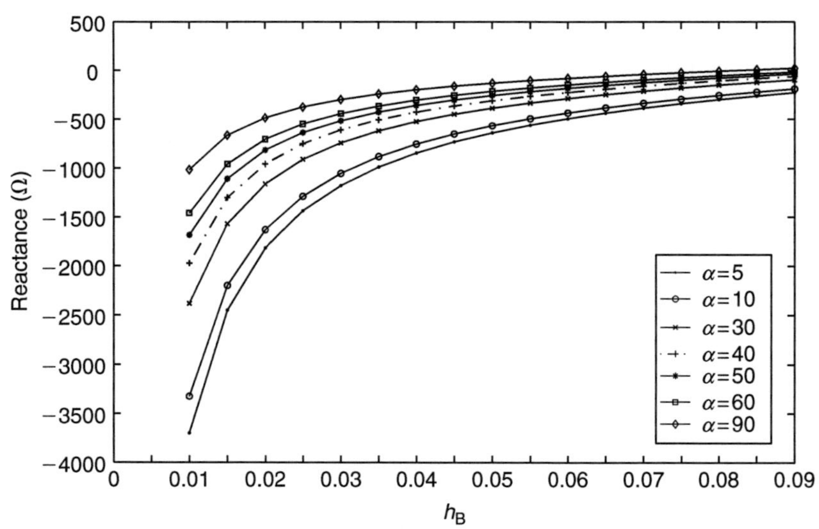

FIGURE 4.22
Reactance of bow-tie antennas of various flare angles (from the expressions in Table 4.6).

TABLE 4.6

Empirical Values for Evaluating Bow-Tie Antenna Circuit Model Parameters

Flare Angle	K_{BR} (Ω)	K_{BL}	K_{BC}
5	7.5	0.2888	0.5275
10	8.5	0.2823	0.5875
30	11.0	0.2605	0.8175
40	11.5	0.2470	0.9875
50	12.0	0.2349	1.1525
60	12.5	0.2250	1.2500
90	15.0	0.2128	1.9000

be designed starting with a more easily analyzed and understood antenna such as the bow-tie antenna.

4.5.4 Bow-Tie Antenna with a Series Tuning Inductor

Considering an approximately credit card size, bow-tie antenna shows that the input impedance is capacitive. This is true for any bow-tie antenna of <90 mm in height. We must therefore seek a suitable matching circuit that can be built into the antenna to match to an RFID chip's input impedance. The simplest possible matching circuit design, which is simple to physically incorporate into the antenna, will generally provide the most suitable antenna design.

Considering the equivalent circuit of the bow-tie antenna, it is possible to contemplate a matching element as illustrated in Figure 4.23 to achieve a match to the RFID chip impedance. The tuning element, L_{Smatch} can be physically incorporated into the antenna design. Such a series inductor may be conceptualized as indicated in Figure 4.24. An outline of the antenna design structure is given in Figure 4.25. The copper strip width, l_w, the size of the gap between the wings, g, and the amount of copper removed, i_c, can then be adjusted to find a suitable inductor value to form a match between the bow-tie impedance and the RFID IC impedance.

Taking a bow-tie antenna 80 mm in height, the properties of the antenna can be calculated as outlined in Table 4.7 where the center frequency of operation is taken as 915 MHz. However, adding the inductive strip will modify the antenna model developed earlier since the added strip will affect the self-capacitance and the inductance of the model. Instead of using an empirical method, simulation results from Ansoft HFSS (finite element method based simulation package) can be used to understand the effect of adding

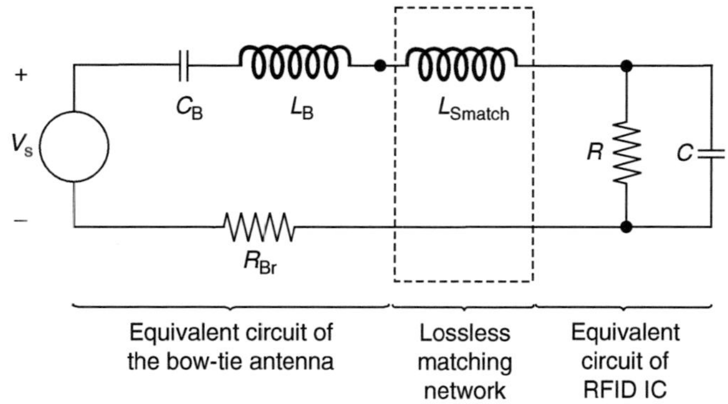

FIGURE 4.23
An RFID tag with a bow-tie antenna and a simple matching circuit.

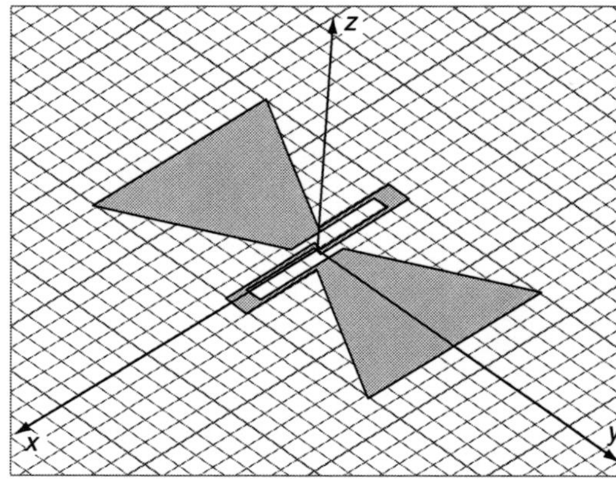

FIGURE 4.24
A bow-tie antenna with a series tuning inductor.

the inductive strip. Figure 4.25 describes the variable parameters used in the simulations, while Table 4.8 outlines the results of simulations performed at 915 MHz.

Increasing the inductance of the strip by altering the strip size increases the reactance of the bow-tie antenna. It can also be observed that the addition of the large inductor not only alters the reactance but also alters the radiation resistance of the antenna.

The reactance contribution to the impedance transformation from the series inductor can be increased by reducing the width of the inductor, i_w, increasing the size of the inductor by reducing i_c, or by reducing the gap, g, between inductance lines. The real impedance of the antenna can easily be adjusted by increasing or reducing the height of the antenna as outlined in the bow-tie antenna model in Section 4.5.2. These results broadly agree with the manner in which a series inductor transforms the antenna impedances as shown in the antenna equivalent circuit in Figure 4.23.

The simulation results suggest two possibilities for constructing the bow-tie antenna as given in Table 4.9. The antenna parameters can be further evaluated for their merits in terms of maximum power transfer to help a designer select the most suitable antenna design for a specific application. The two different physical constructions (as shown in Figures 4.26 and 4.27) were built to evaluate their performance.

Figures 4.28 and 4.29 show the impedance variation of the two bow-tie antenna designs over an operational frequency range of 850–950 MHz. The change in real impedance over

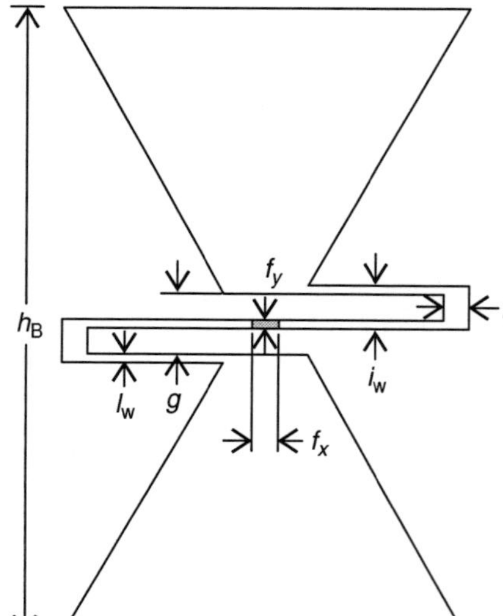

FIGURE 4.25
Bow-tie antenna design structure with a series inductor.

TABLE 4.7

Bow-Tie Antenna Input Impedance Characteristics

R_r	C_B	L_B	h_b	**Input Impedance**
25 Ω	0.8 pF	23.1 nH	80 mm	25 − 84j

TABLE 4.8

Simulation Results

h_b (mm)	l_w (mm)	i_w (mm)	i_c (mm)	g (mm)	f_y (mm)	f_x (mm)	Input Impedance (Ω)
71	1	6	2	9	1	1	17.00 + 122.00j
71	1	7	5	9	1	1	16.47 + 93.54j
72	1	7	5	8	2	1	16.82 + 98.50j
72	1	7	2	8	2	1	18.75 + 142.65j
72	1	7	2	8	2	2	18.97 + 141.61j
72	2	9	2	8	2	1	16.165 + 93.14j
72	1	4	3	8	2	1	17.68 + 80.35j
74	1	6	2	9	1	2	20.35 + 146.10j
75	1	4	3	5	1	2	19.093 + 89.86j
75	1	6	2	9	1	2	21.345 + 154.42j
76	1	3.5	3	4	1	2	19.52 + 79.33j
76	1	4.5	3	4	1	2	20.5 + 102.92j
76	1	6	2	9	1	1	22.732 + 162.33j
76	1	6	3	9	1	1	22 + 148j
78	1	6	4	7	1	1	17.8 + 101.83j
78	1	6	3	7	1	1	18.312 + 114.70j
78	1	6	2	7	1	1	18.95 + 128.34j
78	1	6	5	7	1	1	17.32 + 89.35j
79	1	6	6	9	1	1	18.054 + 90.814j
79	1	6	5	9	1	1	23.929 + 141.78j
79	1	6	2	6	2	1	19.37 + 138.00j
81	1	6	5	7	1	1	26.65 + 152.32j
81	1	6	6	7	1	1	19.433 + 98.55j

the frequency range is similar for both designs and neither design provides an advantage in impedance variations over frequency. Both designs can be expected to maintain a reasonably low-radiation quality factor over a range of frequencies with the added benefit of being able to better mitigate the detuning effects from environmental factors.

TABLE 4.9

Tag Bow-Tie Antenna Configurations

Antenna Name	h_b (mm)	l_w (mm)	i_w (mm)	i_c (mm)	g (mm)	f_y (mm)	f_x (mm)	Input Impedance (Ω)
BowAS	72	1	7	2	8	2	2	18.97 + 141.61j
BowS	74	1	6	2	9	1	2	20.35 + 146.10j

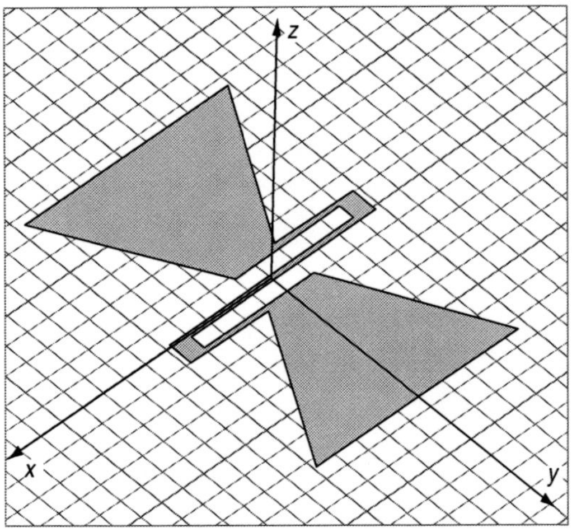

FIGURE 4.26
Bow-tie antenna design, BowS.

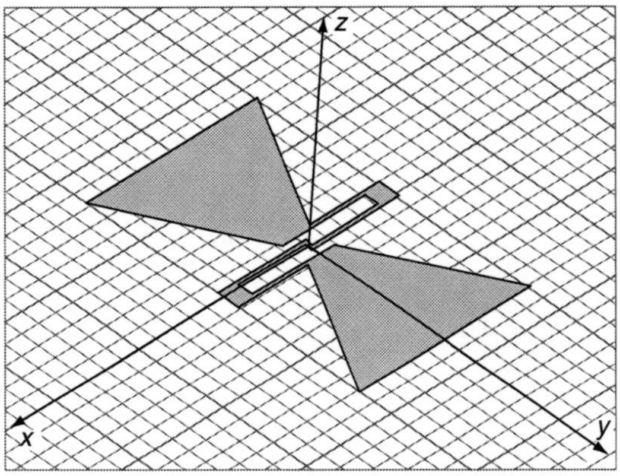

FIGURE 4.27
Bow-tie antenna design, BowAS.

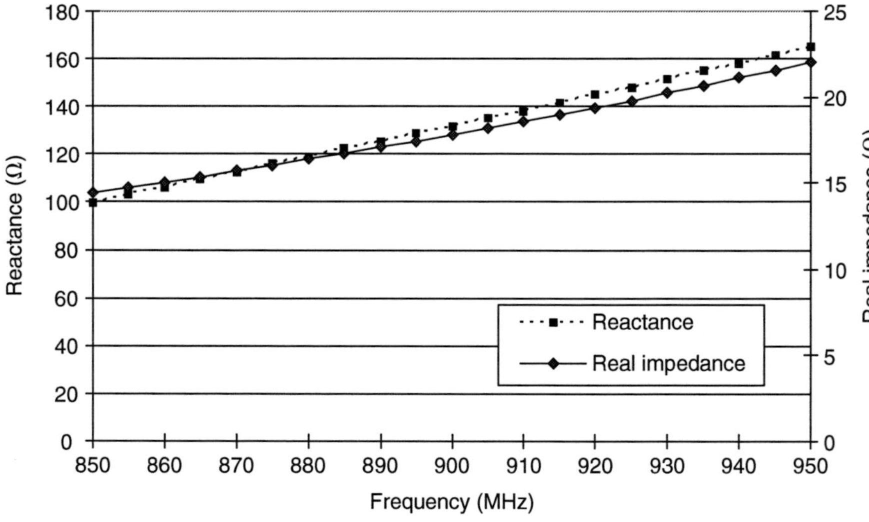

FIGURE 4.28
BowAS impedance variation over a frequency range of 850–950 MHz obtained from simulated results.

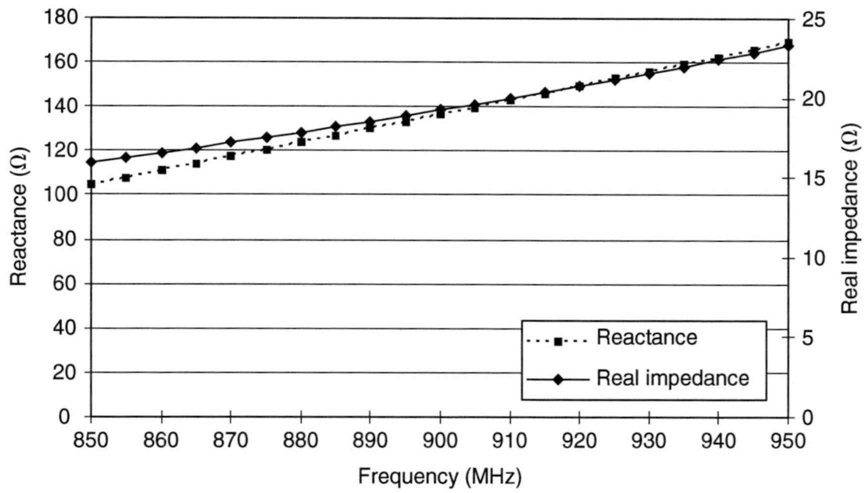

FIGURE 4.29
BowS impedance variation over a frequency range of 850–950 MHz obtained from simulated results.

Figure 4.30 shows the radiation pattern of the bow-tie antenna. As expected, the radiation pattern confirms that the antenna behaves similarly to an electric dipole.

The simulation tools can be used to create a better understanding of the functioning of the bow-tie antenna and confirm expectations. Figure 4.31 shows the surface current distribution on the bow-tie antenna as it is fed with an oscillating current. The diagrams show the current traversing from the feed point, outward toward the end of the bow-tie wings, over a period of $\pi/2$ of the oscillating current.

A practical construction of the tag can be achieved using copper sheets 0.035 mm in thickness or more to the specifications given in Table 4.9 and the description given in Figure 4.25. Then an RFID strap can be attached to the antenna terminals using a conductive adhesive. Figure 4.32 shows one of the antennas constructed in the laboratory for testing purposes. Read range measurements (conducted in an indoor laboratory environment) of

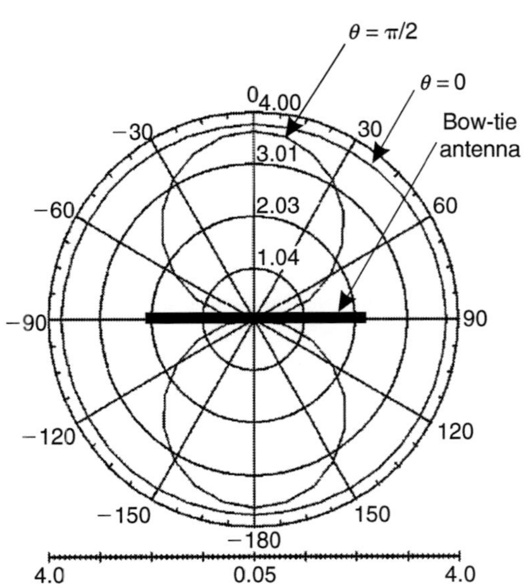

FIGURE 4.30
The radiation pattern of BowS antenna obtained from simulated results.

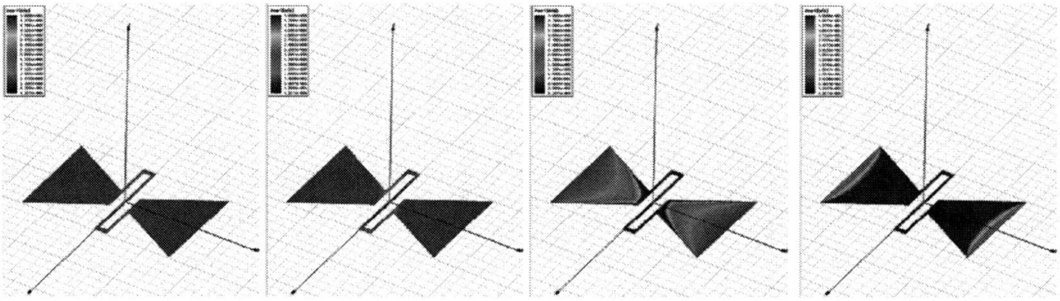

FIGURE 4.31
Surface current distribution plots of BowS.

BowS constructed and placed against polystyrene foam boxes showed a maximum read range of 7.30 m when the tag is favorably oriented with the reader antenna, and where the transmitted power is 1 W, using a 6 dBi gain reader antenna with an interrogator with a monostatic antenna configuration. BowAS showed a maximum read range of 6.70 m. Similar read ranges were obtained when the tags were placed against corrugated cardboard boxes.

If fine-tuning is required for maximum read range, the antenna dimensions can be altered. One such simple adjustment is changing i_c by stripping away small portions of the inductor until maximum read range is obtained at a required frequency. While simulation results are capable of giving a very accurate result, finer adjustments almost always need to be made manually, as described previously, to obtain the optimal tag dimensions for optimum performance.

FIGURE 4.32
A practical construction of the BowS antenna used in laboratory tests.

4.6 Conclusions

This chapter has explored the subject of RFID label antenna design, and illustrated the requirements that a successful label antenna must possess by exploring the nature of RFID label antennas and designing two RFID label antennas for tagging cases. The two different structural designs presented have illustrated the antenna design process outlined in this chapter and produced two successful RFID label antenna designs.

The discussion earlier clarifies that RFID label antenna design limits the designer to planar structures with inductive input impedance due to cost limitations and the nature of the load impedance presented by an RFID chip. Consideration of an adequate size for an antenna involves designing an antenna with an impedance that is a conjugate of the RFID chip's input impedance.

Considering the subject of matching bandwidth, an interpretation of the Bode–Fano theorem provided a theoretical limit to the achievable power transfer to the *RC* load of an RFID label IC. It has been observed that, in practice, if impedance matching is performed over a certain bandwidth, there is a limit to the minimum achievable reflection coefficient. Thus for a given chip impedance (*RC* load), there is a compromise between the maximum matching bandwidth and the maximum power transfer to the load.

The RFID label antennas presented in this section have many advantages. One of the main advantages is the simplicity of the matching network. The bow-tie antenna required only a simple matching network and both empirical and simulation methods were used in designing that network. In addition, because of the small size of the antennas considered, an excessively complex equivalent circuit for the antenna was not required. The antennas are also easily tunable by trimming the size of the inductors. Future work may be used to evaluate the performance of the antennas against various packaging materials.

Acknowledgment

The authors would like to thank Ms. M.L. Ng and Mr. K.S. Leong for assistance offered in the construction of prototype tags shown in Figure 4.32.

References

Alien Technologies, RFID tags, ALC-140-xx RFID transponder IC (January 2005). Available at: http://www.alientechnology.com/products/rfid_tags.php

Ansoft Corporation Web site (2005, August). Available at: http://www.ansoft.com.au

C.A. Balanis, *Antenna Theory: Analysis and Design*, New York: John Wiley & Sons, 1996.

G.H. Brown and O.M. Woodward, Experimentally determined radiation characteristics of conical and triangular antennas, *RCA Review*, 425–452, 1952.

S.-Y. Chen and P. Hsu, CPW-fed folded-slot antenna for 5.8 GHz RFID tags, *Electronic Letters*, 24, 1516–1517, November 2004.

D.K. Cheng, *Field and Wave Electromagnetics*, 2nd ed., New York: Addison-Wesley Publishing, ch. 4, 1989.

R.H. Clarke, D. Twede, J.R. Tazelaar, and K.K. Boyer, Radio frequency identification (RFID) performance: the effect of tag orientation and packaging contents, *Packaging Technology and Science*, 19(1), 45–54, 2005.

P.H. Cole, A study of factors affecting the design of EPC antennas and readers for supermarket shelves, Auto-ID Center workshop, October 2003.

P.H. Cole, D.C. Ranasinghe, and B. Jamali, Coupling relations in RFID systems II: practical performance measurements, Auto-ID Center workshop, June 2003.

K. Eshraghian, P.H. Cole, and A.K. Roy, Electromagnetic coupling in subharmonic transponders, *Journal of Electrical and Electronic Engineering*, 2, 28–35, 1982.

M.B. Eunni, *A Novel Planar Microstrip Antenna Design for UHF RFID*, Master thesis, A.M.A. College of Engineering, Kancheepuram, Madras University, May 2004.

R.M. Fano, Theoretical limitations on the broadband matching of arbitrary impedances, *Journal of the Franklin Institute*, 249, 57–83, January 1950.

FCC Regulations, Title 47, Telecommunications, ch. 1, Part 15, Radio frequency devices (August 2005). Available at: http://www.fcc.gov

M. Hirvonen, P. Pursula, K. Jaakkola, and K. Laukkanen, Planar inverted-F antenna for radio frequency identification, *Electronic Letters*, 40, 848–850, July 2004.

Imping RFID Technology Series, *The RFID Antenna: Maximum Power Transfer*, Technical report, 2005.

J.D. Kraus and R.J. Marhefka, *Antennas—For All Applications*, 3rd ed., New York: McGraw-Hill, 2002.

K.S. Leong, M.L. Ng, D.M. Hall, and P.H. Cole, A small passive UHF RFID tag for livestock identification, *IEEE 2005 International Symposium on Microwave, Antenna, Propagation and EMC Technologies for Wireless Communications*, 1, 67–70, August 2005.

G. Pope, M.Y. Loukine, D.M. Hall, and P.H. Cole, Innovative systems design for 13.56 MHz RFID, *Proceedings of the First Annual Wireless and Portable Design Conference*, p. 240, September 15–18, 1997.

D.C. Ranasinghe, *New Directions in Advanced RFID Systems*, PhD thesis, School of Electrical and Electronic Engineering, The University of Adelaide, 2007.

D.C. Ranasinghe, D.M. Hall, P.H. Cole, and D.W. Engels, An embedded UHF RFID label antenna for tagging metallic objects, *Proceedings of the 2004 Intelligent Sensors, Sensor Networks and Information Processing Conference*, Melbourne, Australia, pp. 337–342, December 14–17, 2004.

D.C. Ranasinghe, K.S. Leong, M.L. Ng, and P.H. Cole, Small UHF RFID label antenna design and limitations, *IEEE International Workshop on Antenna Technology: Small Antennas and Novel Metamaterials*, New York, March 2006.

K.V.S. Rao, P.V. Nikitin, and S.F. Lam, Antenna design for UHF RFID tags: a review and a practical application, *IEEE Transactions on Antennas and Propagation*, 53(12), 3870–3876, 2005.

W.L. Stutzman and G.A. Thiele, *Antenna Theory and Design*, 2nd ed., New York: John Wiley & Sons, 1988.

Q. Xianming and Y. Ning, A folded dipole antenna for RFID, *Proceedings of IEEE Antennas and Propagation Society International Symposium*, 1, 97–100, June 2004.

5

Contemporary RFID Reader Architecture

Behnam Jamali and Peter H. Cole

CONTENTS

5.1 Software-Defined RFID Reader

Wireless communication is becoming softer and softer. This means that radios, terminals, and networks are becoming reconfigurable and programmable. Software and cognitive radios are becoming one of the big trends in this field. Although these concepts have been studied for decades, many challenges remain making this vision a reality. In this work we study several aspects of a software-based RFID reader, including experimentation, implementation, and theoretical work.

5.1.1 Introduction

RFID systems are demanding more data, faster logging, and higher interrogation rates, so they require a powerful system that can manage information from a variety of sources, store the data, and transmit it reliably and continuously for long periods of time to other readers or a host PC. In addition, data logging is an extremely important part of RFID because it is the best method for determining what happens during the tag interrogation and for testing software strategies and protocol efficiencies. In the near future, software will do much more than running a fancy graphical user interface (GUI). More and more of the functionality of traditional radio systems will be implemented in software.

5.1.2 Problem Statement

When one hears the litany of reliability and readability woes that plague RFID in the supply chain arena, it is possible to wonder if anyone is making any progress. The problems with RFID range from nonfunctioning tags to environmental conditions, such as temperature, humidity, and radio frequency interference, often from other readers. All these effects can render tags unreadable. The challenge continues in a shipping and handling process, in which tags can be thrown out of alignment (detuned) so that they do not receive the reader signal at their resonant frequency. And the list goes on. Average read rates are still under 80% (Lewis, 2004). Some of this is the result of tag failure, but other important problems are incompatible tag and reader combinations as well as reader-to-reader interference resulting in ghost tag reading (Engels, 2003).

Traditionally, RFID systems have been designed with only a single-reader scenario in mind. The increasing use of RFID in multiple industries and also increasing deployment of mobile RFID readers results in situations where readers are to operate in close proximity with each other, leading to interference that in turn may result in incorrect or slower operation.

The reader collision problem within an RFID network plays an important role in ubiquitous RFID implementation (Engels, 2002). An anticollision protocol needs to be developed for a mobile ad hoc network of RFID readers. One of the challenges of RFID development is to make the tags as simple as possible even though doing so adds extra complexity to the reader. In a case where multiple readers try to read the same tag, the tag cannot select a particular reader to respond to. Therefore, passive tags, in which the collision may take place, are not able to take part in the collision resolution. Reader–reader collision not only decreases throughput of tag identification, but also increases the bandwidth usage.

5.1.3 Solution

The software-defined RFID reader, from here on referred to as SDLR, was developed to solve this interoperability problem. Traditional radios use hardware circuits, fixed at the time of manufacture, to perform the high-speed signal-processing tasks that convert back and forth between user data and the radio waveform. SDLR exploits advances in components such as digital signal processors (DSPs) and field-programmable gate arrays to make the hardware generic, and move all of the waveform-specific tasks into software. One SDLR device can support a variety of communications standards, just as one PC can run a variety of software applications. SDLR has a number of benefits in addition to improving interoperability.

SDLR has significant advantages over traditional RFID readers, that is why many Auto-ID labs are supporting its development.

Desirable characteristics include:

1. Receive and transmit various modulation methods using a common set of hardware.
2. Alter functionality of the system by downloading and running new software.
3. Possibility of adaptively choosing an operating frequency and a mode best suited to prevailing conditions.
4. Opportunity to recognize and avoid interference with other RFID readers and communication devices in a high-density reader environment.
5. Elimination of analog hardware and its cost, resulting in simplification of data logging reader architectures and improved performance.
6. Chance for new experimentation and development of new RFID protocols.
7. Support of multiple modulation formats.
8. System can be simulated *Exactly*.
9. Flexible bandwidth selection and management.
10. Ability to dynamically join or leave an ad hoc network formed by other readers in proximity.
11. Ability to transmit neighbor information to other readers or a server along with a request to transmit.
12. Ability to scan the status of its neighbors and respond accordingly.
13. On collision with other readers, reducing its power level by a predefined factor.

Although SDLR offers benefits as outlined above, a few obstacles remain to its universal acceptance. Those include:

1. Difficulty of writing software for various target systems
2. Need for interfaces to digital signals and algorithms
3. Poor dynamic range in some of the designs

5.1.4 Novel Properties of SDLR

SDLR is capable of adjusting its software to suit the particular RFID environment in use. The following sections give an overview of two novel applications that can be used to demonstrate the advantages of integrating the software radio into an RFID reader.

5.1.4.1 Automatic Change of Modulation Scheme

A framework for changing modulation schemes on a per packet basis is given in a paper by Bose and Hu (2001). In this case, the modulation scheme is determined by a packet header that identifies the modulation scheme. With proper signal processing it is possible to determine the modulation scheme used at the transmitter (Keith Nolan and Linda Doyle, 2001). A novel approach would be for an RFID reader to detect the modulation scheme of the incoming signal and then reconfigure its signal-processing stack accordingly. This powerful feature can be used to provide flexibility to readers not only in a multitag environment but also in a high-density reader environment. An FPGA device could dynamically load a new modulation scheme and become part of a new reader network without changing the hardware.

A layer in the communication stack can be used to act as a modulation detector and load the appropriate op-code into the FPGA device to demodulate the incoming signal. The modulation detection layer must perform analysis of the incoming signal to determine the modulation scheme used. This information can be added to attributes of the message block as it is passed to the next layer in the stack. The DSP, then, makes sure that the appropriate demodulation scheme is available to demodulate the incoming signal. If it does not have the appropriate demodulation scheme then, it raises an alarm, notifying the operator. Or alternatively it can send a request to the host computer asking for the latest demodulation scheme.

5.1.4.2 Peer to Peer Component Sharing

Tuttlebee (1998) describes Over-The-Air Reconfiguration (OTAR) for a software radio. Using this concept, a remote device can be reprogrammed by the transfer of new software into the device. Up to now, hardware restrictions have meant that OTAR has only been used for one-off projects such as satellites.

Now, advances in hardware allow OTAR to be used in ad hoc networks. An ad hoc network can be a connected set of RFID readers without any centralized or hierarchical structure. In an ad hoc environment, readers might be communicating with tags using many different protocols, modulation schemes, and location-specific parameters. The ability of nodes to share information about network conditions would be of key importance to ensure reliable communication is happening.

In such a situation, readers can go beyond just sharing information by sharing their network layer component as downloads. These network layers, for example, a new demodulation scheme, can also be downloaded from a central computer. In the absence of or unavailability of such a central computer system, the component sharing method can be used as a backup or alternative procedure to keep the system running. For example, a reader installed in a new location could contact its nearest operational reader using this scheme to download the most popular communication environment variables used in that area. These settings would include information such as available frequency allocation, bandwidth, average number of tags per unit of time, number of readers and their locations, noise conditions, etc. This information can be used to optimize communications. This scheme therefore can tremendously increase the flexibility of an RIFD system. Such functionality is not available currently in any RFID system.

5.1.5 Software versus Hardware

A traditional radio receiver requires many analog components, as shown in Figure 5.1. First the radio signal is received by the antenna and is then amplified by an LNA (low noise amplifier). These parts are shared by all the radio channels (Brannon, 1995). After these components one receiver is required for each channel. The RF signal is down converted to baseband by a mixer and a local oscillator. At each stage analog filters are used to discriminate the signal outside the frequency band. The last step before the digitization

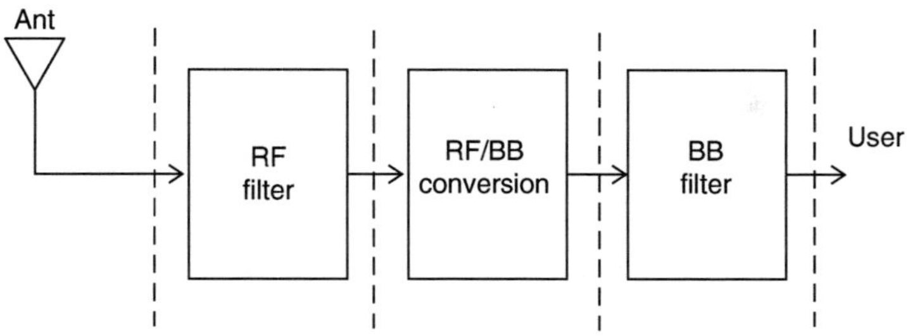

FIGURE 5.1
Simplified hardware chain of a traditional RFID reader.

is to decompose the signal into inphase and quadrature components. These signals are then converted in a narrow-band analog-to-digital converter (ADC). In UHF, for example, FCC regulations limit the total bandwidth to 25 MHz while the channel bandwidth is about 500 kHz. A reader therefore can handle more than 50 channels.

If the ADC is moved closer to the antenna, more components can be shared for all channels and more of the signal processing can be done in software (Figure 5.2). This means that the hardware cost can be substantially reduced. An ideal software radio would consist of an antenna, one ADC that samples directly on the antenna signal, one digital-to-analog converter (DAC) that generates the outgoing antenna signal in the transceiver and a DSP. All the signal processing should then be done in software in the DSP. This ideal situation, however, is not feasible or very expensive with today's technology.

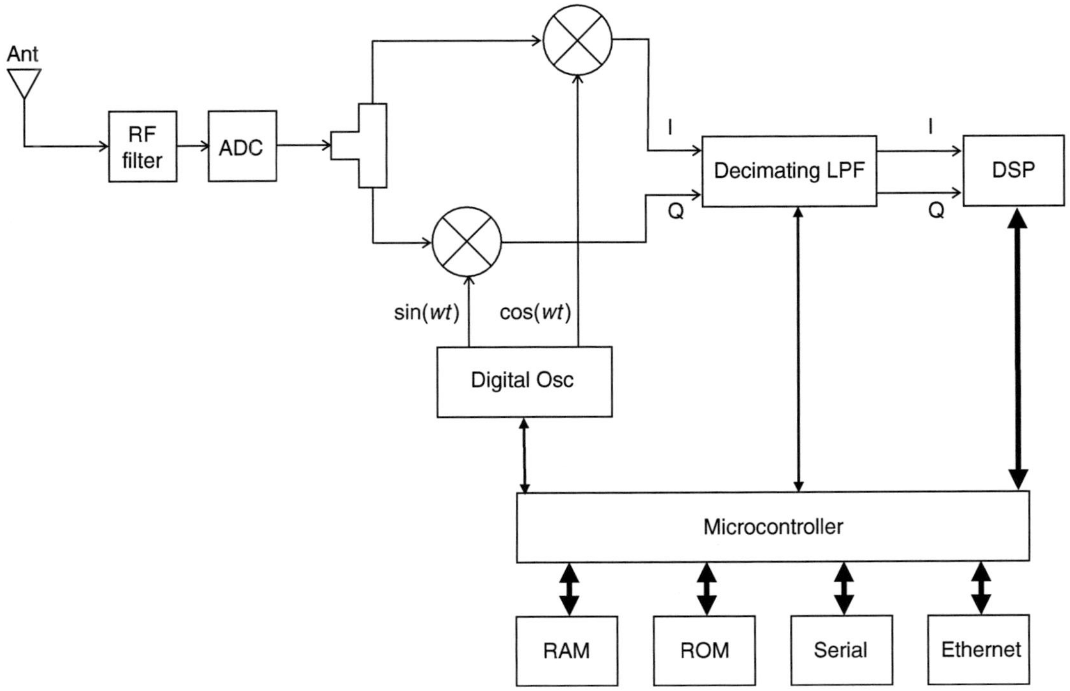

FIGURE 5.2
Simplified hardware and software chain of the SDLR receiver.

There are other advantages with a software radio architecture, besides the hardware reduction. A software radio reader could be reconfigured without replacing any hardware. The same hardware could also be used for different systems, for example, different tag generations and different classes, since all the signal processing is performed in software. The same software can be recompiled for a different processor and can be run on different platforms such as in handheld terminals. Instead of implementing several RFID receivers for different physical regions, one receiver could be used for all systems, and different software packages could be used to switch between systems.

The requirements on the ADC in a software radio architecture are very high, as the signal received by the antenna, for instance, does fluctuate in amplitude by a large amount. When such a wide dynamic range signal is converted in the ADC the dynamic range must be high enough so that a weak tag reply can be separated from the harmonics and the strong carriers. The SNR (signal to noise ratio) must also be high so that weak replies can be seen above the noise floor.

5.1.6 Summary

This section has demonstrated a flexible approach to implementation of an SDLR using dynamic communication schemes. This platform provides a powerful solution for rapidly building and testing new RFID communication systems and protocols, and it forms a basis for developing highly flexible RFID infrastructures.

The SDLR is an inexpensive and versatile solution to the collection of temporal and spatial field data for numerous research and educational endeavors. Defining an open architecture and implementing readers compatible with this architecture further enhances interoperability. It can be integrated with a large array of available sensors. In the next section we will discuss the SDLR in more detail.

5.2 SDLR Architecture

RFID systems are demanding more data, faster logging, and higher interrogation rates, so they require a powerful system that can manage information from a variety of sources, store the data, and transmit it reliably and continuously for long periods of time to other readers or a host PC. In addition, data logging is an extremely important part of RFID because it is the best method for determining what happens during the tag interrogation and to test software strategies and protocol efficiencies.

5.2.1 Introduction

Today's continuously changing technology in RFID brings the need to build futureproof RFID readers. If the functions that were formerly carried out by hardware can be performed by software, new functionality can be deployed easily by updating the software. With the existing stringent requirements of RFID spectrum, increasing traffic rates, and the need to adhere with the regulations on spectrum usage, this requires even more sophisticated signal-processing algorithms that can only be implemented on a software-based RFID reader system.

The software-based RFID reader system will also allow the addition of new functionalities with a short time-to-market. The SDLR includes excellent multipath and antenna diversity performance, resulting in superb tag-reading performance. User-specific functions can be configured easily to the performance required for different environments, and only the software needs to be upgraded rather than a completely new hardware design. By integrating everything into software, fewer external components are required, which further reduces the cost. This allows innovative new features and a rapid development cycle.

The SDLR can be part of a large distributed and dynamic system in which each reader is responsible for the management of its own local population of tags that is changing dynamically. In such a system, the reader acts as a gateway between the low-cost simple tags and a very sophisticated distributed information system that can interface to enterprise software applications.

The fundamental physics of antennas and radio frequency propagation properties at different frequencies cause devices operating at various bands to have different benefits and functionality trade-offs. Therefore, their use will remain a reality for a foreseeable future. The SDLR must be designed around this notion as a modular system and be able to support multiple frequency bands.

Furthermore, there is a need for flexibility in RFID design because the specifications are still changing, and even within a single specification, the tags can be asked to reply at different frequencies. For instance, tags working within the EPC Global Generation 2 specifications can be asked to reply at a subcarrier frequency ranging from 40 up to 620 kHz. Therefore, the reader must be able to be easily reconfigured to support frequency bands and protocols of different geographical regions and those that will become available in the future.

5.2.2 Hardware Design

This section concerns the design and development of the hardware for a software-based data logging reader. The prototype instrument developed for this project is called software data logging reader (SDLR). It features a ColdFire network processor, which enables it to perform high-speed data processing. It runs on an embedded Linux operating system and comes with a number of PC-like functions built in that enable it to run additional applications simultaneously. Additionally, it supports standard network protocols including DHCP, UDP/IP over Ethernet, 820.11x (Wi-Fi), HTTP, and SNMP, and its superior network adaptability enables its software to be easily implemented from remote sites.

The general setup of the implemented SDLR transceiver is shown in Figure 5.3. In the following sections we discuss various considerations that led to its particular design. A detailed description of the overall system and its individual components are also given.

5.2.3 Microcontroller

The microcontroller block is based on ColdFire, MC5407C3 Integrated Processor from Freescale Semiconductor. The MC5407C3 core processor is identical to the MC68EC000 microprocessor and features full compatibility with the MC68000 as well. It operates at 162 MHz. It also provides a UART, SPI, LCD controller, Timer/PWM, and parallel I/O.

The microcontroller runs on uCLinux, which is an operating system for a microcontroller without a memory management unit (MMU). uCLinux is derived from the 2.0 Linux kernel. It is a multithreaded real-time operating system for embedded applications. uCLinux is a mature, robust operating system that already supports a large number of devices, file systems, and networking protocols. It gives the developers complete visibility of the source code. Bug fixes and new features are constantly added, tested, and refined by a large community of programmers and users.

The microcontroller block consists of an Ethernet controller, a UART Serial port, RAM, and ROM. The system provides a boot strap mode function that allows system initialization as well as program/data download from ROM or via the UART (for debugging purposes). The microcontroller uses its parallel interface to communicate and load DSP

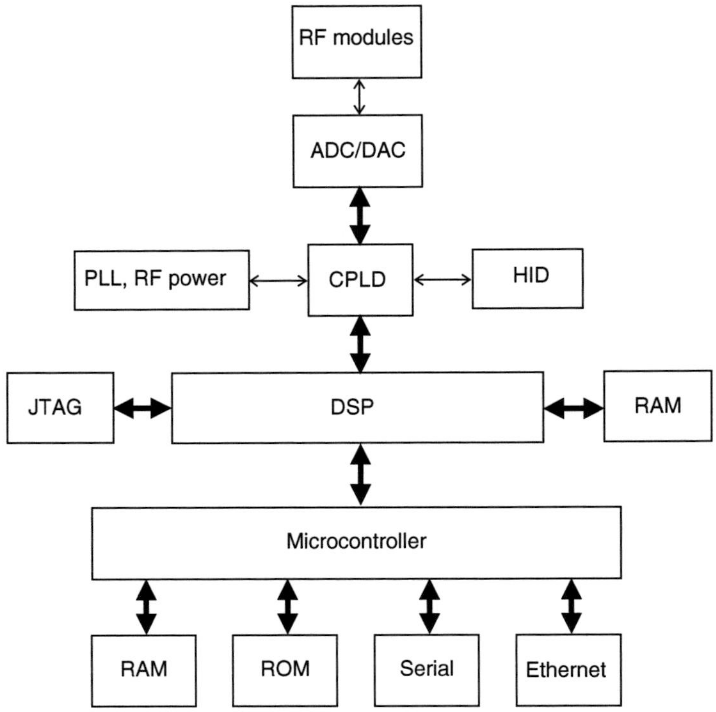

FIGURE 5.3
Simplified and abstract level hardware block diagram of SDLR.

firmware into the DSP chip. The microcontroller is responsible for high-level tasks such as system management, self-test, database maintenance, and serving requests from other readers or host PCs through a TCP/IP network, whereas the DSP handles low-level computationally intensive processing tasks, such as FFT and filtering.

5.2.4 Digital Signal Processors

This system uses the TMS320VC5416–160 fixed-point, DSP chip from Texas Instruments, operating at 160 MHz. The chip includes 256 kB of RAM and 32 kB of ROM, which is used for both program and data storage. A six channel DMA, 16 bit host port interface (HPI), and three multichannel asynchronous serial port (McBSP) are included. It operates with a core voltage of 1.6 V, whereas the IO supply voltage is 3.3 V.

All the low-level signal processing such as FFT, filtering, and waveform shaping are done within the DSP chip. It operates in Microcontroller Mode, allowing it to boot from the microcontroller board. It can be reset manually by power cycling or through the software control of the microcontroller. At power up the DSP waits for the host microcontroller to upload the DSP firmware into its internal RAM through its HPI. After the upload completes the DSP starts its boot cycle. More detail of the software architecture of the DSP subsystem is provided later in this chapter.

5.2.5 Host Port Interface

HPI is an 8/16 bit parallel port used to interface a host processor or a device to the Texas Instruments TMS320C5416 DSP (Texas Instruments Incorporated, 1997, 1999b). The HPI enables glue-less interface with host processors containing single or dual data strobe in addition to separate or multiplexed address and data buses. This section presents a hardware interface and the accompanying software protocol that is involved in communicating between the host and target unit through the 8 bit HPI.

5.2.5.1 HPI Boot Process

The TMS320C5416 contains 4 k-words of on-chip ROM. A portion of this ROM is used to store the bootloader code. The MP/MC bit of the processor is sampled at reset and its value partially determines the configuration of the DSP. If MP/MC is set low, then the C5416 is set to microcomputer mode and the bootloader will start execution following reset. The ROM code bootloader is located at memory address 0xF800.

There are two methods to signal the bootloader that HPI boot is active: INT 2 and memory location 0x007F. The bootloader checks to see if the INT 2 pin is set to one (active); if it is, then HPI mode is selected. The bootloader also clears the memory location 0x007F and uses it as a software flag to show that HPI boot is complete.

A simple way to achieve HPI boot is to connect pin HINT to INT 2 and wait at least 30 clock cycles for the bootloader to become ready. After this time delay the HPI code transfer process can begin. After completing the code transfer, writing the start-up address to memory location 0x007F will signal the DSP that the boot is complete. The TMS320C5416 will detect the change in the memory location 0x007F and branch to the indicated value in that memory location and start program execution. The HPI boot is now complete and the DSP will run normally.

The picture in Figure 5.4 shows the interconnection between the DSP's HPI and the ColdFire GP processor. As can be seen from the figure, the HPI is used in 8 bit mode and is connected to the upper 8 bit of the data bus. Address lines A[1–4] are used to control the HPI registers.

An 8 bit data bus exchanges information between the host and the target. Due to the 16 bit work structure of C5416, all data transfers with a host must consist of two consecutive bytes. The dedicated HBIL pin indicates whether the first or second byte is transferred.

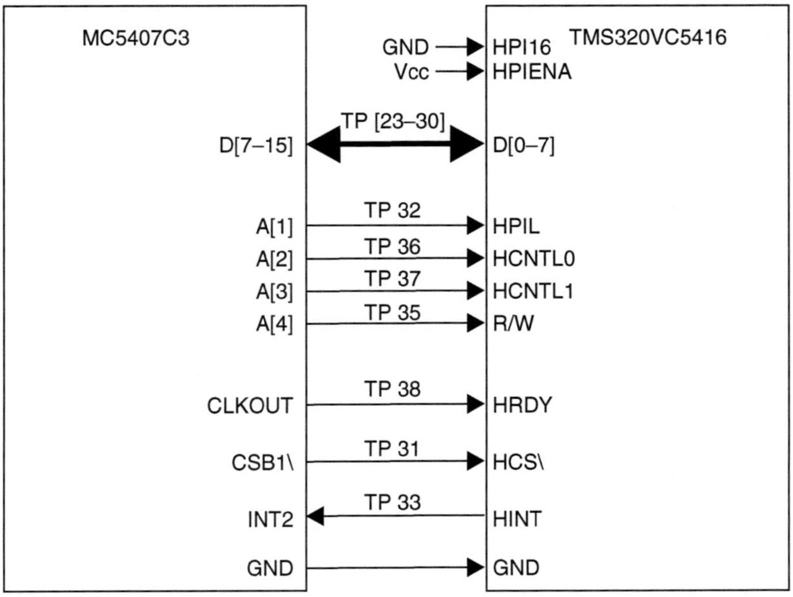

FIGURE 5.4
HPI connection.

The HCNTL0/1 control inputs indicate which one of the three internal HPI registers is accessed, as well as the type of access. In the proposed configuration the host address bus is used to drive these two control inputs along with HBIL and R/W signal.

The host can specify access to the HPIC, the HPIA, or HPID register. The HPIA serves as a pointer to HPI memory. The HPID is used to transfer data to and from the address pointed to by the HPIA. The HPID can also be referenced with the option of automatic address increment. In this mode, a data read causes a postincrement of the HPIA, and a data write causes a preincrement of the HPIA.

The HCS line serves as the enabling input for the HPI and must be low during an access. Following shows the address bits connected to the HPI port and the corresponding simplified C source code used to access the port for reading and writing.

```
/* The signals on the wires are shown in the following table.
  A[4]     A[3]     A[2]     A[1]     A[0]
-----------------------------------------------------------
   1        0        0        0        X        HPIC Read (HI)
   1        0        0        1        X        HPIC Read (LOW)
   0        0        0        0        X        HPIC Write (HI)
   0        0        0        1        X        HPIC Write (LOW)
            0        1                          HPID
            1        0                          HPIA
            1        1                          HPID (Auto-increment)
* /
#define HPI_PORT 0x202000
#define C_SEND 0x00 /* send ctl */
#define A_SEND 0x08 /* send addr */
#define D_SEND 0x04 /* send data, addr auto increment */
#define N_SEND 0x0C /* send data, NO auto increment */
#define HBIL_HI 0x02 /* Hi or Low byte */
#define HPI_R 0x10 /* A[4] is RW\ */
#define BYTE_HI(x) (x & 0xFF00)
#define BYTE_LO(x) (x << 8 & 0xFF00)
/* Writing to HPI */
FPWV (HPI_PORT+C_SEND) =BYTE_HI (data);
FPWV (HPI_PORT+C_SEND+HBIL_HI) =BYTE_LO (data);
/* Reading from HPI */
data= ( (FPWV(HPI_PORT+A_SEND+HPI_R)) & 0xFF00);
data += ( (FPWV(HPI_PORT+A_SEND+HBIL_HI+HPI_R)) & 0xFF00);
```

Host communication with the HPI is dependent on whether the HPI is ready to perform a transfer as indicated by the HRDY output line. When HRDY is high, the HPI is ready for a transfer to occur. When HRDY is low, it reflects that the HPI is busy completing the previous transaction. Since HCS enables the HPI and it is inactive (low) when an access occurs, it may be inferred that HRDY, which is always active high except for the duration of an access, is always active when HCS is active.

5.2.5.2 Software Description

The original kernel and applications are written in a combination of assembly and C source and compiled into a common object file format (COFF). However, to be able to transfer this code to the DSP through the HPI, it is necessary to convert the .out COFF file into a hex format.

A hex extraction utility program that was written for this purpose is called coff2hex. This utility is used to extract hex equivalent of a 16 bit COFF file. During the extraction process, the coff2hex utility also resolves all address references based on the memory map specified in the linker command file. The coff2hex utility is based on an original coff_both.c source for COFF extraction that was published by Texas Instruments (Texas Instruments Incorporated, 1999a).

5.2.6 Complex Programmable Logic Devices

The system used the EPM3256A from Altera class of Complex Programmable Logic Devices (CPLD). This device is based on MAX 3000A family. It contains the electrically erasable programmable read-only memory (EEPROM), which provides instant-on capability and offers 256 macrocells. EPM3256A device supports in-system programmability (ISP) and can be easily reconfigured in the field. Each macrocell is individually configurable for either sequential or combinatorial logic operation.

This device is responsible for the interface between the DSP chip and other peripherals by converting and buffering the signals from the ADC, DAC, PLL, etc., to the McBSP format, which is acceptable by the DSP chip.

The CPLD has various internal registers, which are mapped to its IO pins. These IO pins are connected to PLL, LEDs, and several other peripherals that can be controlled by software.

5.2.7 Analog-to-Digital Converter

The system makes use of an ADS5231 high-speed, dual-channel ADC. The ADS5231 offers 12 bit resolution at sample rates of up to 40 MHz. It is interfaced directly to the CPLD and it drives its clock signal from the CPLD as well.

5.2.8 UHF Module

The UHF module is responsible for interfacing the DSP to the antenna. A simplified block diagram of this module is depicted in Figure 5.5. Some blocks of this design may also be found in Figure 5.3. It operates at 880–1050 MHz UHF band and is compatible with FCC's part 15.247 rules. These rules specify that a maximum output power of 1 W may be exercised in a frequency-hopping system with at least 50 channels. The maximum dwell time for each channel is set to 400 ms at any given frequency. The UHF band module is subject to PLL lock-time and the receiver circuit turn-around time limitations. The lock-time is the time that it takes the PLL to switch from one frequency to another for a given frequency change to a given frequency tolerance. In order to mitigate the effect of PLL lock-time, a two synthesizer design was chosen. This design allows the system to program the second PLL to a new frequency while the first one is still operating and then to switch over to the second PLL when a frequency hop is required. This configuration prevents the dead-time during which the reader RF field would be off and no data can be transmitted. In addition, it reduces the risk of tags' brown-out (when the voltage temporarily drops below the operating voltage level and then recovers).

5.2.8.1 Oscillator

A PSA0965A phase locked loop (PLL) from Z-Communications Inc. is used to generate the operating frequency. It is a small hybrid circuit block based on a National Semiconductor LMX2316 PLL IC. This small module generates an output of 3 dBm with phase noise of −1000 dBc/Hz at 10 kHz from the carrier frequency. The output of the oscillator

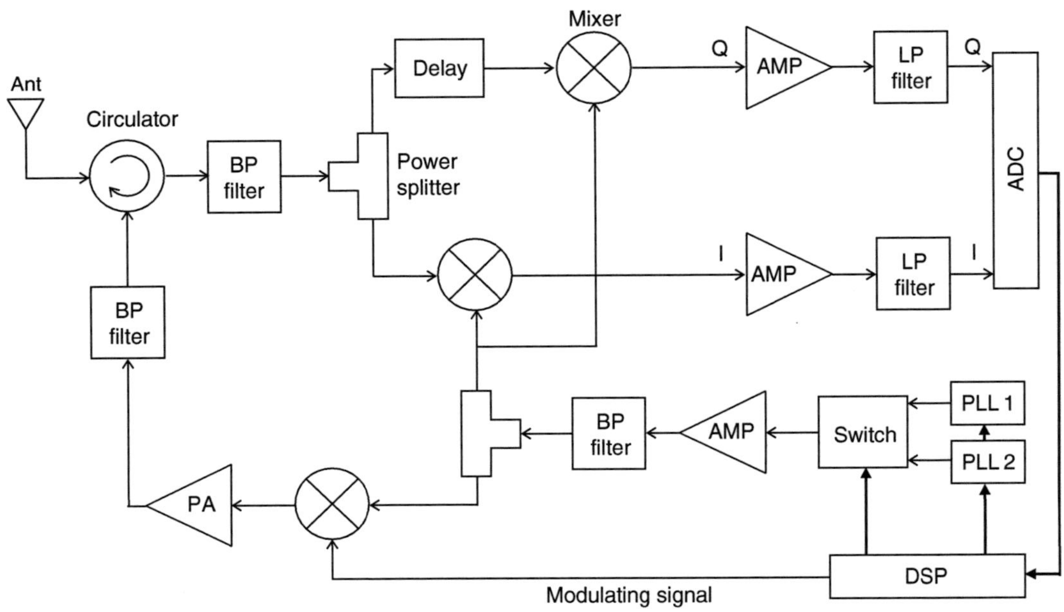

FIGURE 5.5
Simplified hardware chain of the UHF module.

is bandpass filtered and then amplified by 10 dB. The amplifier output is then subjected to a ceramic filter with center frequency of 915 MHz to remove the strong harmonics and other unwanted spurious outputs. The output is then split into two parts, one is fed into the receiver chain and the other into the transmitter chain.

5.2.8.2 Transmit Chain

The transmit chain comprises a PSN0930A PLL as local oscillator and frequency hopping. A 3 bit digitally controlled attenuator is used to adjust the RF output power level. The attenuator controls the gain of the RF power amplifier (an M57785 IC from Mitsubishi Semiconductors). That stage is followed by a ceramic bandpass filter to suppress harmonics and spurious frequencies. The module is capable of delivering up to 4 W of RF power in the UHF band. The modulation is achieved by means of a mixer (MCL SBL-1Z) from minicircuits. The modulating signal is generated using a DAC, which receives its input from the DSP module directly.

5.2.8.3 Receive Chain

A three-port circulator (CN-6) from MECA Electronics Inc. is used to separate the transmitted signal from the weak tag reply. The tag reply is then filtered using a ceramic bandpass filter centered at 915 MHz. It is then split into two parts for I (Inphase) and Q (Quadrature) demodulation. The two phases are then mixed with the local oscillator using minicircuits double balanced mixers. The local oscillator signal for the Q channel is delayed by a quarter of wavelength to generate a 90° lag in the Q path.

The mixed output signals are then amplified and filtered by an antialiasing filter. The outputs are then amplitude limited and then applied to the ADC. Further signal conditioning and processing are performed digitally in the DSP.

5.2.9 Software Design

The system makes use of DSP-BIOS II from Texas Instruments. DSP-BIOS is a scalable real-time kernel, designed specifically for the TMS320C5000 and TMS320C6000 DSP platforms. DSP-BIOS is an integral part of the Code Composer Studio Development Tools from TI. The programming is done using Microsoft Visual Studio C++ and Texas Instruments' Code Composer Studio.

The heart of the SDLR is the TMS320C5416 DSP, which runs a TI DSP-BIOS II real-time kernel and controls the interface in addition to the specific transmit and receive functions. DSP-BIOS provides a convenient multitasking capability through events, software and hardware interrupts, which operate at various levels of priority. Other functions such as tag management and communication to other readers or host PCs are performed in the microcontroller.

The DSP is required to receive commands from the microcontroller, and form packets of symbols, which are embedded into a flexible frame in order to be recognizable by a tag. The data is oversampled and pulse shaped before passing it on to the RF module. Thus, the aims of the DSP implemented in software comprise:

1. Initialization of interfaces and interrupts.
2. Initialization of CPLD, PLL, ADC, and DAC. These are done via McBSP.
3. Reception and transfer of data from/to the microcontroller. This is performed over the HPI.
4. Formation of tag command data frames.
5. Mapping from bytes to symbols. The bit stream is converted into amplitude modulated (AM) symbols, whereby here an AM mode is employed, although other modulation modes are possible.
6. Pulse shaping. The symbol stream is moderately oversampled by a factor of 2, which is required as a minimum by the subsequent mixer hardware, and filtered by a root raised cosine filter.
7. Passing data samples to the RF module. The oversampled and pulse-shaped signal values are passed over to the DAC, and its output is connected to the mixer's input.
8. Reception of I and Q data from ADC. The inphase and quadrature signal values are received from the ADC via the McBSP/CPLD. These samples are oversampled by a factor of 4 compared with the symbol rate to permit sufficient resolution for timing synchronization.

Steps 1 and 2 are performed at boot level. Initialization of PLL, ADC, and DAC needs to be done via the McBSP after the CPLD is initialized and functional. Processing of both the transmitter and receiver functions is performed in interrupt service routines, whereby the processor falls into an idle mode if no interrupts need to be serviced.

5.2.9.1 Device Drivers

A software module that controls how a processor communicates with a device is called a device driver. In SDLR, the interface from the DSP firmware to the hardware is abstracted into a set of device drivers. The device drivers separate the high-level firmware and protocol-specific modules from the low-level input/output (IO) routines of hardware. Device drivers are provided for each of RF's transmit and receive modules as well as for actuators and sensors.

SDLR uses a special abstraction to access device registers independent of the underlying implementation. It hides the mechanisms to access a specific part of the address spaces. The use of this abstraction is in almost all cases as efficient as an assembly language access. These device drivers are designed to provide access to the hardware in a protocol-independent manner that allows the change of communication protocol without the need of changing the rewriting low-level subroutines. The device driver application program interface (API) is a set of C callable functions for writing to or reading from the device or registers. This API is based on the POSIX file IO interface, using `read`, `write`, and `ioctl` functions. A device is made active by calling `open` and released by calling `close`.

5.2.9.2 *open()*

The routine `open()` opens a device as a file for reading, writing, or updating, and returns a file descriptor for that file. The arguments to `open` are the device name and the type of access. In general, `open` can only open preexisting devices and files. Files cannot be created with `open`.

The return value of this call is a file descriptor number, or ERROR if a file name is not specified, the device does not exist, no file descriptors are available, or the driver returns ERROR.

5.2.9.3 *close()*

The routine `close()` closes the specified file and frees the file descriptor. It calls the device driver-specific function to do the work.

The return value is the status of the driver close routine, or ERROR if the file descriptor is invalid.

5.2.9.4 *ioctl()*

The `ioctl` call provides an interface to device-specific configuration function and performs an IO control function on the device. The control functions used by SDLR device drivers are defined in the header file `io.h`. Most requests are passed on to the driver for handling. Since the availability of `ioctl` functions is driver specific, these functions are implemented separately for each specific device.

The return value is the return value of the driver, or ERROR if the file descriptor does not exist.

5.2.9.5 *read()*

The routine `read()` reads a number of bytes (less than or equal to maxbytes) from a specified file descriptor and places them in a buffer. It calls the device driver to do the work.

The return value is the number of bytes read (between 1 and maxbytes, 0 if end of file), or ERROR if the file descriptor does not exist, the driver does not have a read routine, or the driver returns ERROR. If the driver does not have a read routine, errno is set to NOTSUP.

5.2.9.6 *write()*

The routine `write()` writes a number of bytes from buffer to a specified file descriptor. It calls the device driver to do the work.

The return value is the number of bytes written, or ERROR if the file descriptor does not exist, the driver does not have a write routine, or the driver returns ERROR. If the driver does not have a write routine, errno is set to NOTSUP.

The precise meaning of reading or writing from or to a device depends on the nature of the device. For example, writing to an RF driver causes the data to be modulated over the output RF signal, and reading from the RF driver fills an input buffer with samples from an ADC chip. Reading and writing to or from some devices might be illegal. For example, reading from the transmitter and writing to the receiver driver are illegal operations.

5.2.10 EPC Module

The EPC software module is entirely implemented on the DSP. It is based on EPC specifications and supports both EPC Class 0 and Class 1 protocols as well as ISO 18000–6 protocols. It is also designed to accommodate Generation 2 protocol by EPC Global (C1G2) by adding some variables in function calls.

This module consists of functions for generating EPC Class 1 Generation 1 commands such as *Ping, Global Scroll,* and *Masked Scroll* and interpreting the tag reply. It can communicate with EPC 64 bit and 96 bit tag code structure and is capable of handling anticollision protocols.

5.3 Conclusion

By minimizing the hardware requirements for different protocols and implementing software modules that abstract away the differences between protocols, SDLR achieves superior performance to most other solutions in terms of hardware cost and software flexibility.

This possibility has motivated the concept of the SDLR, whereby the digital-to-analog and analog-to-digital conversion are performed as close as possible to the radio frequency. The aim of this work was to extend the digital domain and implement modulation, demodulation, channel coding, and other required processing tasks in software.

Being software based allows the SDLR to be customized to end-user needs. This design has a boot loader feature allowing a user to make changes to the code. It can be upgraded in the field thus lowering operating costs.

References

Bose, V. and Hu, R.M. (2001, February). Dynamic physical layers for wireless networks using software radio. *International Conference on Acoustics, Speech, and Signal Processing, Salt Lake City, USA*, 4, 2045.

Brannon, B. (1995, September). Basics of designing a digital radio receiver, Analog Devices Inc., Greensboro, North Carolina.

Engels, D.W. (2002, February). The reader collision problem. *Auto-ID Laboratories, White Paper Series, Edition 1.*

Engels, D.W. (2003, September). On the probability of ghost reads in an RFID system. *Auto-ID Laboratories, White Paper Series, Edition 1.*

Keith Nolan, Linda Doyle, and Philip Mackenzie (2001, November). Software radio signal space based adaptive modulation for wireless networks. *First Joint IEI/IEE Symposium on*

Telecommunications Systems Research. Network and Telecommunication Research Group, Trinity College, Dublin, Republic of Ireland.

Lewis, S. (2004, January). A basic introduction to RFID technology and its use in supply chain. *Technical Report*, Laran Technologies, White Paper.

Texas Instruments Incorporated (1997, February). *TMS320C54X DSP Reference Set, CPU and Peripherals*. Texas Instruments Incorporated.

Texas Instruments Incorporated (1999a, March). *Extracting Equivalent Hex Values from a COFF File*. Texas Instruments Incorporated.

Texas Instruments Incorporated (1999b, April). *Practical Application of the TMS320C54x Host Port Interface (HPI)*. Texas Instruments Incorporated.

Tuttlebee, W. (1998, September). Software radio: Impacts and implications. *IEEE 5th International Symposium on Spread Spectrum Techniques and Applications*, 2, 541.

6

Progress in RFID Education

Brian J. Garner

CONTENTS

The content of this chapter solely reflects the experience and opinions of the author and does not necessarily represent the views of institutions, companies, and colleagues unless their contribution (research) is specifically cited.

Where reference is made to products and individual applications, the appropriate source typically a Web site or research paper is given. No liability is accepted for errors regarding these citations or to other causes.

6.1 Introduction

The rapid growth in applications using radio frequency identification (RFID) devices, sometimes in association with other mobile (communication) technologies, now warrants significant business interest in extending the options available for automated identification of commercial and industrial items. Unfortunately, education of the respective business, government, and consumer communities has focused in the past on the physics, electronics, and technical features of such devices, rather than stressing the overarching benefits that may be derived from their exploitation in complex and in extreme (e.g., hostile) operating environments.

The author's experience in postgraduate education during the past 25 years regarding emergent technologies, such as electronic data interchange (EDI) and more recently, mobile technologies, strongly supports the need to include the appropriate industrial and consumer marketing perspectives in educational paradigms for emergent and disruptive technologies. While predictive models, in themselves, are helpful in trend analysis and in establishing market potential of new technologies, there are numerous examples of market failure without innovative integration of education and marketing requirements; for example, in the development of effective business process models. A distinction is thus made with training provided by a supplier on the use of a specific product, which may digress from the use of existing educational practices, including e-learning, due to business process integration issues.

The knowledge level of students undertaking advanced courses, typically postgraduate certificate, graduate diploma, or a master degree, requires consideration in the selection and presentation of blended learning material. The increasing use of distance learning, including mobile education options, typically *podcasting*, also needs to be taken into account.

While it is not intended to duplicate training options widely available for students interested in the scope, standards, and industrial/retail use of automated identification technologies, particularly bar codes, a rudimentary knowledge of current business practice and limitations in the use of such technologies is considered desirable to appreciate the rationale and transcendence of RFID.

Appropriate reference sources for remedial education on the EAN bar coding standard for automated identification purposes are provided in the appendix to this chapter. Note, however, that several symbologies are still in use by the business community.

6.2 Overview of Business Drivers, Trends, and Processes

Effective technological forecasting techniques require analytical skills, and in particular, a knowledge of diffusion theories. However, for marketing purposes, knowledge of business trends and processes is also fundamental to understand the current business drivers and engage the attention of business prospects including governments.

Global integration of business logistics, supported by EDI, by universal standards for product identification at the item, package, pallet, and container level, by "just-in-time" manufacturing processes and by government facilitation of customs harmonization (i.e., electronic clearance of goods) has established significant productivity advances in the advanced economies and remarkable improvements in living standards in the newly industrialized economies (NIEs). This productivity drive now underway through business process improvement has nurtured the management discipline of business process modeling (BPM), particularly focused on the supply chain management requirements to achieve *just-in-time* objectives.

Finance and service industries, such as Insurance and Superannuation (in Australia), and more generally, the Financial Planning sector, have also embraced the *productivity imperative* using EDI, including financial EDI and BPM. More recently, cross-industry standards for naming conventions have been adopted in Australia and progress with global (ISO) *semantic web standards* will accelerate the use of *service-oriented architectures* (SOAs) in the establishment of common web application design compatible with *middleware* requirements.

The rising cost of employees, compounded by adverse demographic trends in the advanced economies, has spurred management interest in new electronic processes and technologies that remove people from the supply, administration, and financial chains currently in operation. How may such processes be classified? Three categories have proven useful and germane in an educational, specifically e-learning and RFID context:

1. *Tightly coupled processes*:

   Illustrative examples of these processes would be

   - Electronic business XML processes
   - Supply chains (business-to-business)
   - e-Taxation (government to business/tax agents/individuals)
   - Super industry flow model in Australia (refer below) (*Source*: Public presentation by Dr. Cameron at Deakin University, Melbourne, August 2002.)
   - Real-time automation and process control
   - Disaster recovery planning

Super industry flow model

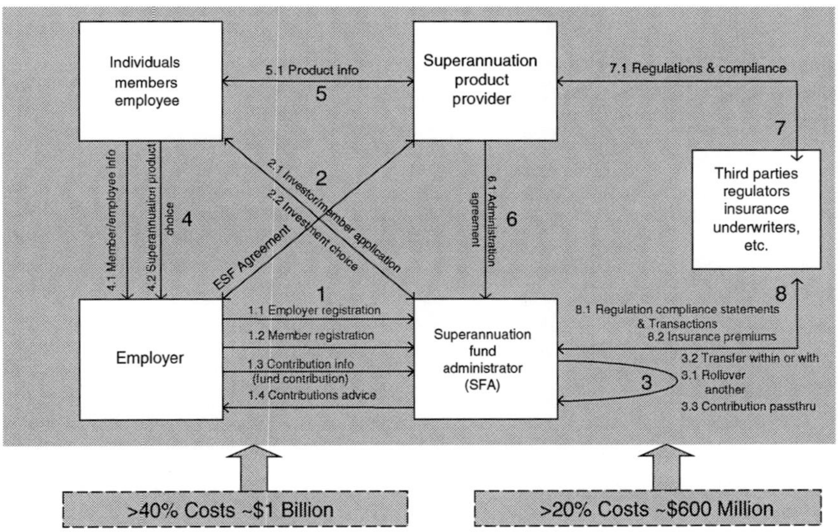

2. *Loosely coupled processes (note that choice of business partner, both B2B and B2C are accommodated)*:

   Illustrative examples would include

   - Travel industry
   - Hospitality and education
   - Entertainment industry
   - Health chains (e.g., telemedicine)
   - ICT Industry
   - Age care networks

3. *Social networks*:

   Formal social networks might include

   - Coordinated welfare agencies
   - Voluntary services for the disabled

- Amateur sporting networks
- Libraries
- Private clubs/associations

The marketing opportunities for RFID typically depend not only on the *productivity imperative* in categories (1) and (2) but also on quality control and continuous audit requirements, focusing in particular on *compliance requirements* (e.g., safety requirements in mining and manufacturing) established by government regulation or by formal business/community contracts, including social compacts.

6.3 Instructional Frameworks for RFID/Smart Labels

The basic instructional framework used successfully by the author at Deakin (e.g., EDI program, computer audit, and in knowledge engineering courses) and at Hong Kong Polytechnic University may be summarized in an RFID context as follows:

1. Introduction
    - Conventional labeling techniques
    - RF Device characterization
    - Progress with standards for RFID
    - Regulatory implications (EPCglobal)
    - Business drivers
2. Current application areas
    - Retail industries (includes jewelry)
    - Transport industries (air, shipping, vehicle)
    - Healthcare/medical developments
    - Gaming (Casino) software support
    - Supply chain integrity (refer www.technologyreview.com)
    - Mobile education tools (e.g., PDA/GPS)
    - Tracking and authentication of individuals (e.g., electronic passports and boarding passes)
3. New developments
    - Process integration (e.g., with biometrics and global positioning systems)
    - Commercial workflow automation (e.g., document tracking)
    - Network security (includes RFID encryption and *smartcard* technologies)
    - Printed and thin film electronics (includes reference to organic substrates)

Given that many of these topics are covered by other authors in this *RFID Handbook*, this chapter will address the educational issues in highlighting the marketing and process advantages of RFID and *smart labels*, namely:

- Solutions using RFID to limitations of existing technologies in automated identification domains
- Novel applications (i.e., unique to RFID solutions)

- Blended learning experience with current RFID implementations
- Summary of educational technologies for RFID instruction

6.4 e-Pedagogy in Learning Process Definition for Domains Requiring Automated Identification Technologies

The importance of leadership and innovation on *pedagogical requirements for blended learning* is the subject of considerable current research into educational practice. Given the broad scope of this subject, and the likelihood that part, at least, of the training program will be delivered through *e-learning*, instructional frameworks and axioms specifically relevant to RFID *e-learning* requirements will be selected and exemplified in the first instance.

6.4.1 Instructional Strategies and Axioms

The essence of an *e-pedagogy* is the relationship between learning objectives and instructional design. In practice, a set of learning objectives would be defined, as is customary in any training environment. However, it should be recognized that the value proposition for the student depends on an explicit link between the assessment (success) criteria for each objective and the design of an appropriate instructional process. A direct correlation needs to be established between the critical success factors (CSFs) associated with the objective and the learning paradigm employed. *Context* is paramount in any training scenario, but in practice the educator is required to consider the motivation, self-assessment, and tacit knowledge appropriate to e-learning context selection [1]. The instructional strategy selected for RFID students, however, depends on their current knowledge level, including business awareness, and on the complexity of the business process/application.

The subject complexity is also a key issue in designing the requisite learning grid, which in simple terms is the educational resource grid available to students, with the attendant web-link network and access authorities. Research by Philp and Garner [2] at Deakin University in knowledge-mediated software engineering identified the need for *context-specific dynamics* in managing the diverse programming support situations that may be required. Students seeking to understand corporate information security management risks in a data hosting organization, for example, would usually be challenged by wireless LAN implementation schematics, such as the referenced context shown in Figure 6.1, without access to associated instructional links (e.g., *virtual private network* instruction).

More generally, the management of subject complexity requires student access to relevant *explanation knowledge* without having to prompt the student. The ability to personalize context selection and dynamic support, rather than rely on the usual *helpdesk* solution, is seen to be a CSF for e-learning management in frameworks that rely on learning grids. This observation and conclusion follows from the overarching importance of effective self-assessment in distance learning, and consequently, effective e-pedagogy may be simplified to the formalization of axioms that strengthen self-assessment requirements with reference to the CSFs. Readers wishing to know more of current processes and analytical tools to assist students in self-assessment may wish to consult Tim Robert's book [3] entitled *Self, Peer and Group Assessment in E-Learning*. While there is, without question, an urgent need to compare and contrast current practices, the emphasis by educators, predominantly academics, to promote a *culture of critique* has largely ignored the contextual significance of content, particularly complexity issues, in self-assessment, such as in training process

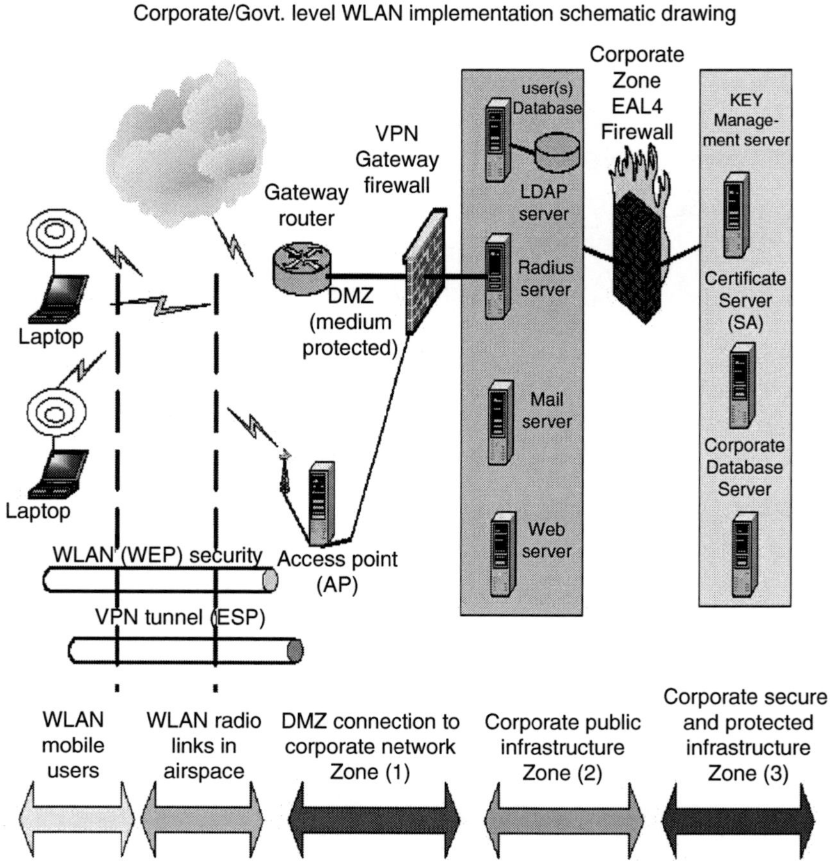

FIGURE 6.1
WLAN (WEP) security schematic.

control operatives, who would typically not have a detailed knowledge of the chemical processes involved or crisis management requirements.

Axioms derived from an analysis by the author of *self-assessment requirements, particularly CSFs in RFID education* may thus be summarized as

- *Conceptualization of roles*, and the evolution of roles in event management, including software module activation, is important in teaching business process dynamics. In disaster recovery management, for example, role uncertainty and constrained resource availability have a major impact on the effectiveness of business recovery.
- *Personalized contexts* are required to enable students to exercise their skills and imagination during a knowledge discovery process. Studies of electronic gaming experiments support the requirement for student involvement in *rule-based scenarios*, and where possible, self-selection of roles, not just in routine problem solving, but in the planning and execution of interim strategies designed to achieve a requested outcome.

- *Situated learning*, in which the effect of a new technology is to be explored, requires an appropriate scenario and communication of the underlying process knowledge, possibly using BPM illustrations and by stressing *group role dependencies*.
- *Knowledge-mediated e-learning* is often required when students demonstrate (using objective tests) that they lack the tacit knowledge to progress using the recommended self-assessment processes. Note that the current trend to semantic enrichment using metadata support for web and grid (service-oriented) architectures does not, at this stage, offer students the required tools to analyze contextual semantics and ambiguities during the context-awareness process.
- *Intelligent human–computer interaction systems* are known to provide effective support to students in context awareness, but are not typically customized to individual student requirements by existing learning management systems.

6.5 Suggested Strategy for RFID Education

The basic framework outlined earlier (p. 112) may usefully be focused and expanded in strategic terms, given the cited importance of learning scenarios, e-pedagogy, self-assessment, and the axioms derived from CSFs.

The recommended strategy drives the selection of appropriate learning scenarios, which may involve BPM (simulation) to define functional RFID objectives. Students should be expected to master the following topics, and to include in their sourcing of technical device specifications, use of *RFID Primers/Newsletters* available from EPCglobal and from manufacturers/consultants:

- Business (value) propositions
- EDI Scenario
- Marketing CSFs in RFID applications
- RF Spectrum use in RFID: international protocol standards
- RFID operational/management requirements, including RFID implementation
- Process technology integration issues

Note: Novice learners would be expected to demonstrate prior competence in *Automated Identification technologies, other than RFID,* using course material available from external reference sources such as those cited in the appendix.

6.6 Business (Value) Propositions

An overarching justification for RFID lies in their role in tightly coupled processes, such as global business supply chains relying on *just-in-time* production. The *Auto-ID Center*, for example, was set up to develop a low-cost RFID system for this important business requirement. Their technology was licensed in 2003 to EPCglobal, a partner in the collaborative *Auto-ID Project*, and six tag classes, including the two protocols produced by Auto-ID, are royalty-free to manufacturers and end users. RFID standards coordinated by EPCglobal and approved by the International Standards Organization (the ISO 18000 series) also depend on

- Object identification using trade identification numbers for global use (GTIN)
- Global location numbers for the identification of locations (GLN)
- Global returnable asset identifier (GRAI) for pallets, and so on

In global supply chains, manufacturers typically seek to minimize inventory, to satisfy their customers *on demand*, and to manage their assets, including financial cash flow to maximal advantage. This exact requirement may be identified from the importance of global logistics linking multiple manufacturing sites, warehouses, and customer locations. The ubiquitous requirement to locate and identify product, irrespective of packaging and distribution process underpins the revolution in automated identification techniques and technologies. The *value propositions* sought by business, as appropriate to this complex set of real-time interactions between suppliers and customers, may be generalized as

- Total visibility of product and logistics across the (global) supply chain
- Effective contingency plans for supply interruptions, including where possible, automated advice to customers of goods locations and revised delivery schedules
- Optimal sourcing of business partners able to guarantee performance in a global process management context
- Reliable automation and near-real-time information updates

Given these demanding requirements, automated identification of product movement without human intervention is essential to maximize financial performance. The role of RFID, supported by effective information and communication systems, is consequently increasing due to the limitations of visual inspection systems, albeit with machine-readable bar codes, or optical character recognition.

6.7 Electronic Data Interchange Scenarios

In the 1990s, a business communications revolution occurred using EDI standards for the computer-to-computer exchange and processing of transaction data. The more common standards are ANSI "X12," which originated in the United States, and EDIFACT, which was developed under UN auspices. EDIFACT was conceived as a comprehensive solution to

- EDI for administration
- EDI for commerce
- EDI for transportation requirement

Subsequent participation by global suppliers of business software, represented by the *OASIS Group*, simplified the EDI standards, particularly for small and medium enterprises (SMEs), by using the "XML" language. This *UN/CEFACT* initiative comprised the following WorkGroups, in which XML has been universally adopted as the de facto standard, particularly for Internet commerce (refer Figure 6.2).

- Business Domain Groups (ewg)
- Business Process Analysis (bpawg)
- Business Process Definitions
- Core Components Discovery Teams (ewg)
- Cross Domain Harmonization (ewg)
- International Trade Procedures (itpwg)
- Reference Models (bpawg/itpwg)
- Trade Facilitation Best Practices (itpwg)

FIGURE 6.2
UN/CEFACT work groups for ebXML.

The *learning scenario* preferred by the author in introducing EDI and automated identification technologies is, thus, a *supply chain management scenario*, using business (modeling) processes in business-to-business (B2B) trading. Logistics management provides the detailed context. This *scenario* has been used as the basis of EDI instruction at Deakin University, and from 1992 to 1997 in India through the All India Management Association.

A useful instructional schematic for the EDI business cycle, and consequently, for training in logistics management and in automated identification requirements, is shown in Figure 6.3. Naturally, such schematics may be extended to accommodate internal processes and logistics, which has happened in the retail sector, but it should also be

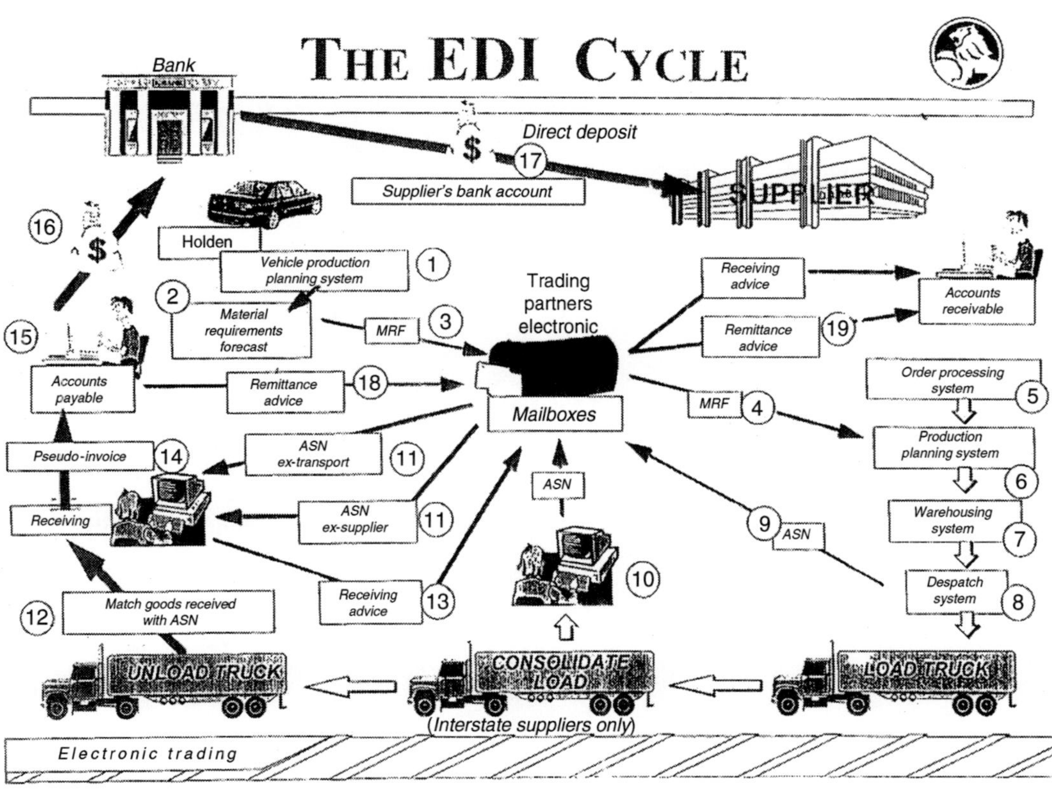

FIGURE 6.3
Supply chain management scenario.

noted that RFID is exciting general business interest in automated identification as a new *business control and continuous auditing strategy*.

6.8 Critical Success Factors in RFID Application Marketing

Stimulating management interest in the adoption of RFID in automated identification has posed a marketing challenge since 2004, originally due to the lack of RFID readers in quantity. The market has developed steadily, however, and novel applications are now reported worldwide. The scope of current applications encompasses:

- Supply chain management, including retail applications
- Management of expensive mining assets (e.g., *RFID tag-based* tracking system in production scheduling and for safety control in minimizing underground truck collisions—BHP Billiton)
- Loss (fraud) prevention (e.g., jewelry items, gaming "chips")
- Tracking sensitive documents in government organizations
- Compliance requirements in workflow process management (e.g., financial and legal sectors)
- Airline and postal security (customer luggage, freight, parcels)
- Personal identification devices (e.g., passports)

While each application will usually have a unique business driver and *value proposition*, a number of CSFs are beginning to emerge and may well influence the future direction of the RFID industry and its associated research programs. CSFs known to be important in RFID application marketing would include

- High reliability and response of read/write RFID volume applications.
- Flexibility in solutions engineering when *off-the-shelf* products are inadequate. For example, integrating RFID with global positioning systems and with BPM solutions may well achieve management performance targets and demonstrate business value.
- Engineering RFID solutions for hostile environments is increasingly required, demanding novel substrates for embedded tags and an adequate signal processing protocol/reader for multiple item configurations, such as parcels or luggage on conveyor belts.
- System integration skills for *near-real-time* database processing when updating/retrieving RFID data on-site.

The ability to understand the financial implications of these CSFs is important in selecting an educational strategy, particularly a *learning reinforcement strategy*.

6.9 RF Spectrum Use in RFID: International Protocol Standards

Students familiar with mobile technologies may nonetheless find it useful to review the allocated RF spectrum use in relation to the ISO/IEC 18000–6 RFID frequency used by the Japanese in the 800/900 MHz band. Shown below are the relative spectrum allocations.

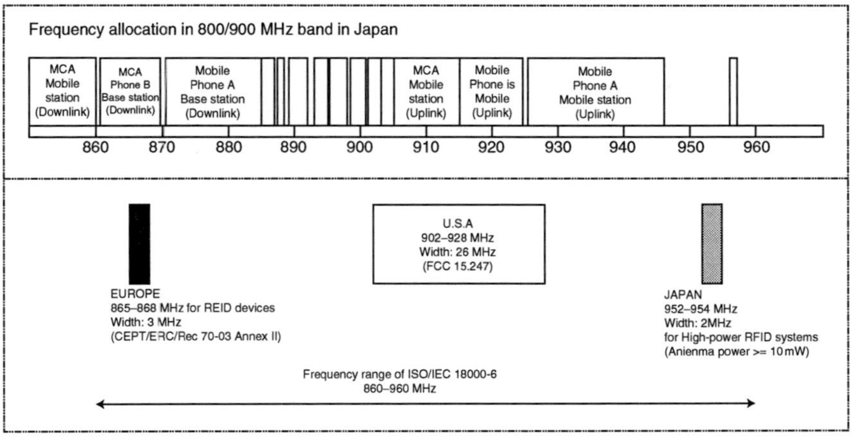

6.10 RFID Operational/Management Requirements, Including RFID Implementations

While the initial uptake of RFID was slower than expected, the evolution of RFID systems since 2005 has, in fact, been quite remarkable due to

- Successful pilot studies by industry groups, coordinated by *EPCglobal*, typically focused on transport, packaging, and warehousing applications (e.g., in Australia).
- Recognition that the RFID readers produced for high volume, online applications (e.g., in 2005) were unable to provide the speeds or reliability required by industry
- Pioneering research in Australia into new protocols for read/write at high speed (refer *Magellan* Web site: www.magtech.com.au)
- Intensive marketing of *Magellan's technology* by system integrators and by companies providing industry-specific software services (e.g., *gaming industry*)
- Rapid development methodologies for novel RFID tags using new organic substrates appropriate to niche requirements and application-specific readers, such as the *Magellan tray reader* for tracking commercial files/folders
- Growing range of security applications, typically involving *smartcards with integrated RFID capabilities*, which may include encrypted data
- Novel RFID application research, supported by government grants (e.g., at Hong Kong Polytechnic University)

To fully appreciate the scope and complexity of current RFID technology, the rapid advances made by *Magellan* using phase jitter modulation (PJM) is worth emphasizing. PJM is specified in the ISI/IEC 18000–3 Mode 2 standard. It provides a superfast high-frequency (13.56 MHz) technology. A brief digression into Magellan technology innovations accepted by the market, particularly the PJM StackTag, will reinforce student enthusiasm for the potential of RFID:

- Faster data rates than VHF and most UHF technologies in use.
- The PJM ItemTag uses tuned coils, as do most item tags, and requires a significant separation for individual item identification.

- The proprietary PJM StackTag is designed to read multiple tags, which touch or are in a stack of some form. Hundreds of tags can theoretically be read, almost simultaneously.
- Excellent anticollision function due to the use of eight channels offering a notional data rate of 848 kbits per second.
- High- and low-power modes for RFID tags.
- Patented system using frequency and time division multiple access.
- Memory capacity up to 10 kbit is available using PJM technology, whereas legacy RFID systems are typically limited to 512 bits.

Implementation of RFID systems may require project management expertise of a high order due to the simultaneous rollout of new business processes. Careful consideration of all stakeholder requirements and an effective risk management strategy are essential to achieve quality outcomes. Standalone applications, without significant BPM issues, may still justify a pilot project to demonstrate the economic viability and technical feasibility of the RFID proposal. High volume and/or fast response applications, which continuous production and distribution lines require, naturally entail a detailed implementation strategy. Successful implementations, based on anecdotal feedback received by the author, include consideration of

- Ownership of the project by the customer, typically facilitated by a *Steering Committee* chaired by a senior manager
- Comprehensive risk management process
- Offline technical assessments by an industry expert to identify CSFs
- Establishment of a *Knowledge Process Team*, including relevant production supervisors, to focus on on-site problems during implementation
- Training strategy that identifies *blended learning* requirements given that production line workers are often fearful when challenged by a changed work environment

6.11 Process Technology Integration Issues

Business integration offers scope for process efficiencies, elimination of duplication, and more effective use of human resources. BPM, supported by an SOA, has been identified as a *value-adding strategy* for responding to global business challenges and to the opportunities presented by technological change.

Less well appreciated are the facilitation, coordination, and knowledge acquisition requirements to integrate new technologies, such as RFID, with process knowledge, particularly when the most important process knowledge is tacit and is shared between several individuals. Multiple roles will require a *Knowledge Process Team* to map existing protocols with the functions and events comprising the new process requirements. This typically requires a collaborative knowledge acquisition methodology for any changes to workflow that affect process quality, and where related compliance/performance objectives are at risk.

The human resource challenges are often compounded by poor software integration quality in the new data capture processes, and probably, in online database update. *Risk propagation scenarios* need to be considered to ensure effective risk management and contingency planning in real-time processes, particularly safety critical environments.

Applications, in which several technologies are rolled out for the first time, *RFID and GPS, and disruptive technologies in general, for example,* pose significant difficulty in education on risk management.

As noted later, the educational challenges are considerable in achieving an effective RFID implementation, as marketing of new technologies typically understate the scope and methods of workforce induction requirements.

6.12 Summary of Educational Challenges in RFID Education

The overarching requirement for most employees, whose roles are affected by RFID, and possibly, RFID augmented with other new technologies such as GPS, is mapping their current process knowledge with the new process requirements. A suggested educational strategy for achieving this objective entails the presentation of training materials on

- Risk management principles and practice.
- Creation of *Knowledge Process Teams* to address specific areas of risk. Naturally, an understanding of the corporate knowledge management systems is required by, at least, one member of each *Team.*
- Adoption of a *blended learning strategy for learning reinforcement* [4].
- Assessment by senior management of compliance with *IT governance principles and practice* in their evaluation of software integration strategies.

Finally, the need for motivational strategies and mentoring of production line employees, when faced with handling technological change should be recognized in the implementation of educational strategies.

References

1. Garner, B.J. (2007): Motivation, self-assessment & tacit knowledge in e-learning context selection, Unpublished paper.
2. Philp, B. and Garner, B.J. (2001): Knowledge mediation in software quality engineering. *Proceedings of ASWEC 2001 Conference,* Canberra, Australia.
3. Roberts, T.S., editor (2006): *Self, Peer and Group Assessment in E-Learning,* Idea Group Inc., London.
4. Garner, B.J. and KcKay, E. (2007): Learning reinforcement strategies for a changing workforce. *Proceedings of WBE 2007,* Chamonix, France.

Appendix

Reference sources for RFID students requiring knowledge of bar codes in automated identification

1. *Keeping Track—An Introduction to Automatic Data Capture* (1995). Videotape, 70 min. Produced by Practical Marketing on behalf of AIM Australia.
2. Richter, G., Hollingworth, M. (1995). *What's in a Number? Facts about the EAN Numbering and Barcoding System.* EAN Australia, Melbourne.
3. Palmer, R.C. (1991). *The Bar Code Book.* Peterborough, NH, Helmers.

Section II

Technology

7

RFID Reader Synchronization

Kin Seong Leong, Mun Leng Ng, Alfio R. Grasso, and Peter H. Cole

CONTENTS

7.1 Introduction

Radio frequency identification (RFID) has received much attention recently as it is widely believed that RFID can revolutionize supply chain management, complementing bar codes

This chapter is based extensively on Leong et al. (2006a,d) with permission © 2006 IEEE and © 2006 Academy Publisher.

as the main object-tracking system. Several major supply chain operators and retailers, such as Wal-Mart in the United States, have deployed RFID systems in some of their supply chains (Roberti, 2004). Initial test runs of RFID deployment show encouraging results (Rendon, 2005), and hence large-scale RFID deployment is planned. However, before any successful deployment can be achieved, some RFID issues have to be resolved. One of them is the RFID reader collision problem, which is the focus of this chapter.

The term ''reader collision(s)'' is discussed extensively in Engels and Sarma (2002) and Carbunar et al. (2005). In this chapter, reader collision is simply defined as the phenomenon where an interrogation signal from a certain reader disrupts the communication between a tag and another reader, and this reader collision problem is potentially magnified in a dense reader environment such as in a warehouse. To visualize a simple reader collision situation, we can imagine a situation as shown in Figure 7.1.

In addition, the band RFID is using is an industrial, scientific, and medical (ISM) band. Normally, there is other electronic equipment, intentional or nonintentional radiators that are operating in this band. Unlike some of the other signaling equipment permitted to use the band, RFID antennas use comparatively intense RF power to energize passive tags within their interrogating zones and hence can interfere with nearby electronic equipment operating in the band of interest. Due to this reason, RFID is subjected to very strict regulations around the world.

Various regulatory and standardization bodies have tried to regulate the operations of RFID readers. There are mainly two different standards on the spectrum management adopted. One is the EN 302 208-1 v1.1.2: electromagnetic compatibility and radio spectrum matters (ERM); RFID equipment operating in the band 865–868 MHz with power levels up to 2 W as introduced by the European regulatory body (ETSI, 2006) and the Title 47, telecommunication, chapter 1, Part 15, radio frequency devices as introduced by Federal Communications Commission (FCC, 2001).

The air interface and command sets between an RFID reader and an RFID tag are standardized by the ''EPC radio-frequency identity protocols class-1 generation-2 UHF RFID protocol for communications at 860 MHz–960 MHz version 1.0.9'' as recommended by EPCglobal.

As will be discussed in more detail in the later part of this chapter, the restrictions from ETSI 302 208 that are put on the operation of RFID readers are very strict, making it quite impossible to have an uncoordinated large-scale deployment of RFID readers. Hence, this chapter introduces the idea of RFID reader synchronization to enable good RFID performance in a dense reader environment while adhering to strict regulations.

The next section introduces the ETSI 302 208, FCC Part 15, and RFID UHF Protocol standards and their impact on RFID reader deployment, especially in the Europe. Section 7.3 explains the concept of RFID reader synchronization and how it adheres to strict regulations. Section 7.4 suggests possible ways in implementing an RFID synchronization system. A case study on RFID reader synchronization is presented in Section 7.5. Ways of fine-tuning RFID reader positioning are discussed in Section 7.6. Variations of possible reader synchronization schemes are presented in Section 7.7. Sections 7.8 and 7.9 offer views

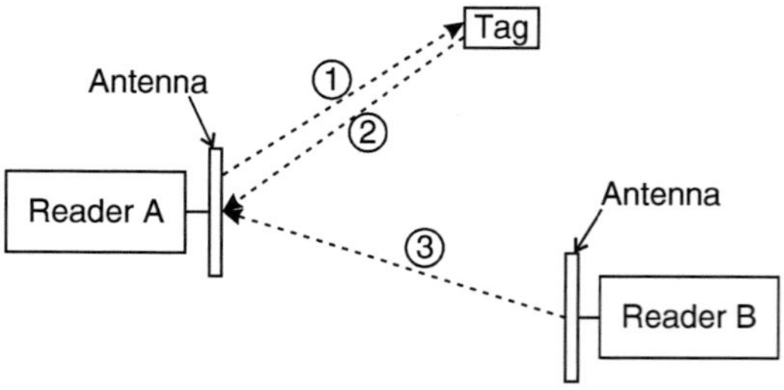

FIGURE 7.1
A simple illustration of reader collision. Reply from tag will be interfered by signal sent from another nearby reader. (From Leong, K.S., Ng, M.L., and Cole, P.H., *IEEE 2005 International Symposium on Microwave, Antenna, Propagation and EMC Technologies for Wireless Communications*, Beijing, China, 2005, © 2005 by IEEE. With permission.)

on reader synchronization in the United States and the current progress of reader synchronization standardization in Europe, respectively, followed by conclusions in Section 7.10.

7.2 RFID Regulations

7.2.1 ETSI 302 208

ETSI 302 208 is a European regulation governing the operation of RFID readers (ETSI, 2006). It allocates the frequency band of 865–868 MHz for RFID deployment. This frequency band is then divided into 15 sub-bands or channels, each spanning a total of 200 kHz. However, when a reader is operating at the maximum radiated power, which is 2 W effective radiated power (ERP), only 10 sub-bands are available, while the remaining 5 are used as guard bands or for lower power readers. ETSI 302 208 also introduces the concept of "Listen Before Talk." An extract from the ETSI 302 208 best describes the essence of "Listen Before Talk." It says "Prior to each transmission, the receiver in the interrogator shall first monitor in accordance with the defined listen time for the presence of another signal within its intended sub-band of transmission. The listen time shall comprise a fixed period of 5 ms plus a random time of 0 ms to 5 ms in 11 steps. If the sub-band is free the random time shall be set to 0 ms" (ETSI, 2006). The threshold to determine the presence of another signal within the intended sub-band is shown in Table 7.1. The measurement method is defined in the same standard.

Furthermore, once a sub-band has been selected, the RFID reader is permitted to use that sub-band for up to 4 s. After use, it must free the sub-band for at least 100 ms. A reader can however, listen to another sub-band for 5 ms and if free use that new sub-band.

7.2.2 FCC

The U.S. FCC Title 47 Part 15.247, with operation within the band 902–928 MHz, uses the technique of frequency hopping spread spectrum (FHSS). FHSS is a method of transmitting radio signals by rapidly switching a carrier among many frequency channels, using a pseudorandom sequence known to both transmitter and receiver. In the context of FCC Part 15.247 for a frequency hopping system operating in the 902–928 MHz band: if the 20 dB bandwidth of the hopping channel is <250 kHz, the system shall use at least 50 hopping frequencies and the average time of occupancy on any frequency shall not be >0.4 s within a 20 s period; if the 20 dB bandwidth of the hopping channel is 250 kHz or greater, the system shall use at least 25 hopping frequencies and the average time of occupancy on any frequency shall not be >0.4 s within a 10 s period. The maximum allowed 20 dB bandwidth

TABLE 7.1

Transmit and Threshold Power

ERP (W)	ERP (dBW)	Threshold (dBW)
Up to 0.1	Up to −10	≤−113
0.1 to 0.5	−10 to −3	≤−120
0.5 to 2.0	−3 to 3	≤−126

Source: From Leong, K.S., Ng, M.L., and Cole, P.H. in *IEEE 2005 International Symposium on Microwave, Antenna, Propagation and EMC Technologies for Wireless Communications*, Beijing, China, 2005, 658–661, © 2005 by IEEE. With permission.

of the hopping channel is 500 kHz. For an RFID system deployed in the United States following EPCglobal RFID protocol (EPCglobal, 2004), which is discussed in detail in the next part, a total of 50 channels are available, each with 500 kHz bandwidth. Hence the occupancy time on any channel cannot be >0.4 s within a 10 s period. In addition, a 1 W limit is allowed for systems employing at least 50 hopping channels with the use of antennas with directional gains that do not exceed 6 dBi, giving a maximum total radiated power of 4 W effective isotropic radiated power (EIRP). FHSS is adopted mainly in North and South America, though with a slight difference in the total bandwidth in different countries.

7.2.3 RFID Protocol

The idea behind standardization of RFID protocols is to have a uniform air interface and command set between an RFID reader and an RFID tag, so that an RFID reader produced by a company can be integrated into an RFID network setup by another company. There are two standards although they are basically the same. The first one is the ISO/IEC 18000-6:2004 Amendment 1: information technology—RFID for item management—Part 6: parameters for air interface communications at 860–960 MHz (ISO, 2004), which is based on the recommendation of the second standard, EPC Class 1 Generation 2 Protocol "EPC Radio-frequency Identification Protocols Class 1 Generation 2 UHF RFID Protocol for Communication at 860 MHz–960 MHz" (EPCglobal, 2004), in short EPC C1G2, by EPCglobal. This protocol outlines the air interfaces and commands between an RFID reader and an RFID tag. It also includes the spectrum management of RFID operation. Frequency hopping or frequency agile systems are the suggested techniques. An allocated frequency band, as allowed by local regulatory body, is divided into sub-bands or channels. A reader will only use a certain channel for communication, not the entire allocated frequency band.

EPC C1G2 covers single, multiple, and dense reader modes; multiple reader mode is for an environment where the number of simultaneously active readers is modest relative to the number of available channels while dense reader mode is for an environment where the number of simultaneously active readers is comparable with more than the number of available channels.

This document only focuses on dense reader mode. In dense reader mode, for narrow bandwidth (European 200 kHz) channels, it is suggested in this protocol that odd-numbered channels should be used for tag backscatter while even-numbered channels will be used for reader interrogation. For a wide bandwidth channel (USA FCC 500 kHz channel (FCC, 2001)), all available channels can be used for reader interrogation as tag backscatter replies will be located at the boundaries of these channels.

7.2.4 Challenges in Dense Reader Environment in Europe

With the implementation of ETSI 302 208 and EPC C1G2, it is clear that when a reader is operating at a certain sub-band or channel, this reader will effectively prevent other readers from using that channel within an unacceptably large area. Leong et al. (2005, 2006b) have presented detailed discussions and analysis on this matter and Table 7.2, as extracted from Leong et al. (2006b), summarizes the minimum distance (calculated using a piecewise path loss model with variable environmental factor) between two antennas connected to readers before one antenna operating at a certain channel will prevent the other antenna from using that channel. It should be noted that these results are obtained using a 0 dB isotropic receiving antenna, and do not represent any real life situation, as a typical RFID antenna will be a directional antenna. Nonetheless, the data presented in the table gives sufficient

TABLE 7.2

Minimum Distance between Antennas

Channel Difference	Antenna Projecting Horizontally		
	Front (m)	**Side (m)**	**Back (m)**
0	1400	350	210
1	180	45	30
2	130	25	15
3	95	20	10

Source: From Leong, K.S., Ng, M.L., and Cole, P.H., *2006 International Symposium on Applications and the Internet (SAINT) Workshop, RFID and Extended Network: Deployment of Technologies and Applications*, Phoenix, Arizona, USA, 2006, © 2006 by IEEE. With permission.

evidence that a low threshold value for the LBT as specified in ETSI 302 208 is severe enough to impede the reader deployment in a dense RFID reader system.

7.3 Reader Synchronization

Under the concept of reader synchronization, all the RFID readers in a certain area, for example all the readers in a warehouse, are networked together through a central control unit. The connection method can be the common Ethernet connection, or equivalent, and will be discussed in the next section.

Since all the readers are linked together, physically or wirelessly, they can be directed to carry commands at the same time. In addition, they can be assigned channels dynamically so that the spectrum management is optimized while the reader collision is minimized.

European regulation allows 10 channels when maximum radiated power, 2 W ERP, is used. Following the recommendation of EPC C1G2, under dense reader mode, five of them, the even-numbered channels, are used for reader interrogation. All the readers are ''Listen Before Talk'' compatible. They are configured to start to ''Listen'' at the same time, and then at the end of the listen period, they can all synchronously start to ''Talk,'' as shown in Figure 7.2. This is due to the fact that according to ETSI 302 208, if there is no signal detected in the intended channel of interest, the ''Listen'' time is fixed. Hence, all the readers, which start ''Listening'' at the same time, will start ''Talking'' at the same time. If a reader is turned on at a different time, or if a reader loses synchronization that reader can be made to start again in synchronism with the rest of the readers, after the last reader has finished its ''Talk'' session.

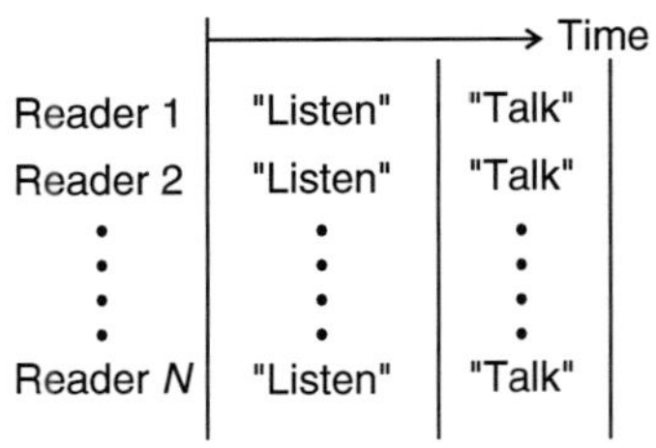

FIGURE 7.2
Synchronization of all readers: all the readers start to ''Listen'' at the same time and finish ''Listen'' at the same time. (From Leong, K.S., M.L. Ng, and P.H. Cole. 2006a. Synchronization of RFID readers for dense RFID reader environments. *International Symposium on Applications and the Internet*. SAINT, Phoenix, Arizona, USA, © 2006 by IEEE. With permission; Leong, K.S., M.L. Ng, A. Grasso, and P.H. Cole. 2006d. *J. Commun.*, 1, 9, © 2006 by IEEE. With permission.)

7.4 Actual Implementation

7.4.1 Connectivity

In actual implementation, RFID readers must be able to communicate with each other to enable synchronization of RFID readers. There are basically two ways in connecting all the readers; either using wired (physical) connection or using wireless connection.

A physically connected system or wired system cannot support mobile readers. In addition, a wired connection may suffer from data latency in the network. Skeie et al. (2001) show that time synchronization in a wired network is possible, but will require additional hardware and system reconfiguration. In the best case, the time difference achievable can be better than 1 ms. A wired system is often considered more reliable and a more secure communication method than a wireless communication.

A wireless system signals through an RF link. It can use one of the five guard bands, mentioned in Section 7.2.1, for sending a synchronizing signal. A synchronizing signal can be a signal with a special pattern. A wireless system can also use any existing wireless protocol such as Bluetooth technology. It supports mobile readers but is inevitably vulnerable to interference (signal integrity problem) and unauthorized signal sniffing (security problem).

Both connectivity methods have their own advantages and disadvantages. The decision in choosing either of these two methods is largely dependent on the positioning of the LBT sensor, which is discussed later.

7.4.2 Positioning of LBT Sensors

An LBT sensor of an RFID reader is responsible for detecting signals in the channel of interest before transmission in that channel. This LBT sensor must have a power sensitivity level better than −126 dBW as specified in ETSI (2006). If not, this LBT sensor will not be able to function efficiently in determining whether there exists a signal with a power level higher than the power level specified in regulations in the channel of interest. An LBT sensor can be the RFID antenna used for transmitting and receiving signals in the communication with RFID tags. An LBT sensor can also be a separate antenna connected to an RFID reader.

In addition, several RFID readers could share an LBT sensor within a close vicinity. This is also known as a centralized system. A localized system is where each and every RFID reader has its own LBT sensor.

A centralized LBT system is as shown in Figure 7.3. The LBT sensor will constantly monitor all the channels allocated for RFID operation, and dynamically assign available channels to all the readers connected to it. The central control system has to be configured

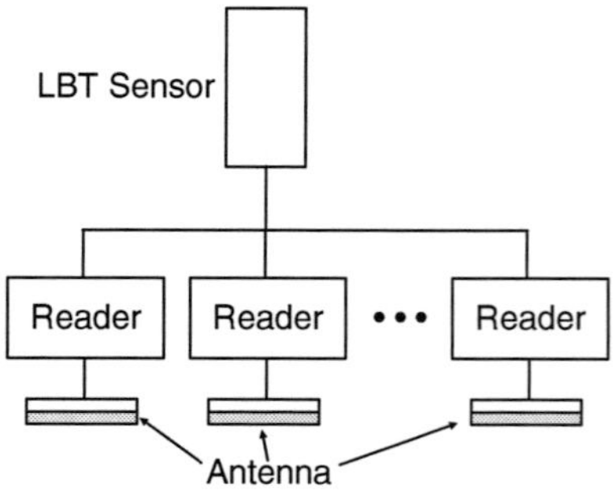

FIGURE 7.3
Centralized LBT system, where readers are connected to one LBT sensor in nearby surroundings. (From Leong, K.S., M.L. Ng, and P.H. Cole. 2006a. Synchronization of RFID readers for dense RFID reader environments. *International Symposium on Applications and the Internet*. SAINT, Phoenix, Arizona, USA, © 2006 by IEEE. With permission; Leong, K.S., M.L. Ng, A. Grasso, and P.H. Cole. 2006d. *J. Commun.*, 1, 9, © 2006 by IEEE. With permission.)

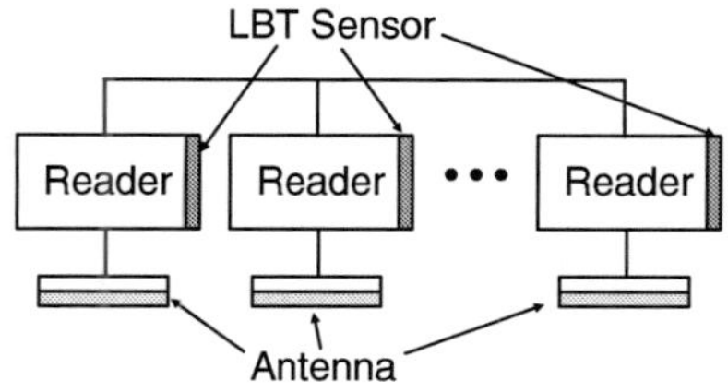

FIGURE 7.4
Localized LBT system, where each reader has its own LBT sensor. (From Leong, K.S., M.L. Ng, and P.H. Cole. 2006a. Synchronization of RFID readers for dense RFID reader environments. *International Symposium on Applications and the Internet*. SAINT, Phoenix, Arizona, USA, © 2006 by IEEE. With permission; Leong, K.S., M.L. Ng, A. Grasso, and P.H. Cole. 2006d. *J. Commun.*, 1, 9, © 2006 by IEEE. With permission.)

during the initial setup of the system. A fine-tuned centralized LBT system offers high reliability. However, it requires additional network hardware to connect all the readers to the LBT sensor. In addition, a centralized LBT system will not be able to be implemented effectively when mobile readers are dominant in the surroundings. Although the readers can communicate with the centralized LBT sensor through a wireless link, it is very difficult for the centralized LBT sensor to estimate the position of mobile readers, and hence is not possible to allocate the best channels for mobile readers. For example, if two mobile readers operate simultaneously in an enclosed area, there is a probability that the two readers move near to each other at some time. The centralized LBT sensor may at that time allocate very nearby channels to those two readers and serious interference between those two readers may occur.

In addition, if two nearby areas are running on different RFID wireless networks and they are uncoordinated, interference with each other will occur, and in the worst case, cause a complete system shut down. The coordination of wireless networks in different premises will be time and cost consuming.

A localized LBT system is as shown in Figure 7.4. Each reader has its own LBT sensor. The LBT sensor can either be a separate antenna (Figure 7.4), or be the same antenna a reader uses to establish communication with an RFID tag within its interrogation zone. As compared with a centralized LBT system, a localized LBT system with wireless connectivity enables relatively easy new reader integration into an existing system, with no additional cabling or setup needed. However, a localized LBT system has the problem with management of channel sharing, signal interference, and possibly creation of unwanted shielding.

In actual fact, the connectivity of readers and the positioning of LBT sensor are closely related. In CISC (2006), a wired system and a centralized LBT are linked together as one configuration, whereas a wireless system and a localized LBT are linked together as another configuration.

7.4.3 Antenna Positioning

The positioning of RFID interrogation antennas depends primarily on the application. Detailed operational considerations for the deployment of RFID system are presented in Leong et al. (2006c). In this chapter, only one example will be given, which is the dock door situation, as it will be used in the case study in the next section. A dock door is usually 2–3 m in length and ~3 m in height. The most effective way to create an RFID interrogation zone is to position two antennas at the sides of the dock door, face-to-face, and with an height elevation, h, as shown in Figure 7.5. The height elevation, h, mainly depends on the average height of objects being shipped through the dock door. A normal choice of h is between 0.5 and 1 m. In addition, antenna A and antenna B will be normally using different channels for tag interrogation.

However, if antenna A and antenna B are operating at the same time, and a tag is located in the middle of the dock door, the tag may be "confused" by the interrogation signals

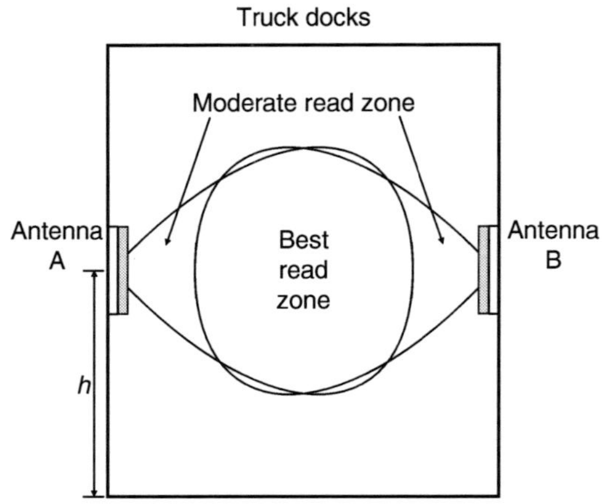

FIGURE 7.5
A typical antenna setting at dock door, with *h* being the height of the antenna from the base of a dock door. (From Leong, K.S., M.L. Ng, and P.H. Cole. 2006a. Synchronization of RFID readers for dense RFID reader environments. *International Symposium on Applications and the Internet*. SAINT, Phoenix, Arizona, USA, © 2006 by IEEE. With permission; Leong, K.S., M.L. Ng, A. Grasso, and P.H. Cole. 2006d. *J. Commun.*, 1, 9, © 2006 by IEEE. With permission.)

from both of the antennas with the result that the tag is misread. This effect is known as the tag confusion problem. The discussion of this issue is outside the scope of this chapter but a simple solution to this is to alternate the operation of antenna A and antenna B every query cycle.

7.5 Case Study

A case study on dense RFID reader deployment at the dock doors of a warehouse is presented here. As shown in Figure 7.6, the dark color rectangles represent trucks loading or unloading goods at the dock doors of a warehouse. Each door is around 3 m in width, and has two RFID antennas facing each other for tag interrogation.

Since all the readers are synchronized in a way described in Section 7.3, they will start "Listening" at the same time and will be assigned a channel for interrogation at the end of "Listen" period. The assignment of channels will be geographically influenced. Two readers assigned to be operating in the same channel will be as far apart as possible. In addition, the neighboring antennas will be using channels as far apart as

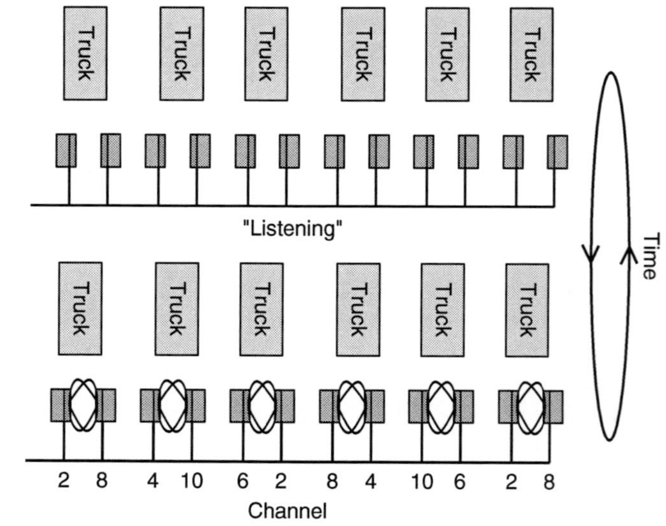

FIGURE 7.6
Alternating of "Listening" and "Talking" mode. (From Leong, K.S., M.L. Ng, and P.H. Cole. 2006a. Synchronization of RFID readers for dense RFID reader environments. *International Symposium on Applications and the Internet*. SAINT, Phoenix, Arizona, USA, © 2006 by IEEE. With permission; Leong, K.S., M.L. Ng, A. Grasso, and P.H. Cole. 2006d. *J. Commun.*, 1, 9, © 2006 by IEEE. With permission.)

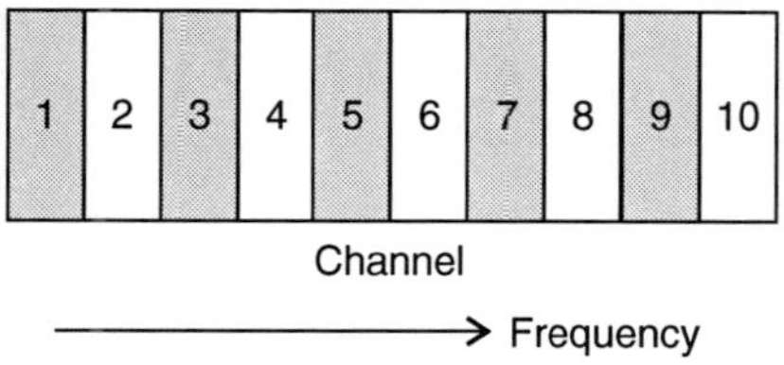

FIGURE 7.7
Channeling of the allocated frequency spectrum. (From Leong, K.S., M.L. Ng, and P.H. Cole. 2006a. Synchronization of RFID readers for dense RFID reader environments. *International Symposium on Applications and the Internet*. SAINT, Phoenix, Arizona, USA, © 2006 by IEEE. With permission; Leong, K.S., M.L. Ng, A. Grasso, and P.H. Cole. 2006d. *J. Commun.*, 1, 9, © 2006 by IEEE. With permission.)

possible. As illustrated in Figure 7.7, the spectrum is split into 10 channels, all 5 of the odd-numbered channels are reserved for tag backscattering, whereas all 5 of the even-numbered channels are assigned for reader interrogation. Figure 7.6 shows how the channel assignment is done. The antenna on the furthest left is using channel 2 for interrogation. The next antenna on its immediate right is using channel 8, which is six channels away. Channel 10, though is the furthest channel away, is not chosen. This is because the arrangement of {2, 8, 4, 10, 6} gives best channel separation between every channel.

7.6 Synchronized RFID System Fine-Tuning

Fine-tuning of a synchronized RFID system, as presented in this section, can be carried out to further reduce the tendency of reader collision. The fine-tuning methods discussed later include the reduction of output power, the reduction of overall reader talking time, the use of external sensors, the use of RF opaque or absorbing materials, and the frequent rearrangement of channel allocations.

7.6.1 Reduction of Output Power

Although up to 2 W ERP can be used in single or small population reader environment, in dense reader populations this higher power may not be necessary. Currently, a state-of-the-art reader can read up to 10 m. However, normal reading operations do not require such a read range. In the case study presented in Section 7.5, the dock doors of the warehouse are around 3 m in width. Since two antennas are positioned facing each other in every dock door, the read range required is also around 1.5–2 m. By reducing the radiated power of readers, the minimum distance between two antennas using the same channel can also be reduced, which is beneficial in a dense reader environment.

Figure 7.8 gives an approximation on the reduction of output power. In the far field region, using the Friis equation, the power received is the inverse function of the square of

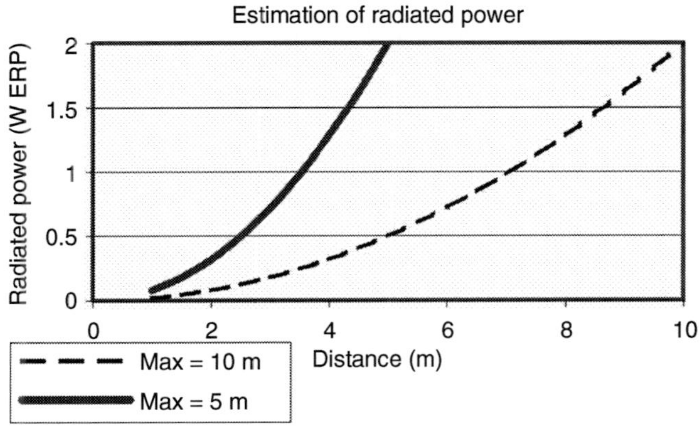

FIGURE 7.8
Estimation of required radiated power given that maximum read range corresponds to maximum radiated power. (From Leong, K.S., M.L. Ng, and P.H. Cole. 2006a. Synchronization of RFID readers for dense RFID reader environments. *International Symposium on Applications and the Internet*. SAINT, Phoenix, Arizona, USA, © 2006 by IEEE. With permission; Leong, K.S., M.L. Ng, A. Grasso, and P.H. Cole. 2006d. *J. Commun.*, 1, 9, © 2006 by IEEE. With permission.)

distance (r^{-2}). If the maximum read range corresponding to maximum radiated power (2 W ERP) of an RFID reader is known, we can compute the required radiated power for a shorter read range. For example, if the maximum read range of a reader is 5 m using 2 W ERP (shown in Figure 7.8), and if a read range of only 2 m is required, the required radiated power can be lowered to 0.32 W ERP. This estimation may not be accurate in real life due to complex electromagnetic propagation phenomena, such as reflection caused by the surroundings objects, but it demonstrates that power reduction is a viable option.

7.6.2 Reduction of Overall Reader Talking Time

While it is possible to talk for 4 s, reader applications should be configured to talk for only the time necessary to capture tag data. There is no optimum talking time. It depends on the application and also the surroundings of the deployment zone. On-site fine-tuning and measurements are needed before the reduction of talking time can be carried out.

7.6.3 Use of External Sensors

Sensors can be used to turn RFID readers on only when tags are approaching to further reduce reader interference in that area. This will free up the channels allocated for those antennas, and also avoid unnecessary interference to other surrounding reader antennas. For example, external sensors can be attached to the dock door in the case study in Section 7.5. When the dock door is not in use, the designated RFID readers would be switched off, as shown in Figure 7.9. Optionally, the central control unit can then dynamically shift the channels assigned for the antennas at door 3 to door 4 as shown.

7.6.4 RF Opaque or RF Absorbing Materials

Another effective, but more expensive, way to reduce reader interference and collision is to use RF opaque or RF absorbing materials to contain the interrogating signal within the designated zone of interrogation. For the case study presented in Section 7.5, the use of such materials is shown in Figure 7.10. Although there will still be some signal leakage through the door openings, it would not have caused much interference. This is due to the fact that the signal strengths at the sides of the antenna are relatively weak as compared with the front of the antenna. According to Leong et al. (2006b), the gain at the side of a typical RFID antenna is ~20 dB less than the gain at the front of the antenna.

7.6.5 Frequent Rearrangement of Channels

Interrogating Channels can be switched around every cycle of "Listen Before Talk." This is to prevent the jamming of the interrogation signal by any external noise. Figure 7.11 shows a simple example on how the switching is done. There are other more complex switching

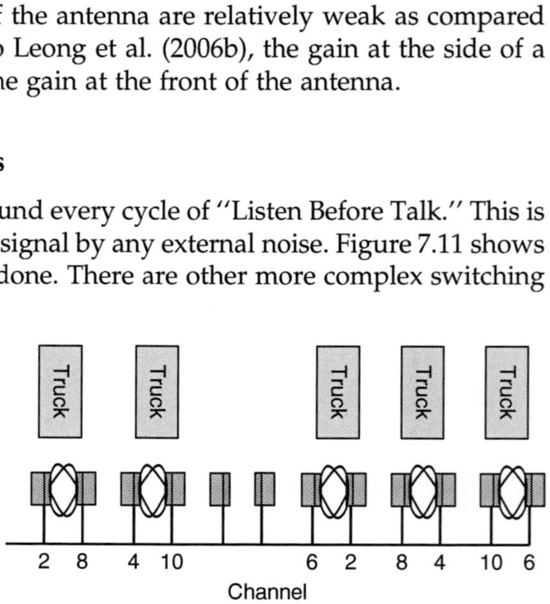

FIGURE 7.9
Using sensors in an RFID system. Both the antennas at dock door 3 are switched off when the absence of truck 3 is detected. (From Leong, K.S., M.L. Ng, and P.H. Cole. 2006a. Synchronization of RFID readers for dense RFID reader environments. *International Symposium on Applications and the Internet.* SAINT, Phoenix, Arizona, USA, © 2006 by IEEE. With permission; Leong, K.S., M.L. Ng, A. Grasso, and P.H. Cole. 2006d. *J. Commun.*, 1, 9, © 2006 by IEEE. With permission.)

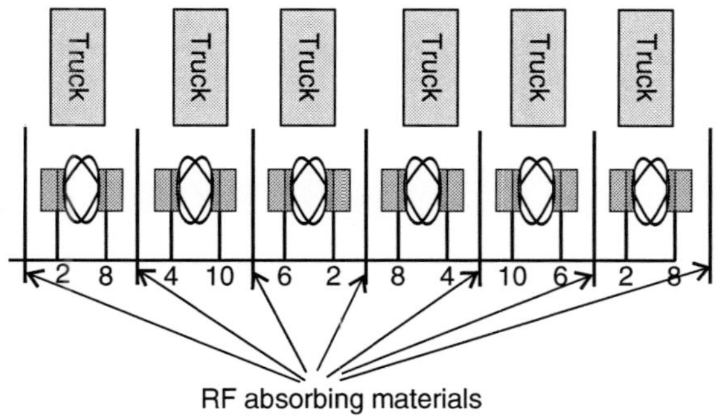

FIGURE 7.10
Use of RF absorbing materials. The antennas facing each other at the same door are at least four channels away. (From Leong, K.S., M.L. Ng, and P.H. Cole. 2006a. Synchronization of RFID readers for dense RFID reader environments. *International Symposium on Applications and the Internet.* SAINT, Phoenix, Arizona, USA, © 2006 by IEEE. With permission; Leong, K.S., M.L. Ng, A. Grasso, and P.H. Cole. 2006d. *J. Commun.*, 1, 9, © 2006 by IEEE. With permission.)

methods involving higher artificial intelligence in the central control unit, depending on the noise received from the surrounding environment, but these await full development.

7.7 Variation of Synchronization

In the previous sections, suggestions on the implementation of a real life RFID reader synchronization system are presented, together with some deployment options, such as the connectivity of all the readers. In addition, fine-tuning methods are presented. In this section, some of the interesting variations of RFID reader synchronization schemes are presented. These variations may not be readily incorporated into the suggested methods mentioned in previous sections, but are presented here for future reference and for completeness.

7.7.1 Separation of Transmitting and Receiving Channels

For the RFID full-power operation (2 W ERP) as governed by ETSI 302 208, only 10 channels are available as shown in Figure 7.7. However, as discussed in Section 7.2, there are actually 15 channels available for RFID in total. Five of the fifteen channels, though used as guard bands, can be used for RFID operation with reduced maximum allowable radiated power. There are 3 channels located lower in frequency than the normal 10 channels, which can only be operated below 100 mW ERP, whereas there are 2 channels higher in frequency than the normal 10 channels, which can be operated below 500 mW ERP. The complete frequency range for RFID operation with respective regulated power level is as shown in Figure 7.12.

The channel numbering system shown in Figure 7.7 is included in Figure 7.12 along with a new channel numbering system to simplify the discussion hereon. Channels 4, 7, 10, and 13

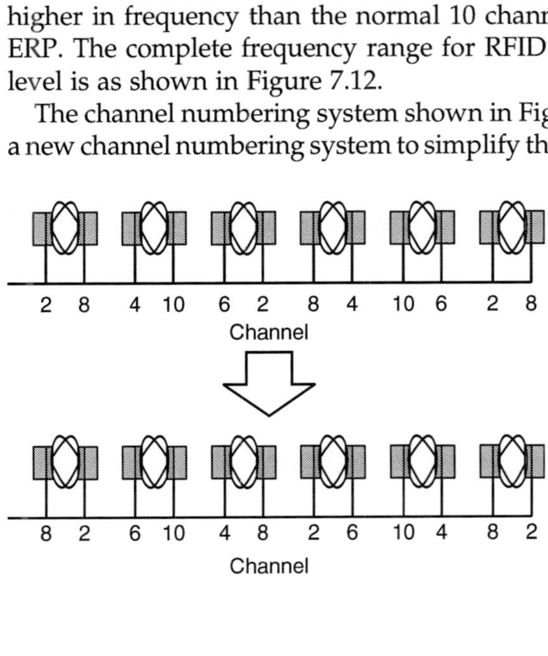

FIGURE 7.11
Channel switching within antennas. (From Leong, K.S., M.L. Ng, and P.H. Cole. 2006a. Synchronization of RFID readers for dense RFID reader environments. *International Symposium on Applications and the Internet.* SAINT, Phoenix, Arizona, USA, © 2006 by IEEE. With permission; Leong, K.S., M.L. Ng, A. Grasso, and P.H. Cole. 2006d. *J. Commun.*, 1, 9, © 2006 by IEEE. With permission.)

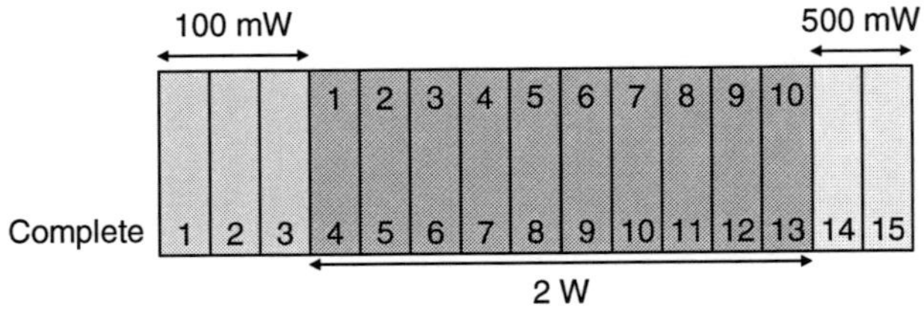

FIGURE 7.12
The complete frequency band allocated for RFID operation as compared with Figure 7.7. (From Leong, K.S., M.L. Ng, and P.H. Cole. 2006a. Synchronization of RFID readers for dense RFID reader environments. *International Symposium on Applications and the Internet.* SAINT, Phoenix, Arizona, USA, © 2006 by IEEE. With permission; Leong, K.S., M.L. Ng, A. Grasso, and P.H. Cole. 2006d. *J. Commun.*, 1, 9, © 2006 by IEEE. With permission.)

are assigned to be the reader transmitting channels while the tag reply channels are the four channels beside the transmitting channels (Impinj, 2006). For example, transmitting channel 4 uses channels 2, 3, 5, and 6 for tag reply (Figure 7.13).

Although the transmitting channels are reduced from a total of five down to four, the transmitting channels are placed two channels away rather than one channel away. From the transmit mask shown in Figure 7.14, an improvement of 5 dB can be obtained. Hence with the reduction of interference between transmitting channels, readers can be placed nearer to each other.

7.7.2 Separation of RFID and Non-RFID Signals

Another variation of synchronization is to differentiate an RFID signal from a non-RFID signal. A method using signal recognition is presented in Intermec (2006). The idea is that all the RFID readers in a certain region can be treated as a single entity in the regulation as outlined in the ETSI 302 208. Hence, it is only required to avoid the signal interference between all the RFID readers and the rest of the short-range devices. If this concept is valid, the interrogation signals of RFID readers are not treated as a signal in a channel when an LBT test is carried out.

The main advantage of this method is that a lot of readers can be deployed in a small confinement area. However, reader antenna positioning can become more challenging, as all the readers can choose any channel for transmission as long as there is no other type of short-range device around.

7.8 Synchronization in the United States

Synchronization cannot be implemented in the United States under Part 15.247 or any countries regulated RFID using FHSS. As discussed earlier, FHSS requires the reader to switch (hop) from channel to channel when a collision is detected. In synchronization, the switching or hopping is not random (or pseudorandom) in nature, and is in violation with the regulations. However, the USA FCC has another clause with Part 15 which does not specify FHSS. This part 15.245 allows an RF transmission to operate at 0.0375 W EIRP on a duty cycle with a 20 dB peak to average ratio.

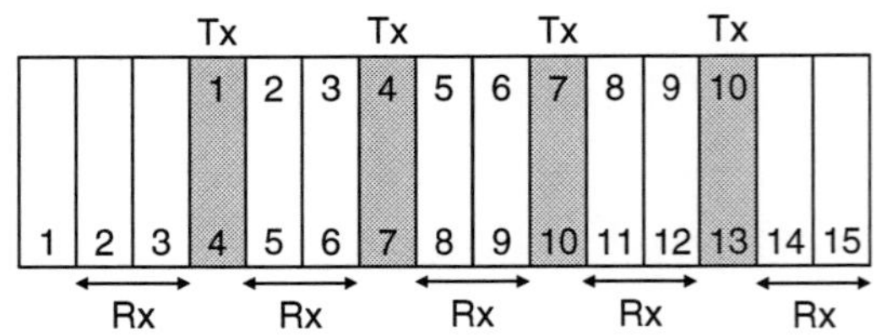

FIGURE 7.13
Variation in the separation of transmitting (Tx) and receiving (Rx) channel of an RFID reader. (From Leong, K.S., M.L. Ng, and P.H. Cole. 2006a. Synchronization of RFID readers for dense RFID reader environments. *International Symposium on Applications and the Internet.* SAINT, Phoenix, Arizona, USA, © 2006 by IEEE. With permission; Leong, K.S., M.L. Ng, A. Grasso, and P.H. Cole. 2006d. *J. Commun.*, 1, 9, © 2006 by IEEE. With permission.)

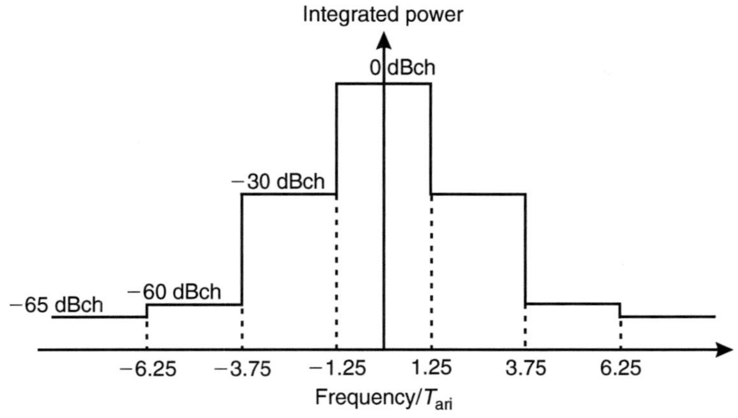

FIGURE 7.14
Transmit mask for dense interrogator environments (EPC radio frequency identity protocols class-1 generation-2 UHF RFID protocol for communications at 860–960 MHz version 1.0.9. (From EPCglobal, EPC Radio-Frequency Identity Protocols Class-1 Generation-2 UHF RFID Protocol for Communications at 860 MHz–960 MHz Version 1.0.9, EPCglobal Standard Specification, 2004, With permission © 2004 by EPCglobal.)

If a synchronized RFID system in European countries, which only occupied a bandwidth of 3 MHz, can perform on par with an FHSS RFID system in the United States, it can be safety deduced that RFID reader synchronization using 15.245 can bring improvement to the RFID operation in the United States. With a broader bandwidth available (50 channels of 500 kHz each as compared with 10 channels of 200 kHz each), a synchronized RFID in the United States can offer more coverage and higher reading speed.

7.9 Updated Progress on Development of RFID Reader Synchronization

The regulation of RFID reader synchronization in the European countries is governed by ETSI. TG34: RFID devices, a technical work group within ETSI has been field-testing late in 2006 in an operational distribution center using a synchronized RFID system (O'Connor, 2006). The testing involved up to 36 adjacent portals operating simultaneously. Using identical pallets comprising 63 "RFID unfriendly" cartons, a read rate of better than 98.5% is recorded. In the near future, RFID reader synchronization would be included in ETSI TR 102 436 "Electromagnetic compatibility and Radio spectrum Matters (ERM); Improved spectrum efficiency for RFID in the UHF Band," and be harmonized with the current EN 302 208. In addition, with the standardization of RFID reader synchronization, the mandatory use of "Listen Before Talk," which is a deterrent to large-scale deployment of RFID system, will be lifted, allowing a better performance and better coverage RFID system in Europe.

7.10 Conclusion

This chapter has identified synchronization of RFID readers as a mechanism to assist in RFID reader deployment in dense reader environments. Some implementation methods and several fine-tuning methods are presented in optimizing the performance of a synchronized RFID system. As compared with conventional unsynchronized RFID systems, a synchronized RFID system can offer more coverage, less reader collision or interference, while strictly following the European regulations and the EPC C1G2 recommendation, and with variation of the normal operating procedure can also deal with the effects of tag confusion. However, these benefits require the use of more complex hardware and hence can marginally increase deployment costs.

References

Carbunar, B., M.K. Ramanathan, M. Koyuturk, et al. 2005. Redundant reader elimination in RFID systems. *Second Annual IEEE Communications Society Conference on SECON*, Santa Clara, California, USA.

CISC. Consolidated proposal on synchronization. 2006. ETSI Electromagnetic Compatibility and Radio Spectrum Matters, Task Group 34, RF Identification Devices, ETSI ERMTG34#12_012.

Engels, D.W. and S.E. Sarma. 2002. The reader collision problem. *IEEE International Conference on Systems, Man and Cybernetics*. vol. 3, 6 pp.

EPCglobal. 2004. EPC Radio-Frequency Identity Protocols Class-1 Generation-2 UHF RFID Protocol for Communications at 860 MHz–960 MHz Version 1.0.9, EPCglobal Standard Specification.

European Telecommunications Standards Institute (ETSI). 2006. EN 302 208-1 V1.1.2 (2006–03), Electromagnetic Compatibility and Radio Spectrum Matters (ERM); Radio Frequency Identification Equipment Operating in the Band 865 MHz to 868 MHz with Power Levels up to 2 W; Part 1: Technical Requirements and Methods of Measurement.

Federal Communications Commission (FCC). 2001. Title 47, Telecommunication, Chapter 1, Part 15, Radio Frequency Devices.

Impinj. RFID Operation in Europe. 2006. ETSI Electromagnetic Compatibility and Radio Spectrum Matters, Task Group 34, RF Identification Devices, ETSI ERMTG34#12_11r1.

Intermec. Smart Listen Before Talk. 2006. ETSI Electromagnetic Compatibility and Radio Spectrum Matters, Task Group 34, RF Identification Devices, ETSI ERMTG34#12_010.

International Standards Organization (ISO). 2004. ISO/IEC 18000-6:2004: Information Technology—Radio Frequency Identification for Item Management—Part 6: Parameters for Air Interface Communications at 860 MHz to 960 MHz.

Leong, K.S., M.L. Ng, and P.H. Cole. 2005. The reader collision problem in RFID systems. *IEEE 2005 International Symposium on MAPE*, Beijing, China.

Leong, K.S., M.L. Ng, and P.H. Cole. 2006a. Synchronization of RFID readers for dense RFID reader environments. *International Symposium on Applications and the Internet*, SAINT, Phoenix, Arizona, USA.

Leong, K.S., M.L. Ng, and P.H. Cole. 2006b. Positioning analysis of multiple antennas in a dense RFID reader environment. *International Symposium on Applications and the Internet*, SAINT, Phoenix, Arizona, USA.

Leong, K.S., M.L. Ng, and P.H. Cole. 2006c. Operational considerations in simulation and deployment of RFID systems. *17th International Zurich Symposium on Electromagnetic Compatibility*, Singapore.

Leong, K.S., M.L. Ng, A. Grasso, and P.H. Cole. 2006d. Dense RFID reader deployment in Europe using synchronization. *Journal of Communications*, ISSN 1796–2021, 1(7), 9–16.

O'Connor, M.C. October 2006. ETSI tests show EPC scaleable in Europe. *RFID Journal*. Available at http://www.rfidjournal.com/article/articleview/2712/1/1.

Rendon, J. 2005. Wal-Mart touts RFID results. Available at: http://searchmobilecomputing.techtarget.com/originalContent/0,289142,sid40_gci1048680,00.html?bucket=NEWS (accessed on April 11, 2007).

Roberti, M. April 2004. Wal-Mart begins RFID rollout. *RFID Journal*. Available at http://www.rfidjournal.com/article/view/926/1/1.

Skeie, T., S. Johannessen, and O. Holmeide. 2001. Highly accurate time synchronization over switched Ethernet. *Proceedings of 8th IEEE International Conference on Emerging Technologies and Factory Automation*, vol. 1, pp. 195–204.

8

Adaptive Tag Anticollision Protocols for RFID Passive Tags

Jihoon Myung, Wonjun Lee, and Timothy K. Shih

CONTENTS

8.1 Introduction

Radio frequency identification (RFID) is an automatic identification system which consists of readers and tags. An RFID reader recognizes an object through reading the identification number (ID) of the RFID tag attached to it [1]. To read tag IDs, the reader sends out a signal supplying instructions to tags. The tag transmits its own ID to the reader, and then the reader consults an external database with the ID to recognize the object. RFID is fast replacing bar code-based identification mechanisms because (1) communication between a reader and a tag is not limited by the requirement of "line-of-sight" reading and (2) each tag is allowed to have a unique ID.

Reader transmissions or tag transmissions lead to collision because readers and tags operate within the same frequency band due to cost considerations. Collisions are divided into *reader collisions* and *tag collisions* [2,3]. When neighboring readers interrogate a tag simultaneously [4,5], reader signals collide and the tag cannot decode any reader signal. On the contrary, when multiple tags transmit IDs to a reader at the same time, tag signals collide and tag collision prevents the reader from recognizing any tag [6]. Collisions make both communication overhead and transmission delay often lose their usefulness. As a result, either the reader may not recognize all objects or retransmissions are required for successful recognition. Especially, since low-functional passive tags can neither detect collisions nor figure out neighboring tags, tag collision gives rise to the need for a tag anticollision protocol that enables the recognition of tags with few collisions, and also executes in real time.

Tag anticollision protocols can be grouped into two broad categories, namely ALOHA-based protocols and tree-based protocols. ALOHA-based protocols [7–21] reduce the occurrence probability of tag collisions since each tag tries to transmit the ID at randomly selected time. ALOHA-based protocols, however, cannot completely prevent collisions, and hence they have the serious problem that a specific tag may not be identified for a long time, leading to the so-called "tag starvation problem." On the other hand, in tree-based protocols such as the binary tree protocol [20–25] and the query tree protocol [26,27], tag identification conceptually forms a tree. Based on the collision resolution algorithms studied in Refs. [28–31], tree-based protocols split a set of tags into two subsets at a time and attempt to recognize the subsets one by one. By splitting until each set has only one tag, the reader can recognize all the tags in the reader's reading range. Tree-based protocols do not cause tag starvation, although they have relatively long identification delay as compared with ALOHA-based protocols.

Based on the analysis earlier, a good tag collision arbitration protocol for passive RFID tags should have the following characteristics:

- The reader ought to recognize all the tags inside its own reading range. Tag starvation problem results in the failure of object tracking and monitoring. Since the reader, however, cannot estimate the number of tags precisely, the guarantee of recognizing all tags must be taken into consideration in the design of the tag anticollision protocol.
- The reader has to recognize tags promptly. Since an object with a tag is potentially mobile, tag identification must keep pace with the object's velocity. If tag identification is carried out slower than the object's velocity, the reader cannot recognize it and the RFID system fails in monitoring or tracking.
- The tag should be recognized while consuming a small amount of resource. Since the passive tag is supplied with power by the reader's signal, tag's available power is limited. In addition, the tag has low computational capability and limited memory. Thus, the tag anticollision protocol must load the tag with the least possible communication and computation overheads.

This chapter introduces tree-based tag anticollision protocols and adaptive splitting protocols [32], an *adaptive query splitting* (AQS) protocol and an *adaptive binary splitting* (ABS) protocol, which are enhanced versions of tree-based protocols by suppressing the occurrence of tag collisions and shortening tag identification delay. For decreasing tag collisions, adaptive splitting protocols adaptively decide the starting point of the tag identification by using information on the previous identification process in an environment where the reader executes tag identification repeatedly for object monitoring and tracking. The reduction in collisions facilitates tag identification with a small delay and few transmissions while recognizing all the tags inside the reader's reading range.

8.2 Tree-Based Tag Anticollision Protocols

Tree-based tag anticollision protocols perform tag identification in units of *reading cycle*. In a reading cycle, a reader transmits a query (or a feedback) to tags and then one or some of tags transmit ID to the reader. Since the passive tag cannot detect collision, the reader detects whether or not tag collision occurs among tag responses and determines the contents of the query (or the feedback) in the next reading cycle according to the result of the detection. On receiving a query (or a feedback) from the reader, the tag decides whether to transmit or not. Only if a single tag transmits in a reading cycle, the reader can recognize it successfully.

In tree-based protocols, the reader recognizes all the tags within its reading range during an *identification frame*, which consists of several reading cycles. The reader attempts to recognize a set of tags in a reading cycle. A set includes tags, which transmit at the same reading cycle. If a set has more than one tag, tag transmissions lead to collision. When tag collision occurs, the mechanisms split the set into two subsets by tag IDs or random binary numbers. After that, the reader attempts to recognize two subsets one by one in the same frame. By continuing the splitting procedure until each set has only one tag, tree-based protocols are capable of recognizing all the tags in the reader's range.

An identification frame in tree-based protocols can be represented by a tree structure as shown in Figure 8.1a. Each node in the tree corresponds to a reading cycle and a number in a node is the number of tag transmissions in that reading cycle. According to the number of

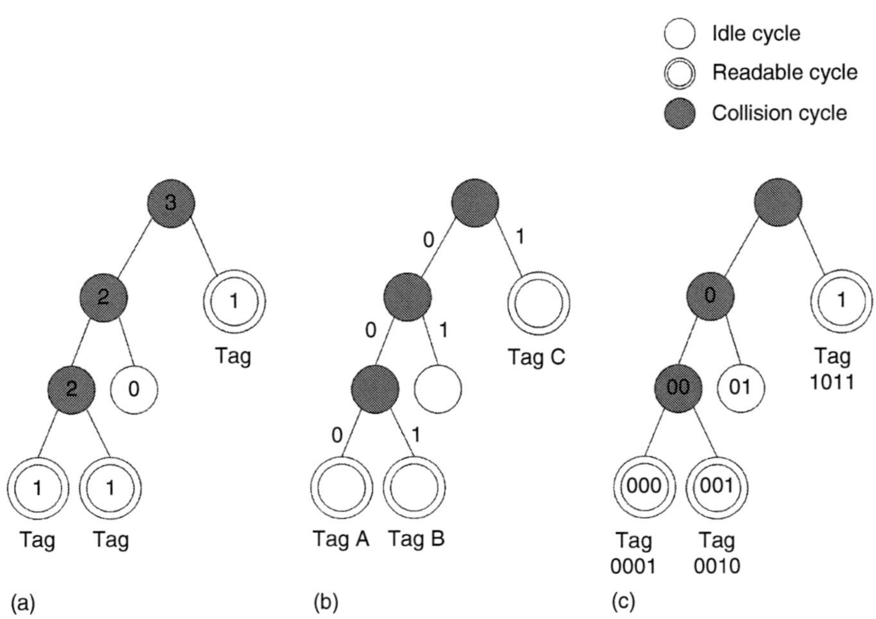

FIGURE 8.1
Tag identification of tree-based protocols. (a) Tree expression of tag identification. (b) Tag identification of the binary tree protocol. (c) Tag identification of the query tree protocol.

tag transmissions in a reading cycle, reading cycles can be divided into three types as follows:

- *Idle cycle*: No transmission is attempted. The idle cycle does not make the reader fail to notice a tag, but it is a source of an unnecessary increment of identification delay.
- *Readable cycle*: Exactly one transmission is attempted. The reader recognizes a tag successfully.
- *Collision cycle*: More than one transmission is attempted. A tag collision occurs and the reader is unable to recognize any tags. The collision cycle defers tag identification and the tag's communication is pure overhead. The reader sends a query (or a feedback) conducting the split of the set including conflicting tags.

In a tree of an identification frame, only a node of a collision cycle has two child nodes because a set is split into two subsets in the collision cycle. Consequently, all intermediate nodes in the tree correspond to collision cycles and all the leaf nodes correspond to either readable cycles or idle cycles. Tag identification in tree-based protocols is coincident with a tree search starting at the root of the tree for finding nodes of readable cycles. The performance of tag identification is influenced significantly by how efficiently it splits the tag set.

8.2.1 Binary Tree Protocol

The binary tree (BT) protocol [20–25] uses random binary numbers generated by colliding tags for the splitting procedure. The tag has a countervalue initialized to 0 at the beginning of the frame. The tag transmits ID when the countervalue is 0. Therefore, all tags, at the beginning of the frame, form one set and transmit concurrently. The reader transmits a feedback to inform tags of the occurrence of tag collision. According to the reader's feedback, all tags change their countervalues. The tag randomly selects a binary number when its transmission causes collision (i.e., the countervalue is 0). By adding the selected binary number to the countervalue, a set is split into two subsets. When tag collisions occur, the tag which is not involved in collision (i.e., the countervalue is not 0) increases its countervalue by 1. When the reader's feedback indicates no collision, all tags decrease their countervalues by 1. The tag infers the successful transmission from the following feedback indicating no collision. The tag recognized by a reader does not transmit any signal until the ongoing frame is terminated. Figure 8.1b shows an example of tag identification of the binary tree protocol and the number by the side of the lines indicates the binary number selected randomly by conflicting tags.

The reader also has a counter to terminate a frame. It initializes the countervalue with 0 in every frame. The countervalue of the reader indicates the number of tag sets which are not yet recognized in a frame. If tag collision occurs, the reader adds 1 to its countervalue since the number of tag sets, which the reader should recognize, increases. Otherwise, it decreases its countervalue by 1. When the countervalue is <0, the reader terminates the frame.

8.2.2 Query Tree Protocol

The query tree (QT) protocol [26,27] uses tag IDs to split a tag set. The reader transmits a query including a bit string. The tag whose first bits of ID equal the bit string of the query responds by transmitting ID. If tag responses of query $q_1q_2 \ldots q_x$ ($q_i \in \{0, 1\}$, $1 \leq x \leq b$, and b

is the number of bits in the tag ID) collide, the reader uses two 1 bit longer queries, $q_1q_2 \ldots q_x0$ and $q_1q_2 \ldots q_x1$ in next reading cycles. The set of tags which match $q_1q_2 \ldots q_x$ is split into two subsets; one is a set of tags which match $q_1q_2 \ldots q_x0$ and the other is a set of tags which match $q_1q_2 \ldots q_x1$. The reader has queue Q for bit strings of queries. At the beginning of the frame, Q is initialized with two 1 bit strings, 0 and 1. The reader dequeues (i.e., removes from queue and returns) a bit string from Q and transmits a query at a time. If tag responses collide, the reader enqueues (i.e., adds to queue) two 1 bit longer bit strings into Q. By expanding the query until either a response or no response follows, all tags are recognized. Figure 8.1c shows an example of tag identification of the query tree protocol and the number inside the nodes indicates the query transmitted by the reader.

Contrary to the binary tree protocol, the query tree protocol imposes simple functions on tags. The query tree protocol is also called a memoryless protocol because tags do not need to have additional memory except ID for identification. However, identification delay is affected by the distribution of tag IDs. For example, as tags have much similar IDs, delay is increased.

8.3 Problems in Tree-Based Protocols

The reader performs tag identification frame repeatedly for object tracking and monitoring.* Let a_x denote tag x. For reader r, let $A_{r,i}$ be the set of tags which dwell inside reader r's range in the ith frame of reader r. To consider the tag's mobility, we classify tags into *staying tags*, *arriving tags*, and *leaving tags*. For the given ith frame of reader r, the staying tag is the tag which was recognized in the last frame by reader r and stays within the reader r's range in the current frame, that is, a staying tag is one of $\{a_x{:}a_x \in A_{r,i-1} \cap A_{r,i}\}$. The arriving tag is the tag which was not recognized in the last frame by reader r and has arrived in the reader r's range before the starting of the current frame, that is, an arriving tag is one of $\{a_x{:}a_x \in A_{r,i} - A_{r,i-1}\}$. The leaving tag is the tag which was recognized in the last frame by reader r and has left from the reader r's range before the starting of the current frame, that is, a leaving tag is one of $\{a_x{:}a_x \in A_{r,i-1} - A_{r,i}\}$. Tag identification should recognize staying tags and arriving tags quickly.

Staying tags have been recognized in the last identification frame and the reader will rerecognize staying tags in the current identification frame. Since the reader already knows information on staying tags, tag collision arbitration can prevent collisions between signals transmitted by staying tags during the current frame. However, the existing tree-based tag anticollision protocols cause collisions between staying tags because they do not take any information on staying tags into consideration. At the beginning of the identification frame, they make one set, which includes all the tags inside the reader's identification range, and start the splitting procedure. To show collisions between staying tags, we measure inter-staying tag collisions through simulations of the binary tree protocol and the query tree protocol. In our simulations, there are 50 tags in an area of 10 m × 10 m, and tags have mobility following the random walk model [33]. A reader, which has the identification range of 3 m, is deployed in the center of the simulation area and recognizes tags repeatedly. To pinpoint an individual tag, we give tags virtual IDs from 1 to 50. To make tag 1 the staying tag, we fix it in the vicinity of the reader. Figure 8.2 shows the number of collisions caused by tag 1 and other staying tags during two consecutive

* Prominent retailers such as Wal-Mart, Target, and Best Buy, as well as logistics companies like UPS and Fed-Ex have made this a requirement of all their operations. Manufacturers are following suite.

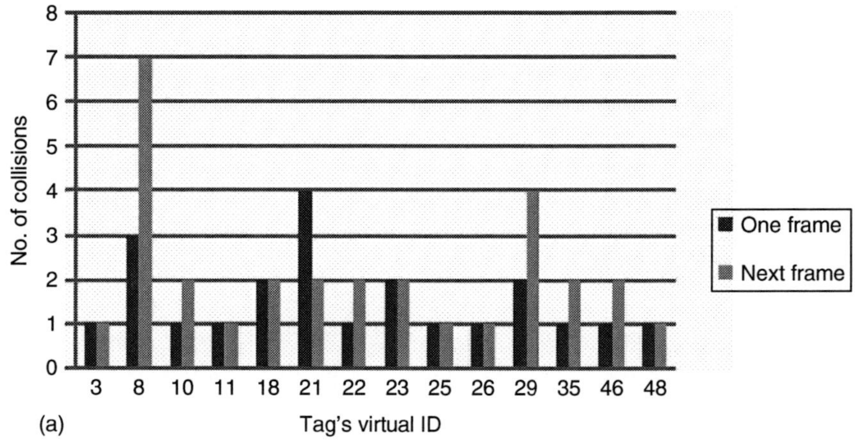

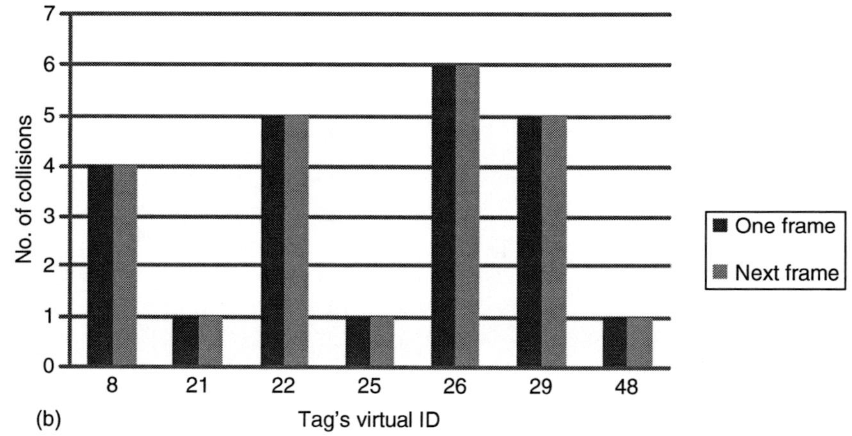

FIGURE 8.2
Collisions between staying tags in tree-based protocols. (a) Collisions by the binary tree protocol. (b) Collisions by the query tree protocol.

identification frames. Under the binary tree protocol, tag 1 suffers from more collisions with some staying tags during the second frame in comparison with the first frame as shown in Figure 8.2a. This is due to the fact that the binary tree protocol adopts the probabilistic approach of exploiting random numbers and does not use information on staying tags. As shown in Figure 8.2b, under the query tree protocol, tag 1 causes the same number of collisions with other staying tags again. The query tree protocol also does not consider collisions by staying tags though it differs from the binary tree protocol in the sense of exploiting tag IDs.

When tag collision occurs in tag identification of tree-based protocols, colliding tags need to retransmit their IDs. Resolution of tag collisions consumes the tag's limited energy and causes additional identification delay. Therefore, eliminating interstaying tag collisions can shorten the total delay for tag identification and reduces the tag's communication overhead. To prevent collisions between staying tags, adaptive splitting protocols start the splitting procedure from several tag sets; each of the sets has one staying tag at most. They are still simple and recognize all tags quickly.

8.4 Adaptive Tag Anticollision Protocols

In tree-based tag anticollision protocols, the tree search causes tag identification delay, and the reduction in identification delay can be accomplished by skipping of collision cycles. However, once a frame is started, the tree searches of the binary tree protocol and the query tree protocol depart from the root or the level 1 nodes of the tree and investigate all intermediate nodes wherein tag collisions occur. The unreasonable starting point of the tree search prolongs identification delay.

The basic idea of adaptive splitting protocols is to adaptively decide the starting point of the tree search with information on tags recognized in the last identification frame. At every identification frame, the tree search of tag identification starts from the nodes which were the leaf nodes of the tree in the last identification frame. Note that these starting nodes were readable cycles or idle cycles in the last frame. To recognize arriving tags, the identification process traces down the path of the tree by inserting two child nodes of the current node into the tree. To handle unnecessary idle cycles induced by leaving tags, the identification process traces up the path of the tree by replacing two leaf nodes with their parent node. The key institution behind this approach is that in most applications employing RFID tags, the set of objects encountered in successive readings from a particular reader does not change substantially and information from one reading can be used for the next.

8.5 Adaptive Query Splitting

AQS uses reader's queries and tag IDs analogous to the query tree protocol. The reader transmits a query including a bit string. The tag responds to a query with its ID, if the prefix of its ID is equal to the bit string of the query, that is, $r_1r_2 \ldots r_x = q_1q_2 \ldots q_x$ where the tag ID is $r_1r_2 \ldots r_b$ (r_i is the ith binary value of the ID and b is the total number of bits of the ID) and the bit string of the query is $q_1q_2 \ldots q_x$ (q_i is the ith binary value of the query, $1 \leq x \leq b$). Tags are memoryless because they do not maintain any information except their own IDs.

The reader has queue Q and candidate queue CQ to make queries. Queue Q maintains bit strings for queries in the current identification frame. Candidate queue CQ maintains bit strings for queries in the next identification frame. The reader uses the bit strings stored in CQ as the starting point of tag identification in the next frame. The starting point of tag identification moves downward (toward descendants in the tree) by the query insertion procedure and moves upward (toward the root of the tree) by the query deletion procedure. At the beginning of the frame, the reader initializes Q with bit strings of CQ and makes CQ empty. If CQ does not have any bit string (e.g., when the reader resets), Q is initialized with two 1 bit strings, 0 and 1. The reader dequeues a bit string from Q and transmits a query at a time. The reader enqueues some of the used bit strings into CQ according to the result of receiving tag responses. The identification frame continues until Q is empty.

8.5.1 Query Insertion

Let $q_1q_2 \ldots q_x$ be the bit string of the transmitted query. According to the number of tags responses, the reader acts as follows:

- Idle query (number of tag responses = 0): The reader enqueues $q_1q_2 \ldots q_x$ into *CQ*.
- Readable query (number of tag responses = 1): The reader enqueues $q_1q_2 \ldots q_x$ into *CQ*.
- Collision query (number of tag responses $\geq$ 2): The reader enqueues $q_1q_2 \ldots q_x0$ and $q_1q_2 \ldots q_x1$ into *Q*.

The reader enqueues idle queries as well as readable queries into *CQ*. Therefore, *CQ* has all the leaf nodes in the tree at the end of the frame. By maintaining *CQ*, the tree search of AQS begins at the leaf nodes in the tree of the last frame and skips nodes whose queries caused tag collisions in the last identification frame. Since some arriving tags may not match any nodes of readable cycles of the last frame, the tree search of AQS starts not only at the nodes of readable cycles but also at the nodes of idle cycles of the last frame. To split a set of conflicting tags into two subsets, the reader uses two queries 1 bit longer than the collision query. By expanding the collision queries, the reader can recognize all tags.

Consider the situation that a reader attempts to recognize three tags whose IDs are 0101, 0110, and 1001, respectively. *CQ* of the reader has no bit string because the reader resets. In this case, tag identification of AQS is equal to that of the query tree protocol. Figure 8.3a shows signals transmitted by the reader and tags for tag identification, and Figure 8.3b is the tree expression of tag identification. Recognizing three tags causes two collisions, and at the end of the frame, *CQ* stores 1, 00, 010, and 011. Thereafter, tag 1101 comes into the reader's reading range, and now the reader attempts to recognize four tags. As shown in

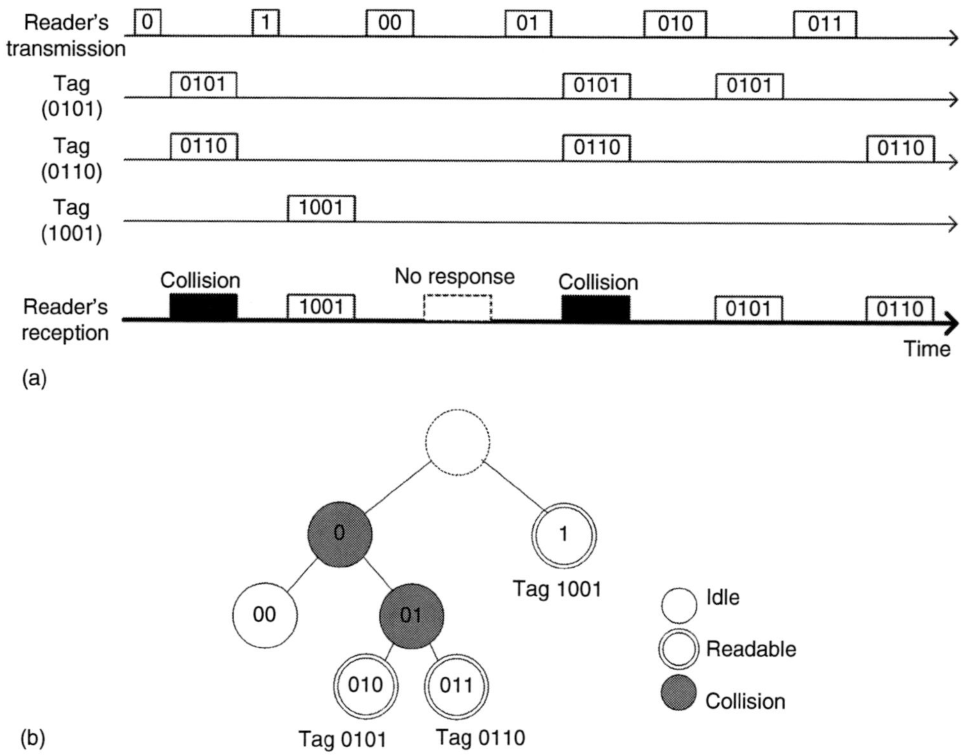

FIGURE 8.3
Tag identification in AQS after the reader resets. (a) Communication between the reader and tags. (b) Tree of tag identification.

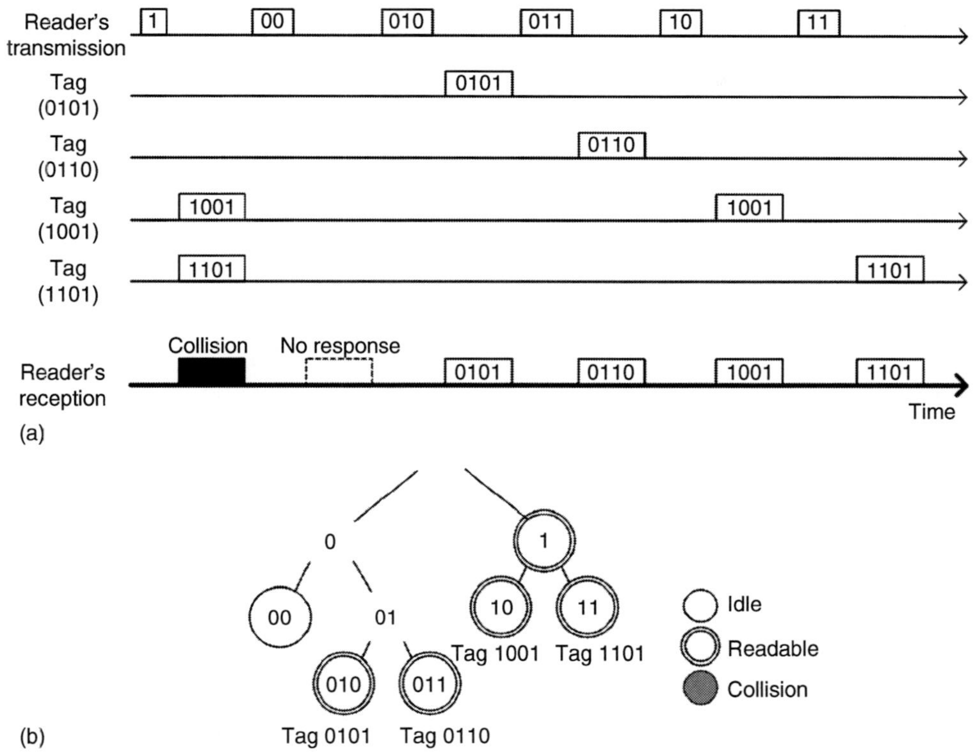

FIGURE 8.4
Tag identification by the query insertion procedure. (a) Communication between the reader and tags. (b) Tree of tag identification.

Figure 8.4a and b, the reader recognizes four tags with only one collision. Since arriving tag 1101 responds to query 1 which staying tag 1001 responds to, query 1 causes a tag collision. By expanding query 1 into query 10 and query 11, the reader recognizes both of two conflicting tags quickly.

8.5.2 Query Deletion

In performing the procedure, the query insertion procedure enables only to expanding the tree and therefore augments the number of leaf nodes in the tree. Since the number of readable queries is the same as the number of tags, idle queries will proliferate and worsen the reader's identification ability. It results from where readable queries are transformed into idle queries by leaving tags. For fast tag identification, the query deletion procedure removes unnecessary idle queries from *CQ* under the condition that *CQ* has bit strings covering all branches in the tree.

Let $\lambda(q_1q_2 \ldots q_x)$ be the number of responses following query $q_1q_2 \ldots q_x$. Since the tag responds to either $q_1q_2 \ldots q_x0$ or $q_1q_2 \ldots q_x1$ when its ID matches query $q_1q_2 \ldots q_x$,

$$\lambda(q_1q_2 \ldots q_x) = \lambda(q_1q_2 \ldots q_x0) + \lambda(q_1q_2 \ldots q_x1).$$

When query $q_1q_2 \ldots q_x$ is a collision query, $\lambda(q_1q_2 \ldots q_x) \geq 2$ since more than one response follows the collision query. The node of collision query $q_1q_2 \ldots q_x$ has two child nodes which are a pair of node types as follows:

- *Two collision nodes*: The node of query $q_1q_2 \ldots q_x$ has two child nodes of collision queries if and only if $\lambda(q_1q_2 \ldots q_x0) \geq 2$, $\lambda(q_1q_2 \ldots q_x1) \geq 2$, and $\lambda(q_1q_2 \ldots q_x) \geq 4$.
- *A collision node and a readable node*: Two child nodes are a node of a collision query and a node of a readable query only if $\lambda(q_1q_2 \ldots q_x0) \geq 1$, $\lambda(q_1q_2 \ldots q_x1) \geq 1$, and $\lambda(q_1q_2 \ldots q_x) \geq 3$.
- *A collision node and an idle node*: Two child nodes are a node of a collision query and a node with an idle query only if $\lambda(q_1q_2 \ldots q_x0) \geq 0$, $\lambda(q_1q_2 \ldots q_x1) \geq 0$, and $\lambda(q_1q_2 \ldots q_x) \geq 2$.
- *Two readable nodes*: The node of query $q_1q_2 \ldots q_x$ has two child nodes of readable queries if and only if $\lambda(q_1q_2 \ldots q_x0) = 1$, $\lambda(q_1q_2 \ldots q_x1) = 1$, and $\lambda(q_1q_2 \ldots q_x) = 2$.

When some tags become leaving tags, a pair of node types can be transformed into as follows:

- *A readable node and an idle node*: $\lambda(q_1q_2 \ldots q_x) = 1$ when $\lambda(q_1q_2 \ldots q_x0) = 0$ and $\lambda(q_1q_2 \ldots q_x1) = 1$, or $\lambda(q_1q_2 \ldots q_x0) = 1$ and $\lambda(q_1q_2 \ldots q_x1) = 0$. Therefore, query $q_1q_2 \ldots q_x$ is not a collision query but a readable query.
- *Two idle nodes*: $\lambda(q_1q_2 \ldots q_x) = 0$ when $\lambda(q_1q_2 \ldots q_x0) = 0$ and $\lambda(q_1q_2 \ldots q_x1) = 0$. Therefore, query $q_1q_2 \ldots q_x$ is also an idle query.

When *CQ* has two bit strings corresponding to the pair of child nodes which have the transformed types after an identification frame, the reader deletes $q_1q_2 \ldots q_x0$ and $q_1q_2 \ldots q_x1$ from *CQ* and enqueues $q_1q_2 \ldots q_x$ into *CQ*. The reader deletes all transformed queries from *CQ* recursively. As the query deletion replaces two bit strings with their common prefix in *CQ*, AQS can recognize all tags with less idle queries.

Consider that tag 0110 moves to the contrary direction of the reader and finally crosses over the boundary of the identification area of the reader after tag identification illustrated in Figure 8.4. In the following frame, query 011 changes into the idle query from the readable query. Figure 8.5 shows the operation of the query deletion procedure after tag 0110 becomes the leaving tag. Since query 010 is a readable query and query 011 is an idle query, query 01 is substituted for queries 010 and 011. The query deletion is implemented once more because query 00 is an idle query and the newly inserted query 01 is a readable query. The reader deletes queries 00 and 01 from *CQ*, and inserts query 0 into *CQ*. Eventually, *CQ* has three queries, 0, 10, and 11, and the reader can recognize tags 0101, 1001, and 1101 with these queries.

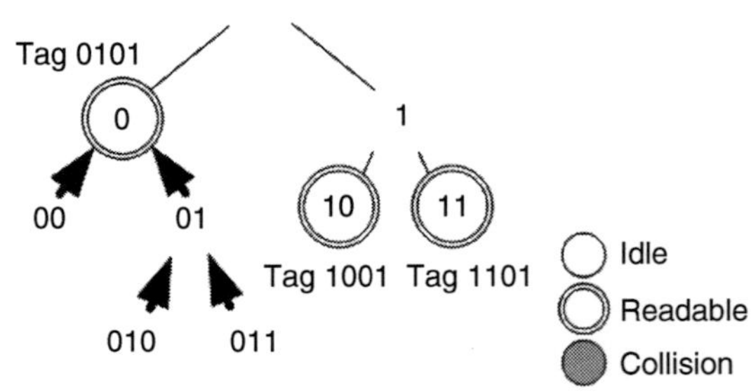

FIGURE 8.5
Query deletion procedure after tag 0110 went out of the reader's reading range.

8.6 Adaptive Binary Splitting

ABS uses random numbers with the aim of splitting a set of tags transmitting at the same cycle like the binary tree protocol but starts the tree search only from the nodes of readable cycles of the last frame. AQS, as described in the previous section, can reduce collisions as compared with the binary tree protocol and the query tree protocol, but it produces idle cycles. To guarantee identification of all tags, the reader uses not only queries of readable cycles but also queries of idle cycles of the last frame. Though the query deletion procedure of AQS eliminates unnecessary idle cycles by leaving tags, it cannot be avoided that the start of the tree search includes some of idle cycles to cover all possible ranges of the tag ID. On the contrary, ABS starts tag identification only from readable cycles of the last frame and uses binary numbers selected randomly in each conflicting tag for the splitting procedure. During tag identification, ABS allocates different cycles to tags by revising the tag's counter into the ranking of recognition. The allocation of cycles to staying tags enables fast identification without collisions between staying tags in the next identification frame. ABS also has a mechanism for deallocating cycles of leaving tags. A transmission of an arriving tag is decided by a random number selected among possible values inside a reader's range. Tag transmissions are aligned in the increasing order of countervalues. ABS achieves fast identification by diminishing not only collisions but also unnecessary idle cycles.

8.6.1 Tag Transmission Control

A tag has two counters, a *progressed slot counter* (PSC) and an *allocated slot counter* (ASC). PSC signifies a numerical order of the readable cycle in a frame. Tags and readers initialize their PSC to 0 at the beginning of the frame and increase by only 1 in the readable cycle. ASC indicates which reading cycle the tag can transmit its ID to the reader. That is, the tag is allowed to transmit when its ASC is equal to PSC. If the tag has ASC less than PSC, it does not attempt the transmission until the completion of the frame, because it has already been recognized in the ongoing frame. To control PSC and ASC, the reader informs tags of the type of the last reading cycle by transmitting a feedback. According to the contents of the reader's feedback, every tag acts as follows:

- Idle (number of tag responses = 0): The tag which has not been recognized in the current frame yet, that is, the tag which has ASC greater than PSC decreases ASC by 1.
- Readable (number of tag responses = 1): All the tags increase PSC by 1. The recognized tag does not react to the reader's feedback and does not transmit its ID to the reader before the start of the next identification frame.
- Collision (number of tag responses ≥ 2): Conflicting tags (tags which have ASCs equal to PSC) randomly select one of two binary numbers, 0 and 1, and then add it to ASC. The tag which has ASC greater than PSC adds 1 to ASC.

During a frame, PSC does not decrease, and an ASC value is not changed in such a way that ASC becomes smaller than PSC. Therefore, a tag recognized by the reader gets to obtain a unique ASC which implies the allocation of the cycle to the tag for the next identification frame. The decrement of ASCs in the idle cycle deallocates cycles corresponding to ASCs of leaving tags. A set of conflicting tags is separated into two subsets by the tag's random binary number selection after tag collision; the first subset includes tags which select 0 and the second subset includes tags which select 1. Since PSC is not changed in the collision cycle, the first subset tries to retransmit in the following cycle. The second subset transmits after the first subset is recognized. The increment in ASC values of unrecognized tags in the collision cycle prevents the second subset from combining with another set (tags which have already had ASC of 1). Preserving ASC at the boundary of two consecutive frames makes it possible that the tree search starts from the readable cycles of the last frame.

8.6.2 Frame Termination

To terminate the frame at once after identifying all tags, the reader of ABS acts as the tag which has the largest ASC. The reader determines the end point of the frame with a PSC and a *terminated slot counter* (TSC). PSC of the reader represents the number of tags recognized successfully. In a readable cycle, the reader adds 1 to PSC. TSC signifies the number of tag sets in the reader's range. If a collision occurs, the reader increases TSC by 1 because the number of tag sets has increased. When a reading cycle of type "idle" is encountered, the reader decreases TSC by 1. This is to reflect the effect of the elimination of idle cycles. As soon as PSC is greater than TSC, the reader concludes that all tags have been recognized and transmits the command terminating the frame to all tags. For fast identification in the next frame, the reader preserves TSC after the end of the frame.

In an environment with multiple readers, an arriving tag can be recognized with its ASC given by other readers. If an arriving tag has ASC less than TSC, it is obvious that the arriving tag has the same ASC with a staying tag or a leaving tag and is recognized by the reader in the following frame. To cope with ASC greater than TSC, the reader supports the TSC value when a frame starts. A tag having ASC greater than TSC changes its ASC to a random number from 0 to TSC. ABS recognizes all tags quickly through scaling ASCs of arriving tags into the range of TSC.

Figure 8.6 illustrates an example of the operation of ABS. Assume that a reader attempts to recognize tags A, B, and C, and there was no tag inside the reader's reading range in the previous frame. Therefore all three tags are arriving tags and the TSC value

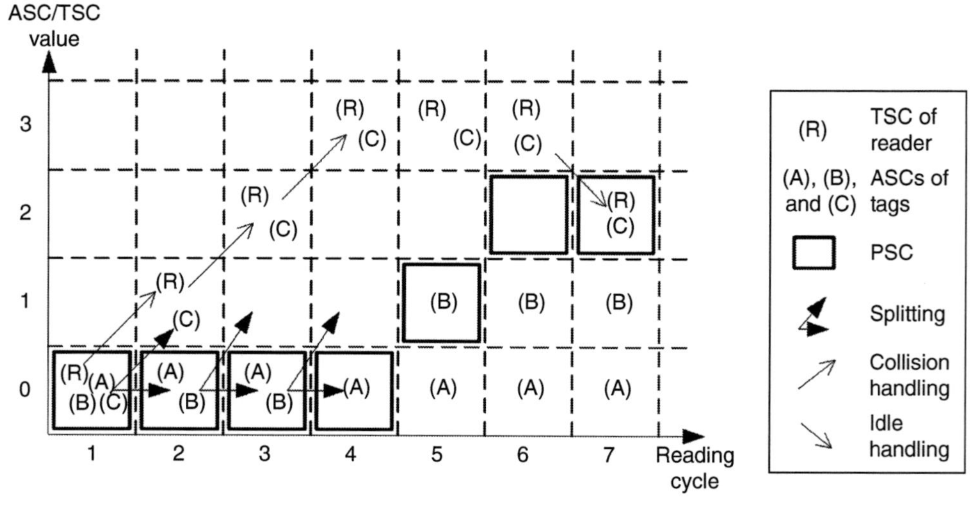

FIGURE 8.6
Tag identification by ABS.

of the reader is 0. In this case, tag identification of ABS is equal to that of the binary tree protocol because all three tags have ASCs of 0. In the first reading cycle, three tags transmit simultaneously and collision occurs. Tags A and B select 0, and tag C selects 1. In the second cycle, a collision between tags A and B occurs and both of them select 0 again. Tag C and the reader increase ASC and TSC by 1, respectively. Tags A and B are split successfully in the third cycle, and hence the reader recognizes tags A and B in the fourth and fifth cycles, respectively. The sixth cycle is an idle cycle because there are no tags which have ASC equal to PSC (=2). ASC of tag C and TSC of the reader are decreased by 1 and the reader recognizes tag C in the seventh cycle. The reader terminates the identification frame after the seventh cycle. By using the values of ASCs and TSC decided in the current frame into tag identification of the next frame, the reader achieves fast tag identification. The arriving tags in the next frame can randomly select their ASCs among numbers ranging from 0 to 2. Especially, when there is no leaving tag and no arriving tag in the next frame, the reader can reidentify these three tags without any collision.

8.7 Performance Evaluation

We evaluate the performance of AQS and ABS compared with the binary tree protocol and the query tree protocol. To measure the efficiency of tag identification in the tree-based protocols, we consider the following aspects:

- *Number of collisions*: We measure the number of collisions between tag-to-reader signals. A collision defers identification and increases power consumption of tags.
- *Number of idle cycles*: The idle cycle is a factor of identification delay.
- *Identification delay*: We measure the total delay for recognizing all tags by the interrogation cycle. Fast identification is the most significant factor in the tree-based anticollision protocols because they do not cause the tag starvation problem.
- *Tag communication overhead*: This metric is the average number of bits transmitted by a tag in a frame. This influences the amount of power consumption. Due to lack of power source in tags, this must be low.

The simulation setup is shown in Table 8.1. To avoid the reader collision problem [4], we place readers in such a manner that their reading ranges do not intersect. To appreciate the impact of tag's mobility, we define meter per frame (MPF). The MPF of tag a_x, MPF(a_x), is given by

TABLE 8.1

Simulation Setup

Parameter	Value
Simulation area	100 m × 100 m
No. of readers	100
Identification range of the reader	3 m
No. of tags	1000
Tag ID	Randomly selected 96 bit ID
Maximum meter per frame (MPF)	2 m/frame

$$\mathrm{MPF}(a_x) = \frac{m_a(t_1, t_2)}{F_p(t_1, t_2)} \text{ (m/frame)},$$

where $m_a(t_1, t_2)$ is the distance that tag a_x moves in the time interval $[t_1, t_2]$ and $F_p(t_1, t_2)$ is the number of frames executed by protocol p in the interval $[t_1, t_2]$. By using MPF, we can ensure that tree-based protocols recognize the same tags in a frame. In our simulations, initial positions and destinations of tags are randomly selected under the simulation area. A tag moves from its initial position toward its destination with MPF, which is randomly selected from 0 to the maximum MPF. We run each simulation 50 times under the earlier parameters and investigate the average results for the performance evaluation.

8.7.1 Impact of the Number of Tags

Figure 8.7 shows the simulation results obtained by changing the number of tags in the simulation area. In the readers' reading ranges, there averagely exist 31.62% of tags. As the number of tags increases, the identification delay gets longer and tag collisions occur

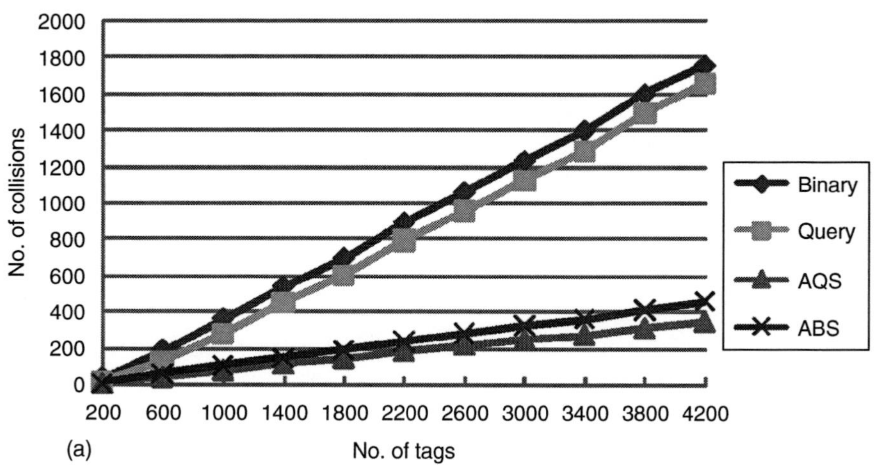

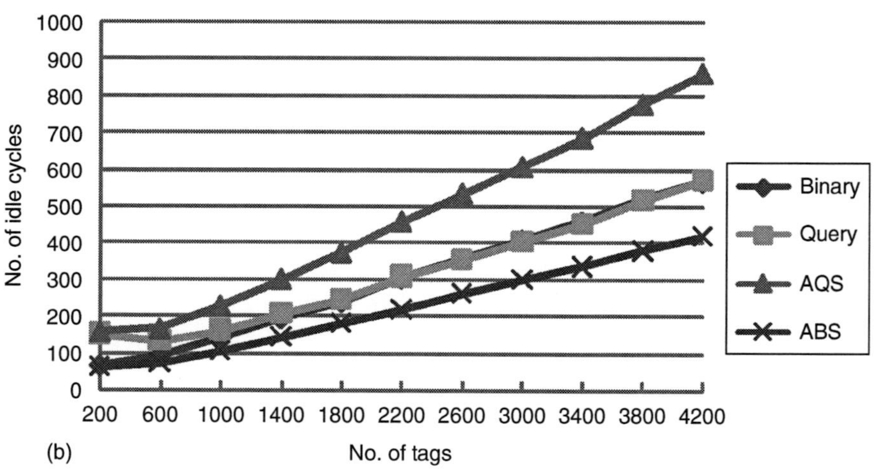

FIGURE 8.7
Performance comparison with varying number of tags. (a) Collisions. (b) Idle cycles.

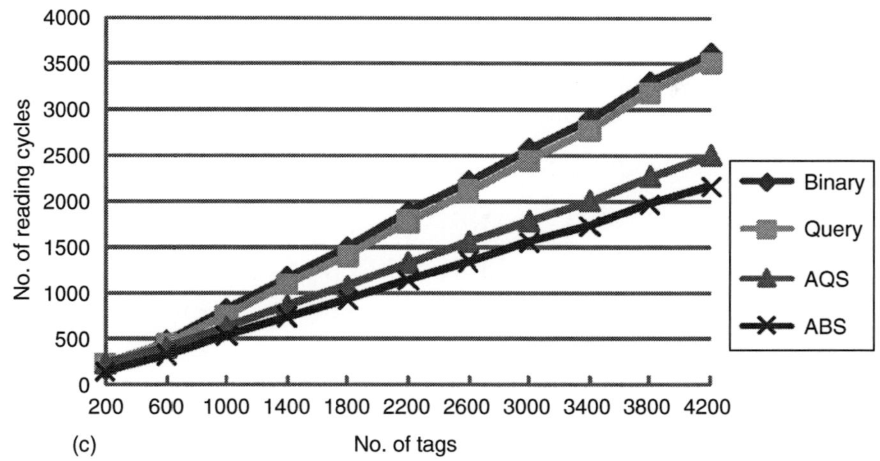

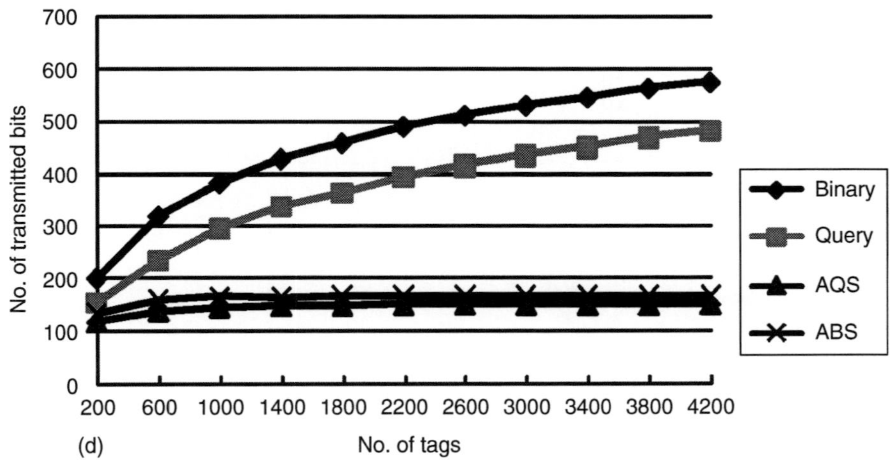

FIGURE 8.7 (continued)
(c) Identification delay. (d) Tag communication overhead.

more often. The binary tree protocol and the query tree protocol show comparable delay curves. The subtle difference in the delay results from the starting points of tag identification. Note that the binary tree protocol departs from the root of the tree and the query tree protocol departs from the level 1 nodes of the tree. The performance of the binary tree protocol can be improved up to the performance level of the query tree protocol by initializing the tag's countervalue to a random binary number instead of the value of 0. By restraining the occurrence of collisions, AQS and ABS have shorter delay than the binary tree protocol and the query tree protocol. Small collisions activate small tag communication overhead. ABS has the shortest delay because it eliminates many idle cycles. AQS generates more idle cycles than others due to additional queries to guarantee recognizing all tags. On the other hand, AQS makes less collision and less tag communication than ABS because idle cycles assist arriving tags to avoid conflicting with other tags.

8.7.2 Impact of Tag Movement

We evaluate the performance by increasing the tag velocity. Figure 8.8 presents the simulation results obtained by varying the maximum MPF. We normalize the measured

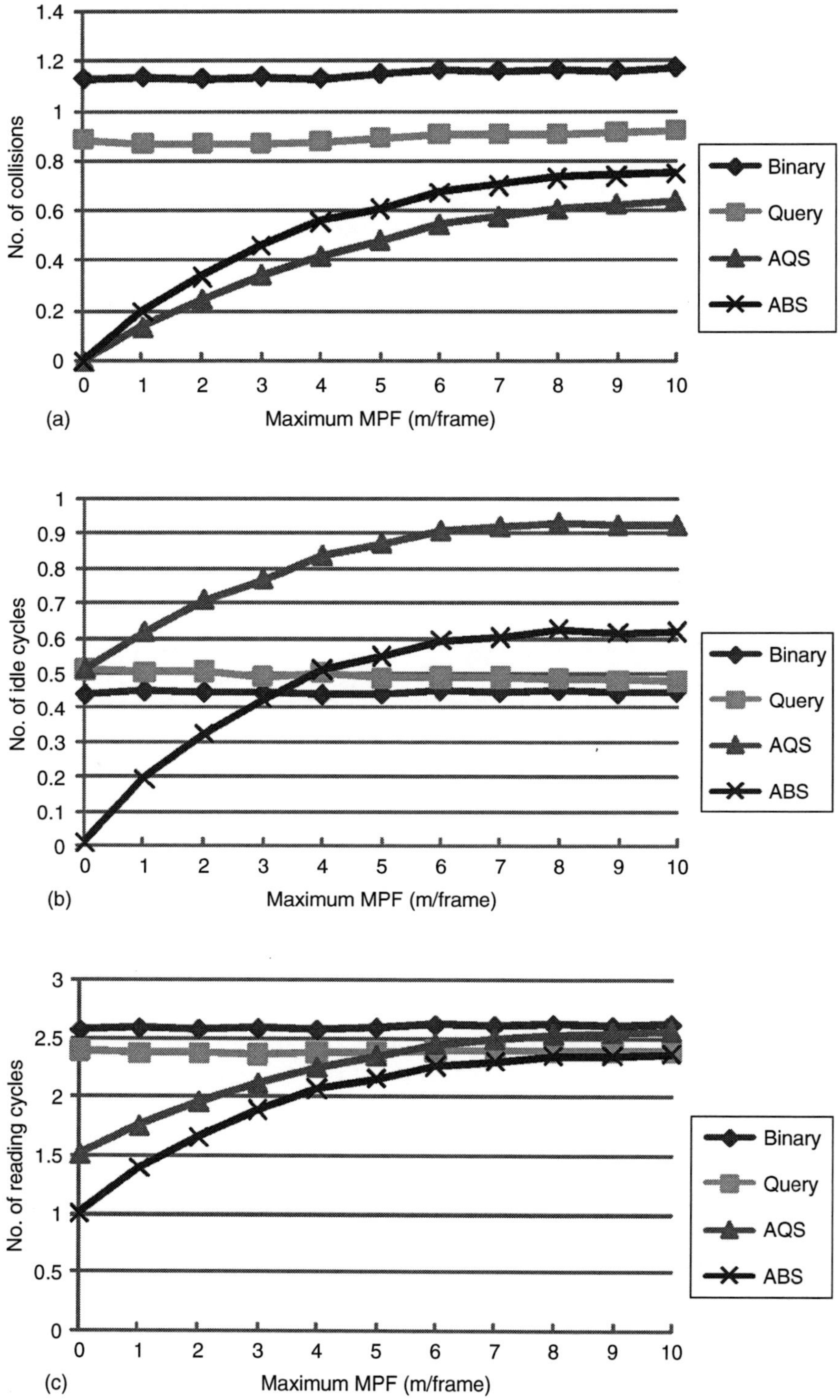

FIGURE 8.8
Impact of tag mobility on tag identification. (a) Collisions. (b) Idle cycles. (c) Identification delay.

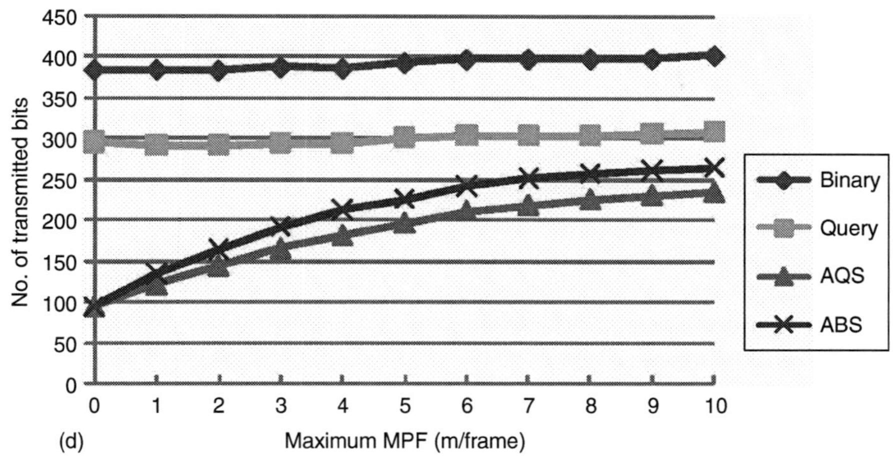

FIGURE 8.8 (continued)
(d) Tag communication overhead.

values by the number of recognized tags. When tags move at low speed, adaptive splitting protocols outperform the binary tree protocol and the query tree protocol considerably. Tags would like to be staying tags, and AQS and ABS block collisions between staying tags completely because they do not allocate more than one staying tag to a set. Especially, they achieve collisionless tag identification when the maximum MPF is 0. As tags move faster, the performance of AQS and ABS deteriorates. When the maximum MPF is >6 m/frame, AQS has longer delay than the query tree protocol. When the tag mobility has high speed, there are few staying tags and collisions between staying tags at the binary tree protocol and the query tree protocol hardly occur. Additionally, AQS and ABS generate idle cycles because leaving tags increase and leaving tags make idle cycles. Hence, AQS and ABS show the performance similar to the binary tree protocol and the query tree protocol at high speeds.

8.7.3 Impact of the Similarity of ID

For the purpose of another comparison, we evaluate the impact of the similarity among IDs. The query tree protocol and AQS may be influenced by the distribution of IDs because they use tag IDs for splitting a tag set. To quantify the similarity of IDs, we define an *identical bit* as the length of the identical prefix all tag IDs have. The tag ID is depicted by $x_1x_2 \ldots x_ax_{a+1} \ldots x_{96}$ (x_i is a binary digit, $1 \leq a < 96$) and all tag IDs have the same $x_1x_2 \ldots x_a$ if the identical bit is a and each tag has a 96 bit ID. Figure 8.9 gives the simulation results for various identical bits from 0 (IDs are completely randomly selected) to 80. We normalize the measured values by the number of recognized tags. As the identical bit increases, the query tree protocol rapidly degenerates as expected. The query tree protocol has the highest communication overhead because the reader transmits all queries causing collisions in every frame. On the other hand, the performance of AQS is not seriously affected by the similarity of IDs. Since candidate queue *CQ* excludes queries of collision cycles of the last frame, AQS uses a collision query only once. However, as the identical bit increases, idle cycles in the trees of the query tree protocol increase. When the identical bit is >48, AQS has longer delay than the binary tree protocol because of an increment of idle cycles. ABS and the binary tree protocol are not affected by the identical bit because they do not use the patterns of IDs. As in the previous scenarios, ABS shows the shortest identification delay.

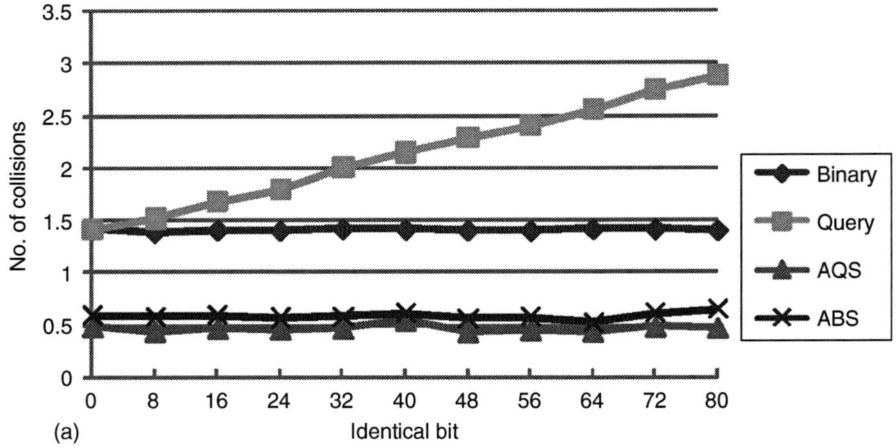

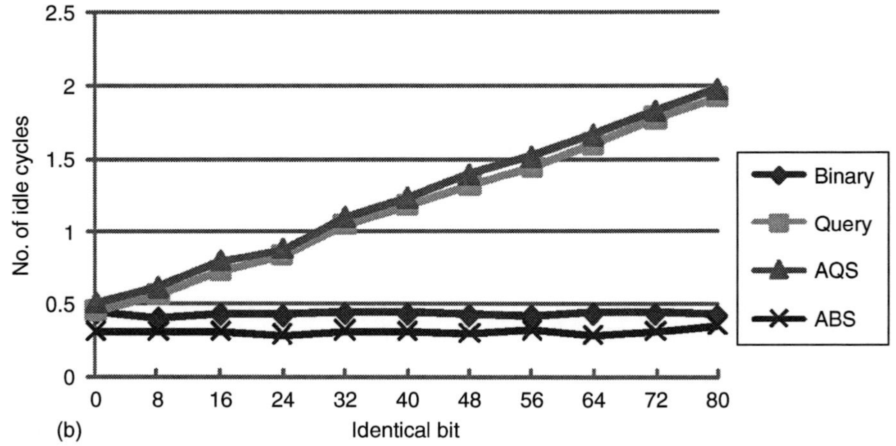

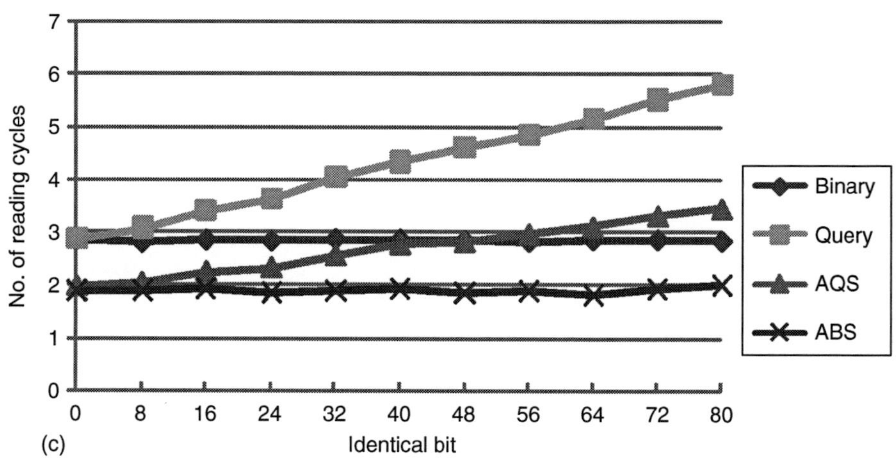

FIGURE 8.9
Impact of ID distribution on tag identification. (a) Collisions. (b) Idle cycles. (c) Identification delay.

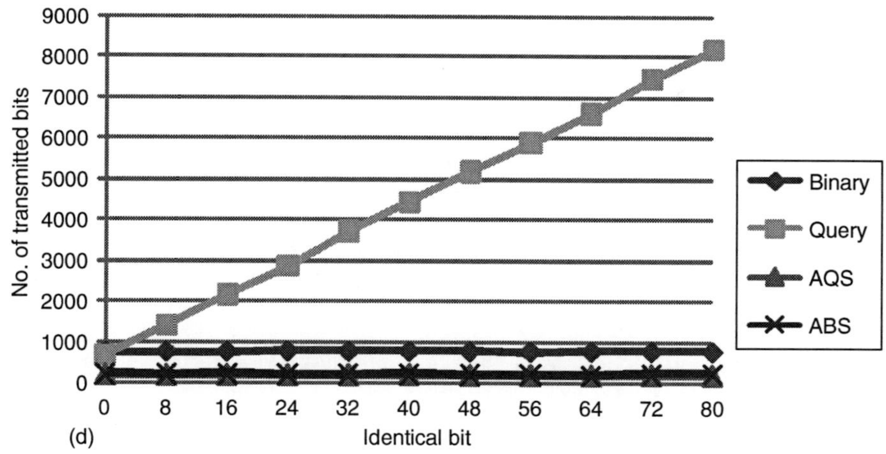

FIGURE 8.9 (continued)
(d) Tag communication overhead.

8.8 Conclusion

A collision caused by tags transmitting simultaneously is a major factor in deferring tag identification of RFID systems. This chapter has described tree-based tag anticollision protocols and adaptive tag anticollision protocols for passive tags. Adaptive splitting protocols are the enhanced tree-based protocols to reduce collisions by exploiting information obtained from the last frame of tag identification. The key institution behind adaptive splitting protocols is that in most applications employing RFID tags, the set of objects encountered in successive readings from a particular reader does not change substantially, and information from one reading can be used for the next. A simulation-based evaluation shows that AQS and ABS significantly reduce delay and communication overhead for the tag reading process.

Acknowledgment

This work was jointly supported by SK Telecom, the Korea Science and Engineering Foundation (KOSEF) grant funded by the Korea government (MOST) (No. R01-2007-000-11203-0) and the Ministry of Information and Communication (MIC), Korea under ITRC IITA-2007-(C1090-0701-0046).

References

1. K. Finkenzeller, *RFID Handbook: Fundamentals and Applications in Contactless Smart Cards and Identification*, John Wiley & Sons, New York, 2003.
2. S.E. Sarma, D. Brock, and D.W. Engels, Radio frequency identification and the electronic product code, *IEEE Micro*, 21(6), 50–54, 2001.
3. S.E. Sarma, S.A. Weis, and D.W. Engels, RFID systems and security and privacy implications, in *Workshop on Cryptographic Hardware in Embedded Systems*, LNCS 2523, pp. 454–470, August 2002.

4. D.W. Engels and S.E. Sarma, The reader collision problem, in *Proceedings of IEEE International Conference on System, Man and Cybernetics*, Hammamet, Tunisie, October 2002.
5. J. Waldrop, D.W. Engels, and S.E. Sarma, Colorwave: an anticollision algorithm for the reader collision, in *Proceedings of IEEE International Conference on Communications*, pp. 1206–1210, May 2003.
6. C. Floerkemeier and M. Lampe, Issues with RFID usage in ubiquitous computing applications, in *Proceedings of the 2nd International Conference of Pervasive Computing*, LNCS 3001, pp. 188–193, April 2004.
7. N. Abramson, The aloha system—another alternative for computer communications, in *Proceedings of Fall Joint Computer Conference, AFIPS Conference*, vol. 37, pp. 281–285, November 1970.
8. L.G. Roberts, Extensions of packet communication technology to a hand held personal terminal, in *Proceedings of Spring Joint Computer Conference, AFIPS Conference*, vol. 40, pp. 295–298, 1972.
9. R. Metcalfe, Steady state analysis of a slotted and controlled aloha system with blocking, in *Proceedings of the 6th Hawaii Conference System Science*, pp. 375–380, January 1973.
10. S. Lam and L. Kleinrock, Packet switching in a multi access broadcast channel: dynamic control procedures, *IEEE Transaction on Automatic Control*, COM-23(9), 891–904, 1975.
11. R. Rao and A. Ephremides, On the stability of interacting queues in a multiple-access system, *IEEE Transactions on Information Theory*, 34(5), 918–930, 1988.
12. V. Anatharam, The stability region of the finite-user slotted aloha protocol, *IEEE Transactions on Information Theory*, 37(3), 535–540, 1991.
13. I.E. Teletar and R.G. Gallager, Combining queuing theory and information theory for multiaccess, *IEEE Journal on Selected Areas in Communications*, 13(6), 963–969, 1995.
14. F.C. Schoute, Control of aloha signalling in a mobile radio trunking system, in *Proceedings of the IEE International Conference on Radio Spectrum Conservation Techniques*, pp. 38–42, 1980.
15. F.C. Schoute, Dynamic frame length aloha, *IEEE Transactions on Communications*, COM-31(4), 565–568, 1983.
16. J.E. Wieselthier, A. Ephremides, and L.A. Michaels, An exact analysis and performance evaluation of framed aloha with capture, *IEEE Transactions on Communications*, COM-38(2), 125–137, 1989.
17. H. Vogt, Efficient object identification with passive RFID tags, in *Proceedings of the International Conference on Pervasive Computing*, LNCS 2414, pp. 98–113, April 2002.
18. J. Zhai and G. Wang, An anti-collision algorithm using two-functioned estimation for RFID tags, in *Proceedings of International Conference on Computational Science and its Applications*, LNCS 3483, pp. 702–711, May 2005.
19. EPC radio-frequency identification protocols class-1 generation-2 UHF RFID protocol for communications at 860 MHz–960 MHz version 1.0.8, EPCglobal, December 2004.
20. Information technology automatic identification and data capture techniques—radio frequency identification for item management air interface—part 6: parameters for air interface communications at 860–960 MHz, Final Draft International Standard ISO 18000-6, November 2003.
21. UCODE, *Philips Semiconductors*, http://www.semiconductors.philips.com, 2005.
22. D.R. Hush and C. Wood, Analysis of tree algorithms for RFID arbitration, in *Proceedings of IEEE International Symposium on Information Theory*, pp. 107, August 1998.
23. M. Jacomet, A. Ehrsam, and U. Gehrig, Contactless identification device with anticollision algorithm, in *Proceedings of IEEE Conference on Circuits, System, Computers and Communications*, pp. 269–273, July 1999.
24. Draft protocol specification for a 900 MHz class 0 radio frequency identification tag, Auto-ID Center, February 2003.
25. S.A. Weis, S.E. Sarma, R.L. Rivest, and D.W. Engels, Security and privacy aspects of low-cost radio frequency identification systems, in *Proceedings of the 1st Annual Conference on Security in Pervasive Computing*, LNCS 2802, pp. 201–212, March 2003.
26. C. Law, K. Lee, and K.-Y. Siu, Efficient memoryless protocol for tag identification, in *Proceedings of the 4th International Workshop on Discrete Algorithms and Methods for Mobile Computing and Communications*, pp. 75–84, August 2000.

27. F. Zhou, C. Chen, D. Jin, C. Huang, and H. Min, Evaluating and optimizing power consumption of anti-collision protocols for applications in RFID systems, in *Proceedings of the International Symposium on Low Power Electronics and Design*, pp. 357–362, August 2004.
28. J.I. Capetanakis, Tree algorithms for packet broadcast channels, *IEEE Transactions on Information Theory*, IT-25(5), 505–515, 1979.
29. G. Fayolle, P. Flajolet, M. Hofri, and P. Jacquet, Analysis of a stack algorithm for random access communication, *IEEE Transactions on Information Theory*, IT-31(2), 244–254, 1985.
30. P. Mathys and P. Flajolet, Q-ary collision resolution algorithms in random access systems with free or blocked channel access, *IEEE Transactions on Information Theory*, IT-31(2), 217–243, 1985.
31. J. Moseley and P. Humblet, A class of efficient contention resolution algorithms for multiple access channels, *IEEE Transactions on Communications*, COM-33(2), 145–151, 1985.
32. J. Myung, W. Lee, J. Srivastava, and T. Shih, Tag-splitting: adaptive collision arbitration protocols for RFID tag identification, *IEEE Transactions on Parallel Distributed Systems*, 18(6), 763–775, 2007.
33. R.A. Guerin, Channel occupancy time distribution in a cellular radio system, *IEEE Transactions on Vehicular Technology*, 36(3), 89–99, 1987.

9

Comparative Performance Analysis of Anticollision Protocols in RFID Networks

Wonjun Lee, Jihoon Choi, and Donghwan Lee

CONTENTS

9.1 Introduction

Radio frequency identification (RFID) is an automatic recognition system that consists of a number of tags and readers. An RFID reader identifies objects by reading the data contained in the tags [1]. RFID can be used in various applications ranging from identifying objects to using memory within its own chip. Attaching tags to wild animals, for instance, makes it possible to track them. For such reasons, the RFID system has been spotlighted as the technology that can replace bar code systems.

Tags can be classified into two types based on the existence of self-electric power: active tags and passive tags. The active tags can transmit data without the aid of a reader because it has its own battery. They also have a more powerful memory than passive tags. On the other hand, it is possible for a passive tag to transmit only when a reader is involved since it does not support self-electric power. Passive tags have constraints in functionality, but they have distinct advantages over active tags. They are small enough to attach to an object easily, and they do not need to consider power consumption due to its dependency on the reader's power. In this chapter, the focus of consideration is to the passive tags. An RFID reader sends out a signal supplying power for tags. The tag extracts energy from the electromagnetic field by charging its capacitor until it is able to operate. When it is charged, communication between the reader and the tag is possible, that is why passive tags can operate without their own battery.

An RFID reader communicates with tags through radio frequency, which is performed in a different manner than the bar code system in which a reader identifies a bar code through the light. Due to these characteristics, the RFID has wider range of identification of tags where tags can be identified even when line of sight (LOS) is not obtained. Both RFID system and bar code system readers are able to identify one object at a time. In a bar code system, one should secure the LOS between a bar code and its reader. It is possible to identify tags as long as they are within the reader's range, but the order of identification should be determined.

There may be multiple tags a reader should identify. All that the tags should do is to respond with the data corresponding to the signal received from a reader. The communication between tags is impossible when passive tags cannot make a decision on whether the channel is busy or not. Since a medium is shared by tags, a collision occurs at the reader's side when two or more tags get transmitted simultaneously [2]. Since collisions make collided signals be retransmitted, it increases delay for identifying tags and consumption of energy. Therefore, the arbitration mechanism is required.

We call the protocol designed for avoiding collisions between a reader and tags an anticollision protocol. The anticollision protocol should have the following characteristics:

- A reader should identify all the tags within its range.
- The anticollision algorithm should have a mechanism which is capable of verifying that all the tags are identified.
- It should minimize the time elapsed for the identification of tags. It lies on the same line as reducing collisions. As the time which is required to identify tags increases, it is more difficult to identify objects moving fast.

We introduce various anticollision protocols including adaptive query splitting (AQS) and adaptive binary splitting (ABS), which we have proposed in our earlier work. We also evaluate their performance.

This chapter is organized as follows. Background material is presented in Section 9.2. We give a detailed description on tree-based and probabilistic anticollision protocols in Sections 9.3 and 9.4, respectively. Section 9.5 presents analytic study on performance of tag anticollision protocols, and is followed by performance evaluation and analysis in Section 9.6. In Section 9.7, we give a summary and discussion. Finally, the conclusions can be found in Section 9.8.

9.2 Background Material

The tag identification process is a continuation of many reading slots, which consist of a reader's request and tags' replies. Figure 9.1 shows data transmission between an RFID reader and tags. We define some terms for detailed description.

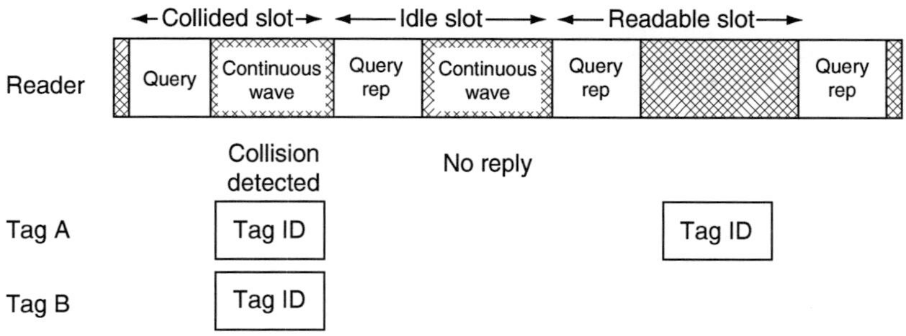

FIGURE 9.1
Data transmission between an RFID reader and tags.

- *Reading slot*: the operation cycle composed of the transmissions of reader's query and replies from the tags.
- *Readable slot*: the reading slot that only one tag replies to. The tag can be identified successfully.
- *Collided slot*: the reading slot that has two or more tags which transmit their replies. It is the case that the result of CRC is failed. The reader cannot recognize any tags.
- *Idle slot*: the reading slot that has no tag reply.

Tag anticollision protocols can be classified into two approaches in terms of the way of determining the point of transmission time: (1) tree-based approach and (2) probabilistic approach. In tree-based tag anticollision protocols, a tag determines the point of transmission on receiving a message from a reader and making a process from the message, that is, a decision of whether to respond to it. On the other hand, probabilistic tag anticollision protocols use the random number generated by a tag to determine the point of transmission. Each tag generates a random number and waits for its transmission time according to the chosen number. Figure 9.2 shows the taxonomy of tag anticollision protocols.

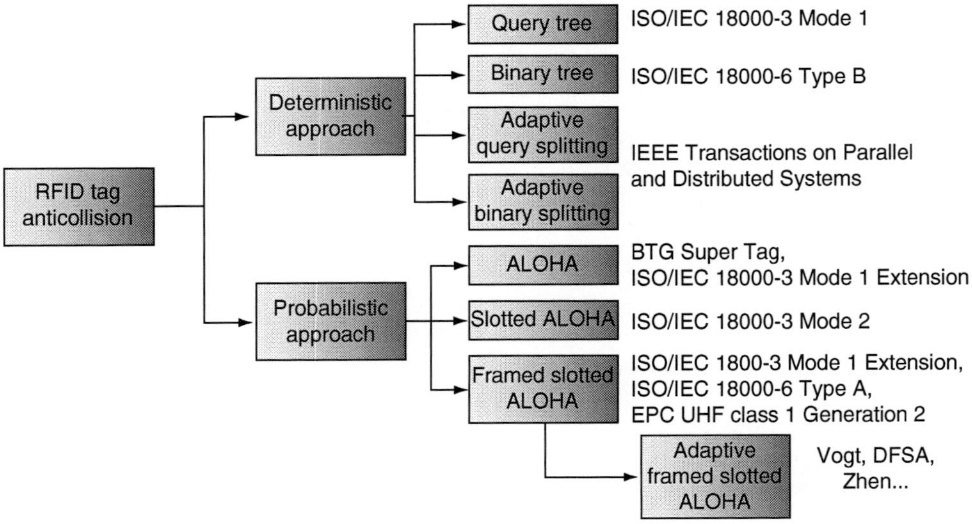

FIGURE 9.2
Taxonomy of RFID tag anticollision protocols.

9.3 Tree-Based Anticollision Protocols

In tree-based tag anticollision protocols, a reader divides tags into two groups. A reader further divides each of them into two groups again. It is required that a reader be able to distinguish each divided group. The process of dividing tags is continued until a group contains only one tag so that each tag could be successfully identified. The dividing process of a group is continued until a reader identifies all the tags.

Tree-based protocols work similar to census. The reader can recognize all tags in its identification range. An identification process of the reader can be illustrated as a process of creating and searching a tree where a node in the tree represents a reading slot. As stated earlier, the result of identification can be classified into three cases: readable, idle, and collided. After the identification cycle is completed, a tree is constructed. A leaf node in the tree corresponds to either readable slot or idle slot, and an intermediate node represents a collided slot.

Both the query tree [3,4] and binary tree [5,6] protocols represent a tree-based RFID tag collision protocol. Although each of them has different dividing algorithms and tree maintenance methods, they are very similar with respect to constructing a searching tree in an identification cycle. The main advantage of tree-based tag anticollision protocol is that all tags in identification range of the reader can be identified. An identification cycle is a process that constructs a tree from root node to leaf nodes. When the next identification cycle begins, information of tree in the previous cycle is initialized. AQS protocol [7] and ABS protocol [7] can improve the performance by using the information of tree in the previous cycle.

9.3.1 Query Tree

A reader transmits a query to tags using a query tree protocol. The query contains the prefixes of the tag identification (ID) codes. All tags within the range of a reader compare the query of the reader with their ID codes and transmit their ID codes to the reader when the result of the comparison is true. This protocol uses a query of reader and prefixes of tag ID codes to divide tags into two groups. Tags in one group transmit their ID codes to the reader while tags in the other group wait for the next query of the reader.

The content of a query is the identifier of each group. The reader repeats dividing tags into two groups until the number of tags in a group is one. When the number of tags in a group is one, the reader is successful in identifying the tag. This identification process can be considered as constructing a searching tree based on tag ID codes. The reader increases the length of the query until the identification cycle is completed.

The operation of the reader can be implemented by using a data structure (e.g., queue or stack). The queries set to 0 and 1 are stored in the data structure initially. When a collision occurs, the reader makes two queries whose lengths are 1 bit longer than the queries which cause the collision by concatenating query and the extra bit (0 and 1). Then the reader inserts itself to the stack or queue. When a readable slot or an idle slot occurs, the reader gets a next query from the data structure without any further processing. In case of using a stack for the data structure, the search is depth first search like the binary tree. When a queue is used, the search is breadth first search.

In Figure 9.3, we give an example of the identification procedure of the query tree protocol. The queue is initialized with null query. At the first reading slot, all tags reply and a collision occurs. It makes reader push query "0" and query "1" into queue. The reader extracts the query "0" from the queue and transmits it to tags. Three tags (001, 010, and 011) which start with "0" reply simultaneously and a collision occurs. The reader

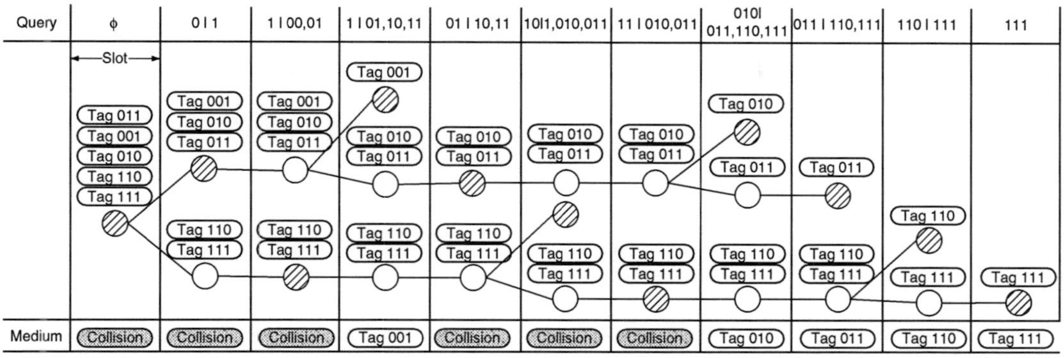

FIGURE 9.3
An example of tag identification using the query tree protocol with five tags.

pushes query "00" and queries "01" into the queue. At the next reading slot, the reader transmits query "1" to tags and two tags (110 and 111) reply. Both queries "10" and "11" are pushed into the queue. A tag with its ID "001" is identified successfully at the next reading slot. At this point, the queue has three queries (01, 10, and 11). The reader transmits query "01" in the next reading slot. Two tags (010 and 011) which start with "01" reply and a collision occurs. The reader pushes query "010" and query "011" into the queue. In this way, tag identification process is continued until the queue is empty. There are five collided slots, one idle slot, and five readable slots under the query tree protocol.

9.3.2 Binary Tree

The binary tree protocol uses the pseudo-random number generator to divide tags into two groups. The counter variable in each tag is used for identifying each group. At the beginning of identification operation, the reader sends a message which notifies the start of its cycle to tags. All tags receiving this message generate random numbers of 0 or 1. Tags set their counter values by adding the generated random number to their counter values. Tags are divided into two groups: one group has 0 in their counter values; the other group has the counter values of 1.

The group with the counter value of 0 tries to transmit and wait for the reply from the reader. If a collision occurs, tags which tried to transmit their ID codes in the previous cycle are divided into two groups by using a pseudo-random number, and tags that did not try the ID transmission increase the value of their counters by 1. If there are no collisions, all tags decrease the value of their counters by 1.

The tags identified successfully set the value of their counters to 0 and wait for the start of a frame message from the reader. Since the binary tree protocol uses the random number generator in branching out, idle slots can occur many times. However, the probability of this is very low. As the information regarding tag ID codes is not used in the identification process as in probabilistic tag anticollision protocols, the performance of the binary tree protocol is not affected by the distribution of the ID codes of tag population.

In Figure 9.4, we provide an example of the identification procedure of the binary tree protocol. Initially, all tags set the value of their counter variables to 0. The tags with the counter value of 0 try to transmit and a collision occurs. Each tag generates random number of 0 or 1. Tag A, tag B, and tag D choose 0 as their counter value and make one group. On the other hand, tag C chooses 1 as its counter value and makes another group. In the next reading slot, the former group tries to transmit and a collision occurs. The reader

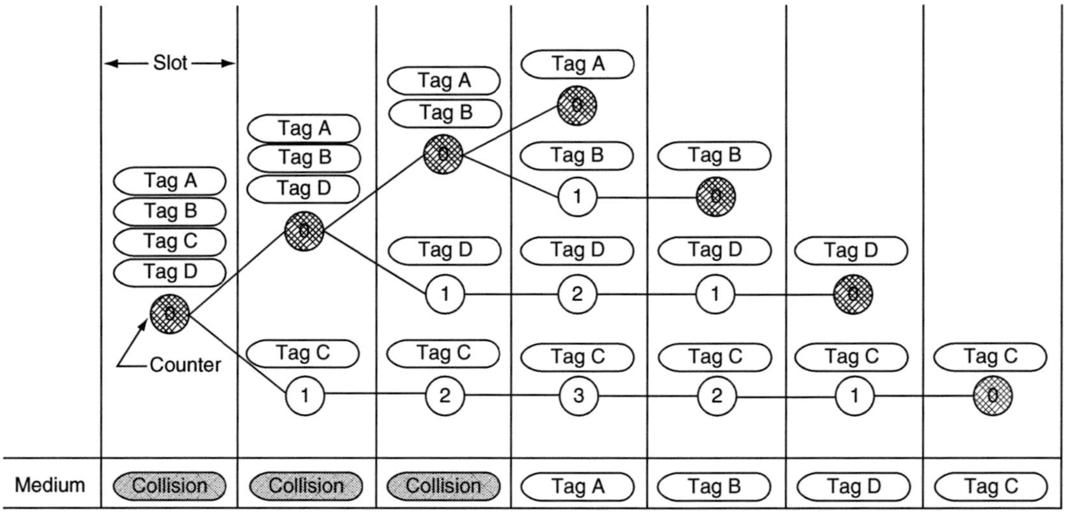

FIGURE 9.4
An example of tag identification using the binary tree protocol with four tags.

informs tags of the result of current reading slot. Tag C which has nonzero value of its counter increases its counter value by 1. Tag A, tag B, and tag D generate random number and add it to their counter value. Since tag A and tag B have the counter's value of 0, they perform transmission and a collision occurs. Tag C and tag D which have nonzero value of their counter increase their counter value by 1. Tag A and tag B generate random number and add it to their counter value. Tag A tries to transmit its data to reader because it has its counter's value of 0. In this case, tag A is successfully identified. The reader notifies the result of current reading slot. All tags decrease their counter value by 1. In this way, tag B, tag D, and tag C are identified in successive reading slots.

9.3.3 Adaptive Query Splitting

AQS protocol is the method that can reduce a search space of a current identification process by using the queries in the leaf nodes of the tree which were constructed in the previous process. Searching from the leaf nodes of the previous tree can decrease identification delay present in the existing search method of query tree protocol that starts from root node of tree while covering all possible search spaces.

The reader maintains a queue Q as well as a candidate queue CQ. There are prefixes of tag ID codes in readable slots and idle slots from the previous identification cycle in CQ. At the start of the identification process, the reader initializes Q with the CQ and empties out CQ.

Figure 9.5 shows the operation of query tree protocol and the AQS protocol. If population of tags is same as in the previous identification cycle, no collisions occur in the current cycle under AQS protocol. If there exists an incoming tag whose ID matches a prefix of readable nodes in the previous cycle, it decreases the number of collision nodes to use prefixes of readable nodes. A new tag, whose ID does not match a prefix of a readable node in previous cycle, is quickly identified with prefixes of idle nodes in previous cycle by the reader.

To prevent the growth of CQ (which carries information of leaf nodes), the query deletion process is used. The query deletion process is the process that merges queries

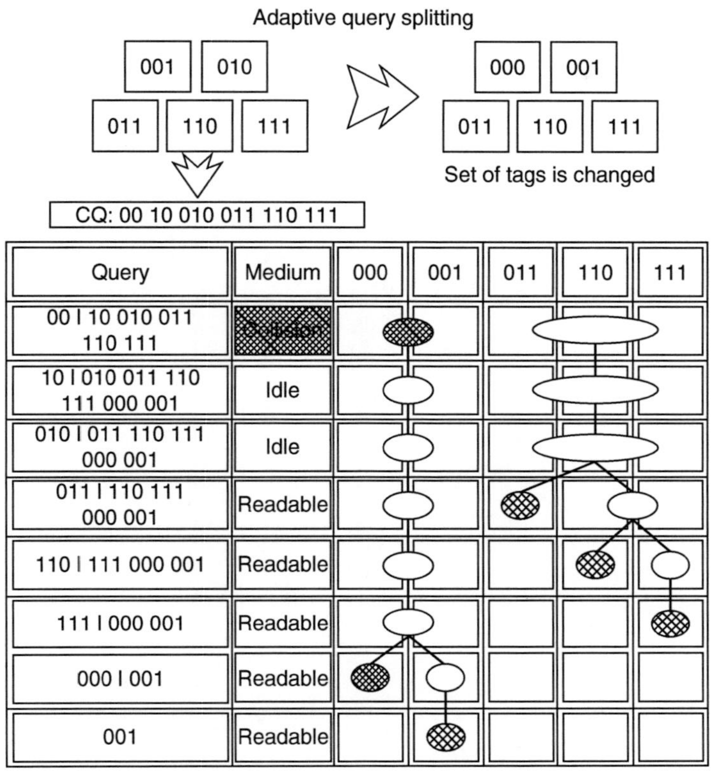

FIGURE 9.5
An example of tag identification using the adaptive query splitting protocol with five tags.

that share the same query content except the last bit. Idle nodes unnecessarily disturb the speed of identification, and CQ can eliminate prefixes of unnecessary idle nodes. For example, if the nodes of the prefix p0p1 ... pi0 and p0p1 ... pi1 contain at least one idle node and the non-idle node is a readable node, CQ stores the prefix p0p1...pi instead of the prefix p0p1 ... pi0 and p0p1 ... pi1.

9.3.4 Adaptive Binary Splitting

ABS protocol is the method by which tags remember their identification order in the previous cycle by adding one to the counters in the tags. A tag maintains a progressed-slot counter (PSC) as well as an allocated-slot counter (ASC). PSC maintains the number of timeslots passed in an identification cycle. At the start of an identification cycle, PSC is initialized to 0. In every readable slot, all tags increase their PSCs by 1. ASC determines whether a tag can transmit its data. If a tag has the same value for both ASC and PSC, the tag can transmit.

A tag can be in one of the following three states:

- *Wait state*: If the tag has ASC greater than PSC, it waits for a command from the reader.
- *Active state*: If a tag has ASC equal to PSC, it transmits its data to the reader.
- *Sleep state*: If a tag has ASC less than PSC, the tag does not transmit any data. This tag waits for the next identification cycle since it has already been identified in the current identification cycle.

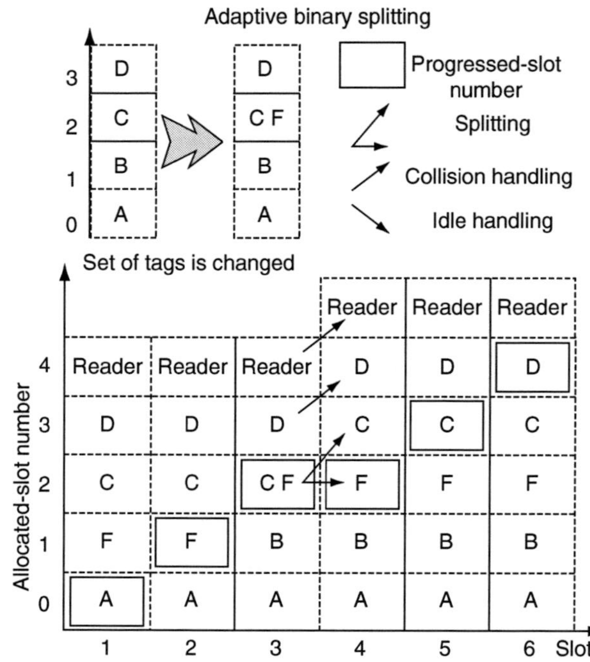

FIGURE 9.6
An example of tag identification using the adaptive binary splitting protocol with four tags.

In a collided slot, the colliding tags, the tags of the active state, add a random number (0 or 1) to ASC. The active tags with the addition of 1 in their ASCs convert their state into wait state. The tags in wait state increase ASC when collision occurs. When the reader sends tags the message which means the idle slot, the tags in the wait state decrease ASC.

Figure 9.6 shows the operations of the binary tree protocol and ABS protocol. If the tag population does not change, neither collision nor idle occurs in the current cycle under ABS protocol. The reader remembers the number of tags in the previous cycle and informs them of it at the start of an identification cycle. If there are incoming tags, they randomly set their counters into a smaller value than the number sent by the reader. Incoming tags can cause collisions with existing tags. When a collision occurs, those collided tags add a randomly selected binary number (0 or 1) to ASC. If there are leaving tags, an idle slot can occur.

9.4 Probabilistic Anticollision Protocols

Probabilistic tag anticollision protocols are based on ALOHA [8]. ALOHA is one of the basic medium access control mechanisms. In ALOHA, each tag generates a random number and waits for its transmission time according to the number chosen. If the data transmitted by a tag is not interfered by other data, the reader can identify the tag. A tag continues to do the same work after its transmission; generating a new random number and transmitting its own data after waiting for random amount of time. If during the interval two or more tags transmit, a complete or partial collision occurs. In order to solve partial collision problems, transmission time is divided into discrete time intervals in the slotted ALOHA [9]. All tags try to transmit their data after random back-off. If there are no partial collisions under the slotted ALOHA protocol, the slotted ALOHA doubles the channel utilization. A framed slotted ALOHA [10–19] groups some slots into a frame, each frame having N slots. In a frame, each tag transmits its data only once. Under the framed slotted ALOHA, collisions caused by backlogged tags can be prevented.

Under ALOHA, slotted ALOHA, and framed slotted ALOHA, the waiting time for a tag is determined by a random function. The important factor which influences performance is the relationship between the number of tags and random space and the maximum value of the back-off timer. If the random space is larger than the number of tags in the reader's range, there exist many collision slots. On the other hand, if it is smaller than the number of tags, there are many idle slots in the frame. It is important to set suitable random space based on the predicted number of tags.

Under the framed slotted ALOHA protocols, the frame size is the size of random space. It is easy to change the frame size at the start of a frame. There have been many proposed protocols which improve framed slotted ALOHA, called adaptive framed slotted ALOHA. Figure 9.7 shows examples of the operation of slotted ALOHA, framed slotted ALOHA, and adaptive framed slotted ALOHA. An optimal frame size happens when the number of tags is the same as the frame size. Many protocols have been proposed that estimate the number of tags, and use the number of readable slots, collided slots, and idle slots. We now introduce three of them. The symbol notions and their descriptions used throughout this chapter are summarized in Table 9.1.

9.4.1 Dynamic Slot Allocation

A readable slot which occurred in the previous frame is the one that contains a single tag. A collided slot has at least two tags. By using these two facts, the lower bound of the number of tags can be estimated. The lower bound of the number of tags can be estimated as

$$\text{Vogt1: } N_{\text{tag}} = S + 2C. \tag{9.1}$$

In Ref. [17], the author proposed a different way to estimate the number of tags. If we know the frame size and the number of tags, we can calculate the expected value of readable slots, idle slots, and collided slots. This expectation is a function of frame size and the number of tags. We can create a vector using the three values earlier. We could also create another vector that consists of the actual value of readable slots, idle slots, and collided slots in the previous frame. By using Chevyshevs inequality, we can make tag estimation function as

$$\text{Vogt2: Estimation function} = \min_{N_{\text{tag}}} \left| \begin{pmatrix} S_{\text{EXP}}(F, N_{\text{tag}}) \\ C_{\text{EXP}}(F, N_{\text{tag}}) \\ I_{\text{EXP}}(F, N_{\text{tag}}) \end{pmatrix} - \begin{pmatrix} S \\ C \\ I \end{pmatrix} \right|. \tag{9.2}$$

We refer to these methods using the author name, Vogt1 and Vogt2, respectively. Vogt1 can easily estimate the number of tags. However, as the number of tags increases, the number of errors increases accordingly. On the contrary, when the number of tags is small, Vogt2 has a large error rate. Vogt2 can estimate the number of tags precisely, but it may have larger computational complexity.

9.4.2 Dynamic Framed Slotted ALOHA

The probability mass function of the number of reacting tags in a slot can be obtained by using a binomial distribution. Given the number of tags and the frame size, we can compute the probabilities of the occurrences of readable, idle, and collision slots. By using these probabilities, we could calculate the collision ratio [18], which is a fraction of the number of collision slots to the frame size. The collision ratio can be obtained by

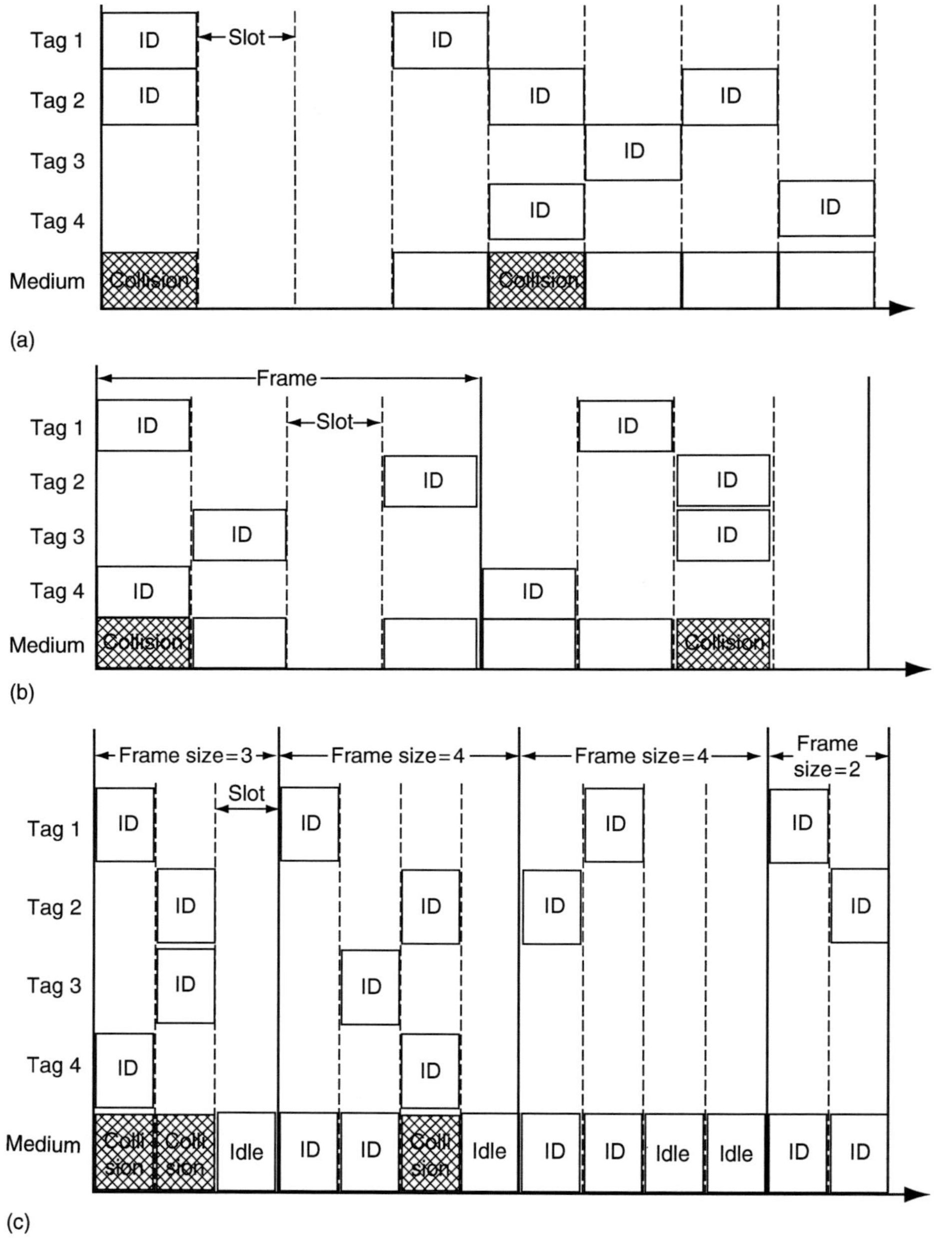

FIGURE 9.7
Examples of the operation of slotted ALOHA, framed slotted ALOHA, and adaptive framed slotted ALOHA. (a) Slotted ALOHA. (b) Framed slotted ALOHA. (c) Adaptive framed slotted ALOHA.

subtracting the summation of the probability of the appearance of readable slots and the probability of the appearance of idle slots from one. We already know the previous frame size and the number of collision slots so that we can estimate the number of tags using the collision ratio. The number of tags can be obtained by

$$\text{DFSA:}\, C_{\text{ratio}} = 1 - \left(1 - \frac{1}{F}\right)^{N_{\text{tag}}}\left(1 + \frac{N_{\text{tag}}}{F-1}\right). \tag{9.3}$$

TABLE 9.1

Notations

Symbol	Description
F	Frame size
N_{tag}	Estimated number of tags
S	Number of readable slots
C	Number of collided slots
I	Number of idle slots
$S_{EXP}(F, N_{tag})$	Expected value of readable slot given frame size and the number of tags
$C_{EXP}(F, N_{tag})$	Expected value of collided slot given frame size and the number of tags
$I_{EXP}(F, N_{tag})$	Expected value of idle slot given frame size and the number of tags

We refer to this protocol as DFSA in this chapter. DFSA must maintain a table to return the number of tags provided that the collision ratio and the frame size are given. It may be burdensome to maintain a table, but DFSA makes possible to estimate the number of tags precisely.

9.4.3 Dynamic Frame Length ALOHA

In Vogtl protocol, we use the number of readable slots and the number of collision slots to estimate the number of tags. The number of tags in a collision slot is at least two. If the number of tags approaches to infinity, we can obtain the expected number of ~2.39 tags in a collision slot [13,19]. We can estimate the number of tags by using the following equation:

$$\text{Zhen: } N_{tag} = S + 2.39C \tag{9.4}$$

We, in this chapter, refer to this estimation method Zhen for convenience's sake. This method can easily estimate the number of tags. This is a less error-prone method that requires only simple calculation.

9.5 Analytic Study

In this section, we analyze the performance of tag anticollision protocols. All four tree-based protocols (binary tree, query tree, ABS, and AQS) use binary search tree as their searching method to identify tags. Therefore, we choose binary tree protocol as the representative of tree-based protocols. On the other hand, adaptive framed slotted ALOHA is mainly researched among the probabilistic tag anticollision protocols. Hence, we compare the performance of adaptive framed slotted ALOHA with the binary tree protocol.

9.5.1 Average Slot Delay Analysis

In general, the performance of anticollision protocols is represented by *average slot delay*. Average slot delay can be defined as the expected number of slots consumed for identifying m tags. In most existing protocols to date which are related with RFID tag, anticollision aimed to minimize average slot delay or maximize average slot throughput. In similar vein, we will present the performance of tag anticollision protocols in slot delay in analytic manner, dividing them into binary tree protocol and adaptive framed slotted ALOHA.

Theorem 9.1
The average slot delay of binary tree protocol approximates to $2.885 \times m$.

Proof
The proof of this theorem follows an analysis presented in Ref. [20], which is based on a research that investigated on tree-based multiple access channel [5]. The average slot delay $\bar{T}(m)$, which is the expected number of consumed slots until identifying m tags, in binary tree protocol is as follows:

$$\bar{T}(m) = \bar{C}(m) + \bar{I}(m) + m, \tag{9.5}$$

where $\bar{C}(m)$ and $\bar{I}(m)$ denote the average number of collided slots and idle slots, respectively. The m in the equation accounts for the m slots corresponding to readable slots. Since the random number generator in a tag for binary tree protocol follows uniform distribution, the probability that k out of m tags try to reply at level L in the tree is given by the binomial distribution as following:

$$P(X = k|m,L) = \binom{m}{k} p^k (1-p)^{m-k}, \tag{9.6}$$

where $p = \frac{1}{2}^L$. Using this, we get the probabilities that a slot at level L of the tree is readable, idle, or collision:

$$P_{\text{read}}(X = 1|m, L) = (1-p)^m, \tag{9.7}$$

$$P_{\text{idle}}(X = 0|m, L) = mp(1-p)^{m-1}, \tag{9.8}$$

$$\begin{aligned} P_{\text{coll}}(X \geq 2|m, L) &= 1 - P_{\text{read}}(X = 1|m, L) - P_{\text{idle}}(X = 0|m, L) \\ &= 1 - (1-p)^m - mp(1-p)^{m-1}. \end{aligned} \tag{9.9}$$

For slots are visited only when their parent experiences collisions, we can write the average slot delay by the summation of the expected number of nodes whose parents are collided slots in all levels:

$$\begin{aligned} \bar{t}_{\text{TS}}(m) &= \sum_{L=0}^{\infty} \sum_{i=0}^{2^L-1} P_{\text{coll}}(X \geq 2|m, L-1) \\ &= 1 + 2\sum_{L=0}^{\infty} 2^L P_{\text{coll}}(X \geq 2|m, L). \end{aligned} \tag{9.10}$$

Substituting from Equation 9.9 gives

$$\bar{t}_{\text{TS}}(m) = 1 + 2\sum_{L=0}^{\infty} 2^L \left[1 - (1-p)^m - mp(1-p)^{m-1}\right] \approx 2.885 \times m. \tag{9.11}$$

In case of framed slotted ALOHA, not like tree-based protocols, there can be diverse variant protocols according to which frame adaptation algorithm is used. However, since the optimal condition of slotted ALOHA channel with given number of nodes has been revealed in the other work [9], we will assume an optimal framed slotted ALOHA protocol.

Theorem 9.2
The average slot delay of optimal framed slotted ALOHA protocol approximates to $e \times m$.

Proof
When the probability that m tags transmit to a slot is p, the successful transmission probability of a tag, S, is given by,

$$S = mp(1-p)^{m-1}. \tag{9.12}$$

Due to the concavity of the equation, we can find the optimal condition through the first derivative with respect to p as follows:

$$\frac{\mathrm{ds}}{\mathrm{d}p} = m(1-p)^{m-1} - m(m-1)p(1-p)^{m-2} = 0. \tag{9.13}$$

Using this, the optimal condition is given by $p = \frac{1}{m}$. When frame size is L, p is represented as $p = \frac{1}{L}$. Therefore, the relationship of L and m when it is in the optimal condition is $L = m$. Under this condition, the $(n+1)$th frame size L_{n+1} in adaptive framed slotted ALOHA is denoted as the following relationship:

$$L_{n+1} = m - \sum_{i=0}^{n} L_i \times S^*, \tag{9.14}$$

where S^* is the optimal utilization of a frame as follows:

$$S^* = \lim_{m \to \infty} P(X = 1|L = m) = \lim_{m \to \infty} \left(1 - \frac{1}{L}\right)^{n-1} = \frac{1}{e}. \tag{9.15}$$

To know the asymptotic property of frame size, n is taken to infinity, then:

$$\lim_{n \to \infty} L_{n+1} = \lim_{n \to \infty} \left(m - \sum_{i=0}^{n} L_i \times S^*\right). \tag{9.16}$$

Intuitively, as time goes by, the number of unidentified tags will decrease. By the optimal condition, frame size will decrease as well. Hence, $\lim_{n \to \infty} L_{n+1}$ converges to zero, and finally we get the following relationship:

$$\sum_{i=0}^{\infty} L_i = e \times m. \tag{9.17}$$

Two theorems in this section allow us to measure the exact performance of tag anticollision protocols. As a result, it seems that adaptive framed slotted ALOHA protocol is little better than binary tree protocol. However, as this result gives us only asymptotic property, it is needed to confirm that in reality. We conducted a Monte Carlo simulation to make sure what we have analyzed. Figure 9.8 depicts the measurement of the average slot delay of the two protocols. The adaptive framed slotted ALOHA protocol with DFSA is assumed as the optimal one in our simulation. The difference between two protocols is not obvious when the number of tags is low, but it clearly increases as the number of tags increases. The reason for this observation is that they converge to the asymptotic average delay values as

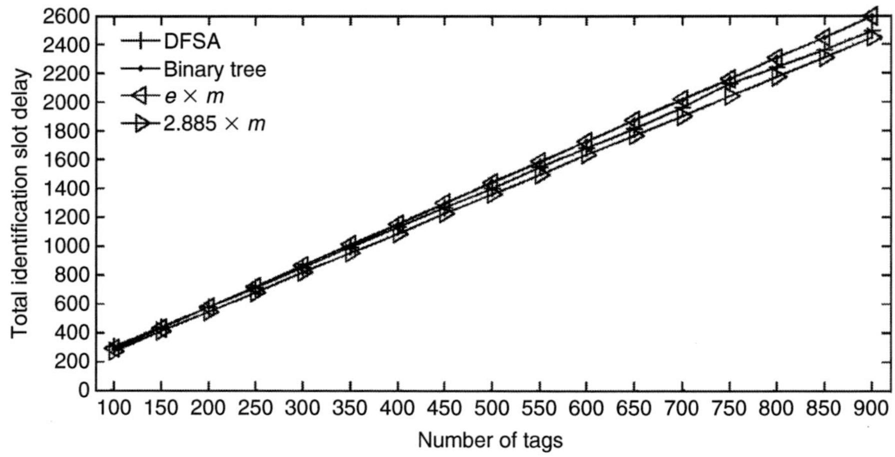

FIGURE 9.8
The average slot delay of DFSA and binary tree protocol.

the number of tags is getting bigger. Nevertheless, we cannot determine which protocol is the best in reality because, in the RFID standards, the frame sizes of probabilistic protocols are formatted to the powers of two, not integer values. Of course, the protocols with the powers of two frames will show the same average slot delay asymptotically, but its converging speed will be fairly degraded.

9.6 Performance Evaluation and Analysis

In this section, we evaluate the performance of tag anticollision protocols examined so far. We make following assumptions for reflecting only the effect on the operational principle of tag anticollision protocols. Reliability in transmission between a reader and a tag is perfectly guaranteed. All the protocols considered make use of the same type of physical functionalities. The transmission rate depends on modulation scheme of physical layer. All protocols have same time for transmitting their ID codes. The reader just targets tag identification and does not perform any additional operations.

The transmitted message format is set based on ISO 18000-6 specification. Table 9.2 shows message formats of tag and reader. Protocols which have fixed-length reader message are frame slotted ALOHA-based protocols, binary tree protocol, and ABS protocol. Query tree protocol and ABS protocol have variable length reader message. Figure 9.9 describes data transmission between a reader and a tag. As we mentioned earlier, the identification operation begins with reader's starting message under frame slotted ALOHA protocols. Since we would like to evaluate the performance of protocols based

TABLE 9.2
Message Format

Message Type	Preamble Detect	Header	Data	CRC
Reader message	400 μs	7 bits	11 bits/variable length	5 bits
Tag response	300 μs	—	96 bits	16 bits

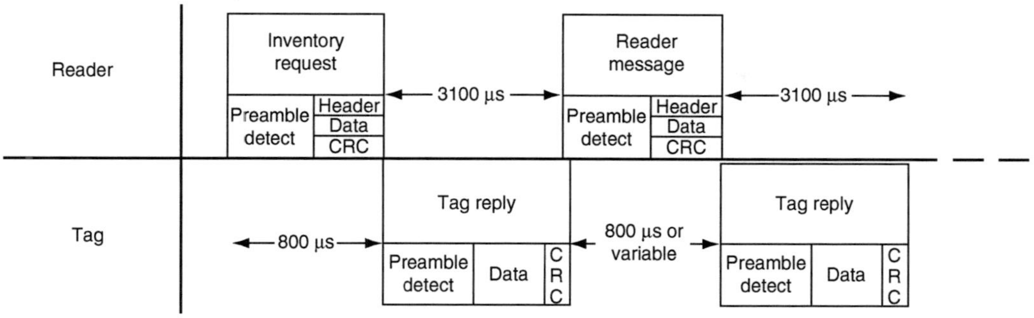

FIGURE 9.9
Data transmission between a reader and a tag.

on transmission slots, we do not consider the computation delay between end point of data receiving and start point of data sending. The reader which receives a data from a tag sends a synchronizing message. Deterministic protocols have also similar data transmission structure. After reader's transmission, tags try to transmit their data to a reader.

9.6.1 Identification of Motionless Tags

Mobility of tags varies accordingly as the kinds of RFID applications are employed. For some applications, one can locate objects in front of a reader, which then are identified. Such a scenario happens when a reader is deployed at an entrance or an exit of a store where products are disposed of. In this case, a user should determine the starting point and wait until all tags are identified. We perform a simulation study to investigate how long a user should wait for in each protocol. As we assume that mobility of tags is not considered, a reader performs identification process without changes on tags' sample during its identification.

Figure 9.10a shows the relationship between the number of tags and the time taken to identify all tags by each protocol. Probabilistic tag anticollision protocols are independent of each round, and they keep three slots: the readable, the collided, and the idle slots, all of which follow the identical probabilistic distribution.

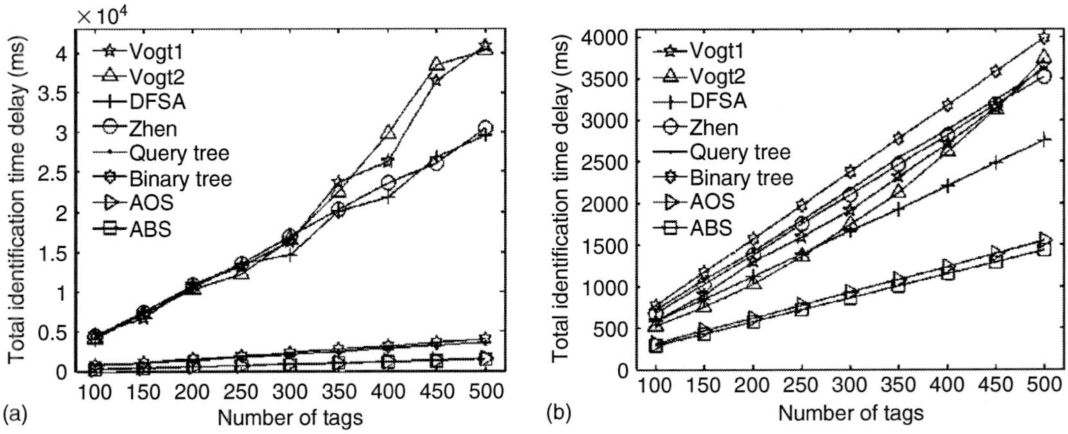

FIGURE 9.10
Total identification delay of motionless tags. (a) Total identification delay (no quiet state). (b) Total identification delay (quiet state).

Query tree and binary tree, which belong to the tree-based protocol, show better performance than the probabilistic protocols. They are, however, somewhat different in total identification delay. There is not much of a difference in the performance of different probabilistic protocols until the sample size becomes >250. The reason why Vogt's methods have larger delays when sample size is over 300 is that they assume that the maximum frame size is 256. In Zhen and DFSA, the predicted number of tags is employed as the frame size, so they outperform Vogt's methods when the number of tags increases. It is expected that both protocols show the similar results provided that Vogt's use the frame size that is the same as the predicted number of tags.

ABS and AQS experience the least amount of delay among the protocols. This is because they already have knowledge of the population of tags since the simulation on them is performed after the identification is done once. Due to the inherent characteristics of ABS and AQS, such as the similar condition with binary tree and query tree, we obtain the same results shown. ABS has the advantage of all slots being configured as readable slots provided that the tag population of previous round is identical to that of the current one. In the same situation, AQS is able to terminate the identification process with only idle and readable slots occurrence in the previous round. Only difference between ABS and AQS is due to the essential fact that ABS does not need to keep additional idle slots.

For applications where there is no mobility of tags and the start point of the identification process is distinct from the end, it is efficient to make use of quiet state of a tag. The tag, which has already been identified successfully, gets into the quiet state, and it does not take part in the process until it gets the next wake-up message from a reader. Figure 9.10b shows the effect on quiet state on the probabilistic protocols. Note that probabilistic protocols have improved performance. Among probabilistic protocols, the performance of DFSA is the best where it can estimate the number of tags more precisely. We can also find that there exists a linear relationship between the number of tags and the identification.

9.6.2 Identification of Moving Tags

This section presents performance of tag anticollision protocols in the applications with high mobility of tags. We consider a situation where objects with an RFID tag move toward a reader through a conveyer belt. There can be >200 tags within its reader's range. We measured performance by changing the velocity of a conveyer belt. Figure 9.11 shows the identified number of objects for each anticollision protocol when 5000 objects get moved. The main purpose of this application is to identify tags passing by a reader.

In such an application, if tags are woken up in advance before an identification process by a reader, we can have the advantage of adapting quiet state of tags because the tag that has been already identified does not need to perform that process again. We show evaluation of both cases in Zhen and DFSA in Figure 9.11a.

As the speed of conveyer belt increases, the interval entering the reader's range is reduced. In other words, the larger the interval becomes, the slower the conveyer belt moves. Among probabilistic protocols, only Zhen and DFSA using quiet state succeed in identifying all tags of 5000 even with the fastest condition. ABS is the second best in adaptability of the speed of a conveyer belt.

The probabilistic tag anticollision protocols adapting quiet state of tags outperform since the number of tags is small in the initial part. The reader performs the identification operation with a small number of tags and makes them sleep. In this case, the size of population is maintained at small size. On the other hand, due to the fact that the tags which were already identified in the pervious identification process rejoin the identification

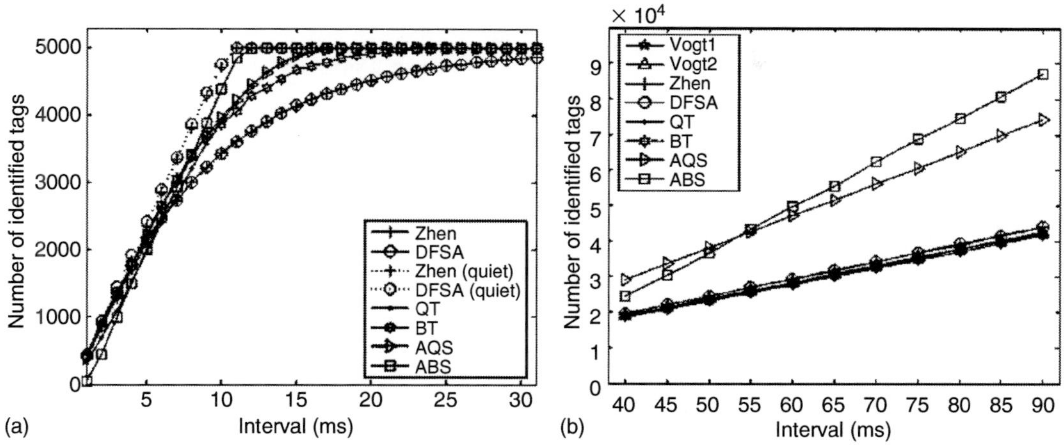

FIGURE 9.11
The number of identified tags and readable slots. (a) The number of identified tags. (b) The number of readable slots.

operation in the current process in deterministic tag anticollision protocols, every process has more tags than the probabilistic tag anticollision protocols adapting quiet state. The average number of tags is ~200.

The probabilistic tag anticollision protocols without using the quiet state of the tag cannot identify 5000 tags with 30 ms interval due to the rejoining of identified tags. AQS and query tree have similar performance, and the performance of the binary tree is not good. However, the binary tree can identify all tags under the interval of 0.026 per incoming tag. It shows that the binary tree can identify all tags when the population is ~200 and 37 incoming tags per second to the reader's range.

The identification delay, the time to identify all tags within a reader's range, is a critical metric. In this chapter, we have defined a new metric as the time taken to identify if a tag left a reader's range after the reader had recognized a tag. This metric can be described as how fast a reader can identify whether a certain tag exists within its reader's range.

In order for an application to make this metric meaningful, we consider an application where the state of the tags are to be checked out in real time. It is meaningless to use the quiet state in such applications, and a method that wakes up tags at appropriate time must be considered. There has been little investigation on such a method, and we compare the probabilistic tag anticollision protocols without quiet state with the deterministic ones.

Accordingly, we can apply the conveyer belt to this. In a conveyer belt, the interval and the moving speed of the tag can be regarded as the speed of the belt and the rate of change of samples, respectively. Figure 9.11b shows the changing rate of samples and the number of readable slots per unit time. The number of readable slots per unit time is the number of readable slots when a reader identifies all tags within its range.

Since even probabilistic protocols without quiet state achieve high identification ratio over 98%, we present the results from 40 ms in interval. ABS and AQS show desirable results while the other protocols are comparable. The reason why the readable slots increase as the speed decreases is that the time taken for tags to pass by increases. Therefore, the identification process is performed during a number of rounds. ABS and AQS achieve good performance compared with other protocols because they can take advantage of the identification information of the previous stage.

TABLE 9.3

Summary of Tag Anticollision Schemes

	Probabilistic Protocols				Deterministic Protocols			
Characteristics	**Query Tree**	**Binary Tree**	**AQS**	**ABS**	**Vogt1**	**Vogt2**	**Zhen**	**DFSA**
The number of variables in tag memory for identification process	0	1	0	2	1	1	1	1
The existence of random number generator in the tag	N	Y	N	Y	Y	Y	Y	Y
Existence of upper bound of the number of tags that can be identified	N	N	N	N	Y	Y	Y	Y
The variety of the reader message length	Y	N	Y	N	N	N	N	N

Note: Y, yes (exist); N, no.

9.7 Discussions and Summary

RFID applications have various requirements. Mobility of tags is a key factor for branching properties of RFID applications. Mobility of tags varies accordingly as what kinds of RFID applications are employed. We divide RFID applications into two cases. The first one is the case that has little mobility of tags. In these kinds of applications, the time for identifying all tags in reader's range is a critical factor for evaluating performance of RFID tag anticollision protocols. We find out that probabilistic and tree-based tag anticollision protocols show similar ability in identifying tags. The second one is the case in which we have to consider mobility of tags for performing identification operation. In case an RFID reader has to recognize tags in reader's range persistently, a reader continues to perform identification process. ABS and AQS achieve good performance than other protocols. They can recognize fast whether the tag is in reader's range or not.

Table 9.3 shows characteristics of tag anticollision protocols. A variable for maintaining a slot number is required in probabilistic tag anticollision protocols. The binary tree protocol should have such a variable for a counter. ABS protocol needs two variables for ASC and PSC. The query tree protocol and AQS protocol require neither counters nor a random number generator. As the length of query varies, the length of reader's message also varies. AQS and ABS are appropriate for satisfying this evaluating factor.

9.8 Conclusions

In this chapter, we have introduced tag anticollision protocols and have presented the performance evaluation results of them. According to the types of RFID applications, we have considered two cases: motionless tags and moving tags. For applications employing motionless tags, a user can determine the start as well as the end of the identification process. When such applications are considered, probabilistic and deterministic tag anti-collision protocols show similar ability in identifying tags. For applications requiring persistent observations on tags, AQS and ABS outperform any other protocols, especially under the situation that tag population varies at low speed. One of the important factors we should consider in evaluating performance of persistent observation is to identify whether a tag exists within its reader's range as quickly as possible. We also have shown that AQS and ABS are appropriate for satisfying this evaluating factor.

Acknowledgment

This work was jointly supported by SK Telecom, the Korea Science and Engineering Foundation (KOSEF) grant funded by the Korea government (MOST) (No. R01-2007-000-11203-0) and the Ministry of Information and Communication (MIC), Korea under ITRC IITA-2007-(C1090-0701-0046).

References

1. K. Finkenzeller, *RFID Handbook: Fundamentals and Applications in Contactless Smart Cards and Identification*, John Wiley & Sons, New York, 2003.
2. C. Floerkemeier and M. Lampe, Issues with RFID usage in ubiquitous computing applications, in *Proceedings of the 2nd International Conference of Pervasive Computing*, LNCS 3001, pp. 188–193, 2004.
3. C. Law, K. Lee, and K.-Y. Siu, Efficient memoryless protocol for tag identification, in *Proceedings of the 4th International Workshop on Discrete Algorithms and Methods for Mobile Computing and Communications*, pp. 75–84, Boston, MA, USA, August 2000.
4. F. Zhou, C. Chen, D. Jin, C. Huang, and H. Min, Evaluating and optimizing power consumption of anti-collision protocols for applications in RFID systems, in *Proceedings of the International Symposium on Low Power Electronics and Design*, Newport Beach, CA, USA, August 2004.
5. D.R. Hush and C. Wood, Analysis of tree algorithms for RFID arbitration, in *Proceedings of IEEE International Symposium on Information Theory*, p. 107, 1998.
6. S.A. Weis, S.E. Sarma, R.L. Rivest, and D.W. Engels, Security and privacy aspects of low-cost radio frequency identification systems, in *Proceedings of the 1st Annual Conference on Security in Pervasive Computing*, LNCS 2802, pp. 201–212, March 2003.
7. J. Myung, W. Lee, J. Srivastava, and T. Shih, Tag-splitting: adaptive collision arbitration protocols for RFID tag identification, *IEEE Transactions on Parallel Distributed Systems*, 18(6), 763–775, 2007.
8. N. Abramson, The aloha system—another alternative for computer communications, in *Proceedings of Fall Joint Computer Conference*, AFIPS Conference, vol. 37, pp. 281–285, 1970.
9. L.G. Roberts, Extensions of packet communication technology to a hand held personal terminal, in *Proceedings of Spring Joint Computer Conference*, AFIPS Conference, vol. 40, pp. 295–298, 1972.
10. EPC radio-frequency identification protocols class-1 generation-2 UHF RFID protocol for communications at 860 MHz–960 MHz version 1.0.8, EPCglobal, December 2004.
11. Information technology automatic identification and data capture techniques—radio frequency identification for item management air interface—part 6: parameters for air interface communications at 860–960 MHz, Final Draft International Standard ISO 18000-6.
12. F.C. Schoute, Control of ALOHA signaling in a mobile radio trunking system, in *Proceedings of the IEE International Conference on Radio Spectrum Conservation Techniques*, pp. 38–42, 1980.
13. F.C. Schoute, Dynamic frame length ALOHA, *IEEE Transactions on Communications*, COM-31(4), 565–568, 1983.
14. J.E. Wieselthier, A. Ephremides, and L.A. Michaels, An exact analysis and performance evaluation of framed ALOHA with capture, *IEEE Transactions on Communications*, COM-38(2), 125–137, 1989.
15. UCODE, Philips Semiconductors, http://www.semiconductors.philips.com, 2005.
16. J. Zhai and G. Wang, An anti-collision algorithm using two-functioned estimation for RFID tags, in *Proceedings of International Conference on Computational Science and its Applications*, LNCS 3483, pp. 702–711, 2005.

17. H. Vogt, Efficient object identification with passive RFID tags, in *Proceedings of the International Conference on Pervasive Computing*, LNCS 2414, pp. 98–113, April 2002.
18. J. Cha and J. Kim, Dynamic framed slotted ALOHA algorithm using fast tag estimation method for RFID system, in *Proceedings of the IEEE Consumer Communications and Networking Conference* (CCNC'06), Las Vegas, USA, January 2006.
19. B. Zhen, M. Kobayashi, and M. Shimizu, Framed ALOHA for multiple RFID objects identification, *IEICE Transactions on Communications*, E88-B(3), 991–999, 2005.
20. M.A. Kaplan and E. Gulko, Analytic properties of multiple-access trees, *IEEE Transactions on Information Theory*, 31(2), 255–263, 1985.

10

Maximizing Read Accuracy by Optimally Locating RFID Interrogators

Lin Wang, Bryan A. Norman, and Jayant Rajgopal

CONTENTS

10.1 Portal Design and Antenna Placement

In recent years there has been a rapid growth in interest in the implementation of RFID technologies to better manage supply chain operations by tracking the movement of products or assets through a system. RFID implementations have also been motivated by mandates from large organizations like the U.S. Department of Defense and Wal-Mart. In typical logistics or supply chain applications, portals have been commonly chosen to streamline the automatic RFID scanning processes so that tags which move through portals are automatically tracked, read, and recorded into a suitable information system. The primary motivation of course, is labor savings since RFID technology can be used to simultaneously read multiple tagged items (typically, pallets or cases today) going through the portal, as opposed to having a human being manually read and record bar-coded information or process RFID tags manually. Portals are currently used by companies at several points within the overall supply chain. These could include shipping (e.g., at a plant or warehouse loading docks), receiving (e.g., at warehouse receiving docks), floor replenishment (e.g., between the backroom/storage area and the retail floor), sales (e.g., at check-out lanes), or packaging materials disposition (e.g., at a box crusher area). In its simplest form a portal is merely a wide doorway with one or more RFID reader antennas mounted at locations along the portal perimeter. In supply chain applications, such portals are designed primarily to read passive tags that receive their power from the reader.

Although their sizes and specific configurations might vary depending on individual applications, RFID-equipped portals are all designed with the objective of maximizing read accuracy as tagged objects move through the portal. Read accuracy is of great importance to the success of RFID applications. A missed read at the pallet or case level gives rise to discrepancies between recorded inventory and physical inventory, which in turn leads to various inefficiencies within the supply chain. For example, suppose 10 tagged cases are received, but only 9 are read and recorded in the system by the receiving dock portal reader. This could lead to nonvalue added activities such as manually reconciling stock for payment purposes or potentially costly outcomes such as inaccurate reorder points within the inventory control system. Missed reads at the item level can be even more worrisome as these could lead to lost sales from stock sitting in a backroom but not being available on the shelf; this directly affects revenue. It is fair to say that despite steadily decreasing costs of tags and readers, a major factor inhibiting the increased adoption of RFID technology in supply chain and logistics applications is the less-than-perfect reliability of the tag-reading process that is prevalent today. Even if the cost of RFID tags falls to the level of barcode labels, their use will not become ubiquitous until read accuracy levels are close to 100%. It is worth noting that although using barcodes for tracking is a slow and manually intensive process, it is rare to miss a read other than by human error. With RFID tags, this human element can be eliminated but unfortunately, the technology usually cannot guarantee 100% accuracy in real-world applications.

Read accuracy at a portal obviously depends on factors such as the quality, reliability, and capabilities of the tags and readers used, as well as the specific application domain. However, when considering a fixed technology (tag/reader) and application, the design of the RFID-equipped portal becomes critical. Tag reads can be missed because of many factors including limitations in the read range, tag orientation, or interference (from water, metal, or other tags). Another complicating factor is that in an automated RFID-scanning process the locations of tags are often not fixed because items are of different sizes and might be moving on a truck, or material-handling equipment such as a pallet, forklift, or conveyor belt. In fact, not only the location but also the orientation of each tag may differ when the interrogation process starts within a portal—this is especially true with item-level applications.

To compensate for the problem of imperfect read-rates, multiple reader antennas are commonly used in portal design, and a major determinant of read accuracy is the correct placement of reader antennas at the portal. Given that tag locations cannot be isolated and fixed, it is important to optimize and fix the locations of multiple RFID reader antennas so that the probability of a tag being read is maximized. The readability of a passive tag is directly related to the amount of power it receives, which in turn, is a function of the distance between the tag and reader antennas as well as their orientations relative to each other. Thus the objective of optimal reader antenna placement is equivalent to maximizing the powering region, that is, the area within the portal where tags can receive the minimal power required for them to be read at almost all likely orientations of the tag.

10.2 Powering Region with a Single Reader Antenna

Historically, read range has been the primary measure used to measure performance in RFID applications. In this section, we begin by showing that performance is not only determined by the relative distance of an RFID tag from a reader, but also by the relative orientation of the tag's antenna with respect to that of the reader. The power received by a passive tag is determined by Friis' equation and includes the effects of orientations and polarizations. Based on Friis' equation, an efficient scheme for calculating the powering region with a single reader antenna is introduced.

10.2.1 Powering Region versus Read Range

Regardless of whether RFID deployment is at the pallet level or at the item level, RF has limited radio range. This is especially true for passive RFID tags, where such distance may range from a few centimeters for near-field inductive coupling to several meters for backscatter. It is also worth noting that the desired read range is case dependent and it is not necessarily always desirable for it to be as large as possible because in some cases, longer read range can be detrimental due to detecting unintended or irrelevant tags.

It is important to realize that read ranges are based on a path loss model that considers that the power received by an RFID tag is only related to the path that the signal traverses. Equation 10.1 shows the free space Friis' transmission equation used in such applications:

$$P_R = P_T \frac{G_T G_R \lambda^2}{(4\pi r)^2} \tag{10.1}$$

In Equation 10.1, P_T is the power from the transmitting (reader) antenna while P_R is the power received by the receiving (tag) antenna. G_T and G_R represent, respectively, the antenna gains of the transmitting and receiving antennas, r is the distance separating the two antennas, and λ is the wavelength (in the same units as r). To use the above formula, both antennas have to be perfectly aligned, that is, at the most favorable orientation. The read range obtainable at some other relative antenna orientation is dependent on the radiation pattern, which might differ based on the specific design used for the antenna. In particular, an antenna that is not omnidirectional is orientation sensitive so that whether a tag with such an antenna can be activated depends not only on its relative distance to the reader but also the relative orientations between the tag and the reader. For example, an RFID tag that claims to have a read range of more than 20 ft could fail to be read at a much shorter distance if the relative orientation between the two antennas is unfavorable. In many RFID applications such as mixed totes and item-level tracking, users might not have full control over the tag orientations, and even in case- or pallet-level applications it is very hard to ensure that the operator will always place the objects in such a way that the tags are oriented in a specified fashion. Therefore, there is a need for a more comprehensive version of Friis' equation in which orientations are included in the coverage calculation. To do this we start with the following formal definitions of read accuracy and powering region.

Definition 1: Given the location of a tag and the locations of a set of reader antennas, *read accuracy* is defined as the percentage of all possible orientations of the tag for which it can be adequately powered by one or more of the reader antennas.

From a probabilistic perspective, read accuracy can be interpreted as the probability that a tag with some random orientation can be read, given its location and the locations and orientations of the reader antennas.

Definition 2: Given some suitably defined fraction α, the *100α% read accuracy powering region* is defined as the space within which a tag can be read with at least 100α% read accuracy.

10.2.2 Friis' Equation and Antenna Gains

It can be shown [1] that the power received by an RFID tag is determined by the version of Friis' equation that is listed below that takes into account the relative positions of the tag and the reader's antennas:

$$P_R = P_T \frac{G_T(\theta_T, \phi_T) G_R(\theta_R, \phi_R) \lambda^2}{(4\pi r)^2} (1 - |\Gamma_T|^2)(1 - |\Gamma_R|^2) |\hat{\mathbf{p}}_T \cdot \hat{\mathbf{p}}_R|^2, \tag{10.2}$$

where

(θ_T, ϕ_T) are the spherical coordinates to define transmitter antenna orientation
(θ_R, ϕ_R) are the spherical coordinates to define receiver antenna orientation
P_R is the received power
P_T is the transmitted power
$G_R(\theta_R, \phi_R)$ is the receiver (tag) gain
$G_T(\theta_T, \phi_T)$ is the transmitter gain
Γ_R is the receiver reflection coefficient
Γ_T is the transmitter reflection coefficient
$\hat{p}_R$ is the receiver polarization vector
$\hat{p}_T$ is the transmitter polarization vector
r is the distance between the transmitter and the receiver
λ is the wavelength

The reflection coefficients Γ_R and Γ_T account for the impedance mismatch between the antenna and circuitry [2] that are introduced in the simple modulation of the backscatter. In an ideal situation, its value is 0, which means no power will be reflected back because of the mismatch. In reality, its magnitude is between 0 and 1 depending on the circuit design. The squared dot product of the polarization vectors $|\hat{p}_T \cdot \hat{p}_R|^2$ is called the polarization loss factor (PLF) and reflects the loss due to the mismatch of the polarizations of a transmitter antenna and a receiver antenna. When readers have a circular-polarized antenna, the PLF is 0.5 no matter what polarization the tag antenna has [3]. Finally, the transmitter and the receiver antenna gains are determined by their orientations as defined by the spherical coordinates (θ_T, ϕ_T) and (θ_R, ϕ_R), respectively.

In Equation 10.2, the antenna gain is not a constant. Rather, it is a function of the antenna's own orientation unless the antenna radiates power isotropically. Antennas of different types differ in their own radiation patterns, which leads to different values of $G_R(\theta_R, \phi_R)$ and $G_T(\theta_T, \phi_T)$ in Equation 10.2. In this section, for purposes of illustration we will use a half-wave dipole antenna for the tag and a patch antenna for the reader. Patch antennas are chosen for RFID readers because they are less sensitive to tag orientations. Similarly, passive backscatter tags with half-wave dipole antennas are common in far-field applications and usually have longer read ranges than inductive-type tags.

Definition 3: The *reader axis* is defined as the line joining the centers of the tag antenna and the reader antenna (which are r units apart).

The following equation may be used to calculate the gain of a half-wave dipole antenna, which is omnidirectional:

$$G_R(\theta_R, \phi_R) = 1.641 \left[\frac{\cos\left(\frac{\pi}{2} \cos\theta_R\right)}{\sin\theta_R} \right]^2 \tag{10.3}$$

In the above formula, we align the z-axis with the direction of the tag's antenna. The spherical coordinate θ_R is defined as the angle between the reader axis and the antenna

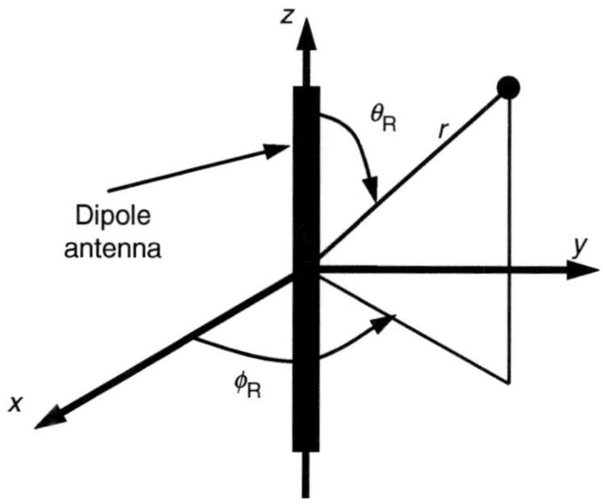

FIGURE 10.1
Dipole antenna angle definition.

direction, while ϕ_R is the angle between the x-axis and the projection of the reader axis on to the x–y plane; these are shown in Figure 10.1. As Equation 10.3 indicates, the gain of a half-wave dipole antenna only depends on θ_R.

The following equation shows the gain of a patch antenna that is used on the reader; the spherical coordinates θ_T and ϕ_T are defined as in Ref. [1] and shown in Figure 10.2:

$$G_T(\theta_T, \phi_t) = 3.136\left[\sin\theta_T \frac{\sin\left(\frac{\pi}{2}\cos\theta_T\right)}{\cos\theta_T}\cos\left(\frac{\pi}{2}\sin\theta_T\sin\phi_T\right)\right]^2 \tag{10.4}$$

RFID antenna design has been undergoing rapid evolution. New antennas have been developed for better read range, smaller size, or better handling of water and metal. Mathematical calculation of antenna gains can become very complicated and in some cases it might not even be possible to find a closed-form expression to describe these. However, an RF anechoic chamber can still be used to measure the antenna radiation pattern or radar cross section in such cases.

Equations 10.2 through 10.4 can be used in conjunction with each other for determining the amount of power received at the tag. It is also worth mentioning that Equation 10.2 is

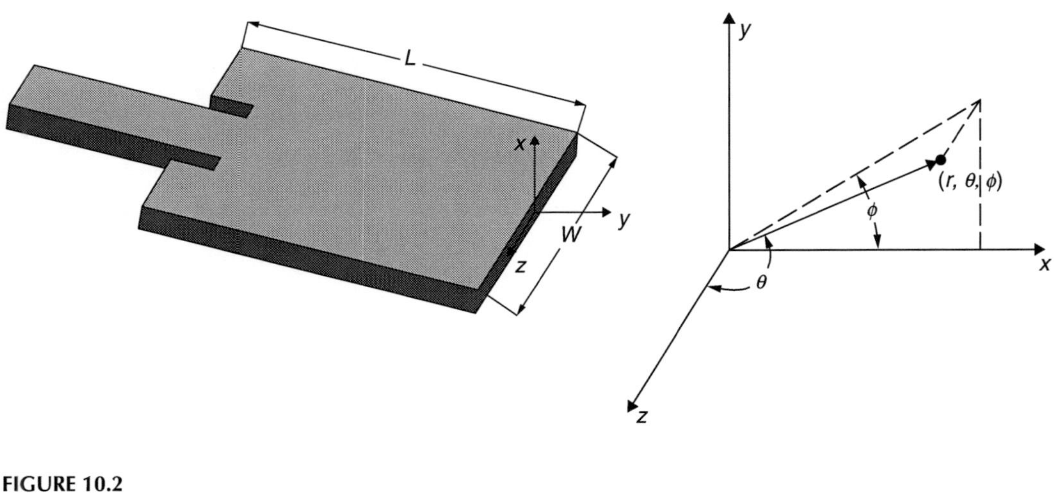

FIGURE 10.2
Patch antenna and its coordinate system.

based on a free space model in which multipath effects and interference are neglected. The received power from the formula is a theoretical value that can only be obtained in an anechoic chamber. When Friis' equation is used for estimating the powering region in a portal, reflection, scattering, diffraction, and shadowing may occur in the signal propagation; it is hard to quantify these since the extent of each may differ on a case-by-case basis. Nevertheless, Friis' equation does provide one with a fundamental tool for power calculations.

10.2.3 Powering Region with a Single Reader Antenna

In the previous section, Friis' equation gives the power received by an RFID tag, given the positions and orientations of the tag and the reader. Suppose that at a fixed location we discretize the set of all possible orientations of an RFID tag into M unit direction vectors; then Friis' equation can be evaluated M times to find the values of power received at that location for all possible orientations. The tag's position is defined as possessing $100\alpha\%$ read accuracy as long as αM of the values computed are greater than a specified value P_{min}, the minimum operational power required to activate the tag.

Although it is true that evaluating the $100\alpha\%$ read accuracy powering region in a portal can be done using Friis' equation, the computational effort could be overwhelming to achieve sufficiently high precision: not only does one need to discretize the orientations in three-dimensional space, but also the portal space. Discretizing a 1 m^3 volume with 1 cm search steps leads to 1 million search points, each of which presumably can take on as many as M different orientations. With a 1° resolution, there will be $M = 64{,}800$ values for the gain for each point (180 × 360°). Therefore, there are $64{,}800 \times 10^6$ evaluations for just a 1 m^3 space.

Greene has developed an efficient algorithm to overcome this computational challenge [2]. With a single reader antenna, we can fix the position and orientation of the reader antenna, and the location and orientation of the tag can be uniquely defined relative to this point of reference. The reader antenna is assumed to be at the origin with its maximum gain direction aligned with the x-axis, and the reader axis as per Definition 3. The monotonicity of power received along the reader axis as a function of distance from the reader yields the following corollary.

Corollary 1: Along the reader axis, if there exists any point with $100\alpha\%$ read accuracy, then there is always a point with $100\alpha\%$ read accuracy such that any point closer to the reader has at least $100\alpha\%$ read accuracy while any point further away fails to meet $100\alpha\%$ read accuracy.

Proof: In Friis' equation, let the transmitter gain along the reader axis be given by G_T. Suppose there are M ($M \rightarrow \infty$) orientations for each tag position along the reader axis; and M corresponding values of the receiver gain G_R at each of these points. Let T be the threshold point with distance r_T such that exactly $100\alpha\%$ of the M orientations at point T can receive power greater than or equal to the minimum activation power. Then for any point T' on the reader axis with distance $r' > r_T$, any of the M orientations will receive power that is greater than the power received from the same orientation at point T. Therefore, at least $100\alpha\%$ of M orientations at point T' can be activated with sufficient power.

Corollary 1 indicates that for the case of a single reader antenna there will be a three-dimensional boundary surface within which lies the $100\alpha\%$ read accuracy powering region. In Ref. [2], instead of evaluating every point in the portal space, an algorithm is developed to obtain such a boundary surface. At any given point on the reader axis for a given tag position, all the parameters in Friis' equation are known except for the receiver gain G_R. Regardless of the mathematical form of the gain function G_R, if substituting the value of the 100α percentile of G_R in Friis' equation gives power that is greater than

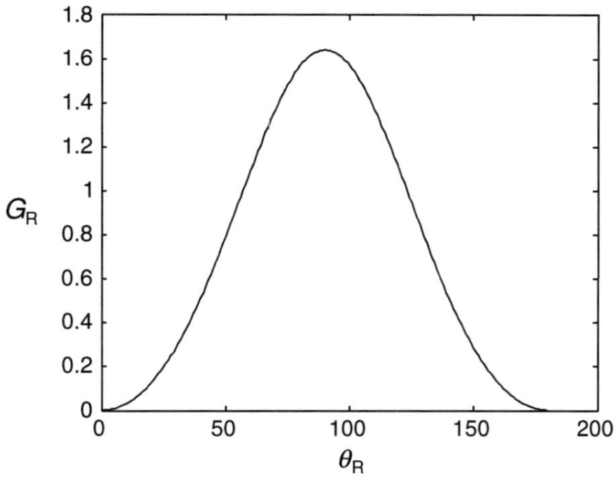

FIGURE 10.3
Half-wave dipole antenna's gain versus θ_R.

the required activation power, then such a point can be read with at least $100\alpha\%$ read accuracy. In other words, a threshold point T on any reader axis is such a point that the received power equals the required activation power P_{min} when G_R assumes the 100α percentile value. Based on the above analysis, we can obtain the following formula:

$$\frac{r^2}{G_T(\theta_T, \phi_T)} = \frac{P_T \cdot G_{R\_100\alpha}\lambda^2}{P_{min}(4\pi)^2}(1 - |\Gamma_T|^2)(1 - |\Gamma_R|^2)|\hat{p}_T \cdot \hat{p}_R| \tag{10.5}$$

In Equation 10.5, $G_{R\_100\alpha}$ is the 100α percentile value of receiver gain while P_{min} is the minimum receiver power to activate a tag. Therefore, the right-hand side of Equation 10.5 is a constant. Let C be the value of the constant in Equation 10.5. Then $f(\theta_T, \phi_T, r) = (r^2/G_T(\theta_T, \phi_T)) = C$ defines the boundary surface of the $100\alpha\%$ read accuracy powering region.

If RFID tags use half-wave dipole antennas, then the antenna gain is a function of θ_R with a period of π and is symmetric about $\theta_R = \pi/2$, as shown in Figure 10.3. This is because the shape of $\cos\left(\frac{\pi}{2}\cos\theta_R\right)/\sin\theta_R$ is similar to that of $\sin\theta_R$.

In the case of a half-wave antenna, the 100α percentile value of the receiver gain is obtained when $\theta_R = 90 \times (1 - \alpha)$ degrees. Figure 10.4 shows an example of a 90% read accuracy powering region when the radio frequency is 915 MHz, transmitter power is 0.5 W, and the minimum operational power of an RFID tag is 50 μW.

10.3 $100\alpha\%$ Read Accuracy Powering Region with Multiple Reader Antennas

In Section 10.2, an efficient algorithm was presented to calculate the $100\alpha\%$ read accuracy powering region of a single reader antenna. From Corollary 1, such a region is bounded by a surface on which each point has exactly $100\alpha\%$ read accuracy. However, when the number of reader antennas placed at a portal is more than one, such a region cannot be obtained by simply merging the $100\alpha\%$ read accuracy powering region from each individual antenna. In this section, we first discuss why merging such regions does not work. After providing a mathematical representation of the problem, we then discuss a methodology for locating readers and related implementation issues.

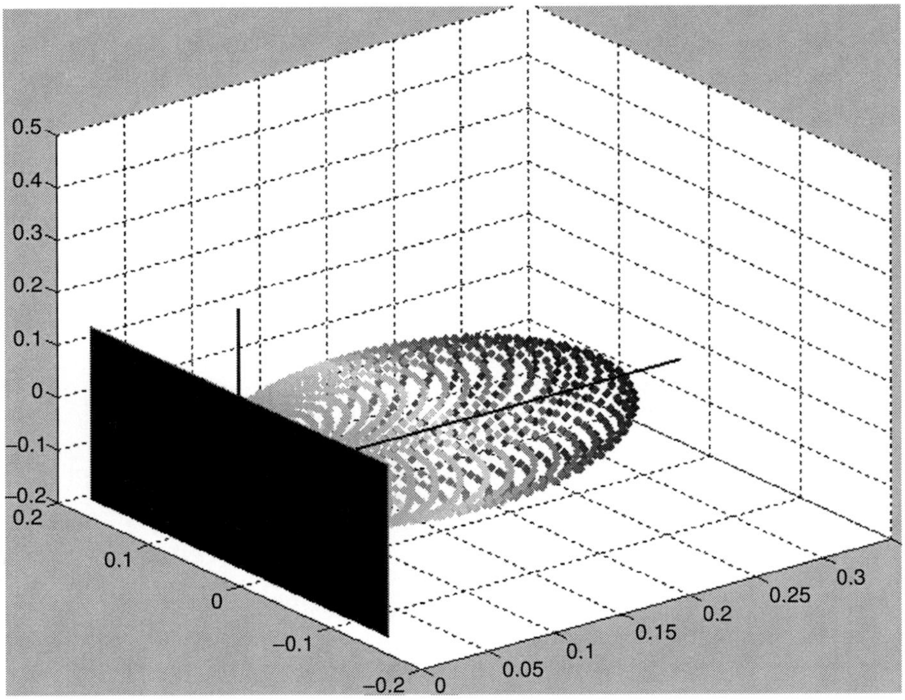

FIGURE 10.4
90% read accuracy powering region.

10.3.1 Merging 100α% Read Accuracy Powering Regions

Given the location of a tag along with its orientation, the tag can be regarded as readable as long as there is at least one reader antenna that provides enough power to activate the tag. The read accuracy with respect to a tag location is thus the percentage of all orientations of the tag at that particular position that can be powered by at least one reader antenna. For a single reader antenna, the 100α% read accuracy powering region is fixed with respect to the reader's location. However, with more than one reader antenna, such a region is not simply the union of the individual 100α% read accuracy powering regions.

Consider Figure 10.5, where a unit ball is used to represent all possible orientations for an RFID tag with its antenna center located at the center of the ball. Note that the surface of the ball does not have any significance by itself—a line joining the center of the ball to any point on the surface represents one possible orientation for the tag's antenna. Now suppose that α is specified as 0.95, and suppose further that each of the two RFID reader antennas shown can individually cover only 90% of all of the orientations. The points on the surface within the conical shadow represent those 10% of the orientations of the tag where it cannot be powered by the corresponding reader. Based on our definition of read accuracy, the current tag position does not meet the specifications for either antenna individually. Thus this point would not lie within the individual football-shaped powering region (Figure 10.4) for either reader. However, because the unreadable orientations shown below are mutually exclusive with respect to each other, orientations missed by one reader antenna are covered by the other; hence the position can actually be read with 100% accuracy using the two reader antennas in conjunction with each other as shown. In a different example, there could be a point that can have 80% of its orientations read by one antenna and 80% of its orientations read by a second antenna, but the union of the two

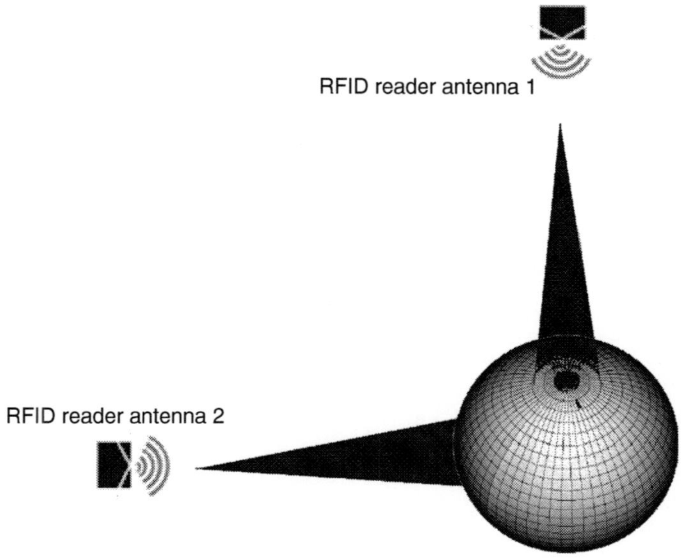

FIGURE 10.5
Orientations for the two reader cases.

antennas results in 95% of the orientations being readable. In this case, the point would not be in either antenna's readable region when the antennas are considered separately but is readable when the antennas are considered jointly. Unfortunately, in determining the coverage area for multiple antennas, Corollary 1 cannot be extended to this case since there is no direct monotonicity result with multiple readers that can be exploited to define the powering region.

Consequently, there is no straightforward way to find the boundary surface of the powering region, which can be disjoint and have some irregular shape. In conclusion (1) union of the $100\alpha\%$ read accuracy powering regions for each individual antenna underestimates the real coverage volume because there might be points that would be considered unreadable by each reader individually, yet would be readable when all of the readers are considered jointly and (2) there is no straightforward process whereby one can define the boundary of the powering region because of the absence of any direct monotonicity result.

10.3.2 Integer Programming Formulation for Multiple Reader Coverage

In this section we formulate the reader antenna location problem as a mathematical program by defining a suitable objective, and translating our read accuracy requirements and the corresponding powering region into mathematical relationships that define the constraints of the math program. Suppose that the region within a portal where a tag can lie is discretized into L distinct points and that each of these is a possible location for the center of the tag's antenna (henceforth referred to as tag position; the entire set of L tag positions will be referred to as the tag space). Suppose further that we are given N potential locations along the portal for the center of any reader antenna (henceforth referred to as reader position), and we have a maximum of n_0 ($<N$) reader antennas available for placement at these positions. Finally suppose that the (infinitely many) orientations that a tag's antenna could take on are also discretized into a total of M distinct orientations. Given a suitably specified fraction α, our objective is to determine the optimal number of readers along with their optimal locations, so as to maximize the number of positions in the tag space that can be powered with at least $100\alpha\%$ read accuracy, that is, for which at least αM orientations can receive the threshold power required to be read.

The reader antenna placement problem can be formulated as the following integer program [4]:

$$\text{Max} \sum_{l=1}^{L} z_l$$

s.t.

$$\sum_{n=1}^{N} x_n \leq n_0 \tag{10.6}$$

$$\begin{aligned} & y_{lm} - \sum_{n=1}^{N} p_{lmn} \cdot x_n \leq 0, \quad \forall l = 1,2,\cdots,L, \quad m = 1,2,\cdots,M \\ & \sum_{m=1}^{M} y_{lm} - \alpha \cdot M \cdot z_l \geq 0, \quad \forall l = 1,2,\cdots,L \end{aligned} \tag{10.7}$$

$$x_n \in \{0,1\}^N, \quad y_{lm} \in \{0,1\}^{L^*M}, \quad z_l \in \{0,1\}^L \tag{10.8}$$

where
- N is the number of candidate reader positions
- L is the number of tag positions in the tag space
- M is the number of discretized orientations considered for each tag position
- n_0 is the maximum number of reader antennas available
- $100\alpha\%$ is the required percentage read accuracy for every point in the powering region
- p_{lmn} is the binary coefficient. $p_{lmn} = 1$ if a tag at point l with orientation m is in range to receive enough power from a reader antenna at location n
- z_l is the binary variable. If $z_l = 1$ then the point l is covered with $100\alpha\%$ read accuracy
- x_n is the binary variable. If $x_n = 1$ then there is a reader antenna placed at location n
- y_{lm} is the binary variable. If $y_{lm} = 1$ then a tag at the point l with orientation m will be covered by at least one reader antenna

Note that the last three quantities are the decision variables. Whether p_{lmn} takes on a value of 0 or 1 is determined by first computing the received power P_R for the location l with orientation m using Friis' equation, and then checking to see whether or not this exceeds $P_{\min}$ (the minimum operational power required to activate the tag). Constraint 1 requires that the number of antennas installed be no more than the number available. Constraint 2 ensures that a tag position with a specific orientation is marked as covered only if the required power is received from at least one of the reader positions. Constraint 3 guarantees that only a point with at least $100\alpha\%$ read accuracy will be counted.

In the above formulation, the number of constraints is $1 + (L^*M) + L$, while the number of binary variables is $N + (L^*M) + L$. The density of the coefficient matrix is determined by the p_{lmn} values, and in general will be much higher than it is for typical 0–1 integer programming problems of this size. The number of binary variables in the problem can be reduced because the structure of the model allows for the y_{lm} to be relaxed and defined as continuous variables with lower and upper bounds of 0 and 1, respectively. Note that the objective attempts to make the z_l values as large as possible, and these are bounded from above by the y_{lm} values in Constraint 3, so that we would also like to make the latter values as large as possible. Thus y_{lm} will be set to 1 in the optimal solution as long as a tag at position l with orientation m can be covered by at least one reader; otherwise Constraint

2 forces y_{lm} to be 0. Therefore, the problem can be reduced to $N+L$ binary variables and $L \times M$ continuous variables in the range [0,1]. However, the number of constraints cannot be reduced.

Unfortunately, the integer program is poorly structured and it is impractical to solve the problem to optimality with a high level of discretization for the portal space as well as the orientations. The fact that the number of constraints is very large and cannot be reduced, and that the technological coefficient matrix problem for the problem is dense makes it a very hard integer programming problem.

10.3.3 Enumeration Scheme for Multiple Reader Coverage

Given that integer programming presents significant computational problems for large problems, an enumeration scheme is examined in this section. Suppose we are given a set of n reader positions. For each reader position and for each of the L points in the tag space, Friis' equation is used for each of M possible orientations to calculate received power. As soon as we find a reader position that can activate more than αM orientations for the tag at that point, we mark the tag position as readable and move on to the next point in the tag space. If no reader can cover the point as required it is marked as unreadable. Thus each tag position could entail as few as αM and as many as nM computations of received power. For the current set of reader positions we use the number of points in L that have at least $100\alpha\%$ read accuracy as the performance measure. This process is repeated for each of the C_n^N choices for the set of reader positions to find the one that yields the maximum coverage across the entire tag space.

Two important issues need special attention with respect to the implementation of the enumeration scheme. First, although it is easy to discretize the portal space into L points, it is not obvious as to how all possible orientations can be evenly discretized into M distinct vectors. Second, it is obvious that the enumeration scheme does not solve the complexity issue in that there are still $C_n^N \times L \times M$ evaluations possible and L rises rapidly with an increase in the portal space resolution. It is important to select reasonable values for L and M, so that the computational effort can be reduced without sacrificing the precision of the final solution.

10.3.3.1 Discretization of Orientations

At the pallet/case level, tags tend to be placed on one of the faces of a case (typically a rectangular shape), although the cases themselves could be oriented in any way relative to the reader. Thus it is not unreasonable to expect that the tags would lie in one of the three Cartesian planes but with an arbitrary orientation within that plane, although this would certainly not be true if cases are not stacked in some uniform fashion as they move through the portal. On the other hand, in item-level applications it is not possible to accurately predict how the product will lie and the tags can thus be expected to randomly take on many different orientations. Even in some pallet-level retailer applications, cases with RFID tags are sometimes carried through the portal by staff, therefore strictly limiting the orientations of tags is not a good idea. To determine the $100\alpha\%$ read accuracy region it is necessary to be able to represent and evaluate the readability of all of these different possible tag orientations. If we discretize all possible orientations into M discrete unit vectors from the location of the tag center, then if there is no bias toward a specific orientation each of these unit vectors should be uniformly distributed on the surface of a unit sphere whose center is coincident with the location of the tag center.

The conventional approach is to discretize uniformly around the latitude and the longitude (e.g., every 3° from 0° to 360°); however, as shown in Figure 10.6, this approach

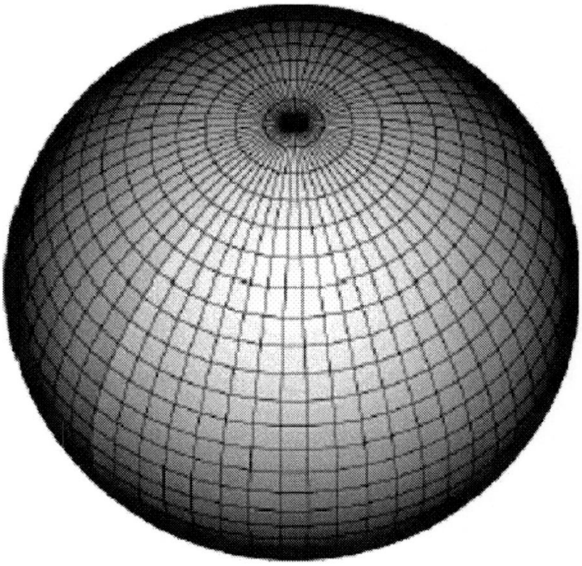

FIGURE 10.6
Longitude–latitude–grid discretization method.

leads to a biased sample that is highly anisotropic and has a stronger concentration of directions pointing toward the poles and relatively few directions pointing toward positions on the equator.

Determining the exact uniform configuration of the orientation vectors is a hard mathematical problem [5,6]. Although a continuous spherical uniform distribution is explicitly defined [7], there is unfortunately no single definition of a corresponding discretized uniform distribution. Researchers in different fields such as geometry, climate modeling, molecular structure, or electrostatics have studied the problem with their own definitions, each of which may lead to some different distribution [5,8]. However, although it may be hard to obtain M vectors uniformly distributed on a sphere for our purposes, such vectors need be computed only once and used repetitively for L points. Therefore, the computation complexity of the discretization algorithms should not be regarded as an important factor in our methodology.

An approximation algorithm by Rusin [9] is used in the numerical examples shown later in the chapter. In this approximation method, a sphere is first cut by a series of evenly spaced horizontal planes, each of which forms a constant-latitude circle on the sphere. On each such circle, points are placed so that the arc distance between each pair of adjacent points is the same. This distance is kept the same for all of the latitude circles. Thus, circles closer to the pole have smaller radii and consequently a smaller number of points on them.

Before listing out the details of the algorithm we review the terminology used. A great circle is defined as a circle around the surface of a sphere that has its center at the same point as the center of the sphere. Great circles which pass through the North and South poles are called *meridians*. The great circle that is perpendicular to the axis (the line joining the two poles) and lies half-way between them is known as the equator, whereas small circles around the surface that are parallel to the equator with centers lying on the axis are called *parallels*. The algorithm may then be described as follows:

Approximation algorithm for creating M uniformly distributed unit vectors on a unit ball:

Begin

$K = \left\lfloor \sqrt{\pi/4 \cdot \mathbf{M}} \right\rfloor$;

Divide a meridian into K equal segments with $K-1$ points $(p_1, p_2, \ldots, p_{K-1})$;

Draw a parallel C_i at each p_i $(i=1,2, \ldots, K-1)$;
For each C_i
Divide C_i into $\left\lfloor 2K \cdot \cos\left(-\frac{\pi}{2}+\frac{i \cdot \pi}{K}\right)\right\rfloor$ equal segments with $\left\lfloor 2K \cdot \cos\left(-\frac{\pi}{2}+\frac{i \cdot \pi}{K}\right)\right\rfloor$ points;
Add two points, one from each pole;
End;

10.3.3.2 Effects of Discretizing the Tag Space and Orientations

The maximum number of evaluations of Friis' equation is $C_n^N \times L \times M$ in the enumeration scheme. In most cases, the number of reader antennas n used in a portal tends to be relatively small, so that the computational complexity is largely determined by the values of N, L, and M. Among these three values, the number of candidate antenna locations N is subject to design issues such as available portal space, the physical size of the antennas, etc. However, the number of discretized points in the portal L and the number of discretized orientations M can be freely selected. In general, larger values lead to better and more reliable final solutions to the placement problem, but at the same time, also lead to increased computational effort. Thus the choice of values for L and M is important because it determines the quality of the solution as well as the amount of computation effort required.

The tag space resolution, which determines the value of L, is a critical parameter because it determines the number of evaluations for each set of reader's antennas. Moreover, a decrease in search resolution (meaning searching more precisely) by a factor of 10 leads to an increase in L by a factor of 1000.

Discretizing the tag orientation and location space is an approximation to what is truly a continuous space and simply looking at the percentage of all points in the tag space that receive coverage (as a measure of performance) can give rise to misleading results as a result of this approximation. This can be illustrated in Figure 10.7, which shows a portal equipped with a single reader antenna. Consider five consecutive grid points in the discretization as shown, and suppose the vertical line represents the actual 90% read accuracy boundary. Then based on the current discretization scheme, the 90% read accuracy powering region covers 3 out of 5, or 60% of the portal space (Points 1, 2, and 3). Suppose now that we use a coarser search resolution where only every other point will be examined, that is, we only have points 1, 3, and 5. Then the coverage actually increases to 66.67% since points 1 and 3 are within the boundary. By the same token if the boundary had been between points 2 and 3 then the coverage would have dropped from 40% to 33.33%. Thus, the percentage coverage attained is not a good comparative measure. In fact, since the coverage percentage should ideally be calculated in continuous three-dimensional space, a higher search resolution is always preferred because the results in such a case are always closer to the actual coverage for the ideal case.

A coarse resolution in the tag space might lead to two different types of errors. First, an optimal set of n antennas might be chosen, but the percentage of the tag space that is covered with 100α% read accuracy might not be correct, as illustrated in Figure 10.7.

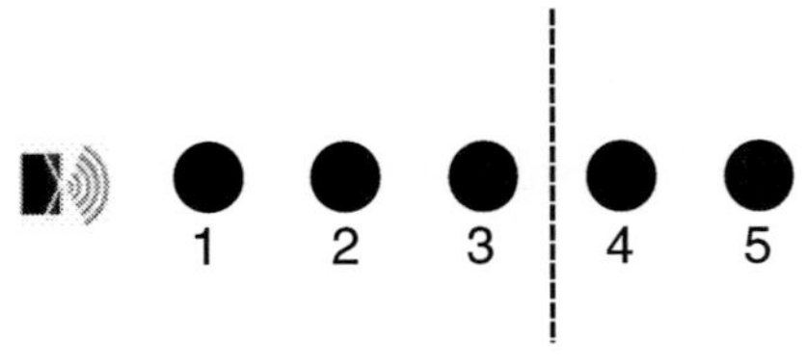

FIGURE 10.7
Coverage percentage calculation with different search solutions.

Second, because the powering region of each set of n antennas is only approximated, a suboptimal set of n antennas might be chosen, which is a much more serious error than the first one. On the other hand, from a computational viewpoint, if a coarse resolution can lead to the same optimal reader placement, then it would be ideal to use such a resolution for determining the actual placement of the readers in the enumeration procedure but a finer resolution can then be used at the end to obtain a more precise estimate of the actual coverage percentage obtained by the placement scheme.

Similar to the tag space search resolution, for every point, the coverage for various orientations should also ideally be calculated in continuous three-dimensional space; therefore a larger value of M is always preferred. A less-than-ideal value of M could give rise to two types of errors. Type 1 error occurs when a point for which more than $100\alpha\%$ of the orientations can be covered with the finest resolution is (mistakenly) classified as not being covered with the smaller value for M. Conversely, a Type 2 error occurs when a point that does not achieve the minimum coverage of $100\alpha\%$ with the finest resolution is classified as being covered with the coarser resolution. Increasing the value of M reduces both types of errors; however, to reduce the computational time, the value of M should be chosen as small as possible without leading to suboptimal or erroneous solutions.

10.3.4 Numerical Examples

In the example shown in Figure 10.8, a portal with dimensions $3 \times 3 \times 3$ m has 18 candidate antenna positions on three walls spaced at 0.5 m intervals. The smaller cube ($2 \times 2 \times 2$ m) inside the portal represents all possible tag locations during the interrogation processes. The tag space is smaller than the portal volume because we assume items move through the central region of the portal. Thus, we eliminate the space that is closest to the wall, floor, or ceiling on consideration. Each of the 18 reader antenna positions has 3 orientations: 45°, 0°, and −45°, respectively. In our tests, we set n_0 to either 2 or 3, and used a value of 90% for the required read accuracy. The transmit power from an RFID reader was 0.5 W with 50 μW needed to activate an RFID tag that operates at 915 MHz.

With a resolution of 0.1 m for the tag space and 450 possible orientations for the tag antenna at each location, the optimal solution, which chooses positions 3 and 16, will cover 71.8% of the 8400 tag locations with 90% read accuracy. When three reader antennas are to be

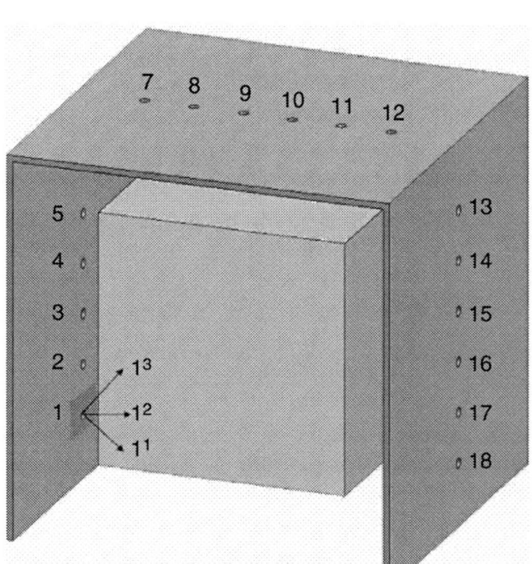

FIGURE 10.8
Portal design with 18 candidate antenna positions, each of which has 3 orientations.

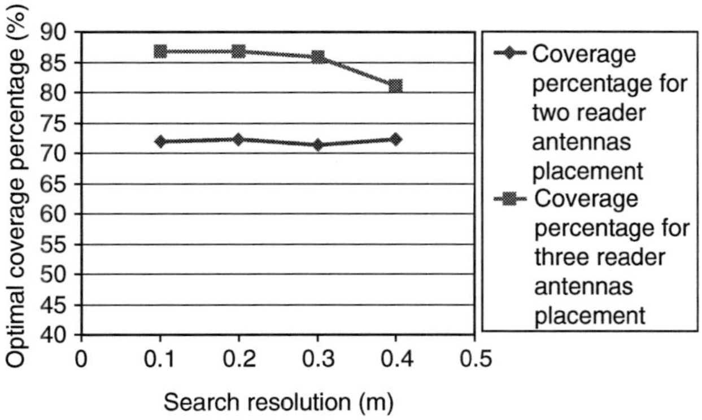

FIGURE 10.9
Optimal coverage percentage for different search resolutions.

placed, the extra one, which is placed at position 9 in the optimal solution, increases coverage by another 14.8%. In other words, the extra antenna can cover approximately 1.18 m^3 more of the tag space with 90% read accuracy. Even though antennas may be placed with different orientations, in the optimal solution they are mounted perpendicular to the walls.

Figure 10.9 displays results from different search resolutions ranging from 0.1 to 0.4 m that were used to find the best antenna placements in the example, but with the coverage reevaluated with the finest discretization resolution used (0.1 m). In all cases, $M = 450$ discretized orientations were used for read accuracy calculations. The coverage percentage is represented by the proportion of discretized points in the $2 \times 2 \times 2$ m portal space that can be read with 90% read accuracy. It can be seen that coarser search resolutions may or may not find the same solution as a finer search resolution. When 0.3 or 0.4 m resolution is used, the enumeration method results in suboptimal solutions.

To evaluate the effect of M, a value of $M = 1916$ is first chosen for calculating the read accuracy of each point, and then the calculation is repeated with smaller M values. Table 10.1 shows the extent of the two types of errors that are caused by smaller M values.

The two types of errors stabilize and quickly converge to a very small value as M increases. In particular, the total classification error is well below 1% once M reaches a value of 450. This point is further illustrated by Figure 10.10, which shows the percentage of tag locations that achieve 90% read accuracy coverage (at the finest tag space discretization) using the optimal solution that is determined by each value of M. For example, the antenna placement found using an M value of 20 results in an actual coverage of about 63%, which is much smaller than the 72% that can be found by using a larger value of M. In the three reader antenna placement case, the optimal placement is found even when M is as small as 20. But for the two reader placement problem, a smaller value of M can result in a suboptimal solution, which covers as much as 10% less than the tag space than the best solution. In the example problems, the optimal two reader placement solution can only be found when the value of M is greater than 100.

Figure 10.11 shows the optimal 90% read accuracy coverage with different numbers of antennas. The marginal benefit is diminishing as n_0 increases. With three readers,

TABLE 10.1
Two Types of Errors for Different Values of M

M	20	44	80	246	450	984	1454
Type 1 Error Percentage (%)	38.18	3.76	11.01	2.67	0.22	0.22	0.44
Type 2 Error Percentage (%)	0.56	3.90	0.29	0.18	0.33	0.18	0.06

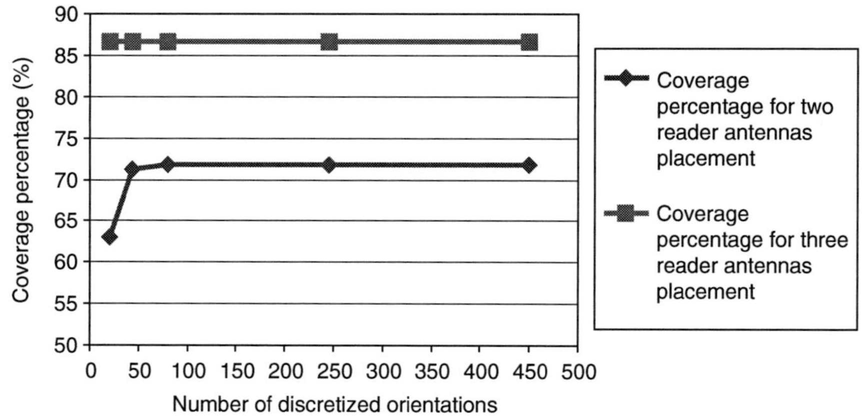

FIGURE 10.10
Optimal coverage percentage for different numbers of discretized orientations.

about 86.5% of the tag space will be covered with 90% read accuracy. The fourth reader brings another 4% of the tag space into the 90% read accuracy region and the covered space becomes saturated as n_0 increases. Due to the limitation in candidate antenna positions in the example, some corners of the portal space will not be covered even with 18 antennas.

10.4 Conclusions and Extensions

Imperfect read rates constitute a major obstacle to the widespread adoption of RFID technology. Without human intervention, it is hard to fix the location and orientation of a tag during the interrogation process. Hence, multiple reader antennas have been used in portals, docks, etc., to improve read accuracy and the powering region. Properly locating these antennas is essential to the success of RFID portal designs. In this chapter, we have addressed the important but difficult problem of determining the location of RFID interrogators to maximize read accuracy.

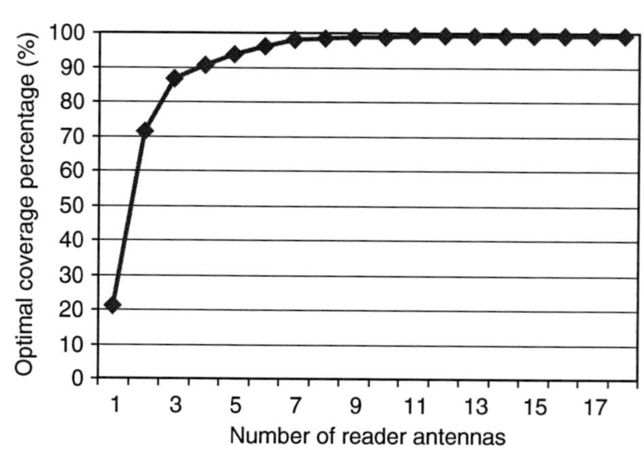

FIGURE 10.11
Optimal coverage percentage for different numbers of reader antennas.

The orientation and polarization of antennas from both readers and tags have direct impacts on the power received and the success of an interrogation process; therefore, the read range parameter, which is defined under ideal conditions, is too simplistic to describe the powering region of RFID interrogators. To fully answer the question of where to locate the RFID antennas requires the use of an expanded Friis' equation and careful definition of the terms read accuracy and powering region.

Although Greene [2] has provided an efficient methodology for determining the powering region of a single reader antenna, merging the powering regions of individual antennas underestimates the real powering region of a multiple reader antenna configuration. An enumeration scheme is developed, where the number of Friis' equation evaluations is $C_n^N \times L \times M$, where N is the number of candidate antenna locations, n is the number of antennas to be installed, L is the number of discretized tag positions, and M is the number of discretized orientations for each tag position. However, the computational effort required for this approach can be overwhelming if these parameters are selected for high resolutions. The effect of varying the values of L and M on the final solution found by the enumeration scheme are studied. Considering the trade-off between computational time and the precision of the solutions, the values of L and M should be carefully chosen so that minimum computational time can be spent without resulting in suboptimal antenna locations. Unfortunately, choosing appropriate values for the enumeration parameters does not totally eliminate the computational problem; in particular when the portal space is considerable in size and the number of candidate antenna positions is not small. In such cases, the number of candidate antenna locations, N, can initially be limited to a relatively small number of points around the portal to facilitate the enumeration process and obtain a solution in a reasonable amount of computational time. The solution could then be fine-tuned by picking the best few locations and refining the search to include points close to these locations.

In this chapter, we assumed that tags can take any orientation and any position within the portal space with equal likelihood. However, in pallet-level RFID applications, tags are more likely to be on cases, which tend to have a limited number of orientations. In addition, cases are usually stacked onto pallet jacks so that tags are likely to be more frequently located in the lower levels of the portal space. In these cases, a weighting factor based on either more probable orientations or more probable tag heights can be added into the enumeration scheme. Use of this weighting scheme may potentially change the final choice of antenna locations.

References

1. Constantine A. Balanis, *Antenna Theory Analysis and Design*, second edition, John Wiley & Sons Inc., 1997.
2. Charles E. Greene, *Area of Operation for a Radio-Frequency Identification (RFID) Tag in the Far-Field*, Ph.D. Dissertation, Department of Electrical Engineering, University of Pittsburgh, February 2006.
3. Klaus Finkenzeller, *RFID Handbook: Fundamentals and Applications in Contactless Smart Cards and Identification*, second edition, John Wiley & Sons Inc., 2003.
4. Lin Wang, Bryan A. Norman, and Jayant Rajgopal, *Optimizing the Placement of Multiple RFID Reader Antennas to Maximize Portal Read-Accuracy*, Technical Report 07-01, Department of Industrial Engineering, University of Pittsburgh, 2007.
5. Hallard T. Croft, Kenneth J. Falconer, and Richard K. Guy, *Unsolved Problems in Geometry*, Problem F17, pp. 165–166, Springer-Verlag, New York, NY, 1991.
6. Edward B. Saff and Arno Kuijlaars, Distributing many points on a sphere, *Mathematical Intelligencer*, 19(1), 1997.

7. Nicholas I. Fisher, Toby Lewis, and Brian J.J. Embleton, *Statistical Analysis of Spherical Data*, Cambridge University Press, 1987.
8. Ali Katanforoush and Mehrdad Shahshahani, Distributing points on the sphere, I, *Experimental Mathematics*, 12(2), 2003.
9. Dave Rusin, Topics on sphere distributions, http://www.math.niu.edu/~rusin/known-math/95/sphere.faq

11

Minimum Energy/Power Considerations

Alex K. Jones, Gerold Joseph Dhanabalan, Swapna Dontharaju, Shenchih Tung, Peter J. Hawrylak, Leonid Mats, Marlin H. Mickle, and James T. Cain

CONTENTS

11.1 Introduction

Typically, the biggest metric of concern for RFID tags and, in particular, the silicon controller devices for the tags has been production cost. In chip fabrication terms, the cost of the device directly relates to the CMOS process chosen for implementation and the area of the device produced. Older CMOS processes such as 0.18–0.35 μm are targeted for RFID as they are relatively cheap to produce than the newer 65–90 nm processes. The design goal is to create a device with the smallest area using one of these older processes. However, one metric that has been held secondary is the power consumption of these devices.

RFID tags generally come in two types, passive and active. Active tags include an internal power source (usually a battery) to power the transceiver used for receiving queries and transmitting responses and the controller that computes the tag's responses to transactions. The controller is typically either an ASIC or a low-power embedded microprocessor. Passive tags do not contain an internal power source. As a result, these tags not only receive information from a query, they also receive energy. This energy is used to power the tag to determine and send a response to the query. Although passive tags are generally cheaper than active tags, they have two major disadvantages: (1) range of passive tags is significantly lower than active tags and (2) complexity of response is significantly reduced over active tags because of the limited energy budget. However, active tags, in addition to being more costly than passive tags, often require battery replacement.

Power consumption affects both passive and active devices. For an active device, the amount of power/energy consumption required by the tag dictates the lifetime that a tag may operate. Ranges and complexity of computation of passive devices for features such as added security capability or access to sensors are directly impacted by power usage. In this chapter, we present techniques and architectures that are applicable to either passive or active tags, or both. These techniques are designed to address the concerns of reducing power in RFID systems without compromising their capability or to extend their capability in a power-efficient manner.

The remainder of this chapter is organized as follows: Section 11.2 presents a technique to significantly increase the memory capacity of a passive tag while minimizing the amount of additional energy required. Section 11.3 describes a power macromodeling technique that works in concert with the RFID design automation flow described in Chapter 13 and allows the effective evaluation of alternate protocol designs. Section 11.4 presents the design and evaluation of a passive active RFID tag that has many of the benefits of an active tag that uses considerably less energy. In particular, we describe a passive switch for an active transceiver called the burst switch and a power-aware packet storage and filtering technique called the smart buffer. Each of these techniques allows us to save power by allowing high energy consumption components of the tag to be powered down.

11.2 Increasing Memory Capacity

Conventional writable memories require some static power to retain their stored values. However, passively powered devices require nonvolatile memories that retain the values stored even when the device is not powered. Nonvolatile memories currently employed on passive tags are typically very small (e.g., <200 byte) due to small power budgets of passive tags.

In this section, a memory architecture is presented that can be employed to expand the memory capacity of fully passive tags and can be employed to lower energy consumption in active tags or hybrid passive and active tags. Our solution is to build a hierarchical low-power memory block based on current state-of-the-art nonvolatile memory blocks.

11.2.1 Memory Block Model

Clock gating has been the conventional solution for low-power operation. However, gating the clock is insufficient for our purposes of developing an ultra-low power nonvolatile memory with capacity many times greater than existing integrated nonvolatile memory blocks. Thus, our memory design is based on preexisting memory blocks with a 200 byte capacity optimized for use in passive RFID systems.

Typical RFID systems are implemented in older CMOS technologies such as 0.18–0.35 μm. Thus, we assume that the power consumed in our system is because of dynamic power, as static power only becomes a dominant effect at much smaller feature sizes [1]. And as such the dynamic power or switching power of the system is governed by the formula shown in the following equation:

$$P = fCV_{dd}^2 \tag{11.1}$$

where

P is the dynamic power
f is the frequency of operation
C is the capacitance of the circuit
V_{dd} is the supply voltage

11.2.2 Memory System Architecture

In our first design, we added a power-enable PMOS device in series with the memory block, as shown in Figure 11.1. This allows only the block that is actually addressed to be powered, while the remaining blocks remain disconnected from power. To manage which bank is activated, an address decoder is required to enable which power-enable signal is asserted. An overview of this architecture is shown in Figure 11.2, where N is the number of memory blocks and M is the number of inputs to the decoder.

The primary benefit of the power-gated implementation is that the static and dynamic powers can both be eliminated from all but the active memory block. The addition of a series power-enable PMOS device does not affect the average power consumed by the device, but reduces the peak power consumed by the device. This technique has been previously applied to I/O buffers to reduce the SSN (simultaneous switching noise) produced on the supply lines when the output buffers switch [2]. The design implication on RFID tags relates to the reduction of peak power consumed by the device, in particular

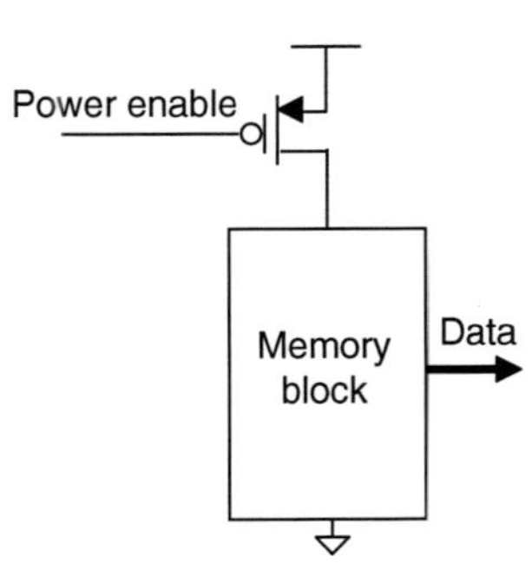

FIGURE 11.1
Memory block with power gate.

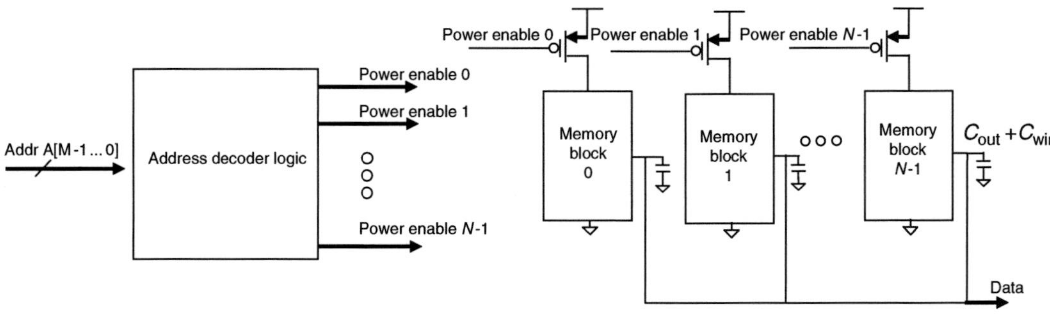

FIGURE 11.2
Flat power-gated memory architecture.

that the peak power requirement in passively powered tags is greatly reduced. A greater peak power requirement can be problematic as the peak power delivered to the tag is limited by the power delivered by the reader. Since RFID tags typically operate at a relatively low frequency (e.g., <500 kHz), the addition of the PMOS device does not have an adverse impact on the speed of the memory.

11.2.2.1 Power-On Reset

The decoder is constructed using standard dynamic CMOS design with a precharge transistor acting as the pull-up network [3]. The precharge signal is driven by a periodic signal related to the clock. In our case, we can use the system clock directly as the precharge signal. For a rising edge triggered memory when precharge is 1, the precharge PMOS transistor is off and the precharge NMOS transistor is on, which allows the pull-down network or evaluation network to pull the appropriate power-enable line to ground, thus turning on the associated memory block. On the back half of the cycle, precharge is 0, turning on the precharge PMOS and disconnecting the power-enable line from ground, thus turning off all the memory blocks.

However, one problem with this power-gating technique that occurs in particular with fully passive RFID tags powered by radio frequency (RF) is due to powering up of the power supply to V_{dd}. Initially, the power-enable lines are low because of the power-up delay of the decoder. This effectively allows each of the memory blocks to power up as the power supply powers up as if they were not power gated, which causes a tremendous amount of power consumption during power-up. This is particularly problematic for passive RFID because the power-up time for V_{dd} from RF energy harvesting is long. Figure 11.3 shows

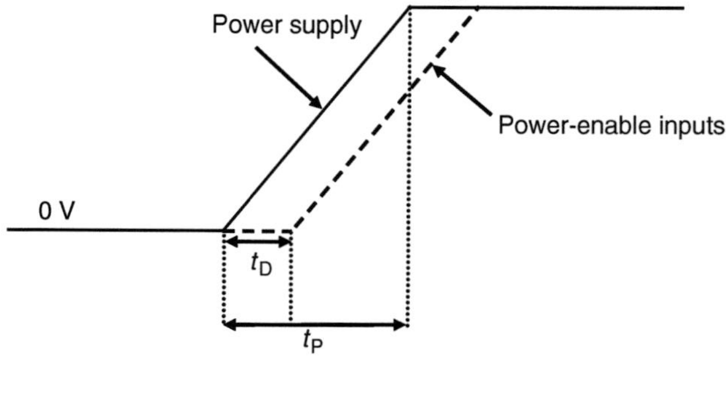

FIGURE 11.3
Simulation for the power-on condition.

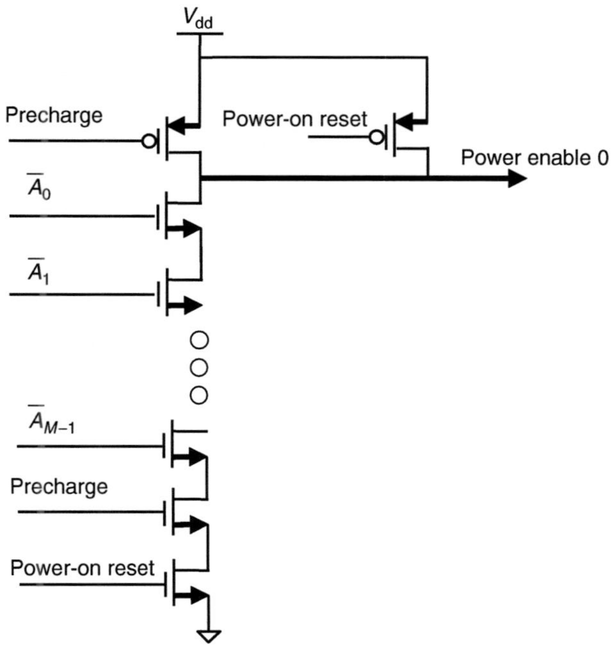

FIGURE 11.4
Precharge for the decoder lines.

the assumptions for our power-on condition of the power supply and the power-enable input lines. The power supply changes to V_{dd} linearly in time t_P and the power-enable lines are delayed by a time t_D, where $t_D \ll t_P$.

To solve this problem, a power-on reset PMOS transistor is added to each of the decoder lines to raise all the power-enable lines high and effectively block any of the memory blocks from charging until V_{dd} has fully powered up. This is shown for the power-enable line 0 in Figure 11.4.

The power-on reset transistor is enabled with the power-on reset signal held low until the power-on reset circuit shuts off the transistor by setting power-on reset high. This is assumed to happen well after the power supply is fully charged, as shown in Figure 11.5, where V_{dd} is activated well after the power supply has powered up. This circuit causes all the power-enable lines to rise at approximately the same speed as the power supply.

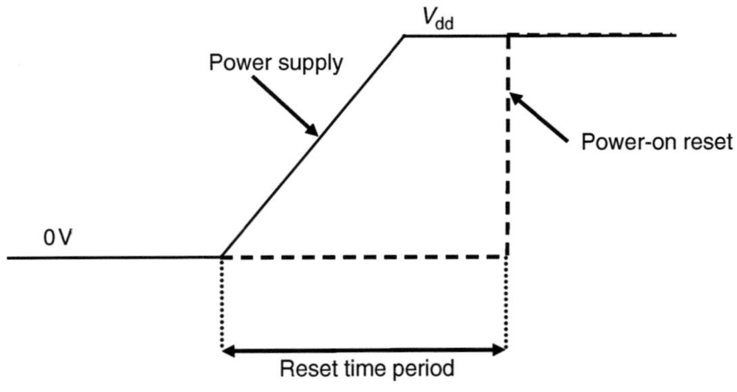

FIGURE 11.5
Simulation for the precharge transistor with power-on reset.

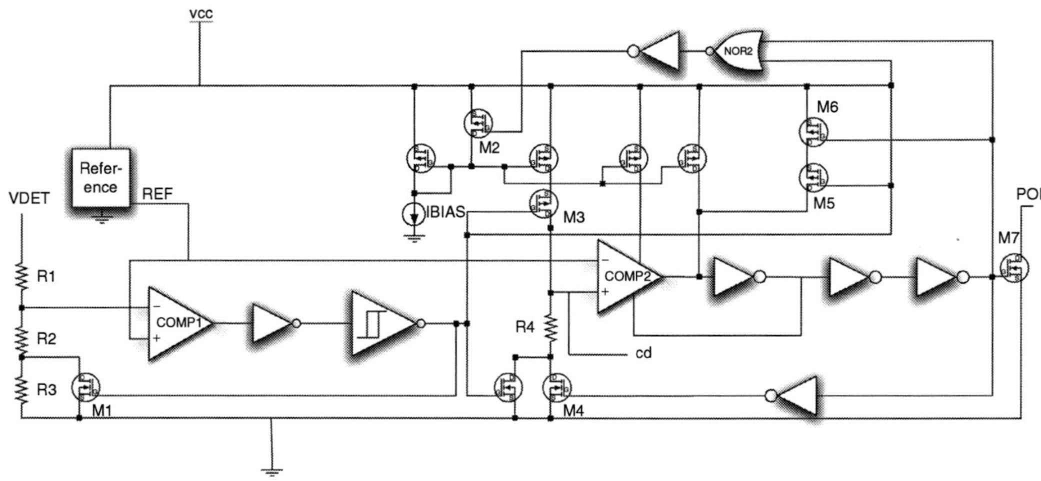

FIGURE 11.6
Power-on reset circuit (Reproduced from Xinquan, L., Weixue, Y., Ligang, and Yu, C., "A low quiescent current and reset time adjustable power-on reset circuit," *Proceedings of International Conference on ASIC (ASIC)*, 559–562, 2005. With permission.)

Thus, the power-enable transistor from Figure 11.1 has a $V_{gs} \approx 0$ during power-up and does not turn on.

Xinquan et al. provide an example of a low-power power-on reset circuit that consumes 5.1 μW, shown in Figure 11.6 [4]. However, if the additional power requirement is problematic for this circuit, a possible solution is to use the burst switch to drive the power-on reset signal. The reader would transmit on its standard frequency to begin powering the tag, and after the reset period had expired, transmit on a different frequency for the burst switch. Thus, the burst switch would activate the power-on reset signal after the requested duration.

11.2.2.2 Banking Memory Blocks

Typically, the data line from each memory block is directly tied together to create a global data line. This is possible because only the active bank drives the line avoiding the potential for multiple drivers. On implementing the architecture from Figure 11.2, the load capacitance for the data lines increases dramatically as each memory block adds its output load capacitance to the system, thus significantly increasing the power required to drive the full memory system. To avoid this, we have grouped the memory blocks into memory banks. The data lines within the banks are tied together, however, between banks they are separated by a multiplexer, as shown in Figure 11.7. This accounts for a significant reduction in required power.

11.2.3 Results

For our fundamental memory building block, we used a model of a 200 byte nonvolatile memory block built in 0.2 μm CMOS operating at 1.5 V with an active power consumption of 5 μW. Using this memory block, the memory block was used to construct a memory that can contain more than 68 KB with a nominal increase in power consumed. This architecture requires approximately 350 memory blocks arranged into banks of 22 blocks each. These specifications are based on conversations with RFID tag manufacturers that include

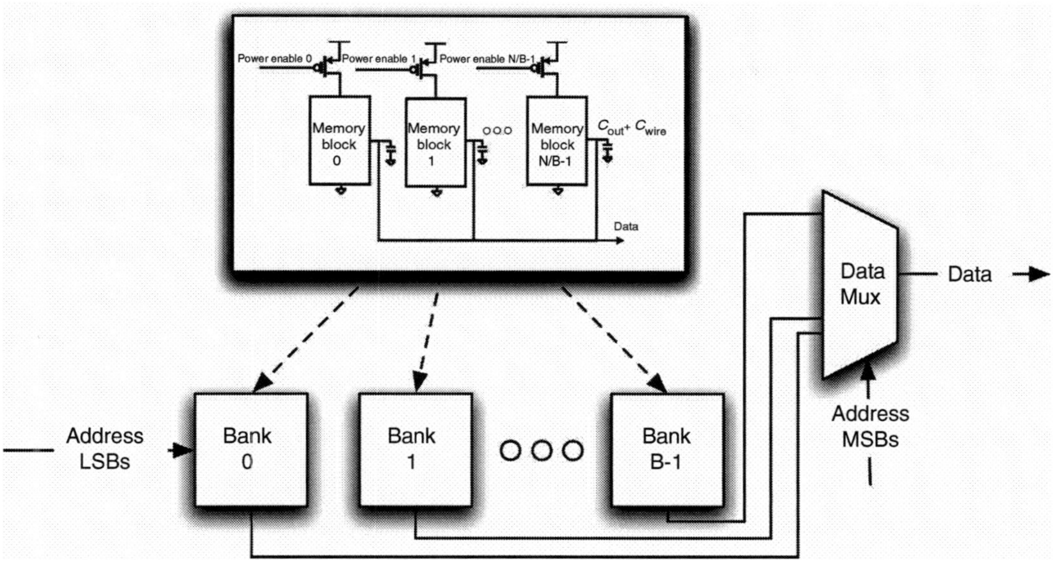

FIGURE 11.7
Banked memory architecture.

nonvolatile memory for tags meeting passive standards such as ISO 18000 Part 6C and have applications that require up to 64 KB of memory.

We model the memory block as an inverter with an approximated load capacitance, C_L, based on details of the memory block architecture, as shown in Figure 11.8a. Based on the memory block specifications, we calculated C_L as 4.44 pF. Any device that allows switching could be used to model the dynamic power of the device; however, the inverter was selected because of its simplified modeling. To this device we added the power-enable PMOS device in series to the system. The circuit diagram is shown in Figure 11.8b.

For a passively powered RFID system, the peak power requirement is the determining factor as to whether the device works correctly. The memory system described in Section

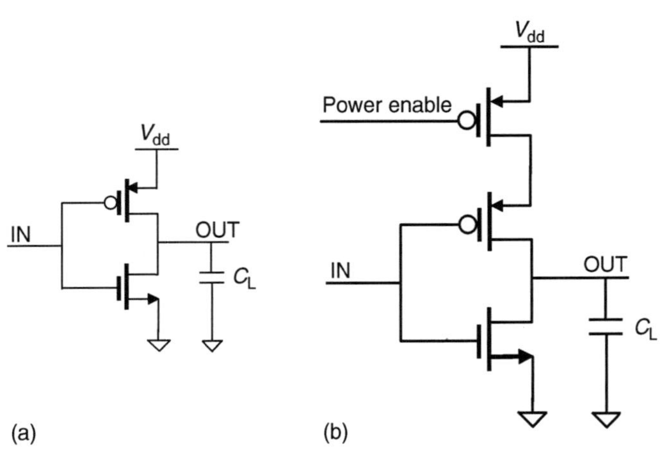

FIGURE 11.8
Dynamic power memory block models. (a) Memory block model. (b) Memory block model with power gate.

TABLE 11.1

Peak Power Reduction with Added Power-Enable PMOS Transistor

	Peak Power (μW)	Average Power (μW)
Without power enable	109.1	5.1
With power enable	57.2	5.1

11.2.2 was implemented in 0.2 μm CMOS technology from TSMC with a combination of tools from Cadence and Micromagic. Power analyses were conducted with HSpice.

The memory block model from Figure 11.8 was examined both with and without the PMOS power-enable transistor. The results shown in Table 11.1 suggest that the power-enable transistor cuts the peak power approximately in half when the device is active as a result of being in series with the pull-up network. This power reduction is due to the peak current attenuation from the power-enable transistor current plot shown in Figure 11.9 from a HSPICE simulation.

When combining all of the memory banks, as shown in Figure 11.2, the power-up time and in particular the delay have a significant impact on the power consumed by the memory. For example, with a power-up time $t_P = 100$ μs and an ideal ramp up of the power-enable inputs $t_D = 0$, average power consumption is approximately 25.5 nW. However, as the power-enable delay increases linearly, the power consumption increases exponentially to reach 25.4 μW for $t_D = 40$ μs based on the simulations as indicated in Figure 11.3. t_D was varied between 26 and 40 μs to study the peak currents in more detail. The results are shown in Figure 11.10. For delays exceeding 30 μs, there are initially spikes

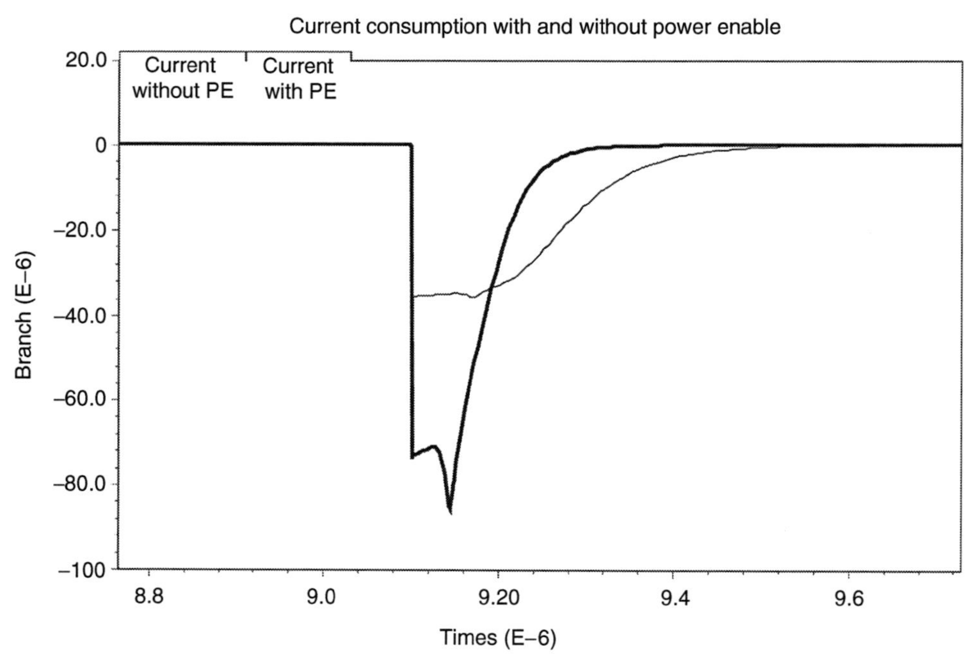

FIGURE 11.9
Peak current reduction from memory banking.

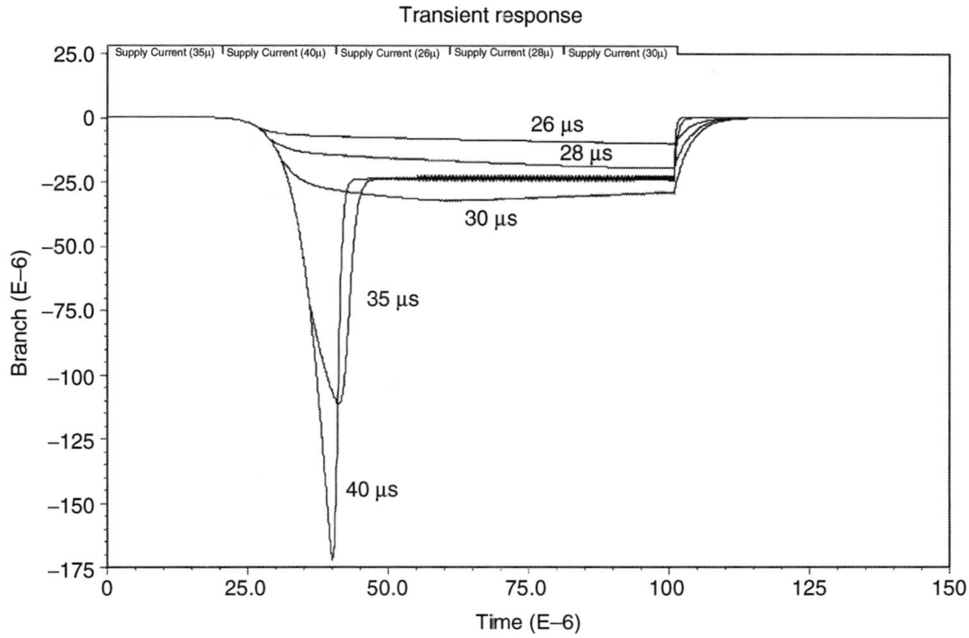

FIGURE 11.10
Supply current versus time for various delays of power-enable signals.

in supply current because the power-enable transistor is turned on causing the internal memory block node capacitances to draw current from the supply. With delays lower than 30 μs, the power-enable transistors do not turn on very much and draw a much lower current.

As noted earlier, the memory blocks were arranged into banks and a multiplexer was used to tie the banks together (see Figure 11.7). In our implementation, we chose 16 banks, each containing 22,200 byte memory blocks, thus the entire system contains 352 memory blocks. Compared with the nonbanked approach, the reduction in average power is significant. The banking approach saves approximately 43% decreasing the power for the memory blocks from 11.8 to 6.7 μW. The overall power savings from the architectural approach is 98.7%, as shown in Table 11.2.

11.3 Power Macromodeling for RFID Protocols

RFID protocols are typically designed without taking into account many of the impacts of their final implementation. For example, the design of a protocol can significantly impact

TABLE 11.2
Power Savings due to Architectural Improvements

Power consumption from naive implementation	1750 μW
Power consumption of banked memories	6.69 μW
Power consumption of address decoder	16.82 μW
Total optimized memory power consumption	23.51 μW
Power savings	98.7%

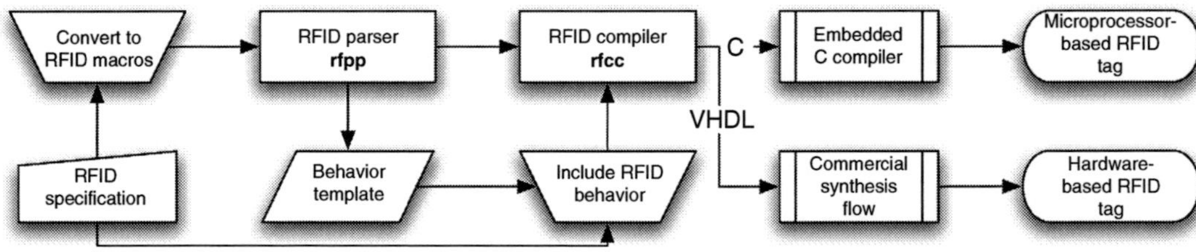

FIGURE 11.11
RFID high-level specification methodology and compilation flow.

the complexity of the protocol realization requiring additional area and cost in the final implementation. Additionally, this complexity can impact the power consumption. Even decisions about primitive opcode encoding can significantly impact power consumption while only minimally impacting area.

With existing design flows, to gain an accurate estimate of power consumed for a protocol implementation, the protocol must be designed, tested for correctness, implemented in hardware, and finally studied for power. This process can take months or years of engineering effort to complete. RFID companies typically do not have this type of man power to dedicate for this purpose.

In this section, we describe a power macromodeling technique that works in concert with the RFID design automation flow described in Chapter 3.

As shown in that chapter, the RFID compiler, as illustrated in Figure 11.11, allows the RFID system designer to design and implement new RFID protocols in a matter of hours. A team of design engineers without this tool would require months or longer. The team would require additional time and effort to examine the power and area impacts of their completed designs to optimize the tag.

However, the generation of the performance and area details of the design requires specialized ASIC synthesis tools such as Synopsys Design Compiler, which must be manually tuned to achieve good results. Performance and area may be estimated at this level but require additional time and computer-aided design effort such as placement, routing, and design rule checking with tools like Cadence SoC Encounter to get a more accurate result. Achieving power consumption statistics requires an additional level of effort by simulating the design and putting it through additional power estimation tools such as Synopsys Nanosim, or HSPICE.

In the following subsections, a tool is described based on a power macromodeling technique that calculates a power estimate at a much higher level during the design automation process of the RFID compiler. The tool generates a behavioral representation of the hardware that generates a custom simulator for the controller design generated by the RFID compiler. Through access to a preprofiled library of blocks in the target CMOS process, the power consumption can be estimated 100 times faster than the fastest ASIC power estimation flows. Thus, the user has near instantaneous feedback about the power consumption of their design without detailed knowledge of ASIC synthesis and power estimation flows.

11.3.1 Power Macromodeling Framework

Power macromodeling was originally proposed by Gupta and Najm as a fast power estimation technique that considers the impact of different input combinations on the circuit [5]. This technique has been shown to be far more accurate than static power estimation methods that do not consider design input values. Power macromodeling has

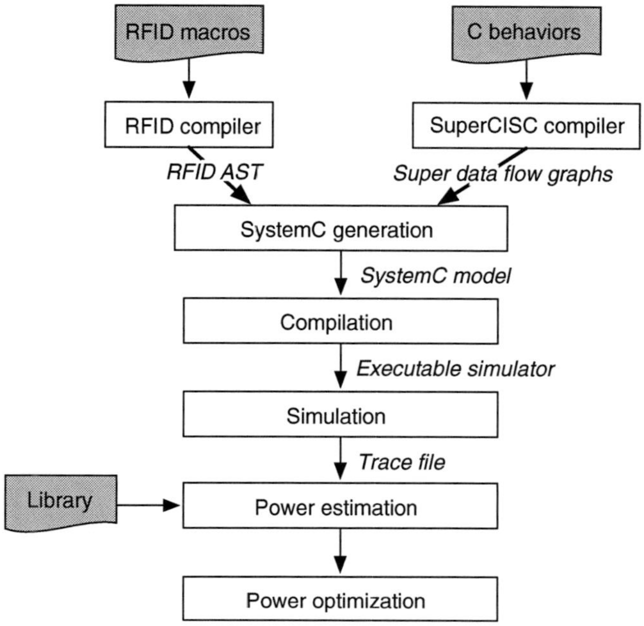

FIGURE 11.12
Power macromodeling flow. Before VHDL code generation in Figure 11.11, this flow generates a SystemC model to allow power analysis. The user changes the specification until an acceptable power result is achieved, and then Figure 11.11 resumes at VHDL generation.

been proposed for a variety of purposes including high-level synthesis of circuits for minimal power consumption [6,7]. Power macromodeling discretizes the components used to build up the circuit into functional blocks that can be implemented and analyzed for their power consumption based on different input stimuli. These results are then compiled into a look-up table of power consumption based on different characteristics of the input stimuli. During a behavioral simulation of the system, rather than computing the power consumption of the block based on the simulated inputs, a table look-up is performed to determine the power value at a significant savings in time and effort.

An overview of our power macromodeling flow used in our RFID compilation flow is shown in Figure 11.12. The SystemC generation occurs just before VHDL generation (Figure 11.11). The RFID compiler reads the RFID protocol description, including the RFID macros and their corresponding behavioral in C and automatically generates a system level simulator for the tag in the SystemC language.

SystemC is a hardware description language built using C++ libraries that include facilities for discrete event simulation, concurrency, fixed-width bit vectors, and block-based design, among others. As it is at heart a C++ program, a SystemC description of the hardware design is compiled into an executable custom simulator for that particular hardware design [8]. SystemC is particularly appropriate for power macromodeling as the power table and estimation behavior can be seamlessly integrated into the simulator through C++ code. Additionally, the final simulator is typically fast as the simulator is a compiled binary rather than an interpreted simulation of other hardware description languages.

After the SystemC simulator is generated, it is compiled into a binary using a software compiler. Probabilistic input test vectors are then automatically generated for use in simulation. The compilation flow executes the simulation and generates annotated trace files with information about all the functional units in the design. These functional units have been preprofiled for power based on input parameters described in more detail in Section 11.3.2. The power estimation of each RFID primitive and the overall tag design are calculated by combining the trace information with the profiles to determine each unit's power, which is aggregated. The resultant power estimates correspond to the actual activity of the tag behavior in hardware.

11.3.2 Library of Power Profiles

Three parameters of input signals can be used to accurately estimate power dissipated in digital circuits, as suggested by other authors [5,9,10]. They are: average input signal probability, p, average transition density, d, and spatial correlation, s. Transition density represents the frequency of bit changes between two or more values in sequence. Signal probability describes the number of 1s to appear within a value. Spatial correlation describes the likelihood for 1s and 0s to appear in groups within the value.

All the types of functional units used by the RFID compiler for hardware generation, such as adders, multiplexers, etc., have been power profiled with different values of p, d, and s. These have been used to construct a library of power profiles.* We used the Markov chain-based sequence generator described by Liu and Papaefthymiou [12], which converts the probabilistic p, d, and s values into an actual sequence of test vectors for use in simulation. Measurements were taken at 0.1 intervals ranging from 0.05 to 0.95 in each probabilistic dimension. The functional units are synthesized using 0.16 μm Oki cell-based ASIC technology. The synthesis was executed with Synopsys Design Compiler and the power was estimated using Synopsys PrimePower.

The power consumption for an adder profiled as described earlier is displayed in Figure 11.13. This chart plots power versus p, d, and s. Power is indicated as a grayscale between black and white where white represents the least power consumed by the device and black indicates the most power consumed by the device. From Figure 11.13, it can be seen that the most power is consumed by the adder when spatial correlation is low and transition density is high. This shows that using transition density, d, alone for dynamic power optimization can be insufficient, although it is often considered the only metric of interest.

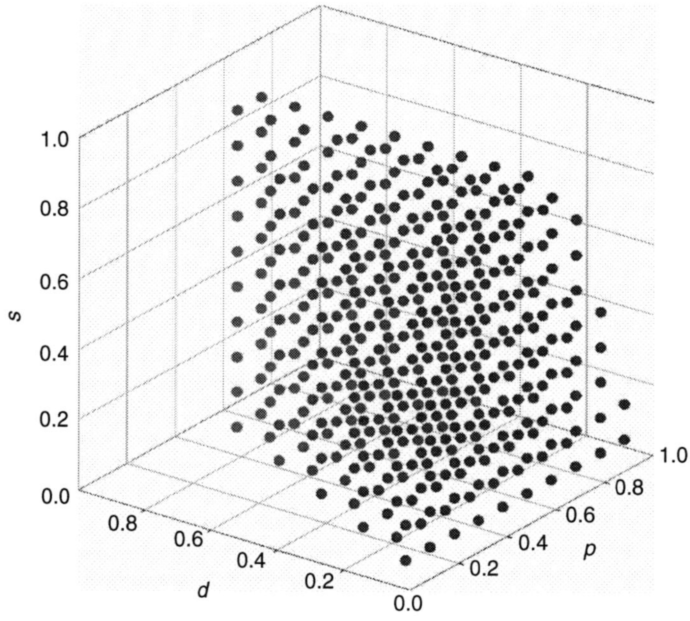

FIGURE 11.13
Four-dimensional plot of p, d, and s versus power for an adder synthesized as 0.16 μm OKI ASIC.

* The power profiles were originally published in Jones et al. [11].

11.3.3 SystemC Simulator Construction

The SystemC simulator construction is completed in two phases. In the first phase, the user-specified C behaviors that correspond to the different RFID primitives are converted into SystemC designs. The compiler translates each super data flow graph (SDFG) into a behavioral SystemC design.* This uses a SystemC abstract syntax tree (AST) for the intermediate representation, which contains data structures that behave according to the SystemC specification.

In the second phase, the compiler generates the simulator framework for the entire tag, which includes the unpacking, decode, and packing logic. The compiler uses the information in the input macros file along with the SystemC AST data structures to generate this. The SystemC hardware blocks generated in the first phase are instantiated in the design for the complete tag simulator. Trace instructions are added to the design, which save information about the functional units associated with changing signal values. The simulator is made by compiling the SystemC design.

11.3.4 Power Estimation

The power macromodeling flow automatically generates probabilistic input test vectors for use in the tag simulation. It uses the Markov chain-based sequence generator, discussed in Section 11.3.2, to generate test vectors with p, d, and s values that are evenly distributed in the three-dimensional space. Since the actual commands issued by the RFID reader may be unknown at the time of simulation, sequences with all possible statistics are applied to the tag simulation. However, while designing tags for a given RFID system, the user may be able to predetermine the expected workloads generated by the RFID reader. The user can use these input vectors instead of the probabilistic vectors for designing tags that are power optimized for the actual workloads.

When the simulation is executed, a trace file is generated with the program execution statistics. If the signal values at a functional unit change, a trace instruction records the unit's identification number and its signal values. During power estimation, each trace instruction is read and the module power is constructed behaviorally. The power estimation is illustrated using a simple example module, as shown in Figure 11.14.

The energy calculation for the module is based on the energy consumed by each of the functional units. In the example, this consists of an adder, an equivalence checker, a selector, and logical not. Consider the adder unit. Based on its inputs, the p, d, and s values

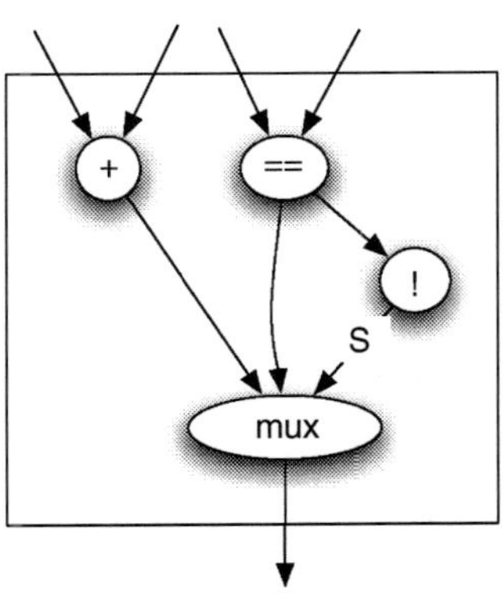

FIGURE 11.14
Example module. Energy of the module is based on the energy of each of the individual functional units addition, equivalence, multiplexer, and logical not.

* An SDFG is an extension of the more common control and data flow graph (CDFG), where basic blocks have been merged by converting control edges into data edges using hardware predication [13,14].

TABLE 11.3

Power Macromodeling versus Traditional Method

Primitives	Collection	Query
Power Macromodeling		
Power (W)	3.42E − 06	1.68E − 04
Time (s)	0.12	0.41
Traditional Method		
Power (W)	2.98E − 06	1.14E − 04
Time (s)	32.12	40.71
Times Speedup	267	99

are computed using the technique described by Liu and Papaefthymiou [12]. For the inputs (0,1), the computed *p*, *d*, and *s* values are 0.017, 0.031, and 0.061. The power consumed by the adder for these values is obtained by looking up the library, and is 1.14×10^{-6} W. The corresponding energy is calculated as the power multiplied by the time spent in executing the add operation. Similarly, the power consumed by the equivalence operator, logical not operator, and the multiplexer units are 1.09×10^{-6}, 1.22×10^{-6}, and 1.14×10^{-6} W, respectively. The energy of the module is constructed by aggregating the individual energies. Energy is averaged over the simulation time to calculate power.

11.3.5 Results

We compared our power macromodeling approach with generating a design and power profiling it with existing tools. The flow we compared with was design synthesis using Synopsys Design Compiler and power profiling by simulating the design in Mentor Graphics ModelSim to generate switching information and power estimation using Synopsys PrimePower.

Table 11.3 shows the run times and power estimated with the power macromodeling technique and the traditional power estimation method for the collection command from ISO 18000 Part 7 and the query command from ISO 18000 Part 6C standards. These commands have similar functionality but use a significantly different protocol. As can be seen from the table, the power consumption is estimated to be very similar between both techniques, and shows the significant advantage of collection over query for power. However, the calculation time was improved by 267 times and 99 times, respectively. For full tag designs, this technique provides an answer in seconds, whereas the other technique can take minutes or hours.

11.4 Combining Passive and Active Technologies

There are many applications that require the capability of an active RFID tag but the application is such that battery replacement is at least inconvenient, if not impossible. To solve this problem we have developed an active tag architecture, as shown in Figure 11.15.

The goal is a tag system with essentially infinite battery shelf life and an active battery life essentially equal to that of the tag. The overall architecture for such a tag is given in Figure 11.15. This architecture is referred to as a passive active radio frequency identification tag (PART). The system contains passive energy receiver called the burst switch connected to the active transceiver. When RF energy is received on the passive receiver it

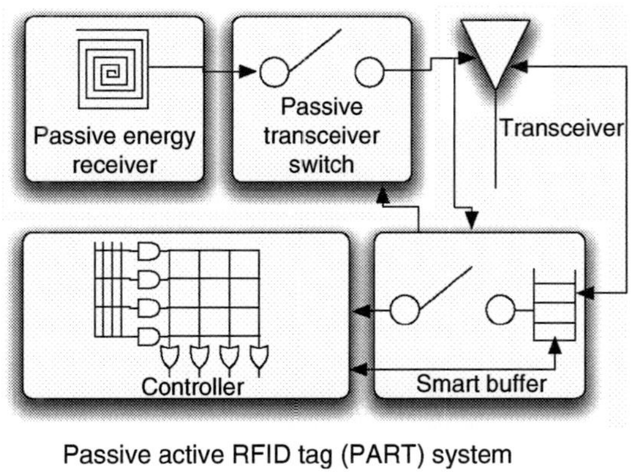

FIGURE 11.15
Overview of the ultra low-power active RFID tag.

activates the active transceiver. This is described in more detail in Section 11.4.1. Once the active transceiver is activated, the smart buffer keeps the main processing controller asleep until it is needed for processing an incoming packet. The smart buffer is described in Section 11.4.2. The controller may be a microprocessor, an ASIC, or even an FPGA, depending on the particular application.

The operation of this architecture is as follows. When the tag is not interrogated the battery power to the transceiver and smart buffer will be disconnected and the controller will be in the sleep mode if processor or SRAM FPGA based or disconnected from power if ASIC or Flash FPGA based. When a group of tags is to be interrogated the passive transceiver burst switch supplies power to the transceiver and smart buffer. If the particular tag is interrogated, the smart buffer awakens or supplies power to the controller and the controller interprets the interrogation and issues the appropriate reply. The controller then returns to the sleep or off mode and power is disconnected from the smart buffer and transceiver.

11.4.1 Burst Switch

There exists a class of devices for converting RF energy into a direct current (DC) potential using a circuit such as that shown in Figure 11.16. In other areas, the circuit has been used to provide DC power to operate remote autonomous devices that have no on-board power supply. In the case of the PART, a battery is used to power the device. The particular application used as an illustration is when the PART is to be attached to the asset for a long

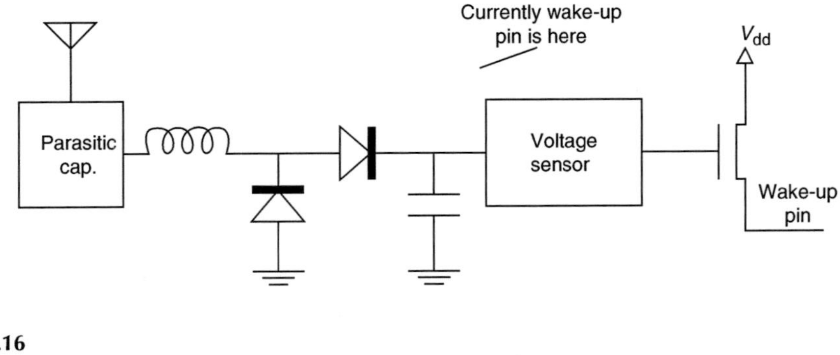

FIGURE 11.16
Simple generic burst switch.

period of time but requires no maintenance. Thus, the battery life must be extended to a maximum—likely the life of the asset being tagged.

Many commercial CMOS (silicon) devices are equipped with a form of sleep circuitry with current draw at a minimum during sleep. An external input signal is used to wake up the device. As an example, the burst switch circuit of Figure 11.16 is used to wake up a microcontroller.

The generic burst switch implementation has been shown to wake up the microcontroller as designed for a prototype of the PART [15]. However, the use of the switch requires considerably more design and analysis to avoid false wake-up states and insure functionality under adverse conditions. Some of these conditions may be analyzed using RF test equipment such as a real time spectrum analyzer (RTSA) from Tektronix. This equipment is in particular very useful in measuring low signal levels such as backscatter from traditional passive ISO 18000 Part 6C tags, as shown in Figure 11.17.

11.4.1.1 Characterization of the Receiver Enable Input Pin

It is important to match the burst switch to the Receiver Enable Circuitry, as seen by the output of the charge pump. The ISO 18000, Part 7 Standard for active RFID allows for a 30 kHz wake-up tone to wake up the tags. In the spirit of this frequency range, the characterization of the Receiver Enable Input (REI) Pin was completed at 50 and 25 kHz. The actual circuitry inside the chip is nonlinear. However, an approximate identification of the circuit was an inductor in the range of 1–10 mH. Other chips can be expected to have variations of this or in fact act as capacitors.

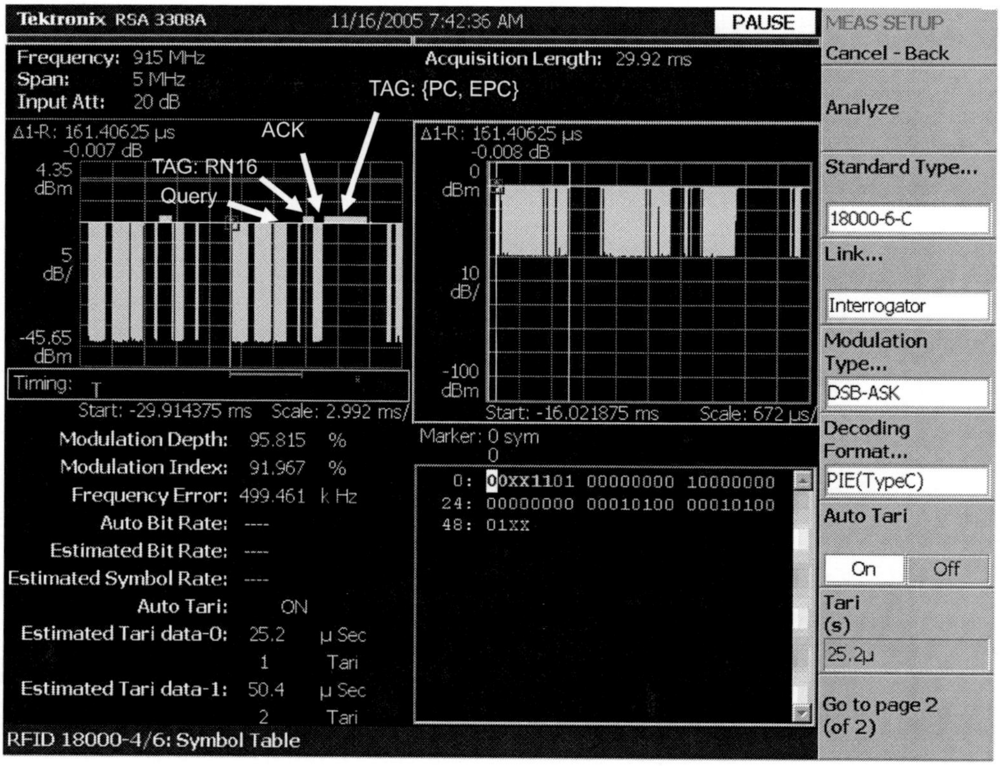

FIGURE 11.17
ISO 18000 Part 6C interrogator/tag communications exchange.

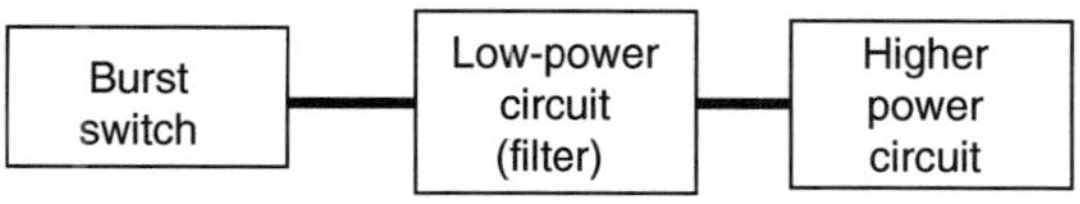

FIGURE 11.18
Low-power filter to avoid false wake-up from RF noise.

The important point here is that it is necessary to first characterize the REI pin as a part of the design procedure. It is believed that the tuning at the input is necessary to essentially provide a resonant tank to the extent possible so as not to deliver any energy to any circuitry on the chip itself, that is, the burst switch is to function as a signaling device, and any energy harvesting that is likely to take place would decrease the response speed and possibly degrade the bandwidth in some manner.

11.4.1.2 Extended Low-Power Logic

One of the difficulties with the simple RF wake-up circuit is that spurious RF energy (noise) could potentially waken the sleeping device. Thus, it may be necessary to interface a low-power or passive circuit (essentially a filter) between the RF switch and the higher power consuming receiver, as shown in Figure 11.17.

The low-power circuit (filter) of Figure 11.18 could be any low-power device that could either operate off of harvested energy or be turned on for a short period of time, increment counters, and go back to sleep. In effect, this device acts like a receiver. A watch dog timer may be used to reset the device after extended noisy periods or after long intervals of inactivity.

An alternative embodiment to the switch is a parallel implementation that will include multiple switches of differing frequencies where logic can be performed with the switches. All of the logical conditions of AND, OR, etc., are candidates with the implementation being the key problem. One such implementation is two switches in series where the combined voltages at two frequencies are required to turn on the device. This particular embodiment is especially interesting as it is totally passive, thus not reducing the shelf life of the battery.

11.4.1.3 Results

To verify the burst switch capability, the burst switch was prototyped on a PC Board and tested using a Samsys RFID reader to generate pulses of RF energy to be recognized by the burst switch [16]. The system was tested using a RTSA from Tektronix, whose output is shown in Figure 11.19. This test was taken with the reader at a distance of 1 m from the tag.

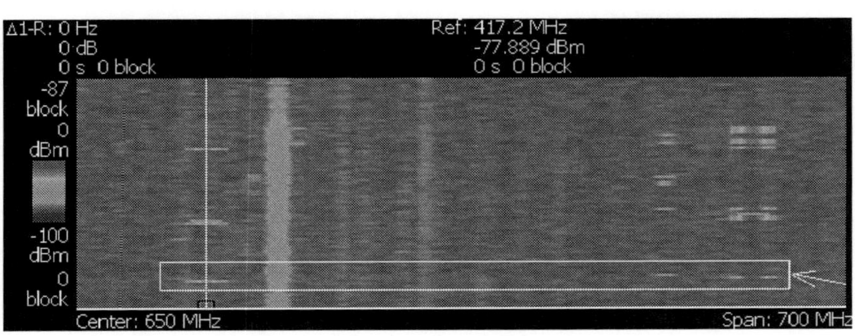

FIGURE 11.19
Burst switch verification with a real-time spectrum analyzer.

In this plot, time is on the *Y*-axis and frequency is on the *X*-axis. The RTSA has a color output and power is indicated by a color gradient. The burst switch activation can be seen in a column near the right at a frequency of approximately 933 MHz. The responses are shown on the left in a column with slightly less intensity at approximately 433 MHz. The white box indicates a burst switch activation that triggered a tag response. The lighter columns are radio interference that occur at low intensity within the lab and are not related to the experiment.

The power consumption savings for the burst switch is highly dependent on the scenario when the tag is used. When the burst switch is used in a hardware-based receiver or software-base receiver [17], the energy consumption of the device is due to the time when the active transceiver is powered. For a *nano Tag*, or a tag whose sole function is when queried to respond with its unique ID, the number of tag accesses determines the lifetime of the battery. For a 100 mA h battery, 1 access per day would last well over 2000 years, 10 accesses per day would last 281 years, and 100 accesses per day would last 28 years [17]. In actuality, the battery would breakdown well before the energy of the battery would be exhausted.

11.4.1.4 Scenarios

The potential impact of the burst switch on energy dissipation in a network was evaluated on a ZigBee sensor network [18]. ZigBee networks are low-power two-way wireless networks intended for embedded consumer electronics [19]. Although this is not an actual RFID communication network, many of the transactions are similar and it can provide insights into how an RFID network can benefit from the burst switch.

The particular ZigBee configuration tested consisted of temperature sensors connected to routers, which in turn were connected to the coordinator. The temperature sensors in the network periodically take a temperature reading and then broadcast that reading to the router controlling them. The router averages the readings and sends the average temperature to the coordinator. The coordinator processes and communicates the temperature information to the outside world. The router can also request the temperature at any time by transmitting the temperature request command.

In this scenario, only the temperature sensors are of interest for evaluating energy consumption. Two cases were considered. Both cases assume that a protocol is setup so that each command or command type has a different length. We used the software receiver to distinguish between the commands [17].

In the first case, a normal active RF receiver is a component of the temperature sensor. In the second case, the receiver consists of a burst switch acting as a front end and a microprocessor that analyzes the signals detected by the burst switch. Since the two commands differ in length, the burst switch can be easily substituted in place of the active receiver.

The temperature sensors are designed such that the only command they receive is the temperature request command. When a burst of RF energy long enough to be the temperature request command is observed the temperature sensors can take appropriate action. In this case, they take a temperature reading and transmit that reading to the router. With an appropriate anticollision protocol, the transmission of multiple temperature readings to the router is regulated to prevent collisions. The sensors were modeled via a Markov process, as shown in Figure 11.20.

This Markov model is identical in structure for both cases, with the energy consumption portion of the model being different. With the models as a basis, the energy consumed by each network for a period of 1 day was determined to be as shown, as in Table 11.4. Using the burst switch receiver in place of the active receiver, the energy consumed by the burst

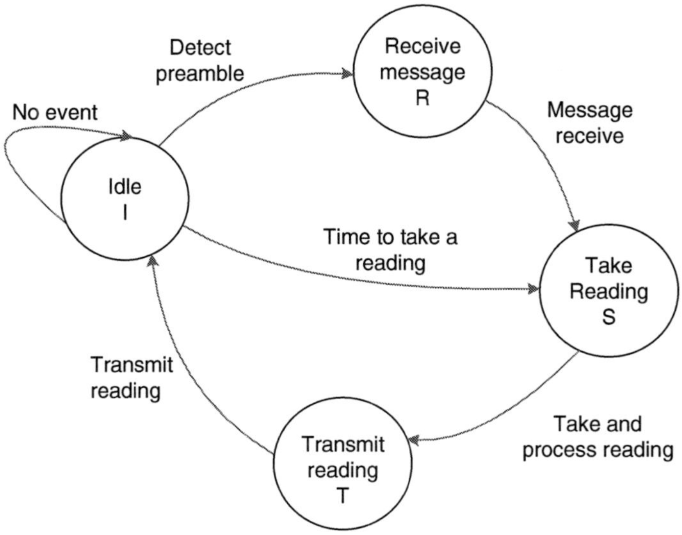

FIGURE 11.20
Markov process of the temperature sensor model.

switch-based sensor was approximately 56% lower than that of the sensor with the RF receiver.

11.4.2 Smart Buffer

Broadcasting is the only communication method between RFID readers and tags. However, RFID technology uses not only broadcast RFID commands but also point-to-point (P2P) commands specified in ISO 18000 Part 7 and ANSI 256 active RFID standards. For broadcast RFID commands such as `Collection` or `Collection with Data` in ISO 18000 Part 7, an RFID reader broadcasts this type of commands to all RFID tags and they wake up and respond associated information back to the reader. For P2P commands such as `Sleep` or `User ID` access in ISO 18000-7, the RFID reader broadcasts this type of command to all RFID tags even though it attempts to access only one of the tags. All tags wake up and process the incoming command. However, only one of the tags needs to respond with the associated information and the rest of the tags ignore the incoming command. For those tags ignoring the incoming command, on-tag controllers dissipate energy for receiving and dropping commands.

To reduce the power consumption of the embedded processor-based tag, it is necessary to decrease the time when the processor is active. Therefore, an RF transceiver coprocessor was created to manage buffering messages to and from the transceiver and activating the processor for responding to RFID primitives when necessary. Thus, the RF transceiver coprocessor or smart buffer (see Figure 11.15) allows the embedded processor to sleep or stay in a low-power idle mode while any nonrelevant packet arriving at the RFID tag is

TABLE 11.4
Energy Consumed by a Sensor Network with and without a Burst Switch

Network model	Energy Consumed (kJ)
Without burst switch	150.9
With burst switch	65.8

ignored, and when valid packets arrive, the smart buffer must also wake up the processor to respond appropriately [20,21].

The amount of power savings allowed by the smart buffer is highly dependent on the scenario. First, the smart buffer must consume less power than the active tag controller, which is supported by the data shown in Section 11.4.2.10. Second, the amount of power savings depends on how many successful accesses the tag receives on average, how long the processor is active, etc. We examine several scenarios and the smart buffer impact in Section 11.4.2.11. As seen in Section 11.4.2.10, the power savings from the smart buffer can be several orders of magnitude for the microprocessor-based tag.

11.4.2.1 Algorithm for RF Transceiver Coprocessor

The conceptual flow in Figure 11.21 shows the mechanism for the RF transceiver coprocessor, that is, the smart buffer. In the first four states on the left of the figure, the smart buffer verifies the message preamble and buffers the incoming packet. The smart buffer then checks to see if the packet was intended for this particular tag. The processor is not used to make the check, it is done in hardware while the processor remains idle or asleep. As a result, only the smart buffer consumes power. If the incoming packet is invalid, the smart buffer will ignore it and go back to the sleep state.

If the incoming packet is identified as an intended packet for the Tag, by matching the tag ID or group ID from the packet, the smart buffer will wake up the processor

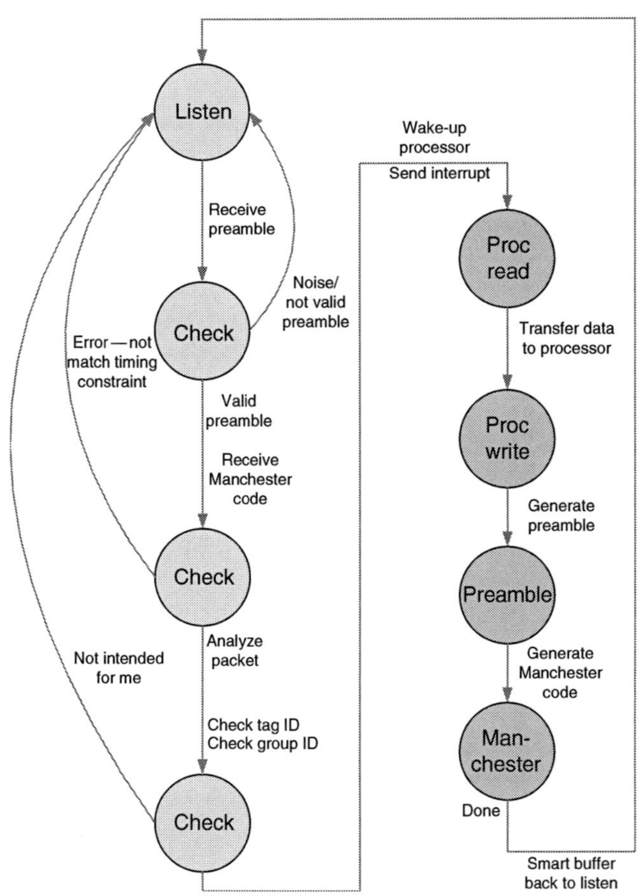

FIGURE 11.21
Conceptual flow of the smart buffer.

to process the packet. The processor reads data from the smart buffer and responds correspondingly.

After the processor writes response data back to the smart buffer, the smart buffer generates a bit stream of data consisting of a preamble signal and the response data with Manchester coding. It is at this point that the processor returns to the low-power mode. On completion of sending the packet response, the smart buffer returns to listening for the next preamble.

Figure 11.22 depicts the top-level diagram of the smart buffer. The smart buffer has four I/O pins to the RF front end circuit. The blocks described in Figure 11.22 are enumerated as follows:

Preamble Removal Unit: Detects the incoming preamble signal and differentiates between signals intended for tags and readers.

Manchester Decoder: Converts Manchester code into binary values.

Preamble Generator: Generates the Manchester code for the tag response.

Packet Analysis Unit: Detects ID flags to determine whether to wake up the processor.

Interrupt Process Unit: Generates an interrupt to the processor.

Processor Control Command Unit: Communicates data to and from the processor.

Air Interface Unit: Communicates data to and from the air interface.

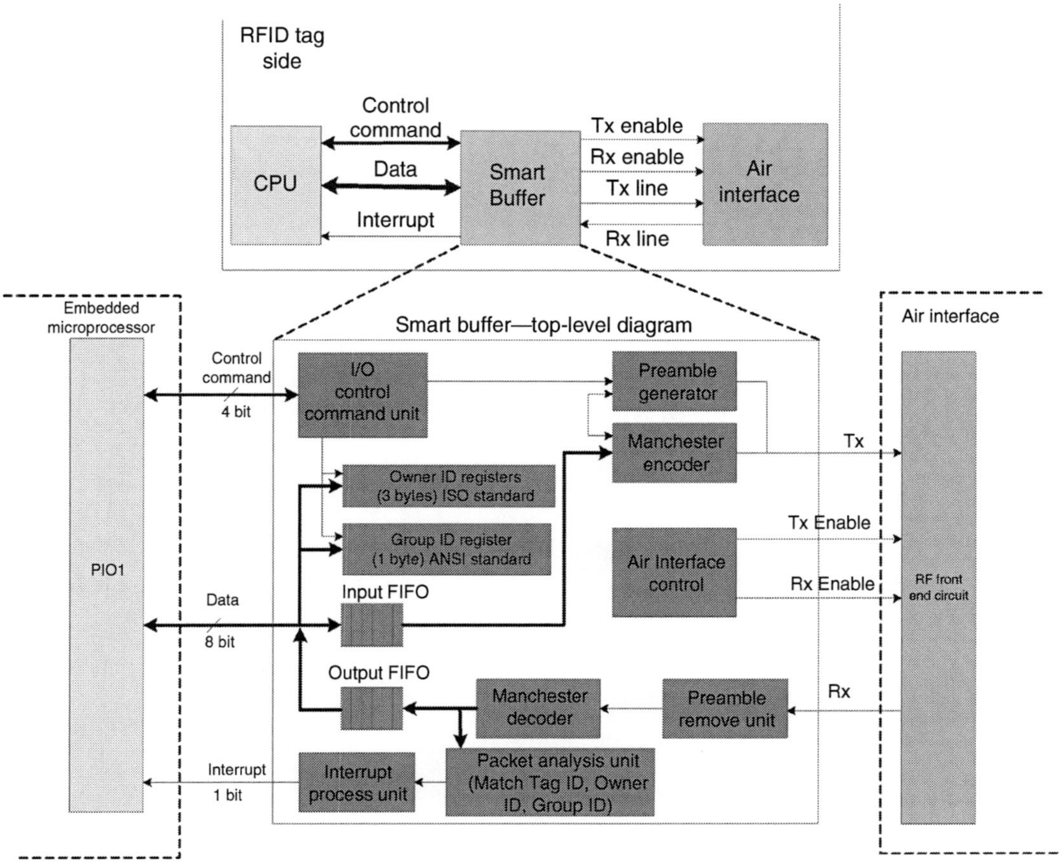

FIGURE 11.22
Top-level block diagram for the smart buffer architecture.

11.4.2.2 Preamble Removal Unit

The preamble removal unit detects a valid incoming preamble signal. It receives bit stream data from the RF front end circuit. These digital bit stream data are converted from an analog signal the RF circuitry receives from the antenna. However, because the input signal is analog, a signal may be due to noise rather than a preamble. Therefore, the smart buffer has the ability to tolerate noise and only recognize the valid preamble signal. The specification of the preamble signal is defined in the ANSI and ISO standards [22,23].

The preamble signal begins with a series of pulses with 30 μs high followed by 30 μs low. Every preamble signal, regardless of whether it has originated from a tag or a reader, has 20 of these regular pulses. These 20 regular pulses are followed by the final sync pulse, which determines whether a tag or reader originated the preamble. If an RFID packet comes from a tag, the final sync pulse is 42 μs high and 54 μs low. If an RFID packet comes from a reader, the final sync pulse is 54 μs high and 54 μs low. The smart buffer can ignore RFID packets that come from other tags by checking the length of the final sync pulse. It will only focus on RFID packets from interrogators as long as the sync pulse is 54/54 pattern. Other packets are not buffered.

The implementation of the preamble removal unit uses four times oversampling within each 30 μs period for each pulse. The design uses counters to count the sampling period. The difference in the sync pulse can be detected by counting for how many samples the final signal is high, 5 for tags and 7 for readers.

11.4.2.3 Manchester Decoder

The Manchester decoder translates the Manchester encoded data immediately following a valid preamble. It is the block for filtering and buffering incoming RFID packets. The decoder extracts a bit stream of nonreturn to zero (NRZ) data from the encoded data.

The Manchester code combines the concept of clock with synchronous data into a single serial data stream, as shown in Figure 11.23. In order to enforce synchronization, Manchester code contains a transition in the middle of each Manchester bit. The Manchester bit

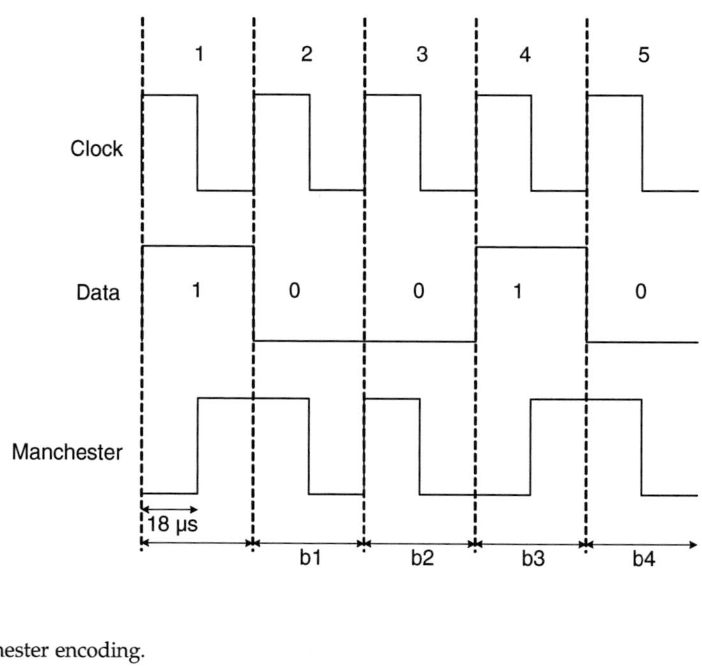

FIGURE 11.23
Example of Manchester encoding.

represents zero (0) NRZ data if this transition is from high to low. Similarly, the Manchester bit represents one (1) NRZ data if the transition is from low to high. By representing data with a guaranteed transition for each bit, slight discrepancies of timing can be tolerated without disrupting the communicated data. The timing specification of the Manchester code is defined in the ISO and ANSI standards.

The Manchester decoder block converts eight serial decoded data bits into a parallel 8-bit datum (e.g., byte). In addition, the Manchester code contains a ninth bit for synchronization, which is always 0. This is called a stop-bit, and is removed during decoding. Each byte is stored in an output FIFO shown in Figure 11.22. When the decoder detects the final bit of an RFID packet, it stops storing data into the FIFO and asserts the end of packet (EOP) signal to the interrupt process unit. Based on the analysis result, the interrupt process unit determines if it is necessary to wake up the processor.

11.4.2.4 Packet Analysis Unit

Based on the header information obtained from the RFID packet, the packet analysis unit attempts to determine the specific characteristics of this particular packet needed to decide whether to wake up the processor for response generation. These elements include (1) the primitive operation code (or opcode), (2) the tag id for which the packet is intended, and (3) some other distinguishing information or id from the packet.

Both the ISO and ANSI standards have their own algorithms for tags to identify RFID packets. Therefore, the packet analysis unit has the ability to switch between these two algorithms seamlessly. For example, if the tag travels to Europe or Asia where the RFID system follows the ISO standard, the smart buffer can perceive that it is being accessed by ISO primitives. However, if the tag returns to the United States, it would switch over to recognizing ANSI commands.

For broadcast commands, both standards allow partitioning of the tags into different bins with a unique identifier where a logical bin can contain an unlimited number of tags. This is accomplished by assigning that unique id to each tag contained within the bin. In ISO this is called the Owner ID and for ANSI the Group ID. For P2P commands, a Tag ID is used to distinguish the destination tag.

The packet analysis algorithms are summarized in Tables 11.5 and 11.6. In both standards, commands are segregated into broadcast commands (B) and P2P commands. If the tag receives an ISO broadcast command, it always responds if no Owner ID is set within the tag. If an Owner ID is set, the tag only responds if an Owner ID is included in the command and it matches the stored Owner ID. For ANSI, broadcast commands are subdivided between 3 opcodes, 30, 16, and 35. If the opcode is either 16 or 35, the tag always responds. If the opcode is 30, the tag only responds if the internal Group ID

TABLE 11.5

Packet Analysis Algorithm for the ISO Standard

Input			Output
B/P2P	**Owner ID field**	**Owner ID in tag**	**Process Command**
B	No	No	Yes
B	No	Yes	No
B	Yes	No	Yes
B	Yes	Yes	If Owner ID matches
P2P	No	No	If Tag ID matches
P2P	No	Yes	If Tag ID matches
P2P	Yes	No	If Tag ID matches
P2P	Yes	Yes	If Tag ID and Owner ID match

TABLE 11.6

Packet Analysis Algorithm for the ANSI Standard

Input	Output
Opcode	**Process Command**
30	If Group ID match[a]
16	Yes
35	Yes
P2P	If Tag ID match

[a] Group ID match always occurs if the currently stored group ID within the tag is zero.

matches the primitive Group ID. ANSI specifies that a stored Group ID of zero (0) always results in a match. For P2P commands, both standards require a Tag ID match. However, ISO requires that the Owner ID must match as well as the Tag ID for P2P commands with an Owner ID present in both the tag and the command.

Once the packet analysis unit verifies that the packet requires a response, it sends a signal to the interrupt process unit to process the packet stored in the FIFO. Because the packet analysis occurs in parallel with the packet buffering, a signal can be sent to the interrupt unit before the entire packet is buffered. If the packet does not require processing it is dropped from the FIFO. This prevents the processor from being powered up unless it is needed to process the packet.

11.4.2.5 Interrupt Process Unit

The interrupt process unit will wake up the processor only when a whole RFID packet has been stored into the output FIFO and the analysis result forwarded by the packet analysis unit is positive. Figure 11.24 shows the state diagram for the interrupt procedure. The interrupt process unit will go back to the idle state after sending an interrupt signal to the processor.

11.4.2.6 Command Control Unit

The command control unit (Figure 11.25) fetches control commands from the processor, such as read and write data to FIFO, update Tag ID or Group/Owner ID, starts the transmission procedure, and so forth. For processor compatibility, it was desirable to minimize the number of lines between the processor and the smart buffer. Thus, four (4) parallel lines are used to communicate the processor to buffer command control.

The processor to smart buffer commands are illustrated in Figure 11.25. After waking up the processor, the smart buffer listens for the processor to initiate commands for data communication. As shown in Figure 11.25, the control unit decodes 4 bit processor commands into five basic operations: transmit, update, push, pull, and null. The double circle represents a potentially multicycle operation.

Based on different control commands, the unit determines the direction of the bidirectional smart buffer, processor interface. In Figure 11.25, `dir = 0` represents that the direction of I/O is from processor to smart buffer and `dir = 1` is the reverse. Because the smart buffer and processor are operated in two different clock domains, a handshaking communication approach is required to push/pull data to/from the FIFOs. Therefore, it is

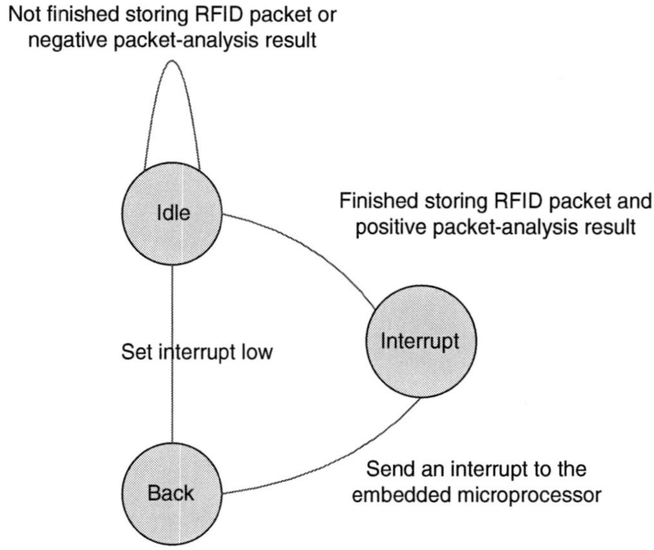

FIGURE 11.24
Interrupt finite state machine.

necessary to dedicate more than one cycle to transmit a single byte of data between the processor and FIFO.

Once the processor has generated a response and completed pushing the response data into the FIFO, it sends the `transmit` command. This signals the smart buffer to generate

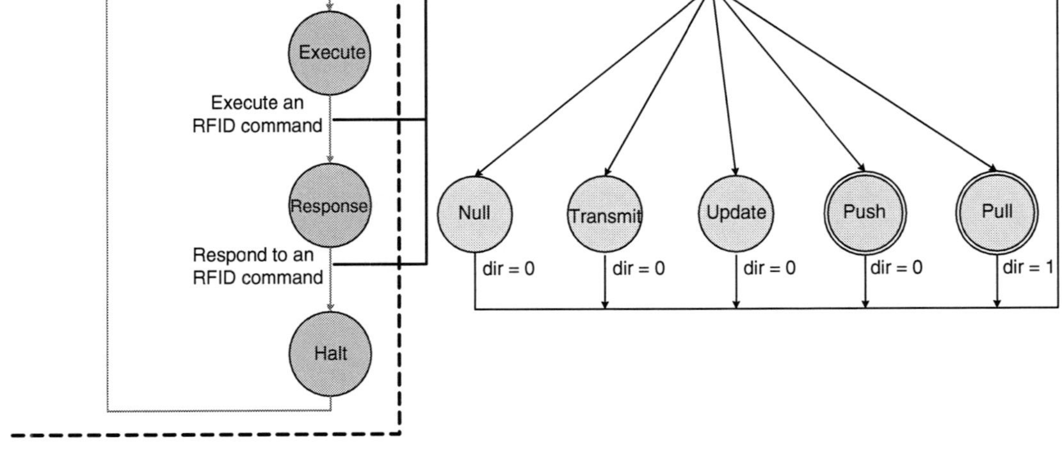

FIGURE 11.25
Fetch and operation diagram for the command control unit.

the preamble signal, convert data in the FIFO to a serial data stream, and encode the bit stream data in Manchester coding.

11.4.2.7 Preamble Generator

According to the ISO and ANSI standards, the preamble generator generates a preamble signal with 20 pulses of 60 μs, 30 μs high and 30 μs low, followed by the final sync pulse. Due to this RFID packet generated from a tag, the final sync pulse is 42 μs high and 54 μs low. The preamble generator uses several counters to trace the period of time for each pulse. These counters are running with the smart buffer system clock at 33 MHz. Therefore, the preamble signal will be skewed less than 1 μs requiring the reader to tolerate an error of +/− 3.3%.

In order to help the receiver filter out an ambient noise in the air, a mark state, or a stable logic low signal for 120 μs is generated and transmitted just before the preamble signal. While the preamble generator creates the mark state and preamble signal, the air interface unit forces the `rx_enable` signal low and raises the `tx_enable` signal.

When the final sync pulse is transmitted, the preamble generator informs the Manchester encoder to begin to output its serial-encoded data immediately. Any significant gap between data and the preamble signal may cause an error, which may not be tolerated in the system.

11.4.2.8 Manchester Encoder

The Manchester encoder starts encoding the data in the input FIFO after it is notified by the preamble generator. This notification occurs during the transmission of the final sync pulse to give the encoder enough time to have the first byte of data ready. First, the Manchester encoder converts the next available byte in the FIFO to eight (8) single serial bits. Those eight (8) bits are individually stored in single-bit shift registers. In addition, a stop bit needs to be appended for each byte of data in the shift registers.

The Manchester encoded data is the output of an NRZ data XOR coding clock shown in Figure 11.23. The NRZ data is synchronized with the coding clock. According to the specification of ISO and ANSI standards, the period of coding clock is 36 μs; 18 μs high and 18 μs low. This clock signal is generated in the Manchester encoder block. The Manchester encoder combines the coding clock and serial data, as shown in Figure 11.26.

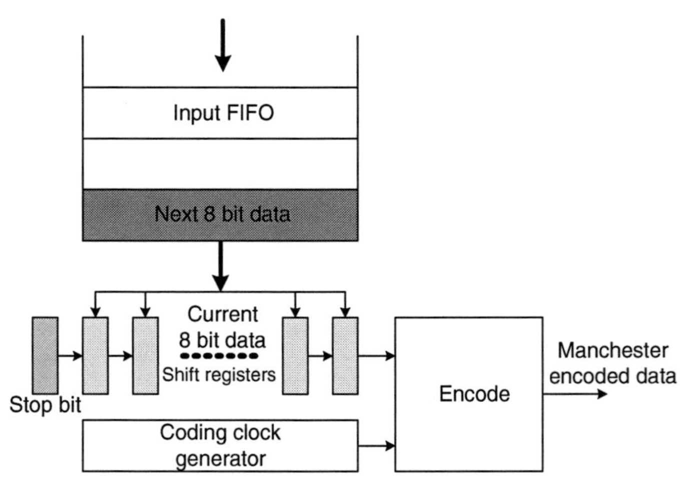

FIGURE 11.26
Top-level diagram for the Manchester encoder.

11.4.2.9 Air Interface Unit

The air interface unit has two output control signals, `tx_enable` and `rx_enable`. Since the smart buffer defaults to listening for incoming signals from the air, the `rx_enable` signal is set high and `tx_enable` signal is set low at all times, except when the smart buffer is ready to transmit an RFID response packet back to readers.

11.4.2.10 Results

The smart buffer was prototyped on the Spartan 3 FPGA as well as studied for ASIC implementation using 0.16 μm OKI standard cells. Although it was possible to implement FIFO blocks on the Spartan 3 FPGAs using Xilinx IP blocks, ASIC versions of these FIFOs were not available. In order to study the impact of having the smart buffer in an ASIC versus the Spartan 3 FPGA, three components were separated from the larger design and synthesized and power profiled independently. These blocks include the preamble detection, a 30 kHz wake-up signal detection, and the Manchester decoder.

The results for power estimation of the ASIC and Spartan 3-based smart buffer components are displayed in Table 11.7. Both power analyses are based on post synthesis simulation with the exact same stimuli. The FPGA power results were computed using Xilinx Xpower, and the ASIC power results were calculated using Synopsys PrimePower. Based on the results from Table 11.7, the ASIC version of the implementation uses orders of magnitude less dynamic power for all three components. Interestingly, the quiescent power alone is nearly three orders magnitude greater than the dynamic power of the ASIC.

11.4.2.11 Scenarios

In order to analyze the power dissipation of each tag in detail, two different Markov process models of an RFID transaction are built: the first shown in Figure 11.27 for a tag with microprocessor only, which is abbreviated as the MP-μP model, and the second shown in Figure 11.28 for a tag with a microprocessor and a smart buffer, which is abbreviated as the MP-μP-SB model. Five states are used in both Markov process models: S_1, sleep; S_2, listen for preamble; S_3, receive command; S_4, process command; and S_5, transmit response. By traversing those five states, both models represent the control flow of an RFID tag for a communication transaction.

During a scenario, when a tag is not participating in a transaction, the tag remains in the sleep state, S_1. After receiving a wake-up tone, it proceeds to state S_2 and listens for the preamble signal. The transition from S_2 to S_3 occurs when a tag receives a valid preamble and is ready to receive the incoming command. After receiving the command, the tag decides whether to continue processing the command (e.g., move to state S_4) or return to the listen for the preamble state (move to state S_2). If the incoming command is intended

TABLE 11.7

ASIC versus FPGA Power Consumption for Portions of the Smart Buffer

Component	Spartan 3 0.16 μm	ASIC
Wake-up signal	0.01 mW	0.001 mW
Preamble detection	0.87 mW	0.005 mW
Manchester decoder	0.95 mW	0.284 mW
Quiescent power	92 mW	0 mW
Total	93.83 mW	0.29 mW

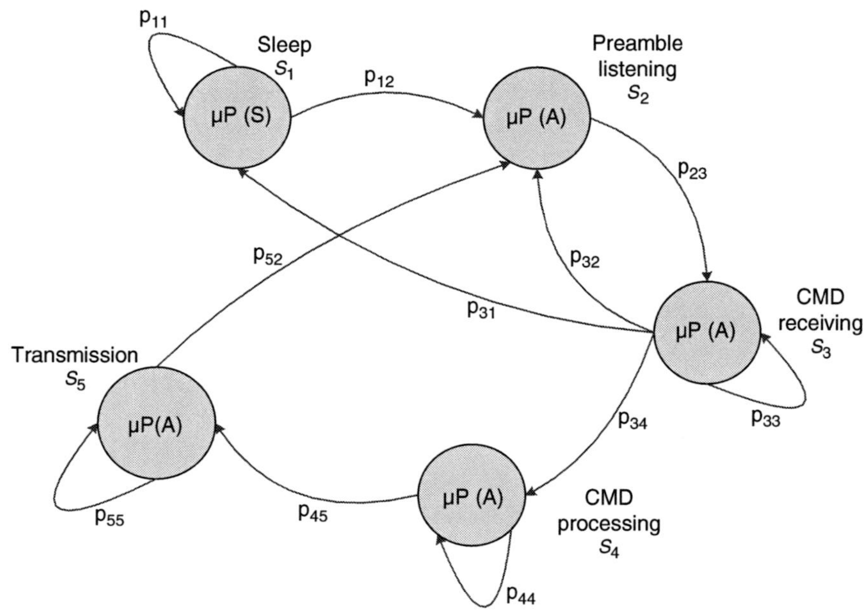

FIGURE 11.27
Markov process model for an RFID tag with an on-tag microprocessor (MP-μP model).

for the tag, it will process the command in state S_4 and transmit the associated information back to the reader in state S_5. Otherwise, the tag will drop the command and move from S_3 to S_2. If the command received is to put the tag to sleep it transitions to the sleep state, S_1.

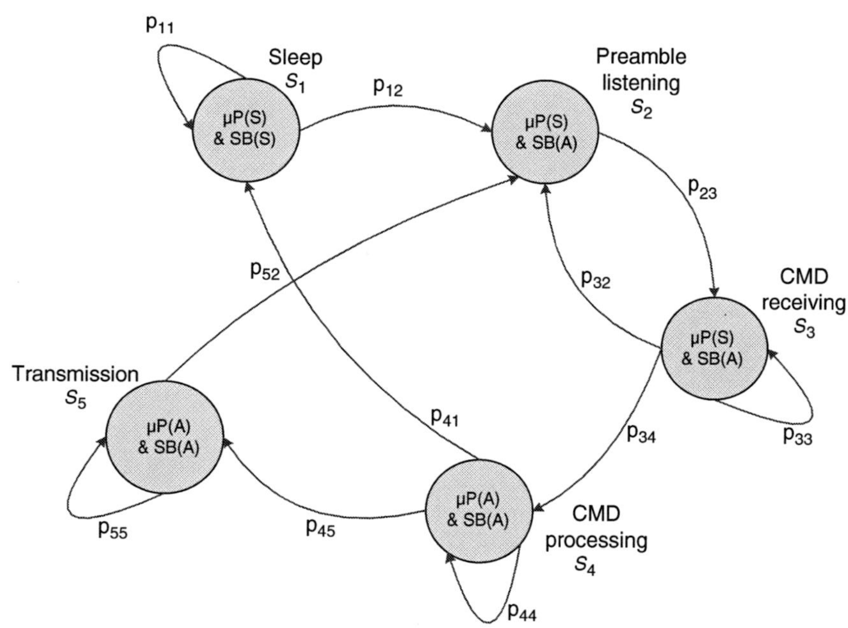

FIGURE 11.28
Markov-process model for an RFID tag with a microprocessor and a smart buffer (MP-μP-SB model).

The only difference between the transitions of the two scenarios is that for the MP-μP scenario the transition to S_1 occurs in S_3 and for MP-μP-SB this transition occurs in S_4. This is because the smart buffer does not currently recognize the sleep command.

The labels μP(A) and μP(S) are used to represent that the microprocessor is in the Active mode and Standby mode, respectively. Similarly the labels SB(A) and SB(S) in the MP-μP-SB model represent that the smart buffer is in the Active mode and Standby mode, respectively.

In the MP-μP model, the microprocessor is in standby in the sleep state, and active in all other states. For many scenarios the on-tag microprocessor will dissipate energy for receiving incoming commands that are not intended for the tag and are eventually dropped. For example, an RFID reader sends N P2P commands equally distributed to N RFID tags in a reachable area. For each tag, the MP-μP model in Figure 11.27 indicates that a microprocessor will stay in a loop between states S_3 and S_2 for $N-1$ communications. Consequently, the microprocessor dissipates energy to process $N-1$ extraneous communications.

In contrast, in the MP-μP-SB model the smart buffer alone becomes Active (SB(A)) while listening for a preamble signal and receiving a command (states S_2 and S_3) while the microprocessor still remains in the Standby mode (μP(S)). Thus, if the incoming command is not destined for this tag, the microprocessor will never become active. This can result in a power reduction of several orders of magnitude (see Section 11.4.2.10).

Based on these two models, an analysis of power dissipation is able to depict the impact of a smart buffer to an RFID tag based on different communication patterns such as broadcast and P2P communication. In all experimental scenarios, a general assumption is that only one RFID reader (R_0) starts the communication to a field of RFID tags (T) containing N tags with a wake-up signal followed by a `collection` broadcast command. The reader then submits a series of k P2P commands. Finally, the reader puts all the tags to sleep using the P2P sleep command. The duration of each experimental scenario is an hour and each experiment is repeated 20 times and the average is taken of all trials.

The result of three different scenarios is shown in Figure 11.29. Scenario one (solid line) sets $k=1$, scenario two (dashed line) sets $k=50$, and scenario three (dotted line) sets $k=100$. Data points with +s are the microprocessor only tag and data points with os have a smart buffer. In this scenario the P2P command selected is the `set owner id` command. The average energy saved by adding the smart buffer is 82%, 88%, and 90% per tag where $N=100$ for scenarios one, two, and three, respectively. As expected, as k increases, the energy consumed by the microprocessor-based tags increases because the microprocessor is processing every packet. In contrast, the smart buffer tags consume approximately the same energy regardless of the number of messages because the microprocessor is rarely activated. The increase in energy consumed by the microprocessor tags as N increases is due to the P2P sleep commands which scales with N. The smart buffer based tags avoid this increase with N.

11.5 Conclusions

In this chapter we have presented several techniques that apply to both passive and active RFID systems that address and reduce power and energy consumption. In particular, we have shown how memory capacities for passive tags can be increased by 350 times with only a 5 time increase in power consumption. This allows a potential for greatly increased capabilities of passive RFID system. We have shown a power macromodeling technique that can reduce the time to determine the power consumption of a protocol by 100 times.

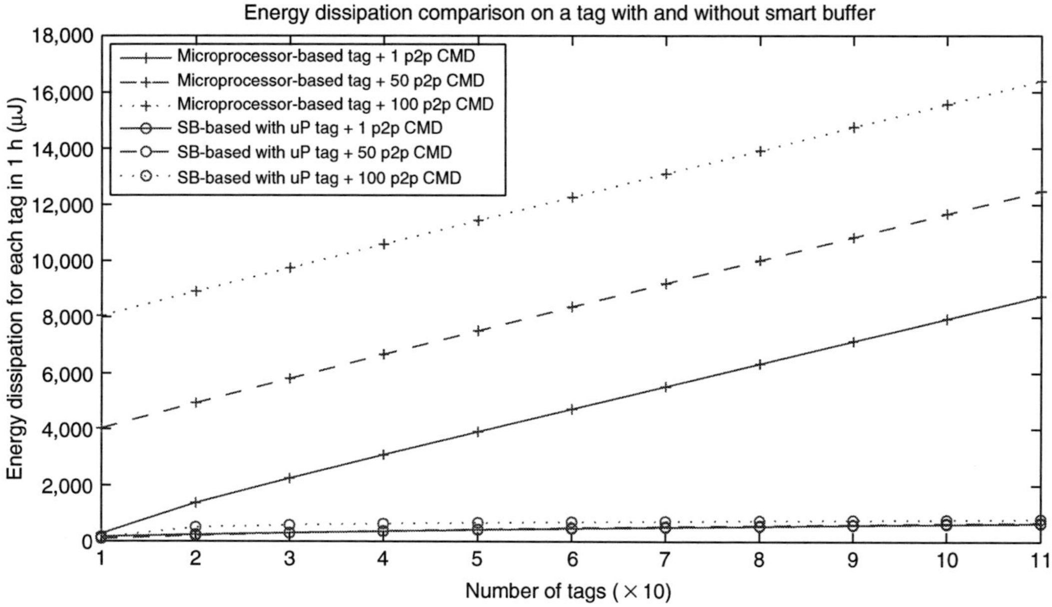

FIGURE 11.29
Energy impacts on a tag for point-to-point communication patterns.

This allows an opportunity for design space exploration of new protocols that considers power as a primary metric. Finally, we have presented a combined passive and active RFID tag that retains many of the capabilities of active RFID tags while having a significantly reduced power consumption through architectural innovation.

This chapter provides only a starting point in the thought process for designing energy-efficient RFID systems. In practice, circuit, antenna, and other various low-level design techniques are and should be applied. The purpose of this chapter is to provide some new directions in the thought process for RFID design. The final product will likely consider these low-level techniques in concert with techniques described here, possibly with new directions yet to be discovered. However, this chapter does show that power is a metric to be considered, and innovation is required for RFID technology to continue to be successful.

References

1. S.G. Narendra and A. Chandrakasan, *Leakage in Nanometer CMOS Technologies*, Springer, 2006.
2. L. Yang and J.S. Yuan. "Design of a new CMOS output buffer with low switching noise," *Proceedings of the International Conference on Microelectronics (ICM)*, pp. 131–134, 2003.
3. J.M. Rabaey, A. Chandrakasan, and B. Nikolic, *Digital Integrated Circuits: A Design Perspective*, 2nd edn., Prentice Hall, 2003.
4. L. Xinquan, Y. Weixue, Ligang, and C. Yu, "A low quiescent current and reset time adjustable power-on reset circuit," *Proceedings of International Conference on ASIC (ASIC)*, pp. 559–562, 2005.
5. S. Gupta and F.N. Najm, "Power macromodeling for high level power estimation," *DAC '97: Proceedings of the 34th Annual Conference on Design Automation*, pp. 365–370, ACM Press, 1997.
6. A.K. Jones, X. Tang, and P. Banerjee, "Compile-time simulation for low-power optimization using SystemC," *IASTED International Conference on Modeling and Simulation*, 2004.

7. X. Tang, T. Jiang, A. Jones, and P. Banerjee, "High-level synthesis for low power hardware implementation of unscheduled data-dominated circuits," *Journal of Low Power Electronics*, 1, 259–272, December 2005.
8. "Overview of the SystemC Initiative." SystemC Website, www.systemc.org.
9. Z. Chen and K. Roy, "A power macromodeling technique based on power sensitivity," *DAC '98: Proceedings of the 35th Annual Conference on Design Automation*, pp. 678–683, ACM Press, 1998.
10. X. Liu and M.C. Papaefthymiou, "A Markov chain sequence generator for power macromodeling," *IEEE Transactions on Computer-Aided Design of Integrated Circuits and Systems (TCAD)*, July 2004.
11. A.K. Jones, R. Hoare, D. Kusic, G. Mehta, J. Fazekas, and J. Foster, "Reducing power while increasing performance with SuperCISC," *ACM Transactions on Embedded Computing Systems (TECS)*, 5, 1–29, August 2006.
12. X. Liu and M.C. Papaefthymiou, "A static power estimation methodology for IP-based design," *Design, Automation, and Test in Europe*, pp. 280–287, March 2001.
13. A.K. Jones, R. Hoare, D. Kusic, J. Fazekas, and J. Foster, "An FPGA-based VLIW processor with custom hardware execution," *ACM International Symposium on Field-Programmable Gate Arrays (FPGA)*, pp. 107–117, 2005.
14. R. Hoare, A.K. Jones, D. Kusic, J. Fazekas, J. Foster, S. Tung, and M. McCloud, "Rapid VLIW processor customization for signal processing applications using combinational hardware functions," *EURASIP Journal on Applied Signal Processing*, Vol. 2006, Article ID 46472, 2006.
15. A.K. Jones, R. Hoare, S. Dontharaju, S. Tung, R. Sprang, J. Fazekas, J.T. Cain, and M.H. Mickle, "An automated, FPGA-based reconfigurable, low-power RFID tag," *Proceedings of the 43rd Design Automation Conference (DAC)*, pp. 131–136, ACM, July 2006.
16. A.K. Jones, S. Dontharaju, S. Tung, P.J. Hawrylak, L. Mats, R. Hoare, J.T. Cain, and M.H. Mickle, "Passive active radio frequency identification tags (PART)," *International Journal of Radio Frequency Identification Technology and Application (IJRFITA)*, 1 (1), 52–73, 2006.
17. P.J. Hawrylak, L. Mats, J.T. Cain, A.K. Jones, S. Tung, and M.H. Mickle, "Ultra low-power computing systems for wireless devices," *International Review on Computers and Software (IRECOS)*, 1 (1), 1–10, 2006.
18. P.J. Hawrylak, *Analysis and Development of a Mathematical Structure to Describe Energy Consumption of Sensor Networks*. PhD thesis, University of Pittsburgh, 2006.
19. ZigBee Standards Organization, "ZigBee Specification." Standard Specification, Document 053474r13, 2006.
20. A.K. Jones, R.R. Hoare, S.R. Dontharaju, S. Tung, R. Sprang, J. Fazekas, J.T. Cain, and M.H. Mickle, "A field programmable RFID tag and associated design flow," *Proceedings of FCCM*, pp. 165–174, 2006.
21. A.K. Jones, R. Hoare, S. Dontharaju, S. Tung, R. Sprang, J. Fazekas, J.T. Cain, and M.H. Mickle, "An automated, FPGA-based reconfigurable, low-power RFID tag," *Journal of Microprocessors and Microsystems*, 31, 116–134, March 2007.
22. American National Standards Institute, "ANSI NCITS 236:2001." Standard Specification, 2002.
23. International Standards Organization, "ISO/IEC FDIS 18000–7:2004(E)." Standard Specification, 2004.

12

Electromagnetic Coupling in RFID

Peter H. Cole and Damith C. Ranasinghe

CONTENTS

12.1 Introduction

Radio frequency identification (RFID) systems are evolving rapidly as a result of (1) increased awareness of the technology; (2) development of improved techniques for multiple tag reading; (3) realization in the business community of the benefits of widespread adoption in the supply chain; (4) adoption by designers of sensible concepts in the arrangement of data between labels and databases; (5) development of efficient data-handling methodologies in the relevant supporting communication networks; (6) appreciation of the need for cost reduction; and (7) development of new manufacturing techniques that will achieve manufacture of billions of labels at acceptable costs. This chapter is designed to provide support to RFID system designers in what is probably the least understood aspect of system design, that of the electromagnetic propagation to and from the label.

12.2 Outline

We will begin with a formal statement of the laws of electrodynamics, and use them to derive boundary conditions that apply where fields interact with materials. Such relations are amplified by introduction of the useful concepts of demagnetizing and depolarizing factors. The retarded potential solutions of Maxwell's equations and their utility in developing near and far radiated fields are developed, especially for infinitesimal electric and magnetic dipoles. A number of field creation structures or near and far fields are illustrated, and label antenna parameters for use in various object geometries are illustrated. Coupling volume theory for electric field- and magnetic field-sensitive antennas is developed at length and the relation to far-field antenna theory is established. Some conclusions about antenna optimization are drawn.

12.3 Notation

The notation and nomenclature used in this chapter for physical quantities will be as defined in ISO 1000 (ISO, 1992). Sinusoidally, varying quantities will be represented by peak (not root-mean-square) value phasors. Lowercase variables will be used for instantaneous values of scalars, and bold script characters for instantaneous values of field vectors. Uppercase variables will be used for phasors representing sinusoidally varying quantities, and bold upright Roman characters will be used for phasors representing sinusoidally varying field vectors. There are some traditional exceptions to these rules where Greek and Roman uppercase symbols do not differ sufficiently.

12.4 Electromagnetic Theory

12.4.1 Complete Laws

The complete laws of electrodynamics were first assembled correctly by Maxwell in the form enunciated in words as follows.

12.4.1.1 Faraday's Law

The circulation of the electric field vector $\mathcal{E}$ around a closed contour is equal to minus the time rate of change of magnetic flux through a surface bounded by that contour, the positive direction of the surface being related to the positive direction of the contour by the right-hand rule.

12.4.1.2 Ampere's Law as Modified by Maxwell

The circulation of the magnetic field vector $\mathcal{H}$ around a closed contour is equal to the sum of the conduction current and the displacement current passing through a surface bounded by that contour, with again the right-hand rule relating the senses of the contour and the surface.

12.4.1.3 Gauss' Law for the Electric Flux

The total electric flux (defined in terms of the $\mathcal{D}$ vector) emerging from a closed surface is equal to the total conduction charge contained within the volume bounded by that surface.

12.4.1.4 Gauss' Law for the Magnetic Flux

The total magnetic flux (defined in terms of the $\mathcal{B}$ vector) emerging from any closed surface is zero.

With the aid of Gauss' and Stokes' laws of mathematics and the definitions

$$\mathcal{D} = \varepsilon_0 \mathcal{E} + \mathcal{P}$$

and

$$\mathcal{B} = \mu_0(\mathcal{H} + \mathcal{M}).$$

These laws may be expressed, when the fields are spatially continuous, in the differential form

$$\begin{aligned} \nabla \times \mathcal{E} &= -\frac{\partial \mathcal{B}}{\partial t}, \\ \nabla \times \mathcal{H} &= \mathcal{J} + \frac{\partial \mathcal{D}}{\partial t}, \\ \nabla \cdot \mathcal{D} &= \rho, \\ \nabla \cdot \mathcal{B} &= 0. \end{aligned}$$

12.4.2 Source and Vortex Interpretation

We remain firmly committed to the source and vortex interpretation (Cole, 2002) of those equations. In that interpretation, the earlier equations stated that the electric field vector $\mathcal{E}$ can have vortices caused by changing magnetic flux; the magnetic field $\mathcal{H}$ can have vortices caused by conduction or displacement currents; the electric flux density $\mathcal{D}$ can have sources caused by conduction charge density; and the magnetic flux density vector $\mathcal{B}$ can have no sources.

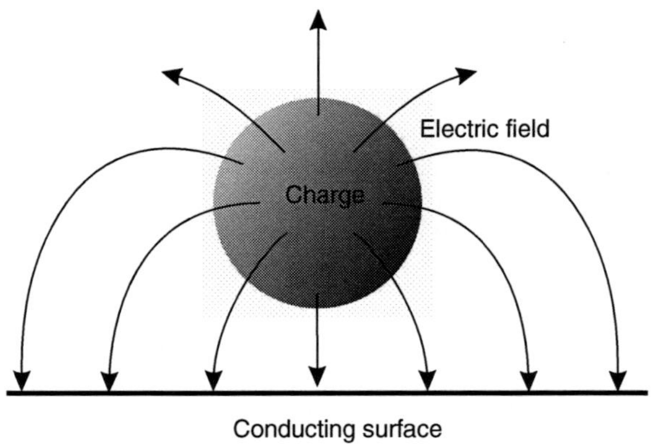

FIGURE 12.1
Electric field near a conducting surface.

In linear media, some of the statements about $\mathcal{D}$ and $\mathcal{B}$ can be extended to $\mathcal{E}$ and $\mathcal{H}$, but when nonuniform fields and boundaries are considered, it can be shown that $\mathcal{E}$, $\mathcal{D}$, and $\mathcal{H}$ can have both sources and vortices, but $\mathcal{B}$ is alone in that it can have no sources.

Figures 12.1 and 12.2 provide archetypical illustrations of the source nature of the electrostatic field and the vortex nature of a magnetodynamic field, as well as illustrations of two of the most important boundary conditions which apply when any electric field $\mathcal{E}$ or a magnetodynamic field $\mathcal{H}$ approaches a conducting surface.

12.4.3 Boundary Conditions

We have already given simplified illustrations for the most important cases of sinusoidal electric and magnetic fields occurring adjacent to conductors.

A full statement of the electromagnetic boundary conditions is provided later. All of those results may be derived from the statement in words of the basic laws given earlier. What we can deduce from those laws, without taking into account the properties on any materials involved, is that the tangential component of $\mathcal{E}$ is continuous across any boundary; the normal component of $\mathcal{B}$ is continuous across such a boundary; the normal component of $\mathcal{D}$ may be discontinuous across a boundary, with a discontinuity being equal to any conduction charge density ρ_{vc} per unit area on the surface; and the tangential component of $\mathcal{H}$ may be discontinuous across a boundary, with the discontinuity being equal to in magnitude and at right angles in direction to a surface current density flowing on the surface.

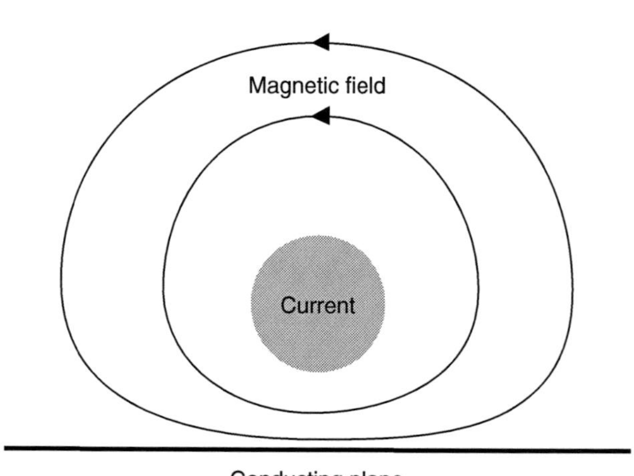

FIGURE 12.2
Oscillating magnetic field near a conducting surface.

When we take into account the restrictions imposed by the properties of the materials which may exist on one or other side of the boundary, we may further conclude that the electric field is continuous across the boundary for all materials and time variations; that there are no electric fields or fluxes, or time-varying magnetic field or flux densities inside a good conductor; that a surface current density can exist only on the surface of a perfect conductor; and that time-varying charge density cannot exist on the surface of a perfect insulator although a static surface charge density can.

12.4.4 Demagnetizing and Depolarizing Factors

The theory of demagnetizing and depolarizing factors (Osborn, 1945) is useful in gaining an appreciation of how the interior fields of an object differ from the fields exterior to that object, and for calculating useful properties of magnetic cores.

When an ellipsoidal shape as illustrated in Figure 12.3 of dielectric or soft magnetic material is introduced into a region in which there was previously a spatially uniform electric field $\mathcal{E}$ or magnetic field $\mathcal{H}$ caused by some distribution of sources (charges or currents), and these sources do not vary as a result, the material becomes uniformly polarized with a polarization vector $\mathcal{P}$ or magnetized with a magnetization vector $\mathcal{M}$. That polarization or magnetization causes on the surface of the shape induced surface charge densities or magnetic pole densities which make an additional contribution $\mathcal{E}^{\mathrm{d}}$ or $\mathcal{H}^{\mathrm{d}}$ to the fields interior to the shape. These fields are in a direction opposite to the original applied fields $\mathcal{E}^{\mathrm{a}}$ or $\mathcal{H}^{\mathrm{a}}$ and tend to depolarize or demagnetize the material, and hence reduce (but not reverse in direction) the polarization or magnetization in the interior of the shape. The internal fields $\mathcal{E}$ and $\mathcal{H}$ are also reduced.

The depolarizing or demagnetizing fields are given by

$$\begin{bmatrix} \mathcal{E}_x^{\mathrm{d}} \\ \mathcal{E}_y^{\mathrm{d}} \\ \mathcal{E}_z^{\mathrm{d}} \end{bmatrix} = -\frac{1}{\varepsilon_0} \begin{bmatrix} N_x & 0 & 0 \\ 0 & N_y & 0 \\ 0 & 0 & N_z \end{bmatrix} \begin{bmatrix} \mathcal{P}_x \\ \mathcal{P}_y \\ \mathcal{P}_z \end{bmatrix},$$

$$\begin{bmatrix} \mathcal{H}_x^{\mathrm{d}} \\ \mathcal{H}_y^{\mathrm{d}} \\ \mathcal{H}_z^{\mathrm{d}} \end{bmatrix} = - \begin{bmatrix} N_x & 0 & 0 \\ 0 & N_y & 0 \\ 0 & 0 & N_z \end{bmatrix} \begin{bmatrix} \mathcal{M}_x \\ \mathcal{M}_y \\ \mathcal{M}_z \end{bmatrix},$$

where N_x, N_y, or N_z are known as depolarizing or demagnetizing factors. These dimensionless constants depend on the shape of the ellipsoid, and are subjected to the condition

$$N_x + N_y + N_z = 1.$$

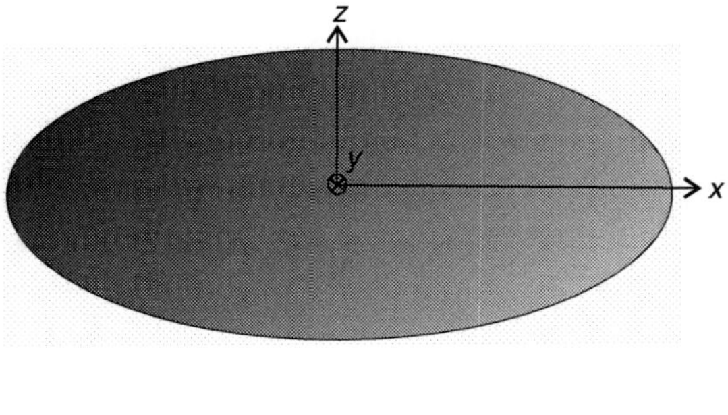

FIGURE 12.3
An ellipsoid of dielectric or magnetic material.

These constraints and relations between N_x, N_y, or N_z for certain symmetrical shapes allow conclusions to be drawn for the cases of the sphere, a long thin rod, or a flat thin desk.

12.4.5 Retarded Potentials

The retarded potentials listed later may be regarded as integral solutions for Maxwell's equations, which are available when charge and current distribution are known.

For the calculation of electric and magnetic fields at a point r_2 caused by a distribution of charge and current at points r_1 over a volume v we may make use of the retarded potentials:

$$\Phi(r_2) = \frac{1}{4\pi\varepsilon_0} \int_v \frac{\rho(r_1) e^{-j\beta r_{12}}}{r_{12}} dv$$

and

$$\mathbf{A}(r_2) = \frac{\mu_0}{4\pi} \int_v \frac{\mathbf{J}(r_1) e^{-j\beta r_{12}}}{r_{12}} dv.$$

The fields in the sinusoidal steady state can be derived from these potentials by the equations

$$\mathbf{E} = -\mathbf{grad}\,\Phi - j\omega\mathbf{A}$$

$$\mathbf{B} = \mathbf{curl}\,\mathbf{A}.$$

These formulae may be used in the calculation of the electromagnetic fields generated by oscillating infinitesimal electric or magnetic dipoles as reported later.

12.4.6 Reciprocity

The integral form of the Lorenz reciprocity relation for two solutions $\mathbf{E}_1$, $\mathbf{H}_1$ and $\mathbf{E}_2$, $\mathbf{H}_2$ of Maxwell's equations in the same region and at the same angular frequency ω is under appropriate conditions

$$\int_S (\mathbf{E}_1 \times \mathbf{H}_2 - \mathbf{E}_2 \times \mathbf{H}_1) \cdot \mathbf{ds} = \int_v (\mathbf{J}_1 \cdot \mathbf{E}_2 - \mathbf{J}_2 \cdot \mathbf{E}_1)\, dv$$

where the integral is over a closed surface S bounding a volume v. In the derivation, which proceeds from Maxwell's equations, it is allowed that the material parameters be characterized by complex (to allow for losses) and possibly tensor (to allow for anisotropy) dielectric permittivity and magnetic permeability, but gyromagnetic behavior of magnetic materials is not permitted.

In certain cases, the right-hand side becomes zero. These cases include those where the surface is a conducting surface, where the surface encloses all sources, where the surface encloses no currents, and where currents flow only by the mechanism of drift (but not by diffusion, as in a semiconductor).

The theorem has in electromagnetic theory, and on the simplification of electromagnetic theory known as lumped circuit theory, some profound consequences. These include the symmetry of impedance, admittance, and suitably defined scattering matrices; the equivalence of transmitting and receiving antennas; the gain of a lossless transmitting antenna being related to effective area of the same antenna used as a receiver; and the propagation loss from an interrogator to label being equal to propagation loss from label to receiver when the antennas have single ports. Use of these results will be made in later sections dealing with coupling between interrogators and labels.

12.5 Field Creation Structures

12.5.1 Near-Field Creation

We first illustrate in Figures 12.4 and 12.5 structures which may be useful in the HF region.

Such structures can generate either electric or magnetic fields, either in the near field, or the midfield, that being the field at the boundary between the near field and the far field.

As a consequence of the value (22 m) of the electromagnetic wavelength at 13.56 MHz, the structures are always in practice electrically small. The first one can be considered as creating in the near field mainly electric field, but as a consequence of the expressions given earlier for dipole fields, which apply to some extent to this situation, the structures will also create some lesser value of near magnetic field.

In the far field, these structures will create electric and magnetic fields in equal proportion, in the sense that for radiated fields in the far field $|\mathbf{E}| = \eta|\mathbf{H}|$.

As HF labels often couple to magnetic fields because they are less easily stopped by conducting materials than are electric fields, there is in fact a more common interest in the creation of strong near-field magnetic fields. The usual structure by means of which this is achieved is a small current carrying loop such as is illustrated in Figure 12.5. This 250 mm square loop has tuning and matching elements at the top end. A strip line transmission line (not visible in the picture) on the underside of the right-hand half conveys the driving signals from the connecting point at the center of the bottom to the driving point terminals at the center of the top. When the loop antenna is made large such as shown for a 1.2 m square loop in Figure 12.6, the capacitive tuning elements are best distributed around the circumference. The balanced drive of Figure 12.5 cannot be implemented in Figure 12.6, but other balanced drive arrangements can be made.

12.5.2 Far-Field Creation

The almost universal choice for the creation of far fields, particularly in the UHF region, is the patch antenna which should be placed against a ground plane of sufficient size to achieve a good front to back ratio.

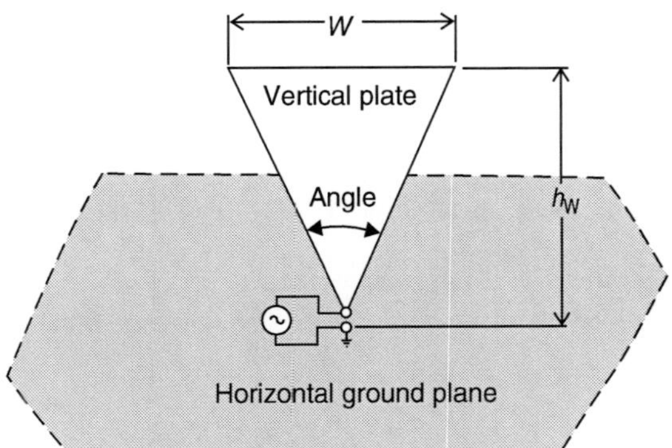

FIGURE 12.4
A monopole plate antenna above a ground plane. (From Cole, P.H., Jamali, B., and Ranasinghe, D., *Coupling Relations in RFID Systems*, 2003 Auto-ID Center White Paper Series, © 2003 by Auto-ID Center. With permission.)

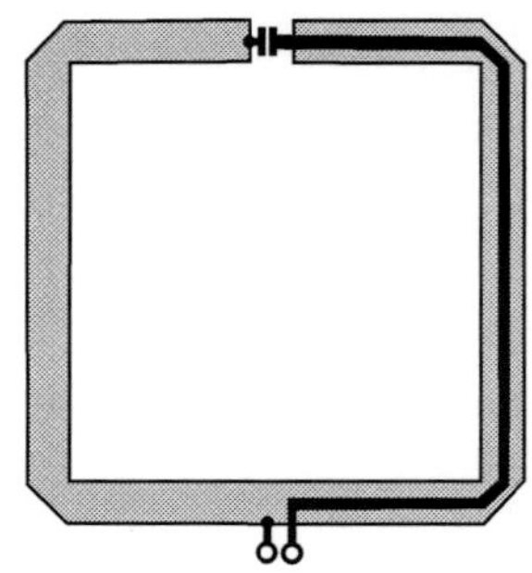

FIGURE 12.5
A small loop antenna.

12.6 Measures of Label Excitation

An appropriate measure of tag excitation in the near field is the oscillating reactive power density per unit volume, and in the far field is power propagating per unit area. At UHF, for calculation of power transfer in the far field, we may use the concepts of gain and effective area as defined in Section 12.7.3. At HF, we will derive power transfer from coupling volume concepts to be defined in Section 12.7.4. In those calculations, we can recognize both electric field-sensitive and magnetic field-sensitive designs of label antennas.

12.6.1 Magnetic Field-Sensitive Antennas

A common example of a magnetic field-sensitive HF label is shown in Figure 12.7. The label is 42 mm wide by 47 mm high. The label is designed to have sufficiently many turns to provide the resonating inductance for the microcircuit input capacitance, as well as a flux-collecting area in the interior which is as large as practicable consistent with the size requirement for the label.

Advantages of working in the near field at HF rather than at LF are that the number of turns required to resonate the microcircuit capacitance is small enough for low-resolution lithography to be used in antenna construction, and no additional external resonating capacitance is required.

Clearly the design illustrated in Figure 12.7 is unsuitable for being placed flat against metal, as the boundary conditions shown in Figure 12.2 will not allow a normal component

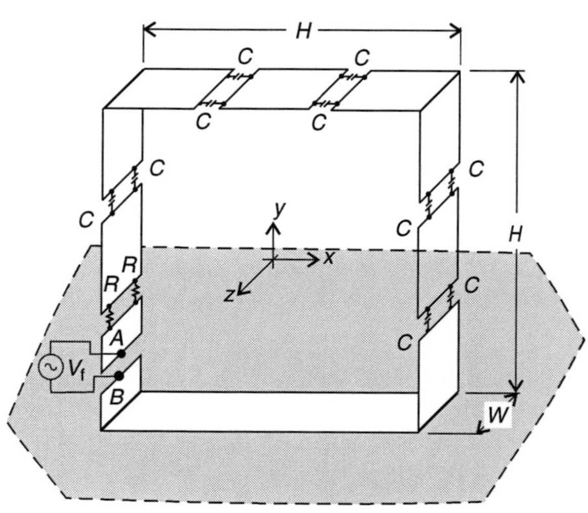

FIGURE 12.6
A large loop antenna. (From Cole, P.H., Jamali, B., and Ranasinghe, D., *Coupling Relations in RFID Systems*, 2003 Auto-ID Center White Paper Series, © 2003 by Auto-ID Center. With permission.)

FIGURE 12.7
A magnetic field-sensitive HF label antenna.

of magnetic flux density at the metal surface. For this situation, the label antenna employing a solenoid with a magnetic core design shown in Figure 12.8 is employed.

Section 12.7.3 describes that without the magnetic core the coupling volume, a concept defined in Section 12.7.4, of a long solenoid is just the physical volume, but when a magnetic core is inserted, the coupling volume increases by factor equal to the effective permeability defined in that section. This behavior may be contrasted with that of electric field labels, in which, in Section 12.7.6, we will find that the inclusion of dielectric material into the interior of the label is not helpful.

12.6.2 Electric Field-Sensitive Antennas

Two varieties of electric field-sensitive antennas are shown in Figures 12.9 and 12.10. Figure 12.9 shows a small bow tie antenna that is intended to be sensitive to electric fields in the horizontal direction. Figure 12.10 shows an electric field-sensitive antenna that is suitable for placement against a horizontal metal plate where it intercepts a downward component of electric flux density.

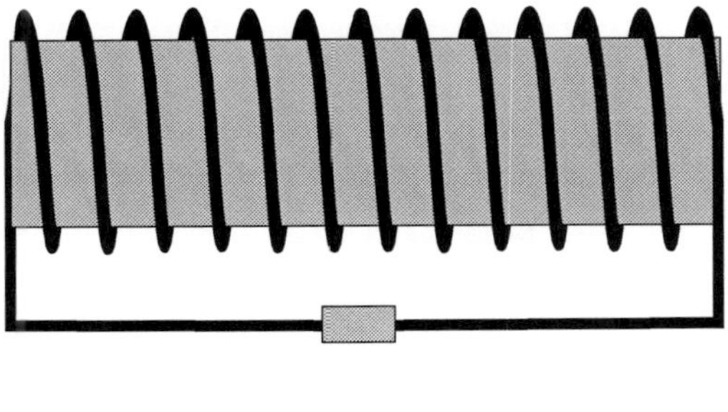

FIGURE 12.8
Antenna for HF operation against metal. (From Cole, P.H., Jamali, B., and Ranasinghe, D., *Coupling Relations in RFID Systems*, 2003 Auto-ID Center White Paper Series, © 2003 by Auto-ID Center. With permission.)

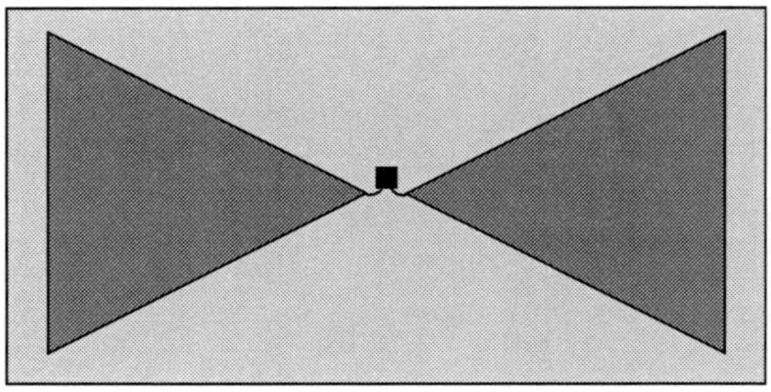

FIGURE 12.9
An electric field-sensitive label. (From Cole, P.H., Jamali, B., and Ranasinghe, D., *Coupling Relations in RFID Systems*, 2003 Auto-ID Center White Paper Series, © 2003 by Auto-ID Center. With permission.)

Analysis of these structures is provided in Sections 12.7.6 and 12.7.7. In the case of the bow tie, the analysis is for the half bow tie (wedge above ground plane) shown in Figures 12.11 and 12.12. The analyses have a common feature that the figure of merit for these antennas, when placed in the energy storage electric field is a coupling volume, in the second case, if no dielectric is present, equal to the physical volume of the structure, and in the first case a volume derived from the label dimensions, even though the antenna itself has no physical volume.

Both of the antennas will also have an effective area, but this area should not be confused with the effective area concept of a radiating antenna or a far-field antenna. The effective area for a near-field electric field-sensitive antenna describes the extent to which the antenna can extract current from the displacement current density of the driving electric field.

12.6.3 Electromagnetic Field Antennas

We consider an antenna as an electromagnetic field antenna on a couple of different bases. Firstly, if the antenna is capable of responding to both electric and magnetic fields we would consider it to be an electromagnetic field antenna. It is almost invariably true that unless the antenna is very small, it does have this property. Proper analysis requires that it is analyzed using the full set of Maxwell's equations, rather than the subset or simplified versions that pertain to electrostatic or magnetostatic problems. A good example of this phenomenon is provided by the electromagnetic field-sensitive antenna shown in Figure 12.13, in which there is no obvious effort to couple to either electric or magnetic field alone.

12.7 Coupling Relations

12.7.1 Near and Far Fields

In the analysis of the performance of RFID systems, it is important to consider whether the labels are placed in the far (propagating) or near (energy storage) fields of the interrogator

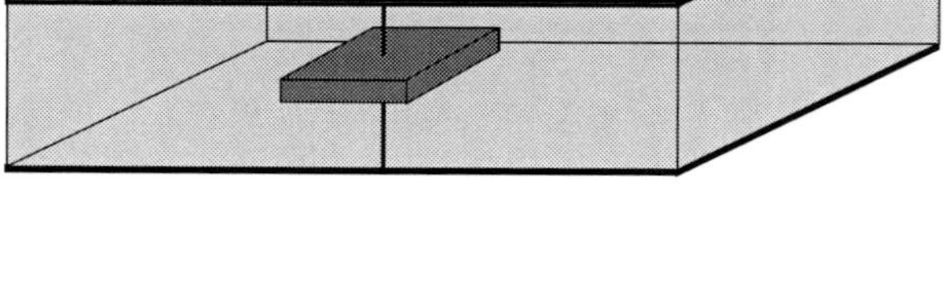

FIGURE 12.10
A parallel plate electric field-sensitive label. (From Cole, P.H., Jamali, B., and Ranasinghe, D., *Coupling Relations in RFID Systems*, 2003 Auto-ID Center White Paper Series, © 2003 by Auto-ID Center. With permission.)

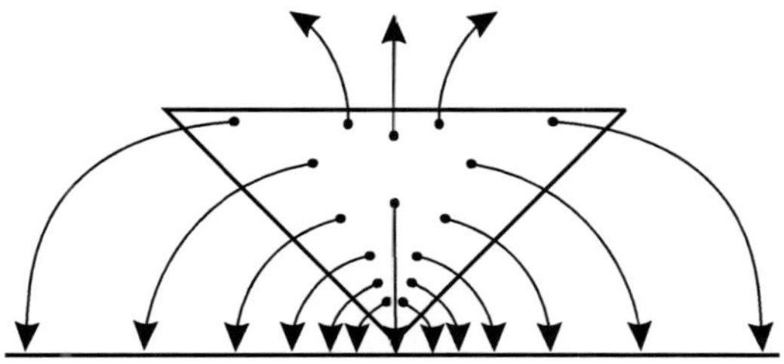

FIGURE 12.11
Field configuration for self-capacitance calculation of half bow tie antenna.

antenna. When that antenna is of small gain, the distance which divides the near and far fields is the size or the radian sphere of radius $r=\lambda/(2\pi)$, where λ is the free space electromagnetic wavelength at the operating frequency.

12.7.2 Field Measures

For a linearly polarized magnetic field described by a peak value phasor H, we may develop the two measures of the exciting field

$$\text{Radial component of Poynting vector } S_r = \frac{\eta|\mathrm{H}|^2}{2} \text{ in W m}^{-2},$$

$$\text{Volume density of reactive power } W_v = \frac{\omega\mu_0|\mathrm{H}|^2}{2} \text{ in V A m}^{-3}.$$

For a linearly polarized magnetic field described by a peak value phasor E, we may develop the two measures of the exciting field

$$\text{Radial component of Poynting vector } S_r = \frac{|\mathrm{E}|^2}{2\eta} \text{in W m}^{-2},$$

$$\text{Volume density of reactive power } W_V = \frac{\omega\varepsilon_0|\mathrm{E}|^2}{2} \text{in V A m}^{-3}.$$

In both cases, the latter expression is ω times the peak value of stored magnetic or electric energy per unit volume. We can easily show that if β is the propagation constant at the frequency used

$$W_V = \beta S_r.$$

The last expression assumes that we are in the far field, that is, there are no near-field reactive energy storage fields to augment W_V without contributing to S_r.

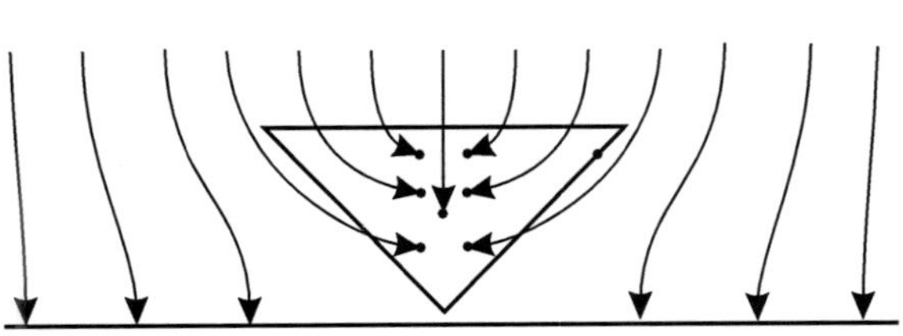

FIGURE 12.12
Field configuration for calculation of effective area of bow tie antenna.

FIGURE 12.13
An electromagnetic field-sensitive antenna.

12.7.3 Far-Field Operation

For calculation of the power coupled in the far field to a label with a lossless receiving antenna, the usual approach is to first calculate the power density per unit area at the label position from the formula

$$\text{Power density per unit area} = \frac{g_t P_t}{4\pi r^2},$$

where
g_t is the gain of the transmitter antenna
P_t is the power it transmits
r is the distance from the transmitter antenna to the label position

In using this formula, we are implicitly assuming that the label has been placed in the direction of strongest radiation from the interrogator antenna.

The power P_r which may be extracted under optimum conditions of tuning and matching by a lossless label antenna placed at the earlier position is given by

$$P_r = A_{er} \times \text{Power flow per unit area},$$

where A_{er} is a property of the label known as its effective area. It is unrelated to the physical area of the antenna (which if it is just a piece of thin wire, does not have a physical area), but has the desirable property that we may imagine the label antenna collects all of the radiated power which flows through that effective area which may be thought of as surrounding the label antenna.

The Lorenz reciprocity theorem of electrodynamics may be used to show that the effective area of a receiving antenna is related to the gain g_r it would have in a transmitting role by the equation

$$A_{er} = \frac{g_r \lambda^2}{4\pi},$$

so we are able to derive for the power transfer ratio the formula

$$\frac{P_r}{P_t} = g_r g_t \left(\frac{\lambda}{4\pi r}\right)^2.$$

12.7.4 Power-Matching Considerations

However we regard our label, we expect to have, in its series equivalent circuit, as well as a reactance, a radiation resistance R_r, a loss resistance R_c, and a load resistance R_L, and possibly a matching reactance. Considerations of maximum power transfer to the load will always require that at optimum, we set

$$R_L = R_r + R_c$$

and the power delivered to the external load is reduced for the value obtained earlier by the ratio

$$\frac{R_r}{R_r + R_c}.$$

12.7.5 Near-Field Operation—Magnetic Field

For near-field operation, the fields which excite the label are basically energy storage fields for which consideration of power flow is inappropriate, and for which pre-Maxwell versions of electrodynamics can give correct calculations. Both the interrogation field creation means and the label field detection means can be considered as weakly coupled inductors of self-inductances L_1 and L_2 and mutual inductance M. When both these coils are tuned to resonance with respective quality factors Q_1 and Q_2, it can be shown that the power P_2 dissipated in the losses of the label coil is related to the power P_1 dissipated in the losses of the "transmitter" coil by

$$\frac{P_2}{P_1} = k^2 Q_1 Q_2 \quad \text{where } k = \frac{M}{\sqrt{L_1 L_2}}.$$

This equation is useful to show the role of the quality factor, Q, of the resonances in both the label and interrogator coils in promoting power transfer, but it is not useful in separately optimizing the properties of those two widely dissimilar elements. For that purpose, we can focus first on the energy storage measure of exciting field, which is the reactive power per unit volume in the field created by the interrogator at the label position. In terms of that measure, we can define (Eshraghian et al., 1982) a figure of merit of a label antenna as the ratio

$$V_c = \frac{[\text{Reactive power flowing in the untuned label coil when it is short-circuited}]}{[\text{Volume density of reactive power created by the interrogator at the label position}]},$$

which clearly has the dimensions of volume, and is for this reason called the coupling volume of the label antenna. For the performance of the interrogator antenna, we can define the companion concept of dispersal volume V_d given by

$$V_d = \frac{[\text{Reactive power flowing in the inductor of the interrogator field creation coil}]}{[\text{Volume density of reactive power created by the interrogator at the label position}]}.$$

When both antennas are tuned, it is possible to show

$$\frac{P_2}{P_1} = \frac{V_c}{V_d} Q_1 Q_2.$$

The benefit of this formulation is that the coupling volume is a property of the label parameters alone, and the dispersal volume is a property of the interrogator antenna parameters alone (and of the label position), and separate optimization becomes possible, whereas k^2 is a complex function of the entire system geometry.

Coupling volumes for various label antennas are readily determined. For an air-cored solenoidal antenna it is approximately the volume of that antenna, and when a magnetic core is in place that volume becomes multiplied by the relative effective permeability

$$\mu_{er} = \frac{\mu_{ir}}{1 + N(\mu_{ir} - 1)},$$

where μ_{ir} is the relative intrinsic permeability of the core material and N is the demagnetizing factor in the direction of the interrogator field.

For a planar coil, which in its idealized state has no physical volume, the coupling volume is given by

$$V_c = \frac{\mu_0 A^2}{L},$$

where A is the flux-collecting area (incorporating by summation of area for each turn) of the coil and L is the self-inductance.

12.7.6 Near Field—Electric Field

In RFID systems, the coupling can be via the magnetic field or the electric field. While in the near field the coupling is almost always chosen to be by way of the magnetic field, it is possible with an appropriate electric field antenna to couple to the electric field, and that can be done whether the tag is in the far field or the near field. The energy transfer is then provided by the electric flux terminating on the antenna surface and inducing a charge on the antenna. The induced charge will oscillate as the field oscillates to produce a current.

The issues of understanding and optimizing coupling to the electric field are important. The coupling volume theory developed here for the electric field case assists in that effort and allows comparisons between different antenna structures for their effectiveness and efficiency in terms of their actual physical volume and the coupling volume. It is also of interest in making comparisons with the magnetic field case.

When we can focus on the energy storage measure of exciting field, which is the reactive power per unit volume in the field created by the interrogator at the label position, we can define a figure of merit of a label antenna as the ratio

$$V_c = \frac{[\text{Reactive power flowing in the untuned label capacitor when it is open-circuited}]}{[\text{Volume density of reactive power created by the interrogator at the label position}]},$$

which clearly has the dimensions of volume, and is for this reason also called the coupling volume of the label antenna. For the performance of the interrogator antenna, we can define the companion concept of dispersal volume V_d given by

$$V_d = \frac{[\text{Reactive power flowing in the capacitance of the interrogator field creation electrodes}]}{[\text{Volume density of reactive power created by the interrogator at the label position}]}$$

When both antennas are tuned, it is possible to show

$$\frac{P_2}{P_1} = \frac{V_c}{V_d} Q_1 Q_2.$$

The benefit of this formulation is that the coupling volume is a property of the label parameters alone, and the dispersal volume is a property of the interrogator antenna parameters alone (and of the label position), and separate optimization becomes possible. Coupling volumes for various label antennas are readily determined, as will be done in the next section.

12.7.7 Coupling Volume of a General Shape

In order to derive the coupling volume of antenna structures, it is important to define a number of concepts. For a given antenna, we can define an electric flux-collecting area as the area in space required by the antenna to generate from the displacement current of the exciting field an oscillating current I when the antenna is placed in an oscillating electric field of flux density D. This area may or may not be equivalent to the physical area. This electric flux-collecting area will be denoted by the symbol A_f and will be referred to as effective area.

The formula for the current I flowing from an antenna placed in an electric field with a flux density D and oscillating angular frequency ω, using the effective area A_f of the antenna structure is

$$I = j\omega D A_f.$$

The reactive power flowing in the antenna with a self-capacitance C when the antenna is open circuit is given by

$$\text{Reactive power} = \frac{|\mathrm{I}|^2}{2\omega C}.$$

The reactive power per unit volume in the field is obtained by

$$W_V = \frac{\varepsilon_0 |\mathrm{E}|^2 \omega}{2}.$$

Thus, the coupling volume of an antenna structure with a self-capacitance C can be obtained as

$$V_c = \frac{\varepsilon_0 A_f^2}{C}.$$

This formula provides the coupling volume of any antenna with a self-capacitance C and an effective area of A_f. It should be pointed out that A_f can easily be determined experimentally by measuring the current flowing from an antenna placed in an oscillating electric field of known strength.

If the antenna takes the form of a parallel plate capacitor and the space between the plates is filled with a dielectric, the coupling volume is reduced by a factor equal to the relative dielectric permittivity.

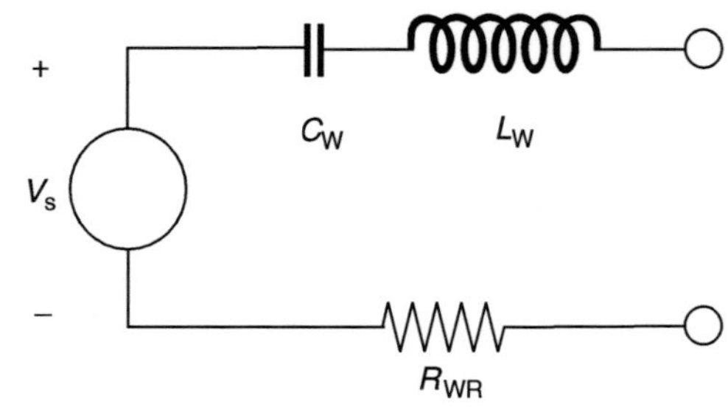

FIGURE 12.14
Three parameter equivalent circuit for a half bow tie antenna. (From Cole, P.H., Jamali, B., and Ranasinghe, D., *Coupling Relations in RFID Systems*, 2003 Auto-ID Center White Paper Series, © 2003 by Auto-ID Center. With permission.)

12.7.8 Properties of the Bow Tie Antenna

Calculating the self-capacitance, as depicted by the field lines in the upper half of Figure 12.11, of the monopole bow tie (wedge above ground plane) antenna by seeking analytical solutions to Laplace's equation presents a difficult problem. A numerical solution is both tractable and much simpler and has been performed to produce results quoted later.

In addition, a detailed study undertaken of results first published by Woodward (Brown and Woodward, 1952) provided the following empirical model for the equivalent circuit shown in Figure 12.14 of a monopole bow tie antenna. The model, derived from Woodward's results (Brown and Woodward, 1952) for the resistance $R_{WR}(\omega)$ and reactance $X(\omega)$ of a bow tie antenna, provides the values for the capacitor, whose value is that of the self-capacitance of the monopole, a value for the inductor placed in the series circuit shown, and a value for the radiation resistance. The following results give the model values for a monopole bow tie (wedge above ground plane) antenna with a flare angle of 90°:

$$C_W = K_{WC}\varepsilon_0 h_W \text{ with } K_{WC} = 7.6,$$

$$L_W = K_{WL}\mu_0 h_W \text{ with } K_{WL} = 0.2135,$$

$$R_{WR} = K_{WR}(\beta h_W)^2 \text{ with } K_{WR} = 30\,\Omega,$$

where h_W is the height (top to ground plane) of the monopole antenna. The significant finding is that, as expected, the low frequency impedance of the antenna is mainly capacitive and this value can thus be obtained by calculating the self-capacitance of the antenna. The radiation resistance obtained has significance in two ways. It allows the amount of radiated power to be calculated for a transmitting antenna, and also provides for a label antenna a means of calculating, using the reciprocity theorem, the effective electric flux-collecting area, as depicted by the field lines in the upper half of Figure 12.12, of the antenna. All of the results from Woodward agree with the results of our own direct measurements of flux-collecting area and our own numerical analysis performed using the method of moments.

12.7.9 Comparison of Formulations

As the Poynting vector-effective area formulation and the coupling volume–dispersal volume formulation are so apparently dissimilar, it is desirable to show that they are equivalent, but useful in different contexts where different approximations may be made. It has been shown, but we have not the space to do it here, that the formulations are indeed equivalent, and that the basic difference between the formulations is whether they emphasize the *radiation resistance* or the *internal losses* of the label antenna.

After considering all issues concerned with the generation of label exciting fields, we are led to conclude that the optimum frequency for operation of an RFID system in the far field is the lowest frequency for which a reasonable match to the radiation resistance of the label antenna can be achieved, at the allowed size of label, without the label or matching element losses intruding.

12.8 Conclusions

The fundamental principles governing electromagnetic coupling between an interrogator and its labels in an RFID system have been outlined, and a set of concepts suitable for describing coupling in the near and far fields, using electric field-, magnetic field-, or electromagnetic field-sensitive antennas, have been defined. Some of the properties of electric field and magnetic field antennas have been derived.

Some important theorems about optimizing antenna sizes and operating frequencies subject to electromagnetic compatibility regulations have been derived.

References

Brown, G.H. and O.M. Woodward Jr. 1952. Experimentally determined radiation characteristics of conical and triangular antennas. *RCA Review*, 13, 425–452.

Cole, P.H. 2002. A study of factors affecting the design of EPC antennas and readers for supermarket shelves. http://www.autoidcenter.org/research/ADE-AUTOID-WH-001.pdf

Eshraghian, K., P.H. Cole, and A.K. Roy. March 1982. Electromagnetic coupling in subharmonic transponders. *Journal of Electrical and Electronics Engineering*, Australia, pp. 28–35.

International Organisation for Standardisation (ISO). 1992. SI units and recommendations for the use of their multiples and of certain other units. International Standard ISO 1000.

Osborn, J.A. 1945. Demagnetising factors of the general ellipsoid. *Physical Review*, 67, 351.

13

RFID Tags for Metallic Object Identification

Mun Leng Ng, Kin Seong Leong, and Peter H. Cole

CONTENTS

13.1 Introduction

RFID is a technology that has existed for many years, but it is only recently that it has experienced the rapid growth that has arisen from application of this technology in various supply chains. From inventory management to theft detection, RFID has been applied in many areas such as in the automotive industry and logistics, as well as in warehouses and retail stores (Angeles, 2005). Potential has also been seen for the application of RFID in capital asset management applications such as keeping track of maintenance tools in the aircraft maintenance sector (Lampe et al., 2004). RFID has the ability of giving a unique identity to each tagged object. Hence, there is a vision to extend this technology to item-level tagging (other than pallet and case-level tagging), to give each specific item-level object a unique identity (Sarma et al., 2001).

The increasing implementation of the RFID technology in supply chains has posed many challenges and one of the biggest is the degradation of the system performance when tagging metallic objects or operating in an environment containing metals. The tagging of objects at pallet, case, and even item levels will most likely involve metallic objects. For example, in the automotive industry, when RFID is used for part tracking, a majority of the parts are made of metal (Strassner and Fleisch, 2003). To meet the challenge posed by metallic objects on RFID and to encourage the full deployment of RFID especially down to item-level tagging, results of research on the design of RFID tags suited for metallic application are presented in this chapter.

Following this introduction, the next section will consider the behavior of electromagnetic waves near metallic surfaces. This will be then followed by a section on the effects of metallic surfaces on RFID tag antennas. Finally, the chapter will present a few examples of passive UHF RFID tags suitable for attaching to metallic structures.

13.2 Electromagnetic Waves near Metallic Surfaces

For a boundary that lies between two media in space with medium 1 characterized by dielectric permittivity ε_1, magnetic permeability μ_1, and electric conductivity σ_1, and medium 2 characterized by ε_2, μ_2, and σ_2, the electromagnetic boundary conditions for a general case can be expressed (in vector form for time varying fields) as follow:

$$\hat{\mathbf{n}} \times (\mathcal{E}_2 - \mathcal{E}_1) = 0, \tag{13.1}$$

$$\hat{\mathbf{n}} \cdot (\mathcal{D}_2 - \mathcal{D}_1) = \rho_s, \tag{13.2}$$

$$\hat{\mathbf{n}} \times (\mathcal{H}_2 - \mathcal{H}_1) = \mathcal{J}_s, \tag{13.3}$$

$$\hat{\mathbf{n}} \cdot (\mathcal{B}_2 - \mathcal{B}_1) = 0, \tag{13.4}$$

where

$\hat{\mathbf{n}}$ is the unit normal vector to the boundary directed from medium 1 to medium 2
$\mathcal{E}$ and $\mathcal{H}$ are the electric and magnetic fields, respectively
$\mathcal{D}$ and $\mathcal{B}$ are the electric and magnetic flux densities, respectively
ρ_s and $\mathcal{J}_s$ are the surface charge density and surface current density, respectively, that may exist at the boundary

If medium 1 is a metallic medium and we assume as a practical approximation that it is a perfect electric conductor with infinite conductivity ($\sigma_1 \to \infty$), there will be no electric field in this medium (i.e., $\mathcal{E}_1 = 0$). Consequently, $\mathcal{D}_1 = 0$, $\mathcal{B}_1 = 0$, and $\mathcal{H}_1 = 0$. Hence, for this case, the boundary conditions become

$$\hat{\mathbf{n}} \times \mathcal{E}_2 = 0, \tag{13.5}$$

$$\hat{\mathbf{n}} \cdot \mathcal{D}_2 = \rho_s, \tag{13.6}$$

$$\hat{\mathbf{n}} \times \mathcal{H}_2 = \mathcal{J}_s, \tag{13.7}$$

$$\hat{\mathbf{n}} \cdot \mathcal{B}_2 = 0. \tag{13.8}$$

Since $\hat{\mathbf{n}}$ is the unit normal vector to the boundary, the expressions can be further written in terms of the corresponding tangential and normal components (denoted with subscripts "t" and "n," respectively) to

$$\mathcal{E}_{2t} = 0, \tag{13.9}$$

$$\mathcal{D}_{2n} = \rho_s, \tag{13.10}$$

$$\mathcal{H}_{2t} = \mathcal{J}_s, \tag{13.11}$$

$$\mathcal{B}_{2n} = 0. \tag{13.12}$$

From here, it can be seen that for oscillating fields there are only perpendicular (normal) components of the electric field at the surface of a perfect electric conductor. There are no

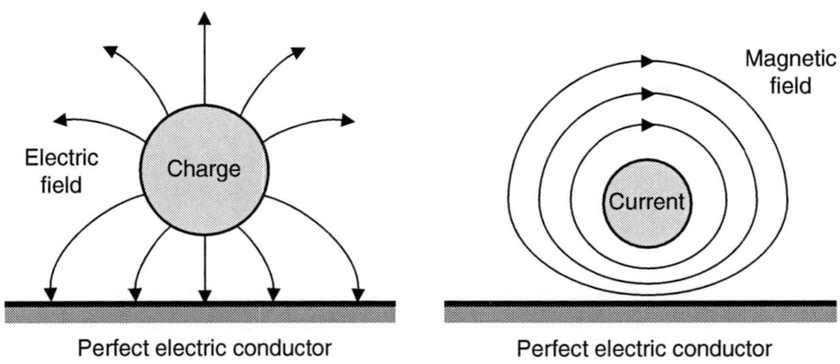

FIGURE 13.1
Electric fields and magnetic fields near a metallic surface. (From Cole, P.H., Jamali, B., and Ranasinghe, D., *Coupling Relations in RFID Systems*, Auto-ID Center White Paper Series, 2003, © 2003 by Auto-ID Center. With permission.)

tangential components of the electric field directly next to a perfect electric conductor. On the other hand, there are only tangential components of the magnetic field directly next to a perfect electric conductor. There are no normal components of the magnetic field to the surface of a perfect electric conductor. Hence, not all components of electromagnetic fields are available near a perfect electric conductor.

There are actually no perfect electric conductors in reality, but since metal has a high conductivity, it is qualified to be close enough as a perfect electric conductor. Figure 13.1 shows simplified illustrations of electric fields and magnetic fields near a metallic surface. Note that only the most relevant to RFID expressions for the boundary conditions have been shown earlier; detailed derivations of those expressions can be found in literatures such as Balanis (1989) and Hayt (1989). The relation of those results to RFID will be discussed in the following section, but we note here that the results provide a preliminary design consideration for one of the RFID tags that is presented in this chapter.

Before proceeding to the next section, another part of the electromagnetic theory that is related to the presence of metallic media in the environment should be mentioned. For a uniform plane electromagnetic wave directed at normal incidence to a boundary formed by having a perfect electric conductor on one side, a phase reversal of the field occurs on reflection from the boundary. Consequently, the total of the incident and reflected electric fields at the boundary will be zero. Further, the total of the incident and reflected magnetic fields at the boundary will be double that of the incident magnetic field. It can be observed that these results satisfy the boundary conditions discussed earlier.

13.3 Effects of Metallic Surfaces on RFID Tag Antennas

Passive UHF RFID tags are able to provide good read ranges for object identification compared with LF or HF RFID tags, and they are also seen as potentially low cost. However, conventional planar passive UHF RFID tags will suffer degradation in performance when attached to metallic objects or structures. A number of commercially available planar and label-like passive UHF RFID tags have been tested against a large aluminum plate (Dobkin and Weigand, 2005). Results from the testing have shown that as the tags are brought closer to the aluminum plate, the read range decreases. It is also shown that the read range has approached zero when the tags are <2 mm from the aluminum plate.

Passive RFID tags obtain their energy from interrogation field from the RFID reader antenna. The energy obtained is then converted to electrical energy within the tag for powering up of the tag chip (Cole, 2002). If there is insufficient interrogation field from the RFID reader antenna reaching the tags, the tags would not be readable. As discussed in the previous section, in the presence of a metallic structure, not all electromagnetic field components are present near the surface of the metallic structure; there are only the normal component of the electric field and tangential component of the magnetic field. Hence, any RFID tag that depends on either the tangential component of the electric field or the normal component of the magnetic field to operate will suffer from serious performance degradation when attached to and close to a metallic surface.

Another issue that arises when placing an RFID tag near a metallic surface is the change of the tag antenna parameters such as the input impedance, directivity, radiation pattern, and the efficiency. Antennas, such as electric dipoles, will suffer a significant change in their impedance when placed near a metallic surface. Plots of impedance change for a basic half wavelength dipole as well as a basic circular loop antenna placed horizontally and located at different distance above a metallic or conducting surface can be found in Balanis (2005). Studies on impedance changes of a folded dipole antenna corresponding to different distances of the antenna from a metallic plate are also presented in Raumonen et al. (2003) and Prothro et al. (2006). The change of the tag antenna impedance will then lead to two issues. First is the perturbation of the resonant frequency of the tag. The resonant frequency of a tuned circuit can be expressed as

$$f_r = \frac{1}{2\pi\sqrt{LC}}, \tag{13.13}$$

where L and C are the inductance and capacitance of the circuit, respectively. This expression shows that the change of the reactive part of the antenna impedance will lead to the change of the resonant frequency. Hence, when the tag is brought close to a metallic surface, the tag antenna impedance will change, which then affects the resonant frequency. This means detuning will occur and the read range (between the RFID tag and the RFID reader antenna) will degrade. The seriousness of the read range performance degradation will depend on the amount of the resonant frequency perturbation and the role that the quality factor of the tag resonance may play in impedance matching.

The second issue caused by the change in the tag antenna impedance is impedance mismatch. The tag antenna is usually designed to have an impedance, which matches as closely as possible to the RFID tag chip impedance. A match between the tag antenna and chip impedances means that one of the impedances is the complex conjugate of the other and hence in theory, a maximum power transfer will occur. When the tag antenna impedance is affected by the presence of metallic structures, the impedance matching will be affected. Hence, the amount of power transfer from the antenna to the tag chip will also change. This will in turn affect the read range performance. In addition, the change of the tag antenna impedance may also affect the bandwidth over which good performance is obtained.

Besides changing the input impedance, the presence of metallic structures may also cause changes to other antenna parameters such as the directivity and the radiation pattern. Antennas such as an electric dipole with an omnidirectional radiation pattern may become directional when placed close to a metallic surface. The reflections caused by the metallic surface may change the concentration of the electromagnetic fields near the antenna and hence, will lead to the change of directivity. Measurements have shown that when an omnidirectional antenna is placed near a cylindrical metallic can at a separation of ~50 mm, the antenna gain has suffered a reduction of as much as 20 dB (in the direction of the antenna nearer to the metallic can) compared with the gain when the antenna is in free space (Foster and Burberry, 1999). The changes in the directivity and radiation pattern will of course depend on the shape and size of the metallic structure and also the separation distance of the antenna from the structure. These have been studied for a folded dipole antenna in Raumonen et al. (2003).

Discussed earlier are some of the effects the metallic structure can possibly cause to RFID tags. Although they are negative effects, and make tagging metallic objects seem difficult, satisfactory solutions may be engineered. One obvious solution for metallic object tagging is to use an antenna, such as a patch antenna, that requires a ground plane to operate. Since the ground plane is a part of the antenna design, this type of antenna will not be affected too much when attached to a metallic object. Another solution is to use a tag antenna design that is able to use the electromagnetic fields that are present near the metallic surface to operate. An increase in the antenna directivity due to the metallic surface may also be obtained. Tag designs corresponding to both solutions above will be discussed further in this chapter.

13.4 RFID Tag Antennas Suitable for Metallic Surfaces

Some of the biggest challenges when it comes to designing RFID tags for metallic objects are the size and cost of the tags. Small tags are most desired, as the tags may be required to fit on smaller size metallic objects or even in tight spaces. However, small tag antennas are usually associated with low radiation resistance, efficiency, and gain. It is also the aim to maintain a low tag cost to increase the feasibility of tagging every item. One direct way to do this is to keep the tag antenna design as simple as possible to reduce the manufacturing cost, and another way is to use if possible cheap materials. In this section, a number of RFID tags suitable for attaching to metallic surfaces will be discussed. Brief design concepts and methods, as well as experimental results, will also be included for some of the tags.

13.4.1 RFID Tag with a Simple Rectangular Patch Antenna

Since patch antennas require a ground plane to operate, they will less likely be affected when attached to metallic surfaces.* As will be shown in the following, we have experimented with an RFID tag consisting of a simple basic rectangular patch antenna that can be used for metallic object identification.

Shown in Figure 13.2 is the top view of the RFID tag. The tag is designed to operate in the Australian UHF RFID band that spans 920–926 MHz. A double-sided copper clad FR4 board material with substrate thickness of $h = 1.6$ mm and relative dielectric permittivity $\varepsilon_r = 4.4$ is used. The dimensions of the rectangular patch are determined first by theoretical calculations using the methods in Balanis (2005) and later verified (with some slight adjustments) by simulations using the Ansoft HFSS simulation program. The patch has a length $L_{patch} = 76$ mm and width $W_{patch} = 99$ mm. Both the substrate and ground plane have the same size and are $6h$ longer on each side compared with the rectangular patch.

The tag design also consists of a simple impedance matching method, whereby the antenna is fed using an inset feed method and the impedance is transformed using a microstrip line to have an overall antenna impedance that is the complex conjugate of the tag chip impedance. The chip used has an impedance of $20-j141\ \Omega$ at 923 MHz. The tag chip is located at the end of the microstrip line between the microstrip line and a small

* This section is based on the work by Ng et al. (2007) © 2007 by IEEE with permission.

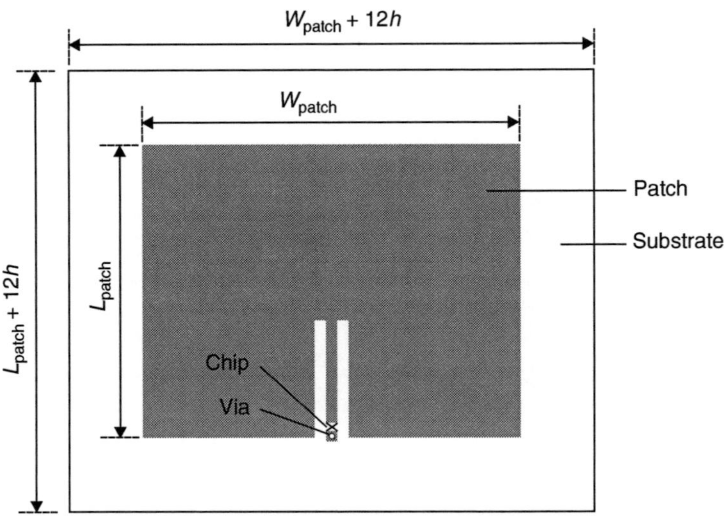

FIGURE 13.2
Structure of the RFID tag consisting of a patch antenna (*top* view). (From Ng, M.L., Leong, K.S., and Cole, P.H., in *IEEE Antennas and Propagation Society International Symposium*, Honolulu, Hawaii, June 10–15, 2007, © 2007 by IEEE. With permission.)

square area that is connected to the ground plane through a via. The inset distance of the feed and the length of the microstrip line are determined using simulations. Further details on the design steps and tag dimensions can be found in Ng et al. (2007).

The tag antenna was fabricated and a practical read range measurement was carried out. An RFID reader (Model ALR-9780-EA) suitable for operation in Australia was used in the measurement. The RFID reader antenna used has a 6 dBi gain and the equivalent isotropic radiated power (EIRP) from the antenna is 4 W. The tag is placed on a $1.5\lambda \times 1.5\lambda$ aluminum metallic plane and with the reader antenna radiating at normal incidence to the metallic plane, the read range measured was 1.44 m.

Since it is always desirable to have a small tag size, further analysis on the tag antenna has been performed by reducing the tag antenna size to find the smallest possible size while still offering acceptable read range performance. The patch width W_{patch} of the tag antenna was reduced in steps of 10 mm from the original size of 99 mm to 19 mm. Changes to the overall input impedance of the tag antenna have been observed through simulations when W_{patch} is reduced. To cater for the impedance changes and maintain a conjugate match of both real and imaginary parts between the antenna and the chip, the inset feed distance and the microstrip line length were adjusted slightly for each decrement of W_{patch}.

RFID tags corresponding to different patch widths were fabricated. Read range measurements of these tags are performed using the same RFID reader and reader antenna as mentioned earlier. The plot of read range versus W_{patch} is shown in Figure 13.3 for the tag in free space and when the tag is attached to a $1.5\lambda \times 1.5\lambda$ aluminum metallic plane. It can be observed from Figure 13.3 that there is a pattern in the reduction of read range when W_{patch} is reduced. The read range of the smallest size tag (with $W_{patch} = 19$ mm) is about half that of the full size tag (with $W_{patch} = 99$ mm). However, despite the read range reduction, the read range for the smallest tag is still acceptable considering the amount of tag size reduction compared with the full size tag. Hence, the smaller tag can be suitable for use in applications that do not require a maximum possible read range but do require a smaller tag size to attach the tag to a limited space or area on the metallic object.

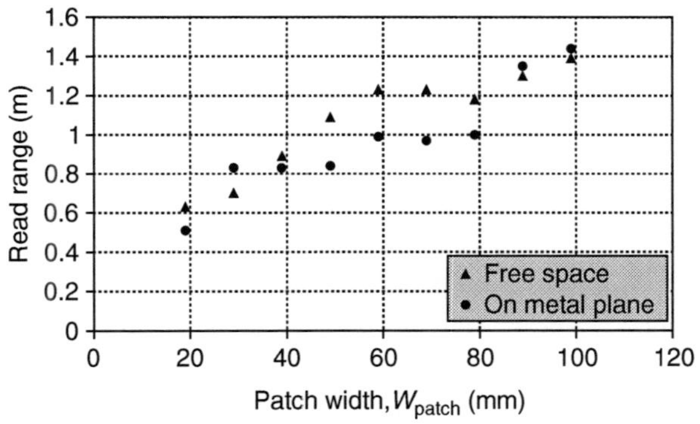

FIGURE 13.3
Practical read range measurement results for tag in free space and tag attached to metallic plane. (From Ng, M.L., Leong, K.S., and Cole, P.H., in *IEEE Antennas and Propagation Society International Symposium*, Honolulu, Hawaii, June 10–15, 2007, © 2007 by IEEE. With permission.)

13.4.2 RFID Tag with a Small Loop Antenna

The RFID tag uses a design approach different from that of the earlier and utilizes the electromagnetic fields that are present near the metallic surface.* According to the theory of electromagnetic boundary conditions, there are only tangential components and no normal components of the magnetic field to the metallic surface. In addition, the magnetic field (tangential component) will be doubled when it is very near the metallic surface. The RFID tag design here exploits this fact by having a loop antenna oriented such that the plane of the loop is perpendicular to the plane of the metallic surface to which the RFID tag will be attached. The idea is to have the rich concentration of the magnetic field near the metallic surface to couple to the loop antenna of the tag.

The structure of the tag, consisting of a rectangular loop antenna and a tag chip, is shown in Figure 13.4. The loop antenna is made of a wide copper strip of width W_{rec}. The physical dimension of the rectangular loop antenna can be adjusted to provide sufficient inductance to be tuned by the combination of the capacitance of the tag chip and the self-capacitance of the inductor. The trade-off of this method is that the small rectangular loop antenna will be able to provide sufficient inductance but not the resistance that corresponds to the tag chip impedance, since it is a characteristic of small loop antennas to have low radiation resistance. However, this method has the benefit of maintaining simplicity and low cost. In Ng et al. (2006), the loop antenna width W_{rec} is varied with the rest of the loop dimensions such as length L_{rec} and height H_{rec} fixed.

Theoretical calculations can be used first to estimate the dimensions of the loop antenna that resonate at the target frequency when tuned with the capacitance of the tag chip. The formulae used for such calculations are the same as for rectangular loop antennas made of

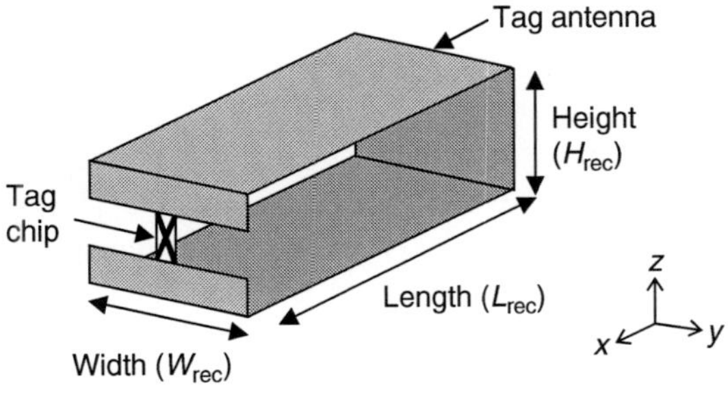

FIGURE 13.4
Structure of the RFID tag consisting of a rectangular loop antenna. (From Ng, M.L., Leong, K.S., and Cole, P.H., in *21st International Technical Conference on Circuits/Systems, Computers and Communications (ITC-CSCC)*, Chiang Mai, Thailand, July 10–13, 2006, © 2006 by ECTI. With permission.)

* The RFID tag shown in Figure 13.4 and this section are based on the work by Ng et al. (2006) © 2006 by ECTI with permission.

circular wire, except that the copper strip of this antenna has to be converted to its equivalent circular wire radius r before calculations are performed. Deriving from the equivalent radius expression given in Balanis (2005), the equivalent radius r corresponding to the wide copper strip of width W_{rec} is

$$r \cong 0.2W_{rec}. \quad (13.14)$$

The rectangular loop antenna with the structure presented here can be approximately represented (assuming it to be lossless, electrically small, and of negligible self-capacitance) by a resistor with radiation resistance R_{rad} and an inductor with inductance L_{ant} in series. Small loop antennas with the same loop area carrying a uniform current have the same radiation resistance. Hence, the radiation resistance and approximate inductance of the rectangular loop antenna are determined using the expressions found in Balanis (2005) that correspond to a small circular loop antenna with loop radius R and circular wire radius r. The expressions are

$$R_{rad} = 20\pi^2(\beta R)^4, \quad (13.15)$$

$$L_{ant} = \mu_0 R\left[\ln\left(\frac{8R}{r}\right) - 2\right], \quad (13.16)$$

where $\beta = \frac{2\pi}{\lambda}$ is the free space propagation constant with wavelength λ and $\mu_0 = 4\pi \times 10^{-7}$ $\mathrm{H\,m^{-1}}$ is the free space permeability. With R fixed depending on the desired loop area, the circular wire radius r corresponding to the required L_{ant} can be determined using these expressions. W_{rec} can then be determined after finding r.

A fabricated tag based on the theoretically calculated dimensions can be checked for its resonant frequency. This can be done by setting a network analyzer to couple to the tag by means of a small untuned loop, and to measure the reflection from that loop over a set frequency range. By observing the reflection pattern on the network analyzer, the approximate resonant frequency of the tag can be found. If the resonant frequency of the tag deviates slightly from the target frequency, the loop antenna width W_{rec} can be adjusted to fine-tune the design. Further details on the fine-tuning steps can be found in Ng et al. (2006).

An example of a tag designed using this method has a rectangular loop antenna with dimensions of $W_{rec} = 15$ mm, $L_{rec} = 25$ mm, and $H_{rec} = 10$ mm (L_{rec} and H_{rec} are fixed from the very beginning of the design process). The tag is designed to resonate at the frequency of around 915 MHz. For the theoretical calculations during the design, a tag chip impedance of 7–j150 Ω at the frequency 915 MHz is assumed. Shown in Figure 13.5 is the simulated directivity patterns of the tag antenna located in both free space and 3 mm above a $1.5\lambda \times 1.5\lambda$ aluminum metallic plane. The small 3 mm gap is required to avoid short circuit of the tag to the metallic plane. As can be observed from Figure 13.5, the shape of the antenna pattern has been significantly changed and the directivity is enhanced with the presence of a metallic plane. It has also been found that a smaller ground plane gives lesser directivity and more backward radiation. As already noted, in terms of impedance matching, there is a reasonable conjugate match to the imaginary part but the real parts maintain somewhat mismatched. The design maintains the advantages of simplicity and economy.

An RFID reader which has the ability to operate over the band 900–940 MHz was set up to measure the read range performance of the RFID tag. The reader has an output peak power of ~250 mW. Taking into consideration of the 8 dBi gain circularly polarized reader antenna used, the total equivalent transmit power is ~1.6 W EIRP. As in the simulation, the

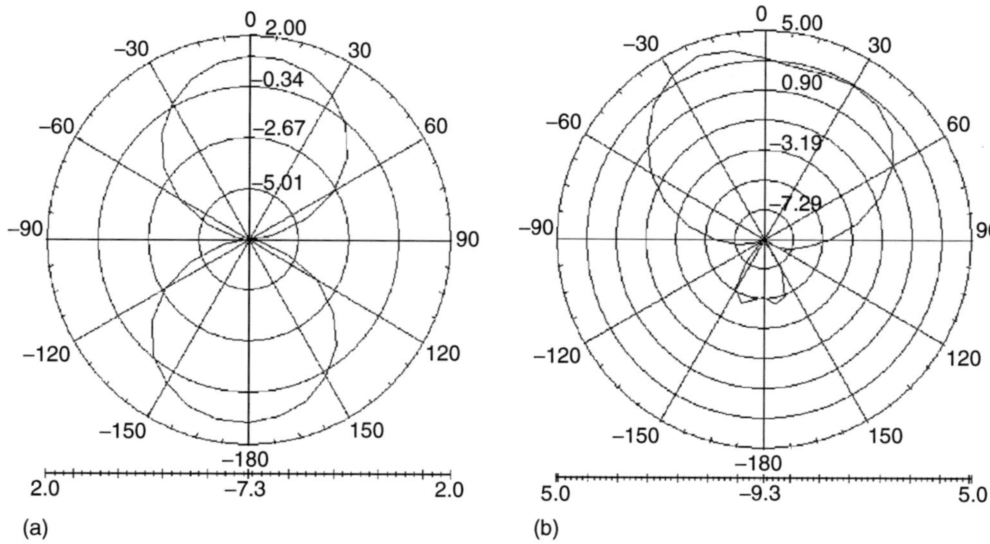

FIGURE 13.5
The *yz*-plane radiation pattern of the RFID tag antenna: (a) tag in free space; (b) tag located 3 mm above a $1.5\lambda \times 1.5\lambda$ metallic plane. (From Ng, M.L., Leong, K.S., and Cole, P.H., in *21st International Technical Conference on Circuits/Systems, Computers and Communications* (ITC-CSCC), Chiang Mai, Thailand, July 10–13, 2006, © 2006 by ECTI. With permission.)

RFID tag is placed 3 mm above a $1.5\lambda \times 1.5\lambda$ aluminum metallic plane. With the reader antenna radiating at normal incidence to the metallic plane, the read range measured at a number of frequencies within 900–940 MHz is shown in Figure 13.6. At 915 MHz, a read range of ~0.83 m is achieved for the above specified total transmit power of the RFID reader. For a total transmit power of 4 W EIRP (e.g., in the United States), the read range is expected to increase by ~1.6 times. Overall, the read range performance of this RFID tag

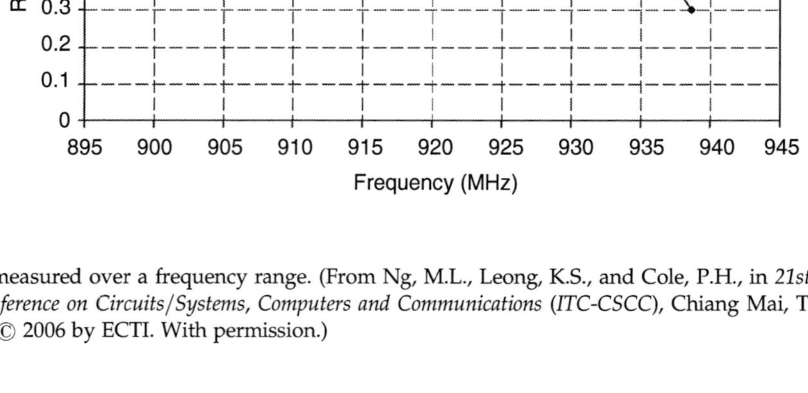

FIGURE 13.6
Read range measured over a frequency range. (From Ng, M.L., Leong, K.S., and Cole, P.H., in *21st International Technical Conference on Circuits/Systems, Computers and Communications* (ITC-CSCC), Chiang Mai, Thailand, July 10–13, 2006, © 2006 by ECTI. With permission.)

is promising, with a good read range achieved over a wide frequency range. The reason for this is that the RFID tag antenna has enhanced performance when it is near a metallic surface, and the structure of this antenna has allowed good coupling to the magnetic components of the interrogation fields from the RFID reader near the metallic surface.

13.4.3 Other RFID Tags

Two RFID tags that are based on rather different design approaches and concepts have been discussed earlier. RFID tags suitable for attaching to metallic objects are, of course, not just limited to these two earlier examples. Other examples, such as presented in Ukkonen et al. (2004a) and Choi et al. (2006), have used a planar inverted-F antenna in their tag designs. There are also different patch antenna designs for RFID tags suitable for metallic objects. In Ukkonen et al. (2004b), a patch antenna with an electromagnetic band gap (EBG) ground plane has been used in the tag design and it has been shown in Ukkonen et al. (2005) that this tag offers a slightly better read range performance compared with a tag consisting of a similar patch antenna but with a conventional ground plane. Yu et al. (2007), Kim et al. (2006), and Son et al. (2006) have also offered different tag designs that use patch antennas.

Although there are different tag designs for tagging metallic objects and they vary in sizes and read range performances, all the tags mentioned earlier have a common problem, that is, they have a certain significant thickness as compared with conventional label-like planar RFID tags. For example, patch antennas will usually have a thickness contributed by the dielectric substrate layer. For the loop antenna discussed in this chapter (loop perpendicular to the metallic surface), the thickness is caused by the area encompassed by the loop. Hence, there is actually no universal choice of tags for metallic object. The choice of the tag antenna used will depend on the type of object to be tagged, the amount of tagging space available, tagging cost limitation, and as well as the read range requirement.

13.5 Conclusion

In recent years, research on RFID involving metallic object identifications has significantly increased. There have been more experiments performed to study the severity of the effects caused by metallic objects toward the RFID tag antenna and the read range performance, some of which have been mentioned in this chapter. With the aid of a few examples, this chapter shows that with proper tag antenna choice and design, it is possible to tag metallic objects with an acceptable level of read range performance. Although the research in the area of metallic object tagging is growing rapidly, more research is still required to encourage wider RFID implementation.

References

Angeles, R. 2005. RFID technologies: Supply-chain applications and implementation issues. *Information Systems Management*, 22(1), 51–65.

Balanis, C.A. 1989. *Advanced Engineering Electromagnetics*. John Wiley & Sons, New York.

Balanis, C.A. 2005. *Antenna Theory: Analysis and Design*, 3rd ed. John Wiley & Sons, New York.

Choi, W., H.W. Son, J.-H. Bae, G.Y. Choi, C.S. Pyo, and J.-S. Chae. 2006. An RFID tag using a planar inverted-F antenna capable of being stuck to metallic objects. *ETRI Journal*, 28(2), 216–218.

Cole, P.H. 2002. *A Study of Factors Affecting the Design of EPC Antennas and Readers for Supermarket Shelves*. Auto-ID Centre White Paper.

Dobkin, D.M. and S.M. Weigand. 2005. Environmental effects on RFID tag antennas. *Microwave Symposium Digest*, 2005 IEEE MTT-S International.

Foster, P.R. and R.A. Burberry. 1999. Antenna problems in RFID systems. *IEE Colloquium on RFID Technology*, London, UK, pp. 3/1–3/5.

Hayt, W.H., Jr. 1989. *Engineering Electromagnetics*, 5th ed. McGraw-Hill, New York.

Kim, S.-J., B. Yu, Y.-S. Chung, F.J. Harackiewicz, and B. Lee. 2006. Patch-type radio frequency identification tag antenna mountable on metallic platforms. *Microwave and Optical Technology Letters*, 48(12), 2446–2448.

Lampe, M., M. Strassner, and E. Fleisch. 2004. A ubiquitous computing environment for aircraft maintenance. *Proceedings of the 2004 ACM Symposium on Applied Computing*, Nicosia, Cyprus, pp. 1586–1592.

Ng, M.L., K.S. Leong, and P.H. Cole. 2006. A small passive UHF RFID tag for metallic item identification. *21st International Technical Conference on Circuits/Systems, Computers and Communications (ITC-CSCC)*, Vol. 2, Chiang Mai, Thailand.

Ng, M.L., K.S. Leong, and P.H. Cole. 2007. Design and miniaturization of an RFID tag using a simple rectangular patch antenna for metallic object identification. *IEEE Antennas and Propagation Society International Symposium*, Honolulu, Hawaii.

Prothro, J.T., G.D. Durgin, and J.D. Griffn. 2006. The effects of a metal ground plane on RFID tag antennas. *IEEE Antennas and Propagation Society International Symposium*, Albuquerque, New Mexico.

Raumonen, P., L. Sydanheimo, L. Ukkonen, M. Keskilammi, and M. Kivikoski. 2003. Folded dipole antenna near metal plate. *IEEE Antennas and Propagation Society International Symposium*, vol. 1, pp. 848–851.

Sarma, S., D. Brock, and D. Engels. 2001. Radio frequency identification and the electronic product code. *IEEE Micro*, 21(6), 50–54.

Son, H.W., J. Yeo, G.Y. Choi, and C.S. Pyo. 2006. A low-cost, wide-band antenna for passive RFID tags mountable on metallic surfaces. *IEEE Antennas and Propagation Society International Symposium*, pp. 1019–1022.

Strassner, M. and E. Fleisch. 2003. *The Promise of Auto-ID in the Automotive Industry*. Auto-ID Center White Paper.

Ukkonen, L., D. Engels, L. Sydanheimo, and M. Kivikoski. 2004a. Planar wire-type inverted-F RFID tag antenna mountable on metallic objects. *IEEE Antennas and Propagation Society International Symposium*, vol. 1, pp. 101–104.

Ukkonen, L., L. Sydanheimo, and M. Kivikoski. 2004b. Patch antenna with EBG ground plane and two-layer substrate for passive RFID of metallic objects. *IEEE Antennas and Propagation Society International Symposium*, vol. 1, pp. 93–96.

Ukkonen, L., L. Sydanheimo, and M. Kivikoski. 2005. Effects of metallic plate size on the performance of microstrip patch-type tag antennas for passive RFID. *Antennas and Wireless Propagation Letters*, vol. 4, pp. 410–413.

Yu, B., S.-J. Kim, B. Jung, F.J. Harackiewicz, and B. Lee. 2007. RFID tag antenna using two-shorted microstrip patches mountable on metallic objects. *Microwave and Optical Technology Letters*, 49(2), 414–416.

14

WISP: A Passively Powered UHF RFID Tag with Sensing and Computation

Daniel J. Yeager, Alanson P. Sample, and Joshua R. Smith

CONTENTS

14.1 Introduction: Ubiquitous Sensing

This chapter describes the wireless identification and sensing platform (WISP), a wireless, battery-free platform for sensing and computation that is powered and read by a standards-compliant ultrahigh frequency (UHF) RFID reader. WISP features a wireless power supply, bidirectional UHF communication with backscatter uplink, and a fully programmable ultralow power 16 bit flash microcontroller with analog to digital converter. This particular point in the sensor platform design space offers some attractive features for ubiquitous sensing, but has not yet been explored very thoroughly. Using the WISP platform, we have implemented the first accelerometer to be powered and read wirelessly in the UHF band, and also the first UHF powered-and-read strain gage. Even without its sensing capabilities, WISP can also be used as an open and programmable RFID tag: with collaborators we implemented the RC5 encryption algorithm on WISP [1], which we believe is the first implementation of a strong cryptographic algorithm on a UHF tag.

One approach to ubiquitous sensing is to use wired sensors. This approach is well-suited to creating purpose-built instrumented environments supporting long-term observation [2]. This approach has the advantage that there is no battery lifetime or battery size constraint. The drawback is the need for wires. In Ref. [2], the authors describe such a purpose-built living space, which includes custom cabinetry to house sensors as well as cables for both power and data. A second approach, favored by the Wireless Sensor Networks community, is to use battery-powered devices that communicate by ordinary radio communication, often in a peer-to-peer fashion [3]. One disadvantage of this approach is the size and lifetime constraints imposed by batteries. Because of the lifetime constraint, it would not be possible to permanently embed a battery-powered device in a building or civil structure such as a bridge. A third approach includes generating power from environmental sources [4], such as vibration, light, or human motion [5], and then communicating the sensor data by ordinary RF transmission. The final class of approaches, within which this work fits, is to deliberately transmit power from a large source device to the sensor platforms, which then harvest this "planted" power, rather than relying on generation of electrical energy from naturally occurring or "wild" sources as in the third approach.

The space of wireless power/data transmission can be subdivided further. Chip-less approaches such as Theremin's cavity resonator microphone [6], as well as more recent examples such as Refs. [7,8], are based on an analog technique in which a quantity to be sensed modifies the frequency or quality factor of a resonant structure. Changes in the resonance can be detected by a "reader" device that is effectively supplying power and collecting analog sensor data. This analog approach is generally limited to a small number of sensors, since the devices cannot be given arbitrarily long unique IDs. Furthermore, these analog sensor devices are not capable of onboard computation, which means the system cannot benefit from channel sharing, error detection and correction, embedded compression and filtering, cryptography, and other capabilities enabled by digital computation and communication.

RFID tags are wirelessly powered digital devices that include a conventional integrated circuit (IC) [9]. Conventional RFID tags are fixed function devices that typically use a minimal, nonprogrammable state machine to report a hard-coded ID when energized by a reader. The Electronic Product Code (EPC) standard operates in the UHF band (915 MHz in the United States), which has substantially improved the range and field-of-view for RFID reading over previous generations of RFID technology. The "EPC Class 1 Generation 1" specification [10] (Gen 1) was the first UHF RFID standard to be widely deployed. The standard's broad adoption enabled a new generation of applications and interoperable products. This standard has been supplanted by a second generation spec of the specification, "EPC Class 1 Generation 2" [11]. As of this writing, WISP supports the Gen 1 but not the Gen 2 specification, although we are developing a Gen 2 WISP.

RFID has been used for sensing in several contexts. Conventional short-range HF RFID tags with a worn RFID reader have been used for activity monitoring in eldercare scenarios [12]; in this case, the RFID read event is in effect a reader-tag proximity sensor. Integrating RFID tags with secondary sensors has been proposed [13] or implemented [14] in various contexts, and a small number of commercially available RFID sensors exist. In almost all cases, these devices are fixed function, and simply report a unique ID and sensor data. Most of the commercially available products are "active tags," meaning that the sensor platform is battery-powered; they use the RFID channel for communication but not power. In addition, existing RFID sensor devices are generally not programmable platforms supporting arbitrary computation. Most recent research in IC-based passive sensor tags has proceeded incrementally to augment the capabilities of RFID tags by integrating additional functional blocks without going beyond a simple digital state machine at the core. Thus these devices are, like ordinary ID-only tags, fixed in function, not programmable. For example, ADCs, specialized logic, and temperature sensors have been integrated into tag ICs with a finite state machine architecture typical of RFID tags [15,16]. WISP represents an alternative design philosophy that focuses on the programmability of a full microcontroller as part of a sensor-enhanced, passive RFID device.

One commercially available fully programmable microcontroller with an RFID interface is described in Ref. [12]. However, this device can only transmit 1 bit of sensor data per read event, and operates at 125 kHz, which limits its range to inches. The Near Field Communication (NFC) standard is also a relevant point of comparison [17]. NFC uses short-range RFID style 13.56 MHz communication to link powered devices. It does not provide any power-harvesting capability.

The general purpose WISPs that are the focus of this chapter were preceded by several less capable devices (also called WISPs) that were described in previous publications. The α-WISP [1] used two mercury switches to mechanically toggle between two commercially produced RFID integrated circuits. Thus if either of the two IDs was detected, it is interpreted as indicating object presence; which of the two IDs is read effectively conveys 1 bit of sensor data. We refer to this technique as "ID modulation." The π-WISP [18] used a microcontroller powered by harvested RF power to activate a GaAs RF switch, which multiplexed two commercially available RFID ICs to one tag antenna, to implement an electronically controlled form of ID modulation [19]. The π-WISP could transmit at most 1 bit of sensor data per query, and used two separate antennae for communication and power-harvesting. In Ref. [20], we presented the first WISP to implement the Gen 1 protocol on its own, without using a commercial RFID tag. This was the first WISP to use a single antenna for power-harvesting, reader-to-WISP data downlink, and WISP-to-reader data uplink. Like its predecessors, this WISP also encoded the sensor data in the EPC ID, but was able to control all bits of the ID, unlike the earlier ID modulation devices. We reported light level measurement using this platform; it also used its onboard processor to compute the CRC of the live light level data [20]. In Ref. [21], we described a next generation of the WISP hardware (known as Rev I), and presented a more detailed discussion of its design. The Rev I WISP used a demodulation technique based on a "miniharvester." This chapter focuses on a next generation of WISP (called Rev 2.3) that is smaller than its predecessors and uses a new demodulator called the "multifunction harvester–demodulator" that reduces demodulator current leakage thus yielding improved range. Figure 14.1 is a photograph of the Rev I and Rev 2.3 WISPs. Instead of a demodulator based on a second parallel harvester (as in the Rev I WISP), the Rev 2.3 WISP's main harvester can be selectively placed in a demodulation mode; when demodulation is not enabled, there is no power leakage through the demodulator.

The remaining sections of this chapter present the WISP design and power budget, followed by experimental results on range, and several representative applications, including several sensors and the RC5 cryptographic algorithm.

14.2 WISP Design

A block diagram of the WISP is shown in Figure 14.2. An antenna and impedance matching circuit precede the analog front end. The power-harvester block rectifies incoming RF energy into DC voltage to power the system. The demodulator follows the envelope

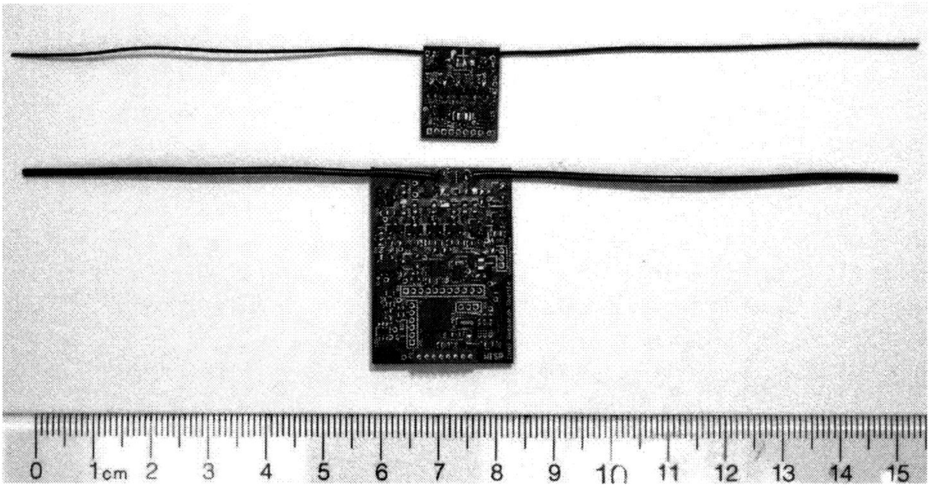

FIGURE 14.1
Photograph of WISP (Rev 2.3, at *top*) and an older WISP (Rev I, *bottom*), with a ruler for scale. The design was shrunk by using both sides of the PCB, using smaller components, and eliminating headers. Both boards include an accelerometer and an external temperature sensor (in addition to one built in to the microcontroller). Additional sensors or LEDs can be added by attaching a daughter board to header pins on the WISP.

of the RF carrier wave to extract the amplitude shift keying (ASK) data stream. This extracted baseband waveform is read by the MSP430 microcontroller (MCU) to receive downlink data from the reader. Uplink data is sent via the modulator circuit, which functions by changing the antenna impedance. Finally, onboard sensors are powered and measured by the MCU.

The Rev I WISP pictured in Figure 14.1 is made of a two-layer FR4 PCB with components limited to the top side. The Rev 2.3 board, also pictured, has components on both sides. A dipole antenna made of solid core wire is visible. The Rev 2.3 WISP in its base configuration (with no sensor daughterboard) has two onboard sensors: a temperature sensor and an accelerometer.

Small header pins expose microcontroller ports for expansion of daughter boards, external sensors, and peripherals.

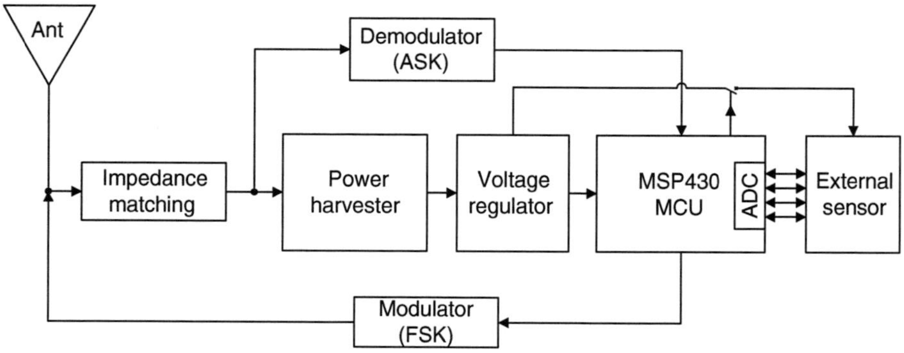

FIGURE 14.2
Block diagram of the WISP platform. (From Sample, A.P., Yeager, D.J., Powledge, P.S., and Smith, J.R., *IEEE International Conference on RFID 2007*, March 26–28, 2007. With permission.)

14.2.1 Analog Front End and Tuning

A schematic of the WISP analog circuitry is shown in Figure 14.3. The WISP analog front end differs slightly in purpose from that of conventional RFID tags. Due to the relatively high power consumption of WISP, the rectifier is designed to supply more current than ordinary tags. This circuit is excited by commercial, EPC Class 1 Generation 1 compliant readers operating at 902–928 MHz with an allowable transmission power of 4 W_{EIRP} (effective isotropic radiated power).

Due to loss in signal strength over transmission distance, there is potentially very little power for the tag. Therefore, efficient conversion of the incoming RF energy to DC power for the tag is an important design consideration. A matching network provides maximum power transfer from the antenna to the rectifier, and a five-stage voltage doubling circuit converts the incoming power to voltage. Low-threshold RF Schottky diodes are used to maximize the voltage output of the rectifier. Finally, this rectified DC voltage is stored in a large capacitor and supplied to a 1.8 V regulator to power the WISP.

To tune the antenna, the two dipole branches were mounted to an SMA connector that was then connected to a network analyzer, and the dipole length was optimized for 902–928 MHz band. Next, an SMA connector was attached to the WISP board in place of the antenna and the WISP was attached to the network analyzer, which was set to sweep from 902 to 928 MHz at a power of 0 dBm. The microcontroller was programmed to remain in LPM4 sleep mode to minimize its power consumption. The WISP's discreet matching network, composed of a series inductor and parallel trimmable capacitor, was tuned until the output voltage of the WISP was maximized. Note that the power-harvester is a nonlinear device, and its efficiency is highly load-dependent. Ultimately, the front end must be tuned to provide maximum output voltage in the presence of the desired load. Optimizing the matching network for the load of the microcontroller in its LPM4 sleep state effectively maximizes read range. To maximize read rate at close range, or power delivered at close range, one would tune the matching network differently.

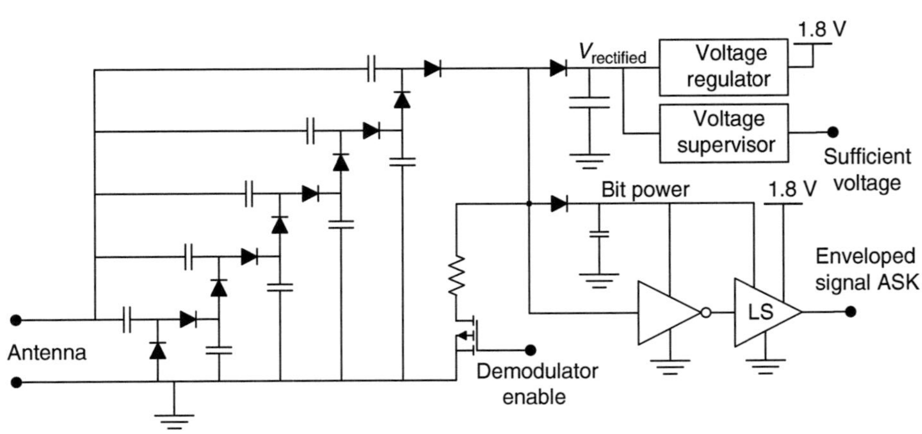

FIGURE 14.3
Schematic of Rev 2.3 WISP. This design improves on the miniharvester of the Rev I WISP. The two rightmost diodes are low frequency, low leakage diodes. The 10 diodes in the voltage doubling ladder are RF Schottky diodes. Instead of using a second miniharvester for demodulation, the demodulator is connected directly to the main harvester, but can be disabled to prevent energy from leaking through the demodulator's pull-down resistor. The component labeled "LS" is a level shifter. (From Sample, A.P., Yeager, D.J., Powledge, P.S., and Smith, J.R., *IEEE International Conference on RFID 2007*, March 26–28, 2007. With permission.)

14.2.2 Demodulation and Modulation

To encode reader-to-WISP data, in the EPC Gen 1 standard the reader amplitude modulates the 915 MHz RF carrier wave it emits. Normally, the carrier waveform remains at a constant amplitude; when bits are transmitted, the amplitude of the carrier drops to ~10% of its normal value. The duration of the low "break" indicates a logical "one" or a "zero." A short break (2 μS) indicates "zero," and a long break (5 μS) indicates "one." The harvester effectively demodulates the 915 MHz carrier, and leaves a baseband data signal—the downlink data—on the order of 70 kHz. Most of this signal is rectified into a DC voltage by the top right diode and large charge storage capacitor. When the demodulator is enabled, some of the signal travels through the demodulator branch. This signal is also rectified to produce a reference voltage for the bit detector, but the filter capacitor after the demodulator diode is much smaller than the power accumulator capacitor. Effectively, it removes the 70 kHz data signal and leaves a slowly varying average power level (i.e., just fast enough so that it can change on the timescale that the tag moves in space, say 10 Hz) that provides a dynamic reference for bit detection. Using this dynamic threshold, the instantaneous output voltage from the voltage doubling ladder is thresholded by a Schmitt trigger inverter, which removes noise and glitches. Finally, a level shifter (labeled LS in the schematic) converts the relative magnitude of the incoming data waveform into a 1.8 V logic level for the MSP430.

Note that the rails of the inverter and level shifter are connected neither to the main unregulated supply (V_{rec}) nor to the regulated supply, but instead are fed by the output of the final, "additional" rectification diode–capacitor pair in the demodulator section. In addition, note that because the final diodes are rectifying a 70 kHz signal instead of an RF signal, we were able to use an ultralow leakage diode for this component instead of the relatively high leakage Agilent HSMS-2852 RF diode used elsewhere.

14.2.3 Uplink

RFID tags do not actively transmit radio signals. Instead, they modulate the impedance of their antenna which causes a change in the amount of energy reflected back to the reader. This modulated reflection is typically called backscatter radiation. In order to change the impedance of the antenna, a transistor is placed between the two branches of the dipole antenna. When the transistor conducts current, it short circuits the two branches of the antenna together, changing the antenna impedance; in the nonconducting state, the transistor ideally has no effect on the antenna, and thus the power-harvesting and data downlink functions occur as if it were not present. WISP accomplishes this impedance modulation with a 5 GHz RF bipolar junction transistor which allows for effective shunting of the 915 MHz carrier wave.

14.2.4 Digital Section and Power Conditioning

As the power available to WISP is extremely limited, careful component selection must be made to minimize current consumption. As advances in IC manufacturing now allow discreet components with <1 μA current consumption and 1.8 V operation, we have shown that it is now possible to construct working, passively powered RFID tags with discreet components.

Most importantly, the general purpose computation abilities of WISP are enabled by an ultralow power microcontroller. The 16 bit flash microcontroller used in the Rev I WISP, the MSP430F1232, can run up to 4 MHz with a 1.8 V supply and consumes ~470 μA when active for this choice of frequency and voltage. (The microcontroller has a 6 MHz 3 V mode which consumes 1800 μA, which was used in Ref. [20]; in Ref. [21], we extended the range over the early results of Ref. [20] by improving the microcontroller's firmware, allowing for operation at lower voltage and clock frequency, and thus longer range.) Of particular interest for low power RFID applications, the MSP430 has various low power modes, and the minimum RAM-retention supply current is only 0.1 μA at 1.5 V. The device provides over 8 kilobytes of flash memory, 256 bytes of RAM, and a 10 bit, 200 kilo-samples per second analog to digital converter (ADC). The low power consumption of this relatively new device is a critical factor in enabling use of a general purpose microcontroller in passive RFID systems.

Another critical design consideration is operation with uncertain power supply conditions. Because the available RF power varies greatly throughout device operation, supervisory circuitry is necessary to wake and sleep the device based on the supply voltage level. The Rev I WISP uses a 1.9 V supervisor and a 1.6 V power-on-reset to control device state and reset the microcontroller, respectively. The supervisor provides roughly 100 mV of headroom on the storage capacitor above the 1.8 V regulator voltage. This serves to buffer the supply voltage from dropping below 1.8 V due to the large power consumption of the microcontroller in active mode.

14.2.5 Software

The onboard MSP430 programmable microcontroller is responsible for implementing EPC Class 1 Generation 1 communication between the WISP and an RFID reader, as well as measuring any attached sensors. Efficient programming for the device is essential in meeting the low power requirements of passive RFID tags. The WISP software can be described on three levels. At the lowest level is the communication code, which generates uplink bits and detects downlink bits. The next level, state and power management, is responsible for managing the device state, including sleep vs. active modes. The third level implements the application layer protocol for encoding sensor data in the tag ID.

14.2.5.1 Packet Decoding and Encoding

The most challenging aspect of programming the MSP430 involves meeting the timing constraints of the EPC protocol while still maintaining a low clock frequency. RFID tags, with custom state machines, are designed at the hardware level to receive and send using the EPC protocol. The general purpose MSP430 must be carefully tuned to perform EPC communication, both in the receiving and transmitting of data. In particular, a mix of C and assembly language is used, where the C code maintains ease of configurability for the firmware for different sensor applications and the assembly code allows fine grained control of the timing of the MSP430 for EPC communication.

The EPC protocol employs ASK modulation to encode data to the tag, representing the data bits 1 and 0 with a long and short gaps in RF power, respectively. To receive data from the reader, the MSP uses the periodic edge of the waveform as a hardware interrupt, and then during the interrupt service routine re-samples the bit line to detect a 1 or 0 during the differentiated part of the waveform. This data is quickly shifted into memory before repeating this process. To detect the end of transmission, a timer is refreshed during each bit. When bits are no longer received, the timer expires, the packet is interpreted, and if appropriate, a response is sent to the reader.

If a valid query is received from the reader, the WISP responds with its current data packet "ID." First, the ID is copied into CPU registers to allow fast access during the transmitting period. Second, the hardware timer is configured for pulse width-modulated (PWM) output. Finally, each bit of the response is read from the CPU registers and used to change the length of the PWM period, in time. Specifically, a zero is represented by a 70 kHz square wave and a one is represented by a 140 kHz square wave. This waveform is sent to the modulator, which creates backscatter radiation. It is important to note that although the signal to the reader is in the form of an amplitude-modulated reflection of energy, the data is encoded as a "higher order" frequency modulation of the "lower order" amplitude modulation. The naming convention of describing the uplink as frequency-modulated is maintained to be consistent with the EPC Gen 1 specification.

14.2.5.2 System State and Power Management Algorithm

Meeting the low power requirements of passive RFID tags requires that the MSP430 dissipate, on average, as little power as possible. With various sleep modes and fast startup time, this processor is well-suited to meet the stringent power requirements. In fact, time is mostly spent in LP4 (low power mode 4) which draws only 0.1 μA, and the running (active) mode current consumption is ~470 μA at 3 MHz and 1.8 V.

Figure 14.4 shows the operational power cycle of the microprocessor. The system is event driven by external interrupts from either the voltage supervisor signal or the bit line communication interrupt. As shown in the diagram, the microcontroller sleeps between events to conserve power.

There are three active mode blocks. The first, designated "Generate Packet," powers and samples an attached sensor and calculates the CRC to complete the EPC-compliant ID. The second block, designated "Receive," is initiated by a communication interrupt from the reader (demodulation circuit). The microcontroller receives the reader command and responds if the query is recognized. The final block, designated "Transmit," involves sending the ID to the reader. Some RFID readers must receive the same ID multiple consecutive times to correctly report the ID. While not shown in Figure 14.4, the same Receive and Transmit sequence is typically repeated three times to ensure that the reader acknowledges the ID.

14.2.5.3 Sensor Data Encoding

To communicate sensor data from WISP to a computer through an RFID reader, the data must be encoded into the tag ID. Using the first byte after the CRC to denote the type

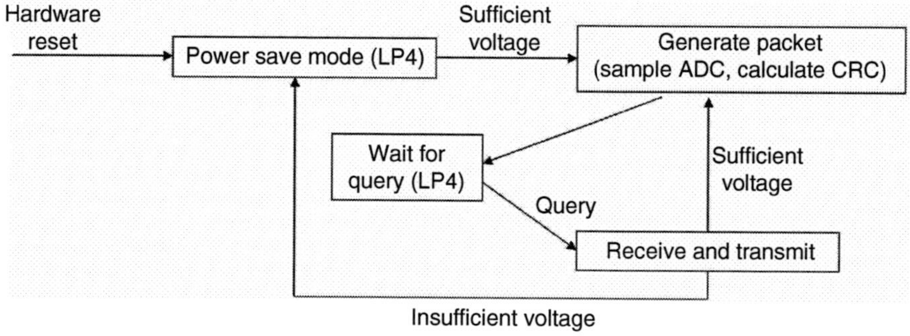

FIGURE 14.4
Operational power cycle. The hardware reset occurs at a lower voltage threshold; the "sufficient voltage" threshold is higher than the hardware reset threshold. When the voltage is between the lower and upper thresholds, the WISP remains in power save mode, which retains state. (From Sample, A.P., Yeager, D.J., Powledge, P.S., and Smith, J.R., *IEEE International Conference on RFID 2007*, March 26–28, 2007.)

of sensor attached, the remaining seven bytes can then be used to encode sensor data. The onboard ADC provides with 10 bit samples, allowing a maximum of five measurements to be transmitted per tag ID. This data is parsed on a computer in real time to display the most recent measurements reported by WISP.

14.3 Power Budget

One of the significant challenges of incorporating microcontrollers, sensors, and peripherals into passive RFID technology is the ability to manage the large power consumption of these devices. For example, the MSP430F1232 running at 3 MHz consumes ~470 μA at 1.8 V. The resulting power consumption is significantly larger than typical passive RFID tags. Under these conditions, the harvester cannot continuously supply power to the WISP during a single reader query.

One method used to overcome this challenge is to use a large storage capacitor (on the order of microfarads) to accumulate charge over multiple EPC queries. This allows for short bursts of power to activate and measure sensors and communicate at long distances where received power is minimal.

If the single query power requirements are not met, the WISP sleeps for several reader transmission cycles. This allows more time for charge accumulation. The approach of duty cycling is often used in low power applications; however, this presents a challenge for RFID networks when the WISP is not necessarily able to respond to each reader query.

The next section examines the issues related to powering the WISP from two perspectives. One is the power required to turn on the device and the other is the energy required for active operation.

14.3.1 Turn On Power Requirement

In order to increase the operational distance of the WISP, the minimum power threshold needed for turn-on is lowered by placing the device in a sleep mode, resulting in a current consumption of 2 μA (MSP430 0.1 μA). Stated another way, the inactive current consumption is minimized, allowing the harvester to rectify the 1.8 V needed for the MSP430 to activate. Given sufficient time, the storage cap will charge to the maximum output voltage of the harvester. Thus, a key parameter for maximizing the read distance of the WISP is not necessarily active current consumption but sleep current consumption. (The other key parameter determining range is the minimum voltage required for operation. The active mode power requirement determines maximum read rate, and is less critical in determining range.) While this strategy can result in slower update rates, it is necessary for powering large sensor loads over long distances. A significant amount of engineering is needed to keep the steady state inactive current consumption under 2 μA.

As shown in Ref. [21], the Rev I WISP requires at least −4 dBm or −0.5 dBm to function, depending on whether the power available to the WISP has dropped from a higher level (in which case, it can function all the way down to −4 dBm) or is rising from a lower level (in which case, it will not start until it has reached nearly −0.5 dBm). (In the latter case, "start-up" energy costs as well as component brown-outs increase the device's effective power requirement.) Using these best- and worst-case input power figures, it is possible to bracket the expected operating distance for the WISP using the Friis equation 14.1 for path loss.

$$P_R = P_T - 20\log\left(\frac{4\pi d}{\lambda}\right) + G_T + G_R, \quad (14.1)$$

where

P_R is the received power
P_T is the transmitted power
λ is the wavelength of the RF carrier wave
d is the distance from reader to WISP
G_T and G_R are the gains of the reader and WISP antennas in dBi

The transmit power of the reader $P_T = 1$ W = 30 dBm. Its center frequency is 915 MHz, corresponding to wavelength $\lambda = 0.33$ m. The transmit antenna gain $G_T = 6$ dBi (this yields an EIRP of ~4 W_{EIRP}, the U.S. regulatory limit for this ISM band). The receive antenna gain $G_R = 2$ dBi using the standard figure for the gain of a dipole antenna. Using the operating thresholds of −4 and −0.5 dBm, the Friis model, which does not model multipath effects, predicts operating range thresholds of 3.3 m (moving from near to far) and 2.2 m (moving from far to near). Experimentally, we have observed longer ranges than predicted by the Friis model enabled by additional multipath power. We have observed EPC packets received up to 4.5 m, which is discussed further in later sections.

14.3.2 Active Energy Consumption

Thus so far, it has been shown that a minimum power requirement needs to be met to rectify enough voltage to turn on the WISP. Additionally, an appropriate duty cycle period is needed to allow the storage capacitor to charge to the turn-on voltage threshold. Since the rectifier cannot supply enough power for continuous operation, it is important to understand the amount of energy that needs to be stored to power the WISP during active periods.

During one EPC Gen 1 reader query, the complete WISP (not just the microcontroller) consumes on average 600 μA at 1.8 V. Only the WISP's active period is considered, which is measured from first bit of the received preamble to 100 μs after the last bit of the response packet is transmitted, totaling 2 ms. For a storage capacitor of 8.5 μF voltage head room needed above 1.8 V is 136 mV, resulting in a total minimum voltage threshold of 1.93 V for a complete packet transmission.

The same methodology of calculating the required stored energy can be used to when selecting sensors to be added to the WISP platform. Sensor tasks and packet generation are generally done before the EPC query. However, it is reasonable to assume that when performing sensor applications the WISP will exhibit similar voltage and current consumption. Inequality (Equation 14.2) expresses an energy feasibility condition for a particular sensor: the energy required to read the sensor must not exceed the usable stored energy. The expression can be used to calculate the voltage headroom required to operate a particular sensor, which in turn determines the range at which the sensor can be operated.

$$V_{dd}(I_S + I_w)T \leq \frac{1}{2}C\left(V_{rec}^2 - V_{dd}^2\right), \quad (14.2)$$

where

I_S and I_w are the current consumption for the sensor and WISP, respectively
C is the capacitance of the storage capacitor
T is the total time of active operation
V_{rec} is the rectified voltage
V_{dd} is the required operating voltage

Assuming that the sensor has the same voltage supply as the WISP, V_{dd} equals 1.8 V. The left-hand side of inequality (Equation 14.2) is a straightforward expression for energy consumed by the sensor and WISP. The right-hand side represents usable stored energy. (Note that the total energy stored on the capacitor is $\frac{1}{2}CV_{rec}^2$, but not all of it is usable since charge stored on the capacitor at a voltage less than V_{dd} cannot operate the WISP.) Inequality (Equation 14.2) makes it clear that the limiting factor when selecting sensors is not only the current consumption (which determines power), but also the total required execution time of the sensor and WISP (energy rather than power). Additionally, it is important that sensors be disabled when not in use to minimize unnecessary energy expenditure.

14.4 Experimental Results

Figure 14.5 shows experimental results of the WISP performance: rectified output voltage, uplink packet error rate, and tag responses per reader query are plotted versus received power (dBm, top scale) and calculated distance (bottom scale). The experimental setup

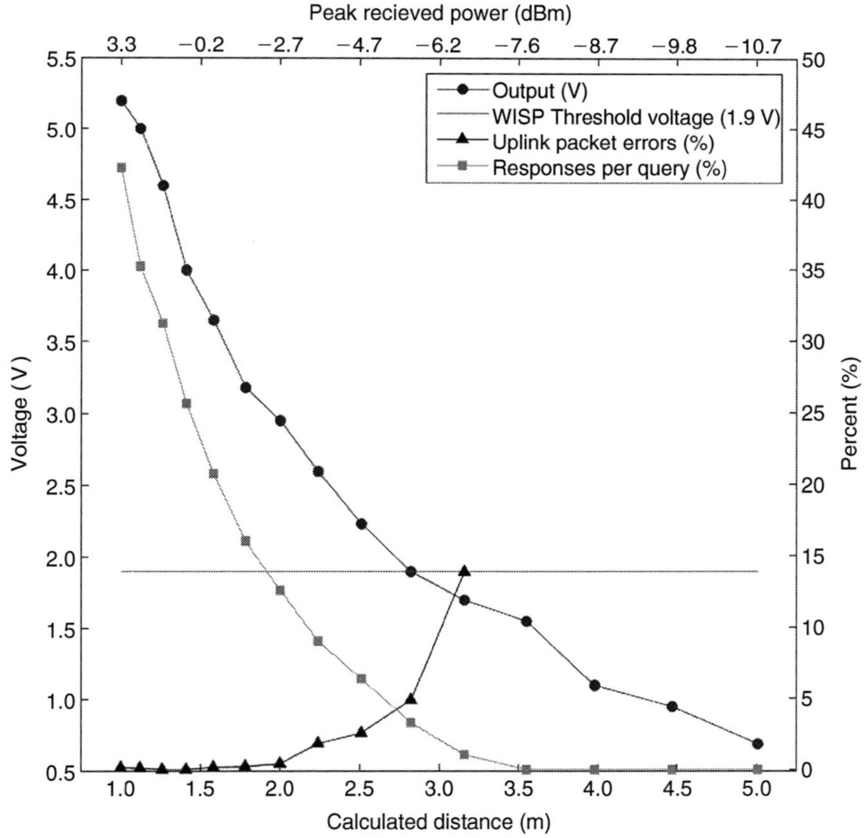

FIGURE 14.5
WISP performance harvested voltage, uplink packet error rate, and "Responses per query" as a function of input power (*top scale*) and calculated distance (*bottom scale*). The input power was controlled with an attentuator and the effective distance was calculated using the Friis transmission formula. The horizontal line with no markers shows the WISP voltage supervisor threshold of 1.9 V. "Uplink packet errors" is the number of failed uplink packets divided by attempted uplink packets, expressed as a percentage. "Responses per query" is the percentage of issued reader queries that return a packet with a valid CRC.

consisted of an EPC Gen 1 RFID reader driving a 6 dBi circularly polarized patch antenna. The reader's antenna and WISP were placed 1 m apart and 1 m off the ground to minimize multipath effects. An adjustable attenuator was used to vary the power delivered from the reader to the patch antenna. Then the Friis path loss equation was used to calculate the loss of the RF signal from the reader to the tag over the 1 m. Thus, the WISP received power is defined as reader transmit power (1 Watt), minus variable attenuator, minus transmission path loss. It should be noted that the 1 Watt source represents peak output power of the RFID reader, while the average output power (not considered here) is highly dependent on reader transmission rate and the specific implantation of the EPC Gen 1 protocol.

To measure rectified output voltage, the WISP is placed in its low power state, and voltage is averaged over a 10 s interval using an oscilloscope. Responses per query shows the number of successful tag responses received by the reader normalized over the total number of queries made. This is roughly equivalent to the operating duty cycle of the WISP, and as expected, is proportional to received power. The response rate drops to 0 at –3 dBm because there is insufficient voltage for operation. Uplink packet errors is the fraction of query responses made by the tag that is not correctly received by the RFID reader. Due to the limited data interface with the RFID reader selected for the experiment, the number of reader-rejected uplink packets is not directly available. To collect this data, the WISP counts the number of query responses it has made and reports the current tally as data encoded in each uplink packet. When the RFID reader application software receives gaps in the running tag response tally, an error is recorded. Figure 14.5 shows that uplink packet error rate increases and duty cycles decreases as available power

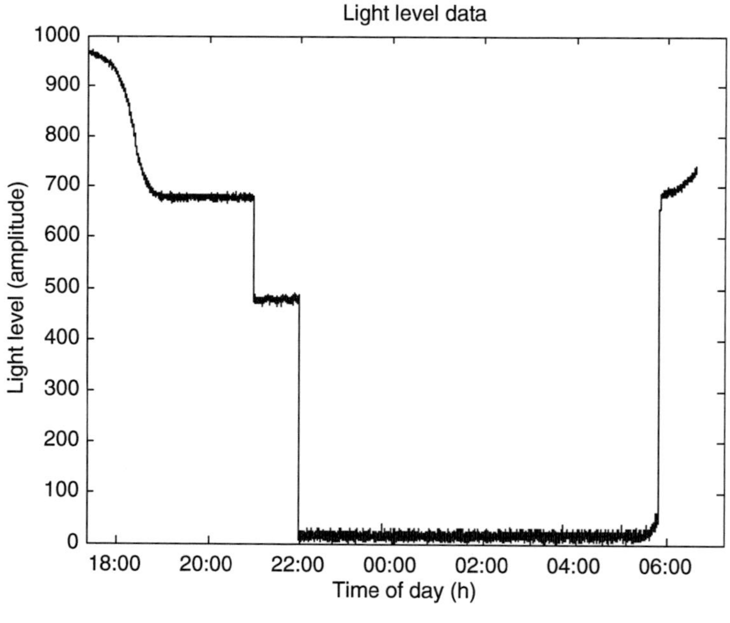

FIGURE 14.6
Light level measured by WISP in a 13 h period. The experiment began at about 5:40 p.m. (17:40 h). The first curve down is sunset. At around 9 p.m. (21:00 h), the measured light level drops substantially, probably because some lights in the laboratory were extinguished. At about 10 p.m. (22:00 h), the light level drops very low, probably because the remainder of the lights in the laboratory were extinguished. Just before 6 a.m. (06:00 h), the beginning of sunrise is visible, and the laboratory lights turn on. (From Smith, J.R., Sample, A., Powledge, P., Mamishev, A., and Roy, S., *Proceedings of Ubicomp 2006: 8th International Conference on Ubiquitous Computing*, Orange Country, CA, USA, September 17–21, 2006. With permission.)

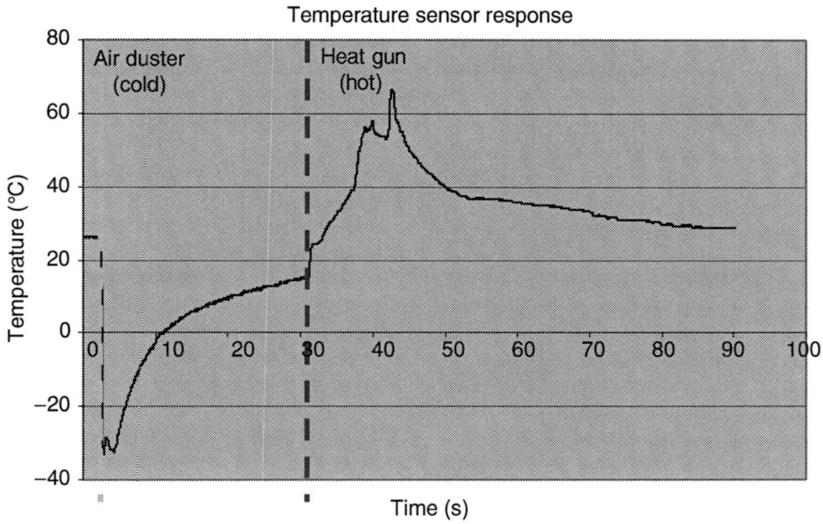

FIGURE 14.7
Temperature measurement with the WISP. A can of compressed air (held inverted) was used to generate a low temperature impulse, visible at about 2.5 s in the figure. The temperature reported by the WISP is allowed to recover toward room temperature for about 30 s, at which time a heat gun is used to generate a high temperature impulse.

decreases. The communication instability is primarily caused by the digitally controlled oscillator, which is used as the system clock, becoming detuned as supply voltage drops below 1.8 V.

14.5 Sensors and Applications

Very low power sensors requiring <50 μA of current are relatively easy to integrate with WISP. Examples include light, temperature, push buttons, and rectified voltage level. Figure 14.6 shows results from a light level measurement made with WISP [20].

We found the accuracy and power consumption of the on-chip MSP430 temperature sensor to be disappointing, but more accurate low power solid state temperature sensors are readily available, and we added an external temperature sensor to the WISP. Figure 14.7 shows a time series of temperature measurements reported by WISP using an external temperature sensor. An inverted can of compressed air was used to generate a low temperature impulse. After the WISP's temperature sensor had recovered for about 30 s, a heat gun was used to generate a high temperature impulse. The WISP continued to function properly even as its temperature sensor was driven from −40°C to +60°C.

Figure 14.8 shows data collected by a triaxial accelerometer on WISP [J.R. Smith et al., submitted]. The accelerometer draws 200 μA at 1.8 V. Due to the relatively high current consumption of these devices, continuously powering them would cripple the range of WISP. To overcome these high power requirements, the sensor is only be powered for a short period of time to take a measurement. Provided that the sensor and conditioning electronics can stabilize sufficiently quickly, this allows for a wide range of sensors to be measured over UHF RFID. Powering this accelerometer, the WISP is able to provide accelerometer measurements at rates of ~10–20 samples per second, depending on

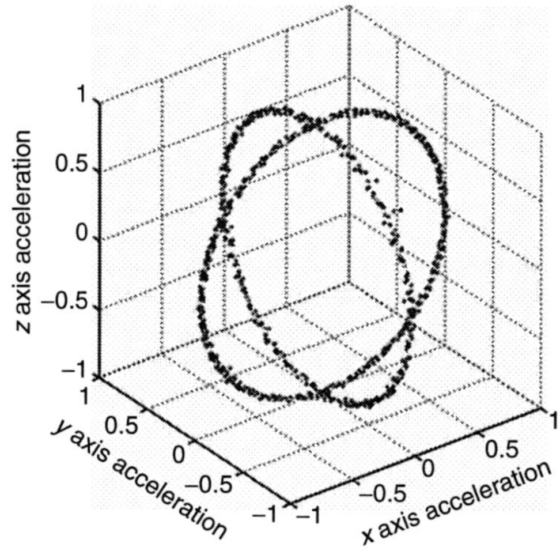

FIGURE 14.8
Data from two of the axes of a wirelessly powered triaxial accelerometer. The WISP was rotated with respect to gravity to generate this figure. As the x- and y-sensitive axes of the WISP change orientation with respect to gravity, the proportion of earth's constant 1 G of gravitational acceleration reported by the changes in x- and y-axes sensors.

range. After the measurement is taken and the data packed into the EPC ID, the WISP calculates the correct CRC. Then the "ID" is reported to the RFID reader, and the information is then decoded in real time by the computer.

Figure 14.9 shows a WISP strain gage device, compared with a wired version of the same transducer, and compared with a more accurate extensiometer device for ground truth measurement [D.J. Yeager et al., submitted]. The power requirements of the strain gage are more substantial than those of the accelerometer: the strain gage is a relatively high current consumption resistive sensor, configured with a Wheatstone bridge and amplification.

Figure 14.10 illustrates an application that makes use of WISP's computing rather than its sensing capabilities. With colleagues, we implemented the RC5 encryption algorithm on the WISP platform [22]. Most previous work on cryptography for UHF RFID assumed that sufficient power for traditional cryptography would not be available on a UHF tag. Therefore, many stripped down (or "minimalist") cryptographic schemes have been proposed for UHF RFID. Unfortunately, many have been broken. WISP shows that traditional "maximalist" cryptography can be implemented on a UHF RFID tag.

14.6 Conclusion

This chapter presented the design of WISP, a programmable, passively powered UHF RFID tag. WISP includes a general purpose 16 bit flash microcontroller, with analog to digital converter. WISP is powered by a read by an ordinary, unmodified standards-compliant UHF RFID reader. After presenting the design, power budget, and power-harvesting performance of WISP, we described several applications: measurement of light, temperature, acceleration, and strain, as well as implementation of the RC5 cryptographic algorithm.

More generally, WISP has proven the feasibility of powering a 16 bit microcontroller and arbitrary low power sensors using only the RF energy from a standards-compliant RFID

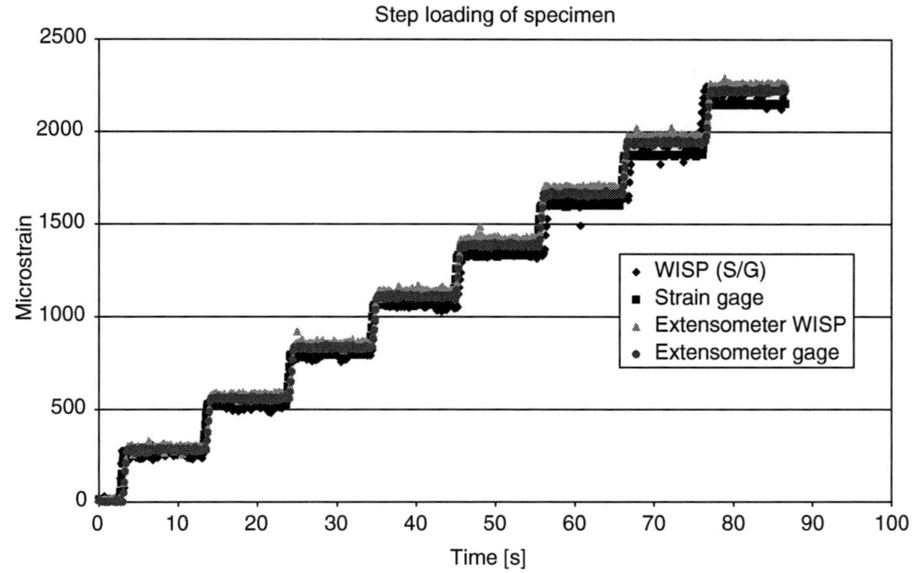

FIGURE 14.9
Strain recorded during the step loading of a specimen. (From Yeager, D.J., Quetin, G.R., Kim, S.H., Duncan, J., Miller, M., Smith, J.R., and Feraboli, P., *Development of a Wirelessly-Powered Strain Gage for Aerospace Applications*, 2007, submitted for publication.) The four traces show data collected by the WISP strain gage "WISP (S/G)," the same transducer powered and read conventionally (strain gage), and a more accurate ground truth sensor measured simultaneously with the WISP (extensiometer WISP) and simultaneously with the conventional strain gage (extensiometer gage).

reader. Furthermore, WISP has demonstrated the communication of sensor data using the EPC Class 1 Generation 1 protocol. The authors believe that WISP is the first of a new class of battery-free, wireless sensing, and computational devices.

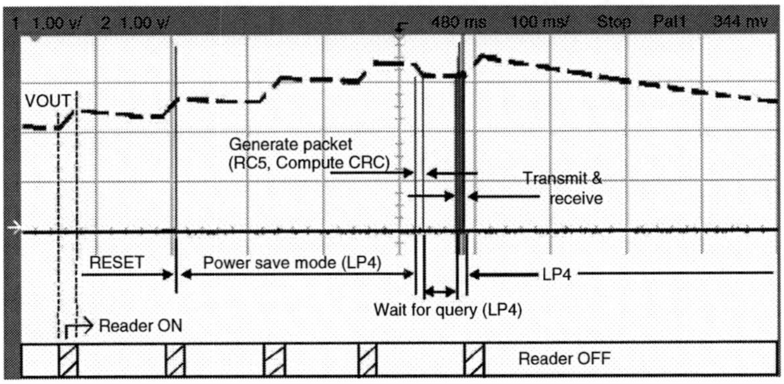

FIGURE 14.10
Trace of VOUT (power supply) before, during, and after RC5 encryption. While the Reader is ON (i.e., sending queries), the WISP's voltage level stairs up. The ramp up from 0 V is omitted on the left. The voltage supervisor wakes up the WISP from LP4 when the voltage level exceeds 3.3 V, and WISP begins its computation (Generate Packet). When the reader receives the WISP's response, the reader stops sending queries, cutting off RF power to the WISP. This is observed as a gradual decline of voltage at the right side of the figure. The WISP first enters LP4 and then resets as the voltage level falls below the minimum operating voltage. The total latency from 0 V until the end of RF response transmission in this case is ~2 s. (From Chae, H.J., Yeager, D.J., Smith, J.R., and Fu, K., *Proceedings of the Conference on RFID Security*, July 2007. With permission.)

References

1. M. Philipose, J.R. Smith, B. Jiang, K. Sundara-Rajan, A. Mamishev, and S. Roy. Battery-free wireless identification and sensing, *IEEE Pervasive Computing*, 4(1), 37–45, 2005.
2. S.S. Intille, K. Larson, E. Munguia Tapia, J.S. Beaudin, P. Kaushik, J. Nawyn, and R. Rockinson, Using a live-in laboratory for ubiquitous computing research. *Proceedings of PERVASIVE 2006*, Vol. LNCS 3968, K.P. Fishkin, B. Schiele, P. Nixon, and A. Quigley (Eds.) Berlin, Heidelberg: Springer-Verlag, 2006, pp. 349–365.
3. D.E. Culler and H. Mulder, Smart sensors to network the world, *Scientific American*, 85–91, June 2004.
4. S. Roundy et al., *Energy Scavenging for Wireless Sensor Networks*, Kluwer Academic Publishers, New York, 2003.
5. J. Paradiso and M. Feldmeier, A compact, wireless, self-powered pushbutton controller. *Proceedings of the 3rd International Conference on Ubiquitous Computing* (*Ubicomp 2001*), Berlin, Heidelberg: Springer-Verlag, pp. 299–304, 2001.
6. Web pages on the cavity resonator microphone can be found at http://www.nsa.gov/museum/museu00029.cfm and http://www.spybusters.com/Great_Seal_Bug.html
7. R. Fletcher, *Low-Cost Electromagnetic Tagging: Design and Implementation*, PhD Dissertation, MIT, Cambridge, MA, 2001.
8. J. Paradiso, K. Hsiao, and A. Benbasat, Tangible music interfaces using passive magnetic tags. *Proceedings of the ACM Conference on Human Factors in Computing Systems: Special Workshop on New Interfaces for Musical Expression* (*CHI 2001*), New York: ACM Press, 2001.
9. K. Finkenzeller, *RFID Handbook*, 2nd ed., John Wiley & Sons, 2003.
10. http://www.epcglobalinc.org/standards_technology/Secure/v1.0/UHF-class1.pdf
11. http://www.epcglobalinc.org/standards/uhfc1g2/UHFC1G2_1_0_9-StandardRatified-20050126.pdf
12. http://www.datasheetcatalog.com/datasheets_pdf/M/C/R/F/MCRF202.shtml
13. R. Want, Enabling ubiquitous sensing with RFID, *Computer*, 37(4), 84–86,2004.
14. K. Opasjumruskit, T. Thanthipwan, O. Sathusen, P. Sirinamarattana, P. Gadmanee, E. Pootarapan, N. Wongkomet, A. Thanachayanont, and M. Thamsirianunt, Self-powered wire-less temperature sensors exploit RFID technology, *IEEE Pervasive Computing Magazine*, 5(1), 54–61, 2006.
15. C. Namjun et al., A 5.1-μW 0.3-mm^2 UHF RFID tag chip integrated with sensors for wireless environmental monitoring, *IEEE European Solid State Circuits Conference* (*ESSCIRC*), Grenoble, France, September 2005.
16. F. Kocer and M.P. Flynn. A new transponder architecture with on-chip ADC for long-range telemetry applications, *IEEE Journal of Solid-State Circuits*, 41(5), 1142–1148, 2006.
17. http://www.nfc-forum.org/home
18. J.R. Smith, K.P. Fishkin, B. Jiang, A. Mamishev, M. Philipose, A. Rea, S. Roy, and K. Sundara-Rajan. RFID-based techniques for human activity detection, *Communications of the ACM*, 48(9), 39–44, September 2005.
19. J.R. Smith, B. Jiang, S. Roy, M. Philipose, K. Sundara-Rajan, and A. Mamishev. ID modulation: Embedding sensor data in an RFID timeseries. *Proceedings of Information Hiding 2005*, LNCS 3727, pp. 234–246, 2005.
20. J.R. Smith, A. Sample, P. Powledge, A. Mamishev, and S. Roy. A wirelessly powered platform for sensing and computation. *Proceedings of Ubicomp 2006: 8th International Conference on Ubiquitous Computing*, Orange Country, CA, USA, pp. 495–506, September 17–21, 2006.
21. A.P. Sample, D.J. Yeager, P.S. Powledge, and J.R. Smith. Design of a passively-powered, programmable sensing platform for UHF RFID systems. *IEEE International Conference on RFID 2007*, March 26–28, 2007.
22. H.J. Chae, D.J. Yeager, J.R. Smith, and K. Fu. Maximalist cryptography and computation on the WISP UHF RFID tag. *Proceedings of the Conference on RFID Security*, Malaga, Spain, July 2007, available online at http://rfidsec07.etsit.uma.es/slides/papers/paper-31.pdf.

Section III

Applications

15

From Automatic Identification and Data Capture to "Smart Business Process": Preparing for a Pilot Integrating RFID

Samuel Fosso Wamba, Élisabeth Lefebvre, Ygal Bendavid, and Louis-A. Lefebvre

CONTENTS

15.1 Introduction

Radio frequency identification (RFID) technology is considered as "the next big thing" in management (Wyld, 2006, p. 154) since the technology enables (1) the optimization of multiple business processes through the improvement, the automation, or even the elimination of existing processes (Strassner and Schoch, 2004; Fosso Wamba et al., 2006) and (2) the emergence of new processes called "intelligent processes" or "smart processes," which are automatically triggering actions or events. The latter point represents one of the most promising benefits from RFID applications and is the focus of this chapter.

Over the last four years, RFID technology has received a great deal of attention which was initially triggered by mandatory requirements from major organizations in the U.S. (e.g., Wal-Mart and U.S. Department of Defense) and in Europe (e.g., Metro AG and Tesco). Since then, the motivations for RFID adoption have moved from mandatory compliance to voluntary undertakings as companies are increasingly exploring the true potential of the technology, especially in the context of supply chains. As stated by Pisello (2006, p. 1), "the network effects of a synchronized supply chain will result in numerous benefits, including improved scan reliability, process automation, and real-time information access."

Recent key developments in technology with respect to hardware (i.e., integrated circuits, readers, antennas, printers) and software (i.e., firmware, middleware) have permitted to overcome some technical limitations of RFID applications. However, if the ability for automatic capture of data has improved substantially, the capacity to efficiently manage this data, and transform it into business intelligence is still limited. As the marketing director of an RFID solution provider involved in our project mentioned "Today, second generation tags are revolutionizing the way data is captured. Companies that initially performed pilots a couple years ago were faced with multiple technological limitations such as limited reading performance. Today, the main focus should be on one core component of RFID systems, that is the middleware." While most supply chain managers have now some knowledge of what RFID technology is about, they do not yet grasp the full implications of the business process redesign entailed by RFID implementation, and their understanding of the required configuration in the middleware to optimize supply chain operations is still limited.

The main objective of this chapter is to propose an approach for configuring and validating business rules in an RFID middleware. The proposed approach, relying on empirical evidence gathered from a detailed field study, will facilitate the dialog and bridge the gap between technical professionals and managers involved in RFID projects. This would in turn allow to better capitalize on the potential of RFID technology, which eventually lead to more successful RFID implementation.

15.2 Background and Context

15.2.1 Middleware as a Key Component of the RFID System

RFID technology is classified as a wireless automatic identification and data capture (AIDC). A basic RFID system is composed of a tag containing a microprocessor, a reader and its antennas, and a computer equipped with a middleware program, in which business rules are configured (Asif and Mandviwalla, 2005). The tag generally attached to a product communicates through radio frequencies with the reader's antennas. The reader sends the location and unique identification of the product to a computer. On the basis of preconfigured rules, the middleware can adjust or initiate business processes automatically.

RFID middleware is considered as one essential intelligence-added component of any RFID system and could be linked with other firm information systems (see Figure 15.1). More precisely, the middleware consists of the operating system, the data repository, and the processing algorithms that convert multiple tag inputs into visible tracking or identification data. The middleware could provide the following: (1) multiple reader configuration, control and monitoring; (2) access coordination to an environment with multiple applications; (3) receiving events generated by a reader; (4) data filtering, smoothing, and aggregating; and (5) data routing to enterprise applications such as a Warehouse Management System (WMS), Enterprise Resource Planning (ERP), Transport Management System (TMS), or a Manufacturing Execution System (MES) (Floerkemeier and Lampe, 2005; Nurminen, 2006).

Driven by the rising number of RFID applications and by mandates from important players such as Wal-Mart, the U.S. Department of Defense, Tesco, Target and Metro AG, and from their own major suppliers, the market for RFID middleware has grown rapidly in

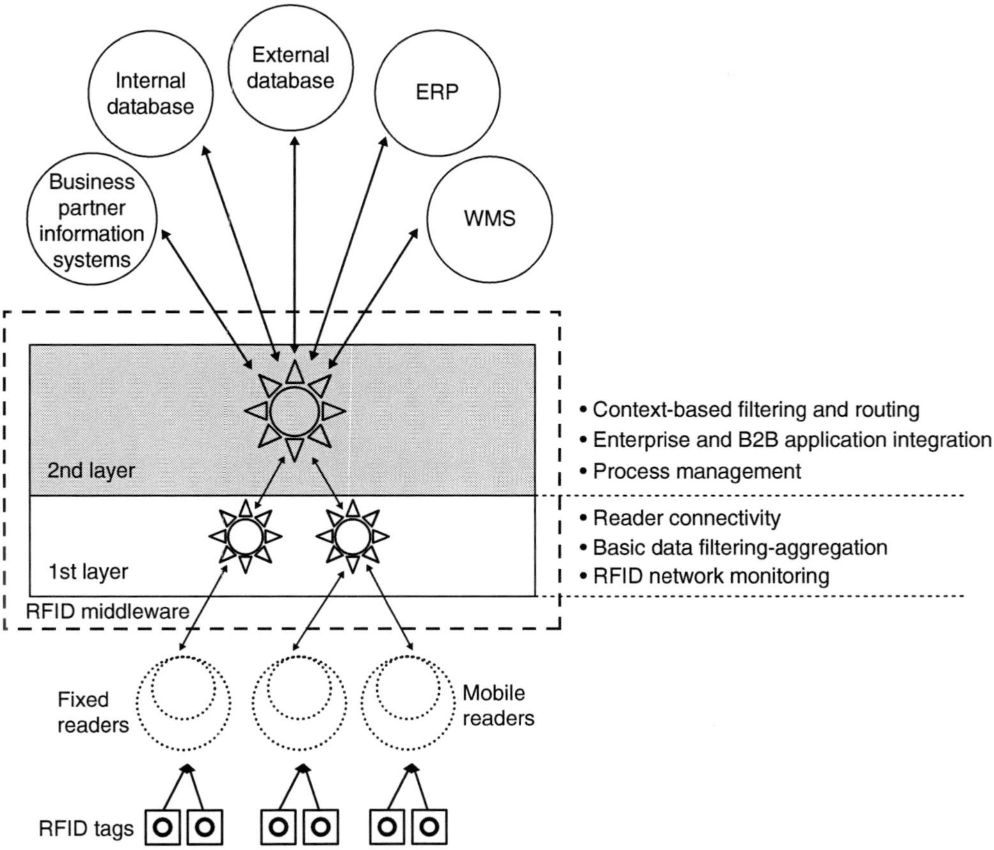

FIGURE 15.1
Positioning RFID middleware in an RFID system.

recent years. O'Connor (2006) estimates that the global market for RFID middleware had expanded from 2004 to 2005 by almost 162%. Although the market for RFID middleware is estimated to reach for $135 million in 2007 or roughly 3% of RFID systems revenues, it is projected to grow to almost $1,557.5 million in 2011 (WinterGreen Research, 2005). The RFID middleware offering is characterized by the presence of numerous firms, ranging from (1) pure-play vendors such as GlobeRanger and OATSystems; (2) application vendors such as Manhattan Associates, RedPrairie, Oracle, and SAP; and (3) integration specialists such as TIBCO Software and Ascential Software, to (iv) platform giants like Sun Microsystems, IBM, Oracle, and Microsoft (O'Connor, 2007).

Some of the major obstacles to widespread RFID adoption tend to disappear. First, the costs of an RFID middleware have decreased drastically, moving from about $125,000 per installed site a few years ago to $5,000 to $20,000 today (Nurminen, 2006). This downward trend positions RFID technology in a more affordable spectrum of potential investments and will eventually facilitate its adoption. Second, the current correct tags reading has reached 100% at the pallet level (Paxar, 2005) and is constantly improving at the item level. Third, chip technology has improved to a point where tags have become much more affordable. The concerns are now more focussed on middleware issues such as the capacity needs in order to manage large number of readers and the huge amount of data these readers generate (Floerkemeier and Lampe, 2005). In addition, the next trends in the RFID middleware will be more and more integrated with other technologies. For example, the integration with wireless technologies, such as Local Area Network (LAN), General Packet Radio Service (GPRS), and Global Positioning System (GPS), could enable real-time tracking and tracing (LogicaCMG, 2004).

In all cases, "installing the software is only the start" (GXS, p. 11, 2005). Indeed, applications at the process level as well as at the technical level need to be configured in the RFID middleware in order to trigger appropriate actions when specific events occur. Finally, the increasing concerns from supply chain members with the accuracy of inventory data and validity of RFID tagged items (O'Connor, 2004) can be effectively addressed by the RFID middleware, which acts as a bridge between the "physical RFID world" and the "software RFID infrastructure." The middleware thus allows the automatic interpretation and the semantic transformation of observations generated from the automatic data collection on tagged items into business logic data prior to their integration into existing information systems (e.g., ERP, WMS) (Fusheng and Peiya, 2005).

Previous work on the role of the middleware in RFID applications within a supply chain context is still scarce but represents an emerging and fast growing area of research. Among the recent conceptual papers, Gunasekaran and Ngai (2005) suggest that RFID technology may facilitate the development of supply chain configurations by acting as an enabler of a build to order (BTO) strategy. In the same line of thought, Pramataris et al. (2005) suggest that RFID technology may constitute a link to more collaborative approaches such as CPFR (Collaborative Planning, Forecasting and Replenishment). Kelepouris et al. (2007) also suggest that RFID technology can act as an enabler of traceability in the food supply chain. While exploring the impacts of RFID technology in a retail supply chain, Lefebvre et al. (2005) also identify the emergence of "intelligent processes" to support RFID-enabled Business-to-Business electronic commerce applications. Finally, Loebbecke and Palmer (2006) also examine the results of a joint RFID pilot project conducted between Kaufhof Department Stores, a leading European retailer and Gerry Weber, a fashion merchandise manufacturer. Results from the pilot study reveals that data derived from RFID technology about "products, processes, product movement, and even customer behavior can be used for proprietary and distinctive capabilities to gain competitive advantage, if turned into understandable and usable "content." The earlier-mentioned work points to the overriding importance of middleware configuration and integration as a key aspect of RFID strategy.

15.2.2 Focus on Business Processes

IT (Information Technologies) and BPR (Business Process Reengineering) are strongly associated. BPR (Hammer and Champy, 1993) is considered as "a critical enabler of new operational and management processes" (Kohli and Hoadley, 2006, p. 41) and could be used for instance as a means to cut nonvalue-added activities and to improve competitiveness (Kohli and Hoadley, 2006). Yet, IT investments represent a major driver for changes in business processes, enhancing informational and coordination capabilities, and thus, leading to cost reductions and better customer services. IT and BPR could therefore be viewed as "complimentary factors and must be changed in a coordinated manner to improve performance" (Kohli and Hoadley, 2006, p. 42). In the particular case of EDI, Riggins and Mukhopadhyay (1994) showed that the alignment of business process and EDI adoption lead to better information sharing, and thus, higher firm performance.

In a supply chain context, Kohli and Sherer (2002) strongly suggest that in order to fully capture the benefits from IT investments, supply chain actors need to conduct major changes in their business processes by adopting a process approach. In fact, "when the process approach is used, other factors that affect the translation of IT assets to impacts are investigated more clearly" (Kohli and Sherer, 2002, p. 7). This process approach is also highlighted by many other authors (e.g., Mooney et al., 1996) when exploring that IT business value and its potential as an enabler of organizational processes and supply chain structure improvement. Finally, the business process approach has been promoted as an appropriate or even an ideal approach to study the impact of IT at a more detailed level by "investigating how IT use in one stage affects a downstream IT and other organizational effects" (Byrd and Davidson, 2003, p. 244).

More recently, some researchers such as Strassner and Schoch (2004), Subirana et al. (2003), Youngil et al. (2006), Lefebvre et al. (2005), and Bornhövd et al. (2004) show that automatic identification technologies such as RFID technology could have a strong impact on business processes. For instance, Lefebvre et al. (2005) used a process mapping methodology and found that RFID technology could be considered as a disruptive technology as it supports a new business model, entails major redesign of existing processes, and fosters a higher level of electronic integration between supply chain members.

Our study builds on the business process approach and focuses on one supply chain in the retail industry.

15.3 Research Design

As the main objective of this study is to improve our understanding of the role of the middleware as an intelligent interface supporting RFID applications, the research design clearly corresponds to an exploratory research initiative (Eisenhardt, 1989) and is grounded in real-life settings. The following sections briefly describe the industry and one of its supply chain, the research activities, the research sites and the data collection methods.

15.3.1 Choice of One Supply Chain in One Industry

The current retail industry is highly globalized and facing fierce challenges: intense competition from powerful mega players (for instance, the ten largest American retailers' accounts for 65% of the U.S. market share), increasingly sophisticated and customized demand from final consumers, and thin profit margins. Retailers have been relying on IT to lower their transaction costs, manage the explosion of the number of Stock Keeping Units (SKUs) within their stores, cope with high volume of daily transactions, and automate manual processes (Fleisch and Tellkamp, 2005). Lately, they have turned to RFID technology and are considered as the lead users of this technology.

As a lead user, Wal-Mart is probably the most cited example in the retail industry: by adopting RFID, it would save annually almost $600 million and would in some cases cut by half its out-of-stock supply chain costs (Asif and Mandviwalla, 2005). RFID deployment at Wal-Mart is increasing at a rapid space. In fact, the number of Wal-Mart stores has increased from 100 stores in 2003 to 1000 stores in 2007 (Cecere and Suleski, 2007). Procter & Gamble represents another convincing example of rather successful deployment of RFID with estimated annual savings of almost $400 million for inventories (Srivastava, 2004) and reductions of out-of-stocks by half in some cases (Johnson, 2007). Yet, results are not as conclusive for all retailers RFID pilots since some have been delayed or even discontinued (Cecere and Suleski, 2007). The retained supply chain under investigation in this paper (Figure 15.2) operates in the beverage retail industry.

The focal firm, called here firm A, is considered as an important player with an overall annual volume of 15 million cases transiting through its Distribution Centers (DC), and

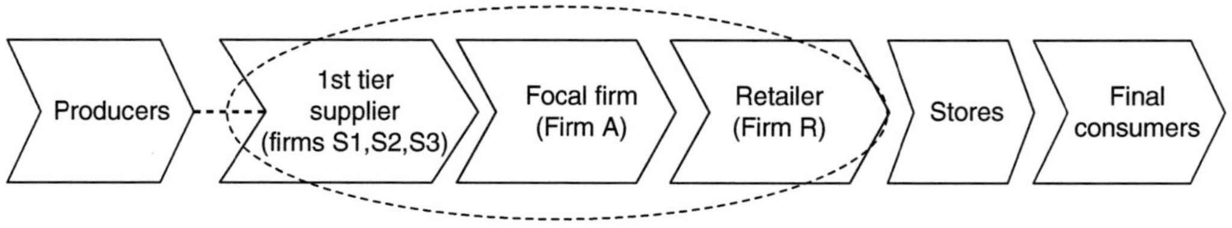

FIGURE 15.2
Focus of the field study in the selected retail supply chain.

with an average of 2.7 million cases passing through its docks and its business partners. The focus of the field study encompasses three layers as indicated in Figure 15.2. Firm A indicated that its primary motivations towards RFID technology were the reductions of warehousing costs (e.g., inventory) and the elimination of inventory discrepancies.

15.3.2 The Field Research and Corresponding Activities

The overall research project was conducted in four phases as indicated in Figure 15.3. Within the scope of this paper, results from phase 2 (i.e., scenario building and validation) and phase 3 (i.e., scenario demonstration) are presented. More specifically, results from steps 10, 11, and 12, which are directly linked to the middleware will be discussed in more detail. All data and information gathered in the previous steps served as an input to the subsequent steps. Although Figure 15.3 seems to indicate that research activities were conducted in a linear manner, a few iterations were actually necessary in order to reach a consensus among the participants.

Twenty four persons participated to steps 10, 11, and 12: seven key executives from the focal firm and its supply chain partners, namely three first-tier suppliers and one retailer, eight professionals and managers from technology firms (including the middleware developers) and nine members of the research teams. The rôle of the researchers ranged from full participants (when elaborating the technological scenario, for business rules configuration and testing in the middleware—i.e., step 10) and to full observers when activities 10, 11, or 12 were concerned.

15.3.3 Research Sites and Data Collection Methods

The field study was carried on-site in the offices and distribution center of the five organizations involved as business partners in the chosen supply chain (see Figure 15.2) and in one university-based research laboratory. Both qualitative and quantitative data were collected. Figure 15.4 summarizes the different data collection methods and their use in the different research sites.

15.4 Results and Discussion

In this study, we have adopted a warehouse perspective in conformity with the choice of the managers of the focal firm A. This choice enables all participants to understand how the work is carried out within one type of a warehouse in order to fully grasp the impacts of implementing RFID technology. Four distinct warehousing activities are usually identified—namely the receiving, the put-away, the picking, and the shipping (Van Den Berg

Preliminary phase: Vision and orientation	
Step a	Choice of test bed based on the partners accessibility openness, readiness, and potential RFID applications
Step b	Vision statement by or with potential industrial and technological partners (focus groups)
Step c	Identification of generic business applications and commitment from strategic business partners to the research project
Phase 1: Opportunity seeking	
Step 1	Determination of the primary motivation towards RFID Understanding the primary motivation to consider the use of RFID technologies (WHY?)
Step 2	Analysis of the product value chain (PVC) Understanding the activities specific to a given product (WHAT?)
Step 3	Identification of the critical activities in the PVC Identification of critical PVC activities (WHICH activities to select and WHY?)
Step 4	Mapping of the network of firms supporting the PVC Mapping the Supply Chain Network to understand the link between the network of firms supporting the product (WHO and WITH WHOM?)
Step 5	Mapping of intra organizational processes for the identified opportunities as they are carried out now («As is»)
Step 6	Mapping of inter organizational processes for the identified opportunities as they are carried out now («As is»)
Phase 2: Scenario building and validation	
Step 7	Evaluation of RFID opportunities in the PVC with respect to the product (level of granularity), to the firms involved in the network and to the specific activities in the PVC
Step 8	Evaluation of RFID potential applications including scenario building and process optimization («As could be»)
Step 9	Mapping and simulating of intra and interorganizational processes integrating RFID technology Selecting specific process for the demonstration
Phase 3: Scenario demonstration	
Step 10	Proof of concept (POC) in laboratory Configuring, testing, and refining business rules in the middleware supporting selected processes
Step 11	Demonstration of retained RFID-enabled scenarios using RFID infrastructure and evaluation of process redesign (e.g., automation, cancellation) at all the supply chain member's levels
Step 12	Demonstration of information system integration (e.g., ERP and middleware)
Step 13	Data analysis and decision to go for the pilot replicating POC scenarios in real-life setting
Phase 4: Real-life implementation	
Step 14	Pilot project in real-life setting
Step 15	Deployment of application and its appropriation by the different organizations involved and their staff

FIGURE 15.3
Steps undertaken in the field study. (Adapted from L.-A. Lefebvre, É. Lefebvre, Y. Bendavid, S. Fosso Wamba, and H. Boeck, *J. Chain Network Sci.*, 5, 101, 2005.)

and Zijm, 1999)—that can benefit from RFID technology (Lefebvre et al., 2005).Within the scope of this paper, only the picking and shipping processes will be discussed. We will first describe the technological infrastructure (Section 4.1) before presenting the underlying logic for the elaboration of the decision rules in the middleware (Section 4.2). Finally,

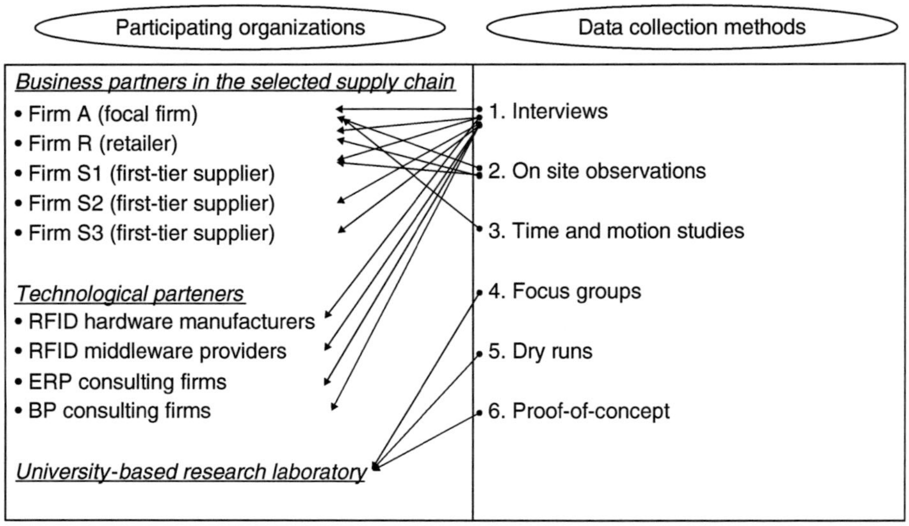

FIGURE 15.4
Research sites and corresponding data collection methods.

some examples of screen shots corresponding to these rules will be illustrated and discussed (Section 4.3).

15.4.1 The Technological Infrastructure

Several RFID-enabled scenarios were tested in a university-based laboratory (Figure 15.5).

The physical flow of products as depicted by the solid black arrows starts with a conveyor equipped with two antennas, one photo eye, and one light stack (top left side of Figure 15.5): this set-up simulates the picking process in the focal firm DC. The function of the photo eye is to automatically detect products equipped with an RFID tag and trigger the activation of the two fixed antennas thus allowing the antennas to be awaken and transmit radio waves when necessary. These two antennas are connected to a fixed reader that captures the information written on the tags and forwards it to the middleware. The stack light which is linked to the reader allows the confirmation of the status of the readings as the products pass on the conveyor belts. When the products reach the focal firm's shipping dock (right hand side of Figure 15.5) they go through another RFID equipped portal with two fixed antennas, two photo eyes and one light stack. Other technological options could be considered such as mounted RFID fork lifts or hand held RFID guns. Informational flows indicated in dotted arrows in Figure 15.5 link the readers to the firm's middleware, which then filters the information to the firm's ERP and-or other systems. The three screens display the different types of information available namely the RFID-enabled business processes, corresponding ERP screens, and the decision rules in the middleware.

15.4.2 The Underlying Logic for Decision Rules in Middleware

Figure 15.6 presents the decision rules that are or will be included in the middleware. The modelization used here corresponds to the EPC (Event-driven Process Chains) formalism, which allows the logical representation of the activities within and between processes. An interesting aspect of the EPC formalism is that it highlights all the events that trigger the

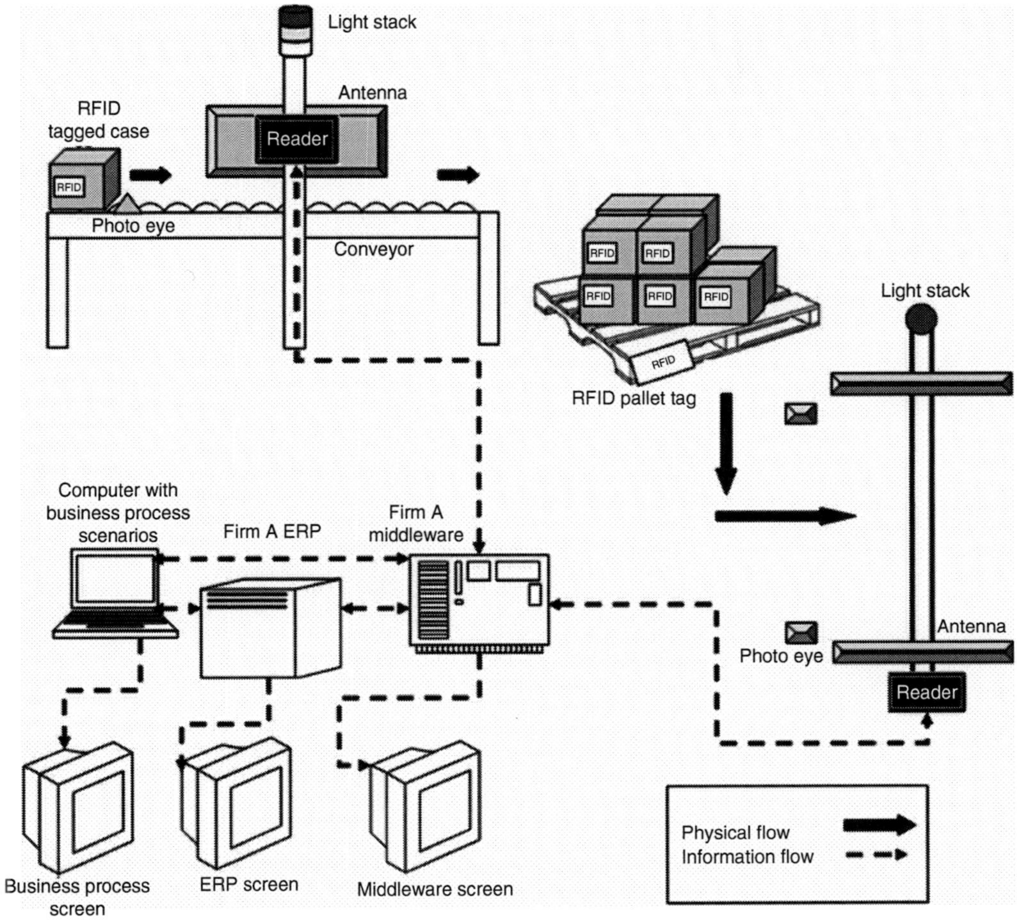

FIGURE 15.5
Technological infrastructure of the laboratory.

activities and the resulting sequence of events. Moreover, the modeling of a business process using EPC formalism uses three types of logical connectors (see bottom part of Figure 15.6) to indicate the workflow between activities and events, mainly the "$\wedge$" (i.e., and), "$\vee$" (i.e., or) "XOR" (i.e., exclusive or).

In addition to the basic representation of a process using EPC formalism, it is possible to assign responsibilities of employees to a specific function, allocate a system which is used to perform the function (e.g., ERP, middleware), specify some business rules, assign them to logical connectors, and quantify their probabilities of occurrence. The use of Business Process Analysis (BPA) tools such as Aris Toolset was therefore required as it supports an extended view of eEPCs.

15.4.3 Examples of Actions Triggered Automatically by Middleware

For the scenario under investigation, the products are tagged at the case and pallet levels in order to ensure the product tracking when products are depalletized (upon receiving) and repalletized during the picking process. More precisely, the following can be observed from Figure 15.7:

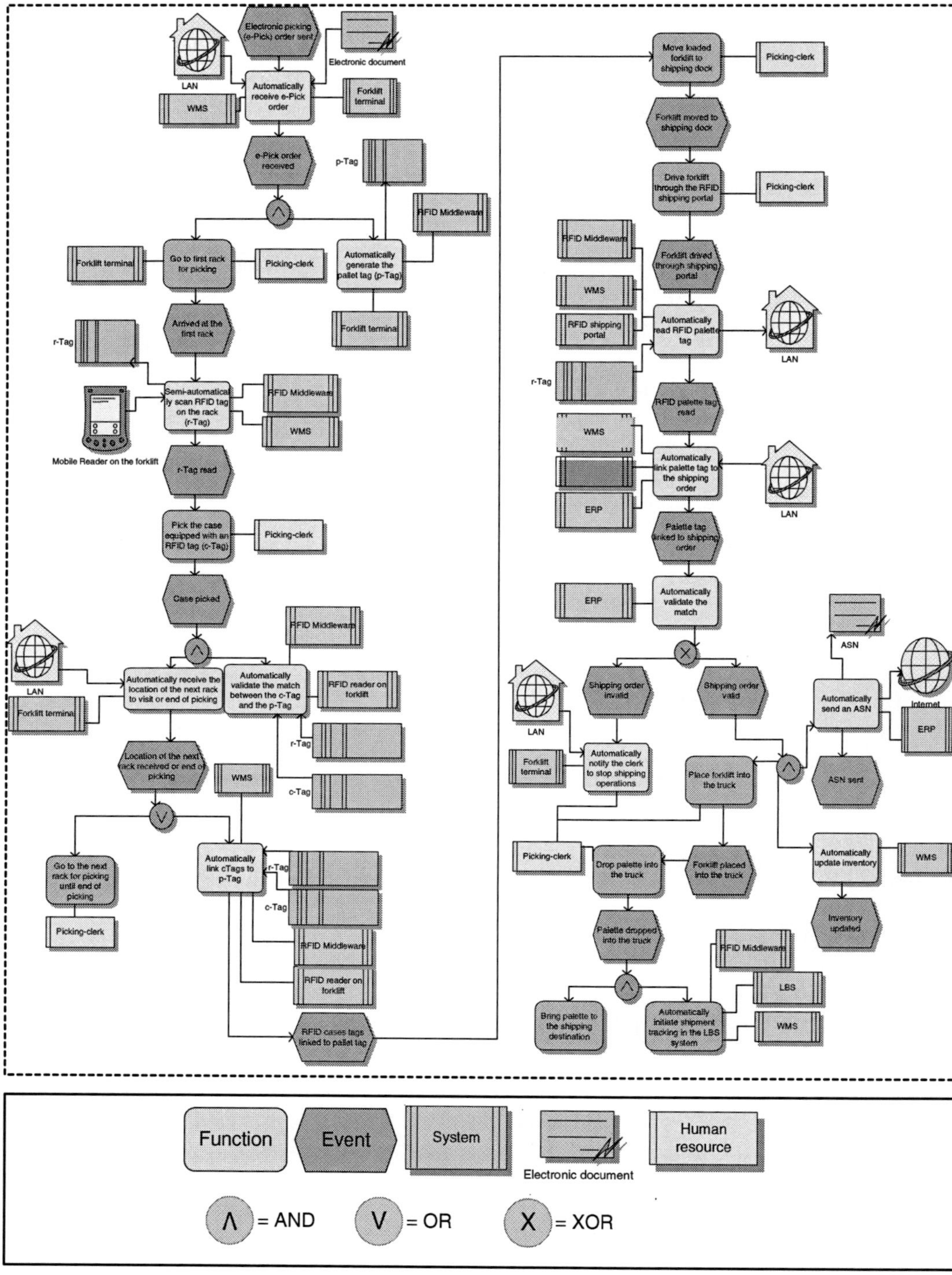

FIGURE 15.6
Decision rules for RFID-enabled picking and shipping processes.

1. The picking process starts with the event ePick order, which is received via an electronic document in the WMS and sent to a forklift clerk via a LAN. Upon reception of this order two functions are performed in parallel, namely (1) the

automatic generation of a pallet tag (p-Tag) based on a number assigned by the middleware (screen shot (b), Figure 15.7) and (2) the assignment of a picking order to a picking clerk (i.e., going to the first rack for picking). At this moment, the pallet status is automatically set to palletize (PLT) in order to specify its status and has all the information related to quantity of cases, types of products, customer ID (screen shot (c), Figure 15.7).

2. The connectors allow the modelization of the sequence of events as the picking clerk moves through the warehouse for the building of his pallet. At the assigned rack, the picking clerk scans the RFID tag (r-Tag) using a mobile RFID reader. As a case is picked, two activities are realized in parallel: (i) the automatic reading and validation of the match between the case tag (c-Tag) and the pallet tag (p-Tag) in the middleware and (ii) the automatic location of the next rack to visit if the end of

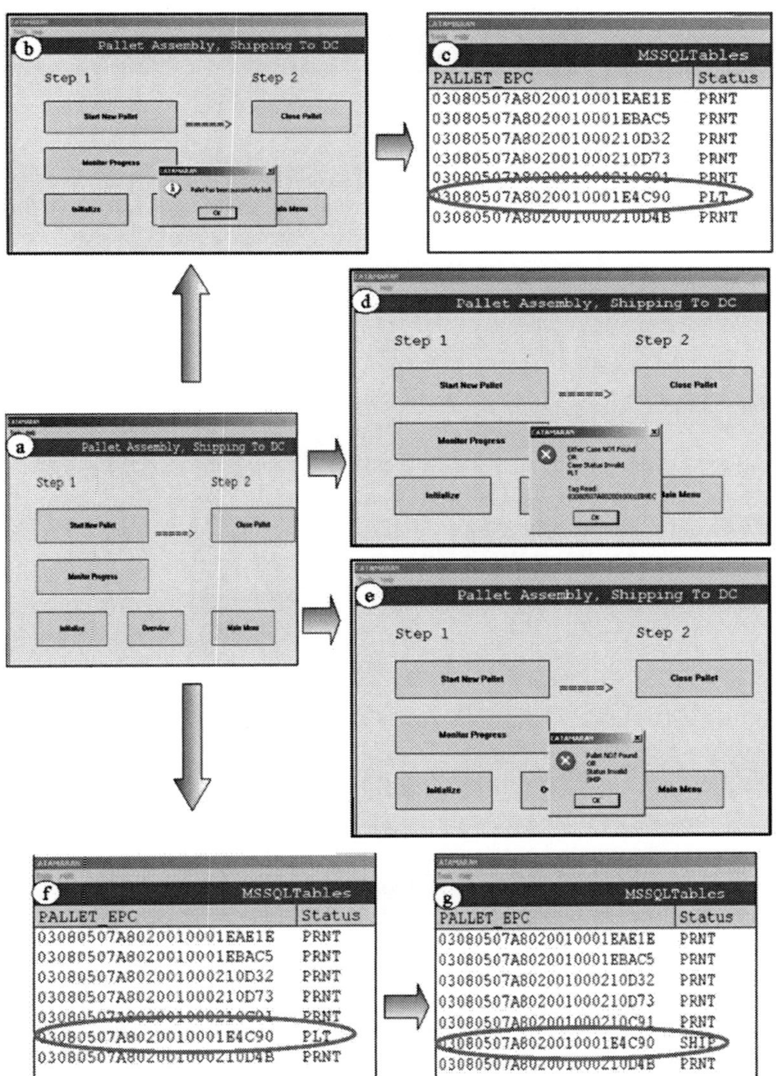

FIGURE 15.7
Business rules validation in the middleware and corresponding screen shots.

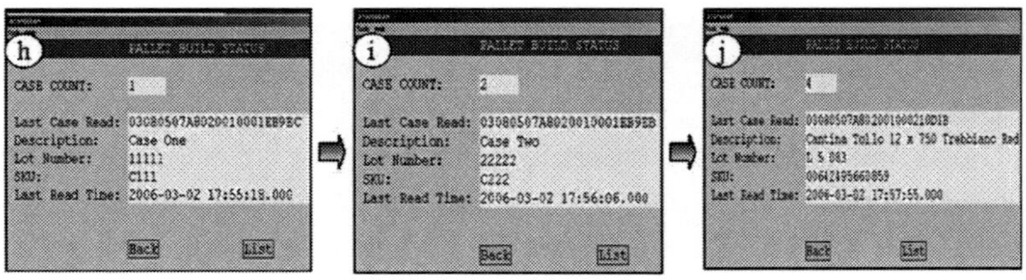

FIGURE 15.8
Automatic update of cases tags (c-Tag) in the palette and corresponding screen shots.

the picking process is not completed. If there is no match, an error message is automatically sent to the picking clerk (screen shot (d), Figure 15.7), resulting in improved picking accuracy and reducing very early in the process the probability of false shipment. Notice, that other configured rules such as horn alarm or indication by a stack light could be used. The conscious choice of mobile RFID reader for supporting the picking process addresses a key concern raised by the stakeholders who wanted to identify any problem at the rack level (and not at the shipping dock level) when using an RFID-enabled portal.

3. When the forklift drives through the shipping portal, the palette tag is automatically read, and thus automatically linked to the shipping order. This in turn triggers automatically the validation of the shipping order by capitalizing on the information system integration (RFID middleware ERP WMS). If it is an invalid shipping order, based on a configured rule in the middleware, an automatic message is sent to the clerk to stop the shipping operation and thus avoid false outbound movement of goods (screen shot (e), Figure 15.7), thus reducing the probability of product discrepancies. In the case of valid shipping order, many other actions can be taken in parallel such as: (i) automatically send an Advance Shipping Notice (ASN) and update inventory; (ii) automatically initiate shipment tracking in the Location Based System (LBS); and (iii) automatically modify the status of the pallet from "PLT: palletized" to fit its new status ("shipped"), allowing real-time tracking of the products. Because RFID tags have unique numbers, the same pallet cannot be shipped twice, since an error message such as "pallet not found" or "status invalid" will prevent such problem (screen shot (e), Figure 15.7).

Moreover, as the picking clerk takes the cases, the quantities are updated automatically (screens shots (h), (i), and (j), Figure 15.8). and when the predetermined number of cases has been taken by the picking clerk depending on the initial circuit, the palette status moved from "PLT" to "closed pallet" meaning that the palette is ready to be shipped (screens shots (f) and (g), Figure 15.7).

The validation of the decision rules for the RFID-enabled picking and shipping processes as simulated in the above mentioned technological infrastructure (Figure 15.5) helped all participants to better understand the challenges and benefits of RFID systems and find a common ground for discussion.

15.5 Conclusion

In this paper, we examined the underlying logic behind the rules configured in an RFID middleware to support "smart business processes" in one retail supply chain. The validation of the retained scenarios integrating the RFID technology in the laboratory settings enable participants to validate the business rules configured in the middleware. Some implications emerge from the field research.

First, it reveals that the University based RFID laboratory can serve as a neutral environment to investigate the real impacts of RFID technology at the firm level and at the supply chain level), thus presenting a "win–win" situation wherein each player in the supply chain is willing to invest on the RFID infrastructure. This can reverse the current situation towards RFID adoption, where the supplier and manufacturers are required to absorb most of the RFID technology costs.

Second, the scenario validation includes the configuration of business rules in the middleware came very late in the redesigning of the processes. In fact, prior to any configuration in the RFID middleware, firms need to conduct upfront homework in terms of identifying inefficient processes, ways to enhance them, redesign the new processes, validate them with key stakeholders (i.e., technology and business partners), and finally translate these processes in business rules to be configured. Moreover, this work has to be conducted at the firm level and also at the supply chain level, suggesting the importance of collaborating with their intra and inter organizational supply chain partner's to agree on business rules. For instance, when considering intraorganizational processes, all the key stakeholders agreed on an RFID-enabled scenario, later "translated" in a set of specific business rules that enabled the integration of selected processes (i.e., picking of an order and its shipping) that are conducted independently. In terms of inter organizational considerations, key stakeholders had to agree on issues such as the ways to organize the information flow between organization and how to ensure the integration of interrelated processes such as the shipping from the focal firm and the receiving at the retailer.

Third, flexibility is a key concern for the middleware configuration. As an example, the intelligence built in basic business rules could enable the same RFID portal to support multiple operational processes such as the "receiving" or the "shipping" of an order. In the laboratory settings, the use of ancillary devices such as photo eyes were used to indicate the presence and the direction of an object. By breaking the photo eye, a message is sent to the reader as an indication to activate the antennas, read the tag number, and send the information to the middleware as an indication to automatically perform a specific action attached to that rule. In real settings, the same logic could be replicated using similar ancillary devices such as motion captor. Moreover, in a warehouse environment other devices including screens and light stack can be used to facilitate the management "RFID transparent processes." A shipping clerk can validate its operation by looking at the visual confirmation sent through these devices, which are in fact the results of processed information and transactions conducted in the middleware.

In terms of investment, the integration of flexibility in RFID infrastructure is the result of a laborious process that highlights the importance scenario building, validation (phase 2) and demonstration (phase 3). It is only by taking the time to assess each scenario and think of ways to include flexibility in the processes that firms can minimize their RFID infrastructure investment, by limiting the installation of costly readers and antennas. On the other hand, an alternative to the use of ancillary devices is the building of more intelligence in the middleware. For example, when a tag is captured at a reading point in the warehouse, if it is not recognized, based on a basic rule (i.e., not created by the internal system) the transaction could automatically be considered as an incoming good and verification against an open ASN could be automatically realized to perform the "receiving" process. While building more intelligence in the middleware is a very interesting way to minimize the reliance on physical infrastructure, it is, however, very demanding in terms of defining, testing and validating the business rules. A cautious step by step approach such as the one undertaken should therefore be considered, starting with simple applications and building on the knowledge gathered from previous iteration to arrive to more complex applications.

References

Z. Asif and M. Mandviwalla, Integrating the supply chain with RFID: A technical and business analysis, *Communications of the Association for Information Systems*, 15, 393–427, 2005.

C. Bornhövd, T. Lin, S. Haller, and J. Schaper, Integrating automatic data acquisition with business processes experiences with SAP's auto-ID infrastructure, *Proceedings of the 30th VLDB Conference*, Toronto, Canada, 2004, Retrieved March 05, 2007, from http://www.vldb.org/conf/2004/IND6P1.PDF

T.A. Byrd and N.W. Davidson, Examining possible antecedents of IT impact on the supply chain and its effect on firm performance, *Information & Management*, 41, 243–255, 2003.

L. Cecere and J. Suleski, What we have learned from three years of retail RFID pilots, AMR Research, 2007, Retrieved May 2007, from http://www.amrresearch.com/Content/View.asp?pmillid = 20358

K.M. Eisenhardt, Building theories from case study research, *Academy of Management Review*, 14(4), 532–550, 1989.

E. Fleisch and C. Tellkamp, Inventory inaccuracy and supply chain performance: A simulation study of a retail supply chain, *International Journal of Production Economics*, 95(3), 373–385, 2005.

C. Floerkemeier and M. Lampe, RFID middleware design—Addressing application requirements and RFID, *Proceedings of sOc–EUSAI 2005 (Smart Objects conference)*, Grenoble, 2005, Retrieved December 2006, from http://www.vs.inf.ethz.ch/publ/papers/floerkem-rfidmi-2005.pdf

S. Fosso Wamba, L.-A. Lefebvre, and É. Lefebvre, Enabling intelligent B-to-B eCommerce Supply chain management using RFID and the EPC network: A case study in the retail industry, ICEC, 2006, pp. 281–288.

W. Fusheng and L. Peiya, Temporal management of RFID data, *Proceedings of the 31st International Conference on Very Large Data Bases*, Trondheim, Norway, 2005, pp. 1128–1139.

A. Gunasekaran and E.W.T. Ngai, Build-to-order supply chain management: A literature review and framework for development, *Journal of Operations Management*, 23(5), 423–451, 2005.

GXS, *Electronic Product Code: RFID Drives the Next Revolution in Adaptive Retail Supply Chain Execution*, 2005, Retrieved December 2006, from http://www.gxs.com/pdfs/whitePapers/WP_RFID_GXS.pdf

M. Hammer and J. Champy, *Reengineering the Corporation: A Manifesto for Business Revolution*, Harper Collins Books, New York, NY, 1993.

J.R. Johnson, P&G achieves "significant" ROI from RFID, RFIDWatch Weekly, 2007, Retrieved May 2007, from http://www.dcvelocity.com/rfidww/?article_id=117

T. Kelepouris, K. Pramatari, and G. Doukidis, RFID-enabled traceability in the food supply chain, *Industrial Management & Data Systems*, 107(2), 183–200, 2007.

R. Kohli and E. Hoadley, Towards developing a framework for measuring organizational impact of IT-enabled BPR: Case studies of three firms, *ACM SIGMIS Database*, 37(1), 40–58, 2006.

R. Kohli and S. Sherer, Measuring payoff of information technology investments: Research issues and guidelines, *Communications of the AIS*, 9(14), 241–268, 2002.

L.-A. Lefebvre, É. Lefebvre, Y. Bendavid, S. Fosso Wamba, and H. Boeck, The potential of RFID in warehousing activities in a retail industry supply chain, *Journal of Chain and Network Science*, 5(2), 101–111, 2005.

C. Loebbecke and J. Palmer, RFID in the fashion industry: Kaufhof department stores AG and Gerry Weber International AG, fashion manufacturer, *Management Information Systems Quarterly Executive (MISQE)*, 5(2), 15–25, 2006.

LogicaCMG, Making waves: RFID adoption in returnable packaging, 2004, Retrieved December 2006, from http://www.logicacmg.com/pdf/RFID_study.pdf

J.G. Mooney, V. Gurbaxani, and K.L. Kraemer, A Process oriented framework for assessing the business value of information technology, *ACM SIGMIS Database*, 27(2), 68–81, 1996.

T. Nurminen, The end of RFID middleware? *RFID Journal*, 2006, Retrieved March 2007, from http://www.rfidjournal.com/article/articleview/2035/1/128/

M.C. O'Connor, RFID users want clean data, *RFID Journal*, 2004, Retrieved January 2007, from http://www.rfidjournal.com/article/articleview/1232/1/14/

M.C. O'Connor, RFID middleware market set for growth, change, *RFID Journal*, 2006, Retrieved January 2007, from http://www.vdc-corp.com/_documents/news/press-attachment-1290.pdf

Paxar Central Europe GmbH, Paxar's Perfect Performance in Metro's SCM, 2005, Retrieved October 2006, from http://www.paxar-emea.com

T. Pisello, The ROI of RFID in the supply chain, *RFID Journal*, 2006, Retrieved March 2007, from http://www.rfidjournal.com/article/articleview/2602/

K.C. Pramataris, G.I. Doukidis, and P. Kourouthanassis, Towards 'smarter' supply and demand-chain collaboration practices enabled by RFID technology, *The Hermes Newsletter*, Eltrum, 31, March–April 2005.

F.J. Riggins and T. Mukhopadhyay, Interdependent benefits from interorganizational systems: Opportunities for business partner reengineering, *Journal of Management Information Systems*, 11(2), 37–67, 1994.

B. Srivastava, Radio frequency ID technology: The next revolution in SCM, *Business Horizons*, 47(6), 260–268, 2004.

M. Strassner and T. Schoch, Today's Impact of Ubiquitous Computing on Business Processes, Institute of Information Management of University of St. Gallen, 2004, Retrieved January, 2005, from www.vs.inf.ethz.ch/publ/papers/Strassner-Schoch-Impact-Ubicomp.pdf

B. Subirana, C. Eckes, G. Herman, S. Sarma, and M. Barrett, Measuring the Impact of Information Technology on Value and Productivity using a Process-Based Approach: The case for RFID Technologies, MIT Sloan, Working Paper, December 2003, Retrieved May 15, 2004, from www.papers.ssrn.com/sol3/papers.cfm?abstract_id=478582

J.P. Van Den Berg and W.H.M. Zijm, Models for warehouse management: Classification and examples, *International Journal of Production Economics*, 59, 519–528, 1999.

WinterGreen Research, RFID Middleware Market Opportunities, Strategies, and Forecasts, 2005 to 2010, 2005, Retrieved November 2006, from http://www.wintergreenresearch.com/reports/RFID_Middleware.html

D.C. Wyld, RFID 101: The next big thing for management, *Management Research News*, 29(4), 154–173, 2006.

K. Youngil, Y. Jung-Woon, and P. Namkyu, RFID Based Business Process Automation for Harbor Operations in Container Depot, Working Paper, Wayne State University, 2006, Retrieved February, 2007, from http://imeresearch.eng.wayne.edu/Proceedings2006/JungWoon.pdf

16

Technological Requirements and Derived Benefits from RFID Enabled Receiving in a Supply Chain

Harold Boeck, Louis-A. Lefebvre, and Élisabeth Lefebvre

CONTENTS

16.1 Introduction

RFID enabled automated receiving optimizes the handoff of products between supplier and client. It consists of receiving products at a manufacturing facility, a distribution center's warehouse or a retail store without manually scanning or verifying the merchandise (O'Connor, 2006). Although current state of the art receiving systems are highly optimized by using barcoding and wireless communications to a central computer, the process can still sometimes be error-prone and time-consuming because of human intervention.

Few studies have been conducted on the receiving process itself (Gu et al., 2007) and even fewer on RFID enabled automated receiving, despite the high profile mandates in this area. In fact, there is a lack of documented results and no common ground for comparing technological RFID scenarios. Furthermore, benefits and the measures used to compare RFID applications differ from study to study and the scope of their benefits has not yet been fully uncovered. The goal of this chapter will be twofold in order to facilitate the adoption of RFID technology in the context of automated receiving: firstly, to present the different technological infrastructures for the RFID enabled automated receiving application in six different organizations and secondly, to develop a more comprehensive list of benefits that can be used to measure the usefulness of such RFID applications.

This chapter focuses on the implementation of RFID enabled automated receiving, which has been identified as one of the quickest profitable SCM (Supply Chain Management) RFID applications. The following section (Section 16.2) briefly outlines the technological and nontechnological issues related to RFID enabled receiving while Section 16.3 offers some information about the detailed field research carried out in six organizations. The next section (Section 16.4) compares the different possible configurations of the application applied to real-life environments, their associated benefits and implications. Finally, Section 16.5 concludes the chapter by highlighting how the application can contribute to building a collaborative advantage in the supply chain.

16.2 Background

16.2.1 Technological Issues

The exponential growth in interest that RFID technology has recently gathered is without a doubt attributable to the highly mediatized compliance mandates from large organizations like Wal-Mart and the U.S. Department of Defense. The objective of these innovative organizations and other early adopters like Marks and Spencer, Tesco and Metro is to use passive Ultra High Frequency (UHF) radio frequency identification for optimizing their supply chain. UHF RFID technology, which is regulated under ISO/IEC 18000-6 and operates in the 860–960 MHz range, has the characteristic over more established Low Frequency (LF) and High Frequency (HF) RFID applications of using far field backscatter communications rather than near-field inductive coupling. The advantage of ISO/IEC 18000-6 and its improved ISO/IEC 18000-6C, commonly referred to as "Gen 2," is that it uses less expensive tags which are able to communicate more quickly, at a greater distance with better anticollision protocols. These characteristics therefore enable companies to identify and track many fast moving products through their supply chain more economically.

Additionally, the Electronic Product Code (EPC) network developed by the Auto-ID Center and managed by GS1 facilitates real-time information sharing between companies belonging to the same supply chain. A supply chain is defined as a group of companies that collaborate together in an effort to bring a product, service, or information from the initial supplier to the final customer. Collaboration among companies that belong to the same supply chain is part of the strategic vision of having their network gain a collaborative advantage by working as a team.

16.2.2 Nontechnological Issues

Innovative firms that have started to use RFID enabled automated receiving have measured and communicated tangible benefits. For example, Paramount Farms a producer of pistachio nuts receives 425 loads of nuts per day (Violino, 2004). It has implemented an RFID system to improve the receiving process of its trailers by affixing a passive 915 MHz tag on each of them. The RFID receiving system automatically gathers the following information: information about the trailer (tare weight, license plate number, owner information), information about its contents (the name of the farmer and the ranch, the location of the specific field where the pistachio nuts were harvested, the method used to harvest the nuts, the merchandise's weight automatically retrieved from the scale house) and information about the receiving process (a date and time stamp of when the shipment was received). Paramount Farms indicates that the automated process speeds data entry and ensures accuracy. Data acquisition has gone from 2 min to instantaneous and transaction time required to initiate a new load has been reduced by 60%. The faster throughput of the trailers at the receiving station also signifies a better utilization of assets. This has in turn reduced leased trailer usage by 30% and allowed Paramount Farms to cancel plans to build a new scale house. At the warehouse level, RFID can potentially redesign the processes (Lefebvre et al., 2006) and create benefits like those observed during a receiving pilot at a Canadian Staples location. RFID has reduced the processing time during receiving from 5.36 to 2.65 min (O'Connor, 2006) and at the store level, the Staples pilot has reduced the processing time from 17.75 to 2.70 min and reduced the number of orders that were delayed thus ensuring that items were available to be sold on time. RFID can thus assist Just-in-Time ordering (Smith, 2005). When combining automated receiving data with retail floor data and POS (Point of Sales) data, automatic replenishment can occur (Roberti, 2005) which can reduce out of stocks by 30% for products selling between 0.1 and 15 units per day (Hardgrave et al., 2006). Similar results are provided by the German grocer Rewe who indicates a 80% reduction in the time required to match deliveries with orders (Wessel, 2007). RFID has the potential to transform the store receiving process which is often manual (Jones et al., 2005) while improving the quality of information (Sellitto et al., 2007).

The retail industry has quickly identified RFID as having the potential to improve collaboration in its supply chain (Jones et al., 2004). It has also indicated that automated receiving is one of the RFID supply chain applications can generate a quick return on investment (ROI) (Roberti, 2007). In 2005, at least 140 Wal-Mart stores used automated receiving (Collins, 2005). In 2007, over a 1000 Wal-Mart locations used RFID and the company has plans to continue increasing the number of distribution centers and stores that use RFID automated receiving (Johnson, 2007). Some concerns have, however, been raised about the added-value of RFID enabled receiving. Many potential adopters of the application therefore see it as having mostly incremental benefits especially when their receiving process is already highly efficient.

16.3 Methodology

As part of a broader research program on RFID applications and deployment within supply chains (SC), special emphasis is placed here on the RFID enabled receiving process, which has been investigated in a detailed field research involving six very different organizations (Table 16.1). These organizations represent different types of SC players, operating in either closed or open loop networks and in different environments. The level of granularity needed for RFID tags also differs from the pallet to the item.

Data collection methods in the six organizations included direct observations, semistructured interviews and analysis of internal documents. Additional valuable data was also collected in a university-based research laboratory following a "living lab" approach (Loeh et al., 2005) whereby RFID enabled processes were modelized and technological scenarios were tested and validated.

TABLE 16.1

Organizations That Participated in the Field Research

Organizations	SC Level	Environment	Network	Volume	Unit Level
M1	Manufacturer	Assembly plant	Open loop	Low	Item
DC1	Distribution center	Warehouse	Open loop	High	Pallet
DC2	Distribution center	Warehouse	Open loop	Very high	Pallet
DC3	Distribution center	Warehouse	Closed loop	Medium	Pallet
DC4	Distribution center	Outdoor storage area	Open loop	Low	Item
R1	Retailer	Store	Closed loop	Medium	Pallet, case

The detailed field research and the living lab approach involved six researchers (two professors and four PhD students). Fifty-two additional people were also involved in the study including key executives and staff members from the six earlier-mentioned organizations and several senior managers from some leading-edge technology-based firms acting as RFID solution providers, including ERP, RFID hardware components, middleware, integrators, and process modelization providers.

16.4 Results

16.4.1 Receiving Process without RFID

The receiving process includes the data and physical handling necessary so that the product is ready to be put-away. It starts once the truck has backed up into the receiving dock and the paper Bill of Lading (BoL) for the merchandise has been handed off to the BoL clerk. Figure 16.1 presents the usual drill down approach for the receiving process from the more aggregate vision to the more detailed one. To simplify the presentation, only two levels are displayed in Figure 16.1.

The information displayed represents the schematized synthesis of the real-life observations in the six investigated organizations. Their respective receiving processes were compared in order to derive a generic receiving process. The process presented here corresponds to a highly efficient environment. These organizations are rather technologically advanced since they use barcoding, optical scanners linked to central computers through a wireless network, specialized software such as a Receiving System and a Warehouse Management System (WMS), which is integrated with an Enterprise Resource Planning (ERP) software. They also place a strong emphasis on the coordination and collaboration between the supplier and the client. The process is presented in the context of a warehouse because they usually have higher volumes and therefore the processes are more sophisticated and complex. By demonstrating that RFID can improve a highly efficient receiving process, we are also implying that it can improve the less efficient ones.

Of course, slight variations to this process exist. The client could require that the supplier already affix its pallets with an SSCC-18 (Serial Shipping Container Code) barcode or another type of identifier. When this is the case, activity "*4.3 Add a SSCC-18 barcode to the pallet*" is no longer necessary. The client could also require that the supplier send an Advanced Ship Notice (EDI transaction # 856) at the time of the shipment. In such a case, the supplier is affixing its pallets with SSCC-18 barcodes as well as electronically sending the information required to interpret the barcode. The electronic document arrives before the physical shipment so that activities from "*2.1. Create BoL in the Receiving System*" to "*2.4. Initiate unloading*" are no longer necessary.

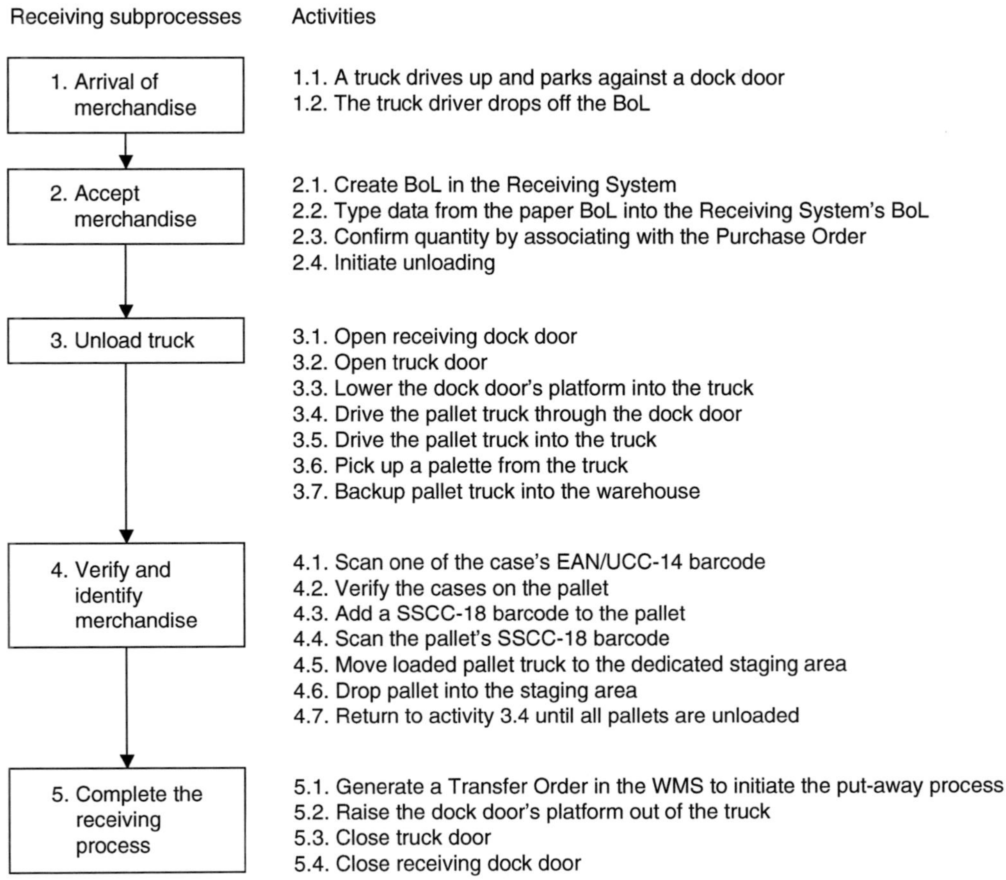

FIGURE 16.1
The current efficient receiving process of a warehouse.

Depending on the quality of the previous shipments, the client may also need to physically or visually inspect the content of the pallet. The pallet may be taken apart. The cases may be opened. An item may be removed for sampling and quality assurance. This can be performed during activity "*4.2. Verify the cases on the pallet.*" The more detailed the inspection, the more expensive it becomes.

16.4.2 RFID Enabled Automated Receiving System

An RFID system depends highly upon its environment and the product that will be tagged. Table 16.2 provides a summary of the different configurations necessary for each RFID system in the six organizations under investigation.

As displayed in Table 16.2, the RFID system differs from one organization to the next, as discussed in the following sections.

16.4.2.1 Location of Data

RFID enabled automated receiving requires replacing the different types of barcodes by smart labels. Smart labels are simply RFID tags glued to an adhesive label. Depending on

TABLE 16.2

Technical Characteristics of the RFID System in the Six Organizations

	RFID Configuration				
Organization	**Data**	**Readers**	**Tags**	**Middleware**	**Main Feedback**
M1	On the network	Mobile readers Handheld readers	Passive UHF tag with a plastic spacer	Pure play	Audible beep Mounted monitor
DC1	On the network	Mobile readers Handheld readers	Passive UHF pallet smart labels Permanent tags	WMS module	Audible beep Mounted monitor
DC2	On the network	Mobile readers Handheld readers	Passive UHF pallet smart labels Permanent tags	WMS module	Audible beep Mounted monitor
DC3	On the tag	Mobile readers Handheld readers	Passive UHF pallet smart labels Permanent tags	WMS module	Audible beep Mounted monitor
DC4	On the tag	Handheld readers	Passive UHF tag with a plastic spacer	ERP module	Audible beep Handheld monitor
R1	On the tag	Fixed readers Handheld readers	Passive UHF pallet smart labels Passive UHF case smart labels	Pure play	Light stacks

the specificities of each organization and of their supply chain, the RFID architecture can differ. The smart labels could possibly contain only a unique identifier such as an EPC if the specific data resides on a shared network. Similar to a barcode, the identifier on the label has no meaning unless one looks up the data to which it is associated in a database. In the case of M1, DC1, and DC2, it is recommended to use the EPC network because the data will be shared more easily between the various business partners who will come in contact with the tag.

Alternatively, if the data resides on the tags themselves (Diekmann et al., 2007) as it is the case for DC3, DC4, and R1, the tags will contain any and all information deemed necessary for the receiving process to be accomplished. The data on the case smart label could contain the following information: EAN/UCC-14 identifier, product description, and lot number. The data on the pallet smart label could contain the following information: SSCC-18 identifier, quantity in pallet, order number, PO (Purchase Order), and BoL. A company may decide to add additional data fields such as the name of the truck driver, the shipment time and destination, storage requirements, etc. In the cold chain, an RFID battery assisted semi-passive tag can record the temperature and various states of the product during transportation. In DC3 and R1, it is recommended to put the data directly on the tag and save the costs associated to using the EPC network because it is a closed loop supply chain. For DC4 it is necessary that the data be directly on the tag even though the business network functions in an open loop because the Internet will not always be accessible in the outdoor environment.

16.4.2.2 RFID Readers

The system requires that RFID readers be in a relatively close proximity to the smart labels during the receiving process. It is recommended that the antennas that are connected to the readers have a circular polarization. Although a linear polarized antenna will be able to read tags further if their orientation can be guaranteed, a circular polarized antenna will provide better read rates in an environment where the tags may not be properly aligned with the reader's antenna. Three different options exist in terms of reader selection. The choice of an option will impact how the RFID enabled automated receiving process will be performed as well as its costs.

16.4.2.2.1 Reader Option #1: Fixed RFID Readers

This option consists of installing fixed RFID readers at the dock doors. The reader's antennas are then placed on each side of the dock door behind protective bollards at the height at which the products will pass. This configuration is referred to as an RFID portal. When the forklift or pallet truck passes through the RFID portal during activity "*3.7. Backup pallet truck into the warehouse*," the tags are automatically scanned. The advantage of this option is that it significantly reduces the activities necessary in the receiving process. It also offers more efficient read rates than the other reader options because it uses more antennas. Its main disadvantage is that it requires more investment than the other two options described later. Additionally, this option may not permit to capture all the case smart labels. If the items within the cases on the pallet contain metal, liquid, or other dielectric components then it may be difficult to obtain a tag read of the cases located in the middle of the pallet because of their influence on RF propagation. Another factor limiting the capacity to read the cases in the middle of the pallets is tag shadowing. Tag shadowing occurs when multiple tags are in close proximity, which can occur with tagged cases on a pallet. Tags are continuously being improved in order to be read more effectively and at greater distances. Paradoxically, this can sometimes be a problem when it causes false positive reads. They occur when an RFID tag is being read by a neighboring reader. RFID portals will therefore be installed with presence of movement detectors and an RF reflective surface like a metal mesh. The presence of detectors ensure that the portal is turned on only when necessary. The metal mesh will isolate the reader's signal to the vicinity of the portal. Figure 16.2 shows an RFID portal near a dock door. A feedback mechanism is installed on the wall to indicate the last pallet smart label to be scanned. A camera can also be mounted on the wall as is the case in Figure 16.2. The camera is positioned right above the feedback mechanism and takes a picture when a tag is scanned for optional auditing purposes. It is recommended that R1 use an RFID portal because it has a limited number of dock doors. The forklift mounted RFID reader alternative is not an option because it does not use forklifts at the store. Although it could install the reader on one of its pallet trucks, an RFID portal ensures that all shipments arriving at the location are scanned even if they are not palletized.

16.4.2.2.2 Reader Option #2: Mobile RFID Readers

This option consists of mounting an RFID reader on the forklifts or pallet trucks instead of near the dock doors. Although the readers are not fixed in this option, it should be noted that every option still requires a fixed reference point. The fixed reference points serves to associate an activity to a location. For the mobile RFID reader option, it is possible to use permanent RFID tags, which will be located near the dock door that the forklift will pass through. When the forklift is in close proximity of the pallet, it will scan the smart label and download the information it contains. It will also scan the permanent RFID tag located near the dock door thereby associating the event with a location.

The advantages of this configuration over RFID portals are numerous. Firstly, fixed RFID readers as their name implies are fixed to a given location. This greatly reduces the visibility of the product in the warehouse to only the dock doors. Because a pallet will always be moved with a forklift or pallet truck, a forklift mounted mobile reader will

FIGURE 16.2
An RFID portal as can be used at R1.

always provide the last location of a pallet anywhere in the warehouse where permanent RFID tags are located. Secondly, a typical warehouse has fewer forklifts than dock doors. Purchasing costs and installation costs of the new system are thus greatly reduced. It is recommended to use the mobile RFID readers at M1, DC1, DC2, and DC3 because it will be more economical than to install RFID portals at every dock door. DC4 should also use a mobile RFID reader because the receiving process is performed outdoors and therefore it is not guaranteed that the arriving shipment will pass through a portal. Although the unit level at DC4 is an item, each individual item is carried on a pallet. Therefore, even though the item is tagged, the forklift can still read it when it takes the pallet.

16.4.2.2.3 Reader Option #3: Handheld RFID Readers

The handheld RFID reader offers the least benefits in terms of process improvements when compared with the two previous options because a certain level of manual intervention is

FIGURE 16.3
Handheld RFID reader.

still required. This option is more of a "semiautomated" receiving process. The handheld reader is a mobile reader carried by an employee as illustrated in Figure 16.3. It is composed of a rugged exterior and contains a keyboard and a terminal. It offers the advantages of portability in remote locations. For example, if receiving is performed in a temporary location or one that is not accessible by an RFID mounted forklift, then the handheld reader is an interesting alternative. Unfortunately, its portability has to be offset by a reduced Effective Radiated Power (ERP) in order to reduce the consumption of the limited battery power supply. Rather than be considered as an alternative to RFID portals and RFID mounted forklift readers, the handheld reader can also complement the other equipment options for exception processing. When a problem shipment needs to be investigated or a pallet needs to be broken down during activity "*4.2. Verify the cases on the pallet*," then the handheld reader is very useful. For this reason, all the organizations studied can also use the handheld RFID reader in their receiving process.

16.4.2.3 RFID Tags

Four types of RFID tags may be used for the automated receiving process. Firstly, the smart labels will be affixed to the pallet and will become the "pallet tag" at DC1, DC2, DC3, and R1. Secondly, permanent RFID tags may be installed near the receiving dock doors or any other location that needs to be automatically identified during the receiving process. M1, DC1, DC2, and DC3 will use permanent RFID tags because mobile RFID readers and dock doors are used at these locations. Some organizations may decide that permanent RFID tags during receiving are optional. Indeed, it may not matter through which receiving dock door the shipment came through as long as its content is properly identified and verified. However, these tags are relatively inexpensive and can provide valuable information that can be used during data analysis such as dock door usage, being able to separate events by location, etc. Thirdly, smart labels may also be affixed on each case within the pallet. They are not necessary in order to perform automated receiving at the pallet level but should an error occur during the receiving process, they will improve exception processing during activity "*4.2. Verify the cases on the pallet.*" Smart labels at the case level are, however, necessary to automate the receiving process at R1 because not all shipments arrive on pallets. Fourthly, smart labels will be affixed at the unit level at M1 and DC4. UHF RFID is not generally used for item level tagging because its far field properties (as defined by ISO 18000-6C) are not designed to isolate individual items. If the objective in automated receiving is not to provide individual tracking nor the capacity to isolate a single item, a UHF smart label can be used for automated receiving at the item level. Special care must be taken when choosing this particular smart label, because the item's composition might contain more metal or water than the corrugated cardboard of the cases and pallets. In the case of M1 and DC4, a plastic spacer was put between the tag and the item to create an air gap. The following text covers this issue in more detail.

RFID inlays are the tags that are used in smart labels. They exist in a variety of sizes, forms, costs, range, memory size, etc. The choice of tag will be highly dependent on the nature of the project and the environment in which they are to be used. Nonetheless, it is possible to give some generic indication of the type of tags to use.

Since it is highly unlikely that the tags' orientation can be guaranteed during the receiving process, it is recommended to use an RFID inlay containing a dual dipole antenna. Additionally, certain inlays are optimized according to the material they will come in contact with. For example, some smart labels are designed to perform efficiently when affixed to corrugated cardboard. The location where the tag or smart label will be placed greatly influences its readability. It is therefore important to consider dynamic and static air gaps in the products or packaging. An air gap behind the tag will offer ideal tag performance. It is best to use a static air gap when possible as this will ensure that an air gap is always behind the tag. The geographical location of the tag is also important as different regions of the world determine the frequency to use. Although these considerations can greatly increase the performance of the system, it is nonetheless crucial to test various types of tags in order to find the most appropriate one.

16.4.2.4 RFID Middleware

The RFID middleware software ensures the bridge between the RFID architecture and the organization's central data repositories such as the Receiving System, WMS, or ERP. The middleware is often referred to as the intelligent portion of the RFID system because it manages and coordinates it. The middleware will associate read items with an activity and a location. It also sends commands to the readers to reprogram the tags. It will clean the raw data to limit false positives and then filter the data before sending it to the receiving system, WMS or ERP. This is a necessary procedure as the latter are not designed to store vast amounts of data gathered by the RFID system. DC4 decided to use a middleware developed by their ERP manufacturer because their IT department has taken the decision to standardize all software on this platform. D1, DC2, and DC3 have opted to use a middleware provided by their highly specialized WMS. M1 decided to use a middleware provided by a "pure play" company that services smaller organizations. R1 has opted for a pure play middleware solution because its existing software does not currently provide an RFID module.

16.4.2.5 Feedback Mechanisms

Feedback mechanisms are an important part of the RFID system but are seldom covered when describing the basic RFID components. Nonetheless, they are required in order to provide a signal capable of being interpreted by a human that will indicate an error in the automation process. This will inform the employee to intervene by processing the exception. Feedback mechanisms can take the form of light stacks, audible devices such as horns or monitors such as the one present on Figure 16.2 for the receiving process at R1. Organizations M1, DC1, DC2, and DC3 use forklift mounted mobile readers, which provide an audible beep when a tag is read and transmits information to a monitor mounted on the forklift. DC4 uses a handheld reader which provides an audible beep and information about the tag on the handheld monitor when it is read.

16.4.3 Derived Benefit

RFID enabled automated receiving can provide a number of benefits to the organizations studied as presented in Table 16.3. These benefits have been regrouped under the following categories: streamlining current processes, improving the quality of information, improving the quality of process execution, enabling business process re-engineering, and creating RFID externalities.

16.4.3.1 Streamlining Current Processes

16.4.3.1.1 Instantaneous Data Acquisition

The nature of RFID is to automate the capture and transmission of data related to a product's identification. In the receiving process presented in Section 16.2 the following activities are therefore eliminated with the RFID system.

TABLE 16.3

Benefits from RFID Enabled Automated Receiving for Each Organization

Benefits	M1	DC1	DC2	DC3	DC4	R1
Streamlining current processes						
Instantaneous data acquisition	+	+	+	+	+	+
Instantaneous verification	+	++	++	+++	+	+++
Reduced paperwork	+	+	++	+	+	+
Reduced errors	+	++	+++	++	+	++
Reduced bottlenecks	+	++	+++	++	+	++
Reduced assets	+	++	++	++	+	+++
Improving the quality of information						
Paperwork gets filled out	+	++	++	+	+++	++
Fewer claims	+	+++	+++	+	+	++
Better informed management decisions	+	+	++	+	+	++
Improving the quality of process execution						
Better management and control of operations	+	++	+++	++	+	+
Quicker shipments or replenishment	+	++	++	++	+	++
Enabling business process re-engineering						
Cross-dock possibility	n/a	n/a	++	n/a	+	n/a
Create smart processes	+	++	++	++	+	++
Creating RFID externalities						
Downstream benefits	+++	++	++	++	+++	+

+ some benefits, ++ relatively high benefits, +++ high benefits.

1.2. The truck driver drops off the BoL
2.1. Create BoL in the Receiving System
2.2. Type data from the paper BoL into the Receiving System's BoL
4.1. Scan one of the case's EAN/UCC-14 barcode
4.3. Add a SSCC-18 barcode to the pallet
4.4. Scan the pallet's SSCC-18 barcode

16.4.3.1.2 Instantaneous Verification

As data from the smart label is captured by the RFID system, the enterprise system can automatically perform the verification activities. When the system is properly configured and integrated, the following tasks are automated.

2.3. Confirm quantity by associating with the Purchase Order
4.2. Verify the cases on the pallet

The benefits from this category will be more significant when many items need to be verified as is the case when the unit level is a pallet at DC1 and DC2 instead of a single item at M1 and DC4. It reaches a peak when the shipment is a mixed pallet of various items like at DC3 and R1.

16.4.3.1.3 Reduced Paperwork

Paper-based information is less efficient than electronically stored data. There is a cost associated with transmitting paper-based information. It has a tendency of being misplaced. It is also bulkier and retrieving it is slower while at the same time requires more effort to copy and manipulate the data it contains. The following activities during the receiving process include paperwork that can be eliminated with the RFID application.

1.2. The truck driver drops off the BoL
4.2. Verify the cases on the pallet

16.4.3.1.4 Reduced Errors

Even though current receiving processes can be very efficient, they are still error-prone because of human intervention. Our field study indicates that errors can occur at several locations especially when the process is not followed as specified. Environments with high volume, open loop networks where an ASN is not available and organizations where work is performed manually are more error-prone and therefore better candidates to RFID benefits.

16.4.3.1.5 Reduced Bottlenecks

An improved movement of goods during the receiving process and fewer errors translate into a higher throughput. The improved throughput makes a bottleneck less likely. When they occur, they potentially halt all other receiving activities and create a queue. Benefits will be more significant for high volume receiving processes where bottlenecks currently occur.

16.4.3.1.6 Reduced Assets

Because the receiving process is performed more quickly, fewer assets involved in the receiving process are required. For example, since throughput is increased, the dock doors used by the trailers have a quicker turnover. Fewer dock doors are now necessary. The same reasoning applies to very coveted warehouse real estate which is freed up: the receiving dock and staging area. Additionally, fewer forklifts and their drivers are necessary to accomplish the same output. Instead of reducing the employee headcount, it can rather mean allocating more employees facing the customer which is one of R1's strategic goals.

16.4.3.2 Improve the Quality of Information

16.4.3.2.1 *Paperwork Gets Filled Out*

Filling out paperwork is often considered as tedious by employees. When it is filled out, it sometimes contains errors. RFID eliminates paperwork because all the necessary documentation is automatically created. This ensures that it is not only gathered, but gathered correctly. This benefit is especially useful in an outside environment where it is less pleasant to fill out paperwork in the rain or in the cold as is the case at DC4. Also, sometimes employees have little time or are less motivated to fill out the paperwork. RFID thus makes the information more available and reliable.

16.4.3.2.2 *Fewer Claims*

A very significant pain point for both suppliers and clients are the numerous claims that occur when a discrepancy in shipments occurs. This is due to shrinkage which includes breakage, misplacements, poor bookkeeping, or theft (Levy et al., 2004). It creates conflict in the relationship between the buyer and seller. It is costly because of the time spent in order to resolve the contentious matter and because of the money spent to compensate the lost goods. RFID during the receiving process can assist in identifying who is responsible for the shrinkage. This is especially useful in open loop networks where there is a high volume of receiving as is the case with DC1 and DC2.

16.4.3.2.3 *Better Informed Management Decisions*

Improved quality of information can lead to better informed management decisions. For example, by knowing which goods have already been received, redundant inventory will not be ordered. This reduced inventory in addition to obviously improving cash flow also has the additional benefit of freeing up warehouse floor space, reducing variable costs such as insurance costs and avoiding depreciation costs on inventory that is not used. It can also improve the management of faster moving items that need to leave the warehouse as soon as possible like at DC2 or at a retail store to reduce Out of Stocks like at R1.

16.4.3.3 Improve the Quality of Process Execution

16.4.3.3.1 *Better Management and Control of Operations*

An efficient process requires that all of the activities be continuously performed in the same optimized manner. Humans do not like repetitive tasks. In our field study we noticed that employees would introduce variety in the receiving process. Sometimes the documented process was not performed as suggested. This observation occurred when comparing different employees performing the same tasks and also when comparing a single employee performing the same task at different occasions at DC1 and DC3. The RFID system can ensure that the process is performed as prescribed. This is especially useful when many employees work on different parts of the same process as is the case at DC2. The system can coordinate the activities and ensure that the process is completed correctly.

16.4.3.3.2 *Quicker Shipments or Replenishment*

Since RFID enabled automated receiving improves the throughput of goods, they have less propensity to arrive late at their destination. This means that the application ensures that Just-in-Time can be executed more efficiently. Environments with faster moving goods can benefit more from this aspect.

16.4.3.4 Enabling Business Process Re-Engineering

16.4.3.4.1 Cross-dock Possibility

Instead of using RFID to create incremental benefits in the receiving process, it is possible to think outside the box and use it to radically change the way the current warehouse is organized. One such way is to use RFID to enable cross-dock activities if the product is not managed under a First In First Out (FIFO) method (Fosso Wamba et al. 2007). In this case, the following receiving activity "*4.5 Move loaded forklift to the dedicated staging area*" is bypassed along with the subsequent put-away and picking process because the goods are moved directly to shipping. This is applicable only in the warehouses of DC2 and DC4.

16.4.3.4.2 Create Smart Processes

The facilitated real-time data acquisition and additional visibility that are provided by RFID can enable new sets of business processes. These smart processes can immediately trigger an event or another process when a specific RFID tag is read in a given situation (Fosso Wamba et al., 2006). These can be, for example, alert notifications when hot items that are in strong demand have arrived and should be treated in priority. Smart processes are still in their nascent phase and have yet to be developed and widely adopted by organizations. Nonetheless the potential they have to offer is interesting.

16.4.3.5 Creating RFID Externalities

16.4.3.5.1 Downstream Benefits

When a receiving process is automated with RFID, it implies that a smart label is affixed to the pallet, case or item. The smart label is still affixed to the unit during the following processes. Therefore the put-away process, picking process, and shipping process can benefit from the infrastructure that was put in place to optimize the receiving process. It is also possible to use the same smart label for the client's receiving process. As the tagged unit moves down the supply chain, it can create additional benefits. The longer the tag remains in use, the more benefits can be derived from it. For example, the tag on the item at M1 and DC4 can be used during its entire lifecycle thereby enabling Product Lifecycle Management (PLM). In R1's retail environment, the pallet or case to which the tag is affixed is quickly disassembled and the tag looses its meaning.

16.5 Conclusion

On the basis of the results of a comprehensive field study carried out in six organizations, this chapter presents the technical considerations that will influence the configuration of an RFID enabled automated receiving application and demonstrates that technical challenges are numerous since such an application requires customization depending on the type of products, the characteristics of the network, the volume of activity, the unit level, and the organization's position in the supply chain. Benefits derived from this type of RFID application are dependent not only on the characteristics of the RFID platform but also on the chosen RFID configuration. A preliminary but rather comprehensive list of benefits derived from the field study represents a useful analysis tool to determine the full benefits that can be derived from the application in a given organization.

Results also point to the overriding importance of managerial considerations related to supplier adoption. In an open loop network, RFID enabled automated receiving clearly impacts the supplier and this represents a major consideration when implementing the application. Client mandates are usually perceived by the supplier as generating additional costs while contributing few benefits. This can create a conflictual situation in the buyer-seller relationship as many reports indicated was the case at Wal-Mart (Fogarty, 2004; Keizer, 2004; Romanow, 2004; Schwartz, 2004). Early adopters agree that the supplier must also see benefits for the application to be adopted by both parties (Schwartz, 2004). In fact, the biggest obstacle to successfully implementing this RFID application is not tied to technical issues but rather to obtaining supplier buy in.

RFID enabled automated receiving application optimizes a process which is shared between buyer and seller and thus should contribute to build a collaborative advantage. In the context of RFID enabled automated receiving, the client shares the mandate requirements and numbering schema to be used. In return, the supplier sends a tagged shipment and electronic information which can be transmitted through the EPC network when the data is not on the tag. In addition to this basic form of collaboration that is required between supplier and client to perform the essential functions of RFID enabled automated receiving, it is also possible for the supplier to capitalize on some benefits from the application. For example, the higher quality and speed of acquisition of the information that has been gathered by the client because of the supplier can be shared back with the supplier in the form of feedback. As an additional incentive, the client could decide to accelerate the payment of the shipment because the possible claims are resolved more efficiently and quickly. Further research on the advantages for the supplier to assist the client in performing automated receiving is clearly necessary and is also expected by the industry.

References

Collins, J. (2005) Hampton unlocks ROI from RFID. *RFID Journal*. Retrieved May 15, 2007, from http://www.rfidjournal.com/article/articleprint/1489/-1/1.

Diekmann, T., Melski, A., and Schumann, M. (2007) Data-on-network vs. data-on-tag: Managing data in complex RFID environments. *40th Annual Hawaii International Conference on System Sciences (HICSS '07)* January 3–6, 2007, Waikoloa, Hawaii (available online at http://www.hicss.hawaii.edu/hicss_40/ apahome40.htm)

Fogarty, K. (2004) RFID: An offer you can't refuse. *Baseline*. Retrieved March 29, 2007, from http://www.baselinemag.com/article2–0,1540,1542214,00.asp.

Fosso Wamba, S., Lefebvre, L.A., and Lefebvre, É. (2006) Enabling intelligent B-to-B eCommerce supply chain management using RFID and the EPC network: A case study in the retail industry. *Proceedings of the 8th International Conference on Electronic Commerce* (*ICEC'06*), 14–16 August, Fredericton, Canada.

Gu, J., Goetschalckx, M. and McGinnis, L.F. (2007) Research on warehouse operation: A comprehensive review. *European Journal of Operational Research* **177** (1): 1–21.

Hardgrave, B.C., Waller, M., and Miller, R. (2006) RFID's impact on out of stocks: A sales velocity analysis. *University of Arkansas Research Article*. Retrieved May 15, 2007, from http://itrc.uark.edu/research/display.asp?article=ITRI-WP068–0606.

Johnson, J.R. (2007) Wal-Mart to enable 400 more stores. *DC Velocity*. Retrieved May 15, 2007, from http://www.dcvelocity.com/rfidww/?article_id=126.

Jones, P., Clarke-Hill, C., Shears, P., Comfort, D., and Hillier, D. (2004) Radio frequency identification in the UK: Opportunities and challenges. *International Journal of Retail & Distribution Management*, 32, 164–171.

Jones, P., Clarke-Hill, C., Hillier, D., and Comfort, D. (2005) The benefits, challenges and impacts of radio frequency identification technology (RFID) for retailers in the UK. *Marketing Intelligence & Planning*, 23, 395–402.

Keizer, G. (2004) Forrester: Most Wal-Mart suppliers won't meet RFID deadline. *Tech Web*. Retrieved May 15, 2007, from http://www.techweb.com/wire/26804096.

Lefebvre, L.A., Lefebvre, É., Bendavid, Y., Fosso Wamba, S., and Boeck, H. (2006) RFID as an enabler of B-to-B e-Commerce and its impact on business processes: A pilot study of a supply chain in the retail industry. *Proceedings of the 39th Annual Hawaii International Conference on System Sciences (HICSS'06)* January 4–7, 2006, Poipu, Hawaii (available online at http://www.hicss.hawaii.edu/ Hicss39/).

Levy, M., Weitz. B.A., and Beattie, S. (2004) *Retailing Management* (Canadian edition), 1st ed. McGraw-Hill Ryerson Higher Education: Toronto, Canada.

Loeh, H., Sung, G., and Katzy, B. (2005) The CeTIM virtual enterprise lab: A living, distributed, collaboration lab. *11th International Conference on Concurrent Enterprising (ICE'05)*, University BW Munich, Germany, 20–22 June, 2005 (available online at http://www.cetim.org/).

O'Connor, M.C. (2006) Staples business depot sees big benefits from RFID test. *RFID Journal*. Retrieved May 15, 2007, from http://www.rfidjournal.com/article/articleprint/2684/

Roberti, M. (2005) Wal-Mart begins RFID process changes. *RFID Journal*. Retrieved May 15, 2007, from http://www.rfidjournal.com/article/articleview/1385–1/1/

Roberti, M. (2007) Collaboration is the key to success. *RFID Journal*. Retrieved May 15, 2007, from http://www.rfidjournal.com/article/view/1327/

Romanow, K. (2004) Wal-Mart RFID supplier conference: Moving ahead, with or without you. *AMR Research*, Retrieved March 29, 2007, from http://www.amrresearch.com/Content/View.asp?pmillid=17365.

Schwartz, E. (2004) Wal-Mart promises RFID will benefit suppliers. *InfoWorld*. Retrieved March 29, 2007, from http://www.infoworld.com/article/04–06/17/HNwalmart_1.html.

Sellitto, C., Burgess, S., and Hawking, P. (2007) Information quality attributes associated with RFID-derived benefits in the retail supply chain. *International Journal of Retail & Distribution Management*, 35, 69–87.

Smith, A.D. (2005) Exploring radio frequency identification technology and its impact on business systems. *Information Management & Computer Security*, 13, 16–28.

Violino, B. (2004) Farm harvests RFID's benefits. *RFID Journal*. Retrieved May 15, 2007, from http://www.rfidjournal.com/article/articleprint/810.

Wessel, R. (2007) Anticipating ROI, rewe expands its RFID deployment. *RFID Journal*. Retrieved May 15, 2007, from http://www.rfidjournal.com/article/articleprint/3107/

17

A Prototype on RFID and Sensor Networks for Elder Health Care

Melody Moh, Loc Ho, Zachary Walker, and Teng-Sheng Moh

CONTENTS

17.1 Introduction

RFID technology has recently become a viable replacement for the Universal Product Code (UPC) technology in many industries. Its fast growth and huge potential benefits have motivated a major move independently taken by Wal-Mart, the world's largest retailer, and the U.S. Department of Defense (DoD), that required their suppliers to install RFID tags by 2005 [1]. In response, several major computer companies, including Intel, HP, IBM, and Sun, have announced their efforts and future plans to support RFID. RFID technology, however, has attracted relatively little attention in the network research community.

Meanwhile, sensors and sensor networks have in recent years been adopted as a major research focus by federal funding agencies. This has resulted in vast amount of research proposals, academic projects, and publications.

In an effort to bridge the gap between industry and academia focuses, we have worked on a prototype that utilizes both technologies and investigated feasibility, technical challenges, and resulting capabilities of their integration.

An RFID system consists of two primary components—a tag and a reader. An RFID tag, like an UPC, is usually attached to a tracking object; a reader is then used to track tagged objects. While sensor network is used to sense and monitor physical, chemical, and biological environments through sensing of sound, temperature, light, and etc., RFID tags allow any objects to be track-able or "sensible" as long as an RFID tag can be attached. Even though RFID technology has limitations, such as low tolerance to fluid or metal environments, tags can extend a sensor network by providing sensing/sensible property to otherwise unsensible objects, thus provide the last-hop connection of a sensor network.

RFID have been used in a number of biomedical and healthcare applications, such as artificial interocular pressure measurement [2], dental implants and molds [3], and hospital workflow including intrahospital patient and equipment tracking [4].

From a recent study, the population of age 65 and older in the United State will grow from 10.6 million in 1975 to 18.2 in 2025, an increase of 72%, while the overall population increase is about 60%. The trend is global; the worldwide population over age 65 will be more than double from 357 million in 1990 to 761 million in 2025 [5]. Longevity has caused expensive age-related disabilities, diseases, and therefore healthcare. To help in addressing this aging population medication needs, we target our prototype on an in-home elder healthcare system. This is a continuation of our work on applying wireless technologies for biomedical applications [6–8].

The rest of this section presents major features of the RFID technology. Related studies on integration of RFID and sensor networks are described in Section 17.2. This is followed by a presentation of the two phases of the prototype system, in Sections 17.3 and 17.4, respectively. Section 17.5 discusses technical challenges and future improvements. Finally, Section 17.6 concludes this chapter.

17.1.1 RFID

Since sensor network has been a familiar topic in academia, we skip its introduction and focus only on RFID. An RFID system, more specifically, includes three components: (1) a tag or transponder located on the object to be identified, (2) an interrogator (reader) which may be a read or write/read device, and (3) an antenna that emits radio signals to activate the tag and read/write data to it.

At its simplest form, a tag is a beacon announcing its presence to a reader. These types of tags are often seen in retail stores used to prevent theft by announcing their presence when taken past a reader. RFID tag capabilities, however, extend well beyond a simple beacon.

Tag can hold a unique identity (UID) of 8 bytes in length and can be used for inventory management at global scale, such as an UPC. More than just a UID, a tag can carry rewriteable persistent storage and accessible via a reader.

RFID tags are classified by its energy source as passive, semiactive (or semipassive), and active. A passive tag has no battery of its own and makes use of the incoming radio waves broadcast by a reader to power its response. An active tag uses its own battery power to perform all operations. A semiactive tag uses its own battery power for some functions but, like the passive tag, uses the radio waves of the reader as an energy source for its own transmission.

RFID readers employ tag-reading algorithms that are capable of identifying hundreds of tags per second. Once identified, a reader may read data from or write to tag memory, depending on the permissions granted by the tag. RFID readers generally fall into two categories—high frequency (HF) and ultrahigh frequency (UHF). Currently HF RFID systems adhere to the ISO (International Organization of Standardization) standard while UHF RFID systems have yet to become standardized globally. Table 17.1 shows a comparison between HF and UHF RFID technology.

17.2 Related Studies

When an RFID tag is given sensing capabilities, the line between RFID and sensor network becomes blurred. Many active and semiactive tags have incorporated sensors into their design, allowing them to take sensor readings and transmit them to a reader at a later time. They are not quite sensor network nodes because they lack the capacity to communicate with one another through a cooperatively formed ad hoc network, but they are beyond simple RFID storage tags. In this way, RFID is converging with sensor networking technology. From the other direction some sensor nodes are now using RFID readers as part of their sensing capabilities. The SkyeRead Mini M1 made by SkyeTek is an example of an RFID reader designed to mate directly with the Crossbow Mica2Dot sensor motes [9].

In the following, we describe several projects and prototypes taken places in industrial and federal research laboratories, as well as some products adopted by companies.

NASA: Sensor Webs—The project has the objective of using readily available technologies to create a wireless network with embedded intelligence [10]. In this way, instead of reporting

TABLE 17.1

Comparison of HF and UHF RFID Technology

	HF RFID	UHF RFID
Frequency	13.56 MHz	902–928 MHz N. America 860–868 MHz Europe 950–956 MHz Japan
Read range	10–20 cm	3–6 m
Read rate	50 tags/s	400 tags/s
Memory size	64–256 bits read/write	64–2048 bits read/write
Power source	Inductive/magnetic field	Capacitive/electric field
Advantage	Low cost Standard frequency	High speed Longer read range

Source: From SkyeTek Inc., SkyeRead Mini, http://skyetek.com/readers_Mini.html, downloaded16 April 2005.

to an external control system, sensed data can be shared throughout the network and be used by the embedded intelligence to act directly on any detected changes. RFID tagged objects, such as firefighters or astronauts, may be sensed and be guided by the intelligent sensor web; or product components and production flow may be sensed and be guided to slow down or to speed up.

Intel Labs: Proactive Healthcare—In addition to leading a major force on sensor network research, Intel Research Labs also initiated an effort to explore technology that can help in caring for the growing elderly population [11]. A joint project called "*Caregiver's Assistant* and *CareNet Display*," developed by Intel Research Seattle and University of Washington, aims to provide elder care by monitoring elders' activities [12]. RFID tags are stuck on household objects. Combined with a sensor network, the system would collect information on which objects are touched and when. These data are used by an artificial-intelligent program, *Caregiver's Assistant*, to fill out a standard Activities of Daily Living (ADL) form.

HP Labs: Smart Rack and Smart Locus—HP opened its U.S. RFID Demo Center at HP Labs on October 2004. Two major research prototypes are Smart Rack and smartLOCUS [13]. Both prototypes gear toward integrating RFID and other types of sensors, such as video cameras or thermal sensors, into "multimodal sensor networks" that use more than one type, or mode, of sensors. Smart Rack uses thermal sensors and HF RFID readers to identify and monitor the temperature of servers sitting in large metal server cabinets. These sensors and readers are networked and the collected data are used to show, in real time, an inventory of the cabinets and temperature profile of each cabinet. It may become a commercial product and offered within HP's OpenView network management system.

Others: DOD and BP Oil—A few other DOD and private sectors are also using RFID with integrated sensor networks. The U.S. Navy, working with Georgia Tech, has developed an RFID sensor network that monitors the temperature, humidity, and air pressure in containers where aircraft parts are stored [14]. The U.S. military's Combat Feeding Program pilot uses active RFID tag-based sensor networks to provide real-time visibility of rations as they move from the manufacturer to units in the filed [15]. The BP oil company uses RFID and sensor network to monitor assets and react quickly to changes in environmental conditions [16].

17.3 Learning Phase—Integrating Off-the-Shelf Sensor Network with Simulated RFID Reader

The first phase of the project is to investigate the capability of sensors and RFID and how they may be integrated. In this section we first illustrate an overview of the phase. Three major components are described in the three subsections. Finally, the last subsection presents performance results.

There are many choices of commercial products, and each costs from hundreds (sensors and HF RFID readers) to thousands of dollars (UHF RFID readers). Before making actual purchase decisions, in the learning phase, we develop a prototype consisting of some hardware and some software simulators.

There are a number of embedded platforms available. Adapting the sensor network platform is the most logical choice. The initial commercially available sensor network platform is the Berkeley mote. The Berkeley mote has been replaced by Mica, Mica2, Mica2Dot, and MCS Cricket manufactured by CrossBow Technology [17]. The Mica2 mote is selected for this phase to determine its capability, feasibility, and integration effect with RFID. Because of expensive hardware cost, in this phase an RFID

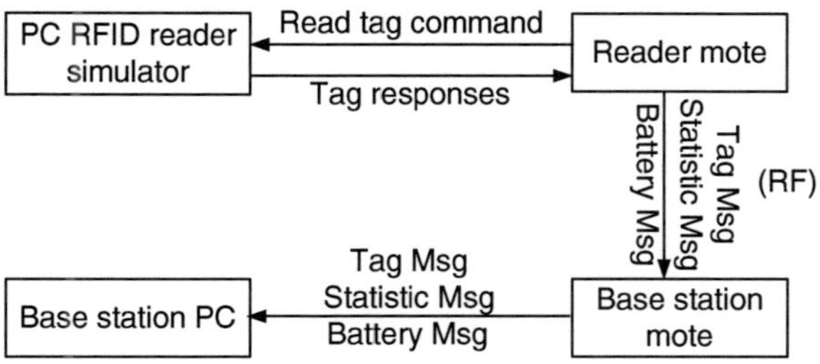

FIGURE 17.1
System component overview.

simulator reader is developed and used. On the basis of the experience learned, in the next phase, the developing phase, an actual RFID readers are used.

In this prototype, there are four system components—two Mica2 motes, one simulated RFID reader, and a base station PC, as illustrated in Figure 17.1. The two Mica2 motes—named RFID reader mote and base station mote—are used for RF communication. The RFID reader simulator is used to simulate an actual RFID reader, communicating via a serial port. The base station PC is also used to perform statistic gathering as well as other required processing. It is connected to the base station mote via a serial port. The message flow of the entire system is also illustrated in Figure 17.1. Each component is described in the following subsections.

17.3.1 Software for RFID Reader Mote and Base Station Mote

The software developed for RFID reader mote is first described in detail, followed by a brief description of the software for base station mote.

The RFID reader mote software is developed using TinyOS and the nesC language [18]. This software interacts with the simulated RFID reader via the mote's serial port. The software module consists of RFID mote (control), RFID reader, battery, and communication modules, as illustrated in Figure 17.2. The control module provides control to all submodules and handles intermodule interactions. The RFID reader module handles interactions with the RFID reader. All RFID-specific details are hidden in the module. The battery module handles battery voltage measurement.

The communication module is divided into three submodules—packet management, serial communication, and RF communication—with an interface module. With limited

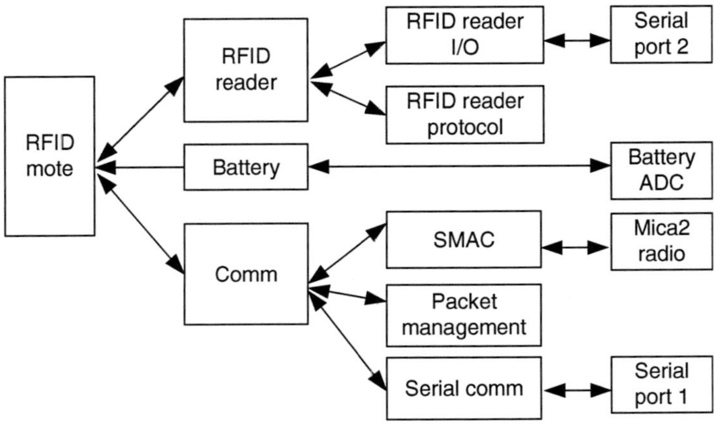

FIGURE 17.2
RFID reader mote software components (with SMAC RF communication submodule).

memory resource, the packet management submodule manages a fixed memory size associated with each communication packet. The serial communication submodule is the TinyOS communication module over serial port. There are a few RF communication modules developed by various members of the open source community. TinyOS has an RF communication module [18]. CrossBow Technology has a mesh RF communication module [17]. An S-MAC module is also available for this platform [19]. The RF communication submodule is designed to allow for easy replacement. All three RF communication submodules are wrapped around a common interface and are selected at compile time.

The RFID reader mote software queries the RFID reader simulator every second for tag messages. In response, the RFID reader simulator sends a set of tag messages to the reader mote. Tag messages are queued by the RFID mote software and transmitted over RF to the base station mote. To allow efficient transmission, up to 12 tag messages are encoded into a single RF message. If there are fewer than 12 tag messages, the RFID mote waits for 300 ms before starting transmission. With a baud rate of 115,200 bps, 300 ms delay is sufficient. (To see this, each tag consists of 19 bytes, but each byte of 8 bits needs 1 stop-bit; thus, transmitting would need to transmit 171 bits. A message of 12 tags needs transmission of 2,052 bits. Given the above baud rate, it needs theoretically 17.8 ms.)

The base station mote software is similar to the RFID mote with run-time behavior changes based on the mote ID (identity). The base station mote software gathers the received packets and forwards them via the mote serial port to the base station PC.

17.3.2 RFID Reader Simulator

A simulator is developed to simulate an HF RFID reader. The HF RFID reader simulator emulates Texas Instrument HF Tag-it protocol [20]. It is written in Java using part of the existing TinyOS serial communication module. When the simulator receives a "Read Transponder Details Command," it sends a series of simulated tag messages to the RFID reader mote via serial port. The number of tags is specified via its command line. The simulated tag ID's are fixed.

17.3.3 Software for Base Station PC

The base station PC is programmed to process data received from the base station mote. It is written in Java, and making use of existing modules from TinyOS. The architecture of this module is designed with component reuse for the next phase of the project. The base station PC software consists of eight modules—RFID Station, RFID Database, RFID Station Packet, RFID Station Statistic, RFID Station GUI (Graphical User Interface), TinyOS Comm, and MySQL Server, as shown in Figure 17.3.

The RFID Station module is the main module and handles all the interactions among various submodules. The RFID Database module handles all database-related interactions. Its main task is to store received tag messages to a persistent storage. The persistent storage is accomplished using MySQL server and interface via Open Database Connection (ODBC). The RFID Station Packet module handles message decoding. The RFID Station Statistic module gathers statistic information based on messages received or statistic message from each mote. With the help of the RF message header, it can determine RF message receive rate as well as lost packet. The RFID Station GUI module (obviously) handles GUI; the GUI screen includes node statistics, list of node details, list of tag details, and command input. Finally, the TinyOS Comm module is the TinyOS reliable communication module. This module handles all serial communications as well as network communication over a serial port.

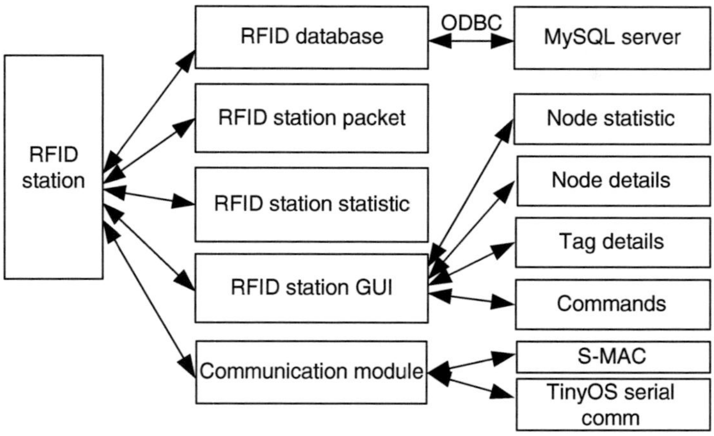

FIGURE 17.3
Base station PC module design overview.

17.3.4 Performance Results

The TinyOS RF communication module has a packet size limited to 29 bytes. This allows only one tag message with overhead to be transmitted in a single RF packet. The CrossBow RF mesh communication module is too slow with too much overhead. It is rated at 1 packet per second. The SMAC RF communication module has a packet size limit of 256 bytes; it is used for bandwidth test described below.

The test setup consists of two motes, an RFID reader simulator, and a base station PC. The reader simulator and base station software both run on the same PC. The two motes are about 2 feet apart, both with battery power. The RFID reader mote (base station mote) communicates with the simulator reader (base station PC) by a USB serial port; the two motes communicate with each other via wireless communication.

The initial test achieved about 10 tag messages of 19 bytes in length. To achieve a better transfer rate, as mentioned earlier, the communication module is modified to queue up to 12 tag messages. This achieves about 25 tag messages or 500 application bytes per second with 100% reliable RF communication. This rate is sufficient in most embedded applications. Most commercial RFID readers can handle between 50 to 100 tags per second. On a pure ID-based application, this usually requires 12 bytes for a tag ID. Thus, this system can handle about 41 (i.e., 500/12) tags per second. If only 8 bytes are required for a tag ID, the support rate goes up to 62 (i.e., 500/8) tags per second.

17.4 Developing Phase—Sensor Network with HF/UHF RFID for Elder Health Care

In this section, we shall describe the development phase, including the details of the application prototype system, results, and future enhancements. As mentioned earlier, we developed a medicine monitoring and notification system for elder patients without personal care assistants. The high-level functionality may be described as follows:

> The system monitors, notifies, and assists an elder patient in taking the accurate amount of his/her medicines at the appropriate time. It notifies the elder patient when it is time to take his/her medicines. A buzzer (Patient Monitor system) mounted on the door of the

patient's room beeps when there are medicines to be taken. When the patient walks to the medicine cabinet (Medicine Monitor system), the system guides the patient in taking the proper type and accurate amount of medicines using a GUI.

The rest of the section is organized as follows. Section 17.4.1 describes the prototype system, including its three subsystems and their interaction. Section 17.4.2 details three application mote software corresponding to these three subsystems. Section 17.4.3 presents the base station software including the GUI module.

17.4.1 Prototype System

The system utilizes the strengths of both HF RFID (lower cost) and UHF RFID (long distance) with three sensor network motes. The system is an extension to a prototype system researched by Intel Labs [21], but completely built from bottom up with newer scale model, 3rd generation sensor network mote, and an additional UHF reader. There is no hardware or software reuse as they are not available to the public.

The system consists of three subsystems: The medicine monitoring subsystem, the patient monitoring subsystem, and the base station subsystem. Together, they use seven components—three Mica2 motes, an HF RFID reader, a UHF RFID reader, a weight scale, and a base station PC. In the following, we first describe each of the subsystems, followed by an illustration of their interactions in the system component configuration.

17.4.1.1 Medicine Monitoring Subsystem

In this subsystem, as shown in Figure 17.4, HF RFID tags are placed on each medicine bottle; each HF tag identifies a medicine bottle. The HF RFID reader (SkyeRead M1 Mini [22]) is used in conjunction to track all medicine bottles within range of the reader.

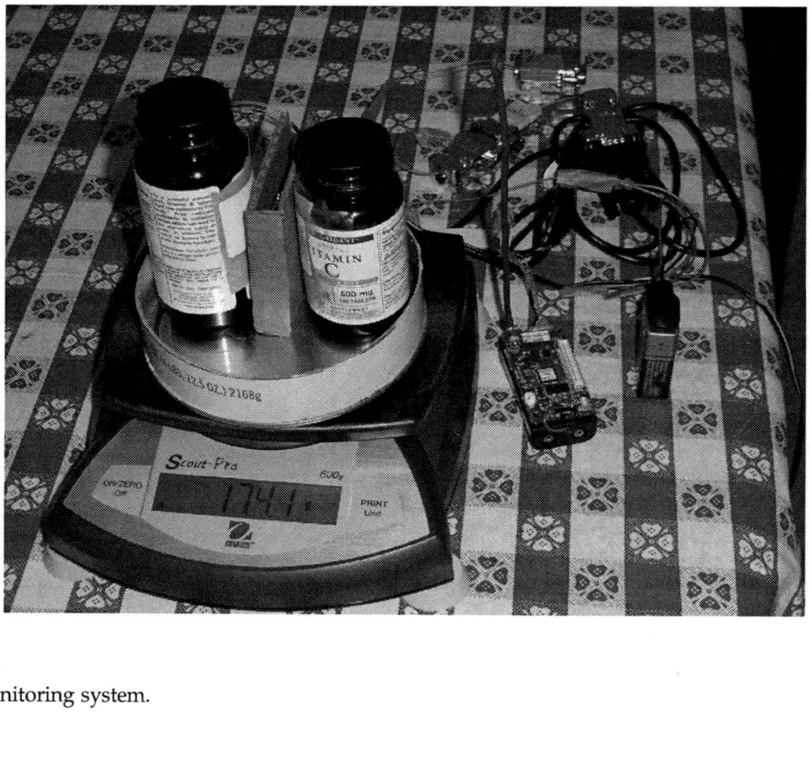

FIGURE 17.4
Medicine monitoring system.

The Medicine Mote communicates with the HF RFID reader and with the scale to monitor HF tags and the total medicine weight. By performing reads of all tags at a regular interval, the system is able to determine *when* and *which* bottle is removed or replaced by the patient. (The short range of the HF reader is actually desirable for this aspect of the application.) The weight scale monitors the amount of medicine on the scale. Combining changes in weight and HF tag event, *which* medicine bottle and the *amount* of medicine taken can be determined when the patient takes their pills. The Medicine Mote software is described in detail in Section 17.4.2.2.

17.4.1.2 Patient Monitoring Subsystem

This subsystem is shown in Figure 17.5. A UHF RFID system including a reader and one or more tags is used to track the elder patient who needs the medicines. This patient wears a UHF tag, which may be detected by the associated RFID reader within 3–6 meters. The AWID UHF RFID reader [23] is chosen for this subsystem. The Patient Mote communicates with the UHF Reader to monitor patient arrival at the door of a room or other areas where the system is installed. Thus, the system is able to determine that the patient is in the vicinity, and alerts the patient to take the required medicines via a buzzer. The Patient Mote software is described in Section 17.4.2.3.

17.4.1.3 Base Station Subsystem

This subsystem is shown in Figure 17.6. The Base Station Mote provides message relay to the Base Station PC. The Base Station PC is a PC running a Linux operation system and it is re-designed from the first phase to accommodate our application. The Base Station software tasks include simulating a display and its GUI for the patient; determining

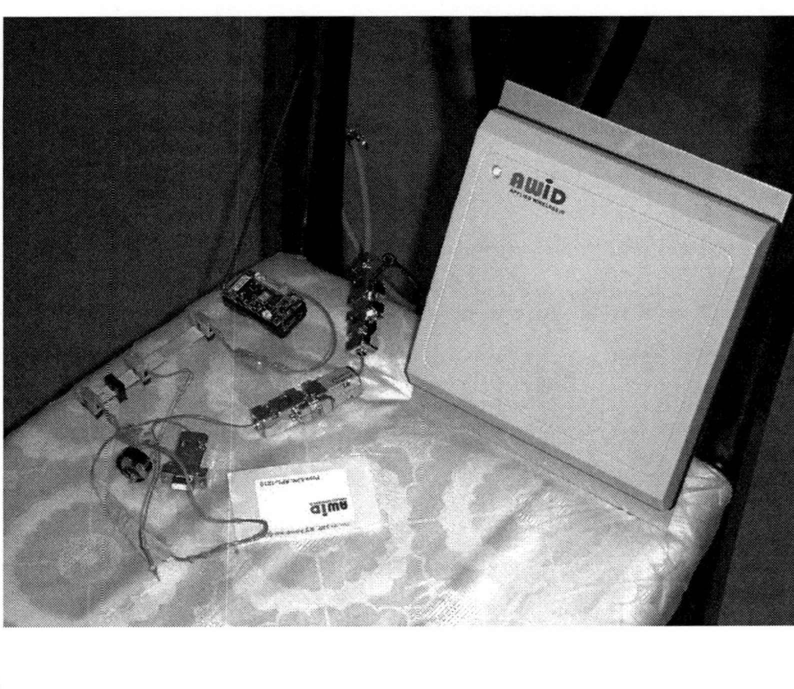

FIGURE 17.5
Patient monitoring subsystem.

FIGURE 17.6
Base station subsystem.

when medicine is required; and maintaining various interactions between the Medicine Mote and Patient Mote. The Base Station Mote software is described in Section 17.4.2.1.

17.4.1.4 System Component Configuration

Next, we describe major interactions among the seven system components, as shown in Figure 17.7. As explained before, motes are mainly used for communicating readers from RFID readers to the control system. The Medicine Mote communicates with the HF RFID reader and weight scale to monitor HF tags and medicines weight. (Recall that each HF RFID tag identifies a medicine bottle.) The Patient Mote communicates with the UHF RFID

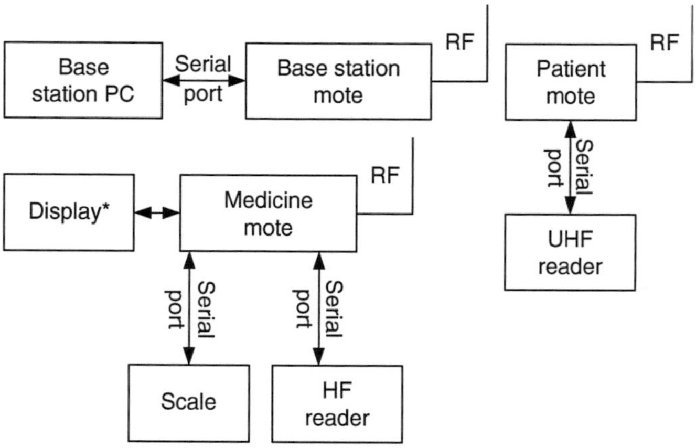

FIGURE 17.7
System component configuration.

reader to monitor patient arrives to room or an area where the system is installed. The Base Station Mote provides message relay to the Base Station PC (the control system).

17.4.2 Application Mote Software

The software for all three motes (Medicine, Patient, and Base Station motes) is identical with compile-time hardware assignment. Figure 17.8 shows the generic mote system software component, enhanced from that in the first phase (Figure 17.2). To support the HF RFID reader, the internal of the RFID reader module developed in the first phase is replaced with the actual protocol interface. The weight scale module is added to handle the scale measurement. The RFID tag and weight data are fused into a single source of information for transmission. A set of data fusion messages is created to indicate the following: weight change, tag no longer detected, tag detected again, and patient detected (details in Section 17.4.3.1). This requires a number of error checking's and handlings to deal with unreliable serial port communications as well as scale weight instability. The UHF RFID reader module is added to communicate with the AWID UHF RFID reader. The scale reader module is added to communicate with the scale.

In correspondence to the three subsystems described in Section 17.4.1, there are also three pieces of mote software: the medicine mote software, the patient software, and the base station mote software. They are described in the following subsections.

17.4.2.1 Base Station Mote Software

The Base Station Mote hardware is a Mica2 mote with break out connectors and wire connectors to connect to the Base Station PC. Its software and functionalities are the simplest of the three motes. Its tasks are to receive wireless messages and relay them to the Base Station PC software application via serial port 1. The Comm module developed in

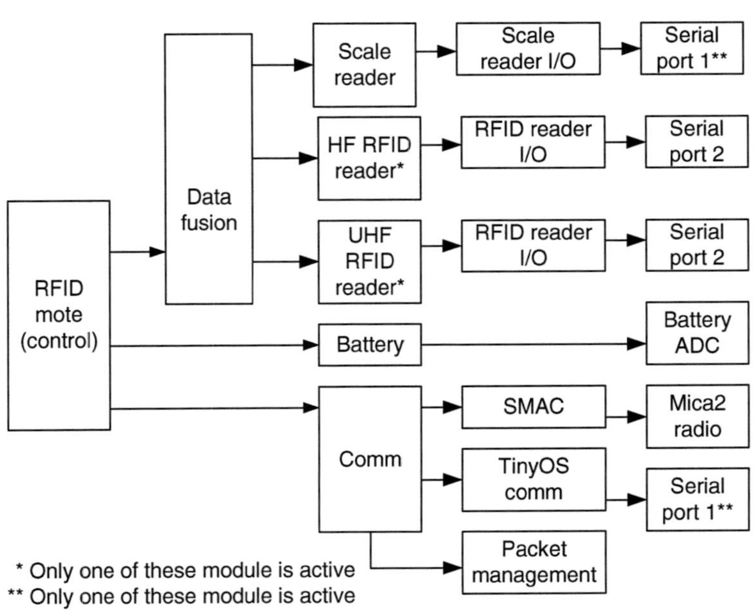

FIGURE 17.8
Mote system software components.

the first phase is sufficient. The application message length and S-MAC packet length changed from 250 bytes to 50 bytes as our maximum message size is 50 bytes. This allows 30 messages to be queued for transmission instead of 12 messages from the first phase of the prototype system.

17.4.2.2 Medicine Mote Software

The Medicine Mote hardware is the same as the Base Station Mote with an HF RFID reader and a scale. As illustrated in Figure 17.8, the scale reader and HF RFID reader modules have its only communication path. The design of the scale reader module encapsulates the scale communication protocol. When a scale reading is available, it notifies the Data Fusion module. Similarly, the HF RFID reader module encapsulates the HF reader communication protocol. All tags detected by the HF RFID reader module are passed to the Data Fusion module for processing.

The most complex module is the Data Fusion module, which performs aggregation on the HF RFID reader and scale data. One of its tasks is intelligent data aggregation, which aims to reduce the number of RF transmission. This process requires an additional message type with five states—medicine detected, medicine removed, medicine placed back, medicine taken, and medicine cleared. The medicine detected state message occurs when a new medicine (or tag) is received. The medicine removed state message occurs when a medicine is not detected after some elapsed interval. The medicine placed back state message occurs when a medicine is detected again after it has been removed. The medicine taken state message occurs when a medicine is placed back and the scale weight is stabilized. The medicine cleared state message is intended to handle the situation where weight changed because of the 100 mg resolution of the scale. In this case, the Base Station PC software will handle this situation by ignoring the changed weight.

The data fusion algorithm used in the Medicine Mote software is described later.

Data Fusion Algorithm (Medicine Mote)

1. On first receipt of each medicine message, perform the following operations:
   a. Record the medicine tag ID
   b. Notify the Base Station PC by sending a medicine detected state message
   c. Set detected counter to 1
2. On subsequent receipt of each medicine message, perform the following operations:
   a. Increment its detected counter by 1
   b. Reset the removed counter to 0
   c. If the medicine is marked medicine removed, notify the Base Station PC by sending a medicine placed back state message and mark the medicine as medicine placed back
3. On every half second timer, perform the following operations:
   a. Reset the detected counter to 0
   b. If the detected counter is 0, increment the removed counter by 1
   c. If the removed counter is greater than a threshold, notify the Base Station PC by sending a medicine removed state message and mark the medicine as medicine removed

4. On each scale weight changed (and stabilized), perform the following operations:
   a. If a medicine is marked medicine removed, determine the scale difference and send a medicine taken state message to the Base Station PC
   b. If no medicine is marked as medicine removed, send a medicine cleared state message to the Base Station PC

To support reliable communication between the Medicine Mote and the Base Station PC application, each message is echoed back to its sender. Message is cleared if the sequence ID matches the expected sequence ID.

Finally, the scale reader module interfaces with the actual scale. At system startup, the scale reader module will initialize the actual scale for proper operation. Every second the scale reader module will send command to query the actual scale weight. This weight is passed to the Data Fusion module for processing. All communication errors with the scale will cause reinitialization.

17.4.2.3 Patient Mote Software

The Patient Mote hardware is the same as the Base Station mote with a UHF RFID reader. The UHF RFID reader is connected to the mote via a broke out board with various connectors and wire connectors. A passive UHF tag is worn on the patient's wrist to provide proper detection by the UHF RFID reader. (Because of limited access to UHF tags and interference of water concentration in human body, the UHF tag is required to be held on the finger for proper operation.) The Patient Mote software uses the same software modules as the Medicine Mote. The UHF RFID reader module is enabled while the scale reader and HF RFID reader modules are disabled via compiler switches. The UHF RFID reader module interfaces with the UHF RFID reader. All protocol specific details are hidden in this module. (For more information on the AWID UHF RFID reader protocol, refer to LR-911 reader manual.) When a UHF tag is detected, it is forwarded to the Data Fusion module for processing. The design of the Data Fusion module is extended to support the Patient Mote functionalities.

The data fusion algorithm for the Patient Mote is described below.

Data Fusion Algorithm (Patient Mote)

1. On reception of a UHF tag ID message from the Base Station PC, enable the UHF RFID reader system. (This indicates medication needs to be taken.)
2. On reception of a cleared tag ID message, disable the UHF RFID reader system. (This indicates no medication required.)
3. On reception of a UHF tag ID from the UHF RFID reader, increment its detected counter if ID matches
4. On every half second timer, perform the following operations:
   a. If counter is greater than 6, set counter to 6 (3 s)
   b. If counter is not zero, enable buzzer and decrement counter by 1
   c. If counter is zero, disable buzzer

17.4.3 Base Station Software

The base station PC software developed in the previous phase (Section 17.3.3, Figure 17.3) is significantly expanded for the application. A data fusion module is added to process data aggregations. All data fusion messages are recorded in the persistent database. In

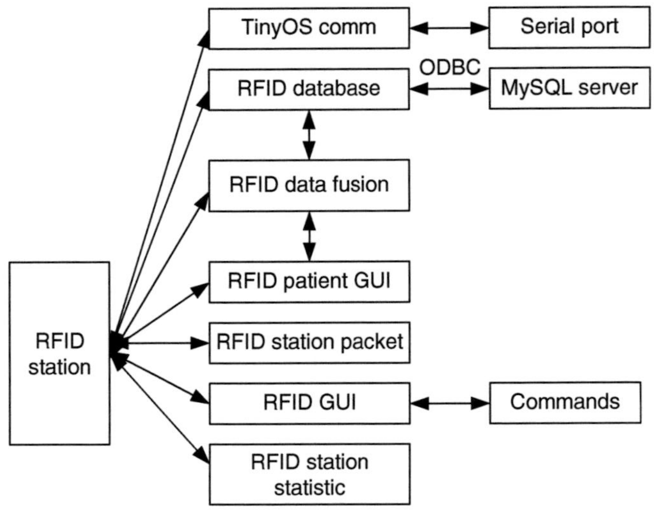

FIGURE 17.9
Application base station software component.

supporting the application, a number of new database tables and schemes are created for recording additional information, patient, medicines, and log of events (from the Patient Mote and Medicine Mote).

The Base Station PC software component is shown in Figure 17.9. The RFID Data Fusion module tasks are to provide a list of medicines that needs to be taken based on the patient data stored in the persistent database and handle response messages from the Medicine Mote and Patient Mote. The list of medicines is queried by the RFID Patient GUI module for display. In the following section, we first discuss details of the Data Fusion module, which is the most complex among all modules; this is followed by the GUI module.

17.4.3.1 RFID Data Fusion Module

The module performs a number of crucial operations, which include interfacing with the database for patient data, computing the required medicines to be taken, handling messages from the Medicine Mote and Patient Mote, providing a list of medicines to be taken to the RFID Patient GUI module, and providing coordination logics to support the various application functionalities.

The RFID Data Fusion module algorithm is described below.

Data Fusion Algorithm (RFID)

1. On system startup, perform these operations:
   a. Query the persistent database for the patient information such as user name and the UHF tag ID
   b. Query the persistent database for a list of medicines to take
   c. Query the persistent database for a list of medicines taken today
   d. Compute the medicine list for the patient based on when a medicine is required to be taken and the current system clock
   e. If there is medicine to be taken, send a UHF tag ID message to the Patient Mote if presents
   f. If there is no medicine to be taken, send a cleared message to the Patient Mote if presents

   g. If there is medicine to be taken, notify the RFID Patient GUI module
   h. If the RFID Patient GUI module receives a notification, it retrieves the list of medicines required to be taken and displays them
2. Perform the following operations once every minute:
   a. Compute the medicine list for the patient based on when a medicine is required to be taken and the current system clock
   b. If there is medicine to be taken, send a UHF tag ID message to the Patient Mote if presents
   c. If there is no medicine to be taken, send a cleared message to the Patient Mote if presents
   d. If the RFID Patient GUI module receives a notification, it retrieves the list of medicine required to be taken and displays them
3. Upon reception of any messages from the Patient Mote or Medicine Mote, mark them as present
4. Upon reception of a Patient Detected state message, log the event
5. Upon reception of the Medicine Detected state message, perform these operations:
   a. If the tag ID matches the tag ID in the database, log the event
   b. If the tag ID does not match the tag ID in the database, notify the RFID Patient GUI to display invalid medicine
6. Upon reception of the Medicine Taken state message, perform these operations:
   a. Log the event
   b. Clear the medicine
   c. Notify the RFID Patient GUI to update its display
7. Upon reception of the Medicine Cleared state message, perform these operations:
   a. Log the event
   b. Add or subtract the additional weight to the last medicine taken record
   c. Notify the RFID Patient GUI to update its display
8. Upon reception of any other messages, write to the persistent storage table accordingly

17.4.3.2 Graphic User Interface Module

To provide user interactions with the system, a GUI is required with a display. An embedded display may be used for this purpose. With limited resource, an emulated display within the Base Station PC software is developed as a replacement. The display emulates and provides a GUI to assist the patient, as shown in Figure 17.10. Note the use of large font size for various medication/vitamins and of different colors for pill quantity; this is to make it easier for old patients. Alternatively, pictures of various medicine brands/bottles may be used to replace medicine names.

17.5 Technical Limitations, Challenges, and Future Improvements

In developing the prototype system, a number of technical challenges and limitations have been encountered. We feel that it is important to discuss them as well as suggest future improvements.

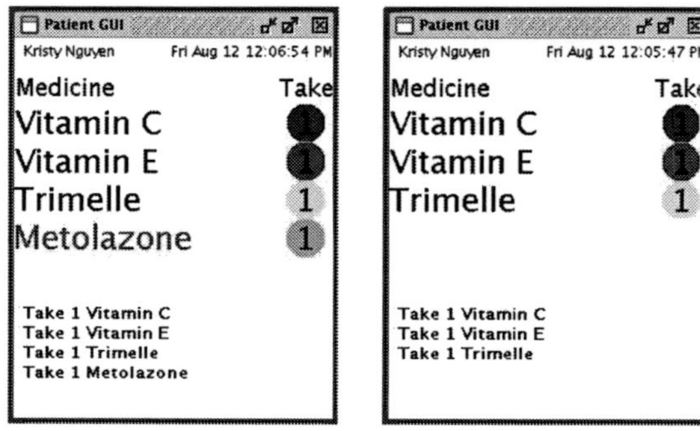

Before Metolazone taken | After Metolazone taken

FIGURE 17.10
Patient GUI.

17.5.1 Technical Limitations and Challenges

The limitations and challenges included struggling with limited availability of hardware, lack of expertise and knowledge of RFID technology, finding and acquiring the proper hardware, immature RFID technology, and the nature of development on an embedded system. Unlike the traditional research of a pure software application or algorithm that requires only a PC or equivalent, our application required seven hardware components, software development, and interaction with each of them. Acquiring the HF and UHF RFID readers were quite difficult. As the UHF RFID reader is relatively new technology with limited availability, a few UHF RFID reader manufacture companies would not sell us their hardware unless we attend their seminars, which could cost thousand of dollars. Although there are only a few UHF RFID readers, we still need to determine which reader would be appropriate for our application. Since we could not purchase all models, we chose the AWID LR-911 model as recommended by UCLA WINMEC [24] as the most reliable model available. For the HF and scale readers, we relied on Intel Labs' prototype as our guide lines. We chose newer models of the same HF RFID and scale readers used by Intel Labs, as newer models are better, smaller, and more accurate.

Because our application involved embedded platform, the development process were quite difficult. After our mote software was written, it was downloaded into the mote for testing. Without an in-circuit debugging capability, the only way to debug any part of the code was via the mote's three LEDs. Message printing capability was added after our mote software functioned properly with basic functionalities. Although the primary way to debug was still via toggling LEDs, we also carefully reviewed the mote source code when the system did not performed as expected.

As the UHF RFID technology is relatively new, we did not find an acceptable UHF tag. The preferred tag is a wrist UHF passive tag that can be worn on the patient's wrist without interference. With the current tag, the patient UHF passive tag is required to hold on the finger for proper detection.

On the HF reader end, the HF reader range was too short. Therefore, an external antenna is required. For this, a custom designed external antenna may be more appropriate. At this stage, it is unlikely that we will acquire an external antenna and integrate in time. It is likely to be added in future phases of the prototype system.

17.5.2 Application Result and Future Improvements

The developed application can monitor and notify the patient to take their medication. The design of the system is fairly flexible. Additional Patient Motes can be added with minimal source changes on the Base Station PC software. As our first application prototype on RFID and sensors, the system is not a very commercially user-friendly. Yet, it has proven the benefit of the integration of sensor network and RFID technologies, which is one major goal for this prototype.

There are a number of improvements and future extensions that can be added to the prototype system in order to be a feasible product. Improvements of the prototype system may include the following:

- Completely battery powered all system components with power management
- Higher precision scale as some medicines require 100 mg resolution
- Better UHF passive tag with less interference
- External HF antenna for more coverage area to detect medicine bottles
- Smaller HF tag in order to place on the bottom of medicine bottles
- An actual embedded display for the patient GUI
- A small form factor PC, such as mini-itx or embedded platform, instead of a laptop
- Molding for all system components for better appearance
- Use or experiment with other MAC protocols

17.6 Conclusion

We have described a project that integrates both sensor network and RFID technologies. It includes an initial learning phase and a development phase. The learning phase investigates technology compatibility and capabilities through a sensor network interacting with a simulated RFID system. Simulating software modules are described as they provide excellent learning experiences, and are needed before hardware purchases are possible. The development phase builds a system that consists of sensors and both HF and UHF RFID components. All the subsystems and the corresponding software modules are presented in detail. The project is targeted for in-home medication monitoring for elder patients. Future work may include, in addition of those improvements listed in Section 17.5.2, extending the prototype from medication monitoring to a broader elder home-care system, from one room to an entire house, featuring more sensors and RFID components distributed at various strategic places and on various household items. Another extension would be adding the capability of notifying family members via email, and networked with an external healthcare center monitor system for any assistance via the Internet.

Acknowledgment

A preliminary version of the work appeared in the following paper: "A Prototype on RFID and Sensor Networks for Elder Healthcare: Progress Report," by L. Ho, Z. Walker, M. Moh, T. Hamada, and C.-F. Su, in *Proc. of ACM Workshop on Experimental Approaches to Wireless Network Design and Analysis*, in conjunction with *ACM SIGCOMM 2005*, Philadelphia, PA, Aug 2005. M. Moh is supported in part by Fujitsu Labs of America and by CSUPERB (California State Univ. Program in Education and Research in Biotechnology). T.-S. Moh is supported in part by CSU Research Grant.

References

1. DOD, DOD RFID website (Official Releases, Memos, etc.), http://www.acq.osd.mil/log/rfid/index.htmhttp://www.dodrfid.org, downloaded 16 April 2005.
2. Finkenzeller, K., *RFID Handbook: Fundamentals and Applications in Contactless Smart Cards and Identification*, Wiley, England, 2003.
3. Ilic, C., Using tags to make teeth, *RFID Journal*, http://www.rfidjournal.com/article/articleview/1206/1/1/, downloaded 22 January 2005.
4. Exavera Technologies, eShepherd overview, http://www.exavera.com/healthcare/eshepherd.php, downloaded 22 January 2005.
5. Hooyman, N. and Kiyak, H., *Social Gerontology: A Multidisciplinary Perspective*, 6th ed., Allyn & Bacon, 2002.
6. Culpepper, B., J.L. Dung, and M. Moh, Design and analysis of Hybrid Indirect Transmissions (HIT) for data gathering in wireless micro sensor networks, *ACM Mobile Computing and Communications Review* (MC^2R), January/February 2004, pp. 61–83, A preliminary version, Hybrid indirect transmissions (HIT) for data gathering in wireless micro sensor networks for biomedical applications, appeared in *Proceedings of the IEEE Computer Communication Workshop*, Laguna Niguel, California, October 2003, pp. 124–133.
7. Moh, M., B.J. Culpepper, L. Dung, T.-S. Moh, T. Hamada, and C.-F. Su, On data gathering protocols for in-body biomedical sensor networks, *Proceedings of the IEEE Globecom*, St. Louis, MO, November 2005.
8. Z. Walker, M. Moh, and T.-S. Moh. A development platform for wireless sensor networks with biomedical applications. *Proceedings of the IEEE Consumers Communication Networks* (CCNC), Las Vegas, NV, January 2007.
9. Crossbow Inc., Mica2Dot Series, http://www.xbow.com/Products/productsdetails.aspx?sid = 73 downloaded 10 April 2005.
10. Colllins, J., NASA creates thinking RF sensors, *RFID Journal*, downloaded April 10, 2005, http://www.rfidjournal.com/article/articleview/1146/1/47/
11. Dishman, E., Inventing wellness systems for aging in place, *IEEE Computer Magazine*, May 2004.
12. Intel Research Seattle, Caregiver's assistant and carenet display: Making eldercare easier, downloaded 10 April 2005, http://seattleweb.Intelresearch.net/Projects/active/Dcdemo/
13. O'Connor, M.C., HP kicks off US RFID demo center, *RFID Journal*, http://www.rfidjournal.com/article/articleview/1211/1/50/, downloaded on 10 April 2005.
14. Roberti, M., Navy revs up RFID sensors, http://www.rfidjournal.com/article/articleview/990/1/, June 2004.
15. Roberti, M., Vendor to foxhole tracking, *RFID Journal*, http://www.rfidjournal.com/article/articleview/847, March 2004.
16. Roberti, M., BP leads the way on sensors, http://www.rfidjournal.com/article/articleview/1216/1/2/, November 1, 2004.
17. Crossbow Inc., Home page, http://www.xbow.com, downloaded 10 April 2005.
18. TinyOS, Tiny OS Community Forum, http://www.tinyos.net, downloaded 4 December 2004.
19. Ye, W., Download S-MAC Source Code for Motes, http://www.isi.edu/ilense/software/smac/#Introduction, downloaded 16 June 2004.
20. Texas Instrument Inc., HF Reader System Series 6000, downloaded 1 January 2005 http://www.ti.com/rfid/docs/manuals/refmanuals/RI-STU-TRDCrefGuide.pdf
21. Fishky, K. and Wang, M., A Flexible, Low-Overhead Ubiquitous System for Medication Monitoring, *Intel Research Technical Report IRS-TR-03–011*, October 2003.
22. SkyeTek Inc., SkyeRead Mini, http://skyetek.com/readers_Mini.html, downloaded16 April 2005.
23. AWID Inc., Home page, http://www.awid.com, downloaded 10 April 2005.
24. WINMEC (Wireless Internet for Mobile Enterprise Consortium), Home page, downloaded 20 February 2005. http://www.winmec.ucla.edu/index.asp

18

Triage with RFID Tags for Massive Incidents

Sozo Inoue, Akihito Sonoda, and Hiroto Yasuura

CONTENTS

18.1 Introduction

In this chapter, a triage system using RFID and mobile devices with a wireless network is proposed, and its advantages are verified through an experiment assuming an incident of massive injured people.

Triage is a procedure used by emergency personnel to ration limited medical resources to massive injured people, in which triage tags are used to (1) classify and transport the injured people effectively and (2) obtain the information about the state and the scale of the

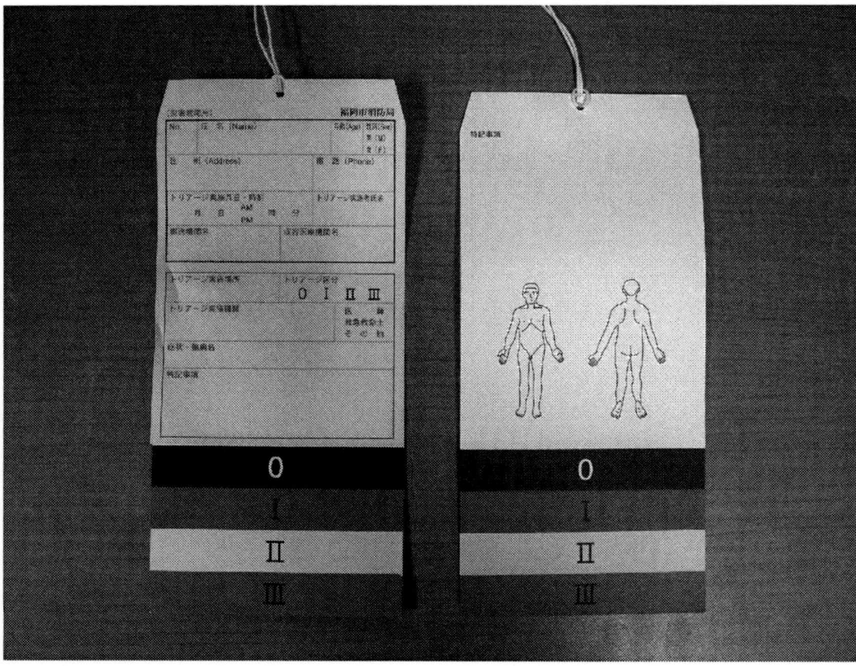

FIGURE 18.1
Triage tag.

casualty incident to publish to the masses or to utilize for decision making such as medical resource procurements. Figure 18.1 is a picture of a triage tag.

So far, triage is operated manually using paper triage tags, tallies, and radiophones. However, manually obtaining the information about the state and the scale of the casualty incident to publish to the masses or to utilize for decision making such as medical resource procurements leads to failure, inaccuracy, and delay in the information transmission while emergency personnel have priority over treatments for injuries, and causes inefficiency in classification and transport of the injured people effectively, which is the essential goal of emergency medical services.

In this chapter, we propose a triage system in which RFID tags, which are silicon chips with their IDs, radio frequency functions, and some additional logic and memory [1,2], are attached to triage tags. Most of the RFID tags are passive, which means the power is supplied through radio frequency communication from external readers. Employed RFID tags in this work are passive and have 1 kb of rewritable memories. Figure 18.2 is a picture of an RFID tag we employed.

Embedding an RFID tag to a triage tag has the following advantages:

1. Terminal that the emergency personnel use can identify the injured person by the unique ID value in each RFID tag.
2. Rewritable RFID tags provide the storage for the information of the injured person, and the emergency personnel can obtain the information when they are in a place where wireless communication is out of service, such as deep in a mountain or in the underground.

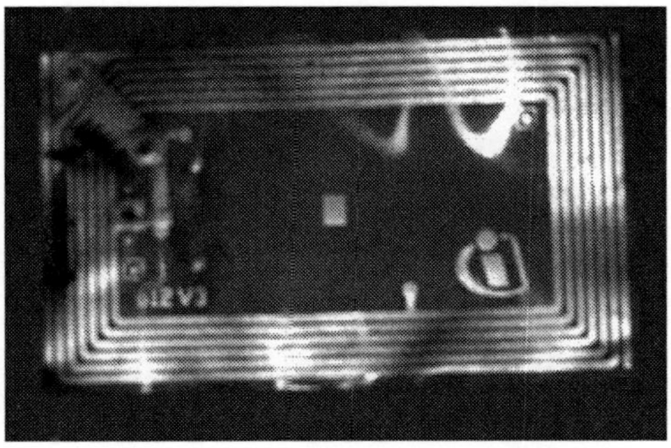

FIGURE 18.2
RFID tag.

3. Time required to input an injured person's information can be independent of the network communication, since the information already input so far about the injured person, which is shown to the emergency personnel when they begin to add information, is not obtained through the network, but from the RFID tag.

This application addresses important challenges for pervasive computing: data integrity, the information that should not be lost after being input; input throughput, the time required to input injured person's information should be as short as possible; availability, that is, emergency personnel should be able to use the system any time, and low latency of communication. These should be as independent of the network status as possible.

In this chapter, we show a realistic solution for the challenges by specializing the network usage in a way that only particular paths are used in particular stages of the workflow by analyzing the workflow and exploiting RFID tags to slim down the possible paths by the following approaches: input throughput and availability are assured by using RFID tags as local buffer; data integrity is assured and latency is improved by defining minimum wireless communication areas in the paths in triage workflow.

To evaluate the effectiveness of the system, we performed two experimental performance of triage assuming a complex car crash of 5 cars and about 82 injured people, where one of the performances is done in the current method, and the other is done using the system. As a result of the experiment, the information collection was accelerated and made more complete by using the system, especially for the information that is utilized in early stages of the triage. The acceleration of the collection also resulted in the acceleration of the transportation of the injured people, which is the most important objective in triage. Moreover, we could estimate that the average acceleration of each input to the terminal by employing RFID tag in the experiment was 14.3 s.

The rest of the chapter is organized as follows: Section 18.2 describes triage and challenges for pervasive computing, Section 18.3 introduces the RFID triage system, Section 18.4 describes the related work, Section 18.5 shows the result of a preliminary experiment to complement the experiment, Section 18.6 shows the result of the experiment to confirm the effectiveness of the system, and Section 18.7 concludes the chapter.

18.2 Triage as a Challenge for Pervasive Computing

We position triage as not only a practical application that requires urgent and continuous improvement since we are facing massive casualty incidents every day, but also a start point for new horizons in pervasive computing concerning unexpected human behavior, rapid deployment, and insufficient computing infrastructure. In this section, we describe the workflow of triage operated by emergency personnel, address technical requirements for triage regarding the paths of the information collected as that of network, and motivate the challenges for pervasive computing hidden in the requirements.

We target on incidents in which 10–100 people are injured in a regional place. Examples of such incidents are a crash/derailment/overturn of a train, a crash/fall of a car, a crash/fire of a plane, and terrorism in a subway/building.

18.2.1 Triage

We call the people who are engaged in triage emergency personnel. Figure 18.3 shows an abstract workflow of triage.

Current triage is done in the following procedure without using information systems by the emergency personnel:

1. Emergency personnel who arrive to the incident site first establish a first-aid area, which is a safe place for first aid close to the incident site, and an operation point, which is a place to command the triage.
2. *First triage*: After the evacuation of injured people once the incident site is secured by firemen, the emergency personnel enter the incident site and perform first triage, in which the personnel distinguish the injury level in about 30 s. During the first triage, they attach triage tags, each of which has a perforated colored label representing the injury level, as the following, to the injured people:
   a. *Black/DECEASED*: The person is severely injured to die and any hospital care does not help survive.

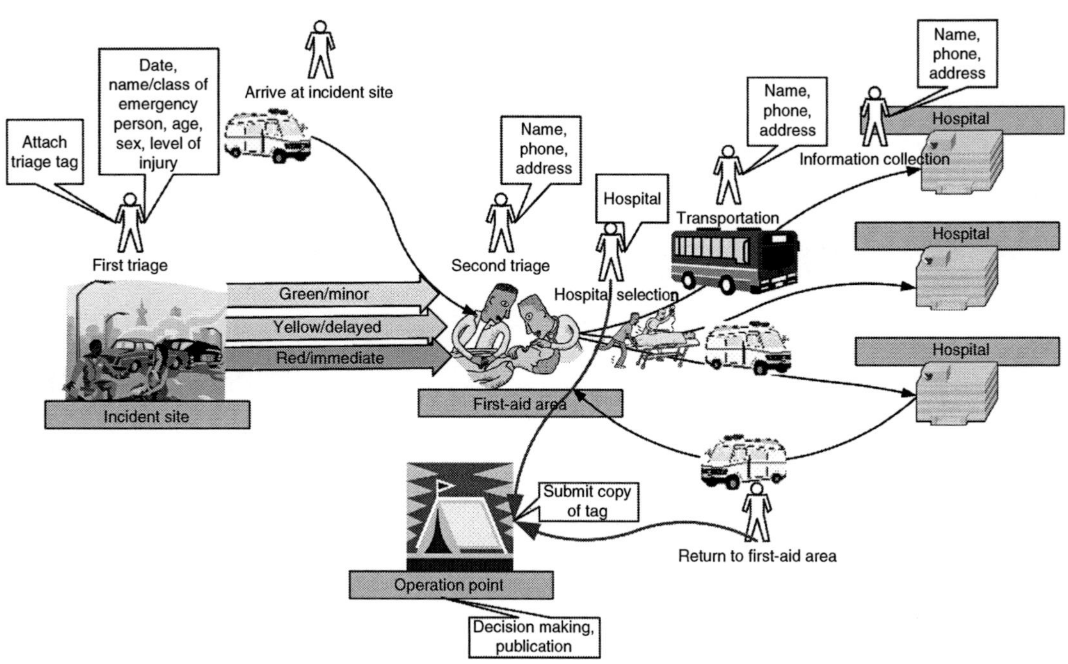

FIGURE 18.3
Workflow in triage.

b. *Red/IMMEDIATE*: The person requires immediate surgery or other life-saving intervention, the person is unconscious, and has first priority to be transported to hospitals.
c. *Yellow/DELAYED*: The person requires hospital care and requires to be watched by trained persons, but the person remains conscious and the condition is stable for the moment.
d. *Green/MINOR*: The person may require a doctor's care but not immediately, and the person is able to walk on her/his own.

Simultaneously, the following information is written by the emergency personnel as much as possible:

a. date of input
b. name of the emergency personnel
c. category of the emergency personnel, such as doctor, emergency medical technician
d. rough age of the injured person
e. sex of the injured person
f. written version of the injury level

3. Injured people are moved from the incident site to the first-aid area.
4. *Second triage*: In the first-aid area, besides treatment of the injured people, second triage is performed if possible, in which the information of the injured people such as

   a. name
   b. phone number
   c. address
   d. updated information from first triage

   are collected as much as the emergency personnel and the injured people can afford to, and written on the triage tags.
5. *Hospital selection*: At the exit of the first-aid area, the hospital to which each injured person is transported is decided and written on the triage tag. A triage tag has carbon copies, and one of them is left to the emergency personnel of the operation point.
6. When an ambulance or a transport vehicle arrives at the first-aid area, the injured people are transported from the first-aid area to the hospital. In the ambulances, the information of the injured people are collected as much as possible and written on the triage tags.
7. Other emergency personnel stand by at each hospital and collect the information of the injured people transported. The ambulance, after transporting injured people to hospitals, returns to the incident site again and repeats the transportation of injured people to hospitals. On returning to the incident site, the ambulance transports the carbon copy of the triage tag at the hospital to the operation point.
8. In the operation point, the information on the carbon copies are collected, and reported to the fire department of the area, which is to be used for decision making, publication, etc.

18.2.2 Requirements

The first mission of triage targeted on in the chapter is to transport injured people as quickly as possible. As for the mission, information collection of injured people affects the latency of transportation in two ways: troublesome operations for information collection by an emergency personnel cause latency by disturbing her/his work, and particular types of input information should be collected in the early stages since they are used for decisions in the transportation. For the latter, information such as the level of injury and hospital to be transported have to be collected within first or second triage before the transportation where these information are used, whereas in the other types they have not been collected since they are used in the afterward. However, in the current triage, all of this information must be collected before transportation for the sake of avoiding loss and latency of the collection after the ambulance has left.

Thus, several technical requirements that can be identified as important challenges for pervasive computing arise regarding the paths through which the information of injured people is collected:

1. *Data Integrity*: The information of injured people should not be lost after being input. However, in current triage, the complete information of injured people cannot be collected when triage tags fail to be cut off, are lost, or left in the pocket of an emergency person. Moreover, the information filled in a triage tag later than the first or second triage stages will be lost when the ambulance does not return to the incident site.
2. *Input Throughput*: The time to input injured person's information should be as short as possible. Especially, it should be so even if the latency of the network communication becomes large.
3. *Availability*: Emergency personnel should be able to use the system. Especially, input operations of injured person's information should be available anytime in the triage, even when the network is unreachable.
4. *Low Latency*: several types of input information should be quickly collected and viewed from the operation point. In current triage, the information filled into a triage tag later than the first or second triage stage is not collected until the ambulance that transported the injured person returns to the incident site, even if they are filled in the ambulance or at the hospital, since additional manual communication from the operation point to hospitals or the ambulance increases the work in the operation point.

18.2.3 Challenges in Pervasive Computing

The requirements listed in the previous section are important challenges in pervasive computing, which aims one in harmony with real world circumstance, including unexpected human behavior, rapid deployment, or insufficient computing infrastructure. In this chapter, we motivate these challenges, and address an improvement in a special but realistic case of pervasive computing, in addition to the improvement of triage by introducing information technology.

Data integrity and latency will be improved by exploiting a wireless network, since the input information can be collected through the network, even if emergency personnel do not directly or indirectly give a triage tag to the person who is in charge of collecting the information. However, this stands on the assumption that information once input can be reached in a destination in the network for more than a while. This assumption does not always hold in pervasive computing. Hence, data integrity and low latency in the semi-reachable network is one of the challenges in pervasive computing.

Input throughput and availability against the network are further challenging requirements in pervasive computing. The real world is not of course composed as a set of packets like TCP/IP, but behaves as the real time system. The interaction between the real world and the network in the interface such as mobile devices, sensors, and RFID tags becomes incoherent if the gap is not absorbed in the network nodes in appropriate layers from physical to human workflows. In these sense, independence of network status for real-world interfaces is also one of the attractive challenges in pervasive computing.

In this chapter, we show a realistic solution for the challenges by specializing the network usage in a way that only particular paths are used in particular stages of the workflow by analyzing the workflow and exploiting RFID tags to slim down the possible paths by the following approaches:

1. Input throughput and availability are assured by using RFID tags as local buffer: users can input data by referring the RFID data already input so far and the input device pushes the input data to the queue that is sent to the destination independent of the user's operation, as well as writing to the RFID tag.
2. Data integrity is assured and latency is improved by providing wireless communication areas at least to the final stages of the paths in triage workflow: hospitals and the operation point. Actually, latency will be more improved if the stages covered with wireless communication areas increases, but we believe it will be better than current triage, as experimented in Section 18.6.

The requirements in environments of insufficient network infrastructure including disasters are also discussed in mobile and ad hoc networks [3–5] with a requirement for quick deployment, but our approach of using RFID tags for input throughput and availability is unique, while most of the RFID applications in the literature use RFID tags as a device to identify objects or people by embedding unique IDs.

18.3 RFID Triage System

In this section, we describe the triage system using RFID Tags. Figure 18.4 shows an abstract workflow in the RFID triage system.

Other than the advantages described in Section 18.2.3, we can employ the following advantages of the system including primitive ones:

1. Using mobile devices with wireless communication, the information of the injured people is collected quickly via the network.
2. Input method using mobile devices provides ease of reading compared with handwriting.
3. Input throughput is improved by automating the information of the emergency personnel and addresses from postal codes, and by reducing the data types necessarily required to be input in the early stages only to the injury level and hospital.

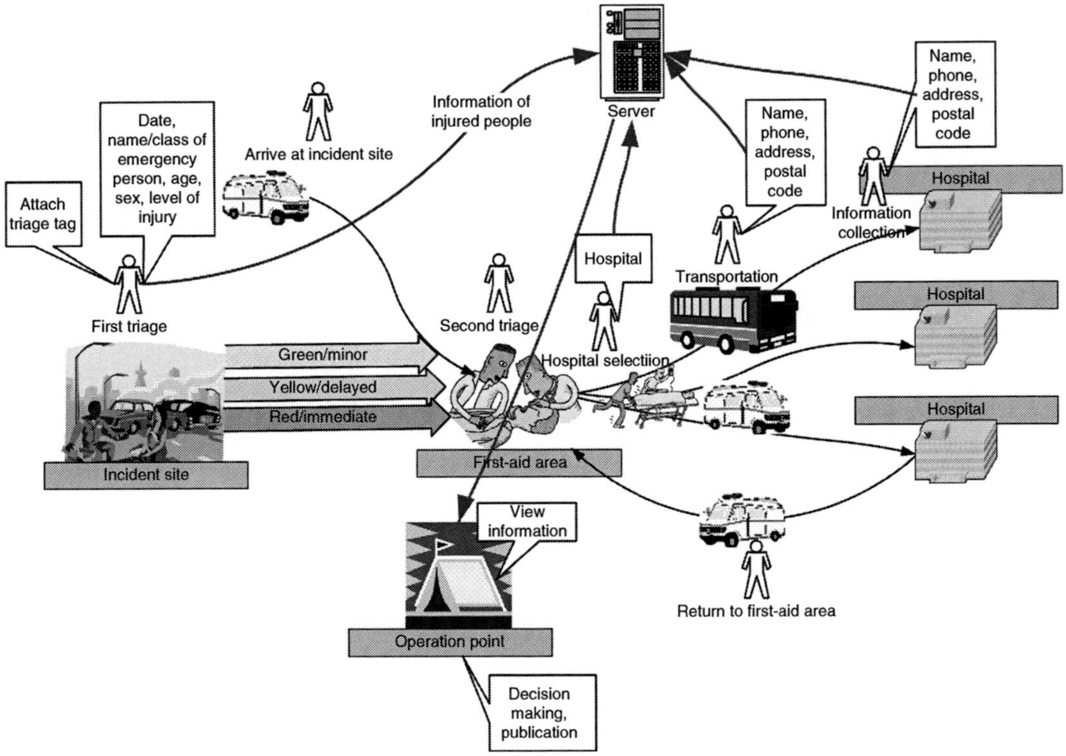

FIGURE 18.4
Workflow in RFID triage system.

18.3.1 System Architecture

The following are the system components, which are shown in Figure 18.5:

1. *Triage Tag*: A tag with input forms for the information of the injured person and an RFID tag with rewritable memory of 1 kb and wireless communication in the frequency of 13.56 MHz. Each RFID tag has a unique ID in the system.
2. *First Triage Terminal*: A mobile terminal device that an emergency person in charge of the first triage stage uses. It is equipped with an RFID reader and a wireless communication interface. An emergency person can input injured peoples' information through the touch panel or buttons. After input, the information is saved to the RFID tag, as well as sent to the server through the wireless network. In case the network is disconnected, the information is stored on the terminal, and resent when the network is connected. Static or automatic information such as the date and name/category of the emergency personnel are input by the terminal automatically. Figure 18.6 is the display of input terminal.
3. *Second Triage Terminal*: Similar to first triage terminal, a mobile terminal device an emergency person in the second triage stage uses.
4. *Hospital Selection Terminal*: Similar to first triage terminal, a mobile terminal device an emergency person in the hospital selection stage uses.
5. *Ambulance Terminal*: A notebook PC equipped with a handy RFID reader and a wireless communication interface, placed in each ambulance. Emergency personnel can input the information of injured people using a keyboard and a mouse. First, several parts of an address can be converted from a postal code.

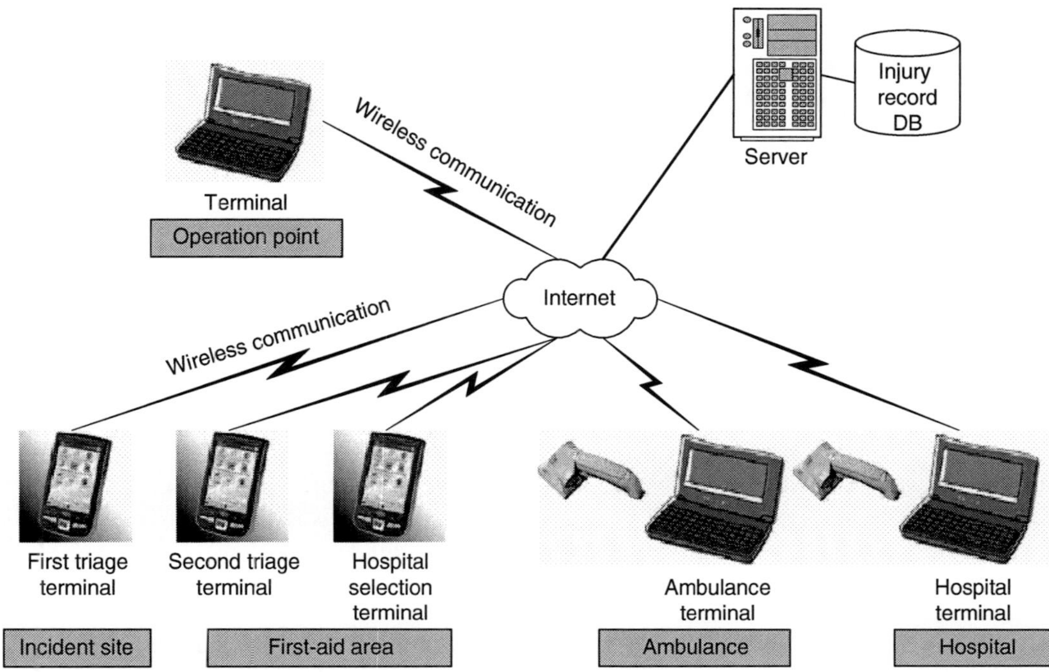

FIGURE 18.5
System and the network architecture.

6. *Hospital Terminal*: Similar to ambulance terminals, a notebook PC emergency personnel in each hospital use.
7. *Operation Point Terminal*: A notebook PC equipped with a wireless communication interface, placed in the operation point. Emergency personnel can browse the information collected to the server through a web browser software.
8. *Server*: Server equipment that has a database of the injured people, placed away from the incident site. The server stores the information sent from the terminals to a database, and responds to the browsing request from the operation point terminals with HTTP protocols. Moreover, the server keeps the information of an injured person up-to-date from those sent by multiple terminals.

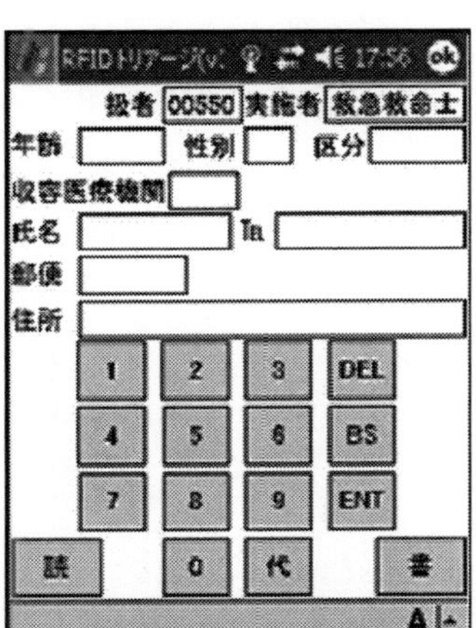

FIGURE 18.6
Display of input terminal in first and second triage stages.

18.3.2 System Flow

Rough workflow of triage does not change when the system is applied. The difference is that instead of writing and collecting by hand, emergency personnel read the information of each injured person from the RFID tag, input the information of each injured person to the terminal, and then write it on the RFID tag. In the following, we describe the usage and the behavior of the system for each role of the emergency personnel.

An emergency person in charge of first triage inputs the information of each injured person as much as possible to the first triage terminal, and then writes it to the RFID tag. Simultaneously, she/he attaches the triage tag with the RFID tag to the injured person, and cuts off the perforated label to the corresponding color. The information of the injured person written to the RFID tag is sent to the server as soon as the network becomes available.

An emergency person in charge of second triage first reads the information of each injured person from the RFID tag through second triage terminal and adds/updates the information to the RFID tag by interviewing the injured person. An emergency person in charge of hospital selection selects the hospital each injured person is transferred to and inputs it to the RFID tag through the hospital selection terminal.

An emergency person in an ambulance or in a hospital does the same operation to the injured person as second triage but does it using a notebook PC and a handy RFID reader. Address is automatically complemented when a postal code is input. Additionally, the emergency person clicks the commit button on the display and records the time of carrying the injured person to the ambulance.

An emergency person in the operation point views the information of the injured people, which is collected to the server, on the operation point terminal using Web browser software, and informs them to the emergency control center in the municipality as she/he needs. The server stores the information sent from the terminals and serves them to the operation point terminal through HTTP protocol. Multiple upload from terminals for the same injured person are sorted in time order and kept up-to-date. This workflow is a natural extension from current triage except the time for inputs, which we evaluated to be trivial in Section 18.6.5.

Table 18.1 shows the types of collected information in each stage in the current workflow or in using the system. From the table, we observe that the information used for the publication and which needs time to input, such as name, phone number, and address,

TABLE 18.1

Types and Stages of Collected Information in Current Workflow or RFID Triage System

Type of Information Stage	Date	Emergency Person	Category of Emergency Person	Age	Sex	Level of Injury	Hospital	Name	Phone Number	Address
First triage	Auto	Auto	Auto	Y	Y	Y	—	—	—	—
Second triage	Auto	Auto	—	—	—	—	—	Y>N	Y>N	Y>N
Hospital selection	—	—	—	—	—	—	Y	—	—	—
Ambulance	—	—	—	—	—	—	Y	Y	Y	Y
Hospital	—	—	—	—	—	—	Y	Y	Y	Y

Note: Y, input by emergency personnel both in current workflow and RFID triage system; Y>N, not input in RFID triage system, while input in current workflow; Auto, automatically or previously input in the system, while input in current workflow; —, not input in both cases.

is input after the hospital selection finishes, and the emergency personnel in charge of the stage not later than hospital selection only have to input the information that is simple to input, such as the level of injury, age, sex, and hospital. Moreover, the name and the category of the emergency personnel, and the date can be input automatically or previously before the incident.

18.4 Related Work

Triage is captured as an application that requires immediate improvement in the field of emergency medicine [6–11] and is realized as one of the important applications in pervasive computing [12].

A variety of information systems can be considered to be useful to support triage, such as wireless and mobile telemedicine systems [13]. Wireless network composition for such an application is discussed [14]. Tiny devices such as sensors to the network are tried to be used for vital sensors for injured people [12]. Active RFID tags are tried to be used for tracking the location of injured people [15].

On the other hand, information systems for supporting disaster management are proposed by featuring several key technologies in pervasive computing. Geographical information systems are used to locate and identify the context of people and objects [16]. Immediate deployment of wireless networks is discussed in [17]. RFID tags are tried to use for planning and guiding the evacuation [18].

Several works have exploited wireless networks, and several exploit RFID tags. However, these works along with networks do not address the requirements for data integrity, input throughput, availability, and latency as we have shown explicitly. Moreover, RFID tags are not used for improving the requirements. Our approach is unique in the sense that we use RFID tags for the alternative of current triage tags, and for improving the requirements by alternating wireless network communication with the communication with RFID tags.

Similar to our approach, U.S. Navy tried to use rewritable RFID tags as triage tags, but they did not try to incorporate wireless network communication and RFID communication [19].

18.5 Preliminary Experiment

Before the massive experiment, we evaluated several points with small number of participants. We measured input operations five times in a room. Figures 18.7 through 18.9 are the results of the measurements.

Figure 18.7 is the comparison of input times of each part in first triage with and without the system. From the figure, the time for an injured person is reduced to about 38 s from about 42 s. The detail shows that the most contribution is because the emergency personnel, with the system, do not have to input the level of injury, the name of the personnel, and the date. On the other hand, the time for preparing and cutting triage tags are increased by having input terminals in the case, and about 2.5 s for writing to triage tags.

Figure 18.8 is the comparison of input times of each part in ambulances or at hospitals with and without the system. From the figure, the time for an injured person increased to about 43 s from about 35 s. The detail shows that the address, the postal code, the phone, and the name of the injured person take longer time with the system than without the

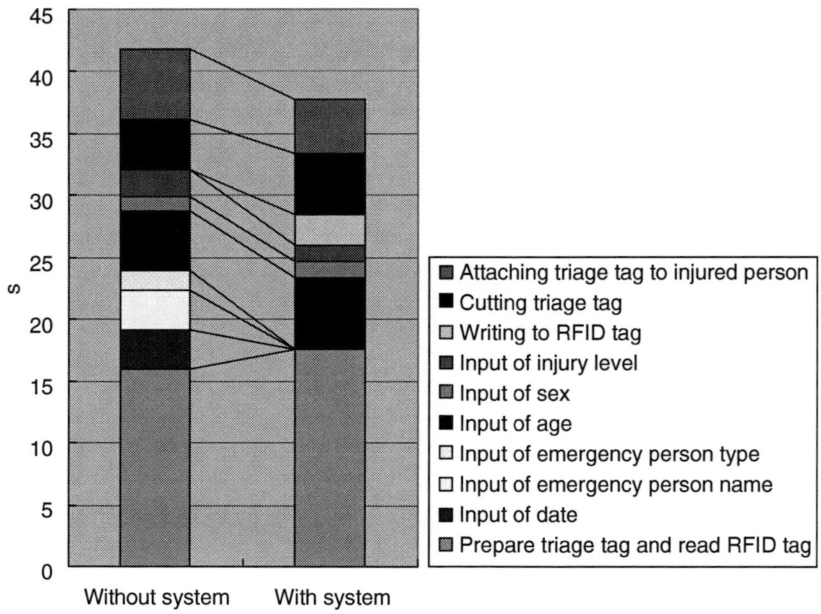

FIGURE 18.7
Input time in first triage.

system. Although the system employs automatic address complements from postal codes, it is still longer than that without the system. Additionally, it needs about 3.5 s for writing on triage tags. It requires further work to reduce the input time for ambulances and hospitals, while there is a problem that, without the system, manual and hastened writing is often difficult to read.

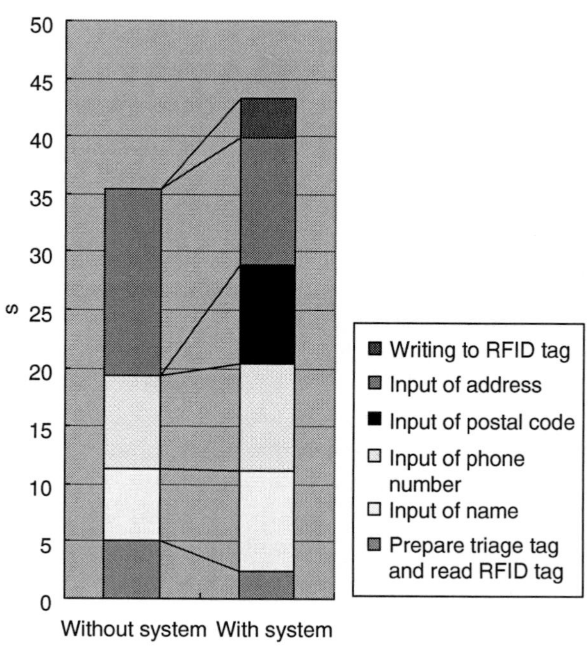

FIGURE 18.8
Input time in ambulance and hospital.

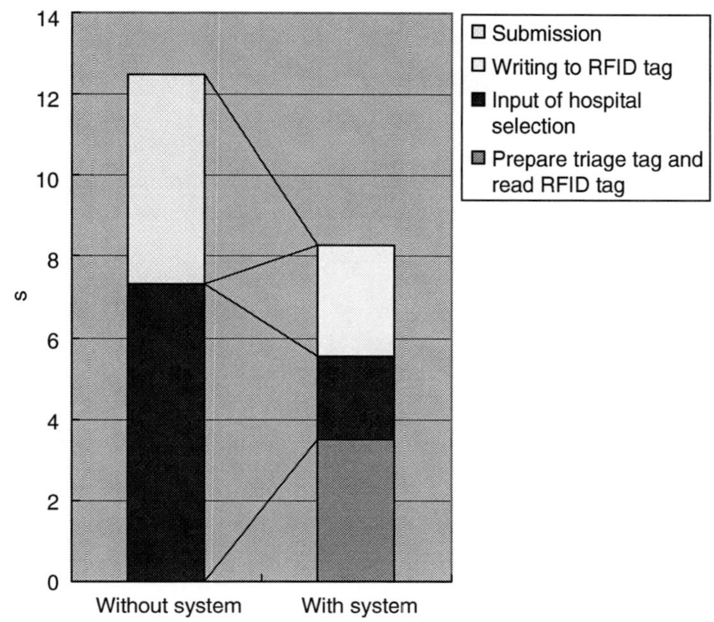

FIGURE 18.9
Input time in hospital selection.

Figure 18.9 is the comparison of input times of each part in hospital selections with and without the system. From the figure, the time for an injured person was reduced to about 8 s from about 12 s. The detail shows that the operations without the system require the emergency personnel to pass the copy of the triage tags to other personnel nearby, whereas they do not with the system. On the other hand, the operations without the system can start the hospital information immediately, whereas in the case with the system, emergency personnel need first to read the information input so far from the RFID tags before input.

18.6 Experiment

In this section, we describe the experiment of using the RFID triage system. The experiment is done in a driving school in a day, assuming a complex car crash of 5 cars and about 82 injured people. On hearing from professionals of emergency operations, this population is considered to be about maximum where current triage is feasible. The goal of the experiment is to evaluate whether and how much the RFID triage system accelerates and completes (1) the transportation of the injured people and (2) the collection of the injured people's information.

18.6.1 Assumption

The experiment spot is a driving school of about 2250 m^2. In the experiment spot, we set hospitals area, which corresponds to three points of hospitals, and set two hospitals about 700 m away from the spot.

We assume that two buses and three cars crash, where the injured people are located and classified as: 5 people as Black/DECEASED, 18 as Red/IMMEDIATE, 31 as Yellow/DELAYED, and 28 as Green/MINOR.

In the experiment, 14 ambulances (3 of them are for mass transportation) and 16 other vehicles such as commander cars and machinery and materials cars are used. The population of emergency personnel is 85, where 4 (2 pairs) for first triage, 1 for second triage, 1 for hospital selection, 3 for each ambulance, and 1 for each hospital.

18.6.2 System Assignment

In the experiment, two first triage terminals, one second triage terminal, and one hospital selection terminal is used as mobile devices. 14 ambulance terminals are used, where 6 of them are equipped with RFID readers and wireless communication interfaces, 3 are equipped with RFID readers but wireless communication interface, and 5 are without both. The ambulance terminals with no RFID readers and wireless communication interfaces are introduced to assume the heterogeneous environment; here not all the ambulances support the RFID triage system but only a usual notebook PC, such that a neighbor municipality helps the triage. The ambulances with no wireless communication interfaces are introduced to assume the situation where communication is not available.

One hospital terminal is placed at each hospital. One operation point terminal is placed at the operation point when the point is made during the experimental performance. The server is set in a data center about 1000 km away from the incident site.

18.6.3 Network

We prepared IEEE 802.11b wireless LAN interfaces for mobile device terminals, and FOMA data communication interface that provides up to 385 kbps by NTT Docomo Inc.

As access point facilities are necessary if we adopt wireless LAN, since otherwise it provides only several 10 m, we had to adopt them since there have been no mobile devices that will be able to be equipped with an RFID reader and any long-range data communication interface of several kilometers. In the future, this cost of access points will vanish when a mobile device with an RFID reader and a long-range data communication interface is used.

18.6.4 Process of Experiment

In the experiment, we first take about 1 h to perform the naive performance, which is the current triage, and then perform the advanced performance, which is the triage using the system, in the same condition.

Each performance starts with the departure of ambulances and fire engines having predefined time differences from the car parking. Each ambulance that transports the injured people to the hospitals inside the experiment spot arrives at the hospitals after driving around inside the spot, and those that transport to the hospitals outside the spot take the predefined route.

The injured people are predefined and informed their level of injury, and reply to the questions from the emergency personnel acting according the defined level. The emergency personnel perform the triage as described in Sections 18.2.1 and 18.3.2.

In the two performances, 7 persons measured the time of the flow of injured people and the flow of the collected information by stopwatches. Moreover, the collected information flow was also logged in the system in the advanced performance.

18.6.5 Evaluation

The result of the experiment is shown in Figures 18.11 through 18.15. The photos of the experiment are given in Figure 18.10.

18.6.5.1 Transportation of Injured People

Figure 18.11 shows the progression of cumulative populations of the injured people who are carried from the incident site to the first-aid area and carried out from the first-aid area to the hospitals in each performance.

The population carried into the first-aid area in the advanced performance always exceeds that of the naive performance (Figure 18.11). This demonstrates that, using the system, the injured people can be carried into the first-aid area faster than the naive performance. This mainly resulted from the fact that multiple injured people could be carried in simultaneously using the system, whereas they must have been carried serially to manually count the people by emergency personnel in order not to lose the information of the injured people in the naive performance. Moreover, the population radically increases at around 2 min after the start. This is the because it was possible to carry in those who are in Green/MINOR level of injury and those who can walk simultaneously as early as possible, whereas they must wait for the people of other levels to be carried in first in the naive performance.

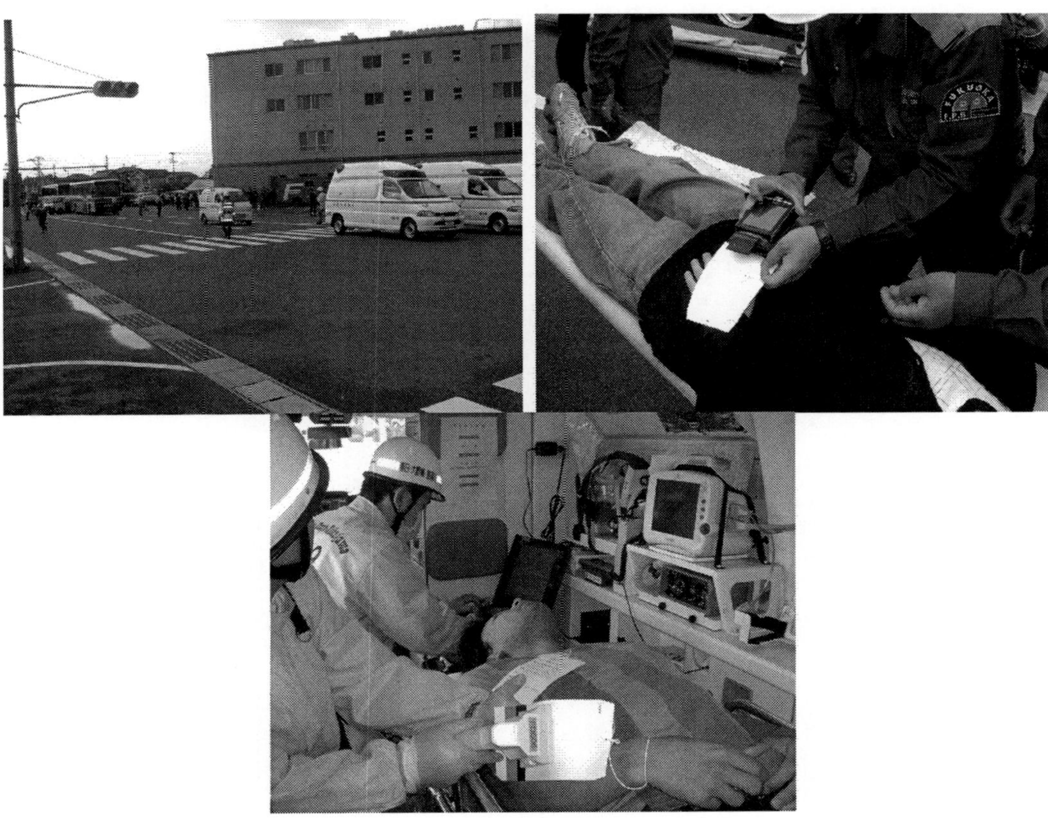

FIGURE 18.10
Photos of the experiment.

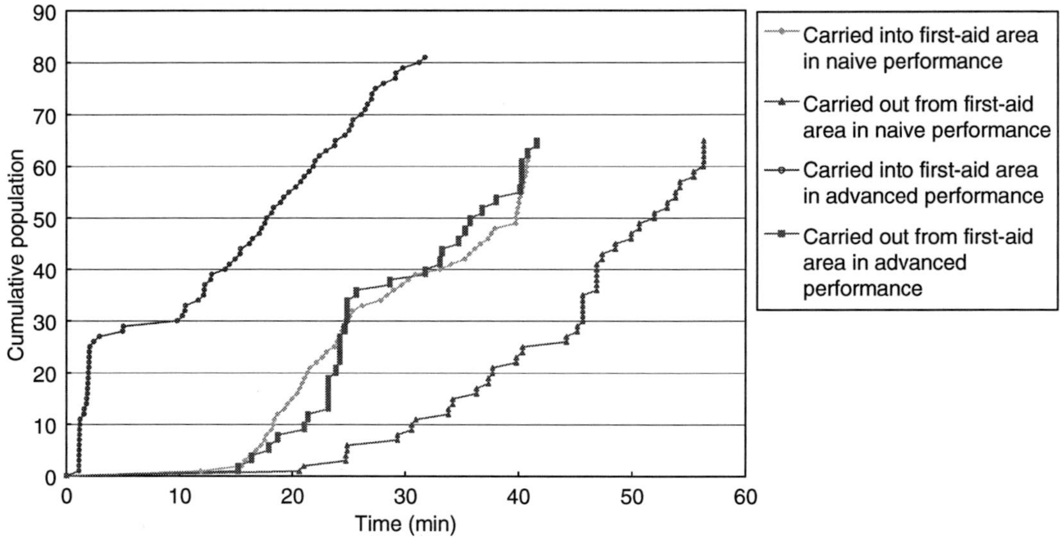

FIGURE 18.11
Transportation time.

Moreover, the population carried out from the first-aid area to the hospitals in the advanced performance (Figure 18.11) always exceeds that of the naive performance. This shows that, using the system, the injured people can be carried to the hospital faster than the naive performance. This is assumed to be because, in addition to the acceleration of carrying into the first-aid area, the hospital selection was effectively performed using a mobile device in the advanced performance. Moreover, the figure shows that plenty of injured people are carried out after 22 min. This is assumed to have resulted from the fact that the injured people of Green/MINOR level could be carried out early followed by the acceleration of being carried in.

The population in naive performance expires before it becomes the total population assumed because of the essential lack of the current workflow: several injured people in Green/MINOR level were dismissed before going to the first-aid area and could not be monitored; the Black/DECEASED people who were not carried could not be monitored, whereas the advanced performance could. Moreover, the population carried out from the first-aid area in the advanced performance exceeds that carried into there. This is because the vehicles for Green/MINOR level could stop at different areas from the ambulances for other levels, and several people of Green/MINOR level happened to be carried out without being carried into the first-aid area.

18.6.5.2 Information Collection

Figure 18.12 shows the progression of the cumulative populations whose information about the injury level is collected to the manual copy at the operation point in naive performance, to the RFID tag of the injured person in the advanced performance, and the server in the advanced performance. Figure 18.13 is about the hospital selection, and Figure 18.14 is about the address. If there are multiple input records for a single type of information, we adopted the first period of the collection. The time in the server was measured in minutes whereas the others were in seconds.

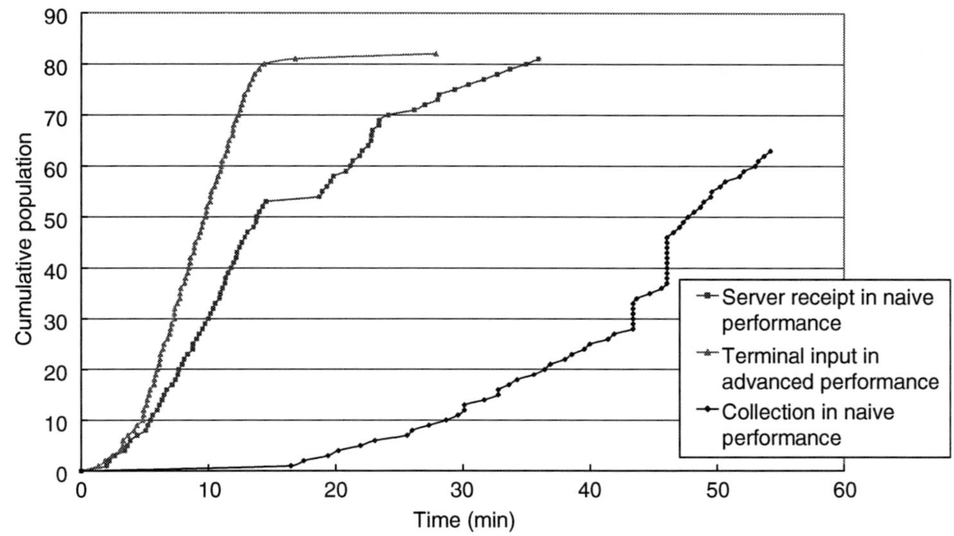

FIGURE 18.12
Time of information collection (level of injury).

In the following, we describe the method for recording and adjusting the record in the naive performance. The level of injury, the sex, and the age were assumed to be collected when the entry in the triage tag is copied to a tally sheet before an injured person is transported from the first-aid area toward a hospital. The name, the address, and the phone number were assumed to be so when they are copied after an ambulance returned to the operation point after the transportation. The records have been made slightly complementary to the deficit in the data obtained from the stopwatch measurement in the following way:

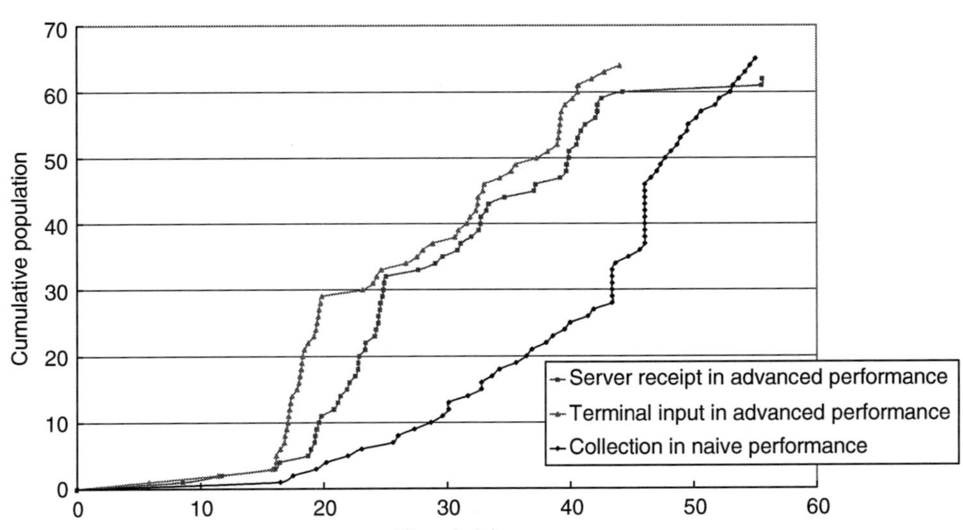

FIGURE 18.13
Time of information collection (hospital selection).

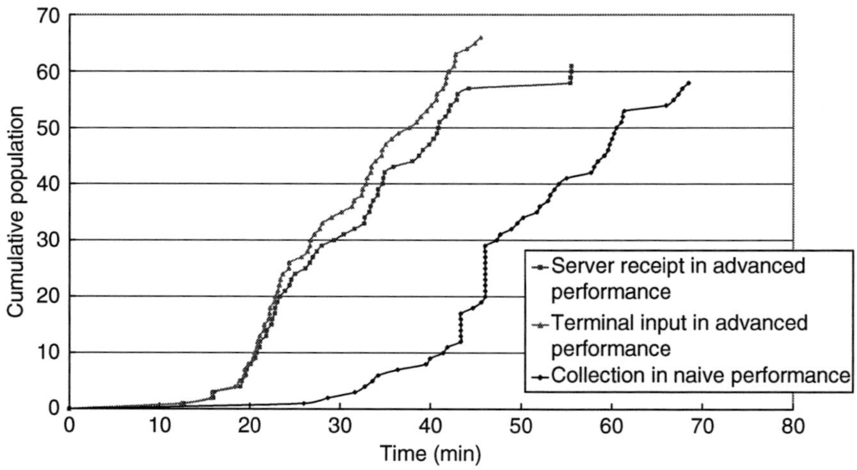

FIGURE 18.14
Time of information collection (address).

1. We adopted the time of particular people for seven lacks of time where the subject injured people could be assumed to be in the same ambulance.
2. We adopted the time of the closest line in the tally sheet for six lacks of time where Green/MINOR people could be assumed to be transported at a time.
3. For two time lacks of the manual copy to the tally sheet after the transportation when the times of collection are known, we adopted the collected times added by the averages of the periods from the collection to the manual copy of other people in the same hospital.
4. For one person whose collection time after the transportation and also that of manual copy lack, we adopted the average time of the times of manual copy for the same hospital.

It is not easy to compare the times of the naive and advanced performance since the methods of measurement are different, but we can see the times for the operation point to view the information for the first time, if we consider that they correspond to the manual copies at the operation point in the naive case, and so to the data arrivals to the server in the advanced case.

From the figures, the information of the injured people using the system could be collected faster than the naive performance in any information. Although we omit the graphs for the other information, there was little difference in the shapes in the graphs of level of injury, sex, and age. Similarly, so was in name, phone number, and address. These mainly owe to the timing of input shown in Table 18.1, and we can observe that the collection of the information in the early stages of the triage could radically be accelerated using the system.

Collecting information of all the injured people is essentially difficult, since that of the Black/DECEASED people are not collected quickly, and the Green/MINOR people tend to walk out on their own. However, Figures 18.12 and 18.14 show that the number using the system is higher than the naive performance, which means that the system provides more accurate and complete information of injured people, especially in the early stages of triage. Although Figure 18.13 does not show the same, this can be assumed since the input time using the system was longer than the naive performance, which we confirmed in

Section 18.5 in which the average input time of the hospital selection in the system was about 8.2 s, whereas 7.3 s without the system, with five trials in each case. Improving the time and complexity of input, there is a possibility of making the information collection of hospital selection accurate.

For data integrity in the advanced performance, all the peoples' data have not been collected in Figure 18.14 since the Black/DESEASED and Green/MINOR peoples' information have not been collected. However, the number of collection is more than that in the naive case. In Figure 18.13, the number of collection in the naive case is better than the advanced case. This can be assumed because the trial with the system needs, except for submission, much time than that without the system, as shown in Section 18.5.

18.6.5.3 Communication Time in the System

From Figures 18.12 through 18.14, we can observe that the network communication time in the system does not affect the triage to the critical delay. In addition, the receipt time on the server can be earlier if the time for communication is reduced, although the result in the advanced performance is better than the naive one. Figure 18.15 shows the distribution of the difference between the time received by the server and that input to the input terminal of each input of the injury information, where the differences of more than 60 s are omitted as a device without wireless communication functionality. The average time of the difference between the time received by the server and that input to the terminal was 17.3 s, and the variance was 35.39.

18.6.6 Discussion

We discuss the implications in network with the requirements discussed in Section 18.2.

Data integrity and availability are apparent to be improved without the experiment, as discussed in Section 18.2.3.

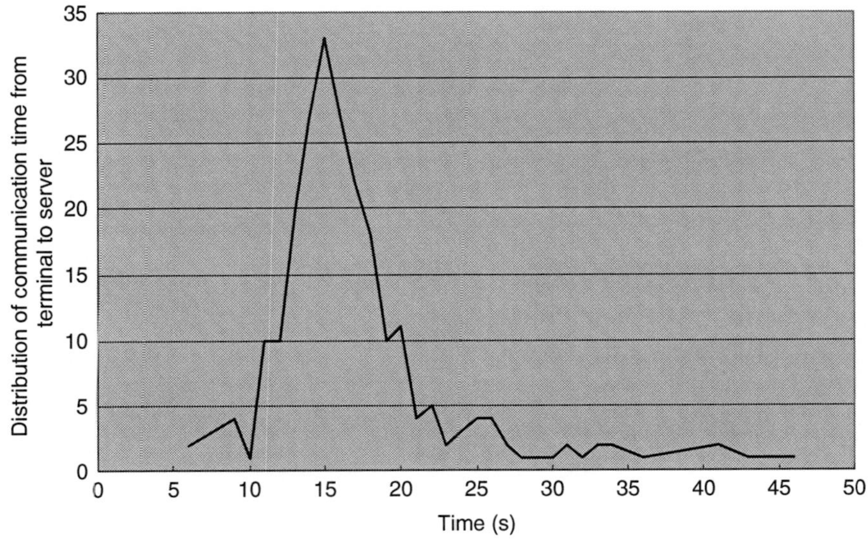

FIGURE 18.15
Distribution of communication time in the system.

As to input throughput, note that the communication time does not affect the input time by the emergency personnel, since they first read the old information from an RFID tag, and finish right after updating the information. The terminal automatically communicates with the server subsequently, but the emergency personnel do not have to wait for the network communication. If the system does not employ RFID tags, the first step of reading from the RFID tag must rely to the network communication. The throughput of an input by discarding RFID tags can be supposed to increase by 14.3 s, since the network communication time is 17.3 s in average and the user-waiting time for the RFID communication is 3 s, as shown above.

Although latency needs further research to clarify the advantage of using RFID tags, we could decrease the latency, especially the required types of information compared with the naive performance, as shown in Figures 18.3 and 18.4, without affecting other factors such as input throughput in triage.

18.7 Conclusion

From the experiment, the RFID triage system is demonstrated to be effective in mass casualty incidents of about 100 injured people. In the future, the system will be more valuable when it becomes applicable for other incidents, such as with smaller number of injured people, or in broad area, such as earthquakes, storms, or floods.

Acknowledgments

This work has been supported by the Grant-in-Aid for Young Scientists (A) No.18680009 of the Ministry of Education, Science, Sports and Culture (MEXT) from 2006 to 2008. We are grateful for the support of the experiment from Toppan Printing Co., Ltd., Fukuoka City, Mizuho Information & Research Institute Inc., Institute of Systems & Information Technologies/KYUSHU, NTT DoCoMo, Inc., NTT West Corp., and Kasuga-Ohnojo-Nakagawa Fire department.

References

1. P. Hewkin, "Smart tags—the distributed memory revolution," *IEEE Review*, 1989.
2. R. Want, K.P. Fishkin, A. Gujar, B.L. Harrison, "Bridging physical and virtual worlds with electronic tags," *Proc. Int'l Conf.* CHI 99, pp. 370–377, 1999.
3. M. Gerla, J. Tsai, "Multicluster, mobile, multimedia radio network," *ACM/Baltzer Journal of Wireless Networks*, 1 (no. 3), pp. 255–265, 1995.
4. G. Zussman, A. Segall, "Energy Efficient Routing in Ad Hoc Disaster Recovery Networks," 22nd Conf. IEEE Computers and Communication Societies (INFOCOM), 10 pages, 2003.
5. T. ElBatt, S. Krishnamurthy, D. Connors, S. Dao, "Power management for throughput enhancement in wireless ad-hoc networks," *Proc. IEEE ICC 2000*, 9 pages, 2000.
6. F. Subash, F. Dunn, B. McNicholl, J. Marlow, "Team triage improves emergency department efficiency," *Emergency Medicine Journal*, 21, pp. 542–544, Sep. 2004.
7. T. Kilner, "Triage decisions of prehospital emergency health care providers, using a multiple casualty scenario paper exercise," *Emergency Medicine Journal*, 19, pp. 348–353, Jul. 2002.

8. S. Goodacre, F. Morris, B. Tesfayohannes, G. Sutton, "Should ambulant patients be directed to reception or triage first?" *Emergency Medicine Journal*, 18, pp. 441–443, Nov. 2001.
9. J. Terris, P. Leman, N. O'Connor, R. Wood, "Making an IMPACT on emergency department flow: improving patient processing assisted by consultant at triage," *Emergency Medicine Journal*, 21, pp. 537–541, Sep. 2004.
10. White, Ann, "Change strategies make for smooth transitions," *Nursing Management*, 35(2) pp. 49–52, Feb. 2004.
11. Parker, Pam, "Move care to a higher level with emergency systems," *Nursing Management*, 35(9), pp. 82–84, Sep. 2004.
12. K. Lorincz, D. Malan, T. Fulford-Jones, A. Nawoj, A. Clavel, V. Shnayder, G. Mainland, M. Welsh, "Sensor networks for emergency response: challenges and opportunities," *IEEE Pervasive*, 3(4), pp. 16–23, 2004.
13. S. Voskarides, C. Pattichis, R. Istepanian, E. Kyriacou, M. Pattichis, C. Schizas, "Mobile health systems: a brief overview," *Proc. SPIE AeroSense* 2002, pp. 124–131, 2002.
14. K. Banitsas, R. Istepanian, S. Tachakra, T. Owens, "Modelling issues of Wireless LANs for accident and emergency departments," *IEEE EMBC Conf.*, vol. 4, pp. 3540–3543, Oct. 2001.
15. http://www.activewaveinc.com/applications_hospitals.html
16. P. Oosterom, S. Zlatanova, S. Fendel, M. Elfriede (Eds.) *Geo-information for disaster management*, Springer Verlag, 1434, 2005.
17. S. Midkiff, C. Bostian, "Rapidly-Deployable Broadband Wireless Networks for Disaster and Emergency Response," 1st IEEE Workshop on Disaster Recover Networks (DIREN '02), 10 pages, June 2002.
18. S. Sharma, S. Gifford, "Using RFID to Evaluate Evacuation Behavior Models," North American Fuzzy Information Processing Society Conference, 5 pages, 2005.
19. J. Yoshida, "Navy to Use RFID Technology in Iraq," *EE Times News*, May, 2003, http://www.embedded.com/showArticle.jhtml?articleID=10700142.

19

RFID Tagging and the Design of "Place"

Anijo Punnen Mathew

CONTENTS

19.1 Background

19.1.1 Problem of "Place"

Architects of the 20th century imagined that their new tools—electricity, steel, concrete, plate glass, mass production and fresh ideas about design—could be used to transform society for the better (Larson et al., 2004). In spite of great promises of the functionalist architects of this era, the problem of "place" still counts as one of the primary needs of humankind. However there has been a significant transformation in the buildings of the 21st century from those of the last. Today's buildings are not just well designed; they are also "intelligent." Computing technology finds pervasive application in many aspects of the modern habitable spaces—environmental control systems, internet based systems for information exchange, cellular systems for instant communication; the list goes on. Our lives today exist in the midst of a complex weave of digital information and communication. Eventually we will all live, work, and play in so-called "smart" environments—environments that not only protect us from the elements but also work *with us* to make inhabitation more efficient and experiential.

All this means that architects of the 21st century have to be concerned about more than the just the "chassis" of the house. Today's designers negotiate with two kinds of computer usage—the first, the use of digital media as a tool *in* the design of spaces, and the second, the use of digital media embedded *within* designed spaces. "Smart" environments offer new opportunities to augment people's lives; not just with good design, but also ubiquitous computing technologies that will provide users with increased communications, awareness, and functionality (Weiser, 1991). With new tools such as inexpensive computing, wireless communication, high performance materials, and new design, fabrication, and supply-chain technologies, we are today equipped with perhaps much more—the ability to transform society in ways unimaginable in the 20th century.

19.1.2 "Disappearing" Computer

Our thinking is beginning to shift from the notion of computers in architecture or computers affecting architectural design to the notion of architecture *as* the computer (Senagala, 2005). "Smart" spaces and interactive environments have found prevalence in architecture with the emergence of powerful mobile computing devices and real time context aware computing (Edwards and Grinter, 2001). As early as 1991, Mark Weiser (Weiser, 1991) coined the term "ubiquitous computing" referring to computers embedded in everyday objects. These arrays of embedded intelligent devices could work invisibly and unobtrusively in the background of our everyday existence. This idea is now referred to as the "disappearing" computer—thousands of "invisible" computers woven into the architectural fabric of our spaces allowing us to connect to a global network of information and workflow.

The disappearance of the computer does not necessarily allude to a physical invisibility; instead it refers to "invisibility through use"—computers that are so embedded into our lives that they work only in the periphery of our senses (Tolmie et al., 2002). An analogous example is that of electricity. When electricity was introduced at the beginning of the 20th century, it was considered a novel technology—one that would never reach the home of the common man. Today, the pervasiveness of electricity is so embedded into our socio-cultural framework that thoughts of life without it form stories of apocalyptic fiction. Not only is electricity ubiquitous and necessary for human society, it is also invisible to us in everyday use. In a similar vein, the basic premise of the disappearing computer is that one day computing too will be both ubiquitous and invisible.

Streitz and Nixon (Streitz and Nixon, 2005) talk about two forms of the ''disappearing computer'': a physical disappearance and a mental disappearance. It is the second category that interests designers—artifacts that may still be large, but not perceived as computers because people discern them as, maybe interactive walls or interactive tables. The image of ubiquitous computing is not any longer that of an omnipresent, servile (or unservile) computer (as science fiction has been telling us for years); instead it is that of an invisible, augmentary appendage to our environments—redefining what we call ''place.''

19.2 Place

19.2.1 Inhabitable Interfaces

Environmental Psychology categorizes ''physical environment'' as 'typically neutral,' only coming into self conscious awareness when individuals form stable and enduring representations of it (Auburn and Barnes, 2006). Place thus can be described as an appropriation of experiences, while space merely the construct that envelopes it. McCullough (McCullough, 2004) presents the idea of place as integral to the idea of architecture—the phenomenological quality of space that enables us to retain memories, weave stories, and describe our experiences. Walls, ceilings, tables, chairs are all simply physical artifacts that aid inhabitants in the development of personal experiences. In short, the design of place (architecture) is the design of interaction between humans and their environment and place itself an interface for this interaction. Thus, architects and designers wield the capability to design artifacts that can influence a person's life—the way they live, move, interact and so on. In short, architects are merely designers of large interfaces for information interchange.

The idea of architecture as an interface is not new; it has found profound use even before the birth of computing. The steeple of a church communicates that it is a place for reflection or religious gathering. The bricks and ivy of Harvard Yard signifies years of scholarly research and education. The solid grey walls of a prison conjure up images of torture and punishment. The bright colors of a playground indicate play and joyfulness. Thus ever since humans have existed, our environments have been acting as interfaces for information interchange. The difference today is that, as computers slowly recede into walls, tables and furniture, we now have in our repertoire new tools of computing and multimedia. Now for the first time users can interact with information—change the way it envelopes their existence; search for more or better information; communicate their needs; and expect a reaction. All of this leads to a realm of experience that was hereinbefore unheard of. Design of interactive (mediated) places will not necessarily change *what* is being designed, only *how* it is designed, and how it is perceived by users. And the onset of ubiquitous computing signals a paradigm shift—of new *places* in unique contexts with their spatial, temporal and material configurations coalescing to form meaningful experiences for its users.

19.2.2 Communication Conundrum

The design of ''smart'' places involves the seamless integration of both the *physical* and *virtual*. As computing initiatives evolve intelligent devices that work in the background of our day to day living, questions arise about how we interact with these devices. Traditional input/output systems such as the keyboard and the mouse become redundant as computation reverts into physical entities such as walls, windows, and furniture. Several ideas are being pursued in universities and research labs around the world—tactile interfaces using gesture recognition and multitouch systems are being evaluated as interactive systems (MERL, 2006; Microsoft, 2007; Synaptics, 2007); and voice recognition systems may one day become our mode of interaction with computers in our homes (Furui, 2000).

However before we can interact with the computers ubiquitously prevalent in our environment, identification becomes critical. As computation becomes pervasive, it is inevitable that there will be an increased demand for more secure and private communication channels. Traditional communication channels such as wi-fi, wi-max, USB, Bluetooth, etc. may prove to be expensive technologies for ubiquitous use. And by themselves, these technologies may not offer the best methods of identification.

This leads us to the communication conundrum—we need an identification technology in "smart" environments and it must be cheap, ubiquitous, and reliable. While several innovative schemes have been proposed, one of the most promising design ideas uses Radio Frequency Identification or RFIDs.

19.3 RFID: State of Art

19.3.1 RFID Technology

Radio Frequency Identification (RFID) technology is an automatic identification method that relies on storing and retrieving data using radio transponders called RFID tags. RFIDs can be of three types—passive (no internal power), semipassive (battery powered microchip) and active (with internal power). Recent developments in RFID technology show that this technology is on its way to becoming smaller (a paper thin microchip was introduced in 2006 by Hitachi), ubiquitous (implantable tags), and cheaper (EPC complaint tags are available at close to 5 cents each) (Morton, 2004; Hitachi, 2006; Roberti, 2006).

RFID tagging techniques have in the last decade demonstrated that identification of both humans and commodities can be connected to elaborate databases without direct manipulation by a separate entity. RFIDs have become state of art in many identification programs—passports, transit cards, payment portals, credit cards, animal identification, etc. RFIDs have slowly started to replace traditional barcode based systems in many database driven entities such as libraries, enterprises, logistics, etc., as significant advance for RFID technology came with Wal-Mart, Target and the U.S. Department of Defense requiring that their suppliers place RFID tags on all shipments in an attempt to improve supply-chain management and the ERP machinery.

As RFID becomes an accepted mainstream technology, it is inevitable that it will find application in many aspects of our daily living. Many companies like Symbol Technologies (now Motorola) have been experimenting with the use of RFIDs in spatial configurations. Very little work however has been done so far using RFIDs in ubiquitous computing systems for the design of "place." As other chapters in this book concentrate on the state of art in RFID technology, this chapter will evaluate how RFID tagging can transform the way we look at "place" and the negotiation and design of "place."

19.3.2 RFIDs in Smart Environments

In a world that has one computer to many, and one computer to one, it is easy to develop interaction methodologies because the interaction is limited. But ubiquitous computing talks of a world where there are many computers to one. In such a scenario, how do we interact with the multiple computers that exist pervasively in our living space? In order for designers to program systems to adapt to the personal demands of an individual, they have to recognize which individual is currently asking for personalization from which computer. Moreover, as ubiquitous computing finds pervasive use in public domains, multiple ids and personalities may need to be recognized—you may have separate preferences for your office, home and play.

All this leads to one of the prime needs of ubiquitous computing–recognition and data structuring. Since as early as 2000, several projects have evaluated the use of RFID technology as a feasible format for identification in "smart" environments. In this half of the chapter we will explore two existing environmental solutions that use RFID technology.

The first system that we will explore is that of utilizing RFID technology in indoor and outdoor environments for navigation and wayfinding.

19.3.2.1 RFID Information Grid for Wayfinding

In spite of great strides in Universal Design concepts, most working environments still remain nonconducive to use by the physically challenged. Environmental limitations restrict most wheelchair bound and visually impaired people from even conducting everyday activities. In the U.S., the percentage of working age blind (for example) who are unemployed remains at 74% in spite of several schemes in universities and corporate systems to encourage physically challenged education and employment. One of the primary reasons for these statistics is a nonconstructive living and working environment which till now has been passive to the needs of physically challenged individuals.

Willis and Helal (Willis and Helal, 2005) of the University of Florida propose a solution—a basic RFID information grid overlaid on the urban spatial framework. Such an RFID information grid can be easily developed from passive, low-cost, High Frequency RFID tags installed within the architectural fabric of a campus (or a city). The tags can be programmed to convey precise location and detailed attributes about the surrounding areas. Because the information about location and spatial attributes is situated within specific tags, it removes the need for an extensive (central) database or a communication infrastructure. People using the space can be provided with interrogators (RFID readers) built into personal devices (PDA's, watches, cell phones) or structured within the framework of mobility devices (wheelchairs, scooters, walking canes).

The beauty of the design lies in its simplicity. RFID tags can be mandated in existing and new architectural projects through Americans with Disabilities Act (ADA) regulations. It can be easily incorporated indoors as proposed by Willis and Helal into the weaving of carpets and/or flooring material. The tags can be programmed with spatial data about both the location and the design of the place, for example, "entry doorway to Museum of Modern Art (MOMA). Two steps up and then glass doorway." It can also parse directional information to electronic wheelchairs or walking canes so that physically challenged users can negotiate architectural elements in their environment.

The second system is a more integrated one; an architectural design that employs RFID identification technology as a core part of its framework.

19.3.2.2 Swiss House

The Swiss House for Advanced Research and Education (SHARE) designed by Huang and Waldvogel (Huang and Waldvogel, 2004) is a novel type of "inhabitable interface" that supports direct and indirect communication and cultural awareness of habitants spread around the globe. The primary role of the Swiss House is to become a physical and virtual environment which fosters closer ties between Switzerland, New England and Eastern Canada in academia, industry and society. Huang and Waldvogel define the basic idea of the space as a convergent architecture—a system that develops the design of the interface elements and software (the "virtual" architecture) in conjunction with the design of the physical environment (the "physical" architecture).

The design of the house is a mesh of physical and virtual interfaces stitched into each other. The house features technologies such as "teleports" which allows users to interact with each other in spite of geographical dislocation; a media wall which extends the perspective of the physical space through the internet to another space at another place; and a knowledge café, which offers the possibility of interaction even while having coffee.

Since the technology was designed to be used by multiple stake holders at many different times, it is important that there is recognition of both user type and personality. And in their design of the house, Huang and Waldvogel wanted the "house" to be aware of inhabitants as they came in and used the technologies. In order to solve this problem they came up with an innovative idea. The close circle of the Swiss scientist community was each sent a personalized Swatch watch containing an RFID tag. The RFID tag has a unique ID that points to the data entry with information about the user on the Swiss House database server. Whenever a user walks into the Swiss House with the RFID Swatch, his/her presence is sensed by the building by RFID interrogators placed around the space. The house once "aware" of the user within its premises is also able to track the use of technologies by this current user. Other (registered) users around the globe are also able to see who is in the house at any time. This system works well because the Swiss House is essentially a public collaborative that accords communication privileges based on community participation. Issues of privacy and security are reduced because the users "expect" to be tracked; and often find it beneficial to their use of the House.

19.4 RFID and Place

In the above examples we saw RFID technology used in two different schematics—one is the conventional system when the interrogator is transient. In such a conventional system, RFID tags on commodities store information that can be retrieved from a stored database when interrogated by an RFID reader. The RFID readers may be portable, which means that they can be carried by an employee, or permanent, built into computational entities, like checkout machines. This concept has evolved through a legacy system from barcode technology which could only expect a passive interaction from the entities that were tagged. Nevertheless such a system is useful in places where there exists large numbers of tagged entities—like a department or clothing store.

The Swiss House on the other hand demonstrates a completely different strategy. In this system, the building itself acts as one big interrogator (albeit through separate readers placed around the space). In short, the place is "aware" of its inhabitant. This is a significant attribute in the design of ubiquitous computing systems. The Swiss House demonstrates that RFID technology coupled with wireless LAN and other connective technology can be used for identification in "smart" spaces. Once a personality has been identified the technology can be extrapolated for interaction between these personalities and the system. Such a system is however useful only when the numbers of entities that need to be identified are limited. It is possible nevertheless to incorporate several layers of interrogators wherein specific ids may trigger specific interrogators based on specific requirements.

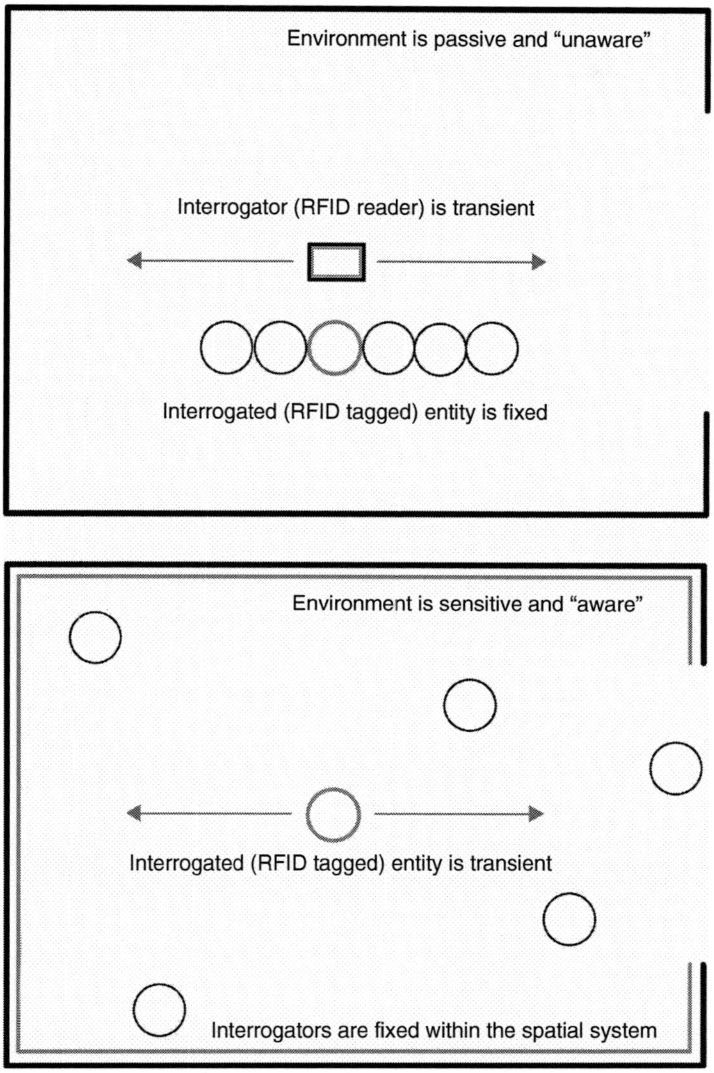

FIGURE 19.1
Differences in the concepts of RFID interrogation in the RFID grid system and the Swiss House.

The critical difference between this method and the conventional one is that of location—in the conventional system, the commodity is fixed and the interrogator is transient; in the spatial system, the interrogator is fixed and the commodity is transient (see Figure 19.1).

The system of the RFID grid for wayfinding follows a conventional RFID schematic:

- Transient:
  RFID interrogators
  - Built into portable devices (PDA, cell phones)
  - Stitched into fabrics (wearable computers)
  - Designed into portables (like shoes, watches, etc.)

- Fixed:

  Individualized RFID tags carrying location information and characteristic. These tags can be

  - Mandated by ADA regulations as
    - Set into existing architecture
    - Designed with new architecture
  - Woven into architectural products (carpets, wallpaper, etc.)

The Swiss House, on the other hand follows a ''spatial'' RFID schematic:

- Transient:

  Individualized RFID tags carrying personal information of user. This tag can be

  - In person (bio-implanted RFID tags)
  - Permanent identification documents (RFID-enabled IDs—passports, university ids, etc.)
  - Temporary identification systems (RFID-enabled watches, RFID-enabled devices picked up from reception, etc.)

- Fixed:

  RFID interrogators within the fabric of architecture in conjunction with wireless LAN and/or other connective technology (Bluetooth, wireless USB) provide for identification and interaction.

From their perspective, spatial systems by default presume that inhabitants will be transient. Hence an identification system that allows for easy movement of tagged entities is what is appropriate for the design of inhabitable interfaces. We can see that both the above systems allow for such interaction.

19.5 Applications

In the following part of this chapter, we will evaluate two other projects (at various levels of conceptualization) using RFIDs in the design and/or negotiation of place. The applications presented here are part of ongoing research projects by the author at the Design Research & Informatics Laboratory (DRIL) at Mississippi State University's College of Architecture, Art and Design. DRIL is a multiplatform platform laboratory for carrying out interdisciplinary research projects and consists of architects, industrial designers and technology experts, internationally recognized for process and product innovations. The exploratory nature of the DRIL enables faculty and students to carry out multiple levels of design research, including research into the use of computing at various levels of architectural design.

While several initiatives at the DRIL look at development of new products and design processes, most projects are exploratory—evaluating the use of existing technology to solve real world problems. Many of these problems have existed before the emergence of the computer; ubiquitous computing solutions are explored in several layers of augmentation to existing solutions. With (almost free) electronics and pervasive information, we aim to develop interfaces that are not just innovative but also help in the society's day to day living. The following are a few examples of how the DRIL puts available innovative technology to common day use.

19.5.1 Library Project

The Library of Congress or Dewey Decimal systems used in public libraries were developed to help in organization of book clusters. Our studies however show that while these systems allows for easy data structuring, it is often inconvenient for novice or occasional end users to comprehend and use these systems. This comes from the inherent lack of users to convert the alphanumeric coding into a tangible search experience.

Problems mostly relate to the user's unfamiliarity with the system and the time taken by them to find a book. Perhaps this is because the current system affords users with a high cognitive load:

1) Memorizing a large alphanumeric code (e.g., NA1469. H43 A4 2003)
2) Having a relative knowledge of the library layout
3) An understanding of the coding

Our studies show that experienced users were able to find books quite easily. This can be easily attributed to their understanding of the library system and layout; in words of one experienced user—"I have been doing it for a long time." However we found that novice or first time users were almost always frustrated by the system. It can become even more of a frustrating experience if the book is misplaced or checked out (a common occurrence in most public libraries).

Libraries began using RFID systems to replace their electro-magnetic and bar code systems in the late 1990s. RFID systems are used primarily in libraries for inventory and tracking (the ability track the movement of a book or a person carrying the book). With RFID-enabled tools, inventory-related tasks can be done in a fraction of the time as with bar code readers. A whole shelf of books can be read by the reader with one sweep of the portable reader which then reports which books are missing or misshelved. Security is another aspect of library operations that may be greatly improved with RFID-based security systems. Rather than purchasing additional tags for security, a single tag can be used for identifying items and securing them. As patrons leave the library, the tags are read to ensure that the item has been checked out. (Ayre, 2004). Because these systems often work at the organizational level, users of the library seldom benefit (directly) from such technology.

Our proposal for improvement suggests the use of existing RFID infrastructure in libraries to decrease cognitive loads, thus making the library search easier. Such a system would not only help in faster book searches but could also be used to add additional layers of information for the particular book that would make the system more efficient for all stakeholders. In the proposed design, a PDA/Pocket PC/Cell Phone or other portable device helps the user to search for books within the library database and also provides information about book and its location. RFID placed on books and the ID cards of the user provide the PDA with information which can help to locate the book from within the library shelves. Proximity sensors on the shelves can detect the RFID on the user ID, enabling the user to locate the appropriate shelf without interpreting the library's coding system.

19.5.1.1 Scenario of Use

To better explain how this design works lets create a scenario of an individual looking for a book in a university library. For the purpose of this study, let's call her Jill. In her first week, Jill receives a research assignment that requires extensive use of the library. In her earlier school, she used an online catalogue to find books. Having found the details of the book on the computer, she would write down the name, title and alpha numeric information of the book on a piece of paper. With this information, she would then go to the library to find the book.

However Jill found that

1) The system was not easy to comprehend especially the first few times.
2) It was time consuming.
3) It is frustrating if the book is misplaced or has been checked out.
4) Or if she forgets the alphanumeric code or loses the piece of paper she has it on.
5) She has to go back and search the online catalogue once again if she wants a different book.
6) She has to wait in a queue to check out the book.

Her current school however has adopted a new RFID assistance system for library search. Jill goes to the front desk and is handed a PDA after the front desk scans her RFID-enabled student ID. On the PDA she uses a search engine linked to the library's database and using keywords, author name or title she is able to find the book she wants. When Jill is ready to get the book, the screen immediately shows her an easy to read graphic layout of the library with information about the floor and the location of the shelf holding her book. Jill follows the directions on the PDA leading to the correct shelf. As she nears the correct shelf, proximity sensors on the shelf detect her ID and a colored LED flashes on the rack. Once near the correct rack, the image on her PDA changes showing her the location of her book in relation to other books on the rack. Noting that the book is on the top right hand corner of the rack, Jill looks for a flashing LED under the book she was looking for. Having found her book, she may choose to look for another book, or check out. On the PDA she sees a tab recommending other books with the same key words or suggested by her professor. These recommended books can be located within the library using the same mechanism. This eliminates the time for Jill to go back to a computer to find the alphanumeric code of the book.

Deciding to check out, she collects all books she has found and moves towards the front desk. She could use her PDA to check out the books; but Jill decides to use a self checkout counter, not unlike the one in her neighborhood supermarket. The counter is placed next to the front desk. She uses her student ID to check out the books at this counter and returns her PDA to the front desk. Having completed her first visit to the library in less time than it would have taken at her last school, Jill leaves to work on her assignment.

19.5.1.2 How the System Works

To locate a book in a library, a database of information has to be stored about that book - Call numbers, ISBN number, title, author's name, abstract, key words. Such information about the book can be used by the described search engine to find the book. This information can be located on the library database accessible through a wireless network in the vicinity of the library. RFID tags transmit this information to a search device—a PDA or cell phones or any other similar device with the appropriate interrogator and software. In addition, the system also recognizes the user who is querying for information. RFID tags on the school ID card (or alternatively on the search device handed at the check in desk) transmit information to interrogators in the library floor and on the shelves. Once the search device attaches itself to a book, it can track and locate the book by locating the RFID information from the database or by sensing proximity.

19.5.1.3 Prototyping and Testing

To test the design, a comprehensive user study was designed and conducted in the Bob and Kathy Luke Library of the College of Architecture, Art, and Design at Mississippi State University. The participants included all stakeholders—faculty members of the school, students as well as library staff ($n = 17$). To verify our claim that the new system will take lesser time than the conventional alphanumeric system, we developed the aforementioned interfaces on a PDA (see Figure 19.2). Owing to constraints of a larger space, the study was conducted in a small library. We believe that the study can be extrapolated to a larger system with similar results.

Some salient results from our tests:

- New system took less time even for experienced users. Fifteen out of 17 users said that the new system made it easier for them to find books in the library.
- Time taken when PDA was used for the second time was much lesser as expected. Results show that even in the first book search the participants took lesser time

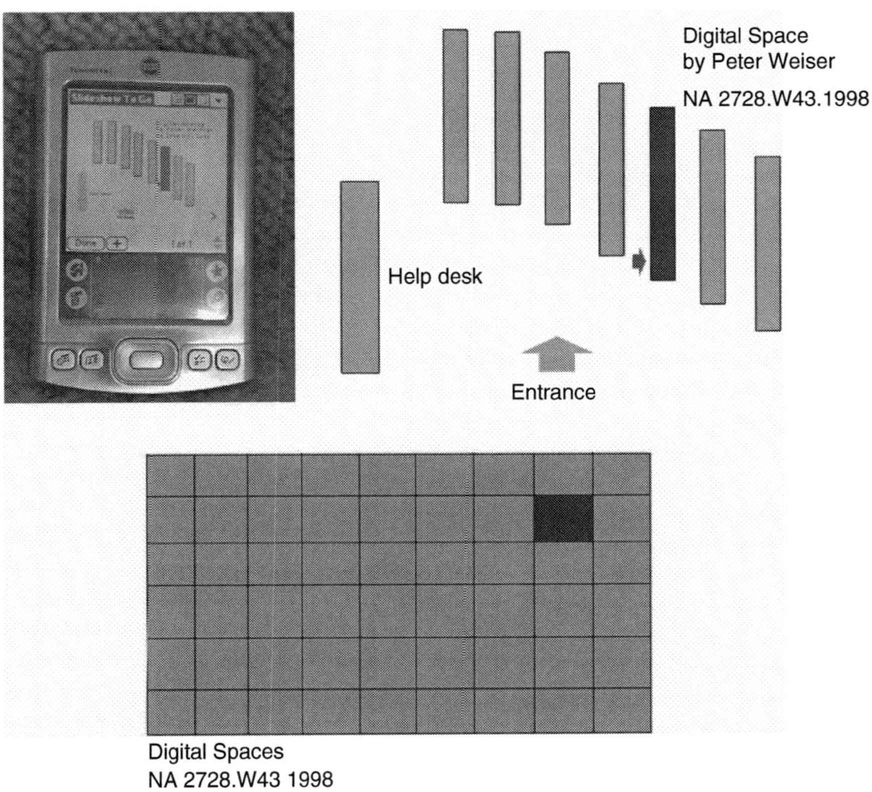

FIGURE 19.2
The PDA used in the library project user study. The image shown in the screen (enlarged) shows the book racks in gray and the desired rack in dark. The final image shows an elevation of a rack in dark marking the section on the shelf where the book is located.

than the conventional method to find the book. Moreover when the participant was asked to search for another book using PDA, it was found that it took much lesser time than the first search indicating a low learning curve.

19.5.2 "Smart" Home Project

Aging in place is the term that is attributed to many adults (especially baby boomers) who wish to grow older without having to move from their own homes. This is verified by several studies—although almost 1.5 million seniors currently reside in nursing homes and one-third of those have been there for more than three years; 77% of American consumers say that nursing homes are a last resort for themselves and their family members. (Boehm et al., 2004). This statistic within itself does not pose a problem except when one considers that as people age, they people face various chronic and temporal illnesses leading to severe limitation of activity in their daily life such as shopping, cooking, answering phone calls, opening doors, paying bills, taking medicine in time and so on. Assisted home living studies show that 50 percent of population, older than 70 years, needs assistance in their daily activities while 11.6 percent of population above 65 years old has severe limitations in self-care (Dominick et al., 2003).

Particularly in rural areas of the United States (like Mississippi) this poses a larger threat because of two issues (Bryden, 2002):

1) A universal tendency for many young people to leave their rural homelands to gain education, training and experience
2) A significant lack of healthcare providers and able physicians

Computer scientists envision that one of the first uses of ubiquitous computing at home will be in the field of proactive healthcare. Eventually these technologies can successfully motivate *long-term* healthy decision making, and thus delay or even prevent the onset of medical problems such as obesity and chronic illnesses, alleviating the pressure on the traditional healthcare system. But as the demography of the U.S. slowly starts becoming older as a population, the power of these technologies will be in the home—where it allows people to age-in-place (Intille, 2004).

At the DRIL, we are evaluating several technologies that can help in the design of a rural "smart" home—*a home that monitors the well being of the resident*. However one of the key characteristics of stitching ubiquitous computing into architecture is to create a nondisruptive environment. Most "smart" home technologies have severe privacy implications. Even the most subtle design may be too intrusive. Using sensors and cameras to monitor a user may not be acceptable by certain communities, even though such a solution would be the most apt for the problem they face. RFIDs can help. RFIDs carry relatively low perception value—which provides them the ability to be relatively invisible. Moreover, used similarly to Willis and Helal's RFID information grid system, a relatively low quality RFID system can be an efficient method of gathering location information.

Location information of an elder can be used in multiple ways:

1) To develop contextual cues—if someone is in the kitchen, it can be assumed that the person is cooking (or at least doing something related to food).
2) To monitor activity—long durations of inactivity may be a sign of a fall or illness.

In our designs at the DRIL, we are evaluating a system which uses RFID technology to look for location clues in "smart" homes.

19.5.2.1 How the System Works

The system uses a hybrid of the two systems that were explained above—interrogators are placed around the house to create an information grid; the user (elder) wears a passive RFID tag. Interrogators can be designed with the house, added at a later stage, or built into carpets or wall paneling systems. The RFID tags can be implanted or worn by the user and may be in the form of non computational products like a ring or embedded into shoes. The RFID tag when attached to the person is read by specific interrogators in the spatial framework which then parses the location information to a central computer. For example when the person enters the kitchen, the specific interrogator in that area reads the presence of the RFID tag and informs the central computer of this status. The central computer then tracks the location of the user as s/he moves around the house based on this simple binary information (present/not present) (see Figure 19.3). The computer compares the information it gathers to historical data to analyze for contextual clues and/or anomalies in activity patterns. Unique identification tags in conjunction with other sensing technologies (pressure sensitive flooring) can also be used to detect the presence of unwelcome guests in the home. Such information can then trigger action—informing a trusted second person (son/daughter or neighbor) in case of pattern anomalies or automatically calling for help (911 or ambulance services) and so on.

The system is simple and effective because of its invisibility. Moreover, the feeling of being "watched" by external entity is reduced because images are not being captured. And because the system works on a relatively simple technology, maintenance issues are minimized—this is especially important in rural areas with limited access to technical help.

19.5.2.2 Prototyping and Testing

The system described above is a very nascent stage of development. At this point we are working with (quasi) rural communities to evaluate perceptions of elderly people towards

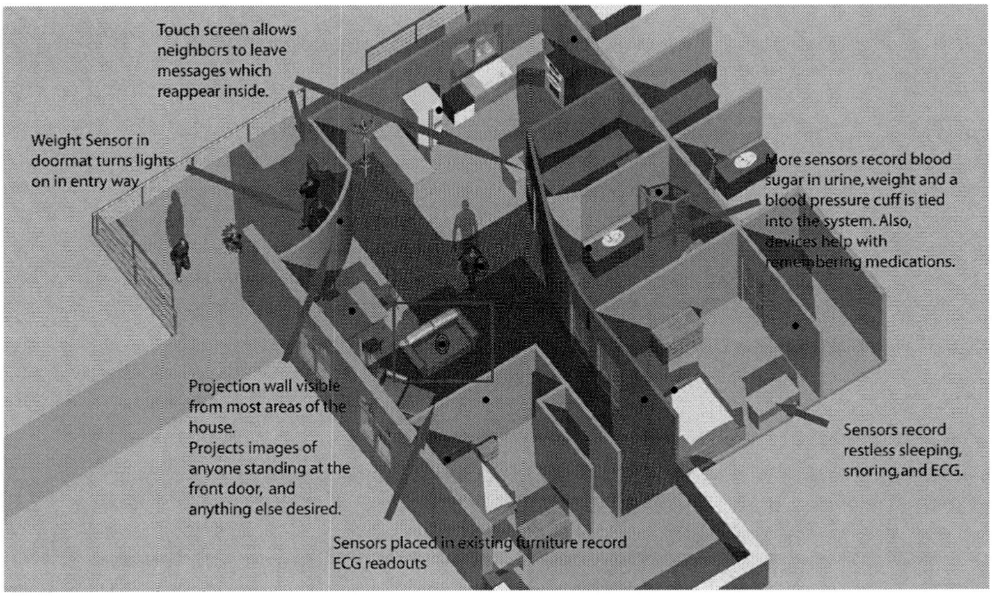

FIGURE 19.3
Design of the "smart" home showing position of RFID interrogators. The interrogator in the couch detects the proximity of the person and parses this information to the central computer.

such technology. Eventually we will work with these communities to develop prototype "smart" homes to test these systems.

19.6 Discussion

In the above sections we saw four applications of RFID technology that concern the design or negotiation of "place." While these examples show the promise of this technology, there are several aspects of it that may not be as rosy as painted. This section of the chapter will look at some of the concerns of using RFID technology within the design of architecture.

19.6.1 Incorporation

All technologies go through what we call a "novelty" phase when first introduced into society. This idea is illustrated by an anecdote—a friend talks of his grandmother who when the radio was introduced would leave windows open so as not to collect radio waves in the house. While we laugh at such anecdotes, this is how our society approaches any new technology. Any new technology traverses through necessary phases of awe, appreciation, critique, incorporation, negotiation and finally acceptance. Rejection of technology (however important and necessary) happens because designers and engineers protect it from this fire branding by society. Mature technology depends on not only on (tangible) efficient performance but also on an (intangible) socio-cultural incorporation.

RFID as a technology is not a new concept; but RFID in the design of place does not find mainstream (real world) applications beyond those in experimental or entertainment settings. Within this perspective, it is important for designers to present novel ideas into mainstream society and allow for it to be critiqued and developed.

19.6.2 Security

One whole section of this book is dedicated to privacy and security as it relates to RFID technology. This alone is an indication of how critical a discussion on security is in this context. However when RFIDs move beyond commodities into our environments, it brings with it much greater implications of security compromises. A system like a "smart" home for example hands over controls of security to a stakeholder that may not even be using the system directly (a son or daughter). In focus group studies with elders in Mississippi, many expressed concerns of what data they wanted to present to even close relatives (even a child of their own). The idea of tracking individuals, in spite of how beneficial, rings in Orwellian nightmares. What implications do implanted biochips carry with respect to personal information in the hand of untrustworthy (or even trustworthy) sources? What happens to an information grid if location information were vandalized by new age (technologically conversant) vandals? What happens to a rural "smart" home if a hacker were to hack into the system and follow the movements of your grandmother?

Designers of technology often assume that the society will use the technology in a particular manner; but users don't. They will always find novel ways to circumvent conventional thinking of the designer and carve out new uses for the solution. Predicting these circumventions may be difficult, but it is important that we pay attention to the use of these systems by actual users to understand not only the working but also the scope of use afforded by the solution. Like other personal information technology (credit cards, social security numbers), RFID is a technology which requires that designers be concerned about issues of privacy and security from the offset of the design. Innovative designs will require a continuous push to safeguard and protect what is valuable to users and the society.

19.6.3 Obsolescence

In a world where information has the ability to travel at the speed of light, innovation happens in weeks and months, not years. Computation technology has demonstrated this quite adequately since its conception. Within the last 50 years, computers have made strides ahead unlike any technology before it and will continue to do so into the future. This trend is likely to continue with current RFID and other communication technology. With regard to this, Huang and Waldvogel (Huang and Waldvogel, 2004) raise an important concern in their paper: what happens when there is a discrepancy between the durability of architectural materials and the rapid obsolescence of technology standards. Unlike a product (cell phones or laptop computers), built environments are designed for decades of use and cannot be refreshed every time there is a significant change in technology.

Huang and Waldvogel suggest that a simple way to address the discrepancy is to separate the elements based on their necessity to be replaced. They claim that in any spatial design, the technology must be designed in layers—each of which can be easily replaced without significantly affecting the other. In addition, systems must be able to scale as well as degrade gracefully. A component that fails should not bring the rest of the system down. A second way to address this discrepancy is to shift the intelligence from the device to the network. Traditional appliances, like telephones or televisions remain relatively unchanged because the intelligence of the system lies in the network and not the device. The home only contains the most simple and minimal "front end" functionality needed to access the network (Edwards and Grinter, 2001). RFID is an ideal technology in this respect because they are cost efficient, easily replaceable, carry a small footprint, and often independent of the front end functional element.

19.6.4 Introspection

Our homes (in an increasingly open world) have become our last bastion of privacy. In many conferences where ideas of ubiquitous computing are presented, the question always arises—do we really need our places to have computers built into them? Are we ready as a society to deal with vandalism and privacy problems at a spatial level (think of spam on the walls of your living room)?

The argument is however (slightly) flawed—as a society; we have always embraced technology into the places where we live, work, and play. Ever since we started living in closed (man made) structures, technology has manifested itself in multiple forms within these structures. New material, fabrication, climatic, communication, surveillance technologies continuously work to improve the way we build our spaces and live our lives in them.

While it is true that computing does have its problems—creases that have to be ironed out; it also brings with it benefits that are impossible to replicate with other technologies. As ubiquitous computing evolves out of its nascent state, it is inevitable that we will face problems. But problems can be solved and we as designers can only do so through continuous innovation and evaluation of these systems in real world applications and designs.

19.7 Closing Conversation

The architect/designer is an important link in the design of "place" because s/he is capable of changing the final product to meet the cultural, ethnic and the socioeconomic requirements of the user. But the architect/designer cannot work alone. Only through a careful mediation of *technical* aspects of design along with the phenomenological and the aesthetic, can we dream of using any technology in our homes. Hence the logical process of designing responsive environments for proactive healthcare is through collaboration. Elegant and usable design can be achieved only through conversations between researchers in various disciplines and through widespread awareness of current literature and research in this realm. This chapter and this book is just a beginning. We have seen in this chapter that, because of low cost and easy interoperability, RFID tagging has the potential to transform the way we design place. As RFID technology matures, its application and acceptance will change. As designers and engineers we have to constantly work to develop design ideas that are not only technically innovative but also socially responsible and useful.

19.8 Acknowledgments

Several acknowledgments are due—to Larry Barrow, Director and Sarah Pittman, coordinator of DRIL and Jim West, Dean of the College of Architecture, Art, and Design at Mississippi State University whose encouragement and mentorship has been most important to my research; the Mayor and people of the City of Meridian and in particular, Sharon Smith for the Aging in Place project; and the MSU Library system and in particular, Susan Hall for the Library project. Most importantly, gratitude is due to my families—first, my professional family which includes students who have been invaluable as research thinkers over these years—Lalatendu Satpathy, Vikash Singh, Landon Voeller, and Sneha Kochak, all of whom have been on the research team or helped with the projects described above. And secondly, my own family—my wife, Shilpi Kumar and my parents, without whom none of this would have been possible.

The research described in this chapter were developed using support from multiple sources: The Hearin Foundation; MSU-OR Research Initiation Grant; MSU- John C. Stennis Institute of Government; the City of Meridian; and the College of Architecture, Art, and Design, Mississippi State University.

References

Auburn, T. and R. Barnes. 2006. Producing place: A neo-Schutzian perspective on the 'psychology of place.' *Journal of Environmental Psychology* 26(1):38–50.

Ayre, L.B. 2004. *Position Paper: RFID and Libraries*, Galecia Group, Petaluma, CA, pp.1-21 [cited. Available from http://www.galecia.com/included/docs/position_rfid_permission.pdf.]

Boehm, E.W., B.J. Holmes, E.G. Brown, L. Bishop, S.E. McAulay, and J. Gaudet. 2004. *Forrester Big Idea: Who Pays For Healthcare Unbound (The $34 Billion Market For Personal Medical Monitoring)* [cited 04-05-2006. Available from http://www.forrester.com/Research/Document/Excerpt/0,7211,16524,00.html.]

Bryden, J.M. 2002. *The Elderly Outside the Metropolis: Myths and Realities*. Lecture for the Haskell Master Class, Arizona State University, Arizona.

Dominick, K.L., F.M. Ahern, C.H. Gold, and D.A. Heller. 2003. Health-related quality of life among older adults with activity-limiting health conditions. *Journal of Mental Health and Aging* 9(1):43–53.

Edwards, W.K. and R.E. Grinter. 2001. At home with ubiquitous computing: Seven challenges. *Paper read at UBICOMP 2001: Ubiquitous Computing International Conference*, September 30–October 2, 2001, Atlanta, GA.

Furui, S. 2000. Speech recognition technology in the ubiquitous/wearable computing environment. *Proceedings of the 2000 IEEE International Conference on Acoustics, Speech, and Signal Processing (ICASSP'00)*, 5–9 June 2000, Istanbul, Turkey.

Hitachi. 2006. *News Release—Worlds smallest and thinnest RFID IC Chip* [cited. Available from http://www.hitachi.com/New/cnews/060206.pdf.]

Huang, J. and M. Waldvogel. 2004. The swisshouse: An inhabitable interface for connecting nations. *Proceedings of the 2004 Conference on Designing Interactive Systems: Processes, Practices, Methods, and Techniques*, 1–4 August 2004, Cambridge, MA, pp. 195–204.

Intille, S. 2004. A new research challenge: Persuasive technology to motivate healthy aging. *Transactions on Information Technology in Biomedicine* 8(3):235–237.

Larson, K., S.S. Intille, T.J. McLeish, J. Beaudin, and R.E. Williams. 2004. Open source building—reinventing places of living. *BT Technology Journal* 22:187–200.

McCullough, M. 2004. *Digital Ground: Architecture, Pervasive Computing, and Environmental Knowing*. Cambridge, MA: MIT Press.

MERL. 2006. *DiamondTouch* [cited. Available from http://www.merl.com/projects/DiamondTouch/.]

Microsoft. 2007. *Microsoft Surface* [cited. Available from http://www.microsoft.com/surface/.]

Morton, S. 2004. Barcelona clubbers get chipped. *BBC News (Technology)*, September 29, 2004.

Roberti, M. 2006. A 5-cent breakthrough. *RFID Journal* [cited. Available from http://www.rfidjournal.com/article/articleview/2295/1/128/].

Senagala, M. 2005. Building is a network for living in: Toward new architectures. Paper Presented at *Smart Architecture: Integration of Digital and Building Technologies; Proceedings of the 2005 Annual Conference of the Association for Computer Aided Design in Architecture*, 13–16 October 2005, Savannah, GA.

Streitz, N. and P. Nixon. 2005. Introduction: The disappearing computer. *Communications of ACM* 48 (3):32–35.

Synaptics. 2007. *The Onyx Concept* [cited. Available from http://www.synaptics.com/onyx/.]

Tolmie, P., J. Pycock, T. Diggins, A. MacLean, and A. Karsenty. 2002. Unremarkable computing. Paper Presented at *SIGCHI Conference on Human Factors in Computing Systems: Changing Our World, Changing Ourselves*, Minneapolis, Minnesota, U.S.A.

Weiser, M. 1991. The computer for the 21st century. *Scientific American* 265(3):94–104.

Willis, S. and S. Helal. 2005. RFID information grid for blind navigation and wayfinding. *Proceedings of the Ninth IEEE International Symposium on Wearable Computers*, 18–21 October 2005, Osaka, Japan, pp. 34–37.

20

Photosensing Wireless Tags for Precise Location and Geometry Queries

Ramesh Raskar, Paul Beardsley, Paul Dietz, and Jeroen van Baar

CONTENTS

20.1 Geometric Context of Wireless Tags

The geometry-rich functionality is achieved by augmenting each tag (Figure 20.1) with a photosensor (Figure 20.2). We achieve optical communication with this composite RF-photosensing tag with modulated light. In this chapter the operations are shown using a projector that is paired with the tag-RF-reader. The projector performs the dual operation of sending optical data to the tag (similar to your TV's IR remote control unit) and also giving visual feedback by projecting instructions on objects. Current tag-readers operate in broadcast mode with no concept of a directional communication but the RFIG tags allow locating of tags within a few millimeters, support selection of individual tags, and create a 2D or 3D coordinate frame for the tags. The system of projector and photosensing tag offers a set of rich geometric operations. It presents a new medium for many of the results from the area of computer vision, with projector and tags replacing camera and image interest points (Raskar et al., 2004).

Our experimental work is based on active, battery-powered radio frequency tags. However, our goal has been to develop methods that can be used with passive, unpowered radio frequency identification (RFID) tags. The key issue in evolving our active tag system to passive tags would be power. In this chapter, we allowed ourselves computation and sensing consistent with only the size and power levels we felt were achievable on a passive RFID system. For example, (a) tags are not photosensing or computing until woken up by the RF reader and (b) we do not have a light emitting diode (LED) on the tag as a visual beacon to a human or camera-based system because it would be power-hungry.

Location tracking using RF received signal strength or time of arrival is popular but requires multiple readers and the accuracy maybe insufficient for complex geometric

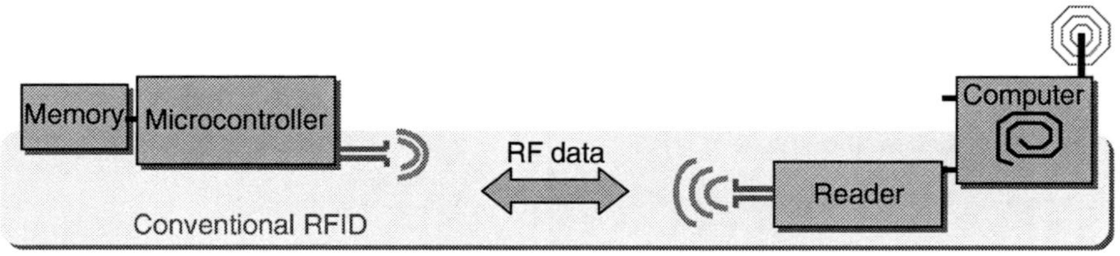

FIGURE 20.1
Conventional radio frequency identity (RFID) transponder communicates with RF reader and responds with the id number stored in the tag's memory.

procedures (Hightower and Borriello, 2001). Previous systems have also married RF tags with optical or ultrasound sensors to improve accuracy. Some systems use active RF tags that respond to laser pointers. The FindIT Flashlight uses a one-way interaction and an indicator light on the tag to signal that the desired object has been found (Ma and Paradiso, 2002). Other systems use a two-way interaction, where the tag responds back to the PDA using a power-hungry protocol like 802.11 or X10 (Patel and Abowd, 2003) and (Ringwald, 2002). CoolTown (The CoolTown Project, 2001) uses beacons that actively transmit devices references but without the ability to point and without visual feedback. The Cricket project (Teller et al., 2003) recovers pose of a handheld device using installed RF- and ultrasound-beacons, and does projected augmentation.

20.2 How It Works

Conventional tag communication works by broadcast from an RF reader, with response from all in-range tags. Limiting the communication to a required tag is traditionally achieved using a short range tag-reader and close physical placement with respect to the tag. In contrast, we can select tags for interaction at long-range using projected light, while ignoring unwanted in-range tags. The handheld device first transmits an RF broadcast. Each in-range tag is awoken by the signal, and its photosensor takes a reading of ambient light, to be used as a zero for subsequent illumination measurements. The projector illumination is turned on. Each tag that detects an increase in illumination sends a response to indicate that it is in the beam of the projector, and is ready for interaction.

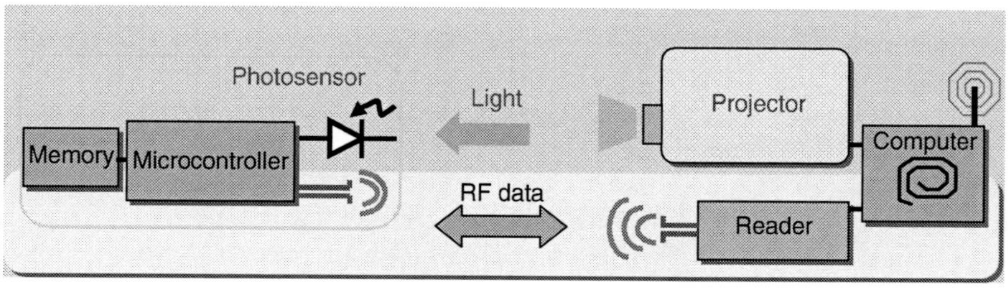

FIGURE 20.2
The RFIG tag: Radio frequency identity and geometry (RFIG) transponder communicates with RF reader as well as spatio-temporal light modulator such as a modulated IR light. For example, with a full-fledged data projector, one can find the stored Id as well as the (x,y) projector pixel location that illuminates the tag.

The handheld device is aimed casually in the direction of a tagged surface. The handheld device sends an RF signal to synchronize the tags, followed by illumination with a sequence of binary patterns, i.e., binary structured light. Each projector pixel emits a unique temporal Gray-code, and thus encodes its position. The tag records the Gray-code that is incident on its photosensor, and then makes an RF transmission of its identity plus the recorded Gray-code back to the RF reader. The projector uses the identity plus the recorded (x,y) location to project instructions, text or images on the tagged object. It is then straightforward to create correctly positioned augmentation on the tagged surface.

20.3 Applications

Several aspects of RFIG have been described in our previous work (Raskar et al., 2004). The work was motivated in terms of the commercially important application of inventory control. But we believe that photosensing tags may have many innovative uses, and in this chapter our goal is to present the new ideas in the context of a few promising examples. We outline broad modes of deployment for geometric analysis. Note these are speculative uses, not work done.

1. Location feedback, e.g., warehouse management (Figure 20.3): Consider the task of locating boxes containing perishable items about to expire. Even with traditional RF tagging with expiry date info recorded in the indexed database, the employee would have to serially inspect boxes and mark the boxes with about-to-expire products. Using RFIG tags, the handheld or fixed projector first locates the queried tags and then illuminates them with symbols such as X and Ok so that the employee has a visual feedback. Note that a second user can perform similar

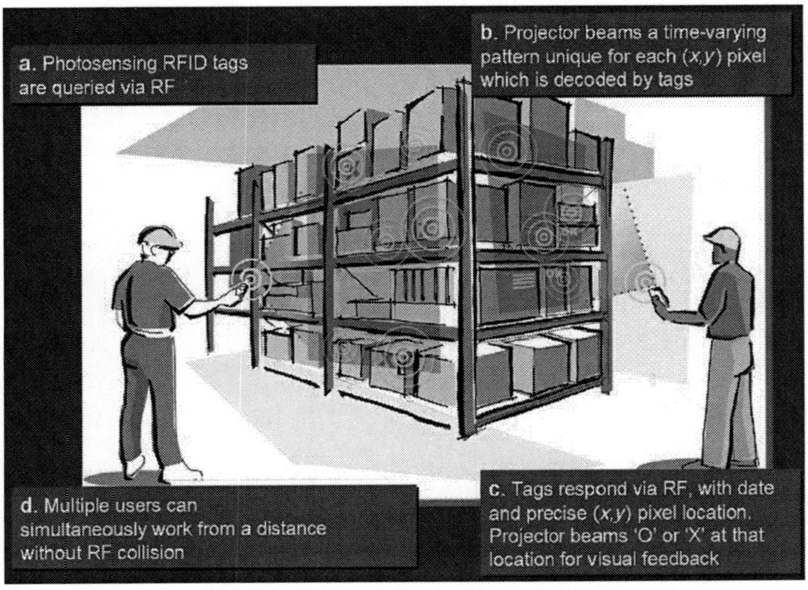

FIGURE 20.3
Application A: Warehouse scenario: Employee locates items about to expire and gets visual feedback. A second user performs similar operations, without conflict in the interaction because the projector beams do not overlap.

FIGURE 20.4
Application B: Detecting obstruction on railway tracks, such as person on tracks near a platform, disabled vehicle at a rail–road intersection or suspicious material on tracks. Finding obstruction with a camera-based system is difficult. The idea is to sprinkle RFIG tags along the track and illuminate them with a fixed or steered beam of temporally modulated light (not necessarily a projector). Tags respond with status of reception of the modulated light. Lack of reception indicates obstruction, which can be relayed to a central monitoring facility where a human observer can carefully observe the scene possibly with a pan-tilt-zoom surveillance camera.

operations, without RF collision with the first reader or the tags because the two projector beams do not overlap.

2. Obstruction detection, e.g., object obstructing a railroad (Figure 20.4): A common computer vision task with camera includes detecting abnormal conditions by performing image processing. One example is detecting obstruction on railway tracks, for example, raising an alarm if a person is on the tracks in a subway station or if there is some suspicious material on tracks. Processing images of videos from camera-based system to detect such events is difficult because the ambient lighting condition can change and several other activities can result in false positives. But one can solve this vision problem by instead sprinkling RFIG tags along the track. One can illuminate these tags with a fixed or steered beam of temporally modulated light (not necessarily a projector), such as 40 Kz infrared beam from a sparse array of light emitters. Then the operation is similar to "beam break" technique commonly used to detect intruders. But a wireless tag based system is ideally suited for applications where running wires to both ends is impractical. Using retro-reflective markers and detecting a return beam is another common strategy to avoid wires but sprinkling a large number of markers creates an authoring nightmare. In case of RFIG, the tags id and location can be easily reported along with the status of reception of the modulated light. Lack of reception indicates obstruction which can be relayed to a central monitoring facility where a human observer can carefully observe the scene possibly with a pan-tilt-zoom surveillance camera.
3. Ordered placement and orientation, e.g., books in a library (Figure 20.5): A common task in libraries, pharmacies, or for facility managers is maintaining a large number of objects in a predetermined order. In a library, if books are RF tagged, it is easy to get a list of books within the RF range. However, without location information it is difficult to find out which books are out of alphabetically sorted order. In addition, without book orientation information, it is difficult to detect books that are placed upside down. With RFIG and a handheld projector, the system gets book title as well as location. Then the system sorts books by title as well as by their 2D geometric location. A mismatch in the two sorted lists indicates that the corresponding book is placed in a wrong position. The system

FIGURE 20.5
Application C: Books in a library. If books are RF tagged, it is easy to get a list of books within the RF range. However, without location information it is difficult to find out which books are out of alphabetically sorted order. In addition, without book orientation information, it is difficult to detect books that are placed upside down. With RFIG and a handheld projector, the system gets book title as well as location. Based on the mismatch in title sort with respect to the location sort, the system gives immediate visual feedback and instructions, shows here as red arrows for original positions.

knows the current location for these books as well their ideal position. The projector display gives immediate visual feedback and instructions, shown in the figure as red arrows from current positions to intended position. A single book can also be tagged with two RFIG transponders, one at the top of the book spine and one at the bottom of the spine. Then comparing the coordinates of these two tags allows one to find if the book has been placed upside down.

4. 3D Path Planning/Guiding, e.g., guiding a robot on assembly line for arbitrarily oriented objects (Figure 20.6): RFIG tags can be used in factories for robot guidance. The idea is similar to other laser guided operations. Suppose a robot is instructed to grab a certain object in a pile moving on a conveyor belt. RFID can simplify the object recognition problem in machine vision but precisely locating

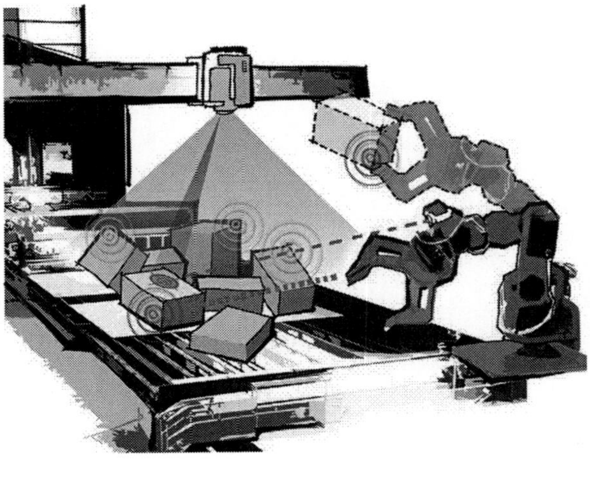

FIGURE 20.6
Application D: Laser guided robot: To guide a robot to pick a certain object in a pile moving on a conveyor belt, the projector locates the RFIG tagged object and illuminates it with easily identifiable temporal pattern. A camera attached to the robot arm locks onto this pattern and allows the robot to home in on this object.

the object will be difficult. The idea is to use a fixed projector to first locate the RFIG tagged object and then illuminate the object with a steady easily identifiable temporal pattern. A camera attached to the robot arm locks onto this pattern by doing pattern matching and allows the robot to home in on this object.

Notice that in a majority of the applications described earlier, the projector behaves similar to devices we are familiar with, remote controls and laser pointers, but with some spatial or temporal modulation of light. The projector is a glorified remote control communicating with a photosensor in the location-sensing phase and a glorified laser pointer in the image projection phase.

20.4 Discussion

Several problems can influence optical communication between the projector and a tag. It can be affected by ambient light. Wavelength division multiplexed communication is commonly used to solve this problem (e.g., TV remote and IR photosensor). The optical communication also gets noisier as projector-tag distance increases, and as the photosensor gets dirty. However, within these limitations, the RFIG method can support very intricate and multipurpose geometric operations with the ambient intelligence provided by wireless tags. The work indicates some of the possibilities for blurring the boundaries between the physical and digital worlds by making the everyday environment into a self-describing wireless data source, a display surface, and a medium for interaction.

References

Hightower, J. and Borriello, G. August 2001. Location systems for ubiquitous computing. *Computer*, 34(8), 57–66.

Ma, H. and Paradiso, J.A. 2002. The FindIT flashlight: responsive tagging based on optically triggered microprocessor wakeup. *Ubicomp*, 160.167.

Patel, S.N. and Abowd, G.D. 2003. A 2-way laser-assisted selection scheme for handhelds in a physical environment. *Ubicomp*, 200.207.

Raskar, R., Beardsley, P., Van baar, J., Wang, Y., Dietz, P., Lee, J., Leigh, D., and Willwacher, T. 2004. RFIG lamps: Interacting with a self-describing world via photosensing wireless tags and projectors. *ACM Trans. Graph.* (*SIGGRAPH*), 22(3), 809–818.

Ringwald, M. 2002. Spontaneous interaction with everyday devices using a PDA. Workshop on supporting spontaneous interaction in ubiquitous computing settings. *UbiComp*.

Teller, S., Chen, J., and Balakrishnan, H. July 2003. Pervasive poseaware applications and infrastructure. *IEEE Comp. Graph. Appl.*, July/August 2003, Canmore, Canada.

The Cooltown Project, 2001. http://www.cooltown.com/research/

21

RFID and NFC on Mobile Phones

Paul Coulton, Omer Rashid, and Reuben Edwards

CONTENTS

21.1 Introduction

Although RFID is establishing itself across a range of business sectors it is likely that it's most significant impact will come through being combined with mobile phones. Mobile phones are often described as a disruptive technology in that they have overturned the dominance of traditional fixed phone service. The incorporation of RFID and NFC onto mobile phone has similar possibilities in terms of micropayments, digital distribution strategies, and interaction with everyday objects, which is likely to radically alter existing consumer practices. In this chapter we will introduce RFID and NFC in relation to mobile phones and provide details of the associated standards and software development tools and environments. Finally, to highlight the potential of these technologies, we will discuss the user experience in relation to three projects which have utilized this technology.

21.2 Overview of RFID/NFC on Mobile Phones

21.2.1 Introduction

While much of the focus on RFID has come through asset tracking, another area that has greatly benefited from RFID is micropayments and prepaid access services for mass transit systems, which has seen huge take up in places such as Japan and Korea and has no doubt fuelled the integration of RFID into the mobile phone feature set.

As this book has highlighted, RFID as a technology covers a range of possible operating frequencies, which in turn affects the nature of the operation of the application of the technology. The particular operating range for mobile phones is generally 13.5 MHz, which limits the range to approx 3 cm or touch as shown in Figure 21.1.

Any discussion of RFID on mobile phones is incomplete without also discussing Near Field Communications (NFC), which is an interface and protocol built on top of RFID and is targeted in particular at consumer electronic devices, providing them with a secure means of communicating without having to exert any intellectual effort in configuring the network [1]. Thus NFC takes RFID beyond the traditional use case scenario of interacting with tags to allow the communication between devices. To connect two devices together, one simply brings them very close together, a few centimetres, or make them touch. The NFC protocol then automatically configures them for communication in a peer-to-peer network. Once the configuration data has been exchanged using NFC, the devices can be set up to continue communication over a longer range technology such as 802.11 or Bluetooth. The other advantage with NFC comes in terms of power saving, and it achieves this by having an Active Mode (AM) and Passive Mode (PM) for communication. In AM both devices generate an RF field over which they can transmit data. In PM, only one device generates the RF field, the other device uses load modulation to transfer the data. This is an ideal scenario for mobile phones as it would allow them to interact with other devices such as laptops while minimizing battery consumption. The data rates available are relatively low, 106, 212, or 424 kbits/s, although for the applications envisaged this should be more than sufficient [1].

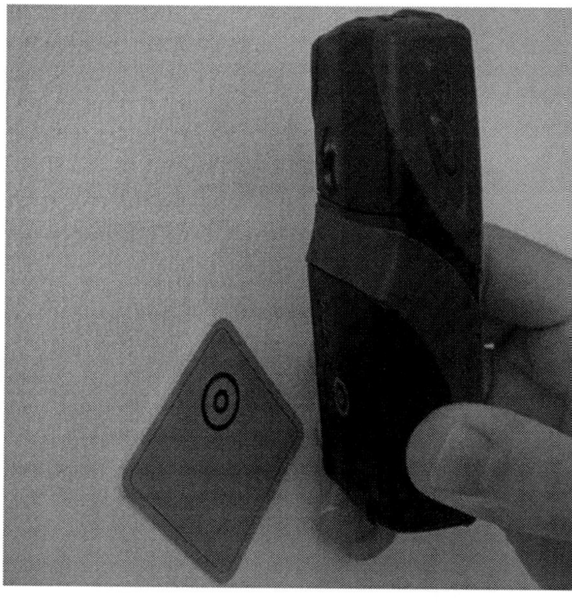

FIGURE 21.1
Phone interfacing with RFID tag.

Nokia was the first to combine mobile phones with RFID/NFC when it introduced clip-on RFID shells for the 5140 and NFC for 3320 and 5140i Series 40 phones (Nokia Xpress-on Mobile RFID/NFC Kits). The RFID/NFC shells can be accessed via J2ME applications running on the phone to trigger defined actions within the application. At the Consumer Electronics Conference in Las Vegas in January of 2007 Nokia demonstrated the 6131 NFC phone with in-built NFC capabilities which is expected to be released in the 2nd quarter of 2007.

These phones were but the first of a growing trend and the Japanese telecommunications giant NTT DoCoMo has reportedly to have shipped more than 5 million RFID enabled mobile phones for use instead of printed tickets in the National Rail Network [2]. Further Sony has started to ship all of its laptops with RFID/NFC technology so that users can download straight to RFID cards or RFID mobile phones [2].

In the following sections we shall discuss some of the important elements of the NFC specifications before discussing programming RFID/NFC by considering both JSR 257 and the Nokia SDK. Finally we shall provide examples of how RFID/NFC has been used for three mixed reality gaming experiences.

21.2.2 NFC Specifications

As has already been highlighted, NFC is already on its way to becoming a part of everyday life and in order to achieve successful consumer adoption of this technology companies involved need to work together closely and applications need to be interoperable. In 2004 the NFC Forum was formed with its main objective to promote the use of NFC technology in consumer devices and services, provide an extensive framework for interoperable applications by developing standards based specifications, and ensure that products and devices claiming to be NFC compliant conform to the forum specifications.

A major step towards the goals of NFC Forum was in 2006 when it announced the first five NFC specifications as follows:

- NFC Data Exchange Format (NDEF)
- NFC Record Type Definition
- NFC Uniform Resource Identifier Service Record Type Definition
- NFC Text Record Type Definition
- NFC Smart Poster Record Type Specification

Further, tag formats based upon ISO 14443 Type A, ISO 14443 Type B, and ISO 18092 were also announced and an NFC Forum compliant device must support these. The technology architecture of NFC is based upon three different modes of operation, i.e., peer to peer, read–write, and card emulation as shown in Figure 21.2.

21.2.2.1 NFC Data Exchange Format Specification

The main objective of NDEF specification [3] is to build a common data format for NFC devices and tags. It does not define any record types as it focuses on data structure of the message to exchange information, record types are defined in separate specifications as listed earlier and they are discussed briefly in the sections below. NDEF is a lightweight binary message format designed to assemble single or multiple application defined payloads into a single message for information exchange between NFC Forum Devices or NFC Forum Device and NFC Forum supported tags. An NDEF message consists of one or more NDEF records which hold the payloads. Figure 21.3 shows the anatomy of an NDEF message [3].

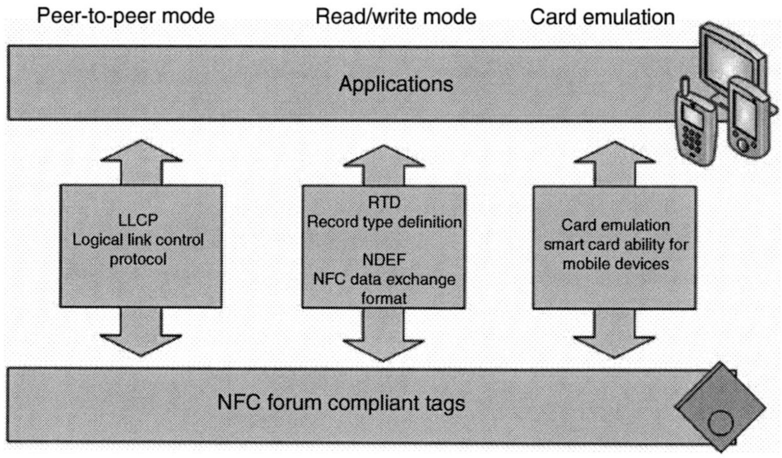

FIGURE 21.2
NFC Modes.

The first record in NDEF message contains the Message Begin (MB) flag and the last record contains the Message End (ME) flag. In figure it can be seen that first record (index 1) has the MB flag set while last record of the message (index *n*) has the ME flag set which also points out that the message head is to the left and tail to the right. If the message is to contain one record only then MB and ME are set at the start and end of the same record.

As the payloads are encapsulated in the NDEF record they can be of different sizes. Each payload is defined by a set of three attributes, i.e., type, length, and an identifier as shown in Figure 21.4. The payload length defines the payload as the number of octets where maximum number of octets in a message being 232-1. Payload Type indicates the kind of data being encapsulated in the record. This can be URIs, NFC Forum specific type format, or MIME types. The key advantages of describing the type of content are that the application receiving this NDEF message can direct it to the application responsible for handling that particular type of data on the user device. Since an NDEF message can comprise of multiple records so payload type of first record should generally dictate the payload type not only for itself but also for the records that follow. It is important to keep in mind that the specification does not define any model for handling of the data based upon type; in fact the processing of data is totally left up to the user application involved in the NFC information exchange. Third payload attribute, payload identifier, is optional and allows the application to identify the payload encapsulated in the record. This is crucial in supporting URI based links. Similar to payload handling the specification leaves the linking mechanism definition to the application.

- *MB*: MB flag (1 bit field) shows the start of an NDEF message when set.
- *ME*: ME flag (1 bit field) shows the end of an NDEF message when set.

	NDEF message				
R_1 MB = 1	...	...	...	...	R_n ME = 1

FIGURE 21.3
Structure of NDEF Message. (From NFC Forum, NFC Data Exchange Format (NDEF) Technical Specification NDEF 1.0, NFCForum-TS-NDEF_1.0, 24th July 2006.)

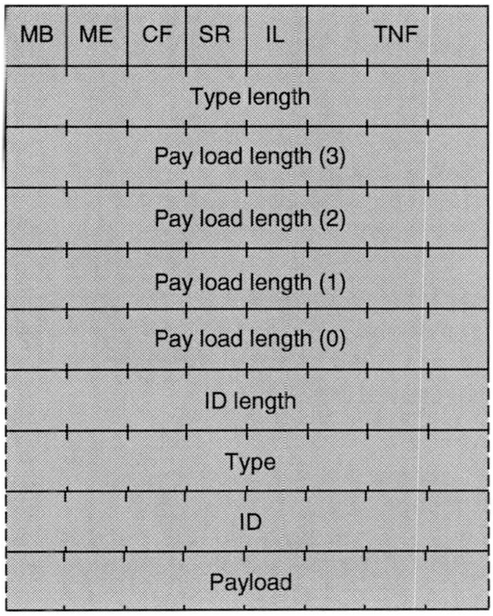

FIGURE 21.4
Structure of an NDEF record. (From NFC Forum, NFC Data Exchange Format (NDEF) Technical Specification NDEF 1.0, NFCForum-TS-NDEF_1.0, 24th July 2006.)

- *CF* (Chunk Flag): 1 bit field which identifies chunked payload.
- *SR* (Short Record): 1 bit field which indicates an SR when set, it indicates that the payload size is within the limits of payload fields, i.e., between 0 and 255 octets.
- *IL* (ID_Length): 1 bit field when set shows that ID_LENGTH filed is present in the header of the record as a single octet.
- *TNF* (Type Name Format): 3 bit field which shows the structure of TYPE field. NDEF specification defines a list of values for this field, e.g., 0×01 means NFC Forum well-known type, 0×03 for absolute URI etc.
- *TYPE_LENGTH*: 8 bit unsigned integer defines the length of TYPE filed in number of octets.
- *ID_LENGTH*: 8 bit unsigned integer which specifies the length in octets for ID field.
- *PAYLOAD_LENGTH*: unsigned integer that indicates the length of PAYLOAD field in octets. If SR flag is set then this filed is set to 1 octet (8 bits) but if SR flag is not set then this field is set to 4 octets (32 bits).
- *TYPE*: identifies the type of payload and must follow the format defined by the value set in TNF field.
- *ID*: contains an identifier in the form of a URI reference which can absolute or relative.

NDEF specification leaves security and internationalization up to implementations. Since the specification supports definition of data types it is up to the application to consider the implications of accepting different data types.

21.2.2.2 NFC Record Type Definition

The NDEF specification discussed earlier defines a common data format for NFC Forum devices but it does not define any record types in detail. Different record types are defined in separate specifications. The RTD specification provides guidelines for the specification of well-known types for inclusion in NDEF messages being exchanged. Each NDEF record contains a record type string which holds the name of the record type referred to as Record Type Name (RTN). RTNs can be specified in several different ways, these can be MIME types, URIs or NFC Forum well-known type names etc.

21.2.2.2.1 NFC Forum Well-Known Type

Well-known type is designed to create primitives for certain common types. It is mainly used when there is no equivalent URI or MIME type available for a certain payload. It is identified within an NDEF message by setting the TNF field of a record to 0 × 01. A well-known type is a URN with a namespace identifier (NID) ''nfc.'' There are two kinds of NFC Forum well-known types; NFC Forum global type and NFC Forum local type. The earlier is solely managed by NFC Forum; other parties are not allowed to define them. The later is available for use within the perspective of another record.

21.2.2.2.2 NFC Forum External Type

This is meant for the organizations who wish to assign a name space to be used for their own purposes. External type is similar to well-known type; however, NSS part is put into another namespace, i.e., ''ext.'' The external type must be formed by taking the domain name of the organization issuing the external type.

21.2.2.3 NFC Text Record Type Definition

As the name suggests Text Record Type defines an NFC well-known type for plain text [4]. It has been created as plain text field which can be used on its own as a solitary record in an NDEF message or in combination with another record type to provide extra description for the content. The specification does raise some security concerns for this record type. These are mainly due to the actual nature of this particular record type. For instance the text on a tag can be over written and the user can be tricked into doing something else than actually desired. To avoid this, the specification recommends the usage of this record type to be for information purposes only.

21.2.2.4 NFC Forum URI Record Type Definition

NFC URI Record Type Definition [5] is used to specify a record to be used with NDEF to retrieve a URI stored in an NFC-forum compliant tag. This specification provides a means to store URIs inside other NFC elements such as Smart Poster [6]. The speciation defines URI service with data model and specifies simple Smart Poster examples. Smart Poster RTD can be considered as an extension to URI RTD. URI is NFC well-known type and is represented as U (0 × 55). Figure 21.5 shows the structure of a URI record [5].

The URI identifier codes are available in the specification, some examples of which are 0 × 04 for https:// and 0 × 03 for http:// etc. If the URI identifier is set as 0 × 04 and the URI filed value is www.mobileradicals.com then the communicating device will receive the URI as https://www.mobileradicals.com. Using Figure 21.6 let us see how a URI will be stored in the record structure provided previously.

Name	Offset	Size	Value	Description
ID Code	0	1 Byte	URI ID Code	ID code of URI
URI	1	N	UTF-8 String	URI

FIGURE 21.5
Structure of URI Record.

Offset	Content	Description
0	0 × D1	SR = 1, TNF = 0 × 01, MB = 1, ME = 1
1	0 × 01	Length of record type
2	0 × 13	Size of payload
3	0 × 55	NFC well-known record type URI (U)
4	0 × 01	URI Identifier (http://www.)
5	0 × 6D 0 × 6f 0 × 62 0 × 69 0 × 6C 0 × 65 0 × 72 0 × 61 0 × 64 0 × 69 0 × 63 0 × 61 0 × 6C 0 × 73 0 × 2E 0 × 63 0 × 6f 0 × 6D	String "mobileradicals.com" in UTF 8

FIGURE 21.6
Example URL as URI record type.

This figure shows a valid URI record type for URL http://www.mobileradicals.com. The NDEF record in this table shows that it is an SR. TNF is NFC well-known type, and since we are not going to use multiple records within this NDEF message so MB and ME have been set in the same record. This is followed by length of record and payload, respectively. Then we declare the record type which is NFC well-known type URI followed by the identifier for URI. Finally we include the string for the URI.

21.2.2.5 NFC Smart Poster Record Type Definition

The Smart Poster record type [5] is an NFC forum well-known record type (well-known type Sp). URI RTD defines a way to store and communicate URIs but there is no functionality to add metadata to those URIs. Smart Poster provides a way to add this meta data to the URIs, for example, consider a poster advertising a new console game, by adding an NFC tag with Smart Poster record we can transform this simple object into a smart object. When users touch the poster with their mobile device (phone or PDA) they will be provided with URI to the advert for the game to be seen on their device along with some text based information for this clip, this information can be the size of clip, total running time, where to buy the game, promotional offers, etc.

Smart Posters are in fact one the key use cases of NFC and is often seen in one way or the other in different documents about NFC, press releases etc. Smart Poster records can also contain certain actions embedded in them as part of the record which can trigger actions on the user device, e.g., launch browser to access the URI etc. Smart Poster payload is an NDEF message which can consist of multiple NDEF records (SR = 0). Depending upon the content being stored the Smart Poster can have none, one, or more of the following components.

21.2.2.5.1 Title Record

This is an optional record representing the service. This record can be used more than once if the Smart Poster is required to support different languages but it is critical not to make duplicate entries for languages being included in the record. This record is an instance of Text RTD.

21.2.2.5.2 URI Record

Since Smart Poster record provides an extension to URI record by adding metadata for the URI, it forms the core of the Smart Poster record and must not be repeated (one URI per smart poster record).

21.2.2.5.3 Action Record

This record is optional and can be used to trigger actions for the URI record, e.g., action can be to launch the browser or simply save the URI as a bookmark for later on. Although this record is optional but not including it can leave the handling of URI to the device and that can result in a different user experience from device to device. It is a good practice to include this record so that appropriate application can be launched to handle the URI.

21.2.2.5.4 Icon Record

This record is optional and can be used to hold one or more MIME type image records so that the device reading the Smart Poster record can include one of those images (depending upon its capabilities) in the URI display.

21.2.2.5.5 Size Record

This record is optional as well and can be used to indicate the size of the content. This can be especially helpful to decide in advance if a device has lesser capabilities to handle that particular object. Using a good combination of size and type records the mobile device can decide whether it can handle the object being referenced or not.

21.2.2.5.6 Type Record

This record is used to declare the type of the external object being referenced through the URI provided in the URI record. This record is optional and as mentioned earlier when used in combination with size record can help determine if the mobile device is able to handle this external object or not.

21.2.3 Overview of JSR-257

The Contactless Communication API, or JSR-257 [7], allows applications to access information on contactless targets, such as RFID tags and visual codes such as QR codes and sema codes [8]. The 2D barcodes are similar to RFID tags in that they contain data often in the form of a URL. However, they hold much smaller amounts of information than the newer RFID tags and they are generally slower and more difficult to read [9].

The primary objective of JSR-257 is to provide easy access to various contactless targets and transfer information to or from them. Before we provide details of programming RFID/NFC applications we shall first consider the mandatory and optional parts of the API, which are divided into five packages as shown in Figure 21.7 [7].

Only `javax.microedition.contactless` is the mandatory package while the rest are optional and can therefore be left unimplemented. A reference implementation must provide a list of target types it supports and it is this list of targets that dictates the packages that must be implemented. If a target type is listed in supported target types an implementation must be provided. All the targets supported by this API implementation are defined in the `TargetType` class. To aid understanding shall briefly consider the functionality provided by each package.

`Javax.microedition.contactless:` This package provides functionality common to all contactless targets supported by this API. The `DiscoveryManager` class sits on top of this API and is the starting point of contactless communication. In order for an

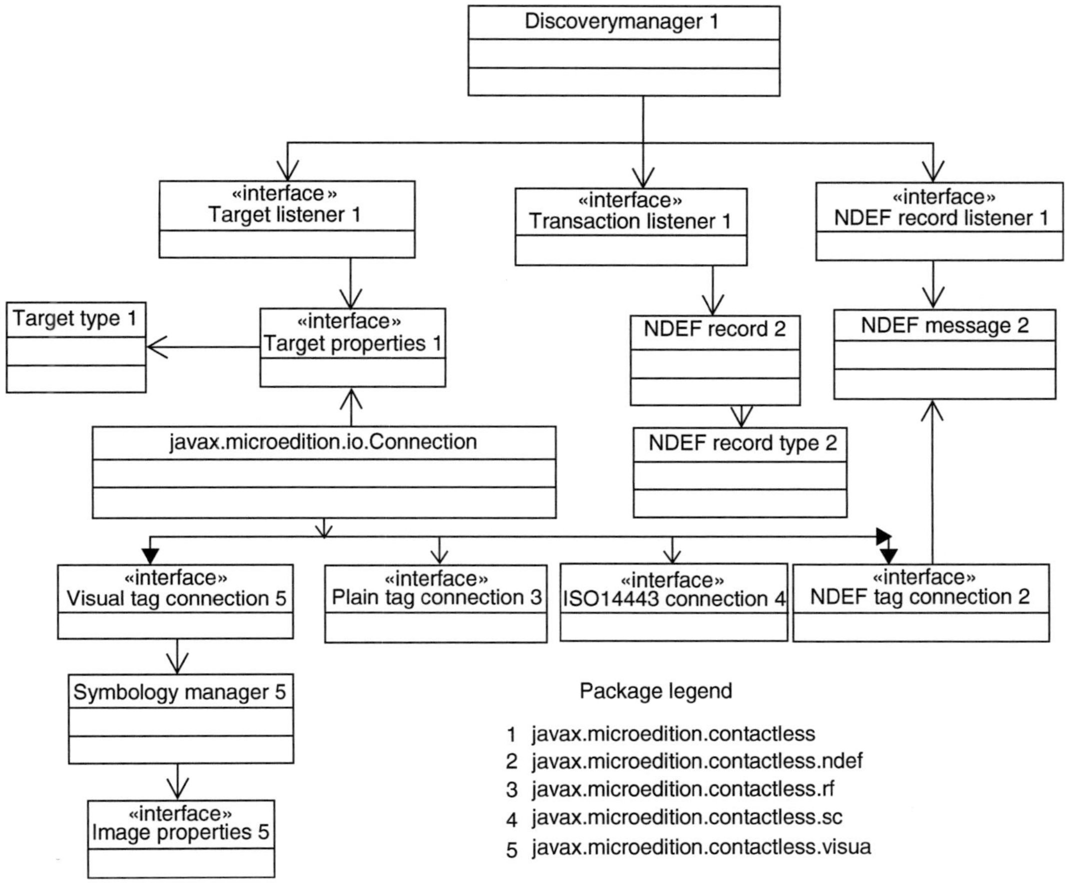

FIGURE 21.7
Overview of JSR 257 Reference Specification.

application to receive notifications about targets in close vicinity an application must obtain an instance of the `DiscoveryManager` class which in return provides the notifications to application about contactless targets. An application can then use subpackages to communicate with different targets depending upon their type. Connections to contactless targets are built on top of Generic Connection Framework (GCF). Each target defines a new protocol to GCF, although only the visual code protocol is visible and all RFID related protocols are hidden due to the nature of communication involved.

The `TargetListener` interface provides notifications back to `DiscoveryManager` instance once a target is discovered by the device hardware. In essence a Java Virtual Machine (JVM) must provide a `TargetListener` for each `TargetType` obtained by calling `getSupportedTargetTypes ()` function of `DiscoveryManager` class. Only one `TargetListener` can be registered at a time as RFID hardware can only handle one physical connection and failing to do so will result in a registration failure and an exception will be thrown.

Class `TargetType` provides all the contactless targets supported by the API while the `TargetProperties` interface provides the properties common to all supported contactless targets. Since the API can support card emulation mode, the interface `TransactionListener` provides information to the application relating activity between secure elements on the device and an external reader.

Javax.microedition.contactless.ndef: NDEF records have been defined by the NFC Forum to exchange data between two NFC devices or an NFC device and a tag. Since these records contain type, type format, identifier, and payload, a connection can be established with any physical target that supports the NDEF data format. To utilize this feature the application must register with an NDEF Listener to receive notifications about tags, or devices, detected. The data from the tag is passed on to the applications registered to receive NDEF notifications.

`NDEFMessage` class represents messages which can consist of a number of NDEF records. This class provides necessary functionality to manipulate these records; i.e., add, delete, or update data payload within the records in the message. Each individual record within the message is represented by the `NDEFRecord` class stored in a byte array. Class `NDEFRecordType` stores the name and format of the record type. Record types are stored as constants within the class and must abide by the rules of NFC Forum RTDs and RFCs for example, MIME types, URL, etc. While other classes handle definition and storage of NDEF messages the `NDEFTagConnection` interface provides the basic connection for exchanging this NDEF data between NFC devices or NFC tag. This interface does not concern itself with physical type of the contactless target; it simply reads and writes the data. The protocol to do achieve this read and write process is extended from GCF.

Javax.microedition.contactless.rf: Although this API implementation aims to use NDEF records as standard, it also provides access to physical RF targets that do not support NDEF. This package contains one interface, `PlainTagConnection`, which encloses functionality to detect RFID tags. This is done to enable comprehensive support for various types of RFID tags since it would be undesirable to have an API to support all different types of tags.

Javax.microedition.contactless.sc: This package is responsible for communication with external smart cards. Although JSR-177 covers communication with smart cards, through an APDU connection, JSR-257 defines an ISO14443 connection interface to access smart cards. Unlike an APDU connection, an ISO4443 connection can read both resident and external smart cards and provide access at a much lower level.

Javax.microedition.contactless.visual: This package provides support for visual tag targets which are detected as contactless targets. This package contains classes and interfaces to read information stored on the visual codes.

21.2.4 Nokia NFC & RFID SDK

Although Nokia were the first to launch the phones equipped with RFID/NFC readers and reference implementation for JSR-257 is available there 6131 NFC will be the first mobile phone on the market that support it. However, the Nokia NFC & RFID API can be used under commercial license to develop MIDlets that utilize Nokia NFC shells or express on RFID shells on the 5140 and 3320 phones. Nokia provides an SDK to develop and test the MIDlets. This SDK consists of the following components:

- *Nokia NFC & RFID API*: enables developers to create MIDlets able to connect to NFC or RFID shells.
- *LI Client API*: enables MIDlets to employ the LI Server, which is also used in Nokia's Field Force Solution. The LI Server is web based and not only supports reporting, but also manages users, mobile phones, tags, and events. It can communicate with a MIDlet using HTTPS over GPRS or via SMS.
- *Nokia NFC & RFID Cover Emulator*: enables development on a PC. It emulates tag touch and can not only emulate both NFC and RFID shells but can also be used to manage and store tags to provide a thorough test and development environment.

21.2.5 Using Nokia NFC & RFID API

Class `ContactlessConnection` sits on the top of this API and provides a connection to NFC or RFID shell and `InterfaceFeature` represents all the capabilities and functionalities of this underlying hardware. `Target` class defines physical target types supported by each shell.

To receive notifications about contactless targets a registration is required with a contactless listener which provides information about contactless events. Contactless events can also be registered with the MIDlet push registry which means that the applications can be auto launched once a certain contactless event occurs, e.g., scanning a tag, etc.

There are certain considerations that need to be taken into account when programming NFC/RFID applications. Because of the nature of the communication taking place it is always a good programming practice to have the listener registered with a separate thread so that the user interface stays responsive. If authentication of the data being read from the tag is to be done from a server, that information should be passed on to the network handler, which should operate under a separate thread. This multithreading is important when programming applications or games such as those discussed in Section 2.

Since all the information about the contactless targets is reported by the listener as events, it is of utmost importance that the application must handle all the events that may arise in. Further, our use case examples will show informative feedback must be given to the user, through visual alerts, vibration and sounds, relating to the contactless event taking place.

21.3 Applications of RFID/NFC on Mobile Phones

To ascertain user experience of RFID/NFC we implemented two different mixed reality games which are games that link the physical and digital worlds to create new experiences. The types of games will often incorporate knowledge of their physical location and landscape, and then provides players with the ability to interact with both real and virtual objects within the physical and digital worlds. These games can incorporate some, or all, of these elements using a rich variety of technologies and are thus an ideal platform for evaluating user experience. In the following sections we will briefly describe the games in questions and the subsequent user experience.

21.3.1 PAC-LAN

PAC-LAN is a version of the video game PACMAN in which human players play the game on a maze based around the Alexandra Park accommodation complex at Lancaster University [9]. One player who takes the role of the main PAC-LAN character collects game pills (using a Nokia 5140 mobile phone equipped with a Nokia Xpress-on™ RFID reader shell), which are in the form of yellow plastic discs fitted with stick-on RFID tags placed around the maze as shown in Figure 21.8. The discs are a direct physical manifestation of the virtual game pills on the mobile screen and are placed at the real location corresponding to the virtual maze.

Four other players take the role of the Ghosts who attempt to hunt down the PAC-LAN player. The game uses a Java 2 Platform Micro Edition (J2ME) application, running on the mobile phone is connected to a central server using a General Packet Radio Service (GPRS) connection. The server relays to the PAC-LAN character his/her current position along with position of all Ghosts based on the pills collected. The game pills are used by the Ghosts, not to gain points, but to obtain the PAC-LAN characters last known position and to reset their kill timer, which must be enabled to allow them to kill PAC-LAN. In this way the Ghosts

FIGURE 21.8
Montage of PAC-LAN Trials.

must regularly interact with the server, which is then able to relay their position to the PAC-LAN. PAC-LAN sees a display with his own position highlighted by a red square around his animated icon while the Ghosts see both a white square highlighting their animated icon and red flashing square around PAC-LAN. These character highlights were added after pretrials revealed players wanted a quicker way of identifying the most important information. The Ghosts can kill the PAC-LAN character by detecting him/her via an RFID tag fitted on their costume (as shown in Figure 21.9), assuming their kill timer has not run-out.

Once PAC-LAN is killed the game is over and the points for the game are calculated in the form of game pills collected and time taken to do so (Figure 21.10). When PAC-LAN eats one of the red power pills, indicated by all ghost icons turning white on the screen, he/she is then able to kill the Ghosts, and thus gain extra points, using the same RFID

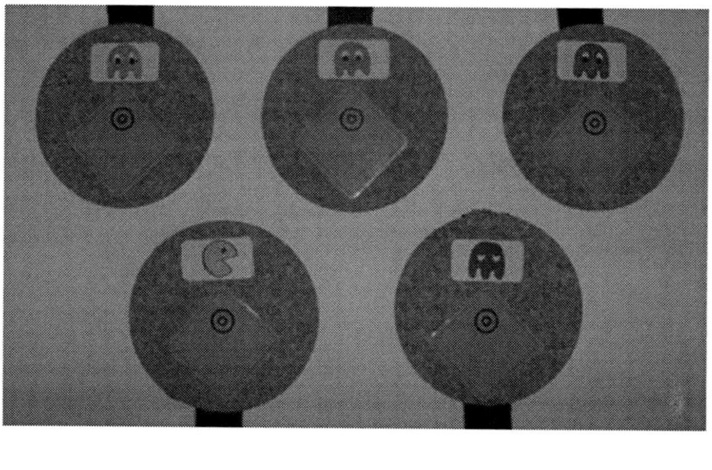

FIGURE 21.9
PAC-LAN kill tags.

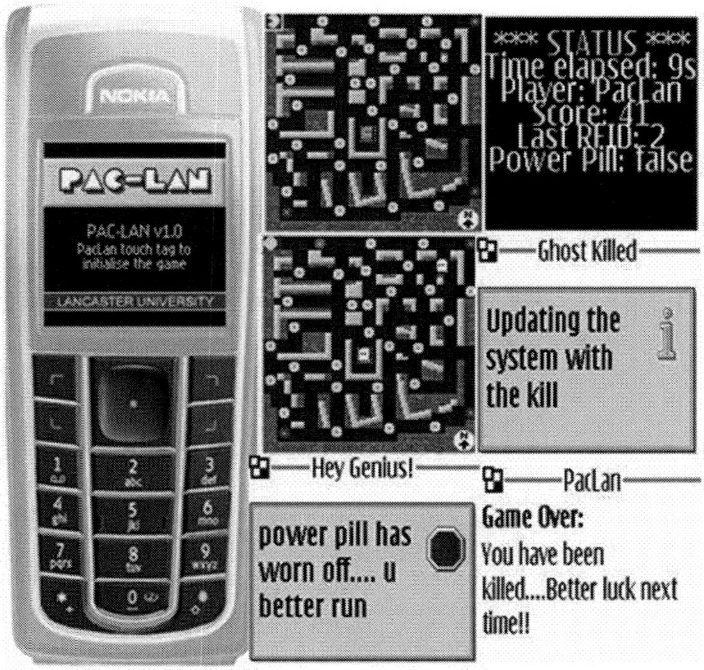

FIGURE 21.10
PAC-LAN Phone UI.

detection process. Dead Ghosts must return to the central point of the game maze where they can be reactivated into the game. Figure 21.4 shows a number of typical screens the PAC-LAN character will experience throughout the game.

The scoring in the game is simple where the PAC-LAN character gets, 50 points for a normal pill, 150 points for a power pill, 1000 points for collecting all the pills, and 500 points for a Ghost kill. The Ghosts get 30 points per pill (this is linked to the length of the kill timer) and 1000 points for killing PAC-LAN. All players lose 1 point per second to ensure they keep tagging.

21.3.2 MobHunt

In terms of operation Mobile Treasure Hunt (MobHunt) utilizes the same combination of mobile application connecting to a server over GPRS as for PAC-LAN. However, it differs from the previous system in that it was designed for flexibility so that new treasure hunts, in new or existing locations, can be created quickly and easily utilizing a simple mobile application and online Web site and utilized Nokia 5140i phones with NFC shells.

There are two mobile clients for MobHunt, one is the basic user client and the other is an administrator client which allows for new MobHunt to be created dynamically in a real world scenario.

The MobHunt user client is a very simple application allowing the user to login before instigating their authentication with the web server. Initially we implemented a standard name and password login solution but decided to simplify the process for the user by providing a unique membership card, as shown in Figure 21.11, that could be kept in a wallet or purse that the user simply touches to their phone to initiate login. This proved to be a much more elegant solution and considerably easier for the user given the restrictions of the mobile phone keypad as an interface [10].

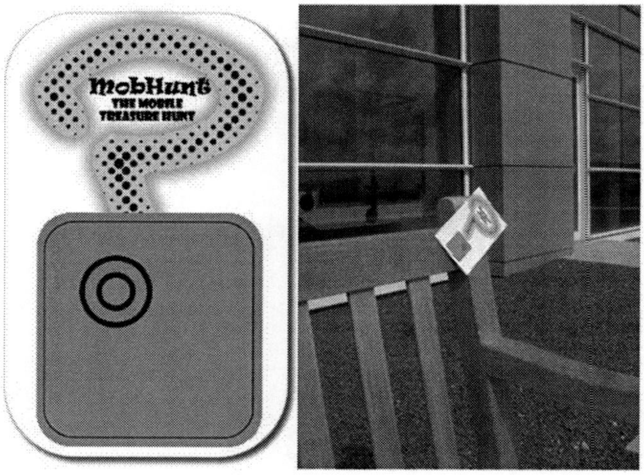

FIGURE 21.11
MobHunt Login/membership card and site marker.

Once the user is logged into the game they simply tag the first RFID tag of the treasure hunt to start the game. A visual place marker was used to make locations or objects used within a game easily visible to players as shown in Figure 21.8. The application also allows the users to browse through the clues as they are collected, this is because initial trials showed users often sought to reconfirm their earlier decisions if they got stuck at some point within the game.

The mobile administrator client allows the creation of MobHunts in the field by scanning tag and entering a description. This information can be subsequently be altered or amended on the web but we felt it would be beneficial to able to place some tags while surveying a site for likely objects or places to include in a game. Figure 21.12 provides some example screen shots for both the administrator and user mobile clients, respectively.

21.3.3 User Experience

The data for PAC-LAN was from eight games played by forty five different players (five per game) and was taken after their first experience of the game. The players were students

FIGURE 21.12
MobHunt User Interface.

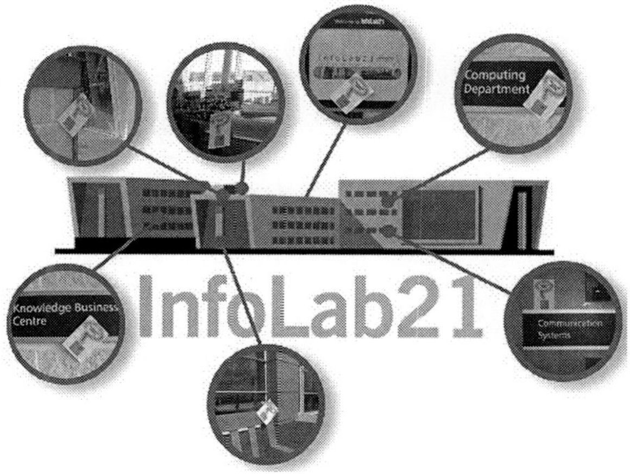

FIGURE 21.13
Infolab21 MobHunt.

from Lancaster University who answered an advert distributed via the University email system. The player groups were selected by the research team and care was taken to select groups from the various faculties across the campus to ensure which did not have a technophile biased sample. Of these forty five players, seven were female and the players were aged between 18 and 24 except six males aged between 25 and 35.

A version of MobHunt was created as tour around the Infolab21 building at Lancaster University and unlike the previous two games was designed to be played indoors rather than outside. The game itself was very simple and primarily provides a tour around the building and its facilities as shown in Figure 21.13 [2].

The game was played by 12 people within Infolab21 and was made up of both University staff and employees from companies within the Infolab21 incubator unit. The ages ranged between 22 and 44 and there were eight males and four women none of whom had any prior exposure to NFC phones.

In terms of learning to use the phone with the tags this proved to be remarkably easy and a very quick demonstration and explanation was all that players required. Interestingly unlike previous research [11] where users expressed concern about the social acceptability of touching tags in public places even the University campus where the trial took place, none of the users in our trials expressed this worry even though they were all conducted in environments open to the general public. When we discussed this further with some of the users they felt that in many ways creating questions in the mind of non players, who were not aware that users were playing a game, added to the sense of fun.

The users found the objects very useful compared with just placing an RFID tag at a location as they found it much easier to see and felt it added to the immersion within the game play.

One of the other aspects we experimented with was related to giving the user feedback after they have successfully read or written from or to a tag. For PAC-LAN we initially created version that had either visual feedback, through a pop-up note, or audio feedback, by playing a short tune. The audio feedback was unanimously preferred as players were often running at speed and the audio feedback was perceived much less intrusive on the game and harder to miss.

Overall the overwhelming user experience across all of the games was that the simple touch interface was very easy to use and as one player put it

"There is something intuitive about simply touching the object you want to connect with."

Further the vast majority of the players expressed a willingness to use and RFID/NFC enabled mobile phone to access other services which is encouraging for the proponents of this technology.

21.4 Conclusions

The use of mobile phones equipped with RFID/NFC not only creates a great method for mobile payments and data distribution but it also allows these experiences to be extended down to object level and they help create Mark Weiser's seminal vision of future technological ubiquity—one in which the increasing availability of processing power would be accompanied by its decreasing visibility, the so-called internet of things.

Further the traditional phone interface is generally too cumbersome for anything beyond dialing numbers and the simple act of touching an object to gain access to information and services is both simple and intuitive for the majority of users.

21.5 Acknowledgments

The authors acknowledge the support of Nokia for the provision of phones, software, and Nokia Forum Pro access to the Mobile Radicals Re-search Group at Lancaster University for the development of the projects featured in this chapter.

References

1. ECMA, Near Field Communication: White Paper, 2004. Ecma/TC32-TG19/2004/1.
2. Coulton, P., Rashid, O., and Bamford, W., Experiencing 'touch' in mobile mixed reality games, *Proceedings of The Fourth Annual International Conference in Computer Game Design and Technology*, 15–16 November 2006, Liverpool, pp. 68–75. ISBN 1-9025-6014-0.
3. NFC Forum, NFC Data Exchange Format (NDEF) Technical Specification NDEF 1.0, NFCForum-TS-NDEF_1.0, 24th July 2006.
4. NFC Forum, Text Record Type Definition Technical Specification RTD-Text 1.0, NFCForum-TS-RTD_Text_1.0, 24th July 2006.
5. NFC Forum, URI Record Type Definition Technical Specification RTD-URI 1.0, NFCForum-TS-RTD_URI_1.0, 24th July 2006.
6. NFC Forum, Smart Poster Record Type Definition Technical Specification SPR 1.1, NFCForum-SmartPoster_RTD_1.0, 24th July 2006.
7. JSR 257 Expert Group, Contactless Communication API JSR 257, Version 1.0, Final Release 17th October 2006.
8. Rashid, O., Mullins, I., Coulton, P., and Edwards, R., Extending cyberspace: Location based games using cellular phones, *ACM Computers in Entertainment*, 4(1), 1–18, 2006.
9. Rashid, O., Bamford, W., Coulton, P., Edwards, R., and Scheibel, J., PAC-LAN: Mixed reality gaming with RFID enabled mobile phones, *ACM Computers in Entertainment*, 4(4), 1–17, 2006.
10. Coulton, P., Rashid, O., Edwards, R., and Thompson, R., Creating entertainment applications for cellular phones, *ACM Computers in Entertainment*, 3(3), 1–12, 2005.
11. Riekki, J., Salminen, T., and Alakärppä, I., Requesting pervasive services by touching RFID tags, *IEEE Pervasive Computing*, 5(1), 40–46, 2006.

22

Applying RFID Techniques for the Next-Generation Automotive Services

Peter Harliman, Joon Goo Lee, Kyong Jin Jo, and Seon Wook Kim

CONTENTS

22.1 Introduction

RFID technology has been used and developed for more than 50 years, first by Harry Stockman in 1948 [1], and its applications have been commonly used around us. Instead of the term RFID, we are popularly using the term wireless applications in their names. Due to this reason, RFID remains unpopular despite its long history. Today most cars are equipped with a remote control to open and lock a door. In Korea, T-Money [2] cards are used for public transportation payments. Although there is no RFID term in their names, both a car remote control and T-Money are in the domain of RFID applications. RFID technology has become more and more widely used in real-world applications even without people realizing it.

TABLE 22.1

RFID and Barcode Comparison

System Parameters	Barcode	RFID
Data quantity (bytes)	1–100	16–64 k
Data density	Low	Very high
Machine readability	Good	Good
People readability	Limited	Impossible
Influence of dirt/dampness	Very high	No influence
Influence of covering	Total failure	Moderate
Data carrier cost	Very low	Medium
Reading electronic cost	Low	Medium
Unauthorized copying/modification	Slight	Almost impossible
Multiple reading	No	Yes
Reading speed	Low	Very fast
Maximum reading distance	50 cm	6 m

Source: From CAEN RFID, About RFID, http://www.caen.it/rfid/about_rfid.php.

RFID tags are often envisioned as a replacement for a barcode technology in the future. The reason to this is that RFID has many advantages over barcodes, such as a sight angle, a distance, a collision, a tag size, etc. [3]. More detailed comparisons between RFID and barcodes are shown in Table 22.1.

Despite its numerous technological advantages over barcodes, the development of RFID technology did not grow as fast as it was expected, especially before the twenty-first century because of a high cost, lack of standards, privacy issues, lack of applications, etc. [4]:

1. *Cost*: RFID readers and tags use more advanced technology than barcode scanners and labels do, and it incurs a cost problem. Table 22.2 compares the prices for both technologies.
2. *Lack of Standards*: Around 10–20 years ago, when RFID applications began to be used in real-world applications, most developers used their own versions of RFID systems. It resulted in lack of standards, which prevented the RFID markets from growing fast.
3. *Privacy Issues*: Privacy is one major issue in RFID deployment. In RFID communication, unlike that of barcode, there is a wide gap in space between a reader and a tag. Some people think that this wide gap can be very vulnerable to data intruders. In February 2005, a team of researchers at Johns Hopkins University Information Security Institute and RSA Laboratories demonstrated that they could capture data from Texas Instruments RFID systems [5]. This issue forced RFID developers to reconsider their design, so it could guarantee enough safety to consumers.

TABLE 22.2

Price Comparisons between RFID and Barcodes

	Barcode	RFID
Reader	U.S. \$100–300	U.S. \$700–2000
Passive Tag	0–3 cent	50–100 cent
Active Tag	—	U.S. \$10–100

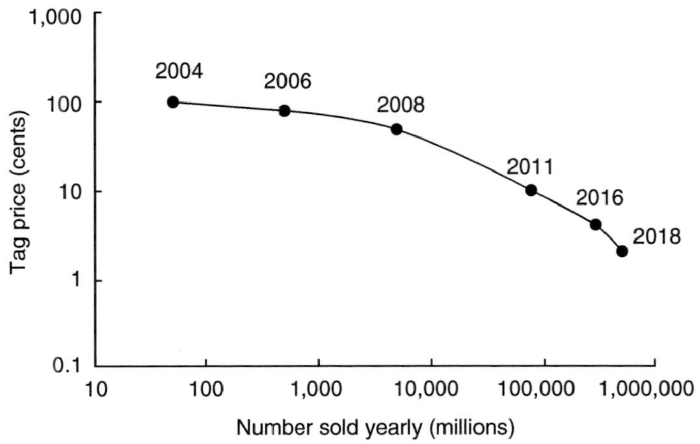

FIGURE 22.1
Tag's price prediction. (From Peter Harrop, The price-sensitivity curve for RFID, http://www.idtechex.com/products/en/articles/00000488.asp, 2006.)

4. *Lack of Applications*: Currently, there are only a limited number of RFID applications that are commercially available. This makes it very difficult for RFID to attract various types of consumers.

Some of those problems have been resolved during the twenty-first century. It has been predicted that the cost of RFID readers and tags will drop quite significantly in the coming years, due to development of semiconductor technology and mass productions, as shown in Figure 22.1. In 2006, one of the major protocols, EPC Gen2, has been already implemented by many commercial RFID reader and tag vendors, and it was adapted as ISO 18000-6 Type C standard [6]. Furthermore, current protocols, especially Gen2, are much safer than the previous ones, which had issues with illegal data capturing. For example, the protocol includes an option to kill a tag. After a tag is killed by a user, the tag will not respond to a reader anymore. This approach alleviates the concern about privacy.

We think that the slow growth problem of RFID comes from the initial concept to introduce the RFID technology as a barcode substitute. Due to this initial purpose, RFID developments always have been driven toward existing barcode applications. Barcode applications generally can be used only in a static environment, for example, by a cashier. In contrast, RFID technology can be used in a dynamic environment, because of its reading distance and reading speed. However, since RFID has so far only been used by the same static environment, we have failed to get the full advantage from RFID technology. Meanwhile, replacing barcode technology requires global effort to change many underlying systems. For this reason, most vendors are not willing to risk replacing a barcode system with RFID. One similar case with this RFID–barcode situation happened in PC–calculator transition around 30–40 years ago, as shown in Figure 22.2. Compared with calculators, PCs are much more powerful for calculations. But in order to be successful in a market, PC makers found many other applications like database, games, etc., in addition to calculations. If PCs were only used for calculation, there would be no chance for them to grow through the market. We should not focus only on barcode's existing applications. Instead, we need to create new applications that obtain maximum benefits from RFID abilities.

In this chapter, we propose innovative ideas for these new applications, with more emphasis on automotive applications. We briefly review several RFID services in Section 22.2 and the currently existing automotive services in Section 22.3. We introduce innovative ideas for the next-generation RFID automotive services in Section 22.4, and discuss an architecture for the new services in Section 22.5. In Section 22.6 we introduce our prototyped system, and finally conclude in Section 22.7.

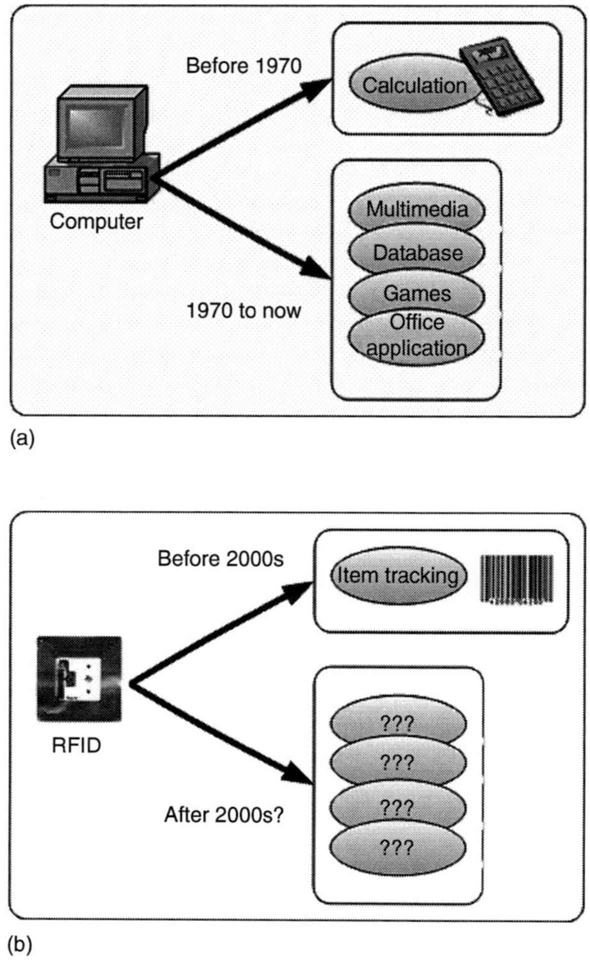

FIGURE 22.2
Analogy between PC-calculator and RFID-barcode cases. (a) PC-calculator. (b) RFID-barcode.

22.2 Background of RFID Services

Current RFID applications can be classified into several categories as follows:

1. *Transportation.* Most countries in the world have used RFID technology for a payment method in public transportations. Figure 22.3a shows T-Money, an RFID smart card that is used for paying transportation fares in Seoul, Korea.
2. *Manufacturing Control.* Some companies have applied RFID technology to track inventories inside their factory. In the application, an RFID system is used internally only by each company. This means that each company can use its own RFID standards without having concern about global standards. Hence, this application is easier to implement than other kinds of RFID applications. For this reason, a manufacturing control is one of the first successful RFID implementations in real-world applications.

(a)

(b)

FIGURE 22.3
Various RFID applications. (a) T-Money RFID card for transports payment. (From Korea Smart Card Co. Ltd., Seoul's New Transportation System, http://www.t-money.co.kr/jsp/newpub/oversea/english/stories/S_story.jsp) (b) RFID tag planted under the skin at Baja Beach Club. (From Robyn Curnow, The price to pay for VIP status, http://edition.cnn.com/2004/TECH/10/05/spark.bajabeach/, 2004.)

(continued)

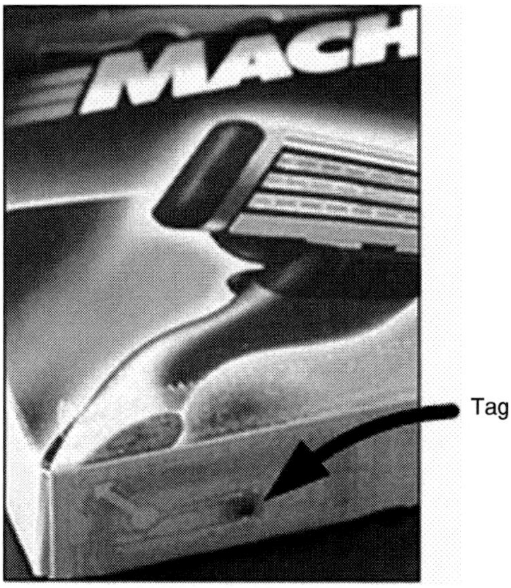

(c)

FIGURE 22.3 (continued)
(c) Gillette Mach 3 embedded with RFID tag in its case. (From Claudia H. Deutsch and Barnaby J. Feder, A radio chip in every consumer product, http://www.uazuay.edu.ec/bibliotecas/cibercultura/A%20Radio%20Chip%20in%20Every%20Consumer%20Product.htm, 2003.)

3. *People Management*. Some companies and educational institutes have added RFID technology to their systems for managing their members. For example, RFID technology is commonly used to manage employee attendance in a company. In 1998, Malaysia became the first country to issue biometric passports, which are passports with RFID tags embedded inside [7]. Since that time, more than 20 countries followed to apply RFID technology in their passports. One disadvantage in this category of application was that IDs can be manipulated by some people. For example, in the case of employee attendance, an employee can easily lend his/her ID card to other employees. For this reason, RFID applications in this category cannot rely solely on RFID. Instead, they combine more than one identification technology inside them. For example, a biometric passport still has its usual physical appearance and a barcode label, in order to preserve the validity of its ID. Another way to overcome this problem is to treat humans like any other item. Rather than embed the tag in a card, it can be inserted beneath the skin. Thus, the validity of an ID is guaranteed because the tag cannot be switched. Baja Beach Club discotheque in Barcelona, Spain, has already planted RFID tags inside its VIP customers [8], as shown in Figure 22.3b. In this application, the RFID tag under a customer's skin is used for an automatic payment method inside the discotheque.
4. *Automatic Payment Method in Stores*. At present, a barcode technology is used in many stores for identifying purchased items. By using RFID, people will not have to pay for them at a counter anymore. Instead, they just need to pass an RFID reader and the payment will be automatically charged online. Gillette moved one step ahead by adding RFID tags inside their Mach 3 Turbo razor blade packaging [9,10], as shown in Figure 22.3c.

22.3 Current Automotive Services

In Section 22.1, we explained that in order for RFID to grow faster, new applications are required. The automotive industry is one of the best development targets for an RFID application.

The automotive world has changed greatly from the way it used to be 10–20 years ago. Surprisingly, the most significant development is not happening in mechanical parts, but in electronic parts [11]. From Figure 22.4, it can be seen that the proportion of electronics cost inside a car has been increasing constantly, and will be increasing even more in the next 10 years. It has been estimated that it will rise to US$36.8 billion in 2005 and will reach US$52.1 billion by 2010 [12]. Some of these electronic parts are used for nonuser-oriented applications (e.g., power train and chassis), but most of them are used for user-oriented applications. For example, cars are often equipped with GPS services, sophisticated in-car multimedia systems and seating, various kinds of sensors, etc. In other words, there have been huge developments in automotive electronics toward user-oriented services.

There are various user-oriented services that can be found inside a modern car, as shown in Figure 22.5, and some are detailed as follows:

1. GPS (global positioning system), was developed by the U.S. Department of Defense, and came into use in 1993 [13]. Nowadays, GPS has been used around the world for many kinds of different applications. The most popular usage is in automotive navigation. Normally, such a GPS device has user-friendly controls and outputs and displays the car's current position on a map. Relying on this GPS navigation, other extended applications have also been developed, such as online road guide systems, emergency rescue systems, and car-tracking services.
2. DSRC (dedicated short range communication) is considered a subset of the RFID technology, since it also uses a radio frequency to communicate [14]. Different from GPS and RFID, DSRC was specifically designed for automotive communication. DSRC technology is known to be fast, stable, and cheap. The DSRC system basically consists of two components, RSE (roadside equipment) and OBE (on board equipment). The RSE communicates and provides useful information

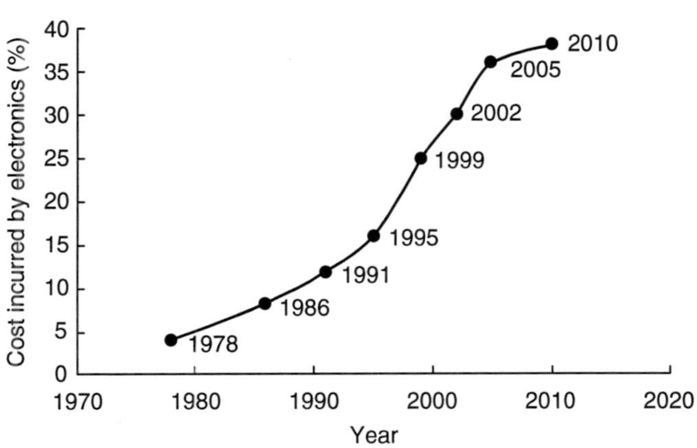

FIGURE 22.4
Proportion of electronics cost inside a car.

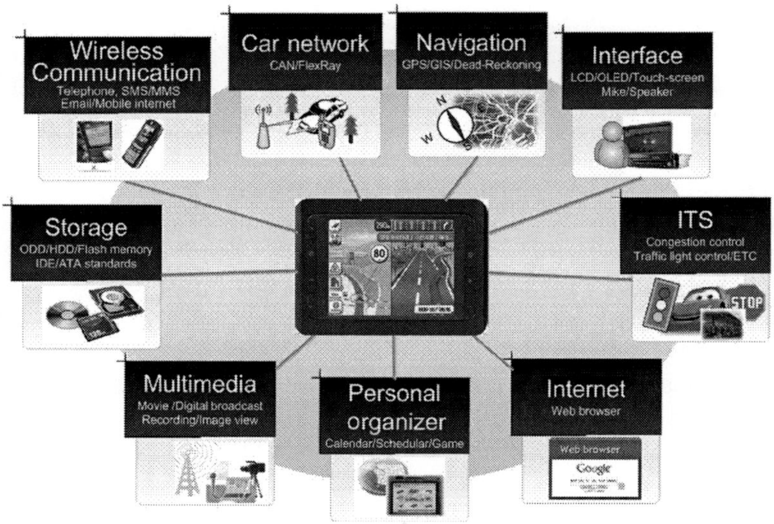

FIGURE 22.5
User-oriented applications in a modern telematics device.

to the OBE. European countries, Japan, Korea, and the United States use the DSRC system for electronic toll collection, which automatically charges a toll payment as a car passes the RSE near a toll gate. This way, the toll gate queuing delay could be avoided. Other applications include an intersection collision avoidance, an emergency vehicle warning system, electronic parking payments, and an in-vehicle signing. Currently, DSRC applications are still very limited compared with GPS applications. However, due to its good performance and its economic efficiency, DSRC is predicted to occupy a great role in modern and future telematic services.

3. In-car Internet access. People are spending more time in their cars than ever before. For this reason, future cars must provide an environment such that a driver can process his/her own job inside a car. In other words, engineers need to expand their concept of a car from a transportation method to a moving office. Some modern cars are already installed with a multimedia tool that can connect to the Internet. This access will lend itself to infotainment (information + entertainment) services. One example of the related services is a multimedia player that can play mp3 files, DVD, and DMB (digital multimedia broadcasting). In order to connect to the Internet in a moving vehicle, a car uses WIFI access points, cellular base stations, satellites, WiBro [15], or HSDPA [16]. Microsoft is developing a new foundry system, which is called Car.Net [17], which enables drivers to use the Internet in their vehicles. By using this system, drivers and passengers can check their e-mails, enjoy various kinds of entertainment, and search for stock market conditions simultaneously.

RFID technology could generate many new user-oriented automotive applications. Some possible automotive RFID applications are shown in Table 22.3, and the following section will discuss some of these ideas in detail.

TABLE 22.3
Key RFID In-Car Appliances

RFID Combined with	Provided Feature
Ignition control	Remote keyless ignition system
Car lock system	Remote keyless entry system
Seat and mirror setting	Automatic customizable seat and mirror setting
Speedometer, odometer, and ignition	Individual authority
Multimedia	Automatic profiled multimedia setting
Internet	Mobile advertising
Road information and multimedia	Advanced road information system

22.4 Innovative Ideas for Automotive RFID Applications

22.4.1 Individual Authority

A car key is one of the most essential components of a car. It is used as a main identification tool that allows a car user to turn on an engine (ignition), open/close a door, open a glove compartment, open a car trunk, open a fuel tank, etc. One limitation in a car key system comes from the fact that a car keyhole is unique. Each car keyhole can only recognize one key, and each key can only be used in one keyhole. If a car is used by more than one driver, the key may be duplicated to give a copy of the key to each user. However, the keyhole is not able to recognize who a current driver is. Consider RFID readers and tags instead of conventional keys and keyholes. An RFID reader can read all tags in its reading zone, and check the validity of the tag's ID. Thus, if we use an RFID tag as an identification tool for a car, we basically move the ID-checking process from a mechanical (like in a keyhole) to an electronic method.

In modern cars, keys have been replaced by remote keyless systems in many functions. More than 70% of currently manufactured cars come with an RKE (remote keyless entry) system, which allows doors to be locked and unlocked without using a key [18]. Meanwhile, some luxurious cars, such as Toyota Prius, Cadillac STS, and Audi A8, are also equipped with an RKI (remote keyless ignition) system, which allows car engines to be turned on without using a key [19]. Instead of using a key, the RKE system uses an electronic key fob. In order to open a door, a user just needs to press a button on the fob within 10–20 m of the car. That typical range is possible since the fob communicates with the transmitter inside the car by using an RF signal. In an RKI system, in case of an ignition, a user just needs to press one button inside the car. However, due to security, most current RKI systems require the presence of the fob inside the car. This technique will prevent an engine from being started by unauthorized users. The car uses RF communication to check whether the fob is in the car or not. The RF transceiver inside the car will broadcast a signal, and if the fob is present, it will reply to the transceiver. It can actually be seen that RKE and RKI systems can be considered as RFID applications, since they use RF communication to check an ID.

Apart from the fact that RKE and RKI systems have more benefits than conventional car keys, there are still some issues that need to be discussed in these systems. For example, current RKE and RKI systems can only work for a single ID. In the case of a car that is shared by multiple users, all possible users need to have the same ID inside their fobs. Hence, the car will not be able to distinguish who the current driver is, since they use the same ID.

TABLE 22.4

Database Example of Single ID without a User Recognition Ability

Name	ID
User 1	10180604
User 2	10180604
User 3	10180604

In addition, the scope of current RKE and RKI systems is very limited. They only focus on services for a door entrance and an ignition. In this application, we can say that RFID technology has not been used effectively. For this reason, in this part, we propose one idea to expand the RKE and RKI systems to a broader scope of applications.

Consider if an RFID reader is put inside a car, and some electronic fobs with RFID tags are developed. Each of the fobs will have its own unique ID. Each possible car user will be given one of these fobs. When a user takes the car, the RFID reader will communicate with the user's fob. As already mentioned, in current RKE and RKI systems, there is only one ID that is recognized as an authorized user. Even if the fobs are duplicated, the IDs for each fob are still the same, as shown in Table 22.4. In our proposed idea, each fob will have a different ID, as shown in Table 22.5. Using these different sets of IDs, the car can recognize an individual driver.

The next step is to use a user recognition ability of a car for creating new applications. Since now the car is able to recognize a current driver, we could connect this user recognition system with some hardware controllers inside the car. A simplified architecture for this idea is shown in Figure 22.6. The RFID reader is connected to a gateway. In this manner, the gateway is defined as an embedded computer that acts as a center of communication between the RFID reader and hardware controllers. This gateway is also the part that implements the user recognition system. It will have some amount of memory to hold a database for all users.

One possible application can be created by connecting a timer and an ignition controller to the gateway. This way, each user will have his own time limit when he drives a car. If the time limit expires, the driver will not be able to turn on the engine anymore. In this idea, the user recognition system will be expanded to keep settings for each user, as shown in Table 22.6. Some possible fields to be controlled are:

1. *Distance*. Each user has his own distance limitation. When a user drives a car more than the distance limit, the car will not be able to run anymore. In order to implement this application, the gateway is connected to a distance counter (an odometer, for example).

TABLE 22.5

Database Example of Multiple IDs with a User Recognition Ability

Name	ID
User 1	10180604
User 2	10216154
User 3	12301240

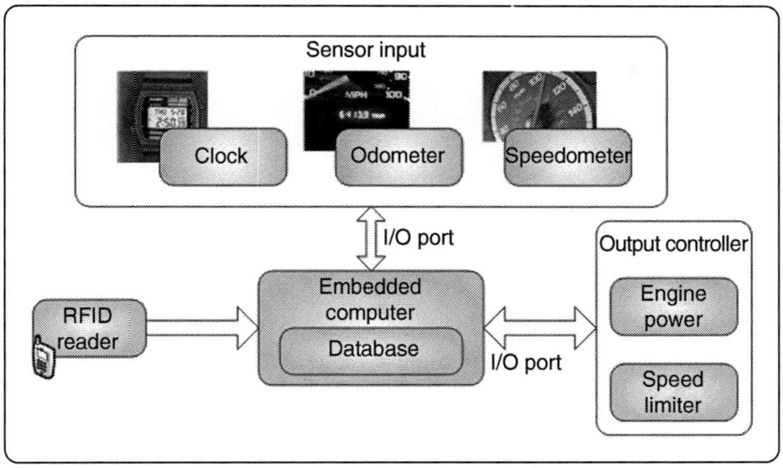

FIGURE 22.6
Individual authority architecture.

2. *Time*. Each driver has his own time limitation, beyond which the car will not be able to run any farther. In order to apply this application, the gateway is connected to a timer.
3. *Speed*. Each user has his own speed limitation. A user cannot drive the car faster than his designated speed limit.

These approaches will have different kinds of benefits. For private cars, a user can limit authorities of other users. Parents who share a car with their children can give a low authority access to them. This will restrict the children to limited boundaries. For company cars, these approaches are also useful since their cars are occasionally used by different drivers. In car rental industries, these approaches can be used to limit use of rented cars.

22.4.2 Customizing a Vehicle for Multiple Drivers

We now expand the database from Table 22.6 to control more items. Here, the idea is not focused on a user's authority as in the previous idea. Instead, it focuses on any item inside the car that can be set for a user's preferences. The architecture for this idea is shown in Figure 22.7.

One example of items to be controlled is car seats, which need to be set according to user's preferences. In conventional cars, the users need to set the seats manually. For a car that is shared by multiple users, each user will need to set the seats whenever there is a new driver or passenger. Some luxurious cars have already used a memory management

TABLE 22.6
Database Example of Multiple IDs with a Customizable Authority

Name	ID	Speed Limit	Duration Limit	Distance Limit
User 1	10180604	15	60	120
User 2	10216154	—	—	50
User 3	12301240	50	30	—

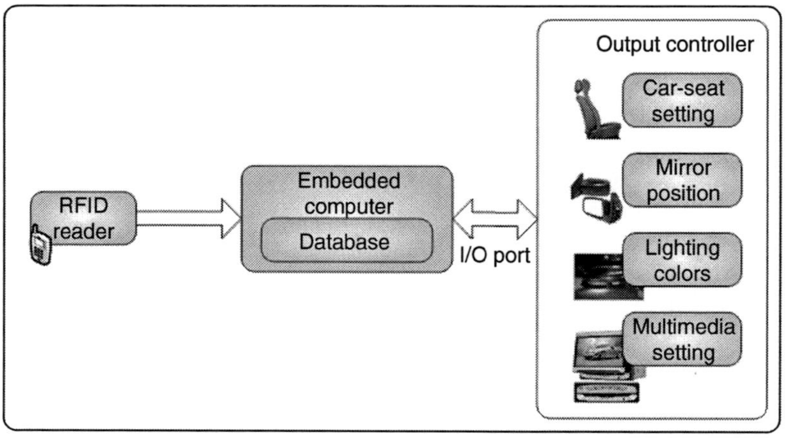

FIGURE 22.7
Customizable setting architecture.

for the car setting. For example, in the Saab 9–5 [20], seat controls have memory settings for three different drivers, including side-view mirror settings. However, they still need to choose among those three settings by pressing some buttons. In our idea, the users do not need to do any manual setting. Similar to the previous idea, the RFID reader will automatically recognize the user from his tag. The gateway then automatically sets the seat according to this user's preference. Other items that might be controlled by using this technique are:

1. *Mirror.* Car mirror settings are very critical for safety. Since they are dependent on driver's height, different drivers will usually require different settings. By using our RFID concept, the settings change automatically when a new driver takes a car.
2. *Multimedia Player Setting.* Drivers will likely want different radio channels and CD player settings. Furthermore, if the car is equipped with in-car Internet access, they also need different settings for Internet (web page, e-mail, etc). With RFID, all settings will be changed automatically by the gateway.
3. *Future Applications.* One of the biggest advantages of using RFID is that it triggers development of new applications. For instance, perfume vendors could develop some kinds of car perfumes that have different aromas, depending on the user's preference. If there is a new user, the perfume will automatically switch its aroma based on new user's preference. It is also possible to develop complex in-car lighting. Teenagers can choose varieties of colors for lighting, while general users can choose conventional lighting.

22.4.3 Automatic Vehicle Management

The condition of car tires needs to be monitored and replaced periodically. Goodyear has embedded RFID tags in their tires for NASCAR race use [21]. Goodyear is the exclusive provider of race car tires for all NASCAR events, supplying about 200,000 tires to racers annually. Normally, participants have to buy their own tires. However, it is not uncommon for a team to use multiple sets of tires in a single race. Hence, the cost for tires could not be affordable for many potential racers. NASCAR came to Goodyear seeking a leasing alternative for drivers who are unable to afford the cost of buying tires for their

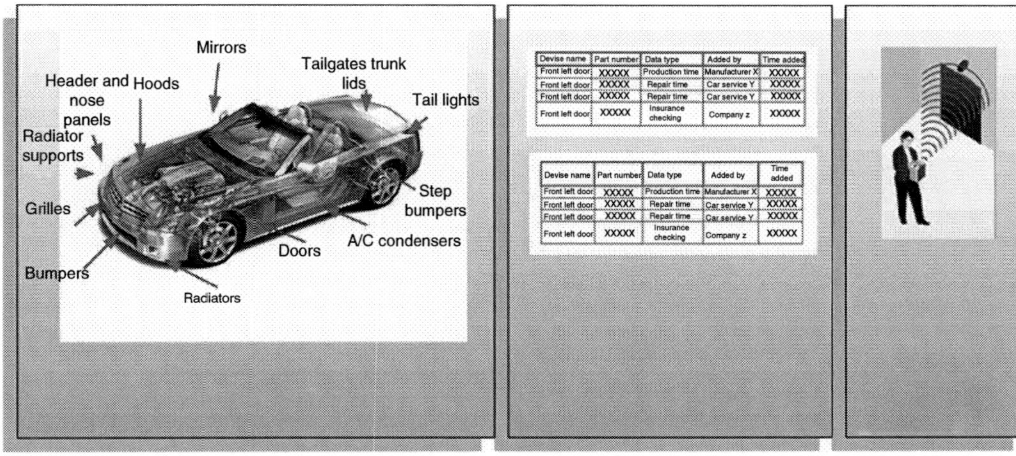

(a) (b) (c)

FIGURE 22.8
Tags are attached to various car parts for an automatic vehicle management. (a) RFID tags attached to various car parts. (b) Information such as historical time data is added to each tag's memory. (c) Allowed person can write/read the info from the tags for their purposes.

vehicles. Goodyear then came with an idea to use RFID tags for managing the leasing of tires. Michelin went further than Goodyear by embedding tags in car tires for public use and more general purposes [22]. We basically extend the RFID tire service from Goodyear and Michelin by attaching tags to other parts of a car. Considering the fact that passive RFID tags costs are low, they can be attached to hundreds of car parts, as illustrated in Figure 22.8. Each tag should hold the part's information. An RFID reader can read information from a tag in milliseconds. Hence, the reader will have no difficulty in periodically reading information from all tags.

Conceptually, we exploit the fact that an RFID tag has some amount of user memory that can be written by an RFID reader. Although the memory size is not big, typically less than 1024 bit, [23], it is enough to store a limited amount of information. One type of information that can be stored in tags is a history timetable for the car engine oil, which needs to be replaced periodically. Assuming that there is a tag attached to the car's oil tank, this tag can hold the projected time for oil replacement. An RFID reader, controlled by a gateway inside the car, will read this information periodically and inform the user when oil needs to be replaced.

In addition, tags will be connected with various kinds of sensors for car management. Related applications include:

1. *Identification.* Tags are used for identification purposes, similar to conventional item tracking.
2. *Operating Conditions.* Sensor devices are attached to tires. For example, a pressure detector is attached to each tire in order to detect when the tires' pressure exceeds a limit. The tag is connected with this sensor device, so that the gateway can read the pressure information through RF signals.
3. *Vehicle Performance.* A sensor for detecting road condition is attached to each tire. The tag connected with this sensor device sends the information to the gateway. The gateway will adjust car's performance according to that road condition.

22.4.4 Advertising with RFID Tags

Many people have said that a multimedia system will be an essential component in cars of the future [24]. For this reason, we need to find new ideas that apply to the future automotive media system, which is predicted to be one of the best places for advertisement. Hence, we present an idea to use RFID technology for advertisement inside cars.

Eric Schmidt, Google's CEO, believes that when he is listening to a radio in his car, an advertisement should personally address his needs [25]. For example, if a person drives by a clothing store, a radio advertisement should remind him that he needs a pair of pants and instruct him to turn left at the upcoming clothing store. For this reason, Google has planned to target on GPS-based in-car personalized advertising in the near future. In October 2006, Viacom Outdoor in London added GPS technology to a number of public buses [26,27]. Each bus has a large digital LED advertising panel, as shown in Figure 22.9. Different from regular LED advertising, here the advertisement depends on the current location of the bus. This means we have a mobile advertising billboard. Furthermore, this kind of a billboard is customizable, which means that we can change the advertisement easily with little time and cost. In both of these cases, GPS is used to track the location of buses, since it is the only currently available technology to track the vehicle position.

The RFID system can be also used to track a vehicle's position. We put some tags at different roadside locations, and an RFID reader inside a bus reads the tag while in transit. Using this approach, we would be able to run a similar advertisement as with GPS. The difference from the GPS approach is that the location detection is done by putting tags in correct positions.

Advertising media can be implemented using in-car multimedia systems. It could be a monitor or a speaker, depending on the output form. Although some people might be willing to listen to any advertising channel, the best way to put an advertisement is by integrating it into some kind of entertainment. For example, the advertisement could be

FIGURE 22.9
Bus with GPS advertising. (From Dody Tsiantar, Getting on board, http://www.time.com/time/magazine/article/0,9171,901060424-1184037,00.html, 2006.)

integrated into some radio channels, so that it will play just like ordinary radio advertising. The only difference is that the advertisement will be different for each listener, depending on which location he currently is. Another method is through an Internet browser. In this method, an advertisement will pop-up on a web page automatically when someone browses the Internet inside a car. The types of advertisement depend on which location the car currently is.

In addition to that, we can also implement a user filtering method for advertisement. In the current advertising method, a person will receive various kinds of advertisement. The problem is that sometimes a person only wants to receive some specific advertisement fields. In this RFID advertisement idea, each advertisement is treated as a single unit of data. For this reason, a driver can select which advertisement fields he is willing to receive. When his car is passing an advertisement tag, an RFID reader inside the car will check whether the tag's advertisement field matches the user preferences. If it does not match, the advertisement data will not be transmitted to the car.

Meanwhile, in order to implement the mobile advertisement source idea that was mentioned before, a tag is attached to a car. This car will then become a mobile billboard. If another car is located near enough to the advertising car, an RFID reader inside this car will communicate with the tag to get the corresponding advertisement.

The architecture for RFID advertising is shown in Figure 22.10. After a tag's ID is read by a reader inside a car, the ID will be sent to an advertisement database center using the Internet wireless access. Using that ID, the database center will search for a correct multimedia advertisement and then send it to the car by using Internet wireless access. Hence, in order to apply for this idea, we need a well-established in-car Internet infrastructure. Although this might sound unreasonable for the current time, in-car Internet infrastructures have been developed suddenly and recently. In the United States, in-car Internet can be seen in some rental cars, although so far it has not been used by many people [28]. Currently, this kind of service is only worthy for a business person who likes to run his business from the car. However, a recent survey shows that the number of car Internet users is increasing [29]. Furthermore, there are many options to get an Internet connection inside a car (GPRS, EDGE, CDMA, WIFI) [30]. For this reason, we believe that the well-established in-car Internet infrastructure, which is required for our idea, will be available in the very near future.

22.4.5 Advanced Road Information System

Some current telematics services have already provided a road information system. In this feature, while passing a road, a driver will be able to obtain information

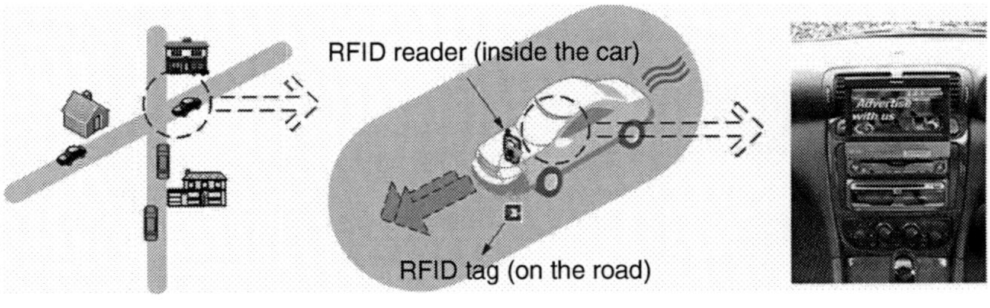

FIGURE 22.10
RFID advertising architecture.

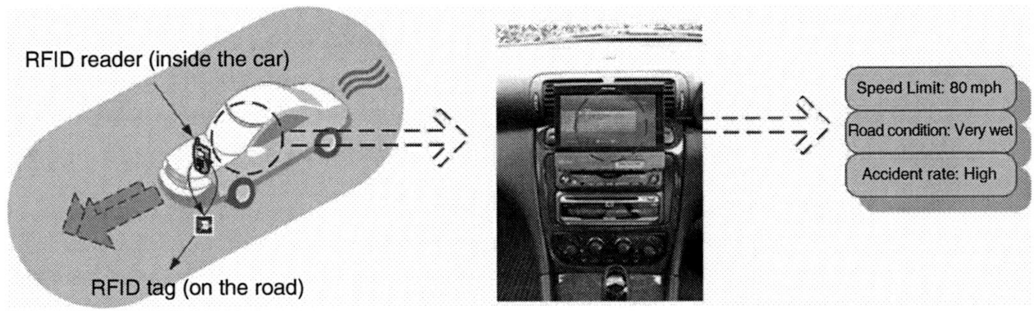

FIGURE 22.11
Advanced road information system architecture.

about the road, including speed limit and road conditions. One disadvantage of the currently used service is that in order to get the road information, the user needs to download and install it. If the road information is changing, the users will need to resynchronize the information. For this reason, a GPS-based road information system is not suitable for supplying such dynamic information as road conditions.

RFID can provide a better solution to implement this road information system. We name this service ARIS (advanced road information system) with its architecture shown in Figure 22.11. In the GPS approach, road information is read from a map data inside a GPS tool (located inside a car). In our approach, the road information will be read from a tag located on the road, assuming that all cars are installed with an RFID reader. In order to change road data, we change the tag's data. After the data has been updated, all cars that pass that tag can read the updated road information. Another option is to combine the existing GPS system with the RFID technology. Here, RFID tags are used as calibration tools to correct GPS errors. This technique will further improve the accuracy of GPS technology.

The main limitation for using RFID in this system is its limited data capacity. Compared with GPS, the data that can be held is relatively small (typically less than 1024 bits). However, this storage size is generally enough, considering two things. First, the amount of information to be sent to the car is small. Second, we could use some protocols to further compress data. Using this way, we could limit the required data to be transferred by focusing only on the information, and leave the decoding process to the reader inside the car. Furthermore, if in-car Internet service is available, a combination with Internet can be used to get more information data. Similarly to RFID advertising, in this idea, the car will read the tag as it runs at a relatively high speed.

22.5 Architecture for Automotive RFID Systems

In order to implement RFID systems with the currently existing automotive systems, the knowledge about the current car networking and telematics systems is needed. Fortunately, current car technologies have already provided flexible network infrastructures, such as CAN [31], ITS [32], and FlexRay [33]. For this reason, automotive RFID architecture can be easily attached onto the current existing car network systems. In addition, we can borrow some ideas from GPS systems, since GPS applications are built as separate add-on options for a vehicle.

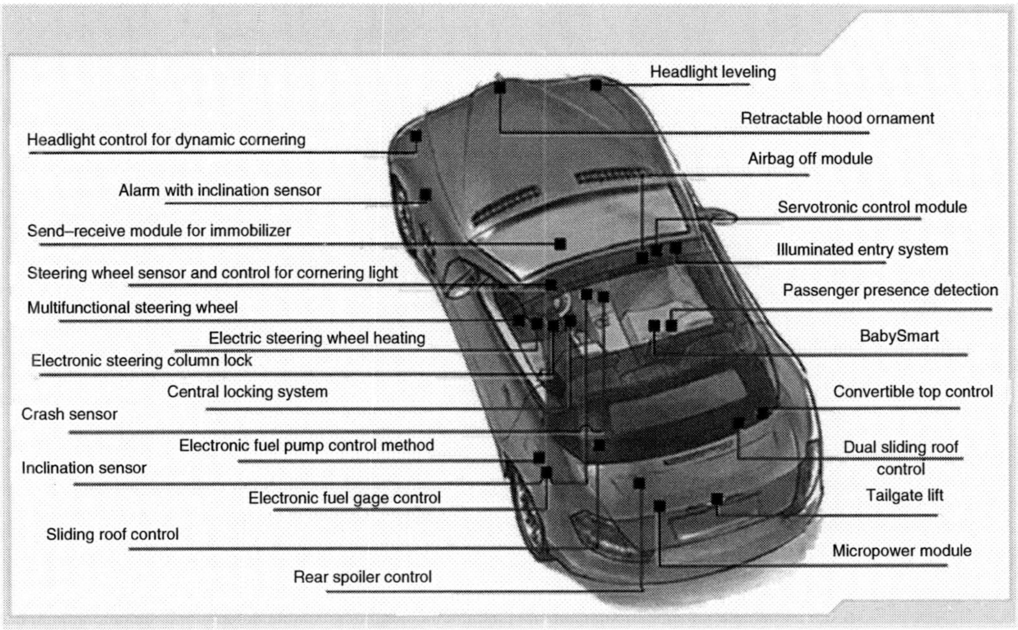

FIGURE 22.12
Electronics complexity inside current cars. (From Helbako, Details that make cars better, http://www.helbako.de/helbako/noflash/sprachen/eng/produkte.htm, 2006. With permission.)

22.5.1 Car Networking Systems

A currently available car contains thousands of circuits, sensors, and other electrical components, as shown in Figure 22.12. Around 30 years ago, communication among those components was handled by point-to-point wire connections. Adding more combinations inside the communication resulted in an enormous increment of wires. Adding more components will increase the complexity even more. This complexity increment would create damaging effects to the car itself, such as increasing its weight, weakening its performance, and reducing its reliability. For a normal car, every extra 50 kg of wiring will result in extra 100 W of power consumption; hence it will increase fuel consumption by 0.2 L for each 100 km [34]. During the 1980s, centralized and distributed networks began to replace this point-to-point communication method, which provided more simple implementation [35].

Controller area network (CAN), which was developed in the 1980s, has been used to connect vehicle sensors and safety systems. Unlike a conventional point-to-point communication, CAN uses a central gateway that controls all communications among car's devices. Hence, this centralized networking system reduces wiring size, weight, and cost compared with conventional point-to-point communications. Furthermore, CAN also provides a standard serial bus that can be used to connect various kinds of devices. This means any vendor can synchronize their devices so that they could be hooked into CAN. This approach eliminated the portability problem that occurred in point-to-point connections, and so increased the productivity of development. More than 100 million CAN nodes were sold in 2000, and currently it is the most widely used vehicle network [34].

Vehicle networks in CAN are commonly classified based on SAE (Society of Automotive Engineers) standards. SAE formally classifies vehicle networks based on their bit transfer rate, as shown in Table 22.7. This classification relies on the fact that each application will need a different requirement of data bandwidth. Using this classification,

TABLE 22.7

Classification of Automotive Networks

Class	Speed	Example Application
A	Less than 10 kbit/s Low speed	Convenience features, e.g., trunk release, electric mirror adjustment
B	10–125 kbit/s Medium speed	General information transfer, e.g., instruments, power window
C	125 kbit/s to 1 Mbit/s High speed	Real-time control, e.g., power train, vehicle dynamics
D	More than 1 Mbit/s	Multimedia applications, e.g., Internet, digital TV

an application that requires huge amount of data, such as multimedia applications, can use a bigger data bus in its communication with a gateway. Meanwhile, another application that does not require a huge amount of data can use a different data bus with smaller capability than the previous bus.

Although CAN includes many kinds of devices inside a car, ITS (intelligent transportation system) is focusing on factors that are visible to consumers. Examples of these factors are safety, transportation times, and fuel costs. In ITS, as in CAN, car networks are handled by a central gateway that controls all communications among various car devices.

FlexRay is one of the newest automotive network communication protocols. It is still under review for protocol specifications. According to its specification, FlexRay will give more benefits than CAN in terms of data rates (10 Mbps), redundancy, safety, fault tolerance, and price. Due to these reasons, FlexRay is predicted to replace other conventional automotive networking protocols in the near future. In 2006, the BMW X5 became the first vehicle to use FlexRay, but limited to its pneumatic damping system. The full use of FlexRay is expected to be accomplished in 2008. Other recent car networking systems include LIN (local interconnect network) [36] and MOST (media oriented system transport) [37].

22.5.2 Modern Telematics Devices

Recently developed modern telematics devices converge many technologies and services. Basically, a telematics device not only has one communication technology, but also numerous communication methods, such as cell-based wireless communications for telephone services, in-car networking for controlling electronics of a car, GPS for LBS (location based service), IP-based wireless communications for Internet services, DSRC for networking with other cars, and digital broadcast communications for a digital radio and TV. It also embeds a huge database for navigation, and has many large storages such as ODD (optical disk drive), HDD (hard disk drive), and flash memory for saving files. A display and a voice/audio system have to be equipped to interface with a user. The device uses TFT-LCD or OLED with various sizes and resolutions for display, voice recognition, and noise cancellation methods for detecting user's voice commands, and speakers with audio DSP or special ASIC for enhancing output sounds. Many brand new technologies are merged into one telematics device. Some major technologies and standards used in a telematics device are shown in Table 22.8.

Figure 22.13 shows that a telematics system includes numerous functional blocks and relevant modules. Typically, a host processor runs an operating system and executes many software applications for the technologies mentioned above.

22.5.3 Proposed Architecture for an Automotive RFID System

In order to provide that standard, the current architecture usually uses a layered system. The layer will act as a gate between two kinds of different networks. In a mobile RFID

TABLE 22.8

Technologies and Standards Used in a Recently Developed Telematics Device

Application	Technologies and Standards
Navigator	GPS (global positioning system), GIS (geographic information system), Dead-Reckoning, NMEA 0183, NMEA 2000 protocol, SiRF, ISO 19100 series, GIS DB, route planning, route guidance, map matching
Interfaces with a user	TFT-LCD, OLED, touch screen, mike, speaker, camera, keypad, vision/audio enhancement, GUI (graphic user interface), voice recognition, noise cancellation, speech synthesis, speaker adaptation
Wireless communication	Telephone, SMS/MMS, e-mail, mobile Internet, video conference, GSM/GPRS/EDGE, CDMA 1x/EV-DO, WCDMA
Internet-based services	Web browser, online market, wireless LAN (IEEE 802.11a/b/g), WiBro/WiMAX (IEEE 802.16), HSDPA
Storage	ODD, HDD, flash memory, IDE/ATA standards, file system, CD-DA, CD-ROM, DVD, DVD-R, DVD+R, DVD-RAM, DVD-RW, compact flash, multimedia card, Secure Digital, eXtreme Digital, memory stick, micro-SD (= T-Flash), mini-SD, secure micro-SD, MMC-micro, RSMMC (reduced size multimedia card)
Car network	CAN (controller area network), FlexRay, RIN, MOST, ISO 11898/11992/11783 series, SAE J1939 series, SAE J2411, SAE J2561
Multimedia	Movie/music play, digital broadcast, recording, image view, multimedia codec solutions (WMV, MPEG, Divx, QuickTime, WMA, MP3, AC3, JPEG), DSP (digital signal processing), DAB, T-DMB, S-DMB, digital radio, DVB-H, multi-tasking
Intelligent Transportation System	Congestion control, traffic light control, electronic toll, collection, emergency warning/rescue, road information providing, DSRC (dedicated short range communication), IEEE 1609 series, WAVE (IEEE 802.11p), LBS
Personal organizer	Calendar, scheduler, address book, word processor, spreadsheet, computer game, operating system, synchronization, software application

reader [38], the layer, which is called HAL (handset adaptation layer), acts as a gate between RFID readers and cell phones. By doing this, both RFID reader vendors and cell phone vendors can develop and produce their own part without having to depend on each other, which surely will accelerate the development for applications. The other advantage in a layered system is that a new system can be added easily.

Figure 22.14 shows our proposed architecture for an automotive RFID system, which consists of three layers: physical layer, protocol layer, and application layer. The physical layer handles communication between reader RF circuits and tags through radio frequency waves. The protocol layer controls the rule of how data transfer should be done, which component handles error checking and collision problems. Last, the application layer is the place where RFID applications should lie.

In the physical layer, more consideration should be put on performance such as the reading distance and the reading speed. The protocol layer is implemented inside a car central gateway. A local database, which holds a database for user IDs and a small amount of information, is also placed here. Next, the application layer should be implemented inside the gateway also. ONS (object naming service) of the EPCglobal Network, which holds a global database for all applications, is also placed in this layer. This ONS will be used to handle requested information from a car. For example, in the case of RFID advertising, ONS will receive the advertisement's ID from a car, find a corresponding advertisement data based on this ID, and then transmit the data back to the car. There is another option to implement the protocol layer separately by using an additional base-band processor, which will require more resources, but will improve the overall network performance.

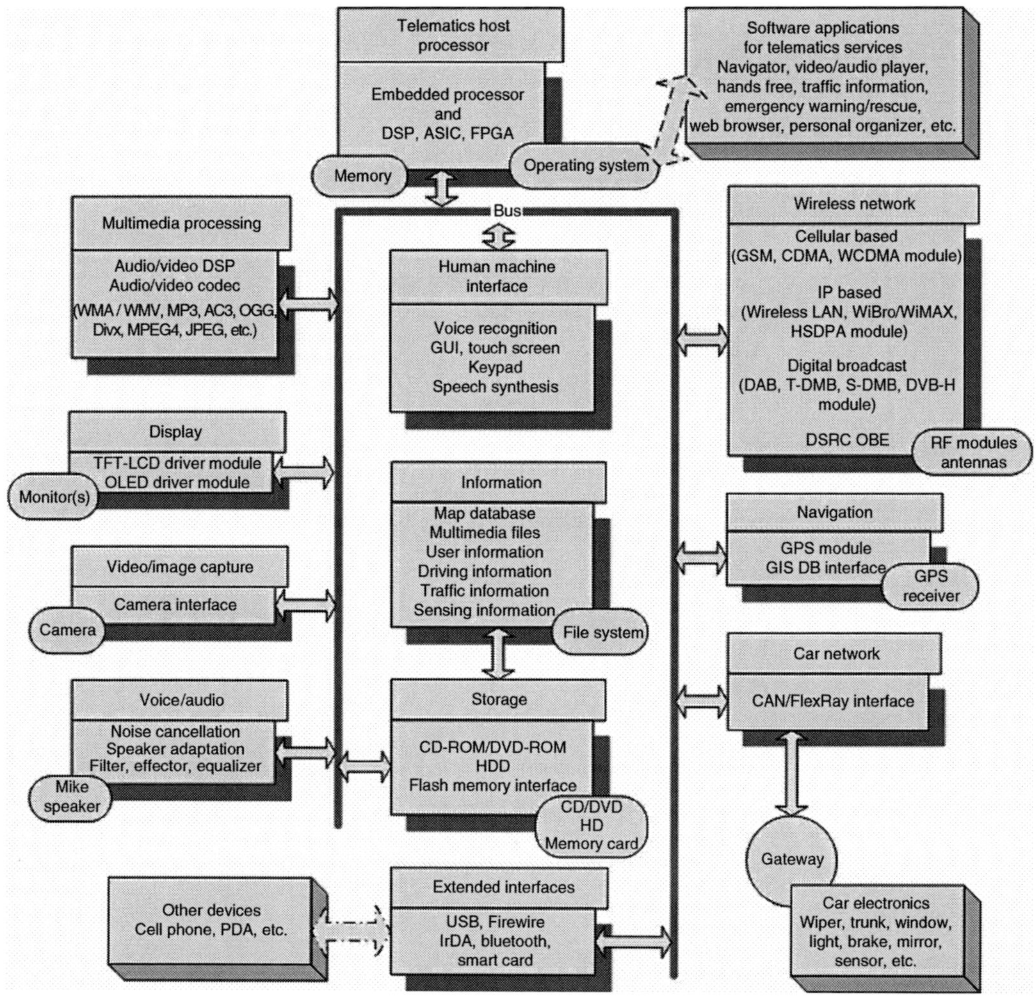

FIGURE 22.13
Typical architecture of a telematics system.

22.6 Experimental Design: TalusRFID

Most current RFID readers are implemented using a general processor. Some researchers have proposed new ideas for implementing an RFID baseband processor. One of examples can be seen in TalusRFID. In this approach, an RFID reader is designed using the Talus architecture [39], which uses Java as its programming language and FPGA technology as its target implementation.

Java has been evolved as one of the most preferable programming languages in the last decade. However, there has not been any RFID reader to be implemented using this language because Java execution speed is a lot slower than any other common programming language. The Talus architecture overcomes this slow-speed problem by using a coprocessor technique, which accelerates the execution on FPGA, hence making it able to implement an RFID baseband processor.

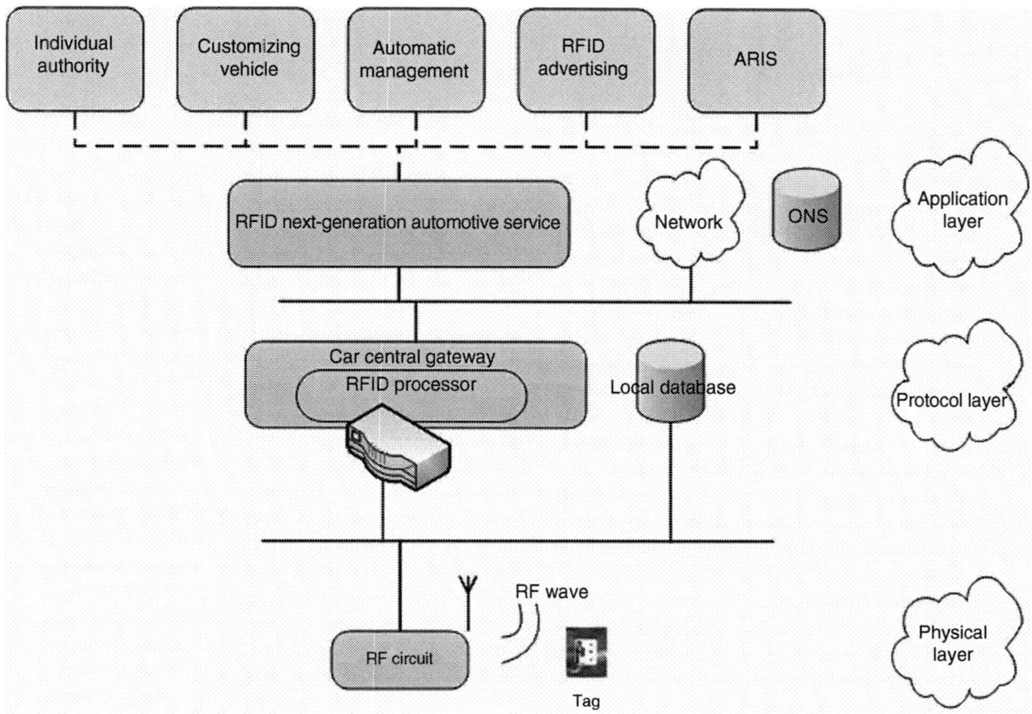

FIGURE 22.14
Proposed architecture for the next generation automotive RFID system.

The TalusRFID reader consists of three different parts:

1. RF circuit
2. Processor part
3. PDA, as the user terminal I/O

As shown in Figure 22.15, the TalusRFID reader can be seen as a prototype for the RFID reader that will be used for the next generation automotive RFID system. The physical layer discussed in Section 22.5.3 is the same as the physical layer in our proposed architecture. The protocol layer is implemented inside a baseband processor by the Java RFID software inside Talus architecture, and the application layer runs identification for users, implemented inside a PDA.

The details of the TalusRFID processor are shown in Figure 22.16, which supports ISO 18000-6 Type B and Type C (EPC Gen2) standards. The protocol layer was implemented in Java bytecodes, and lower layers, such as modulator, demodulator, and filters, are implemented in an FPGA reconfigurable hardware. TalusCore, which is a Java native processor, could only contain 3K instructions in its code memory due to resource limitation. For this reason, it will only execute manually selected time-critical methods while rest of the codes are executed on JVM (Java Virtual Machine), which in Talus is called TVM (Talus Virtual Machine). In order to provide the TVM environment, Embedded Linux is installed in the ARM processor. In order to use hardware parts for RF circuits directly, a software programmer can use some special methods for external hardware control. The Java application can control hardware by invoking them. In this system, hardware demodulator and transmission controller are implemented in hardware. The demodulation of received

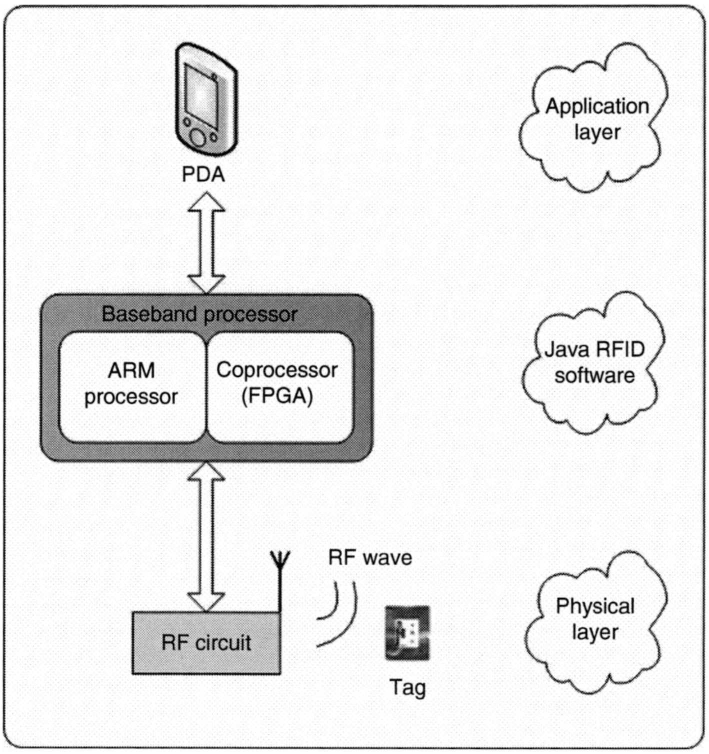

FIGURE 22.15
TalusRFID layer architecture.

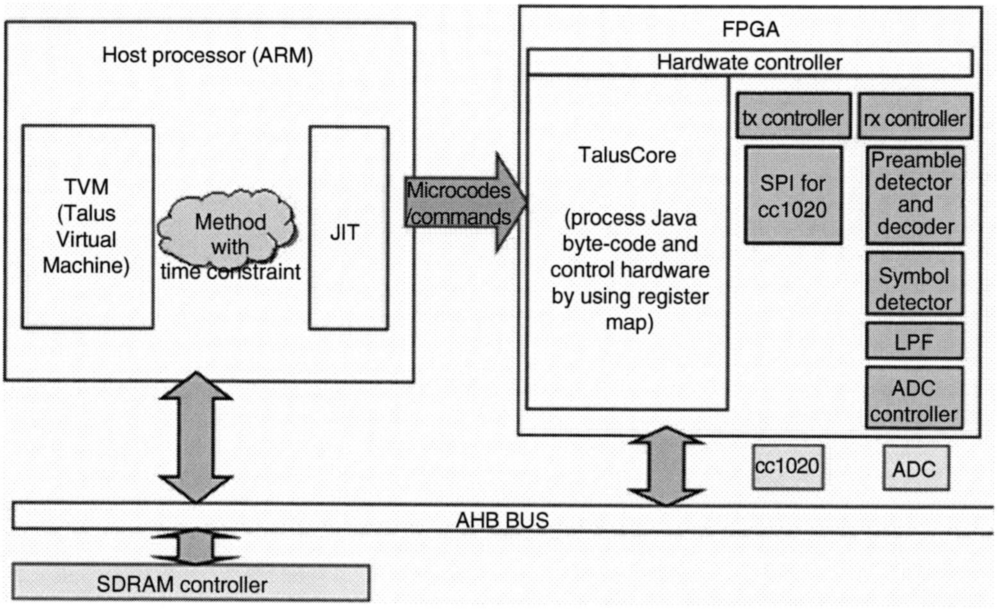

FIGURE 22.16
TalusRFID processor architecture. (From Joon Goo Lee, Peter Harliman, Kyongjin Jo, Sungjea Ko, Seon Wook Kim, and Kwangjoo Choi, TalusRFID: Java-based RFID baseband processor, *Proceedings of the 14th Korean Conference on Semiconductors, 2007.*)

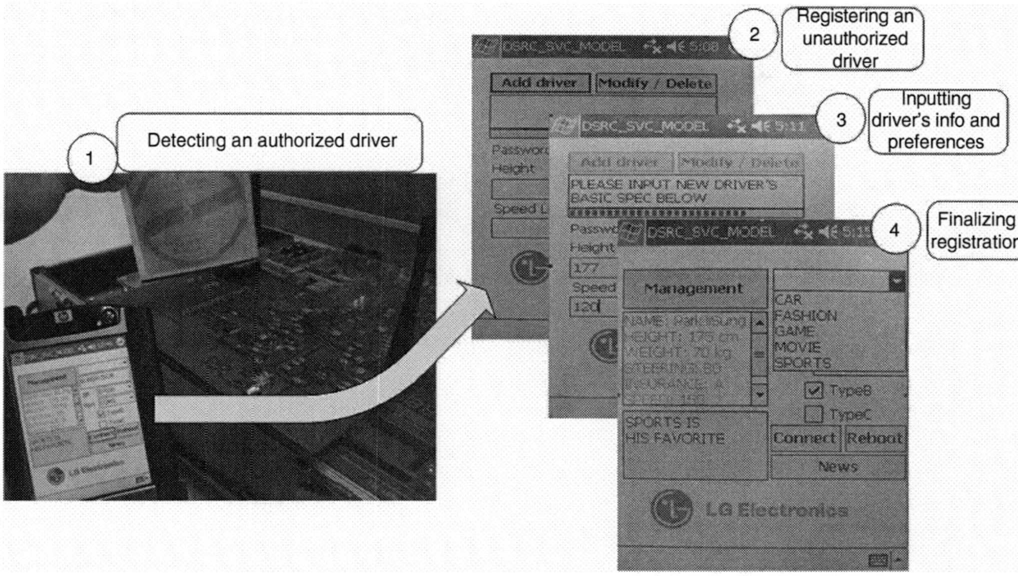

FIGURE 22.17
TalusRFID demo. (From Joon Goo Lee, Peter Harliman, Kyongjin Jo, Sungjea Ko, Seon Wook Kim, and Kwangjoo Choi, TalusRFID: Java-based RFID baseband processor, *Proceedings of the 14th Korean Conference on Semiconductors*, 2007.)

signals in a software manner was almost impossible to satisfy a time constraint for tags' identification in ISO 18000-6 Type C due to overheads of OS. The details of the Talus architecture have been given in detail by Hwang et al. and Lee et al. [39,40].

Figure 22.17 shows the TalusRFID demo with its PDA displays. The demo scenario was designed to illustrate our previous idea in Section 22.4.2 (customizing a vehicle for multiple drivers). In Step 1 of Figure 22.17, the reader detects an authorized tag, and automatically sets car settings based on this tag owner's preference. In Steps 2–4, a car owner wants to add an additional user to the system. In this case, a new tag ID is registered in the system, including the new user's information and preferences. After the registration step is finalized, the system will be able to recognize the new driver as an authorized user.

22.7 Conclusion

Since it was introduced, RFID has been expected to replace barcode technology. For this reason, RFID is always compared with barcode technology in all kinds of parameters. Furthermore, the development of RFID applications has always focused on existing barcode applications. These factors limit the creation of innovative and creative applications for RFID systems. There are unlimited possibilities to generate new ideas for RFID applications. However, it will be easier to develop new ideas that have a visible effect on users. In other words, we should focus on user-oriented applications.

In this chapter, we proposed new ideas for user-oriented RFID automotive applications. For example, RFID technology can be used to provide some limits to drivers, depending on time, distance, speed parameter, etc. In addition, drivers can use RFID technology to automate settings in various devices such as car seats and mirrors. Since the cost of an RFID tag is predicted to drop in the near future, RFID tags can also be attached to numerous car parts in order to relay information to a car management system.

Another possible innovation is to use RFID in advertising. Using RFID technology, a new technique for location-based advertising can be implemented inside a car. In ARIS, RFID is used to improve the accuracy of the existing GPS's road information system.

In order to implement those new applications, current car technologies have already provided flexible network infrastructures, such as CAN, ITS, and FlexRay. For this reason, automotive RFID architecture can be easily attached to the current existing car network systems. However, since automotive RFID applications combine many different types of applications, a fixed standard is required to accelerate the development of applications. With a fixed standard and networking protocol, different developers can focus on developing and producing their own parts without having to worry about compatibility with other parts. In this chapter, we have proposed an architecture for future automotive RFID applications. The architecture was designed so that only minimal modifications are required in integration with the current car networking architecture. Finally, we also discussed our prototyped system, called TalusRFID, which is an implementation of a Java-based RFID reader using the FPGA technology.

Acknowledgment

This work was supported in part by Digital Media Company, LG Electronics.

References

1. Harry Stockman. Communications by means of reflected power. *Proceedings of the Institute of Radio Engineers (IRE)*, 36, 1196–1204, 1948.
2. Korea Smart Card Co. Ltd. Seoul's new transportation system. http://www.t-money.co.kr/jsp/newpub/oversea/english/stories/S_story.jsp
3. Institution of Electrical Engineers (IEE). Radio frequency identification device technology (RFID). http://www.rfidc.com/pdfs_downloads/IEE%20RFID%20Paper.pdf, 2005.
4. John R. Tuttle. Traditional and emerging technologies and applications in the radio frequency identification (RFID) industry. *IEEE Radio Frequency Integrated Circuit Symposium*, 5–8, 1997.
5. Steve Bono, Matthew Green, Adam Stubblefield, Ari Juels, Avi Rubin, and Michael Szydlo. Security analysis of a cryptographically-enabled RFID device. *Proceedings of the 14th USENIX Security Symposium*, 1–16, 2005.
6. Mary Catherine O'Connor. Gen 2 EPC protocol approved as ISO 18000-6C. http://www.rfidjournal.com/article/articleview/2481/, 2006.
7. Iris Corporation Berhad. Malaysia is the first country to use computer chips in passport. http://www.iris.com.my/News/new_detail.asp?id=12, 1998.
8. Robyn Curnow. The price to pay for VIP status. http://edition.cnn.com/2004/TECH/10/05/spark.bajabeach/, 2004.
9. David M. Ewalt and Mary Hayes. Gillette razors get new edge: RFID tags. http://www.informationweek.com/story/IWK20030110S0028, 2003.
10. Claudia H. Deutsch and Barnaby J. Feder. A radio chip in every consumer product. http://www.uazuay.edu.ec/bibliotecas/cibercultura/A%20Radio%20Chip%20in%20Every%20Consumer%20Product.htm, 2003.
11. Klaus Grimm. Software technology in an automotive company: Major challenges. *Proceedings of the 25th International Conference on Software Engineering (ICSE)*, 498–503, 2003.

12. Ariz Scottsdale. Auto electronics market set to exceed US$50 billion by 2010. http://www.instat.com/press.asp?Sku=IN0603375RE&ID=1752, 2006.
13. United States Coast Guard's Navigation Center of Excellence. General information on GPS. http://www.navcen.uscg.gov/gps/default.htm
14. ITS Standards Program. Dedicated short range communications (DSRC). http://www.standards.its.dot.gov/Documents/advisories/dsrc_advisory.htm
15. Seung-Que Lee, Namhun Park, Choongho Cho, Hyongwoo Lee, and Seungwan Ryu. The wireless broadband (WiBro) system for broadband wireless internet services. *Vehicular Technology Magazine*, 44, 106–112, 2006.
16. Javier Gozalvez. Mobile radio—HSDPA goes commercial. *Vehicular Technology Magazine*, 1, 45–53, 2006.
17. Microsoft. Microsoft Car.NET connects motorists to the wireless internet. http://www.microsoft.com/Presspass/press/2000/oct00/carnetpr.mspx, 2000.
18. Maxim Integrated Products, Inc. Remote keyless entry systems overview. http://www.maxim-ic.com/appnotes.cfm/appnote_number/1774, 2002.
19. Warren Clarke. Will your next car have keyless start? http://www.edmunds.com/insideline/do/Features/articleId=106651, 2005.
20. Svenska Aeroplan Aktiebolaget (SAAB). Saab 9–5 features and specifications. http://www2.saabusa.com/95s/features.asp?start=home
21. Goodyear. Goodyear RFID technology—Key to managing your fast-paced world. http://eu.goodyear.com/home_en/sitewides/press/motorshow/rfid/index.jsp, 2006.
22. Michelin. Intelligent tires: Michelin outlines new technology at industry conference. http://www.michelinman.com/difference/releases/pressrelease03092005a.html, 2005.
23. Melanie Rieback, Bruno Crispo, and Andrew Tanenbaum. Is your cat infected with a computer virus? *Proceedings of the 4th IEEE International Conference on Pervasive Computing and Communications*, 169–179, 2006.
24. Rajiv Mehrotra. Telematics Might Steer Your Car into the Future. *IEEE MultiMedia*, 9(3), 9–10, 2002.
25. Donna Bogatin. Google targets GPS-based in-car personalized advertising. http://blogs.zdnet.com/micro-markets/?p=131, 2006.
26. CBS Outdoor. Viacom outdoor launched GPS advertising. http://www.cbsoutdoor.co.uk/web/Current-news/Newspage-UK/Viacom-Outdoor-launches-GPS-advertising-a-global-first-with-Yell.com.htm, 2006.
27. Dody Tsiantar. Getting on board. http://www.time.com/time/magazine/article/0,9171,901060424-1184037,00.html, 2006.
28. Christopher Elliott. Wi-Fi is hitting the road in cars from Avis, but technical and legal bumps lie ahead. http://www.nytimes.com/2007/01/02/technology/02avis.html, 2007.
29. Jeff Goldman. A look at the future of in-car internet access. http://www.wireless-weblog.com/50226711/a_look_at_the_future_of_incar_internet_access.php, 2006.
30. Thomas Nolte, Hans Hansson, and Lucia Lo Bello. Wireless automotive communications. *Proceedings of the 4th International Workshop on Real-Time Networks (RTN'05) in conjunction with the 17th Euromicro International Conference on Real-Time Systems (ECRTS'05)*, 35–38, 2005.
31. Karl Henrik Johansson, Martin Törngren, and Lars Nielsen. Vehicle application of controller area network. *Handbook of Networked and Embedded Control Systems*, pp. 741–766, 2005.
32. Fuqiang Liu and Fengzhong Li. Intelligent transportation system based on the next generation broadband wireless communication. *Proceedings of IEEE International Conference on Service Operations and Logistics, and Informatics*, 41–45, 2005.
33. FlexRay Consortium. FlexRay. http://www.flexray.com/
34. Gabriel Leen and Donal Heffernan. Expanding automotive electronic systems. *IEEE Computer*, 35, 88–93, 2002.
35. Gabriel Leen, Donal Heffernan, and Alan Dunne. Digital networks in the automotive vehicle. *IEE Computing and Control Engineering Journal*, 10, 257–266, 1999.
36. LIN Consortium. Local interconnect network. http://www.lin-subbus.org/
37. MOST Cooperation. Media oriented systems transport. http://www.mostcooperation.com/

38. Joon Goo Lee, Seok Joong Hwang, Seon Wook Kim, Sunshin Ahn, KyungHo Park, Ji Hoon Koo, and Woo Shik Kang. Software architecture for a multi-protocol RFID reader on mobile devices. *Proceedings of the 2nd International Conference on Embedded Software and Systems (ICESS'05)*, 81–88, 2005.
39. Seok Joong Hwang, Peter Harliman, and Seon Wook Kim. Talus: Compiler-assisted Java accelerator. *Proceedings of the 4th International Workshop on SoC and MPSoC Design*, 389–398, 2006.
40. Joon Goo Lee, Peter Harliman, Kyongjin Jo, Sungjea Ko, Seon Wook Kim, and Kwangjoo Choi. TalusRFID: Java-based RFID baseband processor. *Proceedings of the 14th Korean Conference on Semiconductors*, 187–188, 2007.

23

Application of RFID Technologies for Communication Robots

Masahiro Shiomi, Takayuki Kanda, Hiroshi Ishiguro, and Norihiro Hagita

CONTENTS

23.1 Introduction

The development of robots is entering a new stage, focusing on interaction with people in everyday environments. A robot can act as a peer providing mental, communication, and physical support. For example, pet-type robots such as AIBO (Fujita, 2001) and PARO (Shibata, 2004) have been developed to provide mental health care. Humanoid robots have also been developed for the research of human–robot interaction. These robots investigate how humanlike body movements or facial expressions contribute to interaction with people (Breazeal and Scassellati, 1999; Ishiguro et al., 2001; Kozima et al., 2004). In addition, robots have operated in such everyday environments as museums and an expo to investigate the effectiveness of communication robots in everyday environments (Burgard et al., 1998; Nourbakhsh et al., 1999; Siegwart et al., 2003).

For communication robots, person identification functions are very important because such functions enable them to interact with people using personal information such as name and age. For example, name-calling behavior will attract people to interact with a robot. Age information is useful to change the subject when a robot is talking with people; in fact, such behaviors are basic elements of human interaction.

RFID technologies enable communication robots to achieve person identification functions easily. In particular, an active-type RFID tag is very useful for interaction between a robot and people; detection is unaffected by the occurrence of occlusions, the detection area is wide, and the distance between the tag readers and the RFID tag can be roughly estimated. A robot can identify multiple people at the same time by using active-type RFID tags.

Several past works focused on how a system recognizes people and objects in an environment robustly. For example, Schulz and his colleagues proposed a method for estimating the positions of mobile devices such as laptops by using wireless signal strength (Letcher et al., 2005). However, we are focusing on how communication robots interact with people by using personal information from person identification functions. Personal information will enable communication robots to behave more intelligently. We believe that RFID technologies will largely promote the human–robot interaction.

In this chapter, we introduce applications of RFID technologies for communication robots through our two field trials with communication robots and active-type RFID tags at an elementary school and a science museum. In the elementary school, communication robots are used for studying English. Using RFID tags and sensors, these robots identified and interacted with children who came near them. The robots gestured and spoke English with the children, using a vocabulary of about 300 sentences for speaking and 50 words for recognition. Moreover, in the science museum, communication robots were used for providing information to multiple people; the robots explained and guided visitors to exhibits. For this purpose, we installed multiple RFID tag readers and communication robots at the science museum.

23.2 Communication Robots with RFID Tag Technology at an Elementary School

23.2.1 Introduction

In this field trial, two communication robots that had various communicative behaviors interacted with children at an elementary school. The purpose of the trial was to improve the children's ability to speak English through the robot playing with the children and communicating with them in English. We observed scenes of interaction between the robots and the children over the course of 2 weeks.

Our choice of a task for the robots was motivated by the generally poor English language ability of Japanese people. We believe a lack of motivation and opportunities to speak English are major causes of this deficiency. Many children in elementary and junior high school either lack motivation or do not recognize the importance and usefulness of English. In fact, children have no need to speak English in Japan. Even though English teachers speak English during class, children speak Japanese outside of class. In their daily lives, they almost never encounter foreigners who do not speak Japanese. Thus, many children are not motivated to study English.

For this reason, the interactive humanoid robot we developed could only recognize and speak English, and its voice sounded somewhat like that of a child. The robot's utterances were based on recordings of a native English speaker (a professional narrator). In addition, the robot used an RFID tag system to identify each child. With visual, auditory, tactile, and RFID tag information, the robot took the initiative in interacting with children. For example, it called a child's name and initiated interaction after detecting the child from his or her RFID tag.

23.2.1.1 Interactive Humanoid Robot "Robovie"

Figure 23.1 shows Robovie (Ishiguro et al., 2001), an interactive humanoid robot characterized by its humanlike physical expressions and its various sensors. The humanlike body consists of a head, a pair of eyes, and two arms. When combined, these parts can

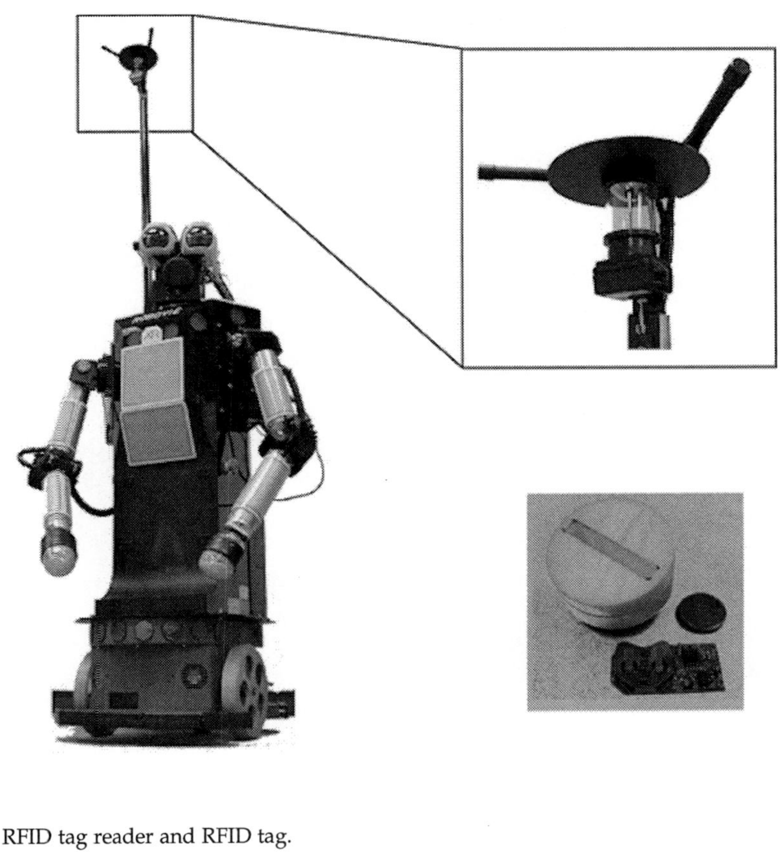

FIGURE 23.1
Robovie with RFID tag reader and RFID tag.

generate the complex body movements required for communication. We decided on a robot height of 120 cm to decrease the risk of scaring children. The diameter was 40 cm. The robot has two 4×2 degrees of freedom (DOFs) in its arms, 3 DOFs in its head, and a mobile platform. It can synthesize and produce a voice via a speaker. We also attached an RFID tag reader to Robovie that enables it to identify the individuals around it.

23.2.1.2 Person Identification

To identify individuals, we developed a multi-person identification system for communication robots using an RFID tag system (SPIDER-IIIA, RF-CODE). In this study, children were given easy-to-wear nameplates (5 cm in diameter) in which an RFID tag was embedded. The tag (shown in Figure 23.1, lower *right*) periodically transmitted its ID to the reader, which was onboard the robot (shown in Figure 23.1, upper *right*). In turn, the reader relayed received IDs to the robot's software system. The RFID tag system can adjust the reception range of the receiver's tag in real time from the robot's software. The RFID tag system provided the robots with a robust means of identifying many children simultaneously. Consequently, the robots could show some humanlike adaptation by recalling the history of interaction with a given person.

23.2.1.3 Interactive Behaviors

Robovie has a software mechanism for performing consistent interactive behaviors (Kanda et al., 2002). The intention behind the design of Robovie is that it should communicate at a young child's level. One hundred interactive behaviors have been developed. Seventy of them are interactive behaviors such as shaking hands (Figure 23.2a), hugging (Figure 23.2b), playing paper–scissors–rock (Figure 23.2c), exercising (Figure 23.2d), greeting, kissing, singing, briefly conversing, and pointing to an object in the surroundings. Twenty are idle behaviors such as scratching the head or folding the arms, and the remaining ten are moving-around behaviors. For the purpose of English education in this study, the situated module could only speak and recognize English. In total, the robot could utter more than 300 sentences and recognize about 50 words.

Several interactive behaviors depended on the person identification function. For example, there was an interactive behavior in which the robot called a child's name if that child was at a certain distance. This behavior was useful for encouraging the child to come and interact with the robot. Another interactive behavior was a body-part game; the robot asked a child to touch a part of the body by saying the part's name.

These interactive behaviors appeared in the following manner based on simple rules. The robot sometimes triggered the interaction with a child by saying "Let's play, touch me please" and it exhibited idling or moving-around behaviors until the child responded; once the child reacted, the robot continued performing friendly behaviors as long as the child responded to it. When the child stopped reacting, the robot stopped the friendly behaviors, said "good-bye," and restarted its idling or moving-around behaviors.

23.2.2 Experiment

We performed the field trial at an elementary school affiliated with Wakayama University. Two identical communication robots were put in the open corridor near the first- and sixth-grade classrooms, and for 2 weeks the two robots interacted with first-grade students and sixth-grade students. The following sections describe the details of the trial.

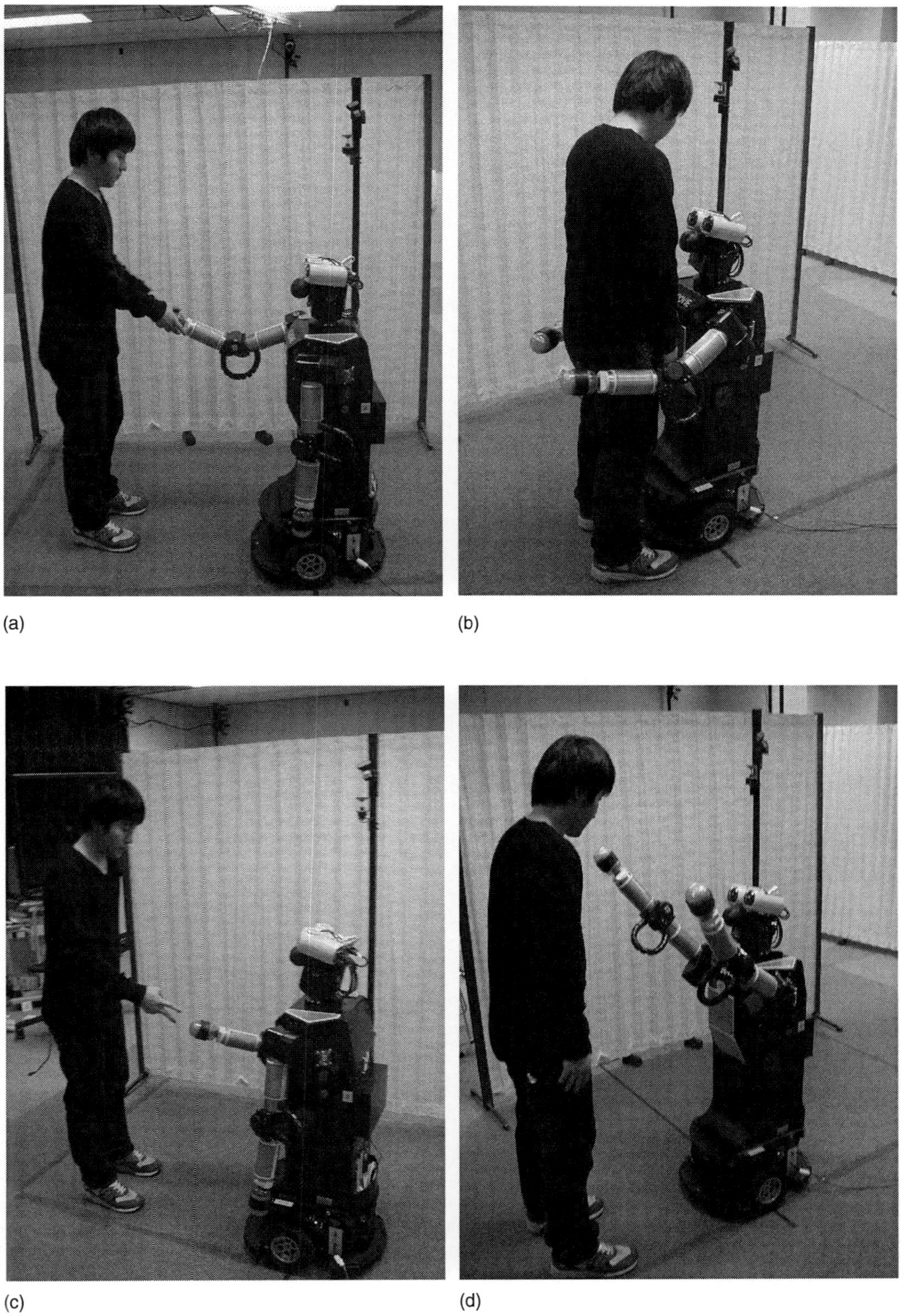

FIGURE 23.2
Scenes of interaction between Robovie and people.

23.2.2.1 Settings

We carried out two sessions, one for first graders and the other for sixth graders. In general, there are six grades in a Japanese elementary school. This particular elementary school has three classes for each grade and about 40 students in each class. There were 119 first-grade students (6–7 years old; 59 males and 60 females), and 109 sixth-grade students (11–12 years old; 53 males and 56 females).

In the three classrooms of the first grade, there are no walls between the classrooms and corridor, so that the corridor (which is called a workspace) is open to every first grader. The first graders' classrooms are located on the ground floor; the sixth graders' classrooms have the same layout as the first graders' and are located on the third floor.

We gave the children safety instructions before the trial. Pictures of the robot were accompanied by messages in Japanese such as "do not treat the robots roughly," and "do not touch the joints because it is not safe." We did not give the children any further instructions.

Both sessions (for first and sixth grades) were conducted for 2 weeks, which is equivalent to nine school days. The two robots were put in the corridor. The children could interact freely with both robots during recess. Every child had a nameplate with an embedded RFID tag (Figure 23.1, bottom *right*) so that the robots could identify the child during interactions.

We did not involve the teachers in the field trial. Two experimenters (university students) looked after the two robots. They did not help the children interact with the robots but simply ensured the safety of the children and robots. For example, when the children crowded closely around the robot, the experimenters would tell them to maintain a safe distance.

23.2.3 Results

23.2.3.1 Children Behaviors

We investigated the microaspects of children's behaviors. Table 23.1 indicates the rate at which children interacted with the robot along with their friends. (Children provided their friends' names on the questionnaire before the experiment, and they were compared with the ID information obtained through the wireless tag system.) Figure 23.3 shows scenes of interaction between a robot and children. In the first grade, 48% of time the children played with the robot along with their friends. In contrast, this was 78% in the sixth grade. There were 11 children who did not interact with the robot at all, who we omitted from the statistical analysis of friend time rate. ANOVA proved a significant difference between the first grade and sixth grade ($F(1,215) = 12.87$, $p < 0.01$). We believe that first-grade children came to the robot to communicate with it, whereas the sixth-grade children used the robot as a method of playing with their friends.

TABLE 23.1

Comparison of Friend-Related Behaviors

Grade	No. of Valid Subjects	No. of Non-Interacting Children	Interaction Time (min)			Friend Time (min)		Friend Time Rate	
			Average	Max.	S.D.	Average	S.D.	Average	S.D.
1	119	0	21.4	162.9	21.7	10.2	12.5	0.48	0.30
6	98	11	17.3	103.6	17.6	11.7	12.0	0.75	0.28

Friend time rate means average of each child's ratio of *Friend time* to *Interaction time*.

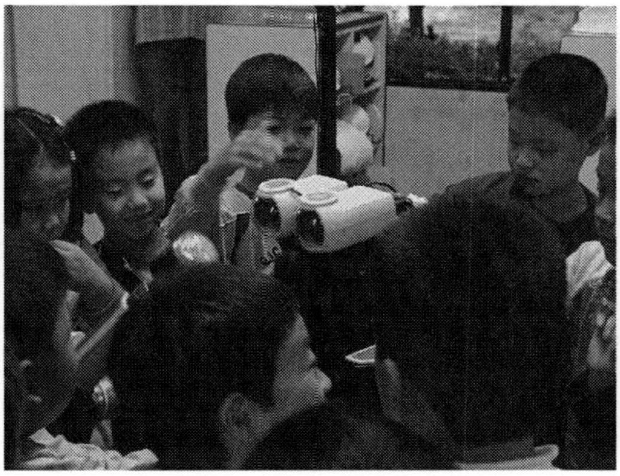

(a)

(b)

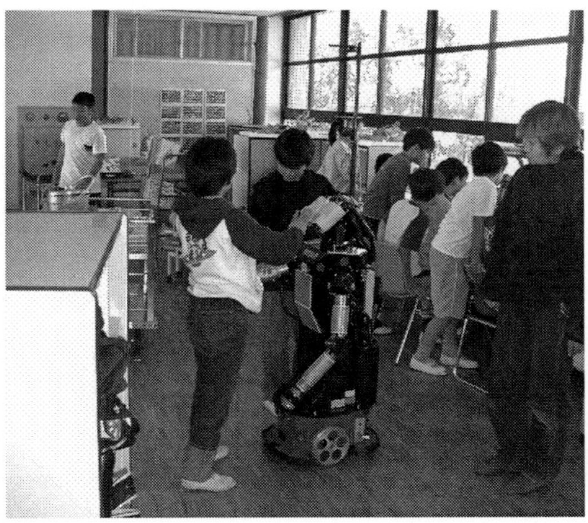

(c)

FIGURE 23.3
Scenes of interaction between Robovie and children. (a) Results of first-grade students. (b) Results of sixth-grade students. (c) Children are interacting with the robot.

By observing their interaction with the robots, we found several interesting cases.

- In Figure 23.3a, a child did not seem to understand English at all. However, once she heard her name said by the robot, she seemed very pleased and began to interact often with the robot.
- In Figure 23.3b, children counted how many times the robot called their respective names. Left child's name was called more often, so left child proudly told other child that the robot preferred left child.
- In Figure 23.3c, a child passed by the robot. He did not intend to play with the robot, but when he saw another child playing with the robot, he joined the interaction.

Those children's behaviors suggest that the robot's behavior of calling names significantly affected and attracted children. Furthermore, observation of successful interaction was related to the desire to participate in the interaction.

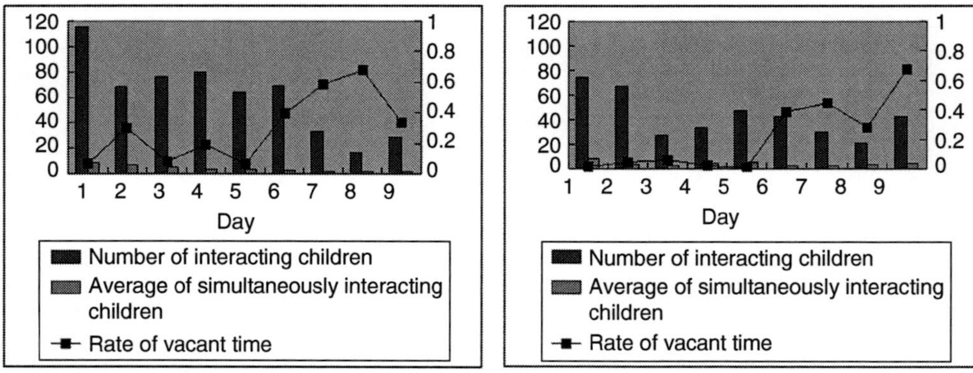

FIGURE 23.4
Transition in number of children playing with robots.

23.2.3.2 Long-Term Interaction

Figure 23.4 shows the changes in relationships among the children and the robots during the 2 weeks for the first-grade and sixth-grade classes. We can divide 2 weeks into the following three phases: (1) first day, (2) first week (except first day), and (3) second week.

1. *First day*: On the first day, as many as 37 first-grade children gathered around each robot (Figure 23.3). They pushed one another to gain a position in front of the robot, tried to touch the robot, and spoke to it in loud voices. Since the corridor and classrooms were filled with their loud voices, it was not always possible to understand what the robots and children said. It seemed that almost all of the children wanted to interact with the robots. There were many children watching the excitement around the robots, and they joined the interaction by switching places with the children around the robot. In total, 116 first-grade students interacted with the robot out of the 119 first-grade students on the first day. Moreover, 75 sixth-grade students interacted with the robot out of the 109 sixth-grade students on the first day.
2. *First week*: The excitement on the first day soon quieted down. The average number of simultaneously interacting children gradually decreased. In the first week, someone was always interacting with the robots, so the rate of vacant time was still quite low. The interaction between the children and the robots became more like inter-human conversation. Several children came in front of the robot, touched it, and watched the response of the robot.
3. *Second week*: At the beginning, the time of vacancy around the robots suddenly increased, and the number of children who played with the robots decreased. Near the end, there were no children around the robot during half of the daily experiment time. On average, there were two children simultaneously interacting with the robot during the second week. This seemed to be advantageous to the robot since it was easy for it to talk with a few of the children simultaneously. The way they played with the robots seemed similar to the play style in the first week. Thus, only the frequency of children playing with the robot decreased.

23.2.3.3 Learning English

In Table 23.2, we describe the main measurements used in this study, that is, the number of minutes each child interacted with the robot in the first and second weeks of the trial, their

TABLE 23.2

Descriptive Statistics and Correlations among Measures

Measures	*M*	S.D.	1	2	3	4	5	6
Interaction time at first week (min)	12.5	14	1	—	—	—	—	—
Interaction time at second week (min)	2.7	5.4	0.27	1	—	—	—	—
Interaction with friends (%)	67	—	−0.11	−0.02	1	—	—	—
Pretest English score	0.69	0.16	−0.02	−0.11	0.08	1	—	—
English score after first week	0.7	0.16	−0.12	−0.13	0.03	0.37	1	—
English score after second week	0.69	0.16	−0.04	0.1	−0.05	0.35	0.4	1

English scores on the pretest, first week test, and posttest, and the amount and percentage of time they interacted with the robot in the presence of friends. In Table 23.3, we show the interaction times with robots of the first and sixth grades during 2 week period.

The analyses we present are analyses of variance (ANOVA) in which the dependent variable is the improvement in each child's English test score from the child's English pretest score. Although many children did not know English at the beginning of the trial, some knew a bit. If they knew any of the phrases on the English test (such as "bye") their improvement might have been small owing to a ceiling effect. Therefore, the appropriate analysis of the effects of the robot on learning is the change from the pretest to the posttest, controlling for the initial pretest score. The main analyses we ran were standard least squares analyses, described as follows:

Model (second week English score − pretest English score) = Intercept + Pretest English score + Week 1 interaction minutes with robot + Week 2 interaction minutes with robot + Percent of interaction time with friends.

The results of this analysis are shown in Table 23.4. This analysis showed the expected significant ceiling effect of pretest English scores on the change in scores from pretest to posttest ($F(1,198) = 86$, $p < 0.001$). That is, the more English the children already knew at the beginning of the trial, the less they learned from the robot. However, the amount of time they interacted with friends and the robots together did not have an impact on the change in the English scores. The amount of time children spent with the robot during the first week also had no effect on their improvement in English by the second week, but the amount of time that children interacted with the robots during the second week did have a significant and positive impact on improvement in English in the second week ($F(1,198) = 5.6$, $p = 0.02$, $d = 0.33$).

One alternative explanation to the improvement in English scores at the end of 2 weeks is that causality was reversed. That is, perhaps those children who were more interested in English and knew more English at the start of the trial were more interested in interacting with the robot. To investigate the possibility that knowledge of English caused the children to interact with the robot more, we ran a regression analysis examining the impact of

TABLE 23.3

Interaction Time with Robots of First-Grade Students and Sixth-Grade Students in 2 Weeks

	Day 1	Day 2	Day 3	Day 4	Day 5	Day 6	Day 7	Day 8	Day 9
First grade ($n = 119$)									
M (min)	7.25	1.85	1.88	2.08	1.6	1.08	0.74	0.13	0.61
S.D.	7.36	3.57	3.14	4.9	3.77	3	2.43	0.51	2.35
Sixth grade ($n = 109$)									
M (min)	3.33	3.09	0.59	1.15	1.3	1.31	0.79	0.2	0.77
S.D.	5.15	5.94	2.01	2.87	2.74	2.64	2.48	0.88	1.37

TABLE 23.4

Analysis of Variance Results for Effect of Interaction Time with Robots on Improvement in English Scores on Posttest (after 2 Weeks)

Source	*df*	*F* Ratio	*p*
Pretest English score	1	85.8	<0.0001
Interaction with friends	1	1.5	ns
Interaction time during first week	1	1.4	ns
Interaction time during second week	1	5.6	0.019
Error	198	—	—

pretest English scores on the first and second weeks of time spent with the robot, controlling for the presence of friends. Pretest English scores did not predict the first week of time with the robot, but there was a marginal positive effect of pretest scores on time with the robot in the second week ($F(1,208) = 2.5$, $p = 0.11$). This suggests that part of the reason for the results shown in Table 23.3 might be the initial ability of some children to understand the robot's English and feel comfortable with it. They might have felt they had something in common with the robot (i.e., the English language).

23.2.4 Discussion and Conclusion

We believe that this field trial provided us with many useful insights that we can apply to the development of future communication robots. The communication robots autonomously interacted with children by using their humanlike bodies and various sensors such as visual, auditory, tactile sensors, and RFID tag system. They also had a mechanism to identify individuals and to adapt their interactive behaviors to them.

The results suggest that the robots did encourage some children to improve their English, and that the robots were more successful in engaging children who already knew at least a little English. These findings support arguments based on previous literature in social psychology on similarity and common ground; they suggest that robots should be designed to have attributes and knowledge in common with their users.

Our robots currently recognize only those who are around them. That is, even if the robot is faced with multiple parties, it does not distinguish the relationships among them. However, as the earlier example indicates, relationships among people might affect the interaction. For example, a child may take a friend to the robot, or someone may take part in the interaction because a friend is playing with the robot. Thus, we believe that a communication robot should also recognize relationships between children (friendship, hostility, etc.).

23.3 Communication Robots with RFID Tag Technology at a Science Museum

23.3.1 Introduction

In this field trial, four communication robots that had various communicative behaviors interacted with numerous people at a science museum. The purpose of the trial was to investigate the effectiveness of communication robots in everyday environments. We performed a 2 month exhibition to gather numerous people's opinions and observe scenes of interaction between communication robots and the people.

We chose to work in a science museum because we expect that many people, with a wide range of ages, would interact with the robots. In a science museum, visitors are going to have the opportunity to interact with the robots and experience the advanced technologies by which they are made, which is the fundamental purpose of a science museum. In addition, we are naturally targeting people who are interested in science and are unlikely to miss the chance to interact with our robots; thus, this field is one of the best choices for collecting feedback and examining the interaction between people and the communication robot in various tasks.

In this field trial, we used an RFID tag system to improve the robots' abilities as with the field trial the elementary school. Moreover, we installed not only RFID tag readers but also other kinds of sensors in the environment. We believe that embedded environment sensors enable communication robots to behave more intelligently.

23.3.2 System Configuration

We used four humanoid robots for interaction with visitors in guidance and giving explanations (the details of the robots are described in Section 23.3.3). In addition, we installed 20 RFID tag readers to record the movements and positions of visitors via their RFID tags on the fourth floor of the Osaka Science Museum. In addition, we installed three infrared cameras and four recording cameras; the details of the embedded sensors are described in Section 23.3.2.3.

23.3.2.1 Science Museum Environment

Seventy-five exhibits were positioned on the fourth floor of the Osaka Science Museum. Figure 23.5 shows a map of the fourth floor of the museum, around which people walk in a counterclockwise direction. Typically, visitors go through the following steps:

1. If a visitor decides to register as part of our project, such personal data as name is gathered at the reception desk (Figure 23.5, point A). The system binds that data to the ID of an RFID tag and automatically produces a synthetic voice for the visitor's name. The visitor receives a tag at the reception desk.

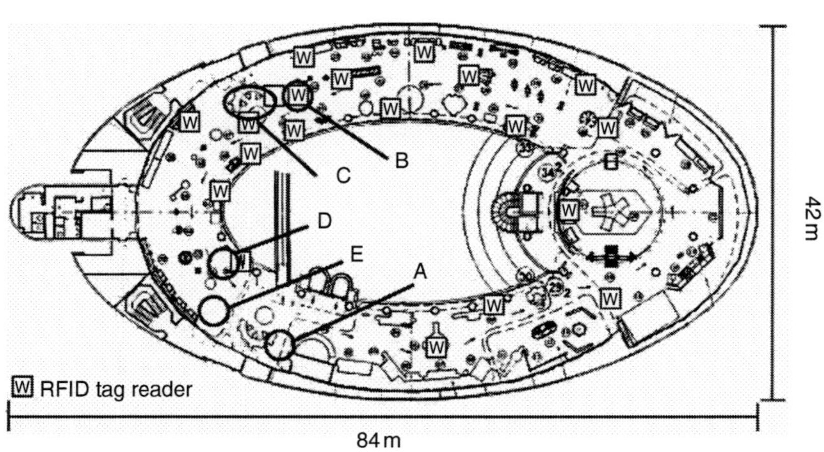

FIGURE 23.5
Map of fourth floor of Osaka Science Museum.

2. When the visitor strolls around the fourth floor wearing the RFID tag, the RFID tag readers detect its signal and the system records the information.
3. Four robots are placed at positions B, C, and D on the fourth floor, as shown in Figure 23.5. After finishing, visitors return their tags at the exit point (Figure 23.5, point E).

23.3.2.2 Humanoid Robots

23.3.2.2.1 Robovie

In this field trial, we used Robovie, as with the field trial at the elementary school. The difference from the past field trial is that we used the Robovies as sensors as well because they contain RFID tag readers. In effect, they became not only interactive robots but also part of the sensor system. Two of the four robots used in this experiment were Robovies.

23.3.2.2.2 Robovie-M

Figure 23.6 shows a "Robovie-M" humanoid robot characterized by its humanlike physical expressions. We decided on a height of 29 cm for this robot. Robovie-M has 22 DOFs and can perform two-legged locomotion, bow its head, and do a handstand. We used a personal computer and a pair of speakers to enable it to speak, since it was originally unequipped for that. The other two robots in this experiment were Robovie-Ms.

23.3.2.3 Embedded Sensors in Environment

On the fourth floor of the Osaka Science Museum, we installed 20 RFID tag readers (Spider-IIIA, RF-CODE), which included the two equipped on the Robovies, three infrared sensors, and four video cameras. All sensor data were sent to a central server database through an Ethernet network. In the following sections, we describe each type of sensor used.

23.3.2.3.1 RFID Tag Readers

We used an active type of RFID tag; it is the same hardware device as in the field trial at the elementary school. This technology enables easy identification of individuals: detection is unaffected by the occurrence of occlusions, the detection area is wide, and the distance

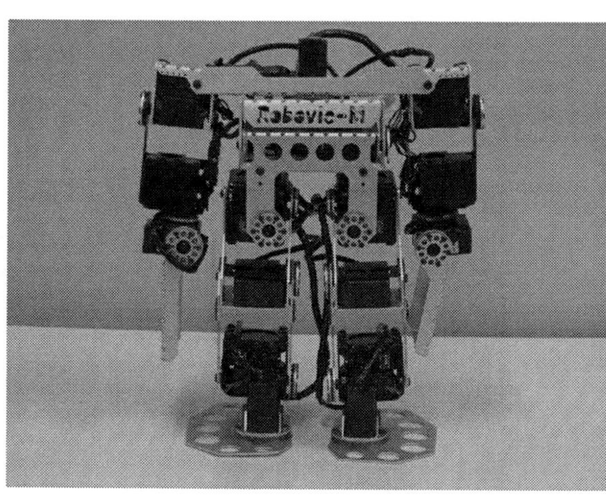

FIGURE 23.6
Robovie-M.

between the tag reader and an RFID tag can be roughly estimated. Such benefits are suitable for large environments.

However, drawbacks include low accuracy over long distances and the inability to detect exact positions. We compensated for these shortcomings by installing many RFID tag readers in the environment.

To achieve approximate distance estimation, we set the RFID tag readers to have eight levels of sensitivity. Detection areas, however, are affected by the position of the RFID tag readers and reflections due to walls. Therefore, we measured each detection area before the experiment. We then attached the tag readers in positions 2 m above the floor, and to successfully detect the tags we had to set the reader sensitivity level to at least five. We placed them around particular exhibits, so that the system could detect whether visitors approached them. Moreover, as a tag reader's detection field has a torus shape using multiple sensitivity levels, the system can estimate the tag position by superposing the circles calculated from the reader outputs.

23.3.2.3.2 *Infrared Cameras*

We placed an infrared LED on top of a Robovie and attached infrared cameras to the ceiling to determine the robot's correct position. The system produces binary images from the infrared cameras and detects bright areas. It calculates absolute coordinates with a reference to the weighted center of the detection area and sends them to the database.

Infrared camera positions are shown in Figure 23.5. The distance between the floor and the ceiling is about 4 m. The width and height of images from an infrared camera are 320 and 240 pixels, respectively. One pixel represents about 1 cm^2 of area.

23.3.2.3.3 *Video Cameras*

The video camera positions are also shown in Figure 23.5. The output images of each video camera are recorded by a PC and used to analyze the data generated by the experiment.

23.3.3 Robots' Behavior

In this section, we introduce the roles and behaviors of the robots. For friendly interaction with visitors, robots need information about them. For example, the field trial at the elementary school shows that children's interest increases when the machines call them by name. Moreover, human interactions are characterized by a shared memory of events.

Acquiring information about visitors, such as their names and memories, is difficult for the robots themselves. However, the embedded sensors enable them to capture this data through, for example, an Ethernet network. This is the way the robots can act more intelligently and overcome the limitations of their features.

23.3.3.1 Locomotive Robot

We used a Robovie as the locomotive robot that moved around in parts of the environment, interacted with visitors, and guided them to exhibits.

23.3.3.1.1 *Interaction with Humans: Childlike Interaction*

The robot can engage in such childlike behavior as handshaking, hugging, and the game of "rock, paper, and scissors" as shown in Figure 23.2. Moreover, it has such reactive behaviors as avoidance and gazing at a touched part of its body, as well as such patient behavior as solitary playing and moving back and forth.

23.3.3.1.2 Interaction with Humans: Using Information from RFID Tags

The robot can detect RFID tag signals around itself by using its RFID tag reader, which allows it to obtain personal data on visitors using RFID tag IDs. It can greet visitors by name or wish them a happy birthday, and so on. In addition, the robot can behave according to the length of the visitors' stays because the system records the time.

23.3.3.1.3 Guiding People to Exhibits: Human Guidance

The robot can guide people to four kinds of exhibits by randomly determining the target. For example, when bringing visitors to the telescope, the robot says, "I am taking you to an exhibit, please follow me!" and approaches the telescope. It suggests that the person looks through it and then talks about its inventor.

23.3.3.1.4 Guiding People to Exhibits: Using Information from RFID Tags

The RFID tags' data are also used for guiding. We used the amount of time that visitors spent near an exhibit to judge whether they tried the exhibit. For example, when an RFID-tagged visitor has stayed around the "a pulley" exhibit longer than a predefined time, the system assumes that the visitor has already tried it. Thus, the robot says, "Yamada-san, thank you for trying 'a pulley.' What did you think of it?" If the system assumes that the visitor has not tried it, the robot will ask, "Yamada-san, you didn't try 'a pulley.' It's really fun, why don't you give it a try?"

23.3.3.2 Robots That Talk with Each Other

Two stationary robots (Robovie and Robovie-M) casually talk about the exhibits as humans do with accurate timing because they are synchronized with each other using an Ethernet network. The topic itself is intelligently determined by data from RFID tags. By knowing the previous visiting course of a visitor, the robots can try to interest the visitor in an exhibit he or she overlooked by starting a conversation on that exhibit.

23.3.4 Robot Bidding Farewell

This robot is positioned near the exit, and after requesting data from their RFID tags, says good-bye to the departing visitors. It also reorients visitors on the tour who are lost by examining the visitor's movement history and time spent on the fourth floor of the Osaka Science Museum, which was recorded by the system. If visitors walk clockwise, they will immediately see this robot at the beginning and will be pointed in the right direction by the robot.

23.3.5 Experiment

We performed experiments to investigate the impressions made by robots on visitors to the fourth floor of the Osaka Science Museum during a 2 month period. As they departed the fourth floor, we asked visitors to complete a questionnaire by ranking five factors on a scale of 1–5, where 5 is the most positive. They were also encouraged to give other opinions on the robots.

By the end of 2 month period, the number of visitors had reached 91,107, the number of subjects who wore RFID tags was 11,927, and the number of returned questionnaires was 2,891. Table 23.5 shows the questionnaire results. It indicates that most visitors had a good impression of the robots and that they would not feel anxiety about robots in the future. Table 23.6 displays the age and gender of subjects who wore RFID tags and returned questionnaires. These results indicate that most questionnaires were returned by women

TABLE 23.5

Results of Questionnaires

	Interesting	Friendly	Effective of Guide	Anxiety about Interaction	Anxiety about Future Robots
Average	4.25	3.86	3.32	2.36	2.46
S.D.	0.90	1.10	1.19	1.36	1.23

and visitors in their 30s and 40s. We think that this trend resulted because only adults were asked to fill out the questionnaires, and the typical visitor group was a mother accompanying her child.

Most opinions were along the lines of "We had a really good time," "I had fun because the robots called me by name," and "We felt close to the robots." The results revealed that visitors held favorable impressions about the presence of the robots. Moreover, visitors described their favorite robot behavior, such as hugging, the calling out of names, and so on. Such behaviors are basic elements of human society.

The freely given opinions of visitors were analyzed and revealed that visitors' opinions of the robots differed according to age (Nomura et al., 2007). For example, younger respondents did not necessarily like the robots more than elder respondents.

23.3.6 Discussion and Conclusion

23.3.6.1 Contributions to HRI Research

More than 90,000 people visited the exhibition, more than 10,000 people interacted with the robots and wore RFID tags, and about 3,000 people returned questionnaires. The results showed that most visitors evaluated the robots highly.

The field trial revealed the advantages of using humanoid robots for interacting with people in a daily environment. In the exhibition, we often observed that people saw the robots, stopped at the robots, started to interact with the robots, and listened to the robot's explanations about exhibits. In addition, after the robot explained a telescope exhibit, one child went to use the telescope (Figure 23.7). When she came back to the robot, another child used the telescope. In other words, the presence of a humanoid robot attracted people to interact with it. This is one of the advantages of humanoid robots.

On the other hand, the lack of speech recognition capability is one important problem. Although it spoke much about the exhibits, the robot could not listen to the questions and comments from the visitors. When a reliable technique of speech recognition in a noisy environment becomes available, the robots will be more useful in this kind of application.

TABLE 23.6

Number of Respondents and Their Percentages Based on Gender and Age

	Male	%	Female	%	Total	%
10s	144	6	205	9	349	15
20s	64	3	118	5	182	8
30s	287	12	822	36	1109	48
40s	215	9	304	13	519	23
50s	19	1	37	2	56	2
60s	31	1	30	1	61	3
70s	17	1	8	0	25	1
Total	777	34	1524	66	2301	100

(a) (b) (c)

FIGURE 23.7
Scenes when visitor had interest in telescope.

Nevertheless, it is interesting that people already accepted and appreciated the robot even without the speech recognition capability.

We believe that this trial demonstrated the possibility of using interactive robots in open environments, which is one of the most important contributions of this work. Human–robot interaction was considered only as a kind of entertainment when research on interactive robots started. Our perspective (and that of many other researchers) is very different. We believe that robots will be part of the fundamental infrastructure of our society and will support a wide range of our daily activities. We believe that it is important to demonstrate possible applications of interactive robots, such as in the field trial at the elementary school, and this work particularly demonstrated the possibility for use in an open environment where both the novelty and interactivity of the robots were appreciated.

23.3.6.2 RFID Tags and Tag Readers

As shown in the scenes of interaction with the robots, the application of RFID technology largely promoted the human–robot interaction, particularly with regard to name-calling behavior. For example, the robot bidding farewell could attract the interest of visitors through the name-calling behavior although the Robovie-M is far cheaper than Robovie and its functionality is very limited, such as its small size, no embedded speech functions (we placed a speaker nearby), and no sensors (an RFID reader was also placed nearby).

However, the information obtained from the distributed RFID tag readers made a relatively small contribution to the system. Robots talked to the visitors about their exhibit-visiting experience, such as "You did not see the telescope exhibit, did you? It is very interesting. Please try it," based on the information from the RFID reader network, but it seemed to be less attractive and impressive to the visitors. Perhaps, robots are too novel for visitors, so they highly appreciate their experience of interacting with the robots while less attention is paid to the detailed services that they offer.

23.4 Conclusion

In this section, we introduced the applications of RFID technologies for communication robots through our two field trials, an elementary school and a science museum. The two trials showed the effectiveness of RFID technologies for communication robots.

In the elementary school, we performed the field trial for 2 weeks using communication robots with first- and sixth-grade elementary school students. In the trial, the robots behaved as English peer tutors for Japanese students. The results suggest that the robot did encourage some children to improve their English. Our findings demonstrate the possibility of having communication robots work in our everyday life, even though the benefits may still be too small to justify practical application. If the communication robots were to acquire a more powerful ability to maintain relationships with humans, we would feel more confident in them serving various roles in our everyday life in the immediate future.

In the science museum, we have developed an interactive robot system that combines communication robots and embedded environment sensors. The system guided visitors through a science museum with humanlike interaction, such as calling their names in a free-play behavior and explaining exhibits with voice and gestures. In a 2 month exhibition, 91,107 people visited the Osaka Science Museum, 11,927 of whom wore RFID tags to participate in the field trial. The results from questionnaires revealed that almost all visitors evaluated these robots highly. In addition, we confirmed the effectiveness of interactive behavior with personal information such as name-calling behavior.

In both trials, RFID technologies enabled communication robots to identify multiple persons simultaneously. That function significantly promoted the human–robot interaction, particularly with regard to the name-calling behavior. In addition, by using a person identification function, communication robots could know detailed information for interacting with people, such as the number of people, personal information of people, approximate distance of people, and moving histories of people. This information is useful for human–robot interaction; already, Kanda and coworkers have proposed a function to estimate friendly relationship by using an RFID tag system (Kanda and Ishiguro, 2006).

By the results of these field trials, we believe that RFID technologies support essential functions of communication robots. For example, robotics researchers should pay attention to long-term interaction between robots and people in the future. To achieve long-term interaction, a robust person identification function is important; RFID technologies can achieve such a function easily. Moreover, they can gather interacting histories between robots and people.

Acknowledgments

We wish to thank the teachers and students at the elementary school for their eager participation and helpful suggestions. We also wish to thank the staff at the Osaka Science Museum for their kind cooperation. This research was supported by the Ministry of Internal Affairs and Communications of Japan.

References

Breazeal, C. and Scassellati, B. 1999. How to build robots that make friends and influence people. *IEEE/RSJ International Conference on Intelligent Robots and Systems*, Kyongju, Korea, pp. 858–863.

Burgard, W., Cremers, A.B., Fox, D., Hähnel, D., Lakemeyer, G., Schulz, D., Steiner, W., and Thrun, S. 1998. The interactive museum tour-guide robot. *Proceedings of the National Conference on Artificial Intelligence*, Madison, Wisconsin, pp. 11–18.

Fujita, M. 2001. Toward the era of digital creatures. *International Journal of Robotics Research* 20(10): 781–794.

Ishiguro, H., Ono, T., Imai, M., Maeda, T., Kanda, T., and Nakatsu, R. 2001. Robovie: an interactive humanoid robot. *International Journal of Industrial Robotics* 28(6): 498–503.

Kanda, T. and Ishiguro, H. 2006. An approach for a social robot to understand human relationships: friendship estimation through interaction with robots. *Interaction Studies* 7(3): 369–403.

Kanda, T., Ishiguro, H., Imai, M., Ono, T., and Mase, K. 2002. A constructive approach for developing interactive humanoid robots. *IEEE/RSJ International Conference on Intelligent Robots and Systems*. pp. 1265–1270.

Kozima, H., Nakagawa, C., Kosugi, D., Kawai, N., and Yano, Y. 2004. A humanoid robot in company with children. *IEEE RAS/RSJ International Conference on Humanoid Robotics*, Santa Monica, California, CD-ROM.

Letcher, J., Fox, D., and LaMarca, A. 2005. Large-scale localization from wireless signal strength. *National Conference on Artificial Intelligence*, Pittsburgh, Pennsylvania, pp. 15–20.

Nomura, T., Tasaki, T., Kanda, T., Shiomi, M., Ishiguro, H., and Hagita, N. 2007. Questionnaire-based social research on opinions of Japanese visitors for communication robots at an exhibition. *Artificial Intelligence & Society* 21(1): 167–183.

Nourbakhsh, I., Bobenage, J., Grange, S., Lutz, R., Meyer, R., and Soto, A. 1999. An affective mobile educator with a full-time job. *Artificial Intelligence* 114(1–2): 95–124.

Shibata, T. 2004. An overview of human interactive robots for psychological enrichment. *The Proceedings of IEEE*, 92(11): 1749–1758.

Siegwart, R. et al. 2003. Robox at Expo. 02: a large scale installation of personal robots. *Robotics and Autonomous Systems* 42: 203–222.

24

Browsing the World with RFID Tags: Design and Implementation of an RFID-Based Distributed Environmental Memory

Alberto Rosi, Marco Mamei, and Franco Zambonelli

CONTENTS

24.1 Introduction

Advances in miniaturization of electronic devices and in wireless communication technologies are making possible to enrich our everyday environments (and the objects in it) with computation and communication capabilities [1]. The result may be an increased capability of interacting with the physical world, by acquiring in a digital form a number of information beyond our normal sensing capabilities. In the next few years, we will no longer simply browse the Web, but will be able to ''browse the world'' around us: we will seamlessly access both data coming form the physical environment via RFID tags, pervasive sensors and data coming form the Web, and we will be able to integrate them as needed in a suitable user-centric perspective.

Starting from the earlier considerations, this chapter presents the design and implementation of a system for tuple-based information access, context acquisition, and coordination in environments enriched with RFID tags [2]. The key idea is that our everyday environments will be more and more populated by RIFD tags, either attached to fixed location of the environment or to objects in it. These tags, while being low cost and battery-free, can make available some small amount of digital memory, in which it is possible to read and write data via wireless access. Thus, it is possible to collectively exploit such physically distributed memory areas as a sort of distributed shared memory: agents in an environment (whether they are humans carrying a mobile device or autonomous robots) can ''access'' such digital memory and, in this way, retrieve useful contextual information and/or coordinate with each other.

The choice of adopting a tuple-based coordination model [3], providing distributed agents with associative content-based access to distributed memory areas (i.e., tuple spaces), suits well modern distributed computing scenarios. In fact, the suitability of tuple spaces abstractions to act both as general repositories of shared contextual information and as suitable means to support uncoupled coordination between agents is widely recognized. Still, while tuple-based coordination has already been extensively adopted for the coordination of distributed agents in distributed computational environments (e.g., high-performance computing [4] and Internet computing [5]), its exploitation in the context of emerging pervasive computing technologies and physical environments has started being explored only recently [6–8].

By bringing together the advantages of RFID technology and the power of tuple-based coordination models—and also thanks to its clean-layered implementation architecture and its simple API—our approach proves itself very flexible, cheap to deploy, other than easy to use and suited for a variety of application scenarios. In addition, our approach clearly distinguishes from related work in the area: on the one hand, RFID technology has been exploited so far only for reading preexisting information on tagged objects and to exploit this information for logistics purposes [9] or to detect activities in an environment [10,11]; on the other hand, systems for pervasive tuple-based coordination proposed so far exploit costlier and more difficult to deploy technologies than RFID [8].

In any case, beside its advantages, we also emphasize that we are aware of a number of current limitations of our system, specifically in terms of reliability and security. These limitations do not generally undermine the usability of our approach (although in critical scenarios, they may limit its role to that of a complimentary support to more reliable and secure technologies), but nevertheless indicate the need for further work and extensions.

The remainder of this chapter is organized as follow. Section 24.2 briefly introduces RFID technology and discusses related work in the area. Section 24.3 presents the design and implementation of our system for RFID tuple spaces, the associated API and the current limitations of our system in terms of reliability, performances, and security. Section 24.4 introduces some application examples to clarify the use of our system and also discusses further potential application areas. Section 24.5 concludes and discusses future works.

24.2 RFID Technology for Pervasive Computing

This section briefly introduces RFID technology and then discusses related work in the area, that is, how RFID technology has been used so far to improve interactions with the physical world.

24.2.1 Overview of RFID Technology

Advances in miniaturization and manufacturing have yielded silicon-based postage-stamp-sized radio transceivers, called radio frequency identification (RFID) tags that can be attached unobtrusively to objects as small as a wristwatch or a pen (Figure 24.1a). Each tag is marked with a unique identifier and is provided with a tiny rewritable memory, up to some KB for advanced models (our test-bed implementation comprises tags with a storage capacity of 512 bit). The memory of tags, organized in cells, can be exploited for storing information in a persistent way. Tags can be purchased off the shelf and can withstand day-to-day use for years. In fact, they are battery-free and do not experience power exhaustion problems.

Suitable devices, called RFID readers, can be used to access the memory of RFID tags for both read and write operations. An RFID reader can be interfaced with any computer-based device, even portable (Figure 24.1b). RFID readers access to RFID tags wirelessly. When in need of reading or writing a tag, an antenna on the RFID reader emits a radio signal that has the twofold function of commanding the tag (e.g., "I want to read your unique identifier," "I want to write the byte 'FF' in the third memory cell") and of bringing energy to it. When receiving a signal, a tag takes power from it, activates itself, and responds or stores data according to the command embedded in the signal. RFID readers divide into short and long range depending on the distance within which they can access RFID tags. Such ranges vary from a few centimeters up to some meters.

Currently, RFID tags are in widespread use as antitheft technology in shops and for the production of personal ID cards. In addition, a growing number of business and logistics information systems are exploiting RFID tags for automating the identification and the tracking of products and merchandises [9]. In these systems, each product in a warehouse or in a production line is uniquely identified by a tag, possibly including some description of the product. The automatic gathering of the data in the tag and its proper integration into an information system allows for better and cheaper management operations.

The trends indicate that, in a few years, most household objects and furniture will be RFID-tagged before purchase (e.g., WalMart already adopts tagging at the level of pallets, and if will not be long before it will adopt it at the level of single package). The increasing

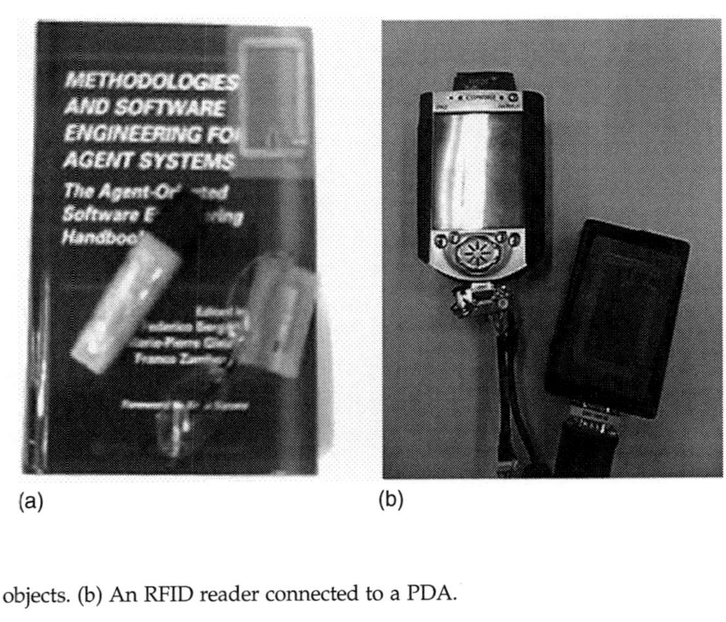

(a) (b)

FIGURE 24.1
(a) Some tagged objects. (b) An RFID reader connected to a PDA.

diffusion of the technology will also make it cheaper and cheaper, and the advent of ink-jet electronics will further reduce the cost of tags and will make it possible to print them nearly everywhere [12]. Moreover, handheld devices will be more and more provided with RFID reading and writing capabilities (e.g., the Nokia 5140 phone can be already equipped with a short-range RFID reader [13].

The future pervasiveness of RFID technology, together with its low cost and wireless data access, makes it an interesting option to explore in pervasive computing scenarios.

24.2.2 Related Work

In this section, we present some related work using RFID technology as the basis of an infrastructure to support pervasive applications. In this section, we will focus on infrastructural issue, and we try to highlight the peculiar characteristics of our proposal. In Section 24.4, instead, we will present some applications that can be realized with RFID technology and our infrastructure that are related to application works in the area.

Several research proposals in the area of pervasive computing exploit—in different ways—RFID tags as a simple and cheap way to enforce some forms of context awareness.

The system described by Satoh [14] assumes that readers are static, embedded in the environment, and connected to a fixed network infrastructure, while tags are attached to objects and mobile devices. This can be used by the network to recognize the presence of specific tagged objects or tagged PDAs into a specific room/location, and to contextualize activities of applications accordingly: if the RFID reader in a room can read a specific tag, this means that the corresponding object is in the proximities.

Our proposal, as well as several other proposals in the area, starts from a different assumption. In particular, we think that more flexibility can be achieved by having mobile devices themselves integrated with an RFID reader, thus having the capability of accessing RFID tags around, as sorts of digital contextual information stores. Such an approach has been used in several proposals with different application goals. However, rather than considering the possibility of storing new information in RFID tags and enforcing coordination through them (as we do), most of these proposals exploit RFID tags only for reading preexistent environmental information and for extracting from it contextual information.

Along these lines, Kulyukin et al. [15] propose embedding tags in an indoor environment and associating location information with them ("I am the tag of the living room"). These can be read by mobile robots carrying on an RFID reader to roughly localize themselves. The work described by Hähnel et al. [11] has similar localization goals, but goes further. There, it is shown that a robot equipped with multiple RFID readers can, while moving in an environment, detect with a very high statistical accuracy its own position as well as the position and motion patterns of other tagged objects/devices.

The system described by Philipose et al. [10] exploits RFID tags not only for localization but also, more generally, for inferring information about contextual activity in an environment. There, users are assumed to wear a special glove instrumented with an RFID reader and connected with a Wi-Fi portable device. Thus, when the user moves and acts in the environment, the type and the sequence of tags read by the glove can suggest what the user is doing, which may be of great use, for example, in elderly monitoring.

None of the earlier proposals exploits RFID tags as environmental memories for writing application-specific information. However, proposals exist in this direction.

The system described by Kawamura et al. [16] is a human memory augmentation system that lies on a distributed environmental memory and enables a user to directly associate experiences with physical objects by using a "touching" operation. However, this work has its focus on discovering memory externalization strategies that performed best, without being really interested in creating a general purpose framework layer. The middleware architecture proposed by Floerkemeier and Lampe [17], instead, definitely shares with our work the motivation of defining a general framework for exploiting RFID tags for both reading and writing application-specific information. There, the authors propose a sort of content-based publish/subscribe mechanisms through which applications can select to which type of tagged data they are interested in, while the middleware service takes care of controlling RFID readers in reading/writing data and in delivering it to applications accordingly. As in our tuple-based approach, the proposal has the advantage of enforcing associative access to tag data. However, the proposal is more oriented to logistics information systems and considers multiple RFID readers embedded in an environment that can be accessed and controlled by a single application agent. Our proposal, instead, is specifically oriented to mobile and pervasive computing scenarios and assumes that each RFID reader is attached to a single device and that application agents running on such device are the only ones to control it.

A previous work of Mamei and Zambonelli [18] has reported about experiments related to exploiting RFID distributed in an indoor environment as a memory infrastructure to store pheromone paths for the sake of enforcing pervasive stigmergic (ant-like) coordination [19]. However, in that work, we neither implemented any specific tuple space abstraction, nor defined a general API to access RFID tags. Conversely, the tuple space system presented in this chapter can be indeed exploited also to deploy distributed pheromone paths.

A proposal that, although not exploiting RFID technology, is worth reporting, is the TinyLime middleware [8]. TinyLime extends the Lime middleware to account for the environmental data collected by a sensor network. In Lime and in TinyLime, each mobile device is assumed to host a local tuple space. When a mobile device such as a laptop or a PDA "walk-through" a network of sensors deployed in an environment, all the data collected by the in-range sensors automatically feeds the local tuple space of the mobile device, which thus can perceive sensorial data collected by sensors simply by reading in the local tuple space. Thus, as far as reading environmental data is involved, TinyLime is functionally analogous to our proposal. However, TinyLime assumes the presence of computer-based sensors, much costlier than RFID tags and exhibiting battery exhaustion problems. In addition, TinyLime misses the possibility of exploiting sensors to enforce coordination between mobile devices, in that it assumes that they cannot write tuples in sensors that are considered as simple data providers.

24.3 RFID-Based Environmental Memory

In this section, we give a general overview of our system, detail its application programming interface, and provide some technical remarks on its current implementation.

24.3.1 Overview

Our system for RFID-based tuple spaces assumes a scenario in which the environment is densely populated by RFID tags, possibly attached to doors, walls, furniture, and other objects of any kind, acting as the digital memory in which tuples are stored. For what said in Section 24.2.1, the presence of tags will be more and more pervasive, mostly avoiding the need of a priori deployment effort. In any case, enriching an environment with tags is a dramatically simple operation, reducing at sticking them around as needed. In addition, our system considers the possibility that an environment can be enriched by tags on-the-fly, whenever needed for application purposes. In other words, if there is a need of environmental memory for users to store information, users themselves can dynamically release a tag in the environment and immediately use it (e.g., by storing a tuple there).

The current EPC standard (www.epcglobalinc.org) adopted by most major retailers uses read-only RFID tags. Moreover, privacy concerns may force tags attached to objects to be deactivated out of the retailer. We think that both these two constraints, that would make our infrastructure unfeasible, are temporary and will disappear in the future. On the one hand, there are a number of researches actively investigating novel writable tags with large storage capacity [20]. On the other hand, research on security and privacy issues in RFID is becoming more and more active, and some solutions addressing these concerns without destroying the tags are emerging (see Section 24.3.5). Ultimately, applications—like the one presented in this chapter—that are made possible by the availability of writable tags will justify the increased costs and/or privacy concerns.

Users, in the form of humans carrying on a PDA connected to an RFID reader (as from Figure 24.1b) or in the form of autonomous robots again connected to an RFID reader, roam in the environment and they can exploit the environmental memory of RFID tags. In particular, application agents running on the PDA or on the robots can control their own RFID reader to access the environmental memory via simple tuple space operations. These operations provide agents the capability of storing new tuples in the environment and of reading/extracting tuples in an associative way. In addition, blocking read/extract operations and asynchronous notification mechanisms enable application agents on different devices to coordinate and synchronize their activities with each other through the environmental memory. In any case, we emphasize that environmental memory may not necessarily be the only coordination mean. For instance, Wi-Fi PDAs or robot could also directly coordinate with each other via TCP/IP. But this is not our focus here.

From the viewpoint of application agents, the perception of the environment is that of a normal tuple space. From a given position in an environment, an agent has access to a locally bounded set of RFID tags, depending on the range of the associated RFID reader (see Figure 24.2a). Thus, performing input tuple space operations implies extracting tuples by applying the associative access mechanism to the aggregate memory of in-range tags. Analogously, inserting a tuple in the tuple space implies storing it in some available memory cells of one of the in-range tags (although we also provide the possibility of storing tuples in specific tags).

In our system, unlike in most of tuple-based middleware, there is no such concept as a single global tuple space [4] or as a multiplicity of localized tuple spaces [5]. In our approach, tuples are spread everywhere in the physical space, without being at all pre-organized in logical containers and rather defining a sort of spatial "continuous" of tuples. The concept of tuple space is logical and subjective, depending on the current position of an agent and on the range of its RFID reader (see Figure 24.2b). The same agent perceives different information in different physical position, whereas two agents in the same position may perceive different information depending on the reading range. Clearly, this implies that coordination between two agents has to be necessarily contextual, that is, spatial. Two agents can synchronize and coordinate only when they access (possibly at different times) to intersecting sets of tuples, that is, when they act in close position of the spatial environment.

24.3.2 Application Programming Interface

The API of our system defines Linda-like Java primitives to write tuples in the environment, read or extract tuples, as well as primitives to subscribe/unsubscribe to events. The

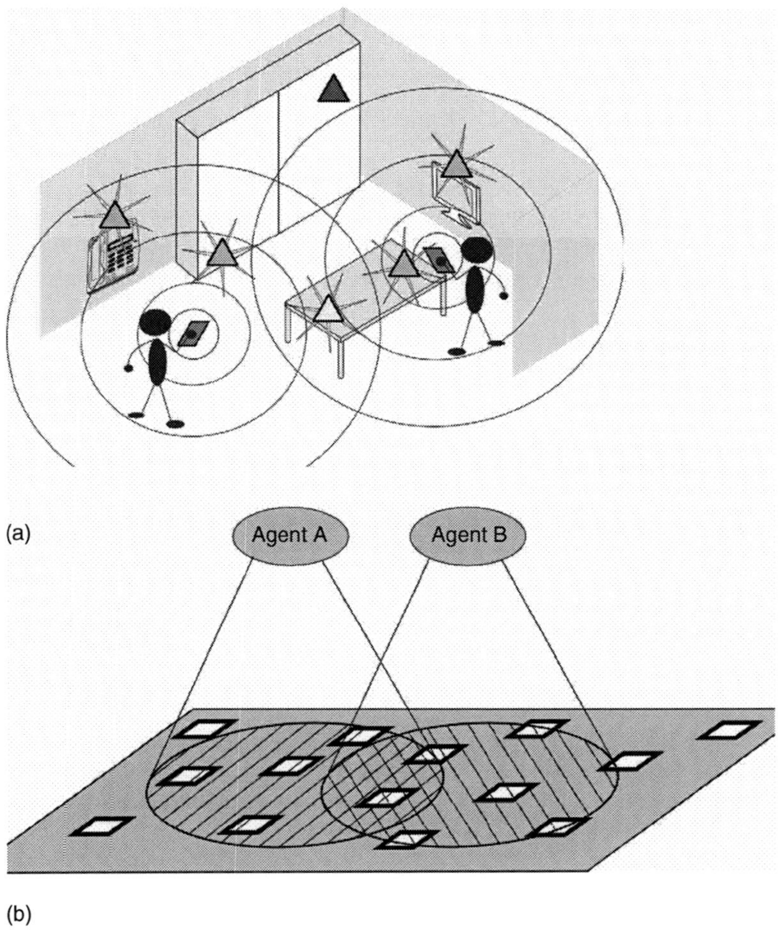

FIGURE 24.2
Overview of the system. (a) Physical overview of two users (or of the same user at different positions) accessing different set of tags in an RFID enriched environment. (b) Logical perception of tuple spaces by two application agents (or by the same user at different positions).

usage of the primitives does not require any a priori bounding to a specific tuple space. Instead, an agent invokes a primitive and this primitive will automatically act on the logical tuple space defined by the currently in-range RFID tags.

Despite the Linda common ground, some important differences have also been necessary to deal with the extremely limited storage capability of RFID tags.

The main difference from the "standard" Linda model is about the definition of a tuple. In our implementation, a tuple is defined by means of a simple name and value pair. Given the limited storage capacity of tags, names and values are coded with just 1 byte each. An application-dependent ontology (i.e., codebook) provides the right correspondence between high-level objects and low-level codes. For example, if an agent wants to create the tuple ("apple," "red"), it will look for the code of the strings "apple" and "red" in the ontology (i.e., "apple" = $0F$, "red" = 10), and then it will actually create ($0F$,10). Once a specific ontology has been selected, both encode and decode operations are invisible from the application point of view. Templates are analogous to tuples, but the value part can be left unspecified (null) to represent wild cards. Both tuples and templates are encoded in a data structure called *TagEntry*.

Taking inspiration from JavaSpace terminology, we defined the *write* primitive to insert a tuple in the environment:

```
boolean write(TagEntry[] tuple, TagEntry[] template);
```

In our implementation, this method can write more than one tuple at once (it writes an array of tuples) and it returns a boolean indicating whether the operation was successful or not. This capability of writing multiple tuples gives to the application developer the illusion of being able to manage complex tuples consisting of several name and value pairs:

```
/* create a tuple*/
TagEntry[] tuple=new TagEntry[2];
tuple[0]=new TagEntry("apple", "red");
tuple[1]=new TagEntry("cost", "10");
// write the tuple in the RFID tag
write(tuple, null);
```

Another remarkable difference with respect to standard Linda operations is the introduction of a template matching process also in the *write* operation. Since in a given environment there may be several in-range tuple spaces (i.e., RFID tags), it may be useful to impose some constraints on where to actually write a tuple. To this end, the second parameter of the *write* method is an array of templates (name–value pairs with wild cards). The write operation first selects among in-range RFID tags, those storing tuples that match *all* the templates in the array (for each template, there must be at least one matching tuple). Then, it writes the new tuple in *one* of the matching tags by a random selection:

```
/* create a tuple*/
TagEntry[] tuple=new TagEntry[2];
tuple[0]=new TagEntry("apple", "red");
tuple[1]=new TagEntry("cost", "10");
/* create a template */
TagEntry[] template=new TagEntry[1];
template[0]=new TagEntry("shop","open");
// this writes the tuple in one tag picked randomly
write(tuple, null);
/* this writes the tuple in one tag picked randomly among those already
containing the tuple "shop=open" */
write(tuple, template);
```

The other Linda methods follow rather simply from these considerations. The *read* and *take* primitives are defined to read or extract a tuple matching a given template from the environment.

```
TagEntry[] read(TagEntry[] template, TagEntry[] tag, int lease);
TagEntry[] take(TagEntry[] template, TagEntry[] tag, int lease);
```

In these methods, there are two template matching processes. The first (second parameter) selects some matching tuple spaces (i.e., tags) among the in-range ones, on the basis of their content, in a process similar to the one described earlier. The second template matching process (first parameter) selects some tuples from the previously selected tuple spaces. At the end, one of the resulting tuples is picked randomly and either read of taken.

Finally, we implemented a *subscribe* primitive to have agents notified whenever a tuple matching a subscription template enters the environment (or, which is the same from the agent viewpoint, when the agent enters a region of the environment in which the tuple can be found). The *subscribe* method returns a unique identifier to be possibly used later to remove that subscription with the *unsubscribe* method.

```
int subscribe(TagEntry[] tuple, TagEntry[] tag, Listener listener);
void unsubscribe(int subid);
```

In conclusion, we want to emphasize that some of the earlier differences from the standard Linda model are likely to disappear when tags with more storage capacity will be available. For example, we expect that both the need to have an ontology (codebook) mapping java objects to bytes and the constraint of having a tuple defined in terms of a name–value pair will be removed when it will be possible to store in tags complete serialized objects.

24.3.3 Implementation

In this section, we describe our hardware test-bed and the implemented software architecture.

24.3.3.1 Hardware

In our implementation, each application agent runs on a HP IPAQ r × 3715, running Pocket PC 2003 Pro and J2ME and provided with a Wi-Fi card and an inside midrange RFID reader. Thus, each agent can be carried around and it can connect to different tags deployed in the environment. In particular, we employed ISO 15693 RFID tags, each with a storage capacity of 512 bits. These tags communicate at 13.56 MHz frequency and the reader can interact with tags within 1.5 m. The information being read is passed to the PDA via a serial cable where it is processed.

In addition, a mobile robot has been realized by installing one PDA connected to an RFID reader onboard of a Lego Mindstorms robot (www.legomindstorms.com). The IPAQ runs an agent controlling both the RFID reader and the robot microprocessor.

24.3.3.2 Software

The software architecture is depicted in Figure 24.3.

- *Driver*. The bottom level consists in the RFID reader device driver. This driver has been realized in terms of a dynamic link library that exports a number of methods to control the reader. In particular, we implemented methods to read and write a specific byte in a specific tag. This is the only layer in our architecture that is hardware-dependent. The driver, in fact, shields top layers from hardware details.
- *RFID Control Logic*. On the top of the driver, we realized an RFID control logic layer. Since RFID communication is error-prone (it happens quite often that an in-range tag fails to respond to the reader), we developed some mechanisms to iterate operations and merge the results accordingly.

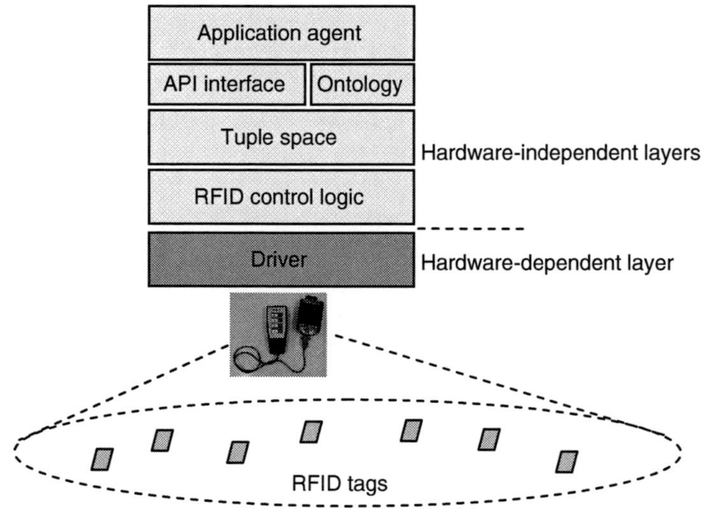

FIGURE 24.3
Software architecture.

- *Tuple Space*. This layer is the core of our tuple space implementation. It is in charge of performing important operations such as pattern matching and notifying agents on triggering conditions (e.g., with regard to event subscriptions and blocking read). For example, on the execution of a blocking read, this layer remains active and ready to notify the application layer on new tuples being inserted in the tags around or new tags coming in-range due to device movements.
- *API Interface and Ontology*. This layer provides the Linda-like methods described earlier and the (application-specific) ontology to map high-level objects in low-level byte values to be stored in the tags. This ontology is integrated with the *TagEntry* constructor (see earlier) to hide the low-level representation from the application point of view.
- *Application Agent*. The top layer of our architecture is the application agent that realizes the application itself. Although nothing would prevent from running more than one agent on the same platform that coordinate using the available tags, our standard application scenario involves multiple devices each with a single agent.

24.3.4 Reliability and Performance Issues

In general, the act of reading and writing RFID tags is something that exhibits high degrees of uncertainty for various reasons (we forward the interested reader to study Refs. [17,21,22] for thorough analysis of these issues).

First, the more a tag is far from the reader, the more the probability that the radio energy that reaches the tag is not enough to activate it. Second, the orientation of the tag with respect to the antenna influences the capability of the reader of providing energy to the tag, so that even a very close tag may happen not to be readable [21]. Third, a metal object in the proximity of a tag tends to absorb the energy of the reader and may prevent a tag from being activated. In addition, the presence of tags close to each other may induce radio interferences in the answers of tags to readers, preventing some of these answers to properly reach the reader [22]. For these reasons, in our system as well as in most systems based on RFID tags, an RFID control layer is introduced (as described in the previous subsection). Nevertheless, there is not any guarantee that a tag in the range of a reader will be properly read (or written) by the reader.

In out system, the impact of the earlier problem is intrinsically limited by the adoption of the tuple space coordination model. The uncertainty in reading tags and the fact that some existing tags may stay undetected (or, which is the same, the fact that some existing matching tuple does not end up in an actual match) does not generally cause, from the viewpoint of the tuple-based coordination model, any application-level critical error. The fact that, among multiple matching tuples existing in the environment, only a few of them may be detected at a given time by a reader, can be perceived by an application agent as part of that nondeterministic process of tuple selection which is promoted by tuple-based coordination. The fact that a specific existing matching tuple *Tx* may remain undetected is not perceived as an error from the application level, but simply as the fact that the requesting agent, from its current position, perceives a projection of the global tuple ensemble that does not include *Tx*. Although some problems may arise if two agents use an undetected tuple *Tx* as a synchronization point, one could deal with the problem by spreading multiple copies of *Tx* around to increase its detection probability.

Getting to performance considerations, we emphasize that the interferences induced when activating multiple tags, together with the limited radio frequency at which tag activation commands have to operate, bound the number of distinct tags that can be read in a timeframe to about 20–100 s^{-1}, depending on the specific technology, the characteristics of the environment, and the positions of tags [17]. Even smaller numbers apply when a tag selection process is included, since the real operation on a specific tag has to be necessarily preceded by read operations to select the tag.

Accordingly, providing performance numbers for our system is a rather useless exercise. In fact, the time required by agents to execute tuple space operations—consisting in a few line of Java code intertwined by invocations of functions on the RFID control layer—is directly proportional to the time required to access RFID tags, and limited by it. In particular, time to perform an operation mostly depends on the amount of reading attempts in a single operation performed within the RFID control layer. In the current implementation, we have set up this number in such a way that tuple space operations take an average of 0.4 s to return to the invoking agents. Such time ensures a good probability of reading all accessible tags in the proximities (enabling, with the hardware in our hands, to read a number of 10–40 distinct tags within a single operation). At the same time, this time is satisfactory from the viewpoint of human reaction times.

24.3.5 Privacy and Security Concerns

The use of RFID technology raises several privacy and security concern. Privacy concerns relate to the fact that anyone with an RFID reader could read information associated with tags in the surrounding, possibly unveiling sensible information. For instance, if we wear or have in our pocket some tagged objects, a user with a reader could take a sort of x-ray picture and discover sensible information (e.g., do I have an expensive cell phone, do I have a pack of cigarettes in my pocket). Security issues arise also because anyone with an RFID reader can modify the data stored in tags (e.g., modifying the expiry date on a tagged bottle of milk). Clearly, such problems affect our system too.

We want to make it clear that our system currently does not integrate specific solutions for preserving privacy and security. Nevertheless, we can at least analyze what possible solutions can be envisioned, and possibly share our bits of experience with those who are currently concerned with extending the technology to account for security and privacy.

In general, the approach of giving users a reader and have it proactively explore the environment puts users in less sensible position with regard to the tracking of personal information: the user is not passively subjected to reading of his own data, and is not forced to wear any tag to exploit the system. This notably contrasts with other proposals for the use of

RFID tags in pervasive computing, where RFID reader embedded in the environment collects information about users by reading tags user wears [14,23]. Of course, this does not solve the problem of having a user being inspected with regard to tagged objects he may wear. The problem of users wearing tagged objects that no one should be allowed to read could be easily solved with the introduction of "read-only-by-user" tags, as proposed by Stajano [24]. In this way, publicly readable (and writable) RFID tags containing no user sensible information could coexist with read-only-by-user ones. The presence of the former ensures the availability of the necessary infrastructure for our memory system to be meaningful and usable. The latter ensures that no sensible information can be accessed. It is worth outlining that, from the viewpoint of users, one user would perceive the overall memory of the surrounding as including both the set of all in-range public tags and the set of his own private ones.

More sensible is the issue of public access to writing of distributed tags, which could affect the possibility of safely interacting across the environmental memory. For instance, with reference to the first-aid rescue scenario, the problem of having some malevolent agent modifying what has been written on the tag of an injured person can have dramatic consequences on his health. A reasonable solution to mitigate the problem could be that of having write-once tags, so as to ensure that a sensible information associated to a tag cannot be maliciously or erroneously modified. It is also worth outlining that the possibility of a single writing operation in some portion of a tag's memory could also be a way to enforce read-only-by-user operations (i.e., by associating to a tag an unmodifiable ID associated to a specific user and have the tag circuitry check for that ID on read operations).

The great current interest in RFID technology, and the increasing number of applications associated with it, will certainly drive the appearance of suitable solutions to privacy and security problems, possibly along the earlier identified lines [25,26]. Our proposal too will benefit from these solutions and, by considering the presence in an environment of read-only-by-user tags to protect private information, of write-once tags to protect by modification critical information, and of normal (read-and-write-by-all) tags to store general contextual information, will become a safe and trusted one to exploit.

24.4 Application Examples

The implementation of tuple spaces within RFID tags enables exploiting the benefits of tuple-based coordination in a wide number of pervasive computing scenarios. Here, we first give some more details about an exemplary application related to agents' coordination in a first-aid scenario. Then, we overview a number of relevant areas to show the wide range of situations that could take advantage of our system.

24.4.1 A First-Aid Rescue Application

Knowing how to monitor and deal with a large number of casualties is critical to disaster response scenarios. If first responders cannot quickly coordinate and triage the injured people, the large numbers could quickly overwhelm emergency field personnel and hospital staff, and prevent them from providing quality trauma care [27]. To avoid such problems, it would be fundamental to provide responders with suitable tools to coordinate their activities at every phase during the disaster response mission.

Let us focus the attention on a possible earthquake scenario. First responders coming to the disaster area have the primary need to coordinate their search for injured people.

To this end, they could use RFID tags likely to be present in the crumbled environment or leave new tags to coordinate efficiently. Specifically, when looking for people, each rescuer could instruct his PDA to simply write the tuple (rescuer = my_id) in the tags around to indicate that the area has already been explored. At the same time, the PDA is programmed to read periodically the tags around and notify to the rescuer if the area had been visited before by another member of the rescue team (see code as follows).

```
while(true) {
/* read tag around */
TagEntry[] template=new TagEnrty[1];
template[0]=new TagEntry(rescuer);
TagEntry[] ts=read(template, null, 0); // non blocking
if(ts contains other rescuer IDs)
  alert("The area has already been explored");
/* write rescuer ID */
TagEntry[] tuple=new TagEnrty[1];
TagEntry[0]=new TagEntry(rescuer, my_id);
write(tuple, null);
}
```

At the same time, rescuers could leave useful context information related to environmental conditions. Such information could be extremely valuable to other team members coming later to the area. For example, first rescuers could leave tuples indicating if the area is suitable for ambulance crossing. Ambulance could be equipped with RFID readers to read such tuples and alert their drivers accordingly.

Tuple-based coordination and RFID tags could also support the workflow in medical operations. First responders, after visiting some injured people (see Figure 24.4a), could stick RFID tags to them and write there tuples with relevant trauma care information (see Figure 24.4b). Adopting the proposed infrastructure, the code to deploy such tuples becomes trivial (see code as follows).

```
/* create a tuple storing: medicated=true, to_be_carried=true,
illness=medium*/
TagEntry[] tuple=new TagEnrty[3];
TagEntry[0]=new TagEntry(medicated,true);
TagEntry[1]=new TagEntry(to_be_carried,true);
TagEntry[2]=new TagEntry(illness, medium);
// write the tuple in the patient RFID tag
write(tuple, null);
```

After that, first responders could move further visiting other persons. In the mean time, nearby paramedics and emergency medical technicians could care for already visited patients by taking advantage of the information stored in their RFID tags (see Figure 24.4c and the code as follows).

```
/* create a template with: to_be_carried=true,*/
TagEntry[] template=new TagEnrty[1];
template[0]=new TagEntry(to_be_carried,true);
/* start a blocking read to find a suitable patient */
read(template, null, true);
/* notify that a patient to be carried has been found */
...
```

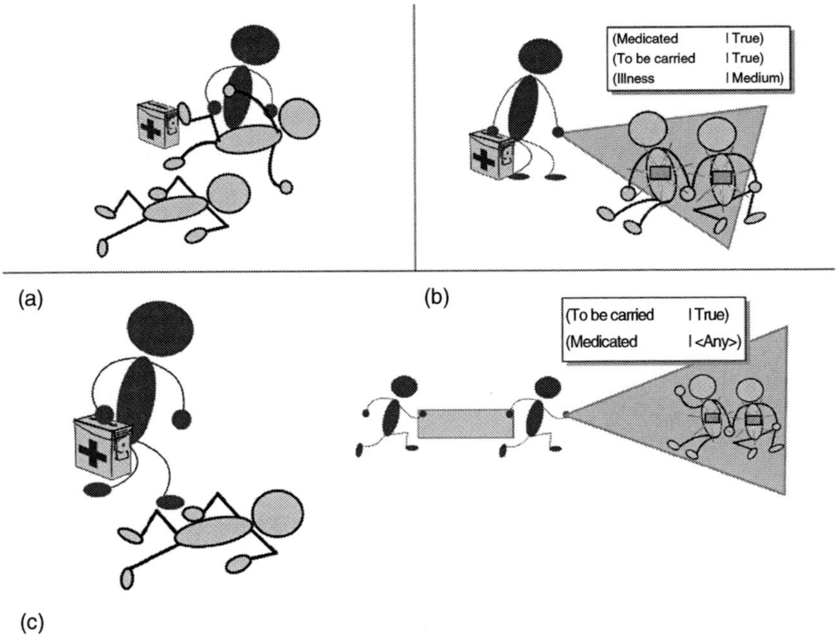

FIGURE 24.4
Tuple-based coordination and RFID in a first-aid scenario. (a) First responder, after providing first aid to injured people, (b) can store in RFID tags attached to them relevant information about their conditions. Following these, (c) paramedics can read those RFID tags to get aware of and treat injured people more quickly and more efficiently.

This could notably reduce latencies and delays and allow to care for several patients at once in parallel. After this first stage, some patients would be carried in the ambulance. During the journey to the hospital, the tuples stored in the RFID tags could be integrated with additional information and be sent wirelessly to the hospital. This would allow doctors there to save time by setting up things in advance (e.g., preparing surgery rooms) to receive the incoming people more quickly and more efficiently.

24.4.2 Post It Localization

As already discussed in Section 24.2, RFID tags can act as a cheap yet effective indoor localization infrastructure [11,14]. The approach we implemented consists of tagging several places in the environment with RFID tags storing a URL where to download a picture of the building map, and their own coordinates within the building. Users are provided with a wireless PDA equipped with an RFID reader. The indoor localization application periodically looks for RFID tags around. If a tag with the coordinates is detected, the map of the building is downloaded and displayed on the user PDA, and the user is localized at the same coordinates of the RFID tags being read (this is consistent with the short reading range of the RFID reader). In addition, localization tags can be used as a virtual "Post It." Tags embedding information about office position can also embed information about people working on it, taking the place of small slip of notepaper temporarily stuck to a door with sentences written above like "I'll be back soon" or "You can temporarily meet me at Professor Zambonelli's office."

The pseudocode of the "Post It" localization application is straightforward (see later), while a screenshot of the indoor localization application is reported in Figure 24.5.

FIGURE 24.5
RFID-based indoor localization.

```
/* Indoor localization tuples are in the form:
TagEntry[] tuple=new TagEntry[3];
tuple[0]=new TagEntry("map", "http://155.185.3.4/map.kml");
tuple[1]=new TagEntry("x", "130");
tuple[2]=new TagEntry("y", "120");
*/
while(true) {
/* read tag around */
  TagEntry[] template=new TagEnrty[1];
  template[0]=new TagEntry("map",null);
  TagEntry[] ts=read(template, null, 0); // non blocking
  If(ts !=null) {
    display(ts[0].value); // map
    localize(ts[1].value,ts[2].value); // x,y
  }
  sleep(1000); sleep 1 second
}
```

24.4.3 Motion Coordination and Events Organizer

Starting from Section 24.2 considerations, and from a previous work of Mamei and Zambonelli [18], we introduce two application examples describing how to obtain self-coordination and self-organization exploiting information spread on environment.

Tuples spread in the environment could enable a group of agents (both humans and robots) to coordinate their respective movements. An exemplary application would be distributed environment exploration. Users could mark already visited areas by writing tuples in the RFID tags around. Later on, other users could decide to explore a specific area if there are not any tuples pointing in that direction (the area is truly unexplored).

Another similar application could be based on spreading information trails, in RFID tags spread in an environment, to be followed as trails of pebbles. Using RFID tags as white pebbles, users can mark footprints of their movements directly in the environment.

Integrating them with contextual and user's behavior information, we can easily obtain a distributed events organizer. We can figure to be walking across a huge park, for example, New York Central Park. A huge park like the one earlier contains several open spaces in which people can self-organize and perform various activities and sports. If along park alleyways we place a line of public RFID tags, with both writing and reading access, we can figure how easy can be organize a soccer match.

A player, having a football in his hands, wants to organize a football match but he is completely alone. While he is walking across alleyways searching for an empty space to transform in a football pitch, he leaves on RFID pebbles as footprint of his passing provided with a small description of the activity he is going to perform, for example, "27th March 2006, 4:10 p.m., arranging a football match" equipped with an implicit message "if you are interested, follow me." Other people, while entering the park, can read RFID tags around to discover interesting activities going on. Then, they can start following the pebbles to reach the location where a selected activity takes place. Now, our lonely player has only to wait for other people to come.

The "Events Organizer" application can be eventually extended gathering information either from devices measuring physical events or from inferencing user's behaviors. For example, wearable accelerometers [28] placed on clothes can indicate if the subject is walking or running (or sitting, standing, shaking hands, etc.), integrating such information over the time we can deduct user's pace and understand the activity he is undertaking.

Likewise, for example, a microphone can capture user's statements indicating aims and intentions supporting the creation of a more distinct profile of the events he is going to perform. Scene classification performed on speech recognition also permits to detect if user is alone, is interacting with another person or with a larger group of people [29]. Information can be taken too observing user habits like suggested by Castelli et al. [30]. A Bayesian system can, for example, infer that our user, every day, uses to have jogging at 07:00 a.m. This information can be stored on RFID pebbles; in this way people interested in running at that time, and with user's pace, can manage to run together.

24.4.4 Context Awareness

Tuples written in the environment can be used to help users (as well as robots) in getting aware of what is in the environment more than their natural and artificial senses can do, by reading the additional information provided by tags. Our RFID approach naturally suits this application in that (1) RFID tags can store—or link to—semantically rich contextual information, (2) context data can be actually written in the environment where it will be most useful. For example, RFID tags stuck on objects or persons could hold information on such objects and persons to be read by people nearby. Reading such tags could be very valuable to assess the application context. As an exemplary application, we developed an RFID-based system to provide tourist information to users. Tourist information related to different art pieces are stored in RFID tags stick to the piece itself (such information can be partially stored locally in the RFID memory, and partially stored in a Web site whose URL is in the RFID). Users can trivially read and display such information on their PDA (see Figure 24.6).

A similar, but more interactive, application we designed is "polite graffiti." It allows users to store short messages or links to content-rich information in RFID tags spread in the environment. Such information can be then accessed and visualized via RFID reader-equipped devices, possibly with an augmented reality interface [31]. The idea of this application is to allow people to express themselves with graffiti-like messages without damaging buildings' facades or monuments. Graffiti could also be displayed on a

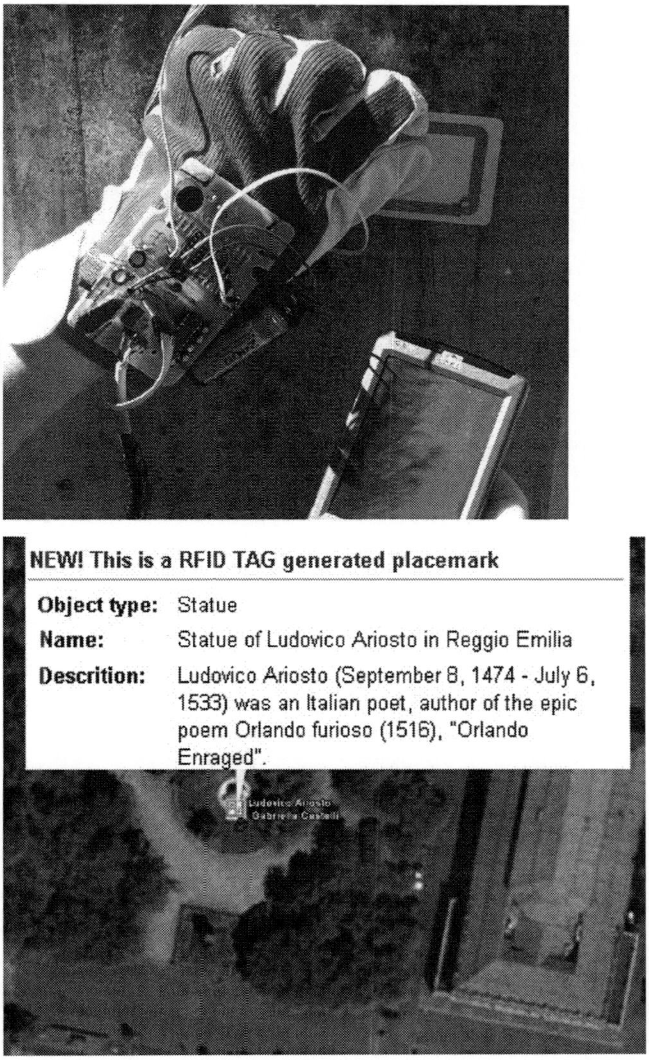

FIGURE 24.6
Context information acquisition by reading information stored in RFID tags.

wide-screen display installed in proximity of the monument and eventually surrounded by advertising messages.

24.4.5 Pervasive Workflow Management

In line with these ideas is the concept of *pervasive workflow management*. Standard workflow management systems are rooted on a centralized engine that keeps track of the status of the workflow being carried on. Workers notify to this engine the tasks being completed and the engine in turn notifies the subsequent tasks that have to be carried on [32]. RFID tags and tuple-based coordination could remove the need for a centralized engine in a pervasive computing environment. Basically, the RFID tags associated to the items to be processed could store a tuple identifying the operations that the item undertook. Workers with RFID readers could simply read the state of the item and process it further. This approach could be employed in traditional manufacturing scenarios as well as in more mundane domestic workflow (e.g., store in the pet's collar a tuple reporting if it has already eaten or not).

24.5 Conclusions and Future Work

In this chapter, we have presented the design and implementation of a system that, by exploiting the physically distributed memory of RFID tags, enables to enforce pervasive tuple-based access to contextual information and distributed coordination activities. The presented system is low cost, easy to deploy and to use, and makes available a simple yet effective application programming interface, suitable for a variety of application scenario.

Our current research work deals with trying to overcome—or at least mitigate—some identified problems of our system related to reliability and security, as discussed earlier in this chapter. In addition, we are exploring the possibility of extending our system to integrate other pervasive computing technologies than RFID tags. In particular, we think that it is possible and useful to integrate into a uniform distributed tuple-based model any embedded device in an environment—from RFID tags to active wireless sensors and wireless computer-based devices—and exploit a single API to access information, control active devices, enforce distributed tuple-based coordination, independently of the specific type of devices existing in the environment.

Acknowledgment

Work supported by the project CASCADAS (IST-027807) funded by the FET Program of the European Commission.

References

1. D. Estrin, D. Culler, K. Pister, and G. Sukjatme, Connecting the physical world with pervasive networks, *IEEE Pervasive Computing*, 1(1): 59–69, 2002.
2. V. Stanford, Pervasive computing goes the last hundred feet with RFID systems, *IEEE Pervasive Computing*, 2(2): 9–14, 2003.
3. D. Gelernter and N. Carriero, Coordination languages and their significance, *Communications of the ACM*, 35(2), 96–107, 1992.
4. S. Ahuja, N. Carriero, and D. Gelernter, Linda and friends, *IEEE Computer*, 19(8): 26–34, 1986.
5. A. Omicini and F. Zambonelli, Coordination for internet application development, *Journal of Autonomous Agents and Multi-Agent Systems*, 2(3): 251–269, 1999.
6. M. Mamei and F. Zambonelli, Programming pervasive and mobile computing applications with TOTA, *Proceedings of the 2nd IEEE Conference on Pervasive Computing and Communication*, Orlando, FL, pp. 415–422, March 2004.
7. C. Fok, G. Roman, and C. Lu, Rapid development and flexible deployment of adaptive wireless sensor network applications, *International Conference on Distributed Computing Systems*, Columbus, Ohio, 2005.
8. C. Curino, M. Giani, M. Giorgetta, A. Giusti, A.L. Murphy, and G.P. Picco, TinyLime: bridging mobile and sensor networks through middleware, *IEEE International Conference on Pervasive Computing and Communications*, Kauai Island, Hawaii, 2005.
9. C. Bornhovd, T. Lin, S. Haller, and J. Schaper, Integrating automatic data acquisition with business processes: experiences with SAP's Auto-ID infrastructure, *International Conference on Very Large Databases*, Toronto, Canada, 2004.
10. M. Philipose, K.P. Fishkin, M. Perkwitz, D.J. Patterson, D. Fox, H. Kautz, and D. Hahnel, Inferring activities from interactions with objects, *IEEE Pervasive Computing*, 3(4): 10–17, 2004.

11. D. Hähnel, W. Burgard, D. Fox, K. Fishkin, and M. Philipose, Mapping and localization with RFID technology, *Proceedings of the IEEE International Conference on Robotics and Automation ICRA*, 2004.
12. G. Collins, Next stretch for plastic electronics, *Scientific American*, 291: 74–81, 2004.
13. Nokia mobile RFID kit, http://www.nokia.com/nokia/0,,55738,00.html
14. I. Satoh, A location model for pervasive computing environments, *Proceedings of IEEE 3rd International Conference on Pervasive Computing and Communications* (PerCom'05), Kauai Island (HW), pp. 215–224, March 2005.
15. V. Kulyukin, C. Gharpure, J. Nicholson, and S. Pavithran, RFID in robot-assisted indoor navigation for visually impaired, *Proceedings of the 2004 IEEE/RSJ International Conference on Intelligent Robots and Systems*, Sendai, Japan, vol. 2, pp. 1979–1985, 2004.
16. T. Kawamura, T. Fukuhara, H. Takeda, Y. Kono, and M. Kidode, Ubiquitous memories: a memory externalization system using physical object, *Personal Ubiquitous Computing*, 11(4): 287–298, Springer-Verlag, London, 2007.
17. C. Floerkemeier and M. Lampe, RFID middleware design—addressing application requirements and RFID, Technical Report, Pervasive Computing Group, ETH Zurich, 2005.
18. M. Mamei and F. Zambonelli, Spreading pheromones in everyday environments through RFID technology, *Proceedings of the 2nd IEEE Symposium on Swarm Intelligence*, Pasadena, CA, pp. 281–288, June 2005.
19. E. Bonabeau, M. Dorigo, and G. Theraulaz, *Swarm Intelligence*, Oxford University Press, Oxford, 1999.
20. K. Greene, Wireless wonder chip, HP's tiny chip could offer a new way for storing and sharing video, audio, and pictures, http://www.technologyreview.com/printer_friendly_article.aspx?id = 17182
21. K.P. Fishkin, B. Jiang, M. Philipose, and S. Roy, I sense a disturbance in the field: unobstrusive detection of interactions with RFID tagged object, *Ubicomp 2004 Conference*, Notthingam, UK, 2004.
22. C. Floerkemeier and M. Lampe, Issues with RFID usage in ubiquitous computing applications, *International Conference on Pervasive Computing*, Vienna, Austria, 2004.
23. S. Hsi and H. Fait, RFID enhances visitors' experience at the exploratorium, *Communications of the ACM*, 48(9): 60–65, 2005.
24. F. Stajano, RDIS is x-ray vision, *Communications of the ACM*, 48(9): 31–33, 2005.
25. G. Avoine, Security and privacy in RFID systems, http://lasecwww.epfl.ch/~gavoine/rfid
26. M.R. Rieback, B. Crispo, and A.S. Tanenbaum, The evolution of RFID security, *IEEE Pervasive Computing*, 5(1): 62–69, 2006.
27. K. Lorincz, D. Malan, T. Fulford-Jones, A. Nawoj, A. Clavel, V. Shnayder, G. Mainland, S. Moulton, and M. Welsh, Sensor networks for emergency response: challenges and opportunities, *IEEE Pervasive Computing*, 3(4), 2004.
28. N. Kern, B. Schiele, and A. Schmidt, Recognizing context for annotating a live life recording, *Personal Ubiquitous Computing*, 11(4): 251–262, Springer-Verlag, London, 2007.
29. N. Kern, B. Schiele, H. Junker, P. Lukowicz, and G. Troster, Wearable sensing to annotate meetings recordings, *Proceedings of ISWC*, Seattle, WA, USA, October 2002.
30. G. Castelli, M. Mamei, and A. Rosi, The whereabouts diary, Technical Report, University of Modena and Reggio, Emilia, Italy, 2007.
31. D. Wagner and D. Schmalstieg, ARToolKitPlus for pose tracking on mobile devices, *Computer Vision Winter Workshop*, St. Lambrecht, Austria, 2007.
32. A. Ricci, A. Omicini, and E. Denti, Virtual enterprises and workflow management as agent coordination issues, *International Journal of Cooperative Information Systems*, 11(3/4): 355–379, 2002.

25

RFID-Enabled Privacy-Preserving Video Surveillance: A Case Study

Jehan Wickramasuriya, Sharad Mehrotra, and Nalini Venkatasubramanian

CONTENTS

RFID-enabled applications are growing at a tremendous rate with uses in a number of different areas such as pharmaceuticals, health care, transportation, retail, defense, logistics and many others. An important aspect of modern radio frequency identification (RFID) technology is the utility it can provide in a variety of different applications. The idea is to enhance existing system components with the benefits RFID provide while still keeping the system modular and efficient. An equally important and overarching concern is that of privacy, which is particularly pertinent when dealing with pervasive applications that deal with potentially personalizing information. The privacy concerns stem from both the issues related to the particular technology (e.g., maintaining anonymity while utilizing RFID for identification purposes) as well as ensuring that the end-to-end system takes them into account too. As an example of how these factors come into play with designing and building an integrated system, this chapter will outline a privacy-preserving video surveillance system that uses RFID hardware for localization and identification. Our focus here is on privacy concerns that pertain to the end-to-end system and more specifically those stemming from interactions between individuals and the environment. In this chapter, we describe the design of a privacy preserving video surveillance system [1] that monitors subjects in an instrumented space only when they are involved in an access violation (e.g., unauthorized entry to a region). In this system, access control policies specify the access rights of individuals to different regions of the monitored space. Policy violations (detected via use of localization sensors such as RFID tags, motion detection, etc.) are used to trigger the video surveillance subsystem. Video manipulation techniques such as masking are used to preserve the privacy of authorized subjects when the surveillance system is turned on. The novel facet of this system is the integration of sensor technology (namely RFID) with a video surveillance subsystem.

25.1 Surveillance for Pervasive Spaces: Utility versus Privacy

Today, even though potential privacy concerns exist, a large number of work spaces and critical infrastructures do not avail the benefits of automation that pervasive spaces offer because of concerns of privacy of individuals. Looking at the other extreme, if the benefits outweigh the concerns, pervasive environments are designed and used with complete disregard to individuals' privacy. Personalizing information is logged and available for (mis)use at the whims of the organizations (and their employees) that collect data. With the heightened consciousness among the public, private, and government organizations for security, surveillance technologies (especially video surveillance) have recently received a lot of attention. Video surveillance systems are being considered/deployed in a variety of public spaces such as metro stations, airports, shipping docks, etc. As cameras go up in more places, so do the concerns about invasion of privacy [2]. Privacy advocates worry whether the potential abuses of video surveillance outweigh its benefits. A fundamental challenge is to design surveillance systems that serve the security needs while protecting the privacy of the individuals.

Example Scenario: We demonstrate the basic functionality of the framework by means of an example that examines security in a hospital setting. Assume the hospital is divided into federated regions that are each covered by video cameras. The RFID sensor information is used in conjunction with the video surveillance subsystem to provide coverage of the monitored regions. Furthermore, access control policies are defined for personnel carrying RFID tags. The enforcement of these policies is used to influence the video surveillance system. Figure 25.1 shows a scenario in which the hospital consists of four regions, each surveilled by a single camera.

The idea is that if a subject is authorized to be in a particular region (which is determined by their RFID tag) he/she may be hidden in the video. This way, subjects' privacy is maintained until they violate their given rights (e.g., enter a region they are not authorized to be in). In Figure 25.1, each region (R1,R2,R3, and R4) is monitored by a video feed that is triggered by a system that processes the information from RFID readers and predefined access policies. A paramedic entering R3 causes the motion sensor to initiate a read on his/her tag that is forwarded to the RFID reader. This causes an access violation and subsequently triggers the video feed for R3. However, medical personnel present in R3 (who are authorized for all regions) will be masked in the feed according to their given privacy level. Additional constraints can also be enforced such as restricting certain patient/doctor pairs or associating a bed (which can also be tagged with RFID) with a particular patient.

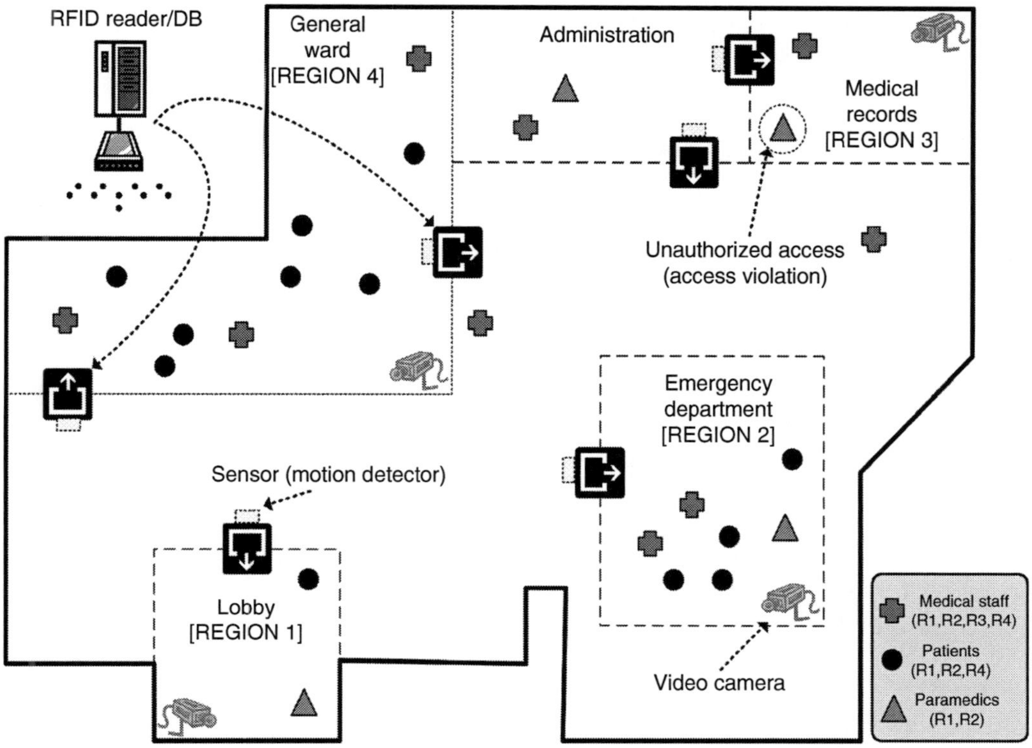

FIGURE 25.1
Instrumentation of a hospital's physical space for privacy protecting video surveillance.

25.2 System Architecture

Figure 25.2 depicts a high-level outline of the system architecture. The infrastructure comprises the following components:

1. *Sensing Module:* Processes data from incoming sensors. More specifically, an RFID control component deals with RF-related messaging arriving at the readers. Data from motion detection sensors are also processed here. A video input module handles the incoming video stream data from the various surveillance cameras.

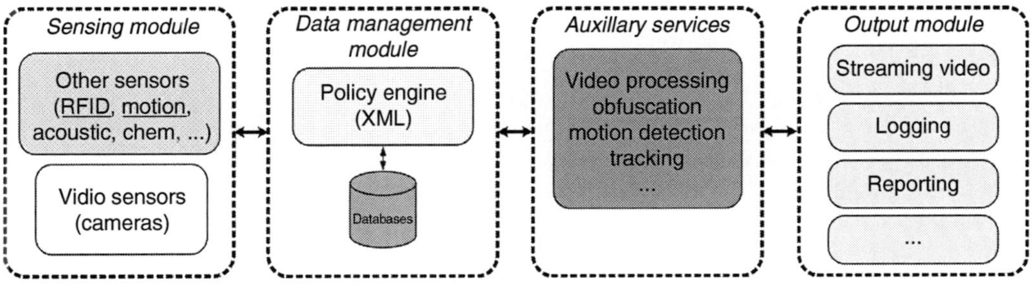

FIGURE 25.2
System architecture.

2. *Data Management Module:* Consists of an XML-based policy engine for access control. This policy engine interacts with a database subsystem consisting of profile and policy databases for the users.
3. *Auxiliary Services:* A service library contains modules that provide auxiliary services on the sensed information (including the incoming video stream(s)). These include obfuscation, motion detection, and object tracking modules. For example, masking may be applied to the video stream before it is passed to the output module, if the subject has been authorized by the policy engine.
4. *Output Module:* Handles customized reporting, logging, and video rendering functionality.

In particular, RFID sensor information is utilized (together with motion detection sensors) for localization of objects within the media space. The system has two major components, the RFID-based localization system and the video subsystem.

25.2.1 Radio Frequency Identification Technology

RFID technology provides the ability to interrogate data content without contact and the necessity of the line-of-sight communication. RFID is a means of storing and retrieving data through electromagnetic transmission to an RF-compatible integrated circuit [3]. A typical system consists of a tag (or a set of tags), a reader/receiver that can read and write data to these tags, and optionally a field generator (for relaying information from a region). The communication occurs at a predefined radio frequency and utilizes a protocol to read and receive data from the tags via inductive coupling.

The system is instrumented as follows; each protected region is equipped with a field generator. The boundaries between adjacent spaces can either be physical, as in a door or wall separating two rooms, or virtual, as in a nonphysical partition used to separate parts of a large room. Since we are interested in entry (and exit) to a region, each field generator is equipped with a motion detector that triggers a read of the region when motion is detected. If there is no tag information associated with the motion, the signal sent to the reader is categorized as unauthorized and video surveillance of the region is triggered. Tags are distributed to personnel, and a database stores the access rights associated with each tag such as desired user privacy levels (which are subsequently mapped to video masking techniques). When entry into a region is detected, the tag information is read (if present) and that information is passed to the RFID control module that forwards an authorization request to the policy engine. The policy decision for the object is then passed to the video-processing module, which uses this information in rendering the video object.

25.2.2 Video Processing Subsystem

A flow chart identifying the main components of the current video-processing subsystem is depicted in Figure 25.3. Our video analysis software is based on detection of moving foreground objects in a static background. The choice of a relatively simple algorithm is motivated by the need for real-time processing. Using a background model, the pixel detection module passes the result to a simple four-connected component algorithm to cluster pixels into blobs. The object tracker identifies objects entering the field of view of the camera and localizes them in the video frame using information passed to it by the RFID control module. Depending on access control policy, the stream is further processed using one of a number of masking techniques (Figure 25.4) before being displayed by the rendering module. The entire process is tuned to maintain a frame rate of approximately

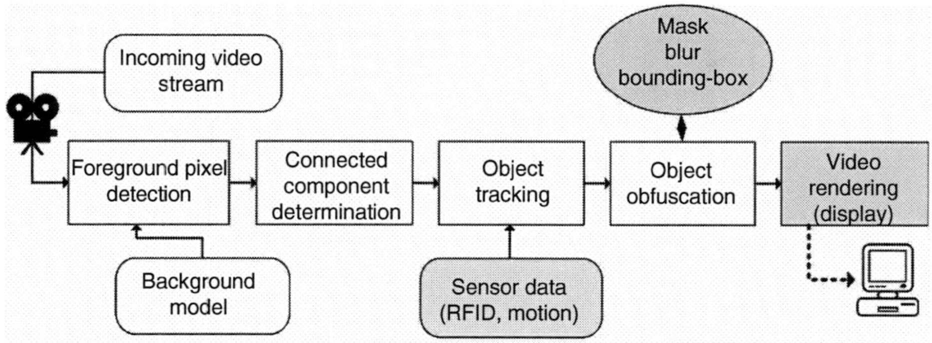

FIGURE 25.3
Flow chart of the video subsystem.

30 fps. It should be noted that the video camera capture facility is motion triggered. This way, video is only captured when there is activity in the region being monitored, making it much easier to focus on events of interest and save storage if video data is being archived.

25.3 RFID-Driven Access Control

An integral part of utilizing the RFID hardware in the system is the ability to specify and enforce security policies based on information gathered via RFID. The access control model is crucial to the system and allows specification of spatial constraints (i.e., which regions a person has access to). It allows the implementer to specify policies that dictate the manner

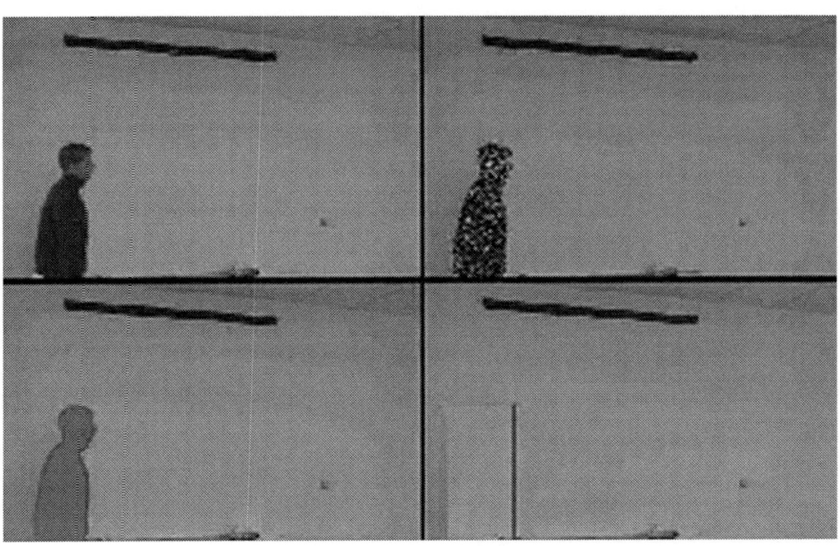

FIGURE 25.4
Privacy-protecting masking techniques for video utilized by our system. They represent different levels of user privacy. A frame of the original video is shown (*top-left*), followed by a noise/blur filter (*top-right*). A pixel-coloring approach is shown (*bottom-left*) followed by a bounding-box based technique (*bottom-right*), which hides details such as gender, race, or even dimensions of the person in question.

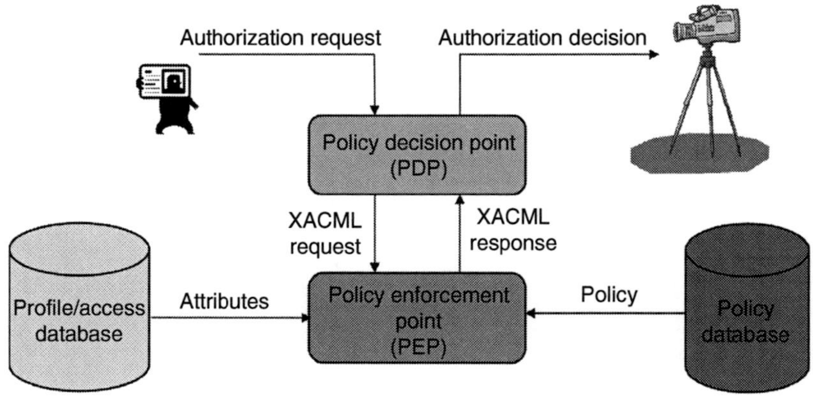

FIGURE 25.5
XACML-based policy framework.

in which video surveillance is conducted in a physical space. In other words, the policy decisions drive the video subsystem. For example, a member of the janitorial staff cleaning an office suite at 2 AM might be considered normal. However, if corporate policy prohibits entry to particular parts of the building after midnight, this event may be considered a potential security breach and needs to be further investigated. The entire policy subsystem is fairly complex and supports both simple and complex predicates; however, for the purposes of this chapter we describe the basic model that supports instantaneous policies (that is to say, the policies can be evaluated using the current RFID-sourced data).

Security policies are specified using the eXtensible Access Control Markup Language (XACML), which are processed by an enforcement engine that provides mediated access to a database. XACML [4] is utilized to define the access policies as well as carry out enforcement on these policies. XACML is a standard, general-purpose access control policy language defined using XML. It is flexible enough to accommodate most system needs, so it may serve as a single interface to policies for multiple applications and environments. In addition to defining a policy language, XACML also specifies a request and response format for authorization decision requests, semantics for determining policy applicability, etc. (Figure 25.5). The components of the access control model are the video-based objects, the potential users, and modes of access that can be modeled as a traditional authorization rule of the form $\langle s, o, m \rangle$, where subject s is authorized to access object o under mode m, where the mode is associated with a particular privacy level. In the following section we give a general description of the type of policies supported by our system and then give specific examples of their specification in XACML. (A simple example of subject specification is shown in Figure 25.6.)*

25.3.1 Access Control Framework

Here we present a framework for access control specification in the context of the system. For the purposes of this chapter, we outline a simplified specification of the framework. Assume a set of regions $R = \{r_1, \ldots, r_n\}$, which are monitored over time, a set of corresponding video streams $V = \{v_1, \ldots, v_n\}$ to these regions. Assume that the region r_i corresponds to the video stream v_i. But note that in a general setting more than one video stream

* In reality, any type of policy specification framework can be implemented and applied to our general set of primitives.

```
<!-- Access request by tag 1001 -->
<Subject>
  <Attribute AttributeId="urn:oasis:names:tc:xacml:1.0:subject:subjectid"
  <Data Type="urn:oasis:names:tc:xacml:1.0:data-type:rfc822Name">
  <Attribute Value>tag1201 </Attribute Value>
  </Attribute>
</Subject>
```

FIGURE 25.6
Subject specification in XACML.

can correspond to a region. There is a set of objects $O=\{o_1, \ldots, o_m\}$, which are being surveilled (e.g., people). These objects can be static (e.g., inventory) or mobile. Each of these objects may have an RFID tag associated with it, which in effect serves as its credential and basis for authorization to a particular region. The mapping between a particular tag and the corresponding object is stored in a profile database. We specify a set of tags $T=\{t_1, \ldots, t_x\}$ and use T as the subject of the authorization rules. Furthermore, each tag also has associated with it a privacy level P, where $P \in L=\{L_0, \ldots, L_N\}$ (Figure 25.4). These privacy levels specify the sensitivity at which video information pertaining to a particular object should be collected or viewed. The privacy level of a tag is determined at the time the tag is issued and can be modified later by an administrator. In our system, we use four levels of increasing sensitivity (L_0 corresponds to no privacy and L_3 to total privacy) and map a masking technique to each of these levels. The various video-masking techniques are discussed in the next section. Finally, to aid in specifying group-based authorization rules, we introduce an object association list, $OA \subseteq O$. This is a useful concept, not in the traditional notion of access groups but for associating inventory (i.e., equipment, etc.) with users authorized to use them. We use the more general notion here that members of OA are simply other objects, however, in practice these objects are of a static nature.

Therefore we can associate with a tag t_i (and therefore an object) a set $Rules := (Ruleset, d)$ which contains a *Ruleset* of authorization rules that allow or deny an action as well as an element d that defines the default ruling of this rule set. Default rulings can be one of $\{+, -, \emptyset\}$, corresponding to allow, deny, and don't care. An $AR_i \in Ruleset$ is a 5-tuple of the form $\langle r, p, oa, ts, res \rangle$, where $r \in R$, $p \in L$, $oa \subseteq OA$, ts is a temporal constraint (e.g., a time range, $[ts_{\min}, ts_{\max}]$ corresponding to the validity of the rule) and $res \in \{+, -\}$ corresponding to allow or deny. A deny result would imply that the object in the corresponding video stream would be unaltered and shown ($p=L_0$), whereas 'allow' would cause the stream to be altered so as to protect the identity of the object in question, depending on p. Additionally, authorization rules are subject to a deny-takes-precedence evaluation strategy. That is if an AR_i, $\langle r, p, oa, ts, - \rangle \in AR'$ exists, all tuples $\langle r, p, oa, ts, + \rangle \in AR'$ are removed.

Some typical examples are given below, following a generic authorization request of the form (tag, region, timestamp), which specifies the tag in question, the region it has entered, as well as the time of the request.

1. *Person with No Tag(s)* $(\emptyset, r_i, ts)$: A person entering a region with no tag will not have any rules associated with him/her aside from the default ruling, which will deny. This corresponds to a privacy level of L_0, meaning the person will be shown (Figure 25.7).
2. *Person with a Valid Tag* (t_i, r_i, ts): A person entering a region with a valid tag will satisfy spatial and temporal constraints and return an allow decision together with the corresponding privacy level, p. $\forall AR, t_i \rightarrow AR_i, \exists AR_i$ such that $r_i=r, \wedge (ts_{\min} \leq ts \leq ts_{\max}) \wedge (res_i=+)$ (Figure 25.7).

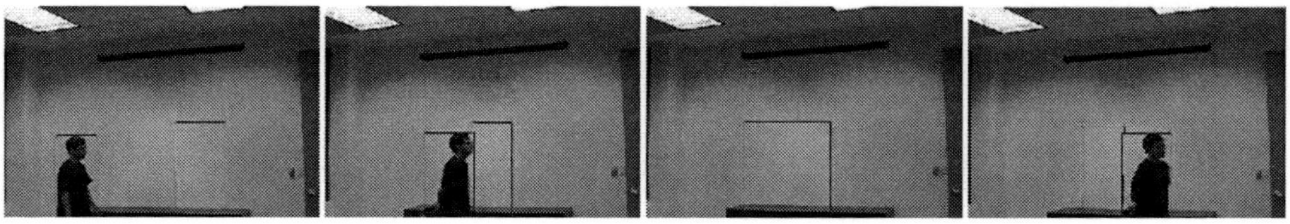

FIGURE 25.7
Interaction between authorized and unauthorized personnel in the space.

3. *Person with an Invalid Tag*: Assuming the tag has been successfully authenticated, two possible violations may cause a deny result from an authorization rule. (1) The access rights associated with the current tag specify that the requesting region is unauthorized, causing a spatial access violation. (2) The access rights associated with the current tag specify that the timestamp associated with the request does not satisfy the time bounds associated with the requesting region.* Figure 25.8 shows an authorization rule that enforces a time constraint on a region expressed in XACML.

```
<!-- Only allow entry into region from 8am to 1pm -->
<Condition FunctionId=" urn:oasis:names:tc:xacml:1.0:function:and">
<Apply FunctionId=" urn:oasis:names:tc:xacml:1.0:function:time-greater-
  than-or-equal"
<Apply FunctionId=" urn:oasis:names:tc:xacml:1.0:function:time-one-and-
  only">
<Environment AttributeSelector DataType=" http://www.w3.
  org/2001/XMLSchema#time"
<AttributeId=" urn:oasis:names:tc:xacml:1.0:environment:current-time"/>
</Apply>
<AttributeValue DataType=" http://www.w3.
  org/2001/XMLSchema#time">08:00:00</AttributeValue>
</Apply>
<Apply FunctionId=" urn:oasis:names:tc:xacml:1.0:function:time-less-than-
  or-equal"
<Apply FunctionId=" urn:oasis:names:tc:xacml:1.0:function:time-one-and-
  only">
<EnvironmentAttributeSelector DataType="http://www.w3.
  org/2001/XMLSchema#time"
<AttributeId=" urn:oasis:names:tc:xacml:1.0:environment:current-time"/>
</Apply>
<AttributeValue DataType=" http://www.w3.
  org/2001/XMLSchema#time">13:00:00</AttributeValue>
</Apply>
</Condition>
```

FIGURE 25.8
Example of a time-bounded authorization condition specified in XACML.

* We adopt two possible approaches to handle a violation of this manner. Either the person is immediately unmasked in the current region or remains masked until subsequent reentry into the region causes reevaluation of the associated access rights.

4. *Group-Based Authorization:* Associate a temporal threshold δ with each tag request (i.e., entry into a surveilled region). If multiple tag events are detected within this threshold, they are treated as a group authorization request. In which case, the respective object association lists are cross-checked.

25.4 System Implementation

Here we describe the setup and hardware for our first prototype implementation of the already mentioned framework, as well as some lessons learned from the initial deployment. Please note that this initial prototyping was done with RFID hardware that is little outdated now, and due to technological advances this may be more streamlined with newer hardware; however, the principles are still valid today.

25.4.1 Hardware and Setup

For our initial experiments we used Canon Optura 20 DV Cameras, capturing video at a resolution of 320 × 240 (typical for a surveillance application). One RFID reader, passive tags, and field generator relays were used for the instrumented regions. All processing was carried out on a uniprocessor Pentium 4 at 2.60 Ghz with 1 GB of RAM equipped with a Pinnacle firewire video capture card under Windows XP Professional. Microsoft's DirectX SDK (8.0) was used for interacting with the camera. The RFID reader was connected to the computer via a four-wire RS-232 interface, and we used medium-range RFID tags that transmit and receive at 916.5 and 433 Mhz, respectively. The TrakTags utilized in the experiments have a range of approximately 275 ft (85 m) and the field generators possess a wake-up range of between 6.5 and 108 ft (2–33 m). The instrumented space is illustrated in Figure 25.9. The range of the field generator relays are hardware adjustable, and were

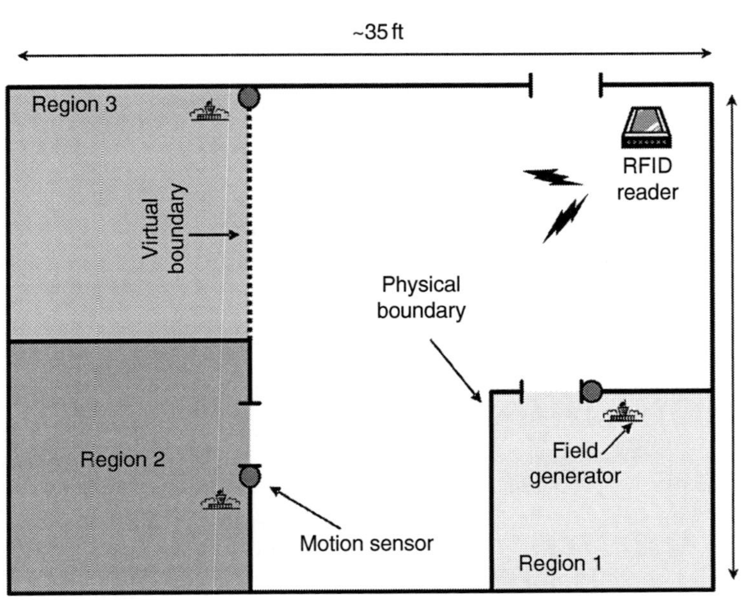

FIGURE 25.9
Experimental setup.

```
long TagEvent(long status, HANDLE funcId,
       rfTagEvent_t* tagEvent, void* user Arg)
{
  if(tagEvent->eventType == RF_TAG_DETECTED) {
    if(PolicyCheck(GenerateXACMLRequest
    (tagEvent->tag->id,tagEvent->fGenerator,enterTime)))
      signalVideoThread(videoSocket, szBuf);
  }
}
```

FIGURE 25.10
RFID event handling.

calibrated to detect activity in the regions of interest. We outfitted an office workspace by creating two regions with physical boundaries (rooms) and one open area that was partitioned by one of the relays via a virtual boundary. Motion sensors were used to monitor activity in and out of the chosen regions. As the motion detector senses movement, it wakes up the field generator which transmits a constant 433 MHz field to detect any tags in the region. This information is sent to the reader. Field generators and motion sensors were positioned at the entry points to the regions being studied. Each region was monitored by a single camera for the purpose of these experiments.

The RFID event handling process involves detection of a tag in a region, checking the associated access control policy, and relaying the result to the video subsystem. On detection of a tag, the tag ID, reporting field generator ID (effectively the zone identification), and timestamp is obtained. A separate thread in the video process waits for this information on a socket and depending on the policy decision renders the object (which may be masked). Figure 25.10 depicts the high-level RFID event handling procedure for (nongroup) events.

25.5 Results and Observation

We demonstrated the functionality of our framework by testing a variety of cases. In the context of each of the scenarios outlined in Section 25.3.1, we also tested the functionality of our tracking and masking algorithms. Interaction between multiple people with varying access rights were tested, especially merging and splitting between masked and unmasked objects. Here we can see the merging/splitting process as they pass each other in view of the camera. In evaluating the performance of our implementation, we investigated the overheads involved in our techniques. We observed that for a single camera the overhead of the various masking techniques (shown in Figure 25.4) were negligible because of the limited field of view of the camera and hence the number of people interacting simultaneously at any given time. We chose a resolution of 320×240 as it strikes a good balance between detail and performance (both in terms of framerate and storage space if the video was being archived). Doubling the resolution to 640×480 affected the framerate significantly as expected, as on average just under 20 fps was achieved. At our target resolution of 320×240, we achieved a constant framerate of 30 fps in all tested cases.

Our video analysis software is based on detection of moving foreground objects in a static background. The choice of a relatively simple algorithm is motivated by the need for real-time processing. Using a background model, the pixel detection module passes the result to a simple four-connected component algorithm to cluster pixels into blobs. The object tracker identifies objects entering the field of view of the camera and localizes them in the video frame using information passed to it by the RFID control module. Depending on access control policy, the stream is further processed using one of a number of masking techniques (Figure 25.4) before being displayed by the rendering module. This entire process is tuned to maintain a frame rate of approximately 30 fps. It should be noted that the video camera capture facility is motion triggered. This way, video is only captured when there is activity in the region being monitored, making it much easier to focus on events of interest and save storage if video data is being archived.

After each frame gets processed into legitimate blobs, the tracker tries to match each person in the list with a candidate blob in the incoming frame using motion vector prediction [5]. In essence, if a person's cog falls within one of the candidate blobs' bounding box in the current frame, then the information is assigned to that particular blob (cog, min/max parameters, area, etc.). If a match is not found between an object and all the blobs in the current frame, the object is removed from the list and assumed to have exited the scene. If a new blob does not have a match in the object list, a new object is created and assigned authorization information (masking level) by the RFID subsystem. During the rendering process, the blob's corresponding masking level is applied to the outgoing video stream. In the case of a merge, the highest masking level among the persons involved in the merge is applied to the corresponding blob. For example, if a person p_1, with privacy level L_1 merges with a person p_2, with privacy level L_2, then the merged blob is masked using L_2.

The PADoC system is being deployed in the CAL-IT^2 building at the University of California, Irvine. The entire building has been instrumented with video cameras and various sensors at the entry and exit points. A general class of policies that are being investigated in the context of this building (in a privacy-preserving manner) are as follows:

1. *Limiting Access to Spaces*: These include triggers such as "Person A is allowed to enter Room X between the hours of 3–5 PM" or "The Robotics group is allowed access to the server room on their own floor."
2. *Protection of Resources*: These types of policies allow control over inventory items. For example, "Members of group A are not allowed to bring/take *K* laptops into/from the conference room."
3. *Monitoring/Flagging of Suspicious Activity*: For example, "Any individual who is present in the lobby area more than 10 times in a 24 h period should be flagged."
4. *Studying Human Interactions*: Policies can be devised for cross-party interaction. For example, "Raise a flag if the number of (spatial) interactions between the Middleware and A.I. groups exceed 25 in a given week." Now this is really a composite trigger similar to that defined for monitoring/flagging. Such a trigger will be translated to a union of two counting triggers, which detect entries by each group into the other's space.

Even though we have realized a fully functional implementation of our framework, the deployment of a such a system should eventually be pushed to the camera level. In this implementation, we tightly coupled the processing capabilities (of the PC) to the camera and processed everything in real time with no archival of the original video data (only the processed video stream is available). Ideally this processing capability should reside at the camera level to make the privacy preservation in media spaces more acceptable to the end users. Optimization of the algorithms used here for object tracking and masking for privacy is a key component of such a realization as well as the possible use of MPEG-4 video. MPEG-4 has a superior compression efficiency, advanced error control, object-based functionality, and fine-grain scalability making it highly suitable for streaming video applications. Recently, MPEG-4 has emerged as a potential front runner for surveillance applications because of its layered representation, which is a natural fit for surveillance tasks as they are inherently object based. It is also desirable to find people moving in the scene, independent of the background. Further, more fine-grained localization via the use of RFID and other low-cost sensors is also something that improves the accuracy of the system. One of the features of these types of integrated applications is that the overall system is as strong as each of the various subcomponents; thus it is possible to improve them somewhat independently with the infrastructure in place.

25.6 Conclusions

In this chapter, we addressed privacy challenges that arose as a result of collecting environmental data for the purposes of surveillance (via RFID and video sensors) in a pervasive space. The fusion of RFID-based localization technology with traditional video surveillance facilitated end-to-end privacy-preserving video surveillance (which would not have been possible without both the video obfuscation and RFID-based localization). Access control policies were used to implement predicates that would serve as triggers to the surveillance system while still protecting the identity of those individuals present in the video feed that were not violating the policies in effect. We view this work as the first step toward a grandiose goal of realizing privacy-preserving pervasive environments. Our work has opened up numerous challenges that we listed in the introduction. Our immediate goals are to improve performance and scalability of our approach as well as to extend our approach to deal with more general classes of predicates. Much of this work requires us to explore formal notions of privacy measures in pervasive environments. Quantifiable measures similar to K-anonymity, uncertainty, entropy but appropriate for pervasive applications will provide us a way to explore trade-offs between privacy and utility including scalability and generalizability of the approach.

References

1. J. Wickramasuriya, M. Datt, S. Mehrotra, and N. Venkatasubramanian. Privacy protecting data collection in media spaces. In *ACM Multimedia 2004*, New York, October 2004.
2. Privacy International. Privacy International: Video Surveillance. http://www.privacyinternational.org/issues/cctv/index.html.
3. F. Heubler, M. Chiesa, and R. Genz. RFID: A week long survey on the technology and its potential. Technical report, Harnessing Technology Project, Interaction Design Institute Ivrea, 2002.
4. S. Godik and T. Moses (Eds.). *eXtensible Access Control Markup Language* (*XACML*) 1.0 Specification Set. OASIS Standard, 2003.
5. A. Lipton, H. Fujiyoshi, and R. Patil. Moving target classification and tracking from real-time video. In *Proceedings of IEEE Image Understanding Workshop*, pp. 129–136, 1998.

Section IV

Security and Privacy

26

Is RFID Technology Secure and Private?

Surinder Mohan Bhaskar

CONTENTS

26.1 Introduction

Radio frequency identification (RFID)[1] chips are used everywhere. A number of examples can be quoted where RFID technology has been implemented—companies and laboratories use them as access keys, to start their cars, and as inventory tracking devices. Drug manufacturers rely on chips to track pharmaceuticals. In the near future, RFID tags are also about to get a lot more personal. Next generation U.S. passports and credit cards will contain RFIDs, and the medical industry is exploring the use of implantable chips to manage patients in an effective manner. According to the RFID market analysis firm IDTechEx, the push for digital inventory tracking and personal ID systems will expand the current annual market for RFIDs from $2.7 billion to as much as $26 billion by 2016 Shadow of RFID chip and antenna when held close to a lamp is shown in Figure 26.1.[2]

During World War II, the British placed radio transponders in Allied aircraft to help early radar system crews detect "good" guys from "bad" guys. The first chips were developed in research laboratories in the 1960s, and by the next decade the U.S. government was using tags to electronically authorize trucks coming into Los Alamos National Laboratory and other secure installations. Commercialized chips became widely manufactured and available in the 1980s, and RFID tags were used to track difficult-to-manage property like farm animals and railroad cars, and so on. But over the last few years, the market for RFIDs has exploded, driven by advances in computer databases and supported by declining chip prices. Now a number of companies, from Motorola to Philips to Texas Instruments, manufacture the chips.[3]

The tags work by broadcasting a few bits of information to specialized electronic readers. Most commercial RFID tags are passive emitters and have no onboard battery:[4] these tags get activated by the reader power. Once activated, these chips broadcast their signal indiscriminately within a certain range, usually a few inches to a few feet. However, active RFID tags with internal power can send signals to hundreds of feet; these are deployed in the automatic toll-paying devices (with names like FasTrak and E-ZPass) that sit on car dashboards, pinging tollgates as autos whiz through.

For protection of information, RFID signals can be encrypted using suitable algorithms. The chips that are used for applications like passports, for example, will likely be

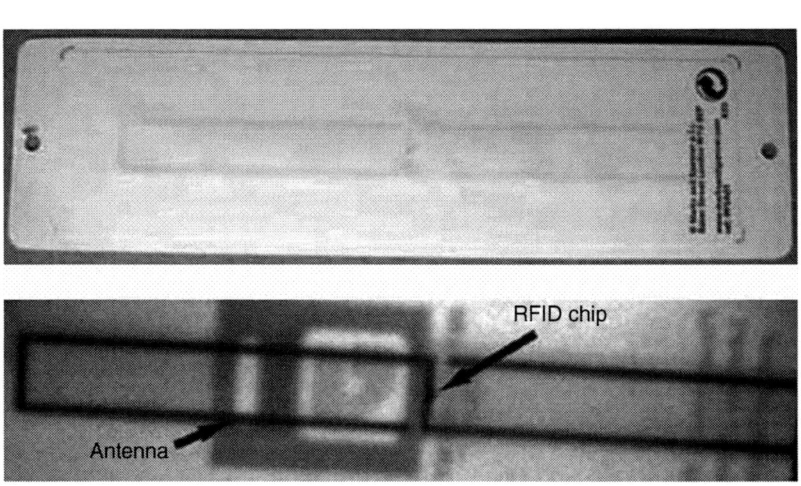

FIGURE 26.1
Shadow of the RFID chip and antenna when held close to a lamp.

coded/encrypted to make it difficult for unauthorized readers to retrieve their onboard information (which will include a person's name, age, nationality, and photo and other sensitive information). But then, most of the commercial RFID tags do not include security as it is very expensive.

This leaves most RFIDs vulnerable to cloning and data tampering, if the RFID chip has a writable memory area. RFID chips that are used to track product shipments or expensive equipment, for example, often contain pricing and item information. These writable areas can be locked, but often they are ignored, either because the companies using RFIDs do not know the working of the chips, or the data fields need to be updated frequently. Either way, these chips are open to hacking or tampering of data.

The world of RFID is like the Internet in its early stages. No one had thought about building security features into the Internet in its early stages, and now we are paying for it in viruses and other attacks by adversaries. The same thing is also true of RFIDs (Figure 26.2).

Hacking of RFID chips is very easy. One can steal the smart card, lift someone's passport, jack someone's car, and even clone the chip embedded in an arm. There are so many accounts of how RFID has been hacked and one such case is shown in Figure 26.3.[5,6]

A wealthy software entrepreneur, James Van Bokkelen, was victimized by a hacker with a laptop. This was not an e-mail scam or bank account hack but something different. An adversary planned to use a cheap, homemade USB device to swipe the office key out of Van Bokkelen's back pocket. He simply got his hand within a few inches of him. As Van Bokkelen approached from the parking lot, the adversary brushed past him. A coil of copper wire flashed briefly in the hacker's palm, then disappeared.

The coil was an antenna for the wallet-sized device known as a cloner, which was concealed up his sleeve. This cloner can elicit, record, and mimic signals from smart card RFID chips. The hacker connected the device to his laptop with a USB cable and downloaded the data from Van Bokkelen's card for processing. Then, once he retrieved

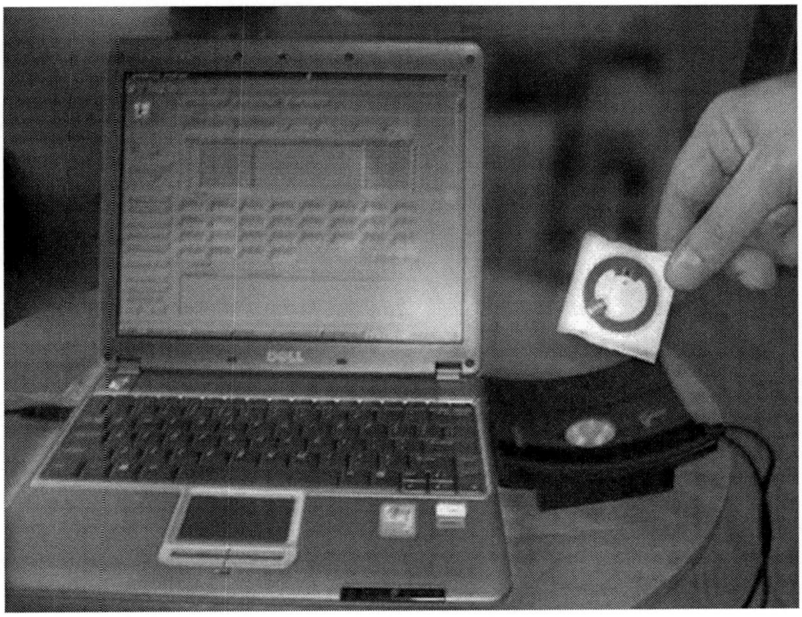

FIGURE 26.2
World's first RFID chip infected with a virus.

FIGURE 26.3
German hacker-cloned RFID e-passport. (From German Hacker Clone e-Passport, http://www.engadnet.com/2006/08/03/german-hackerscolone-rfid-e-passports/)

the code, the hacker switched the cloner from Record mode to Emit. He headed toward the locked door and waved the cloner's antenna in front of a black box attached to the wall. The single red LED blinked green. The lock clicked, and he walked in. Thus, we see how a robbery can be committed by exploiting the information present on an RFID chip. It was so simple and anybody could have very easily walked off with tens of thousands of dollars' worth of computer equipment, and possibly source code worth even more.

In a library, destroying the data on the books' passive-emitting RFID tags is possible by wandering the aisles with an off-the-shelf RFID reader–writer and a laptop. These tags store several writable memory "pages" that store the books' bar codes and loan status, and other information. The RFID-enabled checkout is indeed quite convenient. As the hacker leaves the library, he stops at a desk equipped with a monitor, and shows the books one at a time face up on a metal plate. The titles instantly appear on-screen. A person can borrow four books in less than a minute without bothering the librarian. In one case, a student took the books to his office, where he used a commercially available reader to scan the data from their RFID tags. The reader fed the data to his computer, which was running software that the student had ordered from RFID-maker, Tagsys. As he waved the reader over a book's spine, ID numbers popped up on his monitor. He then found an empty page in the RFID's memory and typed "AB." When he scanned the book again, he saw the bar code with the letters "AB" next to it. This happened because of the Oakland library's failure to lock the writable area. One could even erase the bar codes and then lock the tags. And then the library would have to replace the books.

On the other hand, unlocking the library's tags makes it easier for libraries to change the data in future.

The Future Store in Rheinberg, Germany is the world's preeminent test bed of RFID-based retail shopping. All the items in this high-tech supermarket have embedded RFID price tags, which allow the store and individual product manufacturers—for example, Gillette, Kraft, and Procter & Gamble—to gather near real-time feedback on what is bought. In July 2004, *Wired* hailed the store as the "supermarket of the future." A few months later, German security expert Lukas Grunwald hacked the chips and showed the vulnerability of RFID chips.

Grunwald co-wrote a program called RFDump, which allows access and alters price chips using a PDA (with an RFID reader) and a PC card antenna. With the permission of the store owner, he and his colleagues strolled the aisles, downloading information from hundreds of sensors. They demonstrated how easily they could upload data from one chip onto another. He also showed how he could download the price of a cheap wine into RFDump and then cut and paste it onto the tag of an expensive bottle. The price-switching stunt drew media attention. Today, Grunwald continues to pull even more elaborate pranks with chips from the Future Store in a more sophisticated manner.

Apart from pranks, vandalism, and thievery, Grunwald has recently discovered another use for RFID chips: *espionage*. He programmed RFDump with the ability to place cookies on RFID tags the same way Web sites put cookies on browsers to track returning customers/users. With this, a stalker could place a cookie on his target's E-ZPass, then return to it a few days later to see which toll plazas the car had crossed and at what time. Private citizens and the government could likewise place cookies on library books to monitor who is checking/reading them out.

Despite the increasing popularity of RFID technology, the electronic information it deals with may not be as secure as was once thought. At least that is the story that has emerged from the recent Defcon 2006 network security conference. This 3 day event is a Mecca of sorts for network security experts, programmers, and hackers, who congregate yearly to test their skills against one another, and to show corporations, consumers, and government agencies how vulnerable their networks are, without the risk of criminal prosecution or financial liabilities.

The following sections raise serious concerns about the use of RFID tags. Newer technologies allow everyday objects to beam electronic data to computers equipped with special antennas. Lukas Grunwald successfully demonstrated that electronic passports (e-passports) being introduced in the United States have a major vulnerability in that they could allow criminals to clone embedded secret codes and illegally enter countries with ease. For the demonstration, Grunwald copied personal information stored on an e-passport document and transferred it to another device. He disproved assurances by officials in government and participating private industry that the electronic information stored in e-passports could not be duplicated. When the first RFID passports came out in the United Kingdom, the encryption on the chips was broken within 48 hours.

In 1997, *ExxonMobil* equipped thousands of service stations with SpeedPass, which let customers show a small RFID device attached to a key chain in front of a pump to pay for gas. About 7 years later, three graduate students—Steve Bono, Matthew Green, and Adam Stubblefield—robbed a station in Baltimore using a laptop and a simple RFID broadcasting device. They tricked the system into letting them fill up for free.

The theft was invented by Avi Rubin's computer science laboratory at Johns Hopkins University. Rubin's laboratory is best known for having found massive, hackable flaws in the code running on Diebold's widely adopted electronic voting machines in 2004.

Working with RSA Laboratories manager Ari Juels, the group demonstrated how to crack the RFID chip in ExxonMobil's SpeedPass.[7]

Hacking the tag made by Texas Instruments,[3] is not as simple as breaking into Van Bokkelen's Sandstorm offices with a cloner. The radio signals in these chips, dubbed DST tags, are protected by an encryption cipher that only the chip and the reader can decode. Unfortunately, says Juels, "Texas Instruments used an untested cipher," and this is a case of poor quality control. The Johns Hopkins laboratory found that the code could be broken with a "brute-force attack," in which a special computer known as a cracker is used to try thousands of password combinations per second until it hits on the correct one. With a home-brewed cracker that cost a few hundred dollars, Juels and the Johns Hopkins team successfully performed a brute-force attack on TI's cipher in just 30 min, contrary to experts' estimate that it would take hundreds of years for today's computers to break the publicly available encryption tool SHA-1, which is used to secure credit card transactions on the Internet.[8]

Such stories about hacking the security of RFID tags and many more will keep on adding to the list in the coming years.

26.2 Problems with RFID[5,9–13]

The problems with RFID can be divided into the following three categories:

1. Technology related
2. Privacy and ethics related
3. Security related

26.2.1 Technology-Related Problems

26.2.1.1 Problems with RFID Standards

RFID has been implemented in different ways by different manufacturers; global standards are still being developed and interoperability is a serious issue. It should be noted that some RFID devices are never meant to leave their network as the RFID tags are used for inventory control within a company. This can cause problems for companies.

Customers may also face problems with RFID standards and interoperability issues. For example, ExxonMobil's SpeedPass system is a proprietary RFID system; if another company wanted to use the convenient SpeedPass (say, at the drive-in window of your favorite fast food restaurant), they would have to pay to access it—an unlikely/unwanted scenario. On the other hand, if every company had its own "SpeedPass" system, a consumer would need to carry many different devices, which goes against user-friendliness of the technology.

There are well-developed standards for low- and high-frequency RFID systems, but most companies want to use UHF in the supply chain because it offers longer read range—up to 20 ft under good conditions. UHF technology is relatively new, and standards were not established until recently and are still evolving. Another issue is cost. RFID readers typically cost $1000 or more. Companies would need thousands of readers to cover all their factories, warehouses, and stores. RFID tags are also fairly expensive—20 cents or more—which makes their use for identifying millions of items that cost only a few dollars impractical.

26.2.1.2 RFID Systems Can Be Easily Disrupted

Since RFID systems make use of the electromagnetic spectrum (WiFi networks or cell phones), they are relatively easy to jam by employing energy at the right frequency. Although this would only be an inconvenience for consumers in stores (longer waits at the checkout), it could be disastrous in other environments (e.g., hospitals, battlefields) where RFID is increasingly used.

In addition, active RFID tags (those that use a battery to increase the range of the system) can be repeatedly interrogated to wear the battery down, thus disrupting the system.

26.2.1.3 RFID Reader Collision/Interference

Reader collision occurs when the signals from two or more readers overlap. The tag is unable to respond to simultaneous queries. Systems must be carefully set up to avoid this problem; many systems use an *anticollision protocol* (also called a *singulation protocol*). Anticollision protocols enable the tags to take turns in transmitting to a reader.

26.2.1.4 RFID Tag Collision

Tag collision occurs when many tags are present in a small area; but since the read time is very short, it is easy for vendors to develop systems that ensure that tags respond one at a time by employing suitable algorithms.

26.2.2 Security, Privacy, and Ethics Problems with RFID[12,13]

The following problems with RFID tags and readers have been reported.

26.2.2.1 Contents of an RFID Tag Can Be Read after the Item Leaves the Supply Chain

An RFID tag cannot tell the difference between one reader and another. RFID scanners are very portable; RFID tags can be read from a distance, from a few inches to a few yards. This allows anyone to see the contents of your purse or pocket as you walk down the street. It is also possible that some tags can be turned off when the item leaves the supply chain.

26.2.2.2 RFID Tags Are Difficult to Remove

RFID tags are difficult to remove; some are very small (less than a half millimeter square and as thin as a sheet of paper)—others may be hidden or embedded inside a product where consumers cannot see them. New technologies allow RFID tags to be "printed" right on a product and may not be removable at all.

26.2.2.3 RFID Tags Can Be Read without Your Knowledge

Since the tags can be read without being swiped or obviously scanned (as is the case with magnetic strips or bar codes), anyone with an RFID reader can read the tags embedded in your clothes and other consumer products without your knowledge. For example, you could be scanned *before* you enter the store, just to see whether you are carrying an RFID chip, and so on. You might then be approached by a clerk who knows what you have in your backpack or purse, and can suggest accessories or other items matching the amount you have. This can pose a serious threat to one's privacy and security.

26.2.2.4 RFID Tags Can Be Read at Greater Distances with a High-Gain Antenna

For various reasons, RFID reader/tag systems are designed so that the distance between the tag and the reader is kept to a minimum (see the material on tag collision earlier). However, a high-gain antenna can be used to read the tags from much further away, leading to serious privacy problems.

26.2.2.5 RFID Tags with Unique Serial Numbers Could Be Linked to an Individual Credit Card Number[14]

At present, the Universal Product Code (UPC) implemented with bar codes allows each product sold in a store to have a unique number that identifies that product. Work is in progress on a global system of product identification that would allow *each individual item* to have its own number. When the item is scanned for purchase and is paid for, the RFID tag number for a particular item can be associated with a credit card number.

26.2.3 Security Issues

The security problems surrounding RFID technology can be grouped into several classes:

1. *Data ownership and data-mining techniques*: All methods of data collection involve privacy, data ownership, and the ethical use of data-mining techniques to discover the characteristics of an individual or an organization. For example, customer-loyalty card data could be used to find out private medical information about an individual. This problem predates the use of RFID technology, but the sheer volume of data provided by RFID tags adds a new urgency to these discussions.
2. *Data theft*: For data theft, two things are required: (1) access to a computer system and (2) considerable hacking skills to steal data. RFID tags are made to broadcast information; the possibility of data theft by easily concealable RFID scanners is very real and easy. Chip manufacturers counter this by adding security features such as secure encryption schemes to the chips and data.
3. *Data corruption*: Most RFID tags are rewritable. This feature may be locked (turning the tag into a write-once, read-many device) or left active, depending on application and security sensitivity. For example, in libraries, the RFID tags are frequently left unlocked for the convenience of librarians in reusing the tags on different books or to track check-ins and check-outs. But when tags that should be locked are not locked (e.g., in the supply chain management), the potential does exist for pranksters or malicious users to rewrite the tags with incorrect or fraudulent data.

26.3 Concerns about How RFID Will Be Used[15,16]

Civil liberties groups, among others, have become increasingly concerned, in particular, about the use of RFIDs to track the movement of individuals. For example, passports will soon be required to contain some sort of RFID device to speed border crossings. Scanners placed throughout an airport, for example, could track the location of every passport over time, from the moment you left the parking lot to the moment you got on your airplane.

The Japanese government passed a draft RFID Privacy Guideline that stated the following:

- Indication that RFID tags exist
- Consumer's right of choice regarding reading tags
- Sharing information about social benefits of RFID, and so on
- Issues on linking information on tags and databases that store privacy information
- Restriction of information gathering and use when private information is stored on tags
- Assuring accuracy of information when private information is stored on tags
- Information administrators should be encouraged
- Information sharing and explanation for consumers

There are also concerns about the fact that, even after you leave the store, any RFID device in the things that have been bought is still active. This means that a thief could walk past you in the mall and know exactly what you have in your bags, marking you as a potential victim. A thief could even circle your house with an RFID scanner and pull out data on what you have in your house before robbing it.

Military hardware and even clothing have RFID tags to help track each item through the supply chain. Some analysts are concerned that if there are particular items associated with high-level officers, roadside bombs could be set to go off when triggered by an RFID scan of cars going by. This may pose a serious threat to the internal and national security of a nation as a whole.

There was a recent report revealing clandestine tests at a Wal-Mart store where RFID tags were inserted in packages of lipstick, with scanners hidden on nearby shelves. When a customer picked up a lipstick and put it in her cart, the movement of the tag was registered by the scanners, which triggered surveillance cameras. This allowed researchers 750 miles away to watch those consumers as they walked through the store, looking for related items.

26.4 Various Problematic Situations[10,11]

1. Anyone with an appropriately equipped scanner and close access to the RFID device can activate it and read its contents with ease. Obviously, some concerns are greater than others. If someone walks by your bag of books from the bookstore with a 13.56 MHz "sniffer" with an RF field that can activate the RFID devices in the books you bought, that person can get a complete list of what you just bought. That is certainly an invasion of your privacy, but it could be worse in other cases.
2. Another scenario involves a military situation in which the other side scans vehicles going by, looking for tags that are associated with items that only high-ranking officers can have, and targeting accordingly.
3. The increasing use of RFID devices in company badges is also causing concern. A suitable RF field will cause the RFID chip in the badge to "spill the beans" to whoever activates it. This information is replayed to company scanners, allowing the thief access, and your badge is the one that is "credited" with the access.
4. The smallest tags that will likely be used for consumer items do not have enough computing power to support data encryption to protect your privacy. The most they can handle is PIN-style or password-based protection. This may result in serious security problems.

5. RFID tags inserted in credit cards, passports, and driving licenses being issued is a serious cause for worry. Soon, there will be RFID tags in money. RFID tags are also not very secure. When the first RFID passports came out in the United Kingdom, the encryption on the chips was broken within 48 h.
6. A fact is that without encryption, ''Anyone within range can query a tag and find out what's on them. As we get better performing tags, the longer will be the range over which the tag will transmit.'' Longer range means more potential intruders. RFID tag-to-reader encryption is ''being worked on'' by major players but has been challenging because passive RFID tags are powered by readers, then reflect back a signal communicating their information, with little power left over to set up an encryption channel. But lack of encryption can aid in petty breaches, and might make it possible to corrupt or hack data.
7. Passive RFID tags cannot be read from >20 ft or so. But active RFID tags, which use a battery to broadcast a signal and are used on cargo containers and other large assets, could be read from a satellite if there is little RF ''noise'' (ambient RF energy that causes interference) and the broadcasted signal is powerful enough.

26.5 Other Security Concerns[10,11]

The main security concern surrounding RFID technology is the illicit tracking of RFID tags. Tags that are world-readable pose a risk to both personal location privacy and corporate/military security. Such concerns have been raised with respect to the U.S. Department of Defense's recent deployment of RFID tags for supply chain management. Further more, privacy organizations have expressed concern in the context of ongoing efforts to embed electronic product code (EPC) RFID tags in consumer products.

Cryptography is used to prevent tag cloning. Some tags employ a form of ''rolling code'' scheme, wherein the tag identifier information changes after each scan, thus reducing the usefulness of observed responses. More sophisticated devices engage in challenge response authentications[17] where the tag interacts with the reader. In these protocols, secret tag information is never sent over the insecure communication channel between the tag and the reader. Rather, the reader issues a challenge to the tag, which responds with a result, computed by using a cryptographic circuit keyed with some secret value. Such protocols may be based on symmetric or public key cryptography (asymmetric key cryptography). Cryptographically enabled tags typically require dramatically higher cost and power than simpler equivalents, and as a result, deployment of such tags is much more limited. This cost/power limitation has led some manufacturers to implement cryptographic tags using substantially weakened or proprietary encryption schemes/algorithms, which do not necessarily resist sophisticated attack/penetration. For example, the ExxonMobil Speed-Pass uses a cryptographically enabled tag manufactured by Texas Instruments,[3] called the Digital Signature Transponder (DST),[18] which incorporates a weak, proprietary encryption scheme to perform a Challenge Response Protocol.[17,19]

Still other cryptographic protocols attempt to achieve privacy against unauthorized readers, though these protocols are largely in the research stage. One major challenge in securing RFID tags is a shortage of computational resources within the tag. Standard cryptographic techniques require more resources than are available in most low-cost RFID devices. RSA Security[7] has patented a prototype device that locally jams RFID signals by interrupting a standard collision avoidance protocol, allowing the user to prevent identification if desired. Various policy measures have also been proposed, such as marking RFID-tagged objects with an industry standard label.

26.5.1 Viruses[20]

Ars Technica reported in March 2006 an RFID buffer overflow bug that could infect airport terminal RFID databases for baggage and passport databases to obtain confidential information on the passport holder.

26.5.2 Passports

In an effort to make passports more secure, several countries have implemented RFID in passports. However, the encryption on the U.K. chips was broken in <48 h leaving millions of citizens vulnerable. Since then, further efforts have allowed researchers to clone passport data while the passport is mailed to its owner. Where before a criminal had to secretly open and then reseal the envelope, now it can be done with ease without detection adding significant insecurity to the passport system.

26.6 Controversies[10–13,21,22]

26.6.1 Privacy

> *How would you like it if, for instance, one day you realized your underwear was reporting on your whereabouts?*—California State Senator Debra Bowen, at a 2003 hearing.

The use of RFID technology has engendered considerable controversy and even product boycotts by consumer privacy advocates such as Katherine Albrecht[23] and Liz McIntyre of CASPIAN who refer to RFID tags as "spychips." The four main privacy concerns regarding RFID are as follows:

1. The purchaser of an item will not necessarily be aware of the presence of the tag or be able to remove it, thus leaving him vulnerable to privacy breaches.
2. A tag can be read at a distance without the knowledge of the individual, which is a serious security concern.
3. If a tagged item is paid for by credit card or in conjunction with the use of a loyalty card, then it would be possible to tie the unique ID of that item to the identity of the purchaser. This poses a serious risk to the privacy of the individual concerned.
4. EPCglobal system of tags creates globally unique serial numbers for all products, which enables tracking of the product and the user of this technology.

Most concerns revolve around the fact that RFID tags affixed to products remain functional even after the products have been purchased and taken home, and thus can be used for surveillance and other purposes unrelated to their supply chain inventory functions. This is a breach of security and privacy of the individual.

The earlier concerns may be addressed partly by the use of the clipped tag. The clipped tag is an RFID tag designed to increase consumer privacy. The clipped tag has been suggested by IBM researchers, Paul Moskowitz and Guenter Karjoth. After the point of sale, the consumer may tear off a portion of the tag. This allows the transformation of a long-range tag into a proximity tag that still may be read, but only at short range—less than a few inches or centimeters. The modification of the RFID tag may be confirmed visually. The RFID tag may still be used later for returns, recalls, or recycling.

However, this is not a realistic mitigation considering that the read range is far more a function of the reader than the tag itself. Improvements to technology or simply having readers very close to the tags (such as the security readers at the doors of most major retail stores) will make even supposedly short-range tags easy to read. All the abuses listed previously are therefore still possible.

Another privacy issue is due to RFID's support for a singulation (anticollision) protocol. This is the means by which a reader enumerates all the tags responding to it without mutual interference. The structure of some collision–resolution (Medium Access Control) protocols is such that all but the last bit of each tag's serial number can be deduced by passively eavesdropping on just the *reader's* part of the protocol. Because of this, whenever the relevant types of RFID tags are near readers, the distance at which a tag's signal can be eavesdropped is irrelevant; what counts is the distance at which the much more powerful reader can be received. Just how far this depends on the type of the reader and its power, but in the extreme case some readers have a maximum power output of 4 W, enabling signals to be received from tens of kilometers away. However, more recent UHF tags employing the EPCglobal Gen 2 (ISO 18000-6C) protocol, which is a slotted-Aloha scheme in which the reader never transmits the tag identifying information, are not subject to this particular attack.

The Anti-Collision Scheme of ISO 15693[24] will render this rather implausible. To eavesdrop on the reader part of the protocol—and gather the 63 least significant bits of a UID—would require the reader to send a mask value of 63 bits. This can only happen when the reader detects a collision up to the 63rd bit. In other words, one can eavesdrop on the transmitted mask value of the reader, but for the reader to transmit a 63 bit mask value requires two tags with identical least significant 63 bits. The probability of this happening must be near zero, that is, the eavesdropper needs two virtually identical tags to be read at the same time by the reader in question. This may be good news for some time only.

While dealing with the issues of eavesdropping and skimming, it is important to make a distinction between inductively coupled and radiatively coupled tags. Protocols like ISO 15693[24] use 13.56 MHz radio frequencies and inductive coupling between the tag and the reader. The signal power falls very rapidly to extremely low levels a few antenna diameters away from the reader when inductive coupling is used; so an attacker must be within a few meters to intercept the reader signal, and closer to read a tag. Protocols like 18000-6C, which use 900 MHz signals, usually use radiative coupling between the tag and the reader; a wave is launched, its power falling roughly as the square of the distance. Tag signals can be intercepted from 10 m away under good climatic conditions, and the reader signal can be detected from kilometers away if there are no obstructions.

The potential for privacy violations with RFID was demonstrated by its use in a pilot program by the Gillette Company, which conducted a "smart shelf" test at a Tesco in Cambridge, England. They automatically photographed shoppers taking RFID-tagged safety razors off the shelf to see if the technology could be used to deter shoplifting. This trial resulted in consumer boycott against Gillette and Tesco. While doing business, it is very important to respect the privacy and sentiments of consumers.

In another incident, uncovered by the Chicago Sun-Times, shelves in a Wal-Mart in Broken Arrow, Oklahoma, were equipped with RFID readers to track the Max Factor Lipfinity lipstick containers stacked on them. Webcam images of the shelves were viewed 750 miles (1200 km) away by Procter & Gamble researchers in Cincinnati, Ohio, who could tell when lipsticks were removed from the shelves and observe the shoppers in action at near real-time basis.

In January 2004, privacy advocates from CASPIAN and the German privacy group FoeBuD were invited to the METRO Future Store in Germany, where an RFID pilot project was implemented. It was discovered/uncovered by accident that METRO "Payback" customer-loyalty cards contained RFID tags with customer IDs, a fact that was disclosed neither to customers receiving the cards, nor to this group of privacy advocates.

This happened despite repeated assurances by METRO that no customer identification data was tracked and all RFID usage was clearly disclosed. This is a serious example of privacy violation by METRO.

The controversy was furthered by the accidental exposure of a proposed Auto-ID consortium public relations campaign that was designed to "neutralize opposition" and get consumers to "resign themselves to the inevitability of it" while merely pretending to address their concerns.

During the UN World Summit on the Information Society (WSIS) between 16 and 18 November 2005, founder of the free software movement, Richard Stallman, protested the use of RFID security cards. During the first meeting, it was agreed that future meetings would no longer use RFID cards, and on finding that this assurance was broken, he covered his card in tin foil, and would only uncover it at the security stations. This protest caused the security personnel considerable concern, with some not allowing him to leave a conference room in which he had been the main speaker, and then preventing him from entering another conference room, where he was due to speak.

26.6.2 Human Implantation

The Food and Drug Administration in the United States has approved the use of RFID chips in humans. Some business establishments have also started to *chip* customers, such as the *Baja Beach* nightclub in Barcelona. This has provoked concerns about the privacy of individuals as they can potentially be tracked wherever they go by an identifier unique to them. There is concern that this could lead to abuse by an authoritarian government or to the removal of other individual freedoms.

On July 22, 2006, Reuters reported that two hackers, Newitz and Westhues, showed at a conference in New York City that they could clone the RFID signal from a human implanted RFID chip, proving that the chip is not hack-proof as was previously projected.

26.6.3 Religious Opinion

Critics from the Christian community believe that RFID tagging could represent the mark of the beast (666) mentioned specifically in the *Book of Revelation* (see Revelation 13:16). Katherine Albrecht[23] and Liz McIntyre, authors of *Spychips: How Major Corporations and Government Plan to Track Your Every Move with RFID*, wrote a new book on the subject from a Christian perspective. John Conner, leader of an organization called "The Resistance of Christ," also believes that there is a strong connection. Related subjects include eschatology (last things) and dispensationalism.

26.7 Protection against RFID Interception[25,26]

Various methods can be used to protect against RFID data interception:

- Physically disabling RFID chip: Most RFID chips can be disabled by physical means. For example, the RFID chip inside RFID credit cards can be disabled by a sharp tap of a hammer.
- One can prevent the RFID transponders from receiving power. This is accomplished by obstructing the power supply; one approach is to shield the RFID transponders in a Faraday's cage, intercepting the electromagnetic signal that normally powers them. UHF transponders can be shielded using an antistatic bag. LF and

HF (inductively coupled) transponders can be shielded with conventional aluminum foil.

- By damaging the antenna with larger RFID transponders, one can recognize the spirals of the antenna clearly, using a radiograph. If one splits the antenna circuit, the effective range of the RFID transponder will be greatly reduced.
- Applying an intense electromagnetic impulse to the transponders and antenna can induce high currents, interrupting the circuit and rendering the tag useless. A crude way to do this is to put the RFID tag in a microwave oven. The degree of success may vary, depending on the frequency of the microwave and the shape of the antenna. There is a device in the RFID-Zapper that is built to destroy transponders.
- The system can be blocked by sending a spurious signal in conjunction with the inquiry signal, preferably on the RFID frequency. This blocks the relatively weak signals of the RFID transponder.
- If a simple memory chip is used to confirm the authenticity of the inquiry, then one can record the inquiry and at a later time reverse engineer the signal, allowing replication. For the reader, it appears as if the correct RFID transponder were in the field.
- Many RFID tags include a built-in "kill" function. When provided with the correct passcode, a tag can be either reprogrammed or told to "self-destruct," rendering it useless.
- Newer emerging RFID tags may include some sort of built-in transfer of control and privacy enhancing technologies, using silencing or nonlinkable protocols to ensure that the owner can control and prevent linkage of RFID.

26.8 RFID Shielding

Shielding is another protection mechanism. A number of products are available in the market in the United States that will allow a customer of RFID-enabled cards or passports to shield their data. It is pertinent to note that the U.S. government requires their new employee ID cards to be delivered with an approved shielding sleeve or holder. There are contradicting opinions as to whether aluminum can prevent reading of RFID chips. Some people claim that aluminum shielding, essentially creating a Faraday cage, does work effectively. Others claim that simply wrapping an RFID card in aluminum foil[27] only makes transmission more difficult, and is not completely effective in preventing it. Further research is required for shielding RFID tags.

Shielding is a function of the frequency being used.

- Low-frequency tags, such as those used for implantable devices for humans and pets, are relatively resistant to shielding, though thick metal foil will prevent most reads.
- High-frequency tags (13.56 MHz—smart cards and access badges) are more sensitive to shielding and are difficult to read when within a few centimeters of a metal surface.
- UHF tags (pallets and cartons) are very difficult to read when placed within a few millimeters of a metal surface, although their read range is actually increased when they are spaced 2–4 cm from a metal, due to the positive reinforcement of the reflected wave and the incident wave at the tag. UHF tags can be successfully shielded from most reads by placing them within an antistatic plastic bag.

26.9 Summary

Despite any security concern it does not appear that RFID will be running out of steam in the future. In fact, it is just the opposite and its popularity is increasing. ABI Research (www.abiresearch.com) recently reported that the global market for RFID readers and reader modules grew to >35,500 unit shipments in 2005. Reader unit volumes grew by nearly 14% in the first quarter of 2006 as compared with the first quarter of 2005. The IDC (www.idc.com) market research firm supports these findings, having found similar evidence of a booming RFID market in Malaysia. According to IDC, RFID spending in Malaysia is estimated to grow at a compound annual growth rate of 45.84% from $2.45 million in 2005 to almost $20.94 million in 2010.

New applications such as RFID-enabled self-checkouts, contactless payment systems using credit and debit cards with embedded RFID tags, and payment systems based on finger scans or other biometrics, are also sure to boost the appeal of RFID in upcoming applications—assuming, of course, that the price of RFID tags goes down and that concerns regarding basic privacy and security can be adequately addressed. Thanks to the efforts of experts like Grunwald, some of RFID's inherent weaknesses, such as inadequate security precautions, may be found before widespread deployment—when they are sure to be easier and less costly to fix.

Some vendors are working on combining RFID tags with sensors of different kinds. This would allow the tag to report not simply the same information over and over, but identifying information along with current data picked up by the sensor. For example, an RFID tag attached to the leg of a lamb could report on the temperature readings of the past 24 h to ensure that the meat was properly kept cool. This can be taken as the positive side of RFID usage.

Over time, the proportion of "scan-it-yourself" aisles in retail stores will increase. Eventually, we may wind up with stores that have mostly scan-it-yourself aisles and only a few checkout stations for people who are disabled or unwilling. This will result in added advantage to the users.

Similar to the growth of the Internet, anywhere a security hole exists, some hacker will find and exploit it for fun, profit, or both. The security problems summarized earlier are real and require real, practical solutions. The RFID industry is working on technical solutions to all of the security problems noted earlier and additional progress in security standards in 2007, coupled with increased RFID industry outreach to the general public in the form of press releases and advertising about security features.

And lastly, you cannot ignore the security risks of RFID tags and their impact on the security and privacy of an individual.

References

1. Radio Frequency Identification, SearchNetworking.com RFID, http://searchnetworking.techtarget.com/sDefinition/0,290660,sid7_gci805987,00.html
2. RFID Software a Pandora's box, http://www.in-pharmatechnologist.com/news/ng.asp?n=66894-rfid-hacking-malware
3. Texas Instruments, http://en.wikipedia.org/wiki/Texas_Instruments
4. Klaus Finkenzeller, Giesecke & Devrient GmbH, *RFID Handbook: Fundamentals and Applications in Contactless Smart Cards and Identification*, Second Edition, John Wiley & Sons, New York, 2003.
5. What Are Zombie RFID Tags? http://www.technovelgy.com/ct/Technology-Article.asp?ArtNum=9

6. German Hacker Clone e-Passport, http://www.engadget.com/2006/08/03/german-hackers-clone-rfid-e-passports/
7. RSA Security, http://en.wikipedia.org/wiki/RSA_Security
8. Fake products can bypass quality, safety, http://www.aegis.com/news/Lt/2006/LT060622.html
9. Problems with RFID, http://www.slais.ubc.ca/courses/libr500/04-05-wt2/www/T_Gnissios/problems.htm
10. RFID Is Dead! Or Is It? http://www.b-eye-network.com/blogs/linstedt/archives/2007/01/rfid_is_dead_or.php
11. Evan Schuman, Report: Major RFID Hurdles Ahead, Ziff Davis Internet July 20, 2006, http://www.eweek.com/article2/0,1895,1990814,00.asp
12. RFID & Individual Privacy, http://www.netcaucus.org/events/2005/rfid/one-pagers/cpsr-rfid 2005.pdf
13. Ethical Problems with RFID, http://www.ccsr.cse.dmu.ac.uk/resources/general/ethicol/Ecv14 no5.html
14. Item-Level RFID Tags Cost More than Expected, http://www.eweek.com/article2/0,1895, 1987543,00.asp
15. RFID Privacy: An Overview of Problems and Proposed Solutions, Simson L. Garfinkel, Massachusetts Institute of Technology, Ari Juels, RSA Laboratories, Ravi Pappu, ThingMagic, *IEEE, Security & Privacy*, 3(3), 34–43, 2005.
16. Gildas Avoine, Philippe Oechslin, EPFL, Lausanne, Switzerland, RFID Traceability: A Multilayer Problem, http://fc05.ifca.ai/p11.pdf
17. Challenge Response Protocol, http://en.wikipedia.org/w/index.php?title=Challenge-response_protocol&action=edit
18. Digital Signature Transponders, http://en.wikipedia.org/wiki/Digital_Signature_Transponder
19. William Stalling, *Cryptography and Network Security: Principles and Practice*, Second Edition, Prentice-Hall, New Jersey, 1998.
20. SearchSecurity.com, RFID Virus, http://searchsecurity.techtarget.com/sDefinition/0,290660, sid14_gci1174912,00.html
21. Will RFID Technology Help or Hinder Security? http://searchsecurity.techtarget.com/originalContent/0,289142,sid14_gci1083417,00.html
22. Radio Frequency Identification News and Commentary, RFID Gazette, May 8, http://www.rfidgazette.org/security/, 2007.
23. Katherine Albrecht, http://en.wikipedia.org/wiki/Katherine_Albrecht
24. Anti-Collision Scheme of ISO 15693, http://en.wikipedia.org/wiki/ISO_15693
25. Protecting RFID Tags and Other Discrete Wireless Devices, http://searchsecurity.techtarget.com/topics/0,295493,sid14_tax299848,00.html
26. Rolling Code Scheme, http://en.wikipedia.org/wiki/Rolling_code
27. RFID Shielding, Aluminum Foil Does Not Stop RFID, http://www.omniscienceisbliss.org/rfid.html

27

Privacy and Personal Information Protection in RFID Systems

Yasunobu Nohara, Kensuke Baba, Sozo Inoue, and Hiroto Yasuura

CONTENTS

27.1 Introduction

As RFID systems become popular, more and more people indicate the importance of RFID privacy problems with comments such as, "It is necessary to regard privacy for RFID's success."

At the same time, unfortunately, not many engineers, users, and RFID tag vendors understand the RFID privacy problem comprehensively. RFID is considered to penetrate wide application areas, some of which might be highly sensitive to users' privacy. Although applications in such privacy-sensitive areas have been avoided, and RFID tags are killed before being handed to consumers with products, there are also predictions that RFID's power exists in the areas which involve consumers, that is, users.

Involved as we are in RFID, we must advance our knowledge and understanding of RFID privacy to manage (properly and quickly) the troubles regarding user privacy.

In this chapter, with the help of technology, we try to solve problems with respect to privacy and personal information protection in RFID systems.

27.2 Privacy and Personal Information Protection

First, we would like to distinguish privacy protection from personal information protection. In the literature, information systems, including RFID systems, do not directly manage privacy protection. They manage personal information protection. Although these are now confused worldwide, personal information protection concerns the technology and operation of information systems with respect to personal data management, while privacy is one's right to freedom from intrusion.

Here, it is important to note that personal information protection is not a sufficient condition, but a necessary condition, of privacy. That is, privacy invasion is accomplished after the leakage of personal information. Privacy invasion involves abuse of the information, manipulation of the victim's behavior and actions, and social factors such as legal environments. It would be illogical to state that use of RFID is an invasion of privacy. We cannot expect RFID to manage this essential human right.

Hereafter, we do not use the term "privacy," but "personal information protection," to limit the responsibility of RFID.

27.3 Anonymity and Unlinkability

When we discuss personal information protection in RFID systems, the following two issues are considered as properties of an RFID system [1–3].

1. **Anonymity:** the guarantee that any person using an RFID system is not identifiable from the data in the RFID system. For example, if an adversary can know the correspondence between the ID of a commodity and personal information of a user who bought the commodity, the user and the commodity do not have anonymity against the adversary.
2. **Unlinkability:** the guarantee that any past record of the behavior of a person using an RFID system is not traced from the data in the RFID system. For example, if an RFID tag of a commodity outputs the fixed value, the behavior of the person bringing the commodity can be traced by an adversary. Therefore, the user and the commodity do not have unlinkability against the adversary.

If anonymity against an adversary is broken, the adversary can link past records of a person and the present record of the person by comparing the user's identity. Therefore, anonymity against the adversary is protected if unlinkability is protected against the adversary. On the other hand, anonymity is not always broken if unlinkability is broken. For example, consider the case that the ID of a user is stored in an RFID tag with a suitable encryption. Although the encrypted ID is fixed and linkable against the adversary, anonymity is protected since the adversary cannot know the original ID. Therefore, the achievement of unlinkability is more difficult than that of anonymity.

In a ubiquitous computing environment, a user can be identified by methods like "the person who is in front of me" even if we don't know the user's name. Therefore, it is also important to satisfy unlinkability for protecting anonymity from this viewpoint.

27.4 Personal Information Protection in RFID Systems

In this section, we introduce personal information protection schemes in RFID systems.

Ideas, operations, and technologies for personal information protection of the RFID system are basically the same as those of normal information systems. However, there are two unique features in RFID systems. The first feature is that an adversary can access an RFID tag easily without notice since RFID uses radio frequency. The second feature is that the restriction to the cost of the tag is very severe in RFID systems. Therefore, it is necessary to achieve personal information protection with lightweight operation in RFID systems. In this paper, we focus on personal information protection schemes that could solve this unique problem.

We classify the protection schemes into (1) physical blocking approach, (2) rewritable tag approach, and (3) smart tag approach. Each approach is explained in detail.

27.4.1 Physical Blocking Approach

The physical blocking approach satisfies anonymity and unlinkability by preventing an adversary from accessing RFID tags physically.

The EPCglobal standard [4] specifies kill command, which disables functionality of the tag. Kill command is protected by PIN to prevent wanton deactivation of tags.

The Faraday cage is an enclosure formed by conducting material, and it blocks out radio frequency. While a user encloses RFID tags with a Faraday cage, the tags do not work well because the cage prevents communication between tags and readers.

Juels et al. [5] propose the blocker tag, which prevents an adversary from reading the ID of the tags which are near the blocker tag. The blocker tag is a cheap passive RFID device that can simulate many ordinary RFID tags simultaneously. Since the blocker tag pretends that all possible tags exist there, an adversary cannot identify the tags that are actually present. A blocker tag can block selectively by simulating only selected subsets of ID codes, such as those by a particular manufacturer.

Karjoth and MosKowitz [6] propose clipped tags, in which a user can physically separate the chip from its antenna. In this system, the user can deactivate the tag by removing its antenna. This separation provides visual confirmation that the tag has been deactivated.

The physical blocking approach, however, has a problem: a user cannot use the RFID services because even a regular service reader cannot access the RFID tags.

27.4.2 Changing Output Approach

In this approach, an adversary can access an RFID tag, and read the output of the tag freely. However, the approach satisfies anonymity and unlinkability by changing the output of the

RFID tag. To satisfy unlinkability, it is necessary to change the output of the tag frequently and prevent an adversary from discerning the relations between the outputs.

The changing output approach can be classified into the (1) rewritable tag approach and (2) smart tag approach.

27.4.2.1 Rewritable Tag Approach

In this approach, a nonvolatile RAM (NVRAM), such as a flash memory, is embedded within each RFID tag. The ID of the tag is stored in the NVRAM and the server can rewrite the ID.

Juels and Pappu [7] and Kinoshita et al. [8] propose the external re-encryption scheme and anonymous-ID scheme. These schemes use a re-encryption scheme, which allows transforming a ciphertext C into a new unlinkable ciphertext C' using the public key only, without changing the plaintext. The tag outputs the encrypted ID which is stored in the NVRAM of the tag. The encrypted ID stored in the tag must be renewed because the tag outputs constant value until the encrypted ID is renewed. The renewing process is as follows:

Step 1: The reader gets the encrypted ID from the tag.

Step 2: The reader re-encrypts the encrypted ID with the public key.

Step 3: The reader rewrites the old encrypted ID with the new encrypted ID.

The reading process is as follows:

Step 1: The reader gets the encrypted ID from the tag and sends it to the server.

Step 2: The server decrypts the encrypted ID using the private key, and obtains the ID of the tag.

Inoue and Yasuura [2] propose the private ID scheme, in which each tag has a ROM and an NVRAM. The permanent ID of the tag is stored in the ROM by a producer, and the user can rewrite the temporary ID stored in the NVRAM. The ROM and the NVRAM are used only exclusively. A user cannot read the permanent ID while the temporary ID is stored in the NVRAM. The user can read the permanent ID only when no value is stored in the NVRAM. Permanent ID is used for public uses like supply chain or recycling. Temporary ID is used for private uses.

In the rewritable tag approach, each RFID tag stores its ID in the NVRAM, and the server updates these IDs periodically. Since the tag does not need cryptographic function, the cost of RFID tag is low. However, the running cost of the system is high because the server has to update the tag's ID periodically. Also since the tag outputs constant value until the next update, unlinkability against an adversary is limited.

27.4.2.2 Smart Tag Approach

In this approach, a cryptographic function and a ROM are embedded within each RFID tag. An RFID tag changes its output every time using a cryptographic function—public key encryption, common key encryption or hash function—on itself.

Let N be the number of RFID tags in an RFID system where the ID id_i of an RFID tag T_i is a string of length L over a finite alphabet Σ for $1 \leq i \leq N$. We assume that if $i \neq j$, then $id_i \neq id_j$ for $1 \leq i, j \leq N$, and $2^L \gg N$. For $s,t \in \Sigma^*$, we denote by $s||t$ the concatenation of s and t.

27.4.2.2.1 Public Key Encryption

Kinoshita et al. [9] propose the internal re-encryption scheme, which uses a public key encryption. In this scheme, a public key encryption function and an NVRAM are embedded within each RFID tag. The encrypted ID stored in the NVRAM is re-encrypted by the public key encryption function on the RFID tag. Since the tag changes its output every time, this scheme provides good personal information protection. However, there is the problem that the tag is expensive because a public key encryption function is complex and costly.

27.4.2.2.2 Common Key Encryption

Kinoshita et al. [9] propose the common key encryption scheme, which uses a common key encryption. In this scheme, a common key encryption function, a ROM, and a pseudorandom number generator are embedded within each RFID tag. The server identifies the tag through the following protocol.

Step 1: RFID tag T_i generates a random number R, and sends $X = E_K(id_i || R)$ to the server.

Step 2: The server decrypts X using the common key K and gets id_i.

The calculation of the common key encryption is smaller than that of the public key encryption; however, it is vulnerable to tampering because the common key must be shared among all tags. The reason why the common key must be shared is as follows. If each tag uses an individual key, the server must know which key to use for decrypting of the encrypted ID. However, how does the server determine the tag's ID before decrypting of the encrypted ID? Therefore, it is difficult to use individual common keys.

An exhaustive search of the key can solve this individual key problem; however, the calculation load of the server is high. We describe the detail of the exhaustive search in Section 27.5 because a hash-based scheme also uses the exhaustive search.

27.4.2.2.3 Hash Function

Hash-based schemes [3,10–16] use a hash function as a cryptographic function. Since the hash calculation is a lightweight operation, the hash-based schemes are suitable for RFID systems, where the implementation cost of an RFID tag must be low. However, the calculation load of the server is high because the server needs to do an exhaustive search. We describe the details of the hash-based schemes in Section 27.5.

27.4.3 Comparison

Table 27.1 compares the personal information protection schemes for RFID systems. A bold font denotes the weakness of the scheme.

TABLE 27.1

Personal Information Protection Schemes for RFID Systems

			Smart Tag		
	Physical Blocking	Rewritable Tag	Public Key	Common Key	Hash
Service	**Not Available**	Available	Available	Available	Available
Anonymity	Satisfied	Satisfied	Satisfied	Satisfied	Satisfied
Unlinkability	Satisfied	**Partly satisfied**	Satisfied	Satisfied	Satisfied
Vulnerability	Tamper free	Tamper free	Tamper free	**Vulnerable**	Tamper free
Calculation on server	Small	Small	Small	Small	**Large**
Cost of tag	Low	Low	**High**	Low	Low

27.5 Hash-Based Scheme

In this section, we describe hash-based schemes and compare these schemes.

A hash-based scheme is one of the schemes using the smart tag approach. A tag changes its output using the hash function, which is embedded on the tag. Since the hash function is a lightweight operation, a hash-based scheme is suitable for RFID systems, where the implementation cost of an RFID tag must be low.

We assume the hash function has one-way and pseudorandom properties. One-way means it is computationally infeasible to calculate the input of the hash function from the output of the hash function. Pseudorandom means the output of the hash function is computationally indistinguishable from a true random number.

27.5.1 Randomized Hash Lock Scheme [3]

In this scheme, a hash function H, a ROM, and a pseudorandom number generator are embedded within each RFID tag.

RFID tag T_i stores id_i in the ROM. The server stores the IDs id_i ($1 \leq i \leq N$) of all tags. The server identifies the tag through the following protocol (see Figure 27.1).

Step 1: RFID tag T_i generates a random number R, and sends $X = H(id_i||R)$ and R to the server.

Step 2: The server finds id_i that corresponds to X by checking $X = H(id_i||R)$ for $1 \leq i \leq N$.

Since R changes every time, $X = H(id_i||R)$ is not fixed. It is computationally infeasible to get id_i from X and R due to the one-way property of the hash function. Therefore, this scheme provides unlinkability against an adversary.

27.5.2 Hash-Chain Scheme [10,11]

In this scheme, two different hash functions H and G, a ROM, and an NVRAM are embedded within each RFID tag.

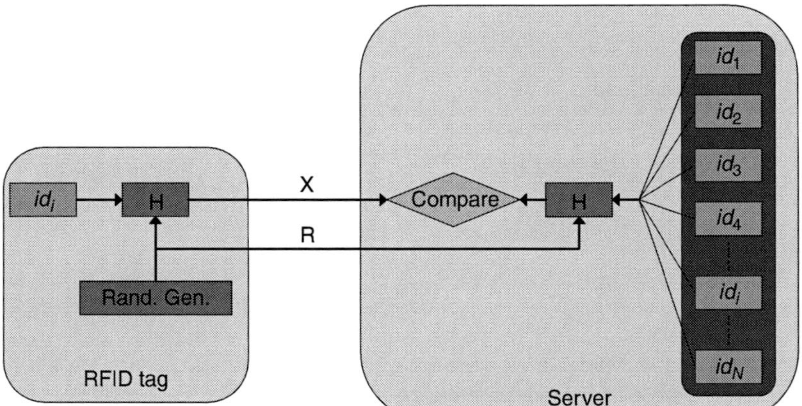

FIGURE 27.1
Randomized hash lock scheme.

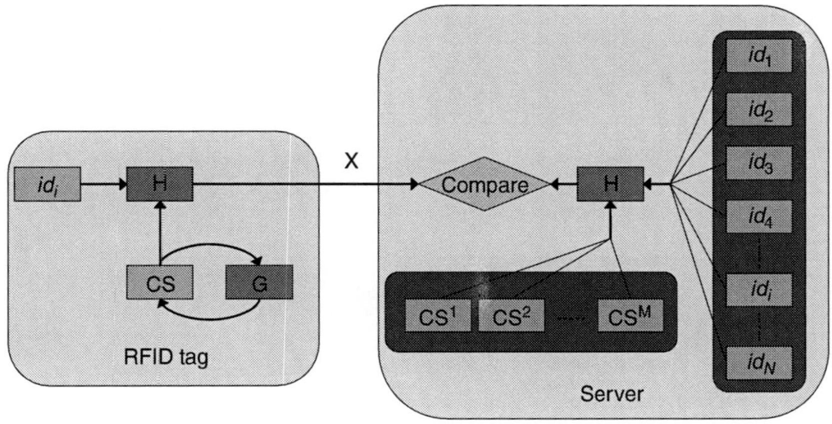

FIGURE 27.2
Hash-chain scheme.

RFID tag T_i stores id_i in the ROM, and stores secret information $cs_i^1 \in \Sigma^{L'}$ in the NVRAM. The server stores the pair $(id_i, cs_i^1)(1 \leq i \leq N)$ of all tags. The server identifies the tag through the following protocol (see Figure 27.2).

Step 1: RFID tag T_i sends $X = H(id_i \| cs_i^l)$ to the server. RFID tag T_i updates $cs_i^{l+1} \leftarrow G(cs_i^l)$.

Step 2: The server finds the id_i corresponding to X by checking $X = H(id_i \| cs_i^l)$ for all $1 \leq i \leq N$ and all $1 \leq l \leq M$ (where M is the maximum length of the hash chain).

Since cs_i^l changes every time, $X = H(id_i \| cs_i^l)$ is not fixed. It is computationally infeasible to get id_i from X due to the one-way property of the hash function. Therefore, this scheme provides unlinkability against an adversary.

Moreover, it is computationally infeasible to get $cs_i^{l'}(l' < l)$ even if id_i and cs_i^l are tampered with. Therefore, the scheme provides forward security, meaning that no RFID tag can be traced from past ID information even if the secret information in the tag is tampered with.

27.5.3 K-Steps ID Matching Scheme [12], Tree-Based Scheme [13]

In these schemes, a hash function H, a ROM, and a pseudorandom number generator are embedded within each RFID tag. These schemes use a tree ID structure. In this section we explain the K-steps ID matching scheme [12].

27.5.3.1 ID Configuration

We use a labeled tree of depth K, such as the tree shown in Figure 27.3. The tree has N leaves, and each leaf corresponds to an RFID tag. Each node has a unique label. ID id_i of an RFID tag corresponding to a leaf node is defined as the sequence of labels from the root node to the leaf node (e.g., a2bX& in Figure 27.3).

In the following, the kth $(1 \leq k \leq S_i)$ label of T_i is denoted by $id_i[k]$, where S_i is the depth of leaf i, and $1 \leq S_i \leq K$.

27.5.3.2 Protocol

In the K-steps ID matching scheme, the server recognizes an ID from the output of an RFID tag through the following protocol.

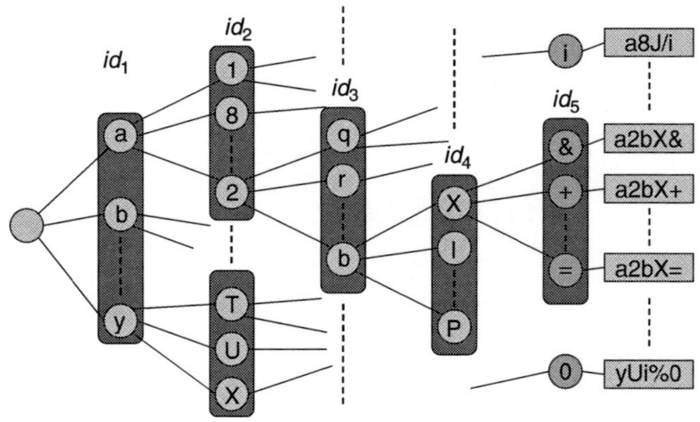

FIGURE 27.3
An ID structure for the K-steps ID matching scheme.

Step 1: RFID tag T_i generates a random number R. T_i then sends $(R, X_1, X_2, \ldots, X_K)$ to the server, where X_k is $H(id_i[k]||R)$ if $1 \leq k \leq S_i$ and a random number R_k if $S_i + 1 \leq k \leq K$.

Step 2: The server operates as follows:

STEP 2-1: let Z be the root of the labeled tree and let $k \leftarrow 1$;

STEP 2-2: find L_i s.t. $H(L_i||R) = X_k$ by computing $H(L_i||R)$ for each child L_i of Z, and update $Z \leftarrow L_i$;

STEP 2-3: output the label corresponding to Z as the ID of the RFID tag if Z is a leaf; otherwise, let $k \leftarrow k+1$ and return to STEP 2-2.

In Step 1, RFID tag T_i sends a random number as X_k for $S_i + 1 \leq k \leq K$, which hides the depth of the leaf S_i to prevent weakening the unlinkability against an adversary.

When $K = 1$, the protocol and the ID structure of the protocol correspond to those of the randomized hash lock scheme [3]. If some procedures of the protocol are changed, it becomes a protocol corresponding to the hash-chain scheme [10,11].

27.5.4 Avoine's Scheme

Avoine et al. [14,15] developed a specific time-memory trade-off that reduces the amount of computation in the hash-chain scheme [10,11]. This time-memory trade-off reduces the hash calculations on the server with the help of precomputation results. However, heavy pre-calculation is needed with Avoine's scheme [14,15].

27.5.5 Yeo's Scheme

Yeo's scheme [16] is one of the hash-chain schemes, and the same ID structure as in K-steps ID matching scheme is used to reduce the server complexity. Yeo et al. propose two types of schemes. One is a scheme without precomputation which uses only a grouping technique. The other is a scheme with precomputation which uses both a grouping technique and a time-memory trade-off technique [14].

27.5.6 Comparison

We compare the hash-based schemes from the viewpoints of security, hash calculation time, the amount of memory needed, and the amount of communication.

TABLE 27.2

Classification of Hash-Based Schemes

	Base Model	ID Structure	Time-Memory
Hash lock [3]	Hash lock	Normal	No
K-step [12], Tree-based [13]	Hash lock	Tree	No
Hash chain [10,11]	Hash chain	Normal	No
Avoine's scheme [14,15]	Hash chain	Normal	Yes
Yeo's without precomp. [16]	Hash chain	Tree ($K=2$)	No
Yeo's with precomp. [16]	Hash chain	Tree ($K=2$)	Yes

27.5.6.1 Classification of Hash-Based Schemes

For our comparison, we classify hash-based schemes with regard to three characteristics:

1. Base model (hash lock or hash chain)
2. ID structure (normal or tree)
3. Introduction of a time-memory trade-off technique [14] (yes or no)

Table 27.2 shows the classification results. There are no proposed schemes with the combination (Hash Lock, Normal, Yes) or (Hash Lock, Tree, Yes) because the responses of RFID tags in a hash lock scheme are randomized, which means a large memory space is needed to apply a time-memory trade-off technique [15].

27.5.6.2 Security

We compare the security of the hash-based schemes with respect to three concerns:

1. Unlinkability
2. Forward security
3. Prevention of replay attacks

27.5.6.2.1 Unlinkability

We analyzed the unlinkability of the hash-based schemes with the *degree of unlinkability* [17]. The degree of unlinkability ranges from 0 to $\log_2 N$ [bit], and unlinkability becomes stronger as the degree of unlinkability increases. When an adversary has no ID information, each degree of unlinkability for the hash-based schemes is $\log_2 N$.

Since anonymity and unlinkability are closely related, this evaluation of unlinkability is also related to that of anonymity.

When an adversary obtains one ID, such as by tampering with an RFID tag, the degree of unlinkability of each scheme differs depending on its ID structure. The degree of unlinkability for the normal ID structure and that for the tree ID structure are given as follows [17]:

$$U_{\text{normal}} = \frac{N-1}{N}\log_2(N-1) \tag{27.1}$$

$$U_{\text{tree}} = \log_2 N + \frac{N-1}{N}\{\log_2(N^{1/K}-1) - \frac{N^{1/K}}{K(N^{1/K}-1)}\log_2 N\} \tag{27.2}$$

The normal ID structure schemes enable user unlinkability, except for the tampered user, but the tree ID structure schemes cannot enable user unlinkability since some users share part of the ID of the tampered user. From Equations 27.1 and 27.2, we can see that the degree of unlinkability with the tree ID structure is lower than that with the normal structure.

However, the tree ID structure schemes provide the same level of unlinkability as the normal ID structure if $\alpha = N^{1/K}$ is large enough. Since the optimized K is much less than 10 even if N becomes 2^{100} [12], the tree ID structure decreases the degree of unlinkability only slightly.

Thus, the decrease in the degree of unlinkability with the K-steps ID matching scheme is only small [17] compared to that with the normal ID structure.

27.5.6.2.2 Forward Security

Forward security is a property that means no RFID tag can be traced from past ID information even if an adversary tampers with the secret information in the tag.

Hash lock schemes, including the K-steps ID matching scheme, cannot provide forward-security because an adversary can easily get a random number R. On the other hand, hash-chain schemes can provide forward security since it is computationally difficult for an adversary to get $cs_i^{l'}$ $(l' < l)$ even if he has tampered with id_i and cs_i^l.

However, Juels and weis have pointed out that hash-chain schemes create a security risk in that an adversary can guess a tag's count number [18]. We discuss this problem in Section 27.5.6.3.

27.5.6.2.3 Prevention of Replay Attacks

A replay attack is one in which a valid data transmission is maliciously or fraudulently repeated. The attack is carried out by an adversary who masquerades as a legitimate user.

Replay attacks must be prevented when a server has to authenticate as well as identify an RFID tag. One way to do this is to use a fresh challenge by the server. Hash lock schemes can prevent replay attacks if a step is added where the server sends a fresh challenge to the tag and includes the challenge in the hash calculations. In K-steps ID matching scheme, the protocol to prevent replay attacks is as follows:

Step 1: The server generates a random number R_s, and then sends R_s to RFID tag T_i.

Step 2: RFID tag T_i generates a random number R_d, and then sends $(R_d, X_1, X_2, \ldots, X_K)$ to the server, where X_k is $H(id_i[k] \| R_s \| R_d)$ if $1 \le k \le S_i$, and a random number R_k if $S_i + 1 \le k \le K$.

Step 3: The server operates as follows:

STEP 3-1: let Z be the root of the labeled tree and let $k \leftarrow 1$;

STEP 3-2: find L_i s.t. $H(L_i \| R_s \| R_d) = X_k$ by computing $H(L_i \| R_s \| R_d)$ for each child L_i of Z, and update $Z \leftarrow L_i$;

STEP 3-3: output the label corresponding to Z as the ID of the RFID tag if Z is a leaf; otherwise, let $k \leftarrow k+1$ and return to STEP 3-2.

Avoine et al. propose a modified hash-chain scheme which prevents replay attacks using a challenge [15]. This technique can be easily adopted in Yeo's scheme without precomputation.

However, the technique of using a fresh challenge cannot be applied directly to Avoine's scheme or Yeo's scheme with precomputation since the randomization of the tag's response prevents the server using a time-memory trade-off (see Section 27.5.6.1). Therefore, the RFID tag must calculate a hash value(s) without a challenge and a hash value with

TABLE 27.3

Comparison of Required Memory and Time

	Hash Calculation on Device	Hash Calculation on Server	Precomputation on Server	Memory on Server
Hash lock	1	N	0	0
K-step, Tree-based	K	$KN^{\frac{1}{k}}$	0	0
Hash chain	2	MN	0	N
Avoine's scheme	2[+1]	$\frac{3^3}{2^3}\frac{M^3\gamma}{c^3\mu^2}[+1]$	$\frac{NM^2}{2}$	cN
Yeo's without precomputation	4	$2M\sqrt{N}$	0	N
Yeo's with precomputation	4[+1]	$\left(\frac{2^5M^6\gamma}{c^3\mu^2}\right)^{1/4}[+1]$	$\left(\frac{2^3c^3N^4\mu^2}{3^4M^2\gamma}\right)^{1/4}\frac{M^2}{2}$	cN

the challenge [15]. The former value(s) enable(s) the server to identify the tag, while the latter prevents replay attacks.

All of the hash-based schemes proposed so far have a countermeasure against replay attacks, and preventing replay attacks increases both the calculation complexity and the communication amount. We discuss this problem in Sections 27.5.6.3 and 27.5.6.4.

27.5.6.3 Comparison of Memory and Time

We compare the different schemes regarding the number of hash calculations on the RFID tag and on the server, the number of precomputations on the server, and the memory required for the precomputation results.

Table 27.3 compares the memory and the time needed for each scheme. In the table, M is the maximum length of the hash chain, μ is the conversion factor, c is the memory size parameter for Avoine's scheme [14], and γ is the rate of successful search parameter in that scheme. For example, the success rate is 99.9% when $\gamma = 8$.

In the table, memory on server denotes the amount of secret information to be stored cs_i^l in hash-chain schemes. Note that the memory amount does not include the space for the ID list, which is required for every scheme. The additional number of calculations for the scheme to prevent replay attacks is given in brackets.

As the table shows, the number of hash calculations on the server in a time-memory trade-off scheme includes M^3 or $M^{1.5}$, while that in the K-step ID matching scheme includes N. Therefore, the K-steps ID matching scheme might be disadvantageous in terms of the required time if M^3 or $M^{1.5}$ is sufficiently smaller than $N^{1/k}$.

The server cannot identify the RFID tag when the tag number is larger than M because it will be outside the search range. In addition, there is a security risk in that an adversary can guess a tag's count number if M is small [18]. Therefore, M must be sufficiently large.

Avoine and Oechslin point out that replacing cs_i^1 by cs_i^k in the database regularly expands the search range of the server [14]. However, the problem of a count number leakage remains, and heavy precomputation (e.g., $M^2 N/2$) is needed for every replacement.

27.5.6.4 Communication Cost

Table 27.4 compares the communication cost for each scheme in terms of the amount of communication data. The costs are shown in each case of preventing or not preventing replay attacks. In the table, r is the length of the random value for the challenge, and h is the length of the hash output.

TABLE 27.4

Communication Cost

	Not Preventing Replay Attacks	Preventing Replay Attacks
Hash lock	$r+h$	$2r+h$
K-steps, Tree-based	$r+Kh$	$2r+Kh$
Hash chain	h	$r+h$
Avoine's scheme	h	$r+2h$
Yeo's without precomputation	$2h$	$r+2h$
Yeo's with precomputation	$2h$	$r+3h$

In tree ID structures, including those of the K-steps ID matching scheme, the communication cost increases in proportion to the tree depth. For the K-steps ID matching scheme, we measured the practical time for the entire execution (including the communication time between the server and the RFID tag), and found it is shorter than that of a naive scheme [3] when N is sufficiently large.

Thus, we expect the communication cost of the K-steps ID matching scheme to be negligible in a practical situation. However, further evaluation is required since we used contact smart cards in our experiment. With contact-less smart cards or RFID tags, the communication cost might increase because of communication failures.

27.6 Conclusion

In this chapter, we have explained personal information protection in RFID systems. First, we explained the distinction between privacy protection and personal information protection, and introduced two properties—anonymity and unlinkability—for personal information protection. Secondly, we surveyed the personal information protection schemes which provided anonymity and unlinkability. Finally, we have described hash-based schemes, which are some of the smart tag approaches.

References

1. ISO/IEC 15408—International Standard Information technology—Security techniques—Evaluation criteria for IT security—Part2: Security functional requirements, 1999.
2. Sozo Inoue and Hiroto Yasuura. RFID privacy using user-controllable uniqueness. In *RFID Privacy Workshop*, MIT, MA, USA, Nov. 2003.
3. Stephan A. Weis, Sanjay E. Sarma, Ronald L. Rivest, and Daniel W. Engels. Security and privacy aspects of low-cost radio frequency identification systems. In *1st International Conference on Security in Pervasive Computing—SPC2003, LNCS*, Vol. 2802, pp. 201–212. Springer, 2004.
4. EPCglobal. http://www.epcglobalinc.org/.
5. Ari Juels, Ronald L. Rivest, and Michael Szydlo. The blocker tag: Selective blocking of RFID tags for consumer privacy. In *10th ACM Conference on Computer and Communications Security—CCS2003*, pp. 103–111. ACM Press, Oct. 2003.
6. Günter Karjoth and Paul A. Moskowitz. Disabling RFID tags with visible confirmation: Clipped tags are silenced. In *Proceedings of 2005 ACM Workshop on Privacy in the Electronic Society*, pp. 27–30, 2005.

7. Ari Juels and Ravikanth Pappu. Squealing euros: Privacy protection in RFID-enabled banknotes. In *Financial Cryptography—FC2003, LNCS*, Vol. 2742, pp. 103–121, Jan. 2003.
8. Shingo Kinoshita, Fumitaka Hoshino, Tomoyuki Komuro, Akiko Fujimura, and Miyako Ohkubo. Low-cost RFID privacy protection scheme. *IPSJ Journal*, 45(8):2007–2021, 2004 (in Japanese).
9. Shingo Kinoshita, Miyako Ohkubo, Fumitaka Hoshino, Gembu Morohashi, Osamu Shionoiri, and Atsushi Kanai. Privacy enhanced active RFID tag. In *Proceedings of 1st International Workshop on Exploiting Context Histories in Smart Environments—ECHSE2005*, 2005.
10. Miyako Ohkubo, Koutarou Suzuki, and Shingo Kinoshita. Cryptographic approach to a privacy friendly tag. In *RFID Privacy Workshop*, MIT, MA, USA, Nov. 2003.
11. Miyako Ohkubo, Koutarou Suzuki, and Shingo Kinoshita. Hash-chain based forward-secure privacy protection scheme for low-cost RFID. In *2004 Symposium on Cryptography and Information Security—SCIS2004*, Vol. 1, pp. 719–724, Jan. 2004.
12. Yasunobu Nohara, Toru Nakamura, Kensuke Baba, Sozo Inoue, and Hiroto Yasuura. Unlinkable identification for large-scale RFID systems. *IPSJ Journal*, 47(8):2362–2370, 2006. Online version: IPSJ Digital Courier, Vol. 2, pp. 489–497.
13. David Molnar and David Wagner. Privacy and security in library: RFID issues, practices, and architectures. In *11th ACM Conference on Computer and Communications Security—CCS2004*, pp. 210–219. ACM Press, Nov. 2004.
14. Gildas Avoine and Philippe Oechslin. A scalable and provably secure hash-based RFID protocol. In *2nd International Workshop on Pervasive Computing and Communications Security—PerSec2005*, pp. 110–114. IEEE Computer Society Press, Mar. 2005.
15. Gildas Avoine, Etienne Dysli, and Philippe Oechslin. Reducing time complexity in RFID systems. In *12th Annual Workshop on Selected Areas in Cryptography—SAC2005, LNCS*, Vol. 3897, pp. 291–306. Springer, 2005.
16. Sang-Soo Yeo and Sung Kwon Kim. Scalable and flexible privacy protection scheme for RFID systems. In *2nd European Workshop on Security in Ad-Hoc and Sensor Networks—ESAS2005, LNCS*, Vol. 3813, pp. 153–163. Springer, 2005.
17. Yasunobu Nohara, Sozo Inoue, Kensuke Baba, and Hiroto Yasuura. Quantitative evaluation of unlinkable ID matching schemes. In *Proceedings of 2005 ACM Workshop on Privacy in the Electronic Society—WPES2005*, pp. 55–60. ACM Press, Nov. 2005.
18. Ari Juels and Stephen A. Weis. Defining strong privacy for RFID. In *IACR Cryptology ePrint Archive Report*, no. 2006–137, 2006.

28

Multilateral Approaches for Reliable Mobile RFID Service Systems

Namje Park and Dongho Won

CONTENTS

28.1 Introduction

Radio frequency identification (RFID) has been recognized as the key technology for the ubiquitous network, which is defined as an environment in which information can be acquired anytime and anywhere through network access service [1,2]. Currently, RFID technologies consider the environment in which RFID tags are mobile and RFID readers are stationary. However, in the future RFID technologies could consider an environment in which RFID tags are stationary and readers are mobile. RFID based on mobile telecommunications services could be the best example of this kind of usage. RFID-based mobile telecommunications services could be defined as services which provide information access through the telecommunication network by reading RFID tags on certain objects using an RFID reader in mobile terminals such as cell phones. RFID tags play an important role as a bridge between off-line objects and online information. The RFID-enabled cell phone was introduced by Nokia in 2004 [2–5].

Furthermore, the RFID tags of the future will evolve as active tags which have networking capabilities, becoming a key component of the ubiquitous network environment rather than the current passive RFID tags. In this stage, RFID tags will need network addresses for communications. For the ubiquitous network, current RFID-related technologies need to be changed to reflect the features of mobile telecommunications services; and additional technologies for RFID-based mobile telecommunications services should be established to provide harmonized operation of services.

A new security technology is required to provide a safe service among mobile RFID tags, terminals, and applications to minimize the threat of personal information infringements and leakage. Firstly, the potential for personal information protection infringement has increased due to the mobility of mobile RFID readers; secondly, information leakage due to mobile communication and wireless internet environment is expected; thirdly, the mobile RFID service can be used illegally; and finally, RFID tag information can be counterfeited or falsified. Therefore, in this chapter, we present an analysis of security and privacy threats, and multilateral security approaches to promoting a globally mobile RFID service. This new technology to RFID will provide a solution that protects absolute confidentiality from basic tags to the user's privacy information.

28.2 Mobile RFID Primer

In this section, we will discuss the mobile RFID technology. We begin with a discussion of the general details of mobile RFID system anatomy, followed by detailed discussions of the components that make up a typical mobile RFID system and the underlying technologies that make them work. The most common of these are described later.

28.2.1 Mobile RFID Technology

RFID is expected to be the base technology for the ubiquitous network or computing, and is likely to be associated with other technologies such as Micro Electro Mechanical Systems (MEMS), Telematics, and Sensors. Meanwhile, it is widely accepted that Korea has established one of the most robust mobile telecommunication networks in the world. In particular, about 78% of the population use mobile phones and >95% of those phones have Internet-enabled functions. Currently, Korea has recognized the potential of RFID technology and has tried to converge it with the mobile phone. Mobile phones integrated with RFID can be expected to create new markets and provide new services to end-users, and as such will be considered as an exemplary technology fusion. Furthermore, it may evolve its particular functions as an end-user terminal device, or a ubiquitous device (U-device), in the world of ubiquitous information technology [2,3,6–8].

Actually, the mobile RFID phone may represent two types of mobile phone device; one is the RFID reader-equipped mobile phone, and the other is the RFID tag-attached mobile phone. Each type of mobile phone has different application domains: On the one hand, for example, the RFID tag-attached type can be used as a device for payment, entry control, and identity authentication, and the main feature of this application stems from the fact that RFID readers exist in the fixed position and recognize each phone, giving the user-specific services like door opening; on the other hand, the RFID reader-equipped mobile phone, to which

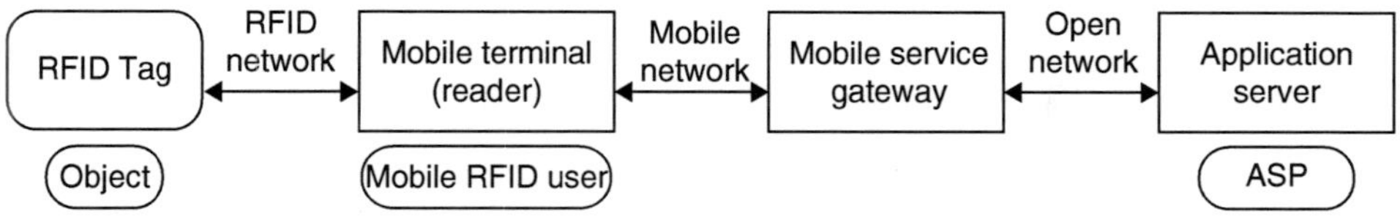

FIGURE 28.1
Model of mobile RFID data communication.

Korea is currently paying considerable attention, can be used to provide end-users with detailed information about the tagged object through accessing the mobile wireless network.

The model of the mobile RFID service as shown in Figure 28.1 defines three additional entities and two relationships compared with those defined in the RFID tag, the RFID access network, RFID reader, the relationship between the RFID tag and RFID reader, and the relationship between the RFID reader and the application server.

Right now, there are several projects concerning "Mobile RFID" as shown in Table 28.1. In particular, Korea's mobile RFID technology is focusing on the ultrahigh frequency (UHF) range (860–960 MHz), since the UHF range may enable a longer reading range and moderate data rates, as well as a relatively small tag size and lower costs [2,3,6]. Then, as a kind of handheld RFID reader, in the selected service domain the UHF RFID phone device can be used to provide object information directly to the end-user using the same UHF RFID tags which have spread widely. Table 28.1 shows a summary of the possible implementations of mobile RFID.

As shown earlier, UHF-band mobile RFID uses 908.55–913.95 MHz and complies with ISO/IEC 18000-6 Types B and C [3,8,9]. From the viewpoint of service deployment, the UHF-band is more profitable according to the following observations:

1. *It has relatively longer range up to 100 cm.*
   - Longer range is favorable for most mobile RFID services, ensuring greater convenience.
2. *Short range is available up to 2 or 3 cm if required;* in the case of the payment system, short range may be supported by reducing the RF strength by application.
3. *Avoiding duplicate investment for the RFID tag.*
   - Most RFID tags in supply chain management (SCM) work in the 900 MHz range, that is, ISO 18000-6 Types A/B/C and EPCglobal.
   - This means that both the SCM and mobile RFID applications can share an RFID tag: thus, a single RFID tag can provide different contents according to its application.

TABLE 28.1
Summary of Mobile RFID Implementations

	Nokia's Mobile RFID	KDDI's Mobile RFID (Passive)	KDDI's Mobile RFID (Active)	NFC (Near-Field Communication)	Korea's Mobile RFID
Radio frequency	13.56 MHz	2.45 GHz	315 MHz	13.56 MHz	860–960 MHz
Reading range	2–3 cm	~5 cm	~10 m		
Compliant standards	ISO/IEC 14443 A			ISO/IEC 18092	ISO/IEC 18000-6 B/C
Feature	HF RF reader	RF reader	Active RFID reader	Tag and reader	UHF RF reader

28.2.2 UHF-Band Mobile RFID Network

Networked RFID comprises an expanded RFID network and communication scope to communicate with a series of networks, inter-networks, and globally distributed application systems, engendering global communication relationships triggered by RFID, for such applications as Business to Business (B2B), Business to Customer (B2C), B2B2C, Government to Customer (G2C), and so on. Mobile RFID loads a compact RFID reader into a cellular phone, thereby providing diverse services through mobile telecommunications networks when reading RFID tags through a cellular phone. Since the provision of these services was first attempted in Korea in 2005, their standardization has been ongoing. Korea's mobile RFID technology is focusing on the UHF range [2–4]. Thus, as a kind of handheld RFID reader, in the selected service domain the UHF RFID phone device can be used to provide object information directly to the end-user using the same UHF RFID tags which have been distributed widely.

The mobile RFID service has been defined as the provision, through the wireless Internet network, of personalized secure services—such as searching for product information, purchasing, verifying, and paying for products—while on the move, by building the RFID reader chip into the mobile terminal [10,11]. The service infrastructure required for providing such an RFID-based mobile service is composed of an RFID reader, handset, communication network, network protocol, information protection, application server, RFID code interpretation, and contents development; the configuration map is as follows.

Figure 28.2 shows the interface structure for the mobile RFID service's communication infrastructure and the types of relevant standards. RFID wireless access communication takes place between the RFID tag and a cellular phone, Code Division Multiple Access (CDMA) mobile communication takes place between a cellular phone and Base Transceiver Station (BTS)/Access Network Transceiver Subsystem (ANTS), and wire communication takes place between BTS/ANTS and a networked RFID application server.

Figure 28.2 represents the entities of the mobile RFID service network architecture. The Object Directory Service (ODS) server plays the role of a Directory Name System (DNS) server which informs the mobile RFID phone of the contents/service server's location, as explained earlier [2,6,12]. The ODS server may be organized in a hierarchical structure similar to that of a DNS server. The Object Traceability Service (OTS) server keeps a record of the tag readings in the RFID readers throughout the life cycle of the objects. Its main purpose is to track objects in the SCM. The Object Information Service (OIS) records the reading of the RFID tag event in the OTS server and may provide additional detailed

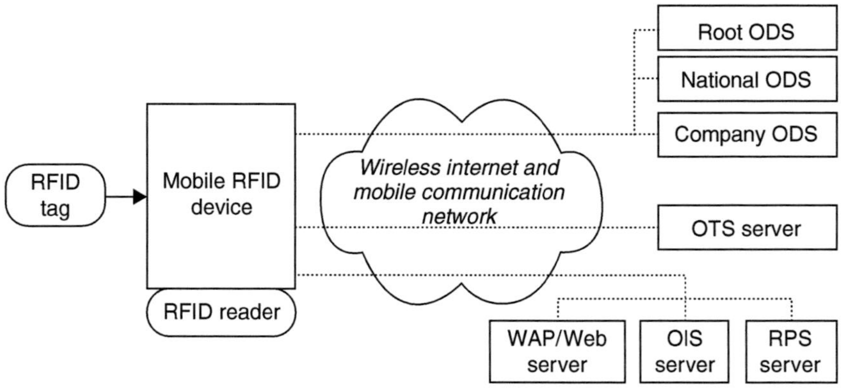

FIGURE 28.2
Conceptual network model for mobile RFID service.

information on an object, such as manufacturing time, manufacturer's name, expiration time, and so on. The RFID Privacy Management Service (RPS) controls access to the information on the object in accordance with the privacy profile put together by the owner of the object. The Wireless Application Protocol (WAP) and Web servers are contents servers that provide wireless Internet contents such as news, games, music, videos, stock trading, lotteries, images, and so on.

The mobile RFID service structure is defined to support ISO/IEC 18000-6 A/B/C through wireless access communication between the tag and the reader; however, as yet there is no RFID reader chip capable of supporting all three wireless connection access specifications so that the communication specification for the mobile phone will be determined by the mobile communication companies [3,12,13]. It will also be possible to mount the RF wireless communication function on the reader chip using Software Defined Radio (SDR) technology and develop an ISO/IEC 18000-6 A/B/C communication protocol in software to choose from the protocols when needed.

The mobile RFID middleware is composed by extending the Wireless Internet Platform for Interoperability (WIPI) software platform to provide RF code-related information obtained from an RF tag through an RFID reader installed in the mobile phone. The networked terminal's function is concerned with the recognition distance to the RFID reader chip built into the cellular phone, transmission power, frequency, interface, technological standard, Personal Identification Number (PIN) specification, Universal Asynchronous Receiver and Transmitter (UART) communication interface, WIPI Application Program Interface (API) extended specification to control the reader chip. RFID reader chip middleware functions are provided to the application program in the form of mobile platform's API. Here, the mobile RFID device driver is the device driver software provided by the reader chip manufacturer.

The mobile RFID network function is concerned with communication protocols such as the ODS communication for code interpretation, the message transportation for the transmission and reception of contents between the mobile phone terminal and the application server, contents negotiation that supports the mobile RFID service environment and ensures optimum contents transfer between the mobile phone terminal and the application server, and session management that enables the application to create and manage the required status information while transmitting the message and the WIPI extended specification which supports these communication services [4,6,10,14,29].

A cellular phone requires a common control interface between the various RFID readers and the application or the middleware; to that end, EPCglobal Inc. and ISO are defining the functions that an RFID reader should commonly support, as well as various common command and standardizing message types. Mobile RFID functions will be extended continuously into standard cellular phone RFID readers, and the RFID supported WIPI extension model using WIPI—the wireless internet standard platform—will define the API required in using the reader suitable for the mobile environment as the API extension of WIPI, while maintaining compatibility among the various devices.

The basic communication scenario for mobile RFID service is as follows: first, a mobile RFID phone reads the RFID tags on an object and fetches the code stored in it [2,4,9]. Second, a mobile RFID phone should execute the code resolution with which the mobile RFID phone obtains the location of the remote server that provides information on the product or an adequate mobile service. The code resolution protocol is identical with the DNS protocol. The ODS server in Figure 28.3 as a DNS server and is similar to EPCglobal's Object Name Service (ONS) server. The mobile RFID phone directs queries on the location of the server with a code to the ODS server, then the ODS server replies by giving the location of the server. Finally, the mobile RFID phone requests contents or a service from the designated server whose location has been acquired from the ODS server.

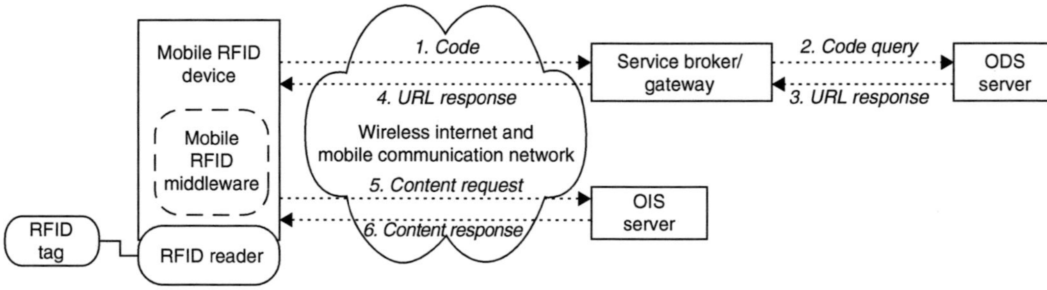

FIGURE 28.3
Block basic communication scenario for mobile RFID service.

Figure 28.4 illustrates the detailed code resolution process. The code store in the RFID tag is formed of a bit string such as ''01001101110...'' and this bit string should be translated into a meaningful form such as EPC, mCode (Mobile RFID Code), uCode, ISO Code, or something else [2,7,8,15]. Given that ''1.2.3.4'' is obtained from a bit string translation and that ''1.2.3.4'' should be converted into a Uniform Resource Name (URN) form as ''urn:mcode:cb:1.2.3.4,'' the remaining code resolution process is the same as the DNS reverse lookup process. Mobile RFID reader requests contents retrieval after code resolution. The RFID application in the mobile RFID phone requests contents from the WAP or Web server returned by the code resolution.

28.3 Security Requirements in Mobile RFID Network

The mobile RFID can be used for entertainment, smart poster service, payment system, and so on. However, the owners of the RFID tags cannot recognize that their RFID tags are identified. In addition, the RFID tags do not store the list of identification. So the people who have a mobile RFID reader can read the information of the RFID tags. Owing to these reasons, mobile RFID service has many security and privacy threats. In this section, we describe security and privacy threats in mobile RFID system. We will discuss vulnerability points in the basic mobile RFID system and privacy threats in mobile RFID

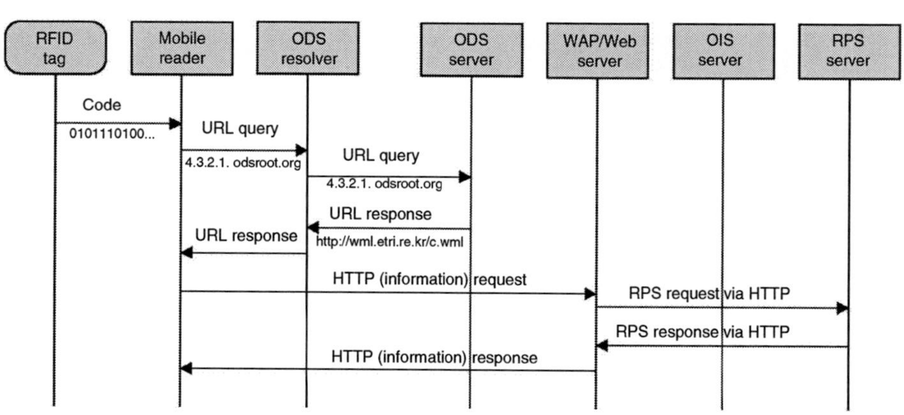

FIGURE 28.4
Detailed mobile RFID's code resolution process.

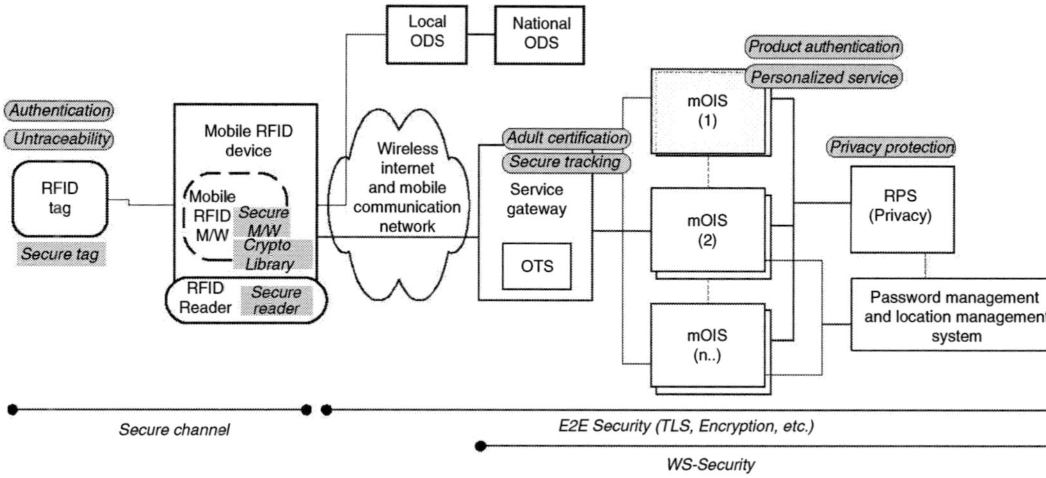

FIGURE 28.5
Conceptual architecture for secure RFID over mobile networks.

service. Then we describe the security and privacy requirements for secure mobile RFID service and its application (Figure 28.5).

28.3.1 Mobile RFID-Oriented Security Threats

The security vulnerability of the mobile RFID consists of potential infringements of the owners' privacy and physical attacks in cyber space [16–19]. The most typical vulnerabilities are as follows:

1. *Privacy vulnerability*: Individual privacy is very likely to be infringed due to the approval of unlimited access to an RFID tag owned by an individual. As such, it is necessary to allow the access to information only for those who need to obtain it in a given application, but to block it for those who have no authority. An individual RFID tag may also become a means by which to track and locate its owner. The infringement of privacy in the internet world results from the collection, storage, and use of customer information by companies, but the problem has become more serious in the mobile RFID world because anyone with an RFID reader can now read any information on anybody who keeps a tag-attached object.
2. *Possible hacking of mobile RFID applications*: It is possible to hack tags, preventing the normal use of tags or obtaining incorrect information from them with a tag information alteration or a tag kill function.
3. *Illegal collection of information*: One may hide a system in a commodity or object for remote communication to wiretap, track, catch information, or personal profile.
4. *Security vulnerabilities* also result from jamming, replay attacks, covert reading, and so on.

Some mobile RFID-oriented privacy threats are summarized later [4,14,16].

Firstly, an RFID tag identifier, the ID, can easily be eavesdropped by intercepting broadcasted radio signals or by actively reading the RFID tag. Accordingly, it is possible to track an RFID-tagged object or monitor a user carrying a specified tag ID by using an invisible rogue RFID reader. Secondly, an RFID tag can contain some important data, such as passwords, IDs, user-specific service data for application, and so on. Consequently, unauthorized tag access can cause denial or misuse of a service, including the permanent disabling of a tag or the illegal modification of tag-stored data. Thirdly, whenever an RFID reader reads a tag ID, its recorded reading history—such as location and time—can be collected without the agreement of the tag user. In particular, if the application of a tag is tightly coupled with an individual, this can cause a violation of privacy due to a leakage of the collected historical context data, such as the user's preference profile. Finally, mobile RFID applications require stricter adult verification. Currently, teenagers and even elementary school students below the age of 10 are using cell phones which are in reality ubiquitous information terminals when in fact they should be private devices; as such, youngsters can access adult contents very easily. A strict and elaborate mechanism for adult verification should be provided to protect young people from adult contents, but adult verification is currently provided within contents at the application layer. That is, the control role is given to contents providers, which means that network operators (called ISPs) cannot control illegal behavior such as the provision of adult contents.

28.3.2 Requirements for Secure Mobile RFID Service

The mobile RFID service structure provides its services by associating the mobile communication network and the RFID application service network based on the RFID tag. The areas to be considered with regard to security are essentially the RFID tag, reader terminal area, mobile communication network area, RFID application service network area, while other security issues such as confidentiality/integrity/authentication/permission/nonrepudiation shall be considered in each network area. Especially, as the mobile RFID service is the end-user service, the issue of privacy protection must inevitably become a serious issue to consider, and as contents accessibility increases due to the off-line hypertext property of RFID, the authentication for adult services is also highly likely to become another important issue for consideration.

1. The mobile RFID service based on the user's ownership of tagged products needs to guarantee the confidentiality of the tag code information or the user's data information to ensure personal privacy protection. In this case, the mobile RFID application service provider (ASP) should ensure the confidentiality of the said information or use other means to prevent personal privacy infringement.
2. The mobile RFID, on behalf of its owner, may need to communicate with ONS, Electronic Product Code—Information Service (EPC-IS) to retrieve the information of a particular tagged item [17–19]. It should identify and authenticate the genuine EPC network and be able to secure the entire transaction, as well as protect the owner's privacy; however, these tasks could create a huge burden on the low-computing and resource-poor mobile device, and are certainly not user-friendly. Therefore, it would be lot easier for the mobile device to securely delegate its work to a nearby, trusted, high-computing, and resource-rich entity, namely the mobile operator. This approach would help in reducing the communication and computational burden on the mobile device.
3. The integrity of the data shall be guaranteed to monitor for counterfeiting/falsification of the data transmitted through the communication path in each section of the mobile RFID service network reference structure. However, additional code-based data integrity—other than the least method (e.g., Cyclic Redundancy Check, CRC) specified in the air interface specification—is not required in the communication section between the tag and reader terminal, given the limit of the calculation

capacity of the tag. However, it is necessary to develop a method of securing data integrity in the tag for the special mobile RFID application service, where personal information is stored in the user data information of the tag and transmitted.

4. Establishing an efficient and convincing trust model is essential if secure transactions, key distribution, and job delegation are to be ensured. With the existence of a trust model, it would be lot easier for the mobile device to delegate its work to the mobile operator.
5. Authentication in the mobile RFID can be divided into device authentication in each network layer and service user authentication.
   - *Device Authentication*: Device authentication refers to the authentication of the RFID reader installed in a cellular phone. Mobile RFID service requires device authentication as it is based on the inter-working service between heterogeneous networks (mobile communication network–RFID application service network).
   - *User Authentication*: User authentication refers to the authentication for mobile RFID service users. It is generally required to enable the reader terminal to access the application server and obtain mobile RFID service contents.
6. The authentication that must be considered in the mobile RFID service structure is as follows [14,16]:
   - *Tag Access Control*: The reader terminal can give various commands to the tag, and the tag will be able to support the access authentication through a password, especially when executing sensitive commands such as Write/Delete/Lock/Kill.
   - *Reader Execution Authorization*: Refers to the function that verifies the validity of the user before executing sensitive reader commands such as Write/Delete/Lock/Kill at the reader terminal. It may be possible to develop the reader execution authorization by developing the reader terminal.
   - *Authorization for Adult Service*: Authorization for an adult service is required as the adult contents provided by a mobile RFID service can be accessed indiscreetly.
   - *User Authorization*: Must provide the access control for each user or the access object when providing different services to each of the users accessing the application server, or when differentiating the access level per user.
7. The mobile RFID application service—including such processes as bill payment between the reader terminal user and the application server—requires nonrepudiation of the data transmitted by the reader terminal user and the application server. In this case, the reader terminal and the application server must be able to execute nonrepudiation.
8. The mobile RFID application service that uses the password to halt the tag or authorize the access to the tag should be able to safely manage such passwords and safely authorize the key to the reader terminal: such functions should be provided by the mobile RFID service infrastructure; for example, the application server or the separate key management server.
9. Since the mobile RFID service is a B2C service using RFID tags for end-users, it is inevitably accompanied by issues of personal privacy infringement and hence must provide solutions for such issues. The personal privacy issue encompasses both location privacy, which relates to the personal identifier role of the RFID tag, and information privacy, which relates to the identification of personal belongings by browsing the tag interface information through the resolution of tag code.

28.4 Multilateral Approaches with Improved Security

The mobile RFID is a technology for developing an RFID reader to be embedded in a mobile terminal and for providing various application services over the wireless networks. Robust mobile RFID security must both protect service network against security threat and shield consumers from privacy intrusions. The keys to robust mobile RFID security are simplicity and a fundamentally secure foundation. This section looks at these security points and recommends an alternative approach to achieving robust mobile RFID security. This section aims at providing secure mobile RFID services, suggesting and analyzing secure mobile RFID service models to solve security issues like security among domains, personal privacy profile, authentication, end-to-end security, and track prevention.

28.4.1 Overview of Secure Mobile RFID Environment

There are many ways to interfere with RFID circumstances, issues which are not only approved theoretically but also possible practically. Besides security vulnerabilities in RFID security like passive signal interception attack on RFID tags and readers, reading of RFID tags by unauthorized readers, falsifying tag or reader identity, use of attack tools against RFID tags, neutralization of RFID tags, and elaborate attack on RFID tags with cryptographic hacking methods, there are also similar vulnerabilities and possible infringement of privacy in mobile RFID circumstances. It requires proper security technologies. Furthermore, needed are some information protection service models that ensure security and privacy protection and management for service providers in practical compliance with present RFID specifications and mobile RFID standards even when tags do not use code algorithms. This chapter suggests and analyzes these mobile RFID information protection service models considering situations mentioned earlier. The provision of secure mobile RFID services needs a combined security framework resolving many security issues like security among domains, personal privacy profile, authentication, end-to-end security, and track prevention. The following is the configuration of the security framework based on the mobile RFID service network.

Generally, in mobile RFID applications such as smart poster, the ASP has the ownership of RFID tags [14,16,20,21]. Thus, mobile RFID users have to subscribe to both the ASP for these kinds of RFID services and to the mobile network operator for the mobile communication service. Namely, there exist three potentially unreliable parties: the user-owned RFID reader, the mobile network operator, and the ASP. Accordingly, a relationship of trust must be established among the three parties to realize a secure mobile RFID service. In particular, when an RFID reader tries to read or change the RFID service data stored in a tag, the reader needs to obtain tag access rights. Additionally, it is important that the new tag access rights obtained whenever some readers access the same tag must be different from the old ones that have already been accessed. Main functions of the proposed mobile RFID information protection service model are the provision of WIPI-based mobile security middleware, tag authentication, tag tracking prevention, reader authentication, message security, and protection of profile-based privacy.

28.4.2 Security-Enhanced Mobile RFID Middleware in the Phone

One of the key problems with mobile RFID technology is how to quickly use the mobile RFID reader and how to integrate it with the application software installed in the mobile device. In the face of numerous existing types of application software, developing an independent mobile RFID middleware layer is a good idea. The mobile RFID middleware layer inhabits the middle ground between the RFID reader and the application logic layer. The mobile RFID middleware layer will manage the RFID readers and server for the application logic layer; so the application logic layer-based mobile RFID technology can focus on implementing commerce logic.

WIPI is required to come into force in Korea in the case of mobile phones as of 2005 to support the interoperability platform for various application software and hardware platforms [22]. Therefore, we chose WIPI as the basic software development platform of the mobile phone: the software architecture and the relationship between each of the software functions are shown as Figure 28.6. The software architecture is composed of REX OS, WIPI Handset Adaptation Layer (HAL) API, WIPI Runtime Engine (WRE), WIPI C API, Phone application, Browser parser, and Phone GUI. Most functions for mobile RFID technology are designed in the WIPI C API: they are Reader control, Tag control, Buffer control and Filter control for interfacing with the RFID reader; and Code Decoder, URN converter, FQDN (Fully Qualified Domain Name) converter, DNS resolver and connect Contents server for communicating with a local ODS server and the contents Web server [30,31].

In the WIPI specifications, the core functions are the functions of the handset hardware, native system software, handset adaptation module, runtime engine, basic APIs, and application programs are the areas of the core functional specifications of WIPI. Actually, in the WIPI specifications, only the handset adaptation and APIs are included, while the other parts of the functions of the wireless Internet platform are considered as requirements of the handset vendors whether they accept it or not. For example, the runtime engine part is required as the mode of the download of the binary code for its maximum performance.

The core functions of the WIPI are the handset adaptation and APIs, which are called the HAL and Application Adaptation Layer (AAL), respectively. The HAL defines an abstract specification layer to support hardware platform independence when porting applications;

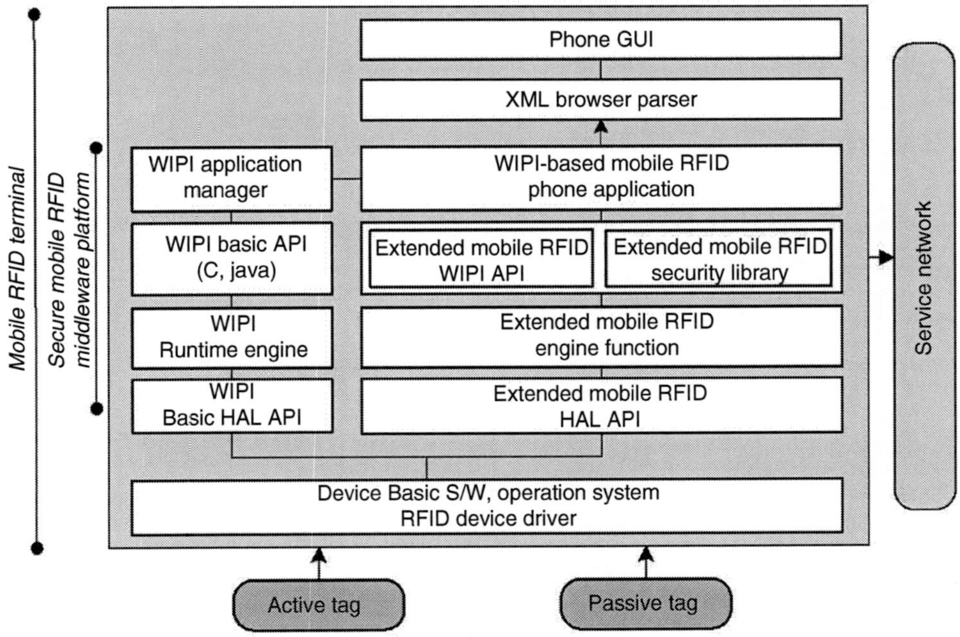

FIGURE 28.6
Mobile RFID's software architecture in the mobile phone.

the AAL defines the specifications for the API of the wireless Internet platform, and supports the C/C++ and Java programming languages.

The mobile RFID middleware is implemented by extending the WIPI platform to provide RF code-related information obtained from the RF tag through the RFID reader installed in the mobile phone [20,21,23,24]. The functions of RFID WIPI C API include RFID reader control, Buffer control, Tag control, Filtering, Networking for Code decoding, URN conversion, FQDN conversion, DNS resolving, and the Content services. WRE software for mobile RFID functions is extended to support RFID WIPI C API, and RFID HAL API. The functions of RFID HAL API include RFID reader control, Buffer control, Tag control, Filtering, Networking for configuring the Internet Protocol (IP) address of the local ODS server. Figure 28.7 shows the middleware functions and software.

The RFID device handler provides the definitions for the functions of starting the platform and transferring the events from the upper layer of HAL to the RFID H/W Reader. The categories of RFID device handle API cover call, RFID device, network, serial communication, short message service, sound, time, code conversion, file system, input method, font, frame buffer, and virtual key. The AAL provides the definitions for the functions of the adaptive, functions for the RFID engine, WIPI C/Java API, Crypto libraries, and RFID security components.

28.4.3 Using Crypto Algorithm in Mobile Terminal Platform

When selecting a suitable mobile RFID service system, consideration should be given to cryptological functions. Applications that do not require a security function would be made unnecessarily expensive by the incorporation of cryptological procedures. On the other hand, in high security applications (e.g., mobile ticketing, payment systems) the omission of cryptological procedures can be a very expensive oversight if manipulated mobile RFID readers are used to gain access to services without authorization.

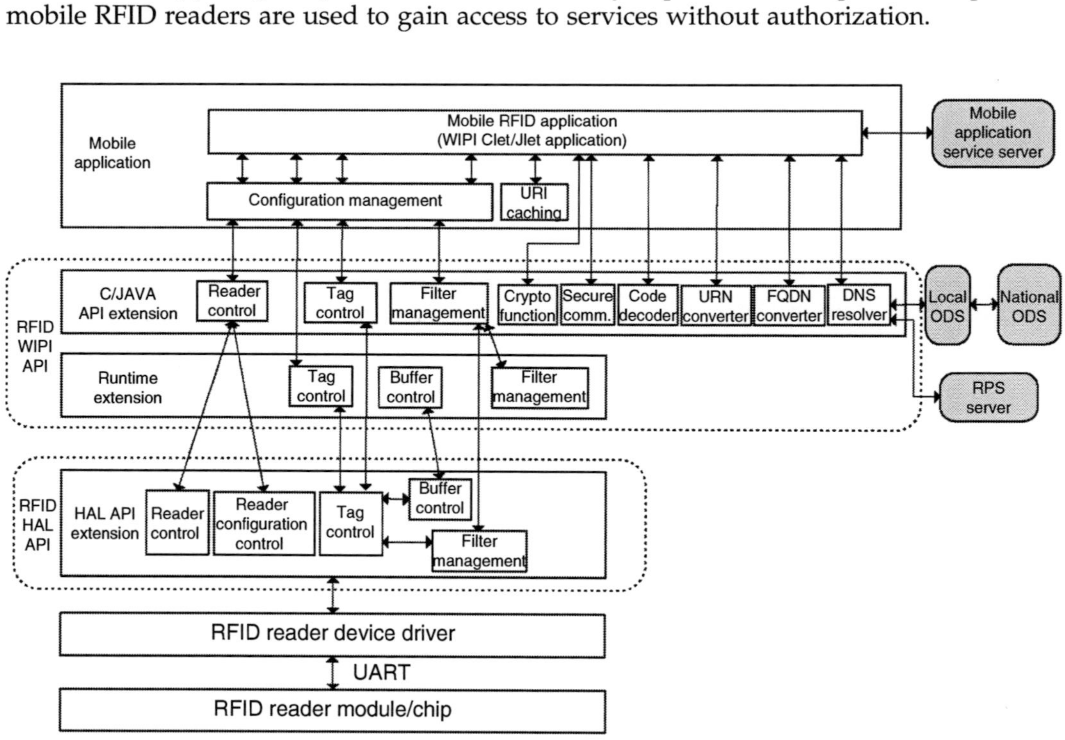

FIGURE 28.7
Security enhanced mobile RFID middleware in the mobile phone.

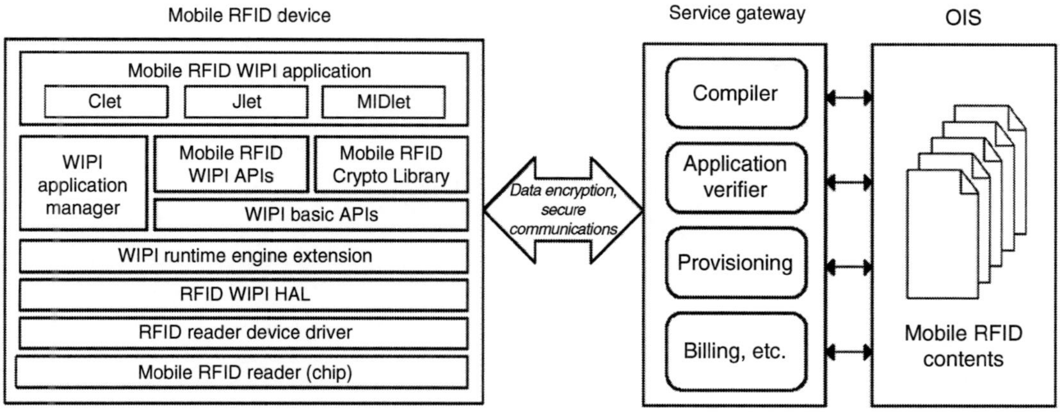

FIGURE 28.8
Service structure of mobile RFID crypto library.

Following is the method of reinforcing RFID data communication security service by using crypto algorithm based on mobile phone terminal platform in mobile RFID service. The mobile RFID crypto library is a crypto library for the efficient processing of the crypto algorithms and security protocols. It provides security mechanisms to the mobile RFID reader and targets the mobile RFID middleware based on the WIPI platform. The mobile RFID crypto library enables the mobile RFID service provider, wireless contents provider, and information security industry to support the information protection service on the mobile RFID middleware terminal platform at a reasonable cost and in a short period of time (Figure 28.8).

Its main features are crypto algorithms (AES, DES, 3DES, SHA-1, HAS160, HMAC, etc.), High-speed Elliptic Curve Cryptosystem (ECC) and Digital signature (ECC, ECDSA, etc.), High-speed Korean standard crypto algorithms and Digital signature (SEED, KCDSA, ARIA, etc.), Secure communication protocol (SSL/TLS, etc.), and Public key crypto standard (PKCS #5, PKCS #8, ASN.1, etc.). Cryptological procedures are used to protect against both passive and active attacks. To achieve this, the transmitted data (plain text) can be altered (encrypted) before transmission so that a potential attacker can no longer draw conclusions about the actual content of the message (plain text). Mobile RFID systems have for a long time used only symmetrical procedures. Because block ciphers are generally very calculation intensive, they play a less important role in mobile RFID service systems.

28.4.4 RFID Tag Access Control and Password Data Management

The mobile terminal platform for the mobile RFID service requires a security function to secure its safety, and this standard defines the security of data on RFID tag information, security of communication with the RFID system—including OIS and ODS, security setup for tag passwords, and security API consisting of approval for execution of the main tag control order [20,21,25]. RFID tag access-control function for secure reader operation must be added to the expanded WIPI API function. The data format of the mobile RFID tag access password based on the WIPI platform is as follows.

This tag access right management method is a method for safely managing by integrating Kill password data and Access password data on WIPI standard platform, which is mobile RFID terminal service environment. Such integrated password information can be safely managed as an expanded security API function on WIPI platform. And let us see the management method for serving by expanding such integrated password information

TABLE 28.2

Data Format of Tag Access Password Based on Mobile Terminal Platform

MIME Type	Data Format
RfidTagPassword	- Tag Password Data = UII Block+\0+APWD+\0+KPWD+\0\0 - UII Block : Tag UII Block (PC + UII Field) - APWD : Access Password, 16 characters - KPWD : Kill Password, 16 characters Ex) UII : FFF0FFF1FFF2FFF3, APWD : FA1B52D1, KPWD : AB1252E1, * RfidTagPassword = "FFF0FFF1FFF2FFF3\FA1B52D1\0AB1252E1\0\0"

in the server system. In the mobile RFID service network environment, the RFID tag password data of private mobile RFID readers are sometimes exposed to hackers because of the lack of awareness of the need for information security. We apply a secret-sharing scheme to solve the RFID tag password data management problems, but the existing scheme is not suitable for the management of large quantities of data because that required operation of large capacity (Table 28.2).

In this section, we recommend to use the secret-sharing scheme in which the right to own the password is weighted in the RFID tag password data management. This method can use security service protocol verified in a theoretical aspect and apply the password management method that is safe between terminal and server.

28.4.5 RFID Tag-Based Adult Certification

The mobile RFID service applications need stricter adult verification. Currently, teenagers and even elementary school students below the age of 10 are using cell phones, which are in reality ubiquitous information terminals when, in fact, they ought to be private devices. As such, they can access adult content very easily. Therefore, a strict and elaborate mechanism for adult verification should be provided to protect young people from adult content. However, adult verification is currently provided within contents at the application layer: namely, the control role is given to content providers, which means that network operators (referred to as ISPs) cannot control illegal behavior such as the unrestricted provision of adult content.

A protocol solution would be a much better means of strictly controlling access to adult content, as ISPs are responsible for clean and well-controlled content provision to users, and the issue of legal responsibility might then arise. Such a protocol-level adult verification could enable service providers to control illegal content provision to young people.

When mobile RFID services provide adult content, adult application data should be included at the user data area of RFID tags. Adult application data shows the content's adult grade. Instead of the current grading method of adapting the user's age, we provide a method of defining grades by dividing them into categories such as word, nudity, sex and language, thus classifying the content in detail. If this grading method can adapt to mobile RFID environments, four values of an item are recorded at user data area of RFID tags and a combined value is selected as final grade for this item. Through this grading method, we can express grades of diverse multimedia content and can control the final grades of various applications by giving different weights to categories, that is, word, nudity, sex and language, and so on. Such RFID tag-based adult authentication service expands the user memory region in the tag to be used and applies the expanded service mechanism to be applied to the relevant application service.

28.4.6 Recoding Scheme of RFID Tag Code Data

This is required to prevent the tracing of the tag owner's location by tag code tracking. There are two methods for satisfying the untraceability requirement. One consists of an authentication between tag and reader. The tag allows the reader to access its ID code only if the reader is authenticated by the tag, so an attacker cannot obtain the tag's ID code without the authentication process. The other consists of an ID recoding technique, which consists of recoding the tag's ID code periodically with a pseudo-ID (or meta-ID), thereby reducing the connectivity of the tag's ID code and the owner.

Tag killing has received so much attention because it has become clear that privacy in item-level tagging will be a hot-button issue in consumer acceptance of RFID. Unfortunately, there are several issues with kill commands. The killing tags prevent all post-point-of-sale uses for RFID tag data. These uses are expected to become more important as the use of RFID tags on retail items spreads. To address these issues, we suggest "recoding of tag code information" as an additional tool for mobile RFID privacy. In recoding, a tag is overwritten with a new ID number when it changes hands. Without knowledge of the map from the old ID number to the new ID, it is impossible to link sightings of the item from before and after recoding. Recoding may occur at point of sale when an item passes from one organization to another.

Both killing and recoding raise infrastructure issues need to be solved before they can become viable privacy protections. In particular, only authorized parties, such as a retailer, should be able to kill or recode tags. In the end, while both are important tools, neither killing nor recoding is the final answer in mobile RFID privacy.

28.4.7 Secure RFID Discovery Service Gateway System

The secure mobile RFID application portal (SMAP) is a secure service portal for various mobile RFID application services. A service provider using SMAP can easily deploy several mobile RFID applications that are guaranteed security and privacy protection. The structure of SMAP framework is as follows.

The secure RFID discovery service gateway is a system that classifies and defines the mobile OIS system as an individual element system in a mobile RFID security application service network, and also supports internal functions comprising each element service system. This gateway system manages the locations and interface of servers and services as registered from the applicable product and service providers that provide product information and contents corresponding to each EPC. Moreover, this system works as a gateway system that seeks appropriate services with the capacity to provide information on any product equipped with a tag containing the applicable EPC or the contents and so forth related to such an EPC, and then provides such services by way of the wired and wireless devices (particularly mobile RFID reader terminals) of users who access the mobile RFID service network.

This portal allows you to find out off-line product information on an EPC, online additional service information, information on authentication for product families or products related to this EPC, and more. This gateway system also has another advantage of applying a privacy policy and security to personal information with regard to inquiries about information, in the interest of creating a secure and reliable environment for making information available. An additional feature of this system is its ability to automatically generate and manage OIS for the mobile RFID. This function allows you to create and use a variety of useful functions, such as a database for managing information corresponding to each EPC by simply setting; security for access control, authentication, encoding, and so on; data exchange for sharing necessary information among OISs; and interface generation and management for the user, and external application to use the OIS.

As a mobile RFID OIS module, mOIS (mobile OIS) offers information on any object, and works for system management or commissioned management and administration. By means of interlocking with the gateway system, the manufacturer's OIS, the application retailer's mOIS, and the mOIS for additional service providers can provide services via the user's RFID reader terminal. The secure OIS is a security-enhanced IS system of EPCglobal. It is capable of strengthening security for events management and tag information provision, and provides security mechanisms such as message security, authentication authorization, and so on. It can also be used as the OIS server of a mobile RFID in the mobile RFID application environment. The method of building a safe server-based information system is carried out by using Web service security technology that is used on the existing wire as a standard and many communication security technologies to expand RFID data security functions. It is suitable to use standardized security technology for compatibility of existing service systems (Figure 28.9).

28.4.8 Policy-Based RFID User Privacy Protection

Widespread deployment of RFID technology may create new threats to privacy due to the automated tracking capability. In the mobile RFID environment, the privacy problem is particularly serious since the RFID reader is contained in a handheld device and many application services are based on the B2C model. The RPS provides mobile RFID users with an information privacy protection service for a personalized tag under the mobile RFID environment [14,16]. When a mobile RFID user possesses an RFID-tagged product, RPS enables the owner to control his backend information connected with the tag, such as product information, distribution info, owner's personal information, and so on.

In academia and industry, most efforts to secure a privacy protection mechanism have focused on the tag/reader authentication protocol. However, these efforts cannot be adapted to the real RFID environment, especially the mobile RFID applications, the environment which the meaningful information of the RFID code value exists in the server at the network domain. Therefore, it is necessary to produce a proper privacy protection mechanism in this mobile RFID environment; a profile-based privacy protection mechanism could be one of the many possible solutions for this environment. The privacy protection framework for mobile RFID services describes the types of privacy infringement in the mobile RFID service environment, the requirements for privacy protection, and the basic structure of RFID privacy protection based on the owner's privacy profile [9,16,17]. A profile-based privacy protection service consists of three service systems as follows (Figure 28.10):

1. *RPS system*: This system incorporates a management function for the owner's privacy policy: it creates a privacy profile for the owner's privacy policy, provides the privacy profile to the service-side system, and manages the event logs from the service-side or RPS system for auditing.
2. *Service-side system*: This system provides information related to the ID code of the RFID tag, and provides an access-control function by the owner-defined privacy profile.
3. *User-side system*: This system has a wireless (or wired) network access function and an RFID reader function. The tag owner accesses the service-side and RPS system via this system.

In order to satisfy the privacy protection requirements of users of the mobile RFID service, the profile-based privacy protection service incorporates the following functions: access

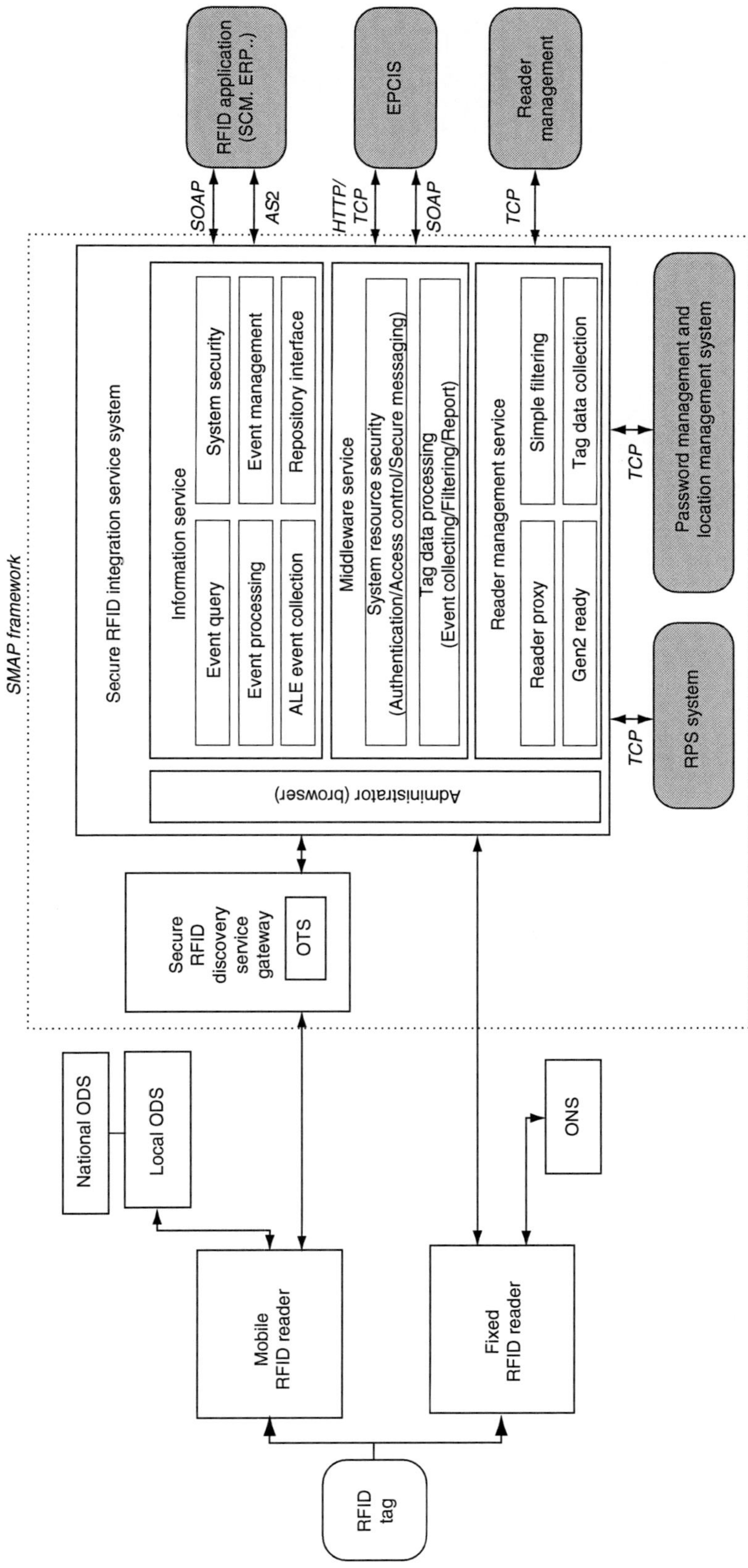

FIGURE 28.9
Structure of security-enhanced mobile RFID's application portal framework.

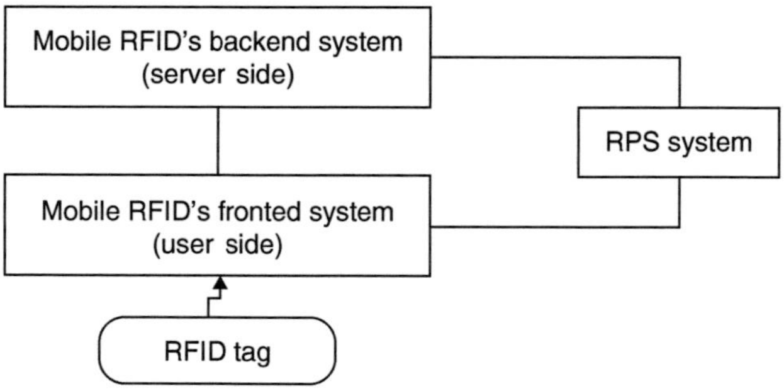

FIGURE 28.10
Service systems comprising the profile-based privacy protection service.

control, registration, privacy profile management, privacy-enhanced log management, obligation notification, and tag data refreshment [14,16,21,26–28].

- Privacy profile management is a core function of the profile-based privacy protection service, which establishes and manages the owner's (default) privacy profile. The RPS system should create and manage a default privacy profile for each application service and an owner-defined privacy profile from the user's privacy protection policy of the registration process. Once created, this privacy profile can be sent to the service-side system if the ASP so requests.
- A privacy-enhanced log management function performs secure event log collection, limitation of event log collection by the owner's privacy profile. Furthermore, it is capable of detecting and analyzing privacy violation elements with regard to the collected event logs.
- The result of obligation that should be performed by the service-side system can be notified to the owner via various means (e-mail, text messaging, etc.). The service-side system should notify the RPS system of the obligation result, while the RPS system should notify the owner of the obligation result.
- The tag data refreshment function provides a mechanism for refreshing the owner's tag data, which should then be written by the user-side system to his/her own tag. Furthermore, the service-side system should reestablish its relationship between the information and the tag's tag data. Thus, the tag data refreshment function would seem to satisfy the untraceability to tag data requirements.

Next is detailed description of the selective RFID privacy protection method in RFID privacy service mechanism [14,16,21,27]. In the RFID tag, data values of privacy level and data values of security level are randomly used in the user memory region to be used in linkage with service protocol described in the previous chapter. If we take a look at the details by each level, they are as follows.

28.4.8.1 *Privacy Level of RFID by Product*

Privacy level refers to a grade of information that the owner of the tagged item can open to the public. For example, Figure 28.11 shows four different privacy levels. Because of people's varying sensitivity to privacy issues, the levels were selected randomly for this paper. The tag set as level 1 provides information about the kinds of product in question. In the case of level 2, information about the product can be served. If the tag is set to level 4, all of the information related to this tag can be presented. When a tag is read by a user, the tag sends the privacy level to the reader. Using the data sent by the tag, the reader can obtain the location of the server and request information customized to the privacy level obtained from the tag.

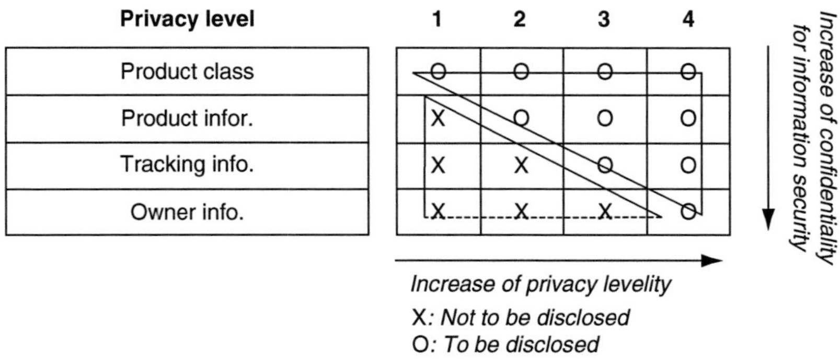

FIGURE 28.11
Privacy level in tag user's product memory bank.

28.4.8.2 Security Level of RFID by Product

The RFID tag memory space containing the level information can be divided as shown in Figure 28.12. The packet from tag to reader includes this information.

The security level has become a key factor in ensuring privacy protection, along with privacy preference. Furthermore, it could be an alternative solution to requests for information that maliciously change the privacy level. Table 28.3 represents the additionally required phases according to the request. A variant of the ways of access control can be added as the importance of the information increases. The sophistication and strength of the enforcement of access control should increase proportionally to the security level.

The main features of this RPS service mechanism consist of establishing and managing the owner's privacy protection policy, providing access control for information associated with the personalized tag by the owner's privacy policy, notifying the obligation result, and executing a privacy audit by audit log management [14,16,21,27]. The brief personal privacy protection process using the earlier functions of the RPS is as follows.

Firstly, the mobile RFID reader reads the tag ID and obtains the network addresses of various types of information, such as the product information integrated into the tag ID through the ODS resolver process. Secondly, it requests from the application server the product information connected to the tag ID. Thirdly, the application receives the personal privacy protection policy in relation to the product information through the RPS. Finally, the product information is protected according to the requirements of the privacy protection policy configured by the individual and is then sent to the reader. The information connected to the tag ID reflecting the personal privacy protection policy through the earlier

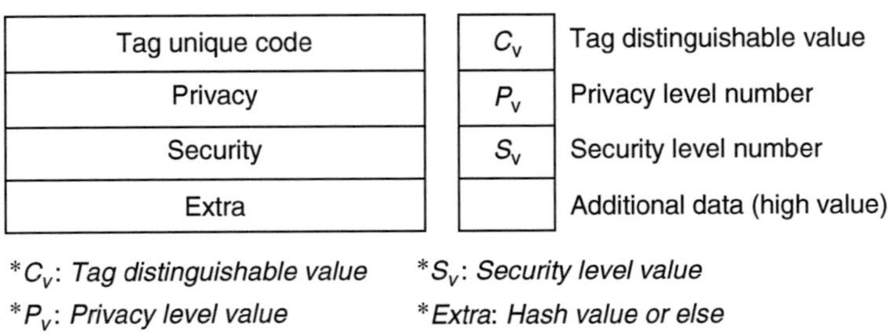

FIGURE 28.12
Data structure in RFID tag.

TABLE 28.3

Security Level for Privacy Based on Mobile RFID

Security Level	[1]	[2]	[3]	[4]
No security	O		X	X
Symmetric key-based security (password-based lock/unlock)	X	O	X	X
Symmetric-based security, protection of location-based traceability	X	X	O	X
Fourth-level heading	X	X	X	O

Note: X: Not to be disclosed; O: To be disclosed.

process is circulated through the network in a process intended to solve the personal privacy infringement issue through the RFID network infrastructure. The detailed service scenario for the profile-based privacy protection service generally arises from the tag personalizing process, such as tagged product purchasing. Figure 28.13 illustrates a basic profile-based privacy protection service scenario of the mobile RFID application [16,21].

1. A consumer reads the ID code from the tagged product with his/her mobile terminal, which is equipped with an RFID reader.
2. The consumer browses the product-related information from the application service network, and then purchases the product with various payment methods. At this moment, the consumer becomes the tag owner.
3. Then, the networked RFID application requests the owner-defined privacy profile from the RPS server, and the RPS server responds by sending the owner-defined privacy profile to the application if the owner's privacy protection policy has been created.
4. The RPS server receives the owner's privacy protection policy for this application service.
5. Anyone requests the information associated with this tag ID to the application.
6. The requestor can browse all the information provided by the application service if the requestor is the owner, but the requestor can only receive limited information, or cannot access any information at all.

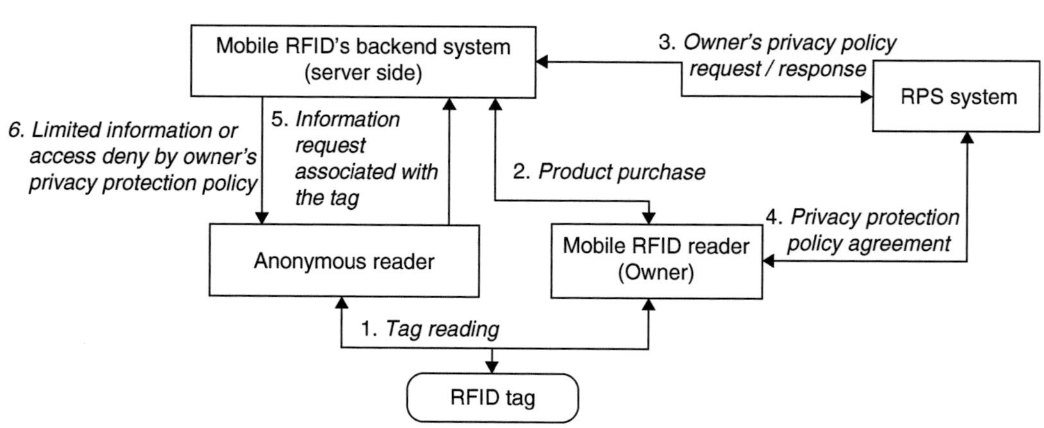

FIGURE 28.13
Service scenario of profile-based privacy protection.

The earlier generic service scenario would seem to be modified according to each specific networked RFID application. In this section, we described the privacy-friendly service mechanism which is hardly focused on in the RFID research area. Most interests are of the privacy of the RFID tag owner. But we are aware of the necessity to protect the privacy of users carrying the mobile reader. To address the reader-side privacy problem, we employ the policy-based service model which is introduced and extend it to mobile RFID service area.

28.5 Conclusion

As mentioned earlier, mobile RFID is an emergent and promising application that uses RFID technology. However, the mobility of reader and its service model—which differs from the RFID service in the retail and supply chain—will give rise to additional security threats.

To address these issues, while both are important tools, neither killing nor recoding is the final answer in RFID privacy. The killing alone is not enough, and new mechanisms are needed for building privacy-preserving RFID architectures. In this chapter, we have tried to introduce the concept of mobile RFID and expose some of the additional security threats caused by it. The frequency band to support the air protocol is allocated from 908.5 to 914 MHz in Korea to comply with ISO 18000-6 for air interface communications at 860–960 MHz. We also describe a way of incorporating the new technology to work with cell phones in particular, both as an external security reading device (replacing 900 MHz) and as an added security service to manage all RFID mobile device mediums. With this purpose in mind, the application areas of this service platform are also briefly presented. By doing so, customized security and privacy protection can be achieved. In this regard, the suggested technique is an effective solution for security and privacy protection in a networked mobile RFID service system.

Acknowledgments

This work was supported by the Korean Ministry of Information and Communication under the Information Technology Research Centre (ITRC) support program supervised by the Institute of Information Technology Assessment (IITA) and by the IT R&D program of MIC/IITA [2005–S-088-03, Development of security technology for secure RFID/USN service]. The authors thank the members of the RFID/USN Research Team at ETRI for formulating many of the concepts discussed in this chapter. The authors also thank Dr. Seungjoo Kim of Sungkyunkwan University and Dr. Howon Kim of ETRI for their encouragement and support.

References

1. T. Tsuji, S. Kouno, J. Noguchi, M. Iguchi, N. Misu, and M. Kawamura. 2004. Asset management solution based on RFID. *NEC Journal of Advanced Technology*, 1(3): 188–193.
2. S. Yoo. 2005. Mobile RFID activities in Korea. Contribution Paper of the APT Standardization Program.

3. J. Chae and S. Oh. 2005. Information report on mobile RFID in Korea. ISO/IEC JTC1/SC 31/WG4 N0922, Information Paper, ISO/IEC JTC1 SC31 WG4 SG 5.
4. W. Park and B. Lee. 2004. Proposal for participating in the correspondence group on RFID in ITU-T. Information Paper, ASTAP Forum.
5. Nokia. RFID Phones—Nokia Mobile RFID Kit. http://europe.nokia.com/nokia
6. Y. Kim and N. Koshizuka. 2006. Review report of standardization issues on network aspects of identification including RFID. ITU-T, Paper TD315.
7. J. Lee and H. Kim. 2006. RFID code structure and tag data structure for mobile RFID services in Korea. *Proceedings of the 8th International Conference on Advanced Communication Technology*. Vol. 2, pp. 1053–1055, IEEE Press.
8. Y. Kim, J. Lee, S. Yoo, and H. Kim. 2006. A network reference model for B2C RFID applications. *Proceedings of the 8th International Conference on Advanced Communication Technology*. Vol. 2, pp. 1049–1052, IEEE Press.
9. B. Park, S. Lee, and H. Youm. 2006. A proposal for personal identifier management framework on the Internet. ITU-T, COM17-D165.
10. L. Sullivan 2004. Middleware enables RFID tests. *Information week*, No. 991.
11. M. Tsukada and A. Narita. 2006. Development models of network aspects of identification systems (including RFID) (NID) and proposal on approach for the standardization. ITU-T, JCA-NID Document 2006-I-014.
12. Y. Sakurai and H. Kim. 2006. Report for business models and service scenarios for network aspects of identification (including RFID). ITU-T, TSAG TD 314.
13. ITU-T, TSAG. 2005. A proposed new work item on object/ID associations.
14. B. Chug et al. 2005. Proposal for the study on a security framework for mobile RFID applications as a new work item on mobile security. ITU-T, COM17D116E, Q9/17, Contribution 116, Geneva.
15. Mobile RFID Forum of Korea. 2005. Mobile RFID code structure and tag data structure for mobile RFID services. Standard Paper. http://www.mrf.or.kr
16. D. Choi, H. Kim, and K. Chung. 2007. Proposed draft of X.rfidsec-1: Privacy protection framework for networked RFID services. ITU-T, COM17C107E, Q9/17, Contribution 107, Geneva.
17. MIC (Ministry of Information and Communication) of Korea. 2005. RFID privacy protection guideline. MIC Report Paper.
18. H. Lee and J. Kim. 2006. Privacy threats and issues in mobile RFID. *Proceedings of the First International Conference on Availability, Reliability and Security*. Vol. 1, pp. 510–514, IEEE Computer Society, Washington, DC, USA.
19. D.M. Konidala and K. Kim. 2006. Mobile RFID security issues. *Proceeding of Symposium on Cryptography and Information Security*. Abstracts, pp. 166, The Institute of Electronics, Information and Communication Engineers. Hiroshima, Japan.
20. N. Park, J. Kwak, S. Kim, D. Won, and H. Kim. 2006. WIPI mobile platform with secure service for mobile RFID network environment. *Lecture Notes in Computer Science*. Vol. 3842, pp. 741–748, Springer-Verlag, Hiedelberg.
21. N. Park, S. Kim, D. Won, and H. Kim. 2006. Security analysis and implementation leveraging globally networked mobile RFIDs. *Lecture Notes in Computer Science*. Vol. 4217, pp. 494–505, Springer-Verlag, Hiedelberg.
22. M. Son, Y. Lee, and C. Pyo. 2006. Design and implementation of mobile RFID technology in the CDMA networks. *Proceedings of the 8th International Conference on Advanced Communication Technology*. Vol. 2, pp. 1033–1036, IEEE Press.
23. Mobile RFID Forum of Korea. 2005. WIPI C API standard for mobile RFID reader. Standard Paper.
24. Mobile RFID Forum of Korea. 2005. HAL API standard for RFID reader of mobile phone. Standard Paper.
25. Mobile RFID Forum of Korea. 2005. Access right management API standard for secure mobile RFID reader, MRFS-4-03. Standard Paper. http://www.mrf.or.kr
26. S. Weis et al. 2003. Security and privacy aspects of low-cost radio frequency identification systems. *First International Conference on Security in Pervasive Computing* (SPC) 2003.
27. B. Lee and H. Kim. Enhanced Security and Privacy Mechanism of RFID Service for Pervasive Mobile Device. *Lecture Notes in Computer Science*. Vol. 3802, pp. 469–475, Springer Berlin/Heidelberg.

28. M. Ohkubo, K. Suzuki, and S. Kinoshita. 2003. Cryptographic approach to privacy-friendly tags. *RFID Privacy Workshop 2003*.
29. ITU-T TSAG RFID CG Deliverable. 2006. Review report of identification based business models and service scenarios.
30. Mobile RFID Forum of Korea. 2005. WIPI network APIs for mobile RFID services. Standard Paper.
31. Mobile RFID Forum of Korea. 2005. WIPI API for mobile RFID reader device. Standard Paper.

29

ONS Security

Benjamin Fabian, Oliver Günther, and Sarah Spiekermann

CONTENTS

29.1 EPCglobal Network

The influential "electronic product code" (EPC) numbering system is about to enhance and finally replace traditional bar codes. It aims to assign a globally unique number to nearly every object equipped with a radio frequency identification chip (RFID tag). This EPC is serving as an identifier for the physical object carrying the tag, which can now be recognized, identified, and tracked by an IT infrastructure [1]. Though the EPC standard is actually a meta framework for different encoding schemes and name spaces, most EPCs have a structure similar to the one shown in Figure 29.1, which depicts an example EPC for one of the most popular standards, the Serialized Global Trade Identification Number (SGTIN) [1].

The consortium EPCglobal, originating from the Auto-ID labs of MIT as well as the former EAN International and Uniform Code Council (both now GS1), places its focus on the development and establishing of global standards for RFID, EPC, and the EPCglobal network. According to their intention, information about an object should in general not be stored on its RFID tag itself, but instead be supplied by distributed servers on the Internet [2]. By using the EPC and the help of an Object Naming Service (ONS) (also called Object *Name* Service in some documents), it will be possible to locate EPC Discovery and EPC

Header 8 bits	Filter 3 bits	Partition 3 bits	EPC manager 20–40 bits	Object class 4–24 bits	Serial number 38 bits
00110000 "SGTIN-96"	001 "Retail"	101 "24:20 bits"	809,453	1734	108,265

FIGURE 29.1
SGTIN electronic product code (EPC).

Information Services (EPCIS), which are remotely accessible data collections about the particular object [3].

One of the advantages the EPCglobal network offers is to let many parties (e.g., manufacturers, suppliers, shops, or after-sale service providers) dynamically register any kind of EPCIS for the objects they are concerned with, thereby creating an open way to exchange product-related information. The designers of the EPCglobal network anticipate that a static list of available services would in general be outdated soon. As a solution, they propose to first ask the ONS (and not yet specified EPC Discovery Services) for a recent list of sources each time fresh information about a particular object is needed. After retrieving this list, one can directly contact all or some of the EPCIS that carry the information itself [4]. This procedure will in most cases not be conducted manually, but in an automated fashion, e.g., by the use of web services [5]. Example application scenarios include supply chain management [5]—increasing efficiency, flexibility, and cooperation—and potentially smart homes [6], where a home IT infrastructure needs to identify objects of the real world to provide services. Since ONS is a critical component of the EPCglobal network, we focus on its security issues. An initial analysis has been given in [7].

29.2 Object Naming Service

From a technical point of view, the ONS [8] is a subsystem of the Domain Name System (DNS) [9,10]. The main design idea is to first encode the EPC into a syntactically correct domain name, then to use the existing DNS infrastructure to query for additional information. This procedure makes use of the Name Authority Pointer (NAPTR) DNS resource record, which is also used for the Session Initiation Protocol (SIP) for Voice-over-IP to map phone numbers into Uniform Resource Identifiers (URI) [11].

The ONS resolution process is described in [4,8]; for a schematic view of the communication procedure see Figure 29.2. After an RFID reader has received an EPC in binary form, it forwards it to some local middleware system. To retrieve the list of relevant EPCIS servers for this particular object, the middleware system converts the EPC to its URI form [1] (e.g., `urn:epc:id:sgtin:809453.1734.108265`). This is handed over to the local ONS resolver, which in turn translates the URI form into a domain name (e.g., `1734.809453.sgtin.id.onsepc.com`) by following a well-defined procedure [8, Section 5]. This name belongs to a subdomain of the domain `onsepc.com`, which is reserved for ONS use.

The current ONS specification states that the serial part (item level, in the example: 108265) of the EPC, which differentiates between objects of the same kind and brand, should not be DNS-encoded as of now, but it leaves room for such a possibility [8, Section 3.2.1]:

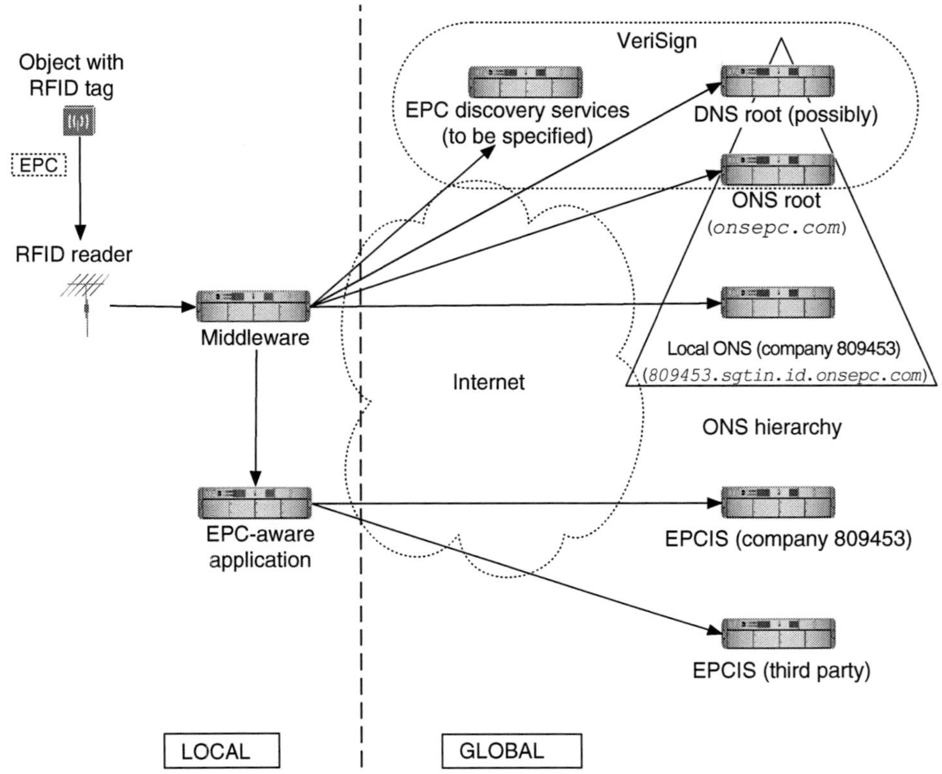

FIGURE 29.2
The EPCglobal network.

> The ability to specify an ONS query at the serial number level of granularity as well as the architectural and economic impacts of that capability is an open issue that will be addressed in subsequent versions of this document. Its lack of mention here should not be construed as making that behavior legal or illegal.

This newly created domain name is now queried for by using the common DNS protocol, possibly involving a recursive query to a local DNS or service provider DNS server, which then queries iteratively accross the Internet [10]. This implies that the (for now partial) EPC moves through the subsequent local networks and the Internet in clear text, and can be read, stored, and analyzed by any interested party on its way through the DNS hierarchy. DNS is an old and central Internet service with a long history of security and configuration issues [12,13] in the protocol itself and in particular implementations. Various vulnerabilities and attacks can be listed by consulting established security sites as CERT [14], SecurityFocus [15], and the SANS Institute's "Top 20 List of Internet Security Vulnerabilities" [16]. A corresponding "Request-for-Comments," RFC 3833: "Threat Analysis of the Domain Name System," was published quite late [17], after two decades of use. Some of the main known threats identified are: Packet interception, i.e., manipulating IP packets carrying DNS information, query prediction by manipulating the query and answer schemes of the DNS protocol, cache poisoning by injecting manipulated information into DNS caches, betrayal by a trusted server that is actually controlled by an attacker, and denial of service, a threat to every Internet service—but DNS itself might be used as an amplifier to attack third parties [17, p. 7].

Besides bugs in DNS software, the fundamental reason for most of these vulnerabilities is the fact that even though DNS is a highly exposed service by definition, it has (in its original and widely deployed form) no way of authenticating a client, the server, or the information that is provided. These DNS weaknesses directly transfer to ONS. However, as a central element of a global information system on physical objects, ONS involves new risks to privacy.

29.3 ONS Risks

29.3.1 Confidentiality and Anonymity

In many situations, the EPC of an RFID tag has to be regarded as highly sensitive information [18]—be it in private [19–21] or in business environments where product and raw material flows constitute valuable market information. Even if the complete serial number of the EPC is not known, the combination of object class and company identifier is enough to determine the brand and kind of object it belongs to. If the use of the EPCglobal network becomes ubiquitous, the eavesdropper could easily add fake serial parts to the captured incomplete EPC and query the corresponding EPCIS servers until a precise match is found. Capturing EPCs can be used to identify assets of an entity, be it an individual, a household, a company, or another organization. If someone happens to own, wear, or carry a rare item, or a rare combination of belongings, identifying and tracking her might be accomplished even without knowing the actual serial numbers, just by using the object classes.

Many different ideas for securing the wireless RFID tag-to-reader communication against unauthorized access and eavesdropping have been proposed, for example confer Refs. [22–25]. However, most solutions to mitigate these privacy problems do not take into account what will happen to the EPC once it is determined by an authorized reading process. To use the information stored in the EPCglobal network about a given EPC, you need to locate the corresponding EPCIS servers first. Even if the connections to these servers are secured by using protocols like Transport Layer Security (TLS), the initial ONS look-up process would have neither been authenticated nor encrypted.

The DNS encoded main part of the EPC, which identifies the asset manufacturer and category, will first traverse every network between the middleware and a possible local DNS server in clear text—this could include an insecure local wireless network. Depending on the configuration of DNS caching and resolution process, this partial EPC will also be transmitted as clear text to additional DNS servers on the resolution path, which could include DNS root servers, servers for `onsepc.com`, and down the corresponding delegation hierarchy [10, Section 2.6 on resolution], until the resolving process finally gets to query a DNS server of the company that serves as main reference for the object in question (usually the manufacturer). All traversed Internet service providers and carriers might capture the partial EPC as well—this includes network taps placed by governmental organizations of countries the packets may cross. It follows that attack trees [26] describing, for example, the profiling of someone's assets will have new branches that represent remote tactics (Figure 29.3)—in addition to those already identified in [21].

There is no simple solution to this problem given the proposed workings of ONS. The main privacy enhancing strategy involves obfuscating the source IP or the real physical orign of the query.

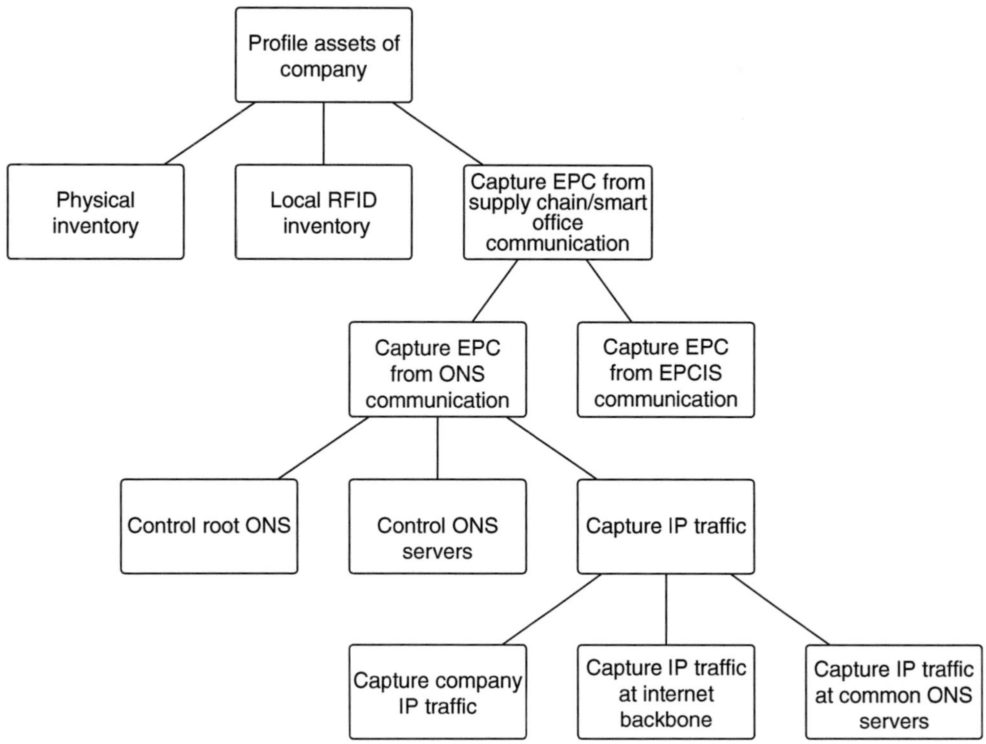

FIGURE 29.3
Asset profiling.

29.3.2 Integrity

Integrity in the ONS context refers to the correctness and completeness of the returned information, i.e., addresses of EPCIS corresponding to the queried EPC. An attacker controlling intermediate DNS servers or launching a successful man-in-the-middle attack on the communication could forge the returned list of URIs and include, e.g., a server under her control. If there are no sufficient authentication measures for the EPCIS in place, the attacker could deliver forged information about this particular or other related EPCs from a similar domain. The corresponding risks will be application-specific: If the query was initiated by a smart refrigerator to order matching ingredients for a cooking recipe, this could result in spoiled meals; if the query was issued by a smart medicine cabinet (as a precursor to an even smarter "home medical advisor" [27, p. 51]) to prevent harmful drug mixes, this could involve more serious risks to personal safety.

29.3.3 Availability

If the EPCglobal network becomes widely accepted, more and more business processes (B2B, B2C) as well as private applications will be able to use it without human intervention. This would leave these processes highly dependable on a working EPC resolution service for finding matching information sources.

ONS will constitute a service highly exposed to attacks from the Internet, if only due to its necessary widespread accessibility. This could include distributed denial-of-service (DDoS) attacks overwhelming the server or its network connection by issuing countless and intense queries, or targeted exploits that shut down the server software or its operating system. Therefore an integration of the EPCglobal network (with ONS as proposed) into core business processes could leave even formerly non-IT related companies dependable on the availability of Internet services. This will most probably increase overall business risk.

29.4 Risk Mitigation

29.4.1 Network Design

Larger enterprises may be able to somewhat reduce risks to EPC confidentiality by using a well-designed internal network structure, especially a carefully planned DNS server hierarchy. All ONS queries from internal machines at any company site could be forwarded—preferably using Virtual Private Networks (VPN)—to a central company DNS server, which in turn does the external resolution process. Even then all the EPCs that are resolved by the company could be intercepted outside of the Intranet borders, but not easily assigned to particular locations, though an attacker might apply a careful analysis of time, possibly combining this information with captured EPCs from region-specific objects. For an attack in a realistic application scenario, consider a company using smart offices with ubiquitous RFID readers, where outsiders might witness the introduction and the actual kind of new items (such as newly introduced laptops of a specific manufacturer) anywhere in the enterprise.

If a company just uses an internal and private version of the EPCglobal network without depending on outside information, e.g., if only self-manufactured items are of interest, on EPC leakage to outsiders would occur, and risks to integrity and availability could be limited likewise to internal attackers. But this special case would deprive the company of the intended advantages of a global and dynamically updated EPCglobal network, as only company-internal data sources about EPCs could be accessed. Another countermeasure could be the prolonging of ONS and EPCIS caching times to reduce the frequency of the EPC crossing the Internet. Depending on the application scenario, the EPCIS dynamics, and the demand for fresh information, risk-reducing caching strategies might be viable.

29.4.2 Cooperative VPN and Extranets

The idea of concentrating ONS queries to prevent an exact locating of the corresponding items could be extended to small networks of trusted business partners (or neighbors in smart homes) by forming a so-called extranet [28, p. 247] (Figure 29.4).

All parties connect to a central ONS resolving server via VPNs, and this server issues the ONS queries to the outside world. Beyond this point, no protection by VPN would be practical; if access to many different "third parties" beyond the borders of the extranet is required, because the possible communication partners are nearly countless and in general not known in advance, the problem of key management for building VPNs to every company that offers a relevant ONS and EPCIS server would render such solutions not scalable.

Apart from issues of trust and administrative overhead, there will be an increased network load for the central party, depending on the scale of RFID reader deployment, caching strategies and the intense of usage of the EPCglobal network by every single partner. The deployment of an extranet could only limit threats to EPC confidentiality, but not to information integrity or ONS availability.

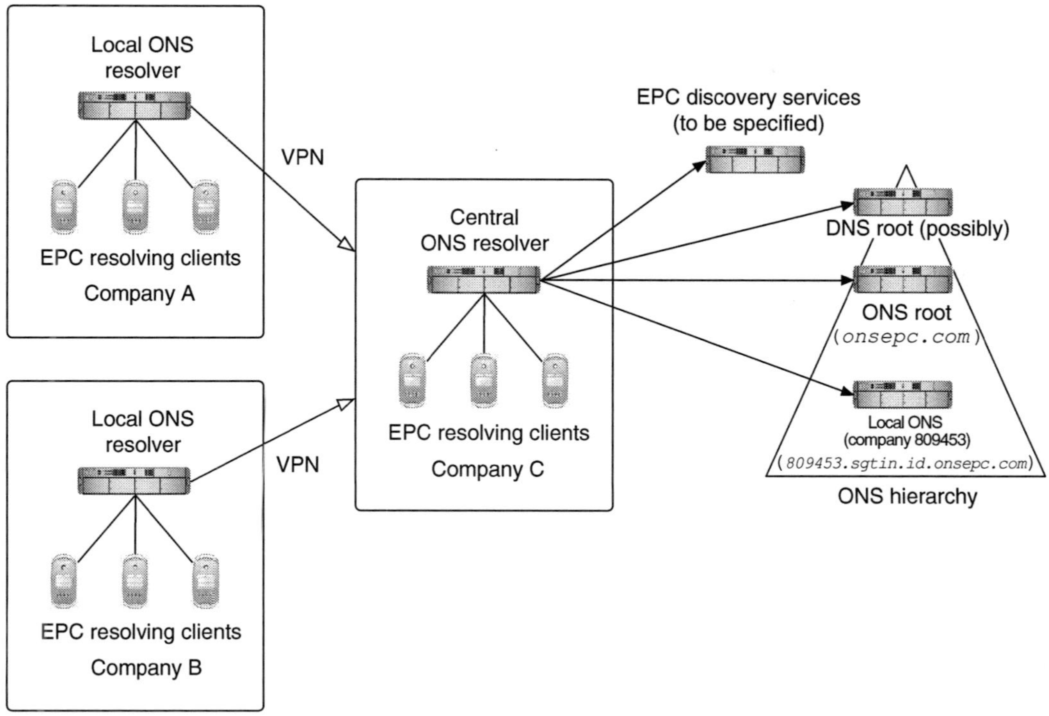

FIGURE 29.4
VPN and extranets.

29.4.3 Mixes and Onion Routing

The culmination of the concentration strategy stated earlier, i.e., collecting ONS queries from different sources to hide the real source IP address, would be the use of so-called anonymous mixes [29], a strategy that might be viable also for private households (Figure 29.5).

There are promising, but still not commonly used approaches such as Onion Routing [30] and especially Tor [31], which cryptographically transform and mix Internet traffic from many different sources, making it highly difficult to match a packet to a particular source. Onion Routing could also anonymize traffic to EPCIS servers, and has the potential to become highly relevant to a privacy preserving use of the EPCglobal network. Again, this approach offers enhanced confidentiality, but does not necessarily increase the integrity of the received messages nor could it do anything about ONS server availability, as any host offering services needs to be somehow addressable and is therefore attackable.

29.4.4 DNS Security Extensions

The main approach to address the security shortcomings of the DNS protocol is called DNSSEC (DNS Security Extensions) [32,33]. In its first version it actually introduced two different and independent procedures, one of these, called TSIG (Transaction Signature [34]), would provide mutual authentication between two DNS servers by using shared secrets, thereby introducing big problems of scalability. The other method provided origin authenticity and data integrity for the delivered DNS information by using public-key cryptography to sign sets of resource records (RR). These signatures are stored in a different RR type. The server's public key could be transferred out-of-band or be stored in an RR of type DNSKEY. The new version of DNSSEC uses four resource

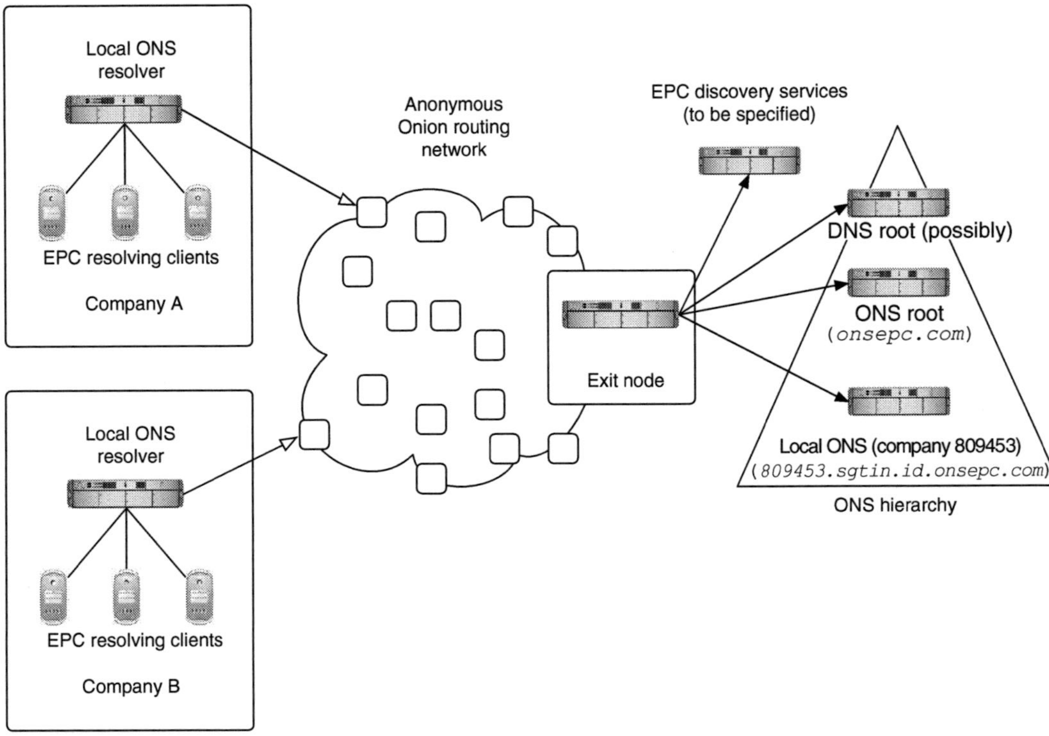

FIGURE 29.5
Onion routing.

record types: Resource Record Signature (RRSIG), DNS Public Key (DNSKEY), Delegation Signer (DS), and Next Secure (NSEC), for details see [32]. To verify an arbitrary DNSKEY, chains of trust down from the root of the DNS would be necessary where each parent DNS server signs the keys of its children after having verified their correctness by some external means.

DNSSEC is a very important approach for securing the Internet at a critical protocol level, but has not been widely adopted so far, except for notable exceptions like the top-level zone of Sweden [33]. Reasons might be the scalability problems of key management, the problem of root zone key control, and difficulties in building chains of trust between servers of many different organizations. Therefore global ONS information integrity could only be assured by DNSSEC in the long run, if the Internet community as a whole adopts it.

However, even if DNSSEC could be widely deployed and even be configured to actually encrypt the DNS information, which is not a stated goal so far [32, Section 4, p. 8], the company prefix of a given EPC could still be guessed by following the sequence of IP addresses the ONS queries are sent to. No additional protection against the availability problem of ONS servers is offered by deploying DNSSEC; on the contrary, signature checking introduces additional load to the involved servers [17].

29.5 Summary

Using the EPCglobal network—as it is designed today—to manage information about objects introduces many new risks. Automated business processes on top of it might induce the same level of dependency on the Internet for traditional businesses tomorrow as for e-commerce companies today. We have focused on the lookup service ONS because of the new and crucial confidentiality issues implied by its current design. We assume the actual EPCIS communication to be more easily securable against third parties, for example, by using standard authentication and encryption by TLS—though trust and integrity problems through improper certificate handling might spoil this assumption, and availability problems do occur likewise. In addition, every single EPCIS constitutes an attractive opportunity for query data analysis, but only for observing a specific user segment of the EPCglobal network (i.e., the customer base of a specific company), unlike, e.g., the global coverage at the ONS root.

If ONS is based on DNS as has been proposed in the specifications, a whole new branch of privacy problems do arise, which could only in part be mitigated by current security technology, and would even then require huge efforts in network design. For companies and individuals alike traffic anonymizers like Tor [31] could present an interesting partial solution to privacy-preserving ONS use and EPCIS access. This approach should be investigated further in relation to scalability, manageability, and adverse effects due to possible authentication measures for accessing the EPCIS. Integrity of ONS information could be dealt with by deploying DNSSEC, though this needs to be set up between all possible business partners and information service providers, which seems very unlikely given the current diverse and complex state of the Internet. Availability of ONS and EPCIS servers is a problem that would have to be approached and dealt with by every company in the resolution path.

Moving from barcode to RFID tags containing an EPC was motivated by saving costs and simplifying supply chains, without taking into account privacy concerns of individuals. The implementation of a global system to store and access heterogeneous information about countless products appears likewise at least in part be motivated by future after sale business. Again, security and privacy measures are no integral part of the original design, but, if at all, an afterthought. Based on a deeper analysis of the multilateral requirements reflecting the security interests of all stakeholders involved, there is urgent need to design an alternative model to ONS along with protocols for its implementation. One interesting research direction is the use of peer-to-peer systems based on distributed hash tables for ONS resolution [35].

References

1. EPCglobal, EPC tag data standards version 1.3, 2006. [Online] Available: http://www.epcglobalinc.org/
2. EPCglobal, The EPCglobal architecture framework version 1.0, 2005. [Online] Available: http://www.epcglobalinc.org
3. M. Harrison, EPC information service (EPCIS), *Auto-ID Labs Research Workshop*, 2004. [Online]. Available: http://www.m-lab.ch/auto-id/SwissReWorkshop/agenda.html
4. Y. Uo, S. Suzuki, et al., Name service on the EPC network, *Auto-ID Labs Research Workshop*, 2004. [Online]. Available: http://www.m-lab.ch/auto-id/SwissReWorkshop/agenda.html
5. K.S. Leong, M.L. Ng, et al., EPC network architecture, *Auto-ID Labs Research Workshop*, 2004. [Online] Available: http://www.m-lab.ch/auto-id/SwissReWorkshop/agenda.html
6. S. Helal, W. Mann, H. El-Zabadani, J. King, Y. Kaddoura, and E. Jansen, The gator tech smart house: A programmable pervasive space, *IEEE Computer Magazine*, pp. 50–60, March 2005.
7. B. Fabian, O. Günther, and S. Spiekermann, Security analysis of the object name service, in *Proceedings of the 1st International Workshop on Security, Privacy and Trust in Pervasive and Ubiquitous Computing (SecPerU 2005), with IEEE ICPS 2005*, Santorini, 2005, pp. 71–76.

8. M. Mealling, EPCglobal object naming service (ONS) 1.0, EPCglobal, 2005. [Online] Available: http://www.epcglobalinc.org
9. A. Salamon. DNS related RFCs. [Online] Available: http://www.dns.net/dnsrd/rfc/
10. P. Albitz and C. Liu, *DNS and BIND*, 4th ed., O'Reilly & Associates, Sebastopol, California, 2001.
11. M. Mealling and R. Daniel, The naming authority pointer (NAPTR) DNS resource record, *Request for Comment—RFC 2915*, September 2000.
12. D. Kaminsky, Explorations in namespace: White-hat hacking across the domain name system, *Communications of the ACM*, 49(6): 62–69, 2006.
13. V. Pappas, Z. Xu, S. Lu, D. Massey, A. Terzis, and L. Zhang, Impact of configuration errors on DNS robustness, in *SIGCOMM '04: Proceedings of the 2004 Conference on Applications, Technologies, Architectures, and Protocols for Computer Communications*, ACM Press, New York, 2004, pp. 319–330.
14. CERT Vulnerability Database. [Online] Available: http://search.cert.org/
15. SecurityFocus Vulnerability Archive. [Online] Available: http://www.securityfocus.com/vulnerabilities
16. SANS Top 20 Internet Security Vulnerabilities. [Online] Available: http://www.sans.org/top20/
17. D. Atkins and R. Austein, Threat analysis of the domain name system (DNS), *Request for Comments—RFC 3833*, 2004.
18. M. Bauer, B. Fabian, M. Fischmann, and S. Gürses, Emerging markets for RFID traces, http://arxiv.org/abs/cs.CY/0606018, 2006.
19. K. Albrecht and L. McIntyre, *Spychips*, Nelson Current, Nashville, Tennessee, 2005.
20. O. Günther and S. Spiekermann, RFID and the perception of control: The consumer's view, *Communications of the ACM*, 48(9): 73–76, September 2005.
21. S. Spiekermann and H. Ziekow, RFID: A 7-point plan to ensure privacy, in *13th European Conference on Information Systems (ECIS)*, Regensburg, May 2005.
22. S.A. Weis, S.E. Sarma, R.L. Rivest, and D.W. Engels, Security and privacy aspects of low-cost radio frequency identification systems, in *Security in Pervasive Computing (SPC 2003)*, ser. LNCS 2802, D. Hutter et al., Eds., Springer-Verlag, Berlin–Heidelberg, 2004, pp. 201–212.
23. S. Garfinkel, A. Juels, and R. Pappu, RFID privacy: An overview of problems and proposed solutions, *IEEE Security and Privacy*, 3(3): 34–43, 2005.
24. A. Juels, RFID security and privacy: A research survey, *IEEE Journal on Selected Areas in Communications*, 24(2): 381–394, 2006.
25. G. Avoine, Bibliography on Security and Privacy in RFID Systems, Massachusetts Institute of Technology, Cambridge, MA, USA. [Online] Available: http://lasecwww.epfl.ch/~gavoine/rfid/
26. B. Schneier, Attack trees, *Dr. Dobb's Journal*, December 1999. [Online] Available: http://www.schneier.com/paper-attacktrees-ddj-ft.html
27. F. Stajano, *Security for Ubiquitous Computing*, Wiley, 2002.
28. W.R. Cheswick, S.M. Bellovin, and A.D. Rubin, *Firewalls and Internet Security*, 2nd ed., Addison-Wesley, 2003.
29. D. Chaum, Untraceable electronic mail, return addresses, and digital pseudonyms, *Communications of the ACM*, 24(2): 84–88, February 1981.
30. Onion Routing Resource Site. [Online] Available: http://www.onion-router.net/
31. R. Dingledine, N. Mathewson, and P. Syverson, Tor: The second-generation onion router, in *Proceedings of the 13th USENIX Security Symposium*, San Diego, California, August 2004.
32. R. Arends, R. Austein, M. Larson, D. Massey, and S. Rose, DNS security introduction and requirements, *Request for Comments—RFC 4033*, March 2005.
33. DNS Security Extensions Resource Site. [Online] Available: http://www.dnssec.org/
34. P. Vixie, O. Gudmundsson, D. Eastlake 3rd, and B. Wellington, Secret key transaction authentication for DNS (TSIG), *Request for Comments—RFC 2845*, May 2000.
35. B. Fabian and O. Günther, Distributed ONS and its impact on privacy, in *Proceedings of the IEEE International Conference on Communications (IEEE ICC 2007), Glasgow*, 2007.

30

Practical Steps for Securing RFID Systems

A. Karygiannis, Bernie Eydt, and Ted S. Phillips

CONTENTS

30.1 Introduction

Like any information technology (IT), radio frequency identification (RFID) presents security and privacy risks that must be carefully mitigated through management, operational, and technical controls in order to realize the numerous benefits the technology has to offer. When practitioners adhere to sound security engineering principles, RFID technology can help a wide range of organizations and individuals realize substantial productivity gains and efficiencies. These organizations and individuals include hospitals and patients, retailers and customers, and manufacturers and distributors throughout the supply chain. This document provides an overview of RFID technology, the associated security and privacy risks, and recommended practices that will enable organizations to realize productivity improvements while safeguarding sensitive information and protecting the privacy of individuals. While RFID security is a rapidly evolving field with a number of promising innovations expected in the coming years, these guidelines focus on controls that are commercially available today.

Each RFID system has different components and customizations so that it can support a particular business process for an organization; as a result, the security risks for RFID systems and the controls available to address them are highly varied. The enterprise and interenterprise subsystems in RFID installations involve common IT components such as servers, databases, and networks and, therefore, can benefit from typical IT security controls for those components. Implementing the recommendations presented in this publication should help organizations improve the security of their RFID systems.

30.1.1 Tailoring Security Controls for RFID Systems

Personnel responsible for designing RFID systems should understand what type of application an RFID system will support so that they can select the appropriate security controls. Each type of application uses a different combination of components and has a different set of risks. For example, protecting the information used to conduct financial transactions in an automated payment system requires different security controls than those used for protecting the information needed to track livestock. Factors to consider include the following:

- The general functional objective of the RFID technology. For example, does the system need to determine the location of an object or the presence of an object, authenticate a person, perform a financial transaction, or ensure that certain items are not separated?
- The nature of the information that the RFID system processes or generates. One application may only need to have a unique, static identifier value for each tagged object, while another application may need to store additional information about each tagged object over time. The sensitivity of the information is also an important consideration.
- The physical and technical environment at the time RFID transactions occurs. This includes the distance between the readers and the tags, and the amount of time in which each transaction must be performed.
- The physical and technical environment before and after RFID transactions take place. For example, human and environmental threats may pose risks to tags' integrity while the tagged objects are in storage or in transit. Some applications require the use of tags with sensors that can track environmental conditions over time, such as temperature and humidity.
- The economics of the business process and RFID system. The economic factors for RFID systems are different than those for traditional IT systems. For example, many RFID tags offer few or no security features; selecting tags that incorporate basic security functionality significantly increases the cost of tags, especially if encryption features are needed. Also, the operational cost of some basic IT security controls, such as setting unique passwords and changing them regularly, may be higher for RFID systems because of the logistical challenges in managing security for thousands or millions of tags.

30.1.2 Risk Management

For RFID implementations to be successful, organizations should effectively manage their risk. RFID technology enables an organization to significantly change its business process to increase its efficiency and effectiveness. This technology is complex and combines a number of different computing and communications technologies. Both the changes to business process and the complexity of the technology generate risk. The major high-level risks associated with RFID systems are as follows:

- *Business process risk.* Direct attacks on RFID system components potentially could undermine the business processes the RFID system was designed to enable. For example, a warehouse that relies solely on RFID to track items in its inventory may not be able to process orders in a timely fashion if the RFID system fails.
- *Business intelligence risk.* An adversary or competitor potentially could gain unauthorized access to RFID-generated information and use it to harm the interests of the organization implementing the RFID system. For example, an adversary might use an RFID reader to determine whether a shipping container holds expensive electronic equipment, and then target the container for theft when it gets a positive reading.
- *Privacy risk.* The misuse of RFID technology could violate personal privacy when the RFID application calls for personally identifiable information to be on the tag or associated with the tag. As people possess more tagged items and networked RFID readers become ever more prevalent, organizations may have the ability to combine and correlate data across applications to infer personal identity and location and build personal profiles in ways that increase the privacy risk.
- *Externality risk.* RFID technology potentially could represent a threat to non-RFID networked or collocated systems, assets, and people. For example, an adversary could gain unauthorized access to computers on an enterprise network through Internet Protocol (IP) enabled RFID readers if the readers are not designed and configured properly.

Organizations need to assess the risks they face and choose an appropriate mix of management, operational, and technical security controls for their environments. These organizational assessments should take into account many factors, such as regulatory requirements, the magnitude of each threat, and cost and performance implications of the technology or operational practice.

An RFID security checklist has been included in Appendix A (Tables A.1 through A.6) of this chapter. These tables summarize the security.

30.2 RFID Security Controls

This chapter discusses security controls that can potentially mitigate the business risks associated with RFID systems. As previously discussed, RFID implementations are highly customized. As a result, the security controls listed are not all applicable or effective for all RFID applications. Organizations need to assess the risks they face and choose an appropriate mix of controls for their environments, taking into account factors such as regulatory requirements, the magnitude of the threat, cost, and performance.

This chapter covers security controls applicable to most RFID implementations. It does not address the security of RFID-enabled smart cards and payment systems. This section also does not discuss security controls related to general IT systems, such as network infrastructure, databases, and Web servers because these are already covered by other security requirements and guidelines. For example, EPCIS servers, which can be accessed by trading partners through the Internet, should be protected by the same types of controls that would be used for any other Internet-facing system (e.g., encryption of sensitive communications, access control to prevent unauthorized access to data and systems) to ensure the security of the data collected by the RFID system. Guidelines on topics such as IT server, application, database, and network security are available from many sources, including NIST's Computer Security Resource Center (CSRC).*

RFID security is a rapidly evolving discipline. Although promising research is noted when applicable, this section focuses on controls that are presently commercially available.

The RFID security controls discussed in this section are divided into three groups:†

- *Management*. A management control involves oversight of the security of the RFID system. For example, the management of an organization might need to update existing policies to address RFID implementations, such as security controls needed for an RF subsystem.
- *Operational*. An operational control involves the actions performed on a daily basis by the system's administrators and users. For example, RFID systems need operational controls that ensure the physical security of the systems and their correct use.
- *Technical*. A technical control uses technology to monitor or restrict the actions that can be performed within the system. RFID systems need technical controls for several reasons such as protecting data on tags, causing tags to self-destruct, and protecting wireless communications.

30.2.1 Management Controls

Management controls are typically involved in risk assessment, system planning, system acquisition, as well as security certifications, accreditations, and assessments. The sections below discuss management controls for RFID systems in more detail.

30.2.1.1 RFID Usage Policy

Control: An RFID usage policy describes the authorized and unauthorized uses of RFID technology in an organization and the personnel roles assigned to particular RFID system tasks. Federal agencies should follow FIPS Publication 199, *Standards for Security Categorization of Federal Information and Information Systems*, when developing the RFID usage policy.

The usage policy also should be consistent or integrated with the organization's privacy policy, which addresses topics such as how personal information is stored and shared. The RFID usage policy should also address privacy issues associated with the

* The CSRC is located at http://csrc.nist.gov/publications/nistpubs/index.html. This site also has links to NIST publications that address general security issues and provide guidelines for the configuration of specific technologies that might be of use when securing an RFID system, including the computing devices in the enterprise subsystem.

† For more information on security controls, 6 see R. Ross, S. Katzke, A. Johnson, M. Swanson, G. Stoneburner, and G. Rogers, *Recommended Security Controls for Federal Information Systems*. NIST Special Publication 800–53 (as amended), December 2006.

tag identifier formats and the potential disclosure of information based on solely on the tag identifier format selected.

Applicability: This control applies to all organizations that use or are considering using RFID technologies.

Benefits: The policy establishes the framework for many other security controls. It provides a vehicle for management to communicate its expectations regarding the RFID system and its security. It enables management to take legal or disciplinary action against individuals or entities that do not comply with the policy.

Weaknesses: The existence of a policy does not ensure compliance with the policy. A policy needs to be coupled with the implementation and enforcement of appropriate operational and technical controls to be effective.

30.2.1.2 IT Security Policies

Control: IT security policies describe the approach to achieve high-level security objectives of the usage policy. The IT security policies related to RFID should cover each RFID subsystem, including network, database, and application security in the enterprise and interenterprise subsystems; they should not just be limited to security of tags and readers in the RF subsystem.

IT security policies for RFID systems should address the following:

- Access control to RFID information, especially records contained in RFID analytic system databases
- Perimeter protection, including port and protocol restrictions for network traffic between the RF and enterprise subsystems and between the enterprise subsystem and a public network or extranet
- Password management, particularly with respect to the generation, distribution, and storage of tags' access, *lock*, and *kill* passwords
- Management system security for readers and middleware, including the use and protection of SNMP read and write community strings*
- RFID security training for system administrators and operators
- Management of associated cryptographic systems, including certification authorities and key management

Applicability: All RFID implementations, particularly those with enterprise subsystems or interenterprise subsystems.

Benefits: Well-crafted security policies govern the mitigation of business risks associated with the use of RFID technologies. The policies provide requirements and guidelines for the individuals designing, implementing, using, and maintaining RFID systems. For example, IT policies help the personnel designing RFID systems or procuring system components to make appropriate decisions. Similarly, they help system administrators correctly implement and configure software and related network components.

---

* *SNMP community strings* are passwords that provide anyone with an SNMP management client and network access the ability to manage the associated systems. Knowledge of the *read community string* provides the holder the ability to view the system configuration and track system behavior. Knowledge of the *write community string* provides the holder the ability to reconfigure system components.

Weaknesses: The existence of a policy does not ensure compliance with the policy. A policy needs to be coupled with the implementation and enforcement of appropriate operational and technical controls to be effective.

30.2.1.3 Agreements with External Organizations

Control: When data associated with an RFID system needs to be shared across organizational boundaries, formal agreements among the participating organizations can codify the roles and responsibilities, and in some cases the legal liability, of each organization. These formal agreements are usually documented as a Memorandum of Agreement (MOA) or Memorandum of Understanding (MOU). The MOU or MOA specifies the network connections and authentication mechanisms to be used, the data to be shared, and the manner in which data should be protected both in transit and at rest. It may also address controls on vendors, subcontractors, and other third parties to the extent they have access to the system.*

If the interenterprise application requires tag passwords to be shared across organizations, then the MOU or MOA should specify how these passwords will be generated, stored, and shared. The memorandum may specify IT security controls such as methods of authentication, access control, or encryption that participating organizations shall implement to protect the passwords.

Applicability: Any RFID system involving more than one organization, which is most common in supply chain applications.

Benefits: Having an MOA or MOU significantly reduces the potential for subsequent misunderstandings and security breaches. They enable signatories to communicate their respective security requirements while also realizing the benefits of the business partnership that led them to collaborate in the development and use of the RFID system.

Weaknesses: Monitoring an external organization's enforcement of an agreement is difficult without full access to its systems and personnel, which is unlikely. As a result, violations may occur without detection. This risk can be mitigated with independent audits if signatories agree to hire third parties to conduct such audits.

30.2.1.4 Minimizing Sensitive Data Stored on Tags

Control: Instead of placing sensitive data on tags, the data could be stored in a secure enterprise subsystem and retrieved using the tag's unique identifier.

Applicability: Applications that use tags with on-board memory and process data that is either considered sensitive or that could be combined with other data to infer sensitive information.

Benefits:

- Adversaries cannot obtain information from the tag through rogue scanning or eavesdropping.
- Data encryption and access control is often more cost effectively performed in the enterprise subsystem than in the RF subsystem.

* For additional information on agreements with external organizations, see NIST Special Publication 800–47, *Security Guide for Interconnecting Information Technology Systems*, which can be found at http://csrc.nist.gov/publications/nistpubs/800–47/sp800–47.pdf

Weaknesses:

- Adversaries can often obtain valuable information from the identifier alone. For example, knowledge of the EPC manager ID and object class bits in certain EPC formats may reveal the make and model of a tagged object concealed in a container. An adversary might target containers based on the perceived worth of their contents.
- Placing data in the enterprise subsystem makes the availability of that data contingent on the availability of the network. Retrieving data over a network also introduces a small delay, which could be unacceptable for some applications.

30.2.2 Operational Controls

There are several types of operational controls, which are given below:

- Physical access controls restrict access to authorized personnel where the RFID systems are deployed.
- Proper placement of RF equipment helps avoid interference and reduce hazards from electromagnetic radiation.
- Organizations can destroy tags after they are no longer useful to prevent adversaries from gaining access to their data.
- Operator training can help ensure that personnel using the system follow appropriate guidelines and policies.
- Information labels and notice can inform users of the intended purposes of the RFID system and simple methods users can employ to mitigate risk.

The following sections discuss operational controls for RFID systems in more detail.

30.2.2.1 Physical Access Control

Control: Physical access controls include fences, gates, walls, locked doors, turnstiles, surveillance cameras, and security guards. When the objective is to limit radio communication over a short distance, room walls or partitioned stalls might provide adequate protection if they are opaque to the relevant radio frequencies that the RF subsystem uses.

Applicability: All RFID implementations except those in which RFID tags or other system components are in public areas.

Benefits: Physical access controls limit the ability of an adversary to get close enough to RFID system components to compromise RFID data security or to modify, damage, or steal RFID system components. Physical security applies to all RFID subsystems. In the RF subsystem, the primary objective of the control is to prevent unauthorized radio communications. In the enterprise and interenterprise subsystems, the primary objective is to prevent physical access to system components.

Examples of risks that are mitigated by physical access controls include the following:

- Unauthorized reading and writing of tag data
- Rogue and cloned tags
- Reader spoofing
- Denial of service resulting from radio interference or unauthorized commands
- Targeting

- Physical destruction of RFID equipment
- HERF/HERO/HERP

Weaknesses:

- Physical access controls are not a countermeasure for radio interference from legitimate radios located within a perimeter designed to block external emissions.
- The effective range of RF signals may be much longer than stated operating ranges, thereby allowing many attacks to occur using customized directional antennas and other technologies.
- Physical access controls do not protect against attacks by insiders (i.e., those granted access to the area).
- HERF/HERO/HERP still exists with respect to radiation emitted within the physical perimeter.
- Physical controls may fail to contain radio signals as expected if ductwork or other openings allow radio signals to escape.

30.2.2.2 Appropriate Placement of Tags and Readers

Control: RFID system equipment can be placed to minimize unnecessary electromagnetic radiation. Tags and readers can be kept away from the following:

- Fuel, ordnance, and other materials that could *cause harm* if exposed to electromagnetic radiation
- Humans and sensitive products (e.g., blood, medicine) that *might be harmed by* sustained exposure to RF subsystem radiation
- Metal and reflective objects that can modify and amplify signals in unintended and potentially harmful ways
- Legitimate radios with which the RF subsystem communication will cause interference

Applicability: All environments in which the organization deploying RFID systems determines the location of the RF equipment (which excludes many consumer and supply chain applications).

Benefits:

- Reduced risk of interference with legitimate radios
- Reduced risk of eavesdropping and unauthorized RF subsystem transactions
- Mitigation of HERF/HERO/HERP

Weaknesses:

- Tag location cannot always be controlled, such as when tags are used to track mobile items (e.g., hospital cart) or items in transit (e.g., pallet on a truck).
- Radio interference may persist even if the tags or readers are placed in a new location that is still sufficiently close to other radios.*

* In this situation, a panel or wall of grounded wire fencing between the two RF sources is a possible alternative means to reduce interference.

30.2.2.3 Secure Disposal of Tags

Control: Secure disposal involves physically or electronically destroying tags, as opposed to just discarding them, when they are no longer needed to perform their intended function. Physical destruction may involve manual tearing or shredding using a paper shredder. Electronic destruction can be accomplished by using a tag's kill feature or using a strong electromagnetic field to render a tag's circuitry permanently inoperable. When a tag supports an electronic disabling mechanism, it usually is the preferred way to disable a tag before it is disposed because it can be accomplished without touching each tag, thereby reducing the cost of the effort.

Applicability: RFID applications in which the continued operating presence of a tag after it has performed its intended function poses a business intelligence or privacy risk (e.g., an adversary can subsequently use the presence of the tag to track items or people).

Benefits: Destroying or disabling tags:

- Eliminates the possibility that they could be used later for tracking or targeting.
- Prevents access to sensitive data stored on tags.

These benefits apply to both business intelligence and privacy risks.

Weaknesses:

- Even if minimal, the effort it takes to destroy a tag increases the tag's life cycle cost, which is a concern if very low costs are required to justify an RFID-enabled business process.
- Destruction of a tag precludes the ability to use it for future value-added applications such as post-sale product support, targeted recalls, receipt-free returns, expiration date monitoring, and sorting assistance for recycling.

30.2.2.4 Operator and Administrator Training

Control: Operator and administrator training provides personnel with the skills and knowledge necessary to comply with RFID usage, IT security, and privacy policies, as well as agreements with external organizations. In most RFID implementations, personnel will perform various roles, which might require different training materials for each role. For example, an administrator of middleware might need different information than an operator of a mobile reader. Appropriate security and privacy training addresses at least three points:

- What constitutes unauthorized use
- How to detect that unauthorized use might be occurring
- To whom to report violations

If HERF/HERO/HERP risks are present, appropriate security training covers mitigation techniques, such as safe handling distances.

If tags are destroyed or recycled, training should cover how to perform these functions. For example, operators might be trained how to clear tag memory before reuse.

Applicability: All RFID implementations.

Benefits: Operator training helps ensure that the system is used and maintained properly. Training also helps operators identify security violations and take appropriate actions to prevent their reoccurrence.

Weaknesses: Training alone cannot ensure proper operation of the system or compliance with policy.

30.2.2.5 Information Labels/Notice

Control: A written message is affixed to or distributed with each tag or is posted near readers. The notice may inform users of the purposes of the RFID system or advise users on how to minimize privacy or other risks (e.g., place an RFID-enabled access card or transponder in metal foil or a sleeve that shields RF radiation when the card or transponder is not in use).

Applicability: All applications in which there is a risk that could be mitigated with simple informational messages. The control is particularly relevant to consumer applications in which privacy is a concern.

Benefits: Information labels or notices can communicate basic information about risks that might otherwise be left unknown by users that are able to take simple steps to mitigate the risk (e.g., remove a tag or place it in a shielded sleeve).

Weaknesses: Distributing a notice is no guarantee that it will be read or understood. Notice is not an appropriate communications medium for complex concepts or instructions that may require formal training.

30.2.2.6 Separation of Duties

Control: RFID system duties are distributed among various personnel roles to minimize the damage resulting from an inadvertent or malicious activity of a single person. The general principle of the control is that malicious collusion between two or more authorized users is much less likely than one person engaging alone in inappropriate behavior.

One example of separation of duties is having different personnel (1) attach tags to objects and (2) read the tags. If an individual performed both functions, the individual could intentionally put the wrong tag on an object to circumvent the objectives of the business process. For example, a store clerk could affix tags intended for low-priced items on high-priced items, and then later work the checkout scanner while the clerk's accomplice purchased the items. The system would not know that the tags had been switched, but if another person performed the checkout, he or she might be suspicious of the checkout total, which could uncover the plot.

Applicability: RFID applications in which an insider might have a motive to perform unauthorized RFID transactions. This scenario is most likely to occur when tags support commercial transactions, especially those related to high-value objects.

Benefits: Separation of duties helps to reduce fraud and malicious damage, because any user attempting to engage in such activities would be forced to collude with at least one other user. Separation of duties also reduces errors, because a second operator will often catch mistakes made or missed by the first.

Weaknesses: Multiple employees still could collude to commit fraud or violate the RFID usage policy. Also, organizations with a limited staff may not be able to perform complete separation of duties.

30.2.2.7 Nonrevealing Identifier Formats

Control: RFID tags are assigned identifiers using identifier formats that do not reveal any information about tagged items or the organization operating the RFID system.

01-005FC5B-0001A3-026A45E29

Header	EPC manager	Object class	Serial number
8 bits	28 bits	24 bits	36 bits

FIGURE 30.1
Example 96-bit EPC.

Nonrevealing identifier format options include serially assigning identifiers and randomly assigning identifiers.*

In contrast, if an adversary reads an identifier that is encoded with a standardized format, such as the EPC format, that adversary may be able to discern the manufacturer or issuer of the item, as well as the type of item. For example, all cans of a soft drink from a certain manufacturer will have the same EPC manager ID and object class bits if their identifiers are encoded in an EPC identifier format. Figure 30.1 shows an example 96-bit EPC and how it can be parsed into the four aforementioned, individual fields.

Tags must have programmable identifiers to support the control. Even tags that are designed to support standard tag formats can still be assigned non-standard identifiers in the field. However, some tags have factory-initialized identifiers that cannot be modified after manufacture.

Applicability: Any applications in which the implementing organization determines that the revelation of a tag's identifier is a business intelligence risk.

Benefits: Adversaries cannot obtain information about tagged items from the identifier alone.

Weaknesses:

- The use of nonrevealing identifier precludes an organization from realizing benefits that come from standard identifier formats that reveal organization and item type information. For example, standard identifier formats are particularly advantageous when designing and maintaining distributed databases in interenterprise systems. Lookup and query functions are much easier in such databases when the identifiers provide information on where item data is located.
- If identifiers are assigned randomly, then a potential exists that two tags may be assigned the same identifier. The likelihood of such an event is very small, but it could lead to errors in the supported business process.†
- If there is logic in how the identifiers are assigned, an adversary may uncover the method that is used, which would defeat the control. For example, an adversary knows that an identifier was assigned to a certain item and that all items of that

* A related control is rotating identifiers. Auto-rotating tags store a list of identifiers and cycle through the list when queried. To support multiple identifiers, databases in the enterprise subsystem must associate each identifier in the list to the particular item. The benefit of rotating identifiers is that organization can make it more difficult to identify and track particular items as well as hide the type of item. Random and serialized identifiers, on the other hand, may not reveal information about the type of item, but since these identifiers are fixed, once they are revealed that particular item can be tracked. One weakness to rotating identifiers is that a rogue reader can easily obtain the complete list of identifiers through repeated queries. Therefore, this control is more appropriate when the primary threat is eavesdropping. While research is being conducted on the concept of rotating identifiers, it is not specified in any RFID standard and proprietary designs are not widely commercially available.

† When two tags are assigned the same identifier, the event is called a collision. If identifiers are randomly assigned, a collision is expected after approximately the square root of the total number of possible identifiers. Therefore, in the case of a 96-bit EPC, a collision is expected after approximately 2^{48} tags, which is an enormous number not likely to be encountered in most RFID applications.

type were assigned sequentially, then the adversary may be able to deduce the approximate range of identifiers that correspond to items of that type. Similarly, when identifiers are serialized, the adversary may be able to deduce the approximate time of the assignment based on the identifier.

30.2.2.8 Fallback Identification System

Control: A fallback identification system provides an alternative means to identify, authenticate, or verify an object when the RFID system is unavailable or an individual tag is inoperable. Options include text labels and AIDC technology such as bar codes.* The fallback may consist of just an identifier, or it may also include additional data about the tagged object. The fallback system is accompanied by standard operating procedures and operator training to ensure that personnel know when and how to use it.

Applicability: All RFID applications.

Benefits: Duplicating tag identifiers and data on a label provides a fallback in case of malicious or accidental tag damage, reader malfunction, or enterprise subsystem network outage. The redundant data can also be used to verify that tag data has not been altered improperly.

Weaknesses: This control has several potential weaknesses, including the following:

- Damage to the tag could render both the stored data and the printed data unusable. Similarly, many enterprise subsystem outages that would affect the RFID system would also affect its fallback alternative.
- The data stored on the label is visible, so it may be easier for unauthorized parties to gain access to it than it would be to read the data from the tag.
- The text label or bar code might not provide the same data capacity as RFID memory, although two-dimensional bar codes can encode at least as many bits as standards-based tag identifiers.
- Text labels and AIDC technologies are static, so they do not provide a complete fallback solution for applications in which tag data changes over time. However, some identification information is still likely to be better than none in most applications.

30.2.3 Technical Controls

There are a number of technical controls currently available for RFID systems, and many others are under development in industrial and university research labs. This section focuses on technical controls that are commercially available as of the publication date of this document. Supplementary information on selected emerging security technologies is provided in footnotes. Many of the technical controls listed are specified in standards, while others are available only in proprietary systems.

Many technical controls related to a tag require the tag to perform additional computations and to have additional volatile memory. Accordingly, a tag that uses such technical controls requires a more sophisticated microchip than those that do not use such controls. In the case of passive tags, the tags may also need to be closer to readers to obtain the required power to perform these computations. Alternatively, readers may need to operate

* If the RFID application's objective is to provide security or authentication, then a fallback technology such as holograms or other optical security features may be used.

at greater power levels, although this may not be feasible or permitted in many cases. These inherent characteristics of passive tags can limit the use of certain technical controls in some environments.

Technical controls exist for all components of RFID systems, including the RF, enterprise, and interenterprise subsystems. This section focuses on technical controls for the RF subsystem. Many controls also exist for the enterprise and interenterprise subsystems, but these typically apply to IT systems in general rather than to RFID systems in particular. The general types of RF subsystem controls include controls to perform the following:

- Provide authentication and integrity services to RFID components and transactions.
- Protect RF communication between reader and tag.
- Protect the data stored on tags.

Examples of each of these types of controls are discussed in the following sections.

30.2.3.1 Authentication and Data Integrity

While a wide variety of authentication methods exists for IT systems, the most common techniques for the RF subsystem of RFID systems are passwords, keyed-hash message authentication codes (HMAC), and digital signatures. In some cases, the primary objective of the authentication technology is to prevent unauthorized reading from or writing to tags. In other cases, the objective is to detect cloning of tags. Authentication techniques based on cryptography often provide integrity services for data included in the authentication transaction; in other words, an adversary cannot modify data in the transaction without the reader or tag detecting the change.

30.2.3.1.1 Password Authentication

Control: A tag does not permit password-protected commands to be executed unless they are accompanied by the correct password. Protected commands may include those that support reading and writing of tag data, memory access control, and the kill feature.

Organizations properly implementing this control will develop a password management system to support it. The password management system addresses all stages of the password, including generation, conveyance, and storage. From a security perspective, effective password generation involves random selection of each password.* Whenever possible, the passwords are assigned to each tag in a physically secure environment to reduce the likelihood of eavesdropping. Tags should not share passwords, although this may not be administratively feasible in some environments, such as those in which the reader is not expected to have access to networked database of tag passwords. In inter-enterprise applications such as supply chains, multiple organizations may need to access databases that contain tag identifiers and passwords. Authenticating external entities likely will require additional security systems. While in traditional IT systems, passwords are often changed on a periodic basis (e.g., every 90 days), in RFID systems, such changes may be infeasible, especially if the tags are not always accessible to the organization assigning the passwords.

Applicability: Any application where authorized execution of a particular command represents a business process, business intelligence, privacy, or externality risk.

* For additional information on proper random number generation, see E. Barker and J. Kelsey, *Recommendation for Random Number Generation Using Deterministic Random Bit Generators*, NIST Special Publication 800–90, June 2006.

Benefits: The likelihood that tags will be used for unauthorized purposes is greatly reduced.

Weaknesses:

- Password management for RFID systems is complex, particularly if the application deploys large number of tags or if passwords must be shared across organizational boundaries as might be the case in supply chains.
- Adversaries can intercept passwords transmitted over the air and then use them at a later time to perform unauthorized transactions.*
- If the application environment precludes access to an on-line tag password database (e.g., mobile readers in remote locations), then the implementing organization may need to take simplifying measures, such as assigning the same password to multiple tags. In cases such as these, the compromise of a single password could compromise the integrity of the entire system.
- RFID passwords can be obtained through brute force methods (i.e., cycling through all possible passwords) when the tag technology is limited to short passwords.†
- RFID passwords can be revealed through power analysis attacks on some types of passive tags.‡

30.2.3.1.2 Keyed-Hash Message Authentication Code

Control: Both the reader and the tag share a common secret key that can be used in combination with a hash algorithm to provide one-way or mutual authentication between tag and reader. When HMAC is applied to messages, it also ensures the integrity of data in the messages. HMAC is specified in FIPS Publication 198.§ HMAC supports any cryptographic hash algorithm, but Federal agencies must use one of the secure hash algorithms

---

* RFID passwords are often transmitted "in the clear" (i.e., without cryptography to hide them), which makes them particularly vulnerable to eavesdropping. The cover-coding technique described in Section 5.3.2.1 mitigates this risk for tags that support cover-coding, but this technique is not without its own limitations.

† For example, EPC Class-1 Generation-1 UHF tags support a maximum password length of 8 bits, which enables only 256 possible passwords. An adversary can cycle through 256 passwords in a matter of seconds. EPC Class-1 Generation-2 tags support 32-bit passwords and, therefore, 2^{32} possible passwords, which are sufficient if the passwords are randomly generated. However, if the binary password is based on American Standard Code for Information Interchange (ASCII) characters, then the actual number of possible passwords may be much smaller. For example, the ASCII representation of a 4-digit decimal number (a common length for personal identification numbers) is 32 bits, but results in only 10,000 possible combinations, a number certainly vulnerable to brute force attacks. Tags typically do not lock-out readers after a certain number of incorrect guesses, which means a determined adversary can continue to guess the password as long as the tag remains within the operating range of the adversary's reader.

‡ The power analysis attack (also called a side channel attack) is based on the fact some passive tags use different levels of power depending on how close the password provided is to the actual password. For instance, if the first bit in a password is incorrect, the tag uses less energy than it would if the eighth bit is incorrect, given how the algorithm is hard-coded into the tag's circuitry. These power differences are detected in the backscatter to the reader, but it requires that the adversary be reasonably close to the tag to get effective measurements. If such measurements are possible, an adversary can determine the password much more quickly than by using a brute force method. Lab experiments proved that someone could crack the 8-bit password protection found on EPC Class-1 Generation-1 tags in one minute. For more information, see Y. Oren and A. Shamir, "Power Analysis of RFID Tags," discussed at the *Cryptographers Panel of the Fifteenth RSA Conference*, San Jose, 2006.

§ The FIPS HMAC is a generalization of HMAC described in H. Krawczyk, M. Bellare, and R. Canetti, "HMAC: keyed-hashing for message authentication," Internet Engineering Task Force, Request for Comments (RFC) 2104, February 1997 and American Bankers Association, "Keyed Hash Message Authentication Code," American National Standards Institute (ANSI) X9.71, Washington, D.C., 2000.

(SHA) specified in FIPS Publication 180–2.* HMAC is not specified in any RFID standard, but it is available in proprietary designs.

Applicability: Applications in which passwords are considered to offer an inadequate authentication mechanism, perhaps because the risk of eavesdropping is high. Applications that require evidence of a tag's authenticity.

Benefits: The advantages of HMAC relative to password authentication include the following:

- Provides evidence of tag's authenticity.†
- Provides integrity protection.‡
- Does not transmit secrets over the air, which eliminates the risk of eavesdropping inherent with clear text passwords.

Weaknesses:

- The management of HMAC keys provides similar challenges to those of password management and may not be practical if mobile readers do not have reliable access to an HMAC key management system.
- The authenticity claims associated with HMAC authentication only hold when the HMAC key remains secret. If an adversary has physical access to a tag and can obtain its HMAC key, then the adversary can clone the tag. This attack, however, assumes that that the adversary has some level of expertise, both in reverse engineering the HMAC-capable tag and in producing a reasonable facsimile.
- When HMAC keys are shared across organizations, authenticity claims rely on an implicit trust between the organizations that may not be present in practice.
- HMAC requires greater computing power than password comparisons, and, therefore, requires more complex tag designs to support it.

30.2.3.1.3 Digital Signatures

Control: Readers digitally sign tag identifiers, timestamps, and related event data to provide for nonrepudiation of tag transactions. The resulting signatures are stored on tags for subsequent verification, although recording signatures in enterprise subsystem databases provides additional assurance of the tag's chain of custody.

Digital signatures are based on *asymmetric cryptography*, also commonly referred to as *public key cryptography*. Federal agencies implementing digital signature technology must

---

* The specified algorithms are SHA-1, SHA-256, SHA-384, and SHA-512. While SHA-1 offers the lowest level of assurance and is not recommended for use in digital signatures beyond 2010, it is likely most applicable to RFID systems due to its greater computational efficiency relative to the other algorithms. See NIST Special Publication 800–57, *Recommendation on Key Management*, Part 1 for additional information.

† The evidence of tag or item authenticity is provided by authenticating a tag to a reader, which can be accomplished when the tag computes and returns an HMAC using a random challenge provided by the reader. Mutual authentication is also possible if both tag and reader provide challenges to each other. Passwords, on the other hand, typically only are used to authenticate readers to tags, thereby protecting the tag against rogue commands. If the tag were to authenticate itself to a reader using a password, an adversary could simply use a rogue reader to obtain the password and the re-use with a legitimate reader. HMAC provides an effective countermeasure to this attack because it never reveals the secret key during any of its transactions.

‡ Integrity protection is when either tag or reader computes an HMAC using as input the data for which integrity protection is desired. Any change in the data results in a different value of the HMAC, which would be detected by the receiving entity.

comply with FIPS Publication 196, *Entity Authentication Using Public Key Cryptography*. The use of digital signature technology in the context of RFID systems is also referred to as *authenticated RFID*. It typically works as follows:

1. The tag has a permanent unique identifier than cannot be modified after manufacturer.
2. The reader generates a public/private key pair and obtains a corresponding public key certificate.
3. The reader uses a specified hash algorithm to compute a message digest of the tag's identifier and possibly other transaction-related data, encrypts the message digest with its private key to create a digital signature for the transaction, and stores the resulting signature on the tag.
4. Other readers read the signature, decrypt it with the first reader's public key, and compute the identical message digest to determine if a match exists. If the message digests match, then verification procedure provides assurance of the authenticity of the earlier transaction. If the message digests do not match, then either the transaction data has been altered or an unauthorized device created the digital signature.
5. The other readers can store their own event transactions on the tag or record them in enterprise subsystem databases for later queries regarding the tag's chain of custody.

Applicability: Applications that require more robust evidence of authenticity than provided by HMAC technology, including authentication of multiple chain of custody events. Applications that require verification of authenticity without network connectivity.

Benefits: Digital signatures offer several advantages relative to HMAC authentication, including:

- Digital signature systems do not require tags to store cryptographic secrets. Instead, readers maintain private keys. In password and HMAC authentication, both the tag and the reader must share a secret for the system to function, but there are no shared secrets in the public key cryptosystems that support digital signatures. Tags are typically much more vulnerable to compromise than readers, so eliminating the need to store secrets on tags enhances overall system security. One private key and one or more public key certificates are on the reader. Integrity is needed for the certificates, but not confidentiality.
- In many cases, digital signatures do not require network connectivity to successfully perform the authentication function. In password and HMAC authentication, a reader is unlikely to have the memory to store the passwords or keys for large numbers of tags. With digital signatures, a reader may only need to store the public key certificate of the entity that initialized the tags or perhaps a relatively small number of readers. In interenterprise systems, each participating organization only has to share the public keys of its readers rather than provide its partners reliable network access to a password or secret key database.
- Digital signatures are compatible with existing RFID tag standards. HMAC requires tags to support hash algorithms and to implement a challenge-response protocol, neither of which are included in existing RFID standards. On the other hand, in authenticated RFID systems, tags can receive, store, and transmit digital signatures with existing read and write commands because the complexity resides in readers or middleware.

Weaknesses:

- A system of digital signatures requires a public key infrastructure (PKI), including registration and certification authorities, revocation functions, and associated policies and practice statements. Successfully implementing and operating a PKI requires careful planning and considerable expertise. In addition, readers or middleware need to support digital signature and other PKI functionality that is not commonly found in current RFID technology.
- Digital signatures systems require more memory than found on many current tags. For example, NIST recommends that RSA signatures have a length of 1024 bits, and a length of 2048 bits after 2010.* Additional memory is required to store identifying information related to the transaction. Providing chain of custody evidence requires storing a digital signature and related identifying information for each transaction.
- Digital signatures that are not generated by the tag are subject to replay attacks. An adversary could query a tag to obtain its evidence of authenticity (i.e., the digital signature created by a previous reader) and then replicate that data on a cloned tag.
- The use of digital signatures to support authentication of readers to tags would require tags to support relatively complex cryptographic functions beyond the capacity of most common tag designs. Consequently, password or symmetric key authentication systems likely will support tag access control, as opposed to tag authenticity verification, for the foreseeable future.

30.2.3.2 RF Interface Protection

Several types of technical controls focus on the RF interface to tags, including:

- Cover-coding can be used to obscure the content of messages from readers to tags.
- Data can be encrypted prior to its transmission.
- Shielding can be installed to limit eavesdropping and rogue scanning.
- The selection of an operating radio frequency can be used to avoid interference from other sources or achieve certain operating characteristics such as the ability to propagate through metals, liquids, and other materials that are opaque to many frequencies.
- Reader and active tag transmission characteristics can be tuned to reduce the likelihood of eavesdropping and help mitigate interference and the hazards from electromagnetic radiation.
- The RF interface for tags can be temporarily shut off to prevent unauthorized access when the tag is not expected to be used for authorized purposes.
- The RF interface may be turned off by default until a user takes an action to activate it.
- Readers may periodically poll tags to determine the presence of the tags, assess system health, and acquire environmental data.

These controls are discussed further in the following sections.

---

* Elliptic curve cryptography can reduce the size of signatures. Elliptic curve methods provide comparable assurance to 1024-bit RSA signatures with 163 bits, and to 2048-bit RSA signatures with 224 bits. This approach combined with greater memory on tags may alleviate storage concerns over time.

30.2.3.2.1 Cover-Coding

Control: Cover-coding is a method for hiding information on the forward channel from eavesdroppers.

In the EPCglobal Class-1 Generation-2 standard, cover-coding is used to obscure passwords and information written to a tag using the *write* command. The EPCglobal Class-1 Generation-2 cover-coding protocol works as follows:

1. The reader sends a message to the tag requesting a key.
2. The tag generates a random 16-bit number (i.e., the key) and returns it to the reader.
3. The reader produces ciphertext (i.e., a message unintelligible to an eavesdropper who cannot intercept the key) by applying an exclusive-or (XOR) operation* to the key and the plain text.
4. The reader sends the ciphertext to the tag.
5. The tag applies the XOR operation using the ciphertext and the key it generated to recover the plain text.†

Cover-coding is an example of *minimalist cryptography* because it operates within the challenging power and memory constraints of passive RFID tags.‡ By itself, the XOR operation would be considered a trivial encryption algorithm in traditional cryptography, but it nonetheless mitigates risk to an acceptable level in many RFID environments.

Figure 30.2 illustrates how cover-coding works. As shown in the figure, the passive tag's back channel signal is weaker than the reader's forward channel signal. This will always be the case for a passive tag, which must use the forward channel to power both its computations and the backscattered signal. In the figure, the adversary is able to eavesdrop on the forward channel but not the back channel. So long as this condition holds, the adversary will not be able to learn the random number sent from the tag and, therefore, will be unable to decipher cover coded information.

Applicability: Cover-coding is useful when eavesdropping is a risk that requires mitigation, but adversaries are expected to be at a greater distance from the tags than readers. Intelligible reception of back channel signals from a passive tag requires proximity of less than four meters in most applications. In many applications, an adversary's reception equipment would be conspicuous if it were located within this range. In contrast, reader signals can be detected at distances of a kilometer or more under ideal conditions.

Cover-coding is designed for RF subsystems in which the forward channel carries stronger signals than the back channel, which essentially limits the control to passive tags. EPCglobal Class-1 Generation-2 technologies support cover-coding. Proprietary technologies support similar features.

Benefits: Cover-coding helps prevent the execution of unauthorized commands that could disable a tag or modify the tag's data. Consequently, cover-coding mitigates business process, business intelligence, and privacy risks.

---

* The XOR operation is a binary operation denoted with the symbol "$\oplus$" that works as follows: $1 \oplus 1 = 0$; $1 \oplus 0 = 1$; $0 \oplus 1 = 1$; $0 \oplus 0 = 0$. When the XOR operation is applied to two multi-bit strings, the XOR operation is applied to the first bit of the each string to produce the first bit of the result, the second bit of each string to produce the second bit of the result, and so on. To work properly, the inputs to the XOR operation must be of equivalent length, and the output is also of the same length.

† The XOR operation is symmetric. For instance, given key K, plaintext P, and ciphertext C, if $P \oplus K = C$, then $C \oplus K = P$.

‡ For more information on minimalist cryptography, see A. Juels, "Minimalist cryptography for low-cost RFID tags," in the *Fourth Conference on Security in Communication Networks*, 2004, pp. 149–164.

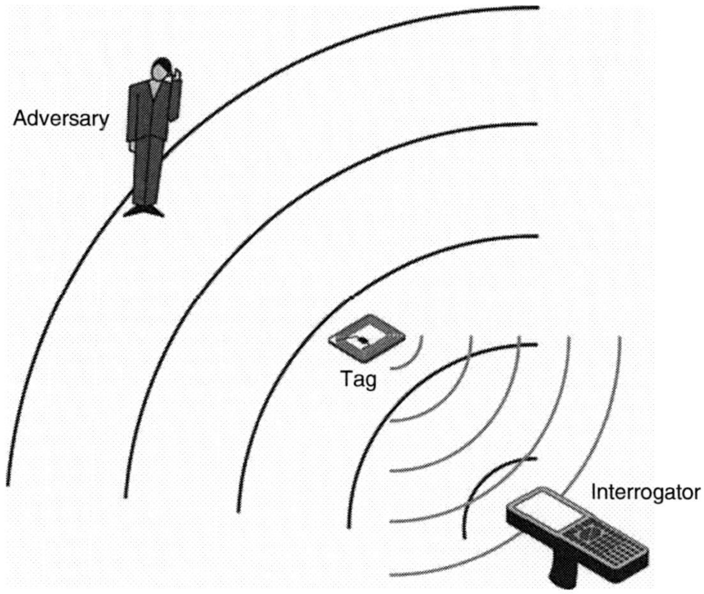

FIGURE 30.2
Cover-coding.

Weaknesses:

- If an adversary can intercept a key distributed on the back channel, the adversary could decrypt any cipher text message generated with that key.
- The effectiveness of cover-coding depends on the performance of the tag's random number generator. If the random number is predictable because of a flaw in the tag's design or cryptanalysis, then an adversary can learn the key and decrypt subsequent communication.

30.2.3.2.2 Encryption of Data in Transit

Control: Data collected or processed by the tag is encrypted prior to over-the-air transmission.

Applicability: Applications that require an effective countermeasure to the threat of eavesdropping and for which cover-coding offers inadequate protections. Tags typically only require on-board encryption capabilities to protect the confidentiality of data in transit if they collect or process data from sensors or other directly connected sources. In these cases, no alternative exists to hide the content of the data over the air because the data originates on the tag.

On-board cryptography for confidentiality is not required for applications in which readers are the only source of data. In these cases, the data can be encrypted in the enterprise subsystem or by a reader before it is written to the tag and then retrieved in its encrypted form from the tag when needed. If the tag never has to perform computations on the data, then it never has to decrypt it, but merely store it.

Proprietary tag designs support encryption for over-the-air confidentiality, but EPCglobal and ISO/IEC 18000 standards do not as of the date of this publication.

Benefits: Encryption of data in transit prevents successful eavesdropping of over-the-air RFID transactions.

Weaknesses:

- Data encryption requires a key management system, which can be complex to manage and operate.
- Cryptographic functions may introduce an unacceptable delay in RFID systems that require very fast read or write transactions.
- Cryptographic functions require additional power to complete, which could impact applications that use passive tags.
- Tags that support onboard encryption currently are more costly than those that do not. One reason for the increased cost is that onboard encryption requires additional logic gates to perform the necessary computations. Most low-cost passive tags do not have enough logic gates to perform complex encryption algorithms.*

30.2.3.2.3 Electromagnetic Shielding

Control: RF shielding encloses an area with a conducting material that limits the propagation of RF signals outside of the shielded area. Shielding can vary in size and form depending on the application.

For example, some RFID-enabled travel documents are protected by a metallic anti-skimming material. This material helps to prevent adversaries from reading the embedded tag when the passport cover is closed. Shipping containers are sometimes shielded to prevent the reading of tags during transit. Shielding is also placed in walls, partitions, or stalls to prevent RF emissions from leaving a confined area. When readers are placed in tunnels on industrial production conveyor belts, the tunnels may be shielded to reduce radio interference. Wrapping a tag in aluminum foil is also an effective means of shielding.

Figure 30.3 shows how shielded partitions can separate collocated readers to prevent interference. The readers near forklift A can operate without inadvertently reading tags on boxes on forklift B due to the shielding in the partition that separates the portals. Shielding may be necessary when middleware is unable to correctly filter duplicate read events from the two portals.

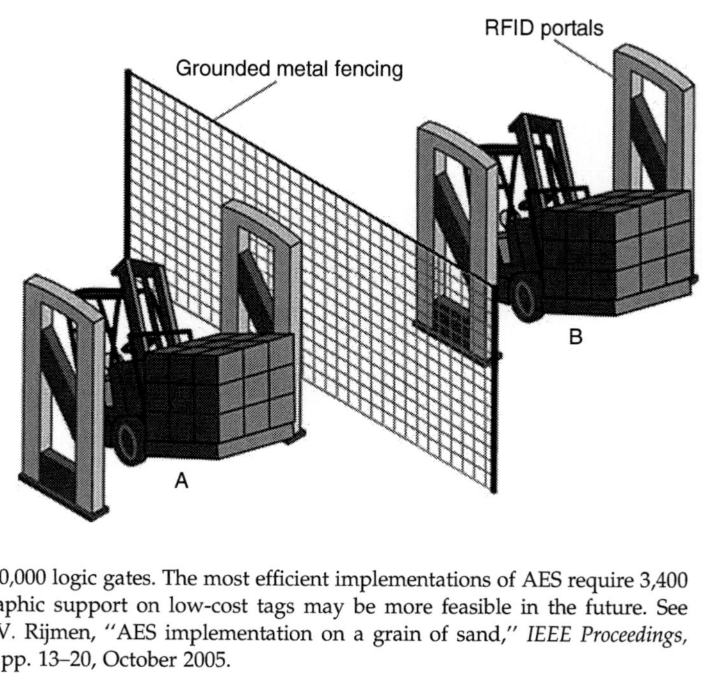

FIGURE 30.3
Grounded metal fencing as shielding.

* Low-cost tags currently have about 10,000 logic gates. The most efficient implementations of AES require 3,400 gates, which suggests that cryptographic support on low-cost tags may be more feasible in the future. See M. Feldhofer, J. Wolkerstorfer, and V. Rijmen, "AES implementation on a grain of sand," *IEEE Proceedings, Information Security*, vol. 152, issue 1, pp. 13–20, October 2005.

Applicability: Shielding is applicable for contexts in which eavesdropping or RF radiation is a concern, and the use of temporary shielding would not stop valid transactions.

Benefits: Shielding can limit the ability of eavesdroppers or unauthorized readers to collect data from an RFID system.

Weaknesses:

- Shielding can prevent or hinder legitimate transactions. For example, shielded containers require objects to be physically removed from the shielding material. This prevents an implementing organization from realizing one of the key benefits of RFID technology, which is to read tags remotely without optical line of sight and additional handling.
- It may still be possible for an adversary to place a radio inside the shielded area. The radio could be used for malicious purposes, such as eavesdropping on RFID transactions or causing interference.

30.2.3.2.4 Radio Frequency Selection

Control: RFID technology can communicate over various radio frequencies, including those in the LF, HF, UHF, and microwave bands. Particular fixed frequencies can be assigned to an RFID application to avoid or reduce the effects of radio interference. Some tag technologies can perform frequency hopping within a limited frequency range, but in these cases, the technique is used primarily to avoid collisions with other tag transactions, not radio interference with different types of radio systems.* In some cases, the implementing organization may need to obtain a license to use a particular desired frequency.

Ideally, an RF site survey will be performed before an RFID system is installed to determine what frequencies are already in use. After the RFID system is installed, site surveys can be conducted to determine if the RF characteristics of the site have changed (e.g., new sources of interference).

Applicability: All implementations whose radio frequency is not determined by other application requirements. Organizations that implement a closed RFID system have more freedom to select an operating frequency because they do not have to interoperate with other organizations. However, if tags are based on a particular air interface standard, the range of potential frequencies will be limited to those supported by the standard.

Benefits: Radio frequency selection permits the avoidance of RF interference with other radio systems that could disrupt the RFID system or other technologies. A particular frequency might be desirable because of radio interference on other frequency bands. Some frequencies also have desirable propagation characteristics, such as the ability to penetrate certain materials.

Weaknesses:

- It may be difficult to identify sources of interference. For example, bug zappers have been found to create interference in passive RFID trials.† Interference can be caused by poorly grounded motors, noisy relays, old fluorescent light ballasts, and

* For example, EPCglobal Class-1 Generation-2 915 MHz UHF systems use frequency hopping techniques. This capability is built into tags complying with the standard. Therefore, organizations implementing RFID systems using EPCglobal Class-1 Generation-2 compliant equipment do not have to configure this capability.

† L. Sullivan, "IBM Shares Lessons Learned From Wal-Mart RFID Deployment," October 15, 2004, http://informationweek.com/story/showArticle.jhtml?articleID=49901908.

other devices that generate unintended RF noise in nearby environments. Each RFID technology deployment should be tested in its intended environment prior to production use to identify these sources of interference.

- New sources of interference can be later introduced at the site.
- When implementing an interenterprise RFID system, all organizations involved in the system will have to agree on a tag type that supports all the frequencies that the organizations collectively intend to use.

30.2.3.2.5 Adjustment of Transmission Characteristics Other Than Frequency

Control: Operators adjust the level of transmitted RF energy from a reader or active tag. The use of particular antenna types and configurations can also determine the direction of transmitted RF energy. Additionally, the duty cycle of a reader can be controlled.

Applicability: All applications for which eavesdropping, radio interference, or hazards of electromagnetic radiation are a concern.

Benefits: Reducing transmitted power can do the following:

- Reduce the likelihood that an adversary can intercept communication.
- Limit radio interference with other legitimate radios.
- Lessen hazards of electromagnetic radiation.

Weaknesses: The drawback of reducing transmission power or the duty cycle is performance degradation, especially with respect to back channel communication from a passive tag. For instance, readers might fail to detect the presence of valid tags. Also, changes in the physical environment or the introduction of new radio equipment can impact the power levels required for consistently successful transactions. Consequently, the benefits of power adjustments based on a site survey can be negated by changes to the environment.

30.2.3.2.6 Temporary Deactivation of Tags

Control: The RF interface on some proprietary tags can be turned off temporarily. Tag manufacturers have different methods of turning their tags on and off. For example, some tags are designed so that the tag is on or off depending upon which end is inserted into a mounting clip. Other tags have replaceable batteries that can be removed to deactivate them.

If the control is implemented, tags would be turned on inside a designated area where the RF subsystem operates. When the tags leave that area, they would be turned off. For example, in a supply chain application, tags may be turned off to prevent unauthorized transactions during shipment. When the tags arrive at their destination, they would be powered on again and managed. Conversely, tags used for in-transit visibility may be turned on for their trip and turned off when they reach their destination.

Applicability: This control is most useful when communication between readers and a tag is infrequent and predictable. For example, a warehouse might store items for an annual event, such as a holiday celebration or parade. In this case, the RFID confers a benefit only for a short period each year, but could remain vulnerable to rogue transactions if left operational for the rest of the year.

Benefits: Deactivating tags temporarily:

- Prevents unauthorized tag transactions during periods of inactivity.
- Extends the battery life of active tags.

Weaknesses:

- If operators or system software fail to reactivate the tag when it is needed, then the missing transactions resulting from the tag's RF silence could adversely impact the supported business process.
- If turning a tag on or off requires human intervention, then this control would result in additional labor expense, which could be significant for systems that process large numbers of tags. The potential increased labor required to operate the system could negatively affect the business case for RFID relative to other AIDC technologies.
- Even if the activation and deactivation process is automated, it introduces a delay that might not be acceptable for many time-sensitive applications.

30.2.3.2.7 Tag Press-to-Activate Switch

Control: The tag remains deactivated by default unless a user or operator takes a positive action, such as holding a press-to-activate switch on the tag to turn it on. When the switch is on, the tag is capable of RF communication, but when pressure on the switch is released, the tag returns to its default deactivated status so that tag transactions can no longer occur.

Applicability: Primarily access control or automated payment applications in which the holder of the tag desires or requires control over when tag transactions occur.

Benefits: A press-to-activate switch provides a user with physical control over when and where the tag can respond to a reader. Consequently, this control mitigates privacy and business intelligence risks by providing a countermeasure to the threat of eavesdropping and the execution of unauthorized tag commands. Eavesdropping would be limited to the immediate vicinity of authorized readers and tracking beyond the immediate vicinity of the authorized readers would not be possible.

This control also provides assurance that a person is knowingly in possession of the tag, and that it has not been intentionally or inadvertently separated from that person. For example, this control could be useful in ticketing or access control applications in which the objective is to get an accurate count of the number of individuals present, and to prevent spare tags in pockets or bags from interfering with the accuracy of the count.

Weaknesses:

- Requiring the user to activate the tag would require some level of instruction, however minimal, which might add a cost or delay in the business process. For example, the user would need to know when and for how long they would need to activate the tag.
- Some users may consider activating a switch to be an inconvenience, which could hinder user acceptance of the technology.
- A press-to-activate switch could distract the user from other functions that the user is performing. For example, a press-to-activate switch is not an appropriate control for an automated toll-payment system because the user needs to have both hands available for driving the vehicle.

30.2.3.2.8 Tag Polling

Control: A reader periodically queries the tag to determine its continued presence and operating status.

Applicability: Process control or asset management applications in which a design objective is periodic or near continuous monitoring. Examples include medical facilities that require real-time inventory of certain medical supplies or systems that collect sensor data. Tag polling also is applicable for high-value business processes that require early indications of system failures or performance problems. This control is most effective in applications in which those with access to the tags are trusted or when detaching the tag is not feasible (e.g., when a tag is embedded in another item such as a poker chip).

Benefits: Operators obtain timely information about system failures, item theft, or unusual environmental conditions that enables them to proactively address problems.

Weaknesses: Tag polling:

- Reduces the battery life of active and semi-active tags.
- May not detect critical events in a timely manner if the polling frequency is too low.
- Is a business intelligence risk if the tag polling enables an adversary to perform traffic analysis, or track or target tags that might have otherwise remained silent.
- Could be circumvented in some cases by detaching the tag, taking the item, and leaving the tag behind so that it continues to signal its presence to readers.

30.2.3.3 Tag Data Protection

Technical controls currently available for protecting tag data include

- Tag memory access control, which can restrict use of tag commands and protect data stored in a tag's memory
- Encrypting the data on tags
- The kill feature, which can prevent subsequent unauthorized use of a tag
- Tamper protection

These controls are described in more detail in the following sections.

30.2.3.3.1 Tag Memory Access Control

Control: Many tags support a password-protected lock feature which provides read or write protection to memory. In some RFID technologies, the lock feature is permanent and in others it is reversible. For example, the EPCglobal Class-1 Generation-2 has five areas of memory, each of which can be protected using the *lock* command.* The memory is either both read- and write-protected, or only write-protected, depending on the parameters issued with the command. The EPCglobal Class-1 Generation-2 UHF standard also has a *permalock* feature. If engaged, permalock will make the lock status (locked or unlocked) permanent for all or part of a tag's memory. ISO/IEC 18000-3 Mode 2 supports both read and write protecting all areas of memory with a 48-bit memory access password. Finally, Mode 2 of the ISO/IEC 18000-3 standard describes a *lock pointer*, which is a memory address. All areas of memory with a lower address than the lock pointer are write-protected, while those areas of memory above the pointer address are not.

* The five areas of memory are registers for the kill password, access password, EPC memory, TID memory, and User memory. When locked, the kill password and access password become both read- and write-protected. If they are locked, the EPC memory, TID memory, and User memory are only write-protected.

Applicability: All applications that store data on tags.

Benefits: A write-protect *lock* command will prevent the contents of a tag's memory from being altered. A read-protect *lock* command will prevent unauthorized users from reading or accessing the data on tags.

Weaknesses:

- The password length on many tags is too short to provide meaningful memory access protection. Even when the technology supports longer passwords, password management is challenging.
- Locking a tag's memory does not prevent data loss from electromagnetic interference or physical tag destruction.

30.2.3.3.2 *Encryption of Data at Rest*

Control: Data stored on a tag is encrypted before it is written to the tag. The control does not require that the tag encrypt or decrypt data. Instead, the encryption is performed by either the reader, middleware, or other enterprise subsystem components.

Applicability: All applications that store additional data beyond an identifier on the tag that needs to be kept confidential on the tag. If the encryption and decryption functions are performed in the enterprise subsystem, then network access is required to read the content of data stored on the tag, which makes the control unsuitable for mobile readers that do not always have real-time network access.

Benefits: Data encryption protects sensitive tag data from being read by individuals with unauthorized access to the tags.

Weaknesses:

- Data encryption requires a key management system, which can be complex to manage and operate.
- Sending tag data to network components for encryption or decryption is a source of network latency, when in conjunction with the time to complete cryptographic functions may introduce an unacceptable delay in RFID systems that require very fast read or write transactions.

30.2.3.3.3 *Kill Feature*

Control: The kill feature permanently disables a tag's functionality using a remote command. The most common implementation of the kill feature is the EPCglobal *kill* command. The EPCglobal Class-1 Generation-2 *kill* command is password-protected using a 32-bit password different from the access password.*

* Several alternative technical controls to the kill feature are under development, but have not yet been fully commercialized. One approach is to disable the tag's antenna in such a way that it can still perform transactions over short distances (e.g., 10 cm or less) but not longer than that. The objective is to greatly reduce the probability that an adversary could track or target someone in possession of the tag after the tag longer serves its primary purpose, but still enable the tag to perform some additional functions, albeit with additional effort. For example, the primary purpose of the tag might be to facilitate a point-of-sale transaction, but using the approach described, the tag could also facilitate a receipt-less return, although the item would need to be placed closer to the reader to complete this post-sale transaction. Another approach is to use multiple control domains as described in the immediately preceding footnote. The objective of both of these approaches is to extend the life a tag to support some residual functionality that would otherwise be eliminated as a result of the kill feature.

Applicability: RFID applications that encounter business intelligence and privacy risks after a tag has moved beyond its intended functional environment (e.g., after a tag moves beyond the supply chain in which it served inventory and checkout functions). EPCglobal tags are the only standards-based tags that support a kill feature.

Benefits: Using the kill feature prevents a tag from being reused improperly. The kill feature was designed and implemented in EPCglobal tags primarily to protect consumer privacy. It also protects improper access to tag data used in business processes. For example, discarded tags that have not been disabled may be read by adversaries to gain access to data, such as which products an organization or individual is purchasing or using.

Weaknesses:

- The existence of a kill feature represents a significant business process threat to an RFID system. If an adversary who learns the kill password improperly disables tags that should remain in operation, the supported application will not function properly because it will not be able to perform transactions on the disabled tags. This risk is particularly salient for organizations that assign the same password to multiple tags because doing so could enable an adversary to disable large numbers of tags with a single compromised password.
- Once killed, a tag cannot be used for any further application involving the asset (e.g., recalls, receipt-less product returns).
- If an organization assigns a weak (e.g., short or easily guessed) password for the *kill* command, unauthorized parties can kill the tag at will.* Moreover, the longer a tag maintains the same password, the more likely it is that the password will be compromised.
- Data stored on the tag is still present in the tag's memory after it is killed (although it can no longer be accessed wirelessly), and, therefore, still may be accessible to someone with physical access to the tag.†
- Although the *kill* command was added to tags as a potential solution for privacy concerns, consumers cannot easily detect whether a tag has been deactivated.‡ Moreover, typical consumers cannot easily kill tags on their own because this action requires a reader and knowledge of the *kill* password.

30.2.3.3.4 Tamper Resistance

Control: Certain RFID tags have tamper resistant or tamper-evident features that help prevent an adversary from altering the tags or removing them from the objects to which they are attached. One simple type of tamper resistance is the use of a frangible, or easily broken, antenna; if a tag of this type is removed, the electric connection with the antenna is severed, rendering the tag inoperable. Other, more complex types of RFID systems monitor the integrity of objects associated with the tags to ensure that the objects have not been compromised, altered, or subjected to extreme conditions.

---

* An EPCglobal Class-1 Generation-2 tag cannot be killed if it has a null password (i.e., one whose bits are all zeros). See EPCglobal, "EPC Radio-Frequency Identity Protocols Class-1 Generation-2 UHF RFID Protocol for Communications at 860–960 MHz Version 1.0.9," January 2005, p. 58.

† Obtaining data from the tag in this circumstance would require an attacker to have specialized equipment and expertise.

‡ This may open a door for future consumer products to test for the presence of passive RFID tags and probe their characteristics. It is hypothesized that cellular phones may be able to provide this service for EPC passive tags since cellular phones already operate in the 860–960 MHz band.

Applicability: Applications in which tags are frequently outside of the direct control of the implementing organization and, therefore, vulnerable to tampering. Tamper resistance and tamper-evident features are currently only available on specialty RFID tags that are designed for tamper resistance to support specific buyer requirements.

Benefits: This control helps to prevent adversaries from breaking the association between a tag and its corresponding object. The more complex tamper-resistant/tamper-evident tags provide health and status monitoring of the attached objects to ensure that they have not been opened, manipulated, damaged, or subjected to extreme temperature, humidity, or shock.

Weaknesses: Sophisticated adversaries may be able to defeat the tamper resistance mechanisms. This is dependent upon the implementation of the tamper resistance feature. For example, a sophisticated adversary may be able to repair a frangible antenna. In addition, tamper-resistance/tamper-evidence technologies do not prevent the theft or destruction of the tag or its associated items.

30.3 Summary

Organizations should use a combination of management, operational, and technical controls to mitigate the business risks of implementing RFID systems. Table 30.1 maps the presented controls to the categories of risks that they mitigate. Because each RFID implementation is highly customized and each organization's requirements are different, the security controls discussed in this section are not all applicable or effective for all RFID applications. Organizations need to assess the risks their RFID implementations face and choose the appropriate controls, taking into account factors such as regulatory

TABLE 30.1

RFID Controls Summary

		Risk Mitigated by Control				
					Externality Risk	
	Control	**Business Process Risk**	**Business Intelligence Risk**	**Privacy Risk**	**Hazards of Electromagnetic Radiation**	**Computer Network Attacks**
Management	RFID usage policy	X	X	X	X	X
	IT security policies	X	X	X		X
	Agreements with external organizations	X	X	X		X
	Minimizing data stored on tags	X	X	X		
Operational	Physical access control	X	X	X	X	X
	Appropriate placement of tags and readers	X	X			X
	Secure disposal of tags	X	X	X		
	Operator and administrator training	X	X	X	X	X
	Information labels/notice	X	X	X	X	
	Separation of duties	X		X		
	Nonrevealing identifier formats		X	X		
	Fallback identification systems	X				

(continued)

TABLE 30.1 (continued)

RFID Controls Summary

	Control	Risk Mitigated by Control				
					Externality Risk	
		Business Process Risk	Business Intelligence Risk	Privacy Risk	Hazards of Electromagnetic Radiation	Computer Network Attacks
Technical	Password authentication	X	X	X		X
	HMAC	X	X	X		X
	Digital signatures	X	X			
	Cover-coding	X	X	X		
	Encryption of data in transit		X	X		
	Electromagnetic shielding		X	X	X	
	Radio frequency selection	X			X	
	Adjustment of transmission characteristics		X	X	X	
	Temporary deactivation of tags	X	X	X		
	Tag press-to-activate switch		X	X		
	Tag polling	X				
	Tag access controls	X	X	X		X
	Encryption of data at rest	X	X	X		
	Kill feature		X	X		
	Tamper resistance	X	X			

requirements, the magnitude of threats, and cost and performance implications of the controls. For example, a remote warehouse may have little need to protect against eavesdropping, but it may require redundant processes in case of system failure. Traditional security controls are often preferable to RFID-specific controls. For example, if RFID data can be stored in an enterprise database rather than on tags, then physical and network security controls for the database server probably is more practical than using tags with cryptographic capabilities.

Appendix

TABLE A.1

RFID Security Checklist: Initiation Phase

No.	Security Practice	Rationale/Discussion	Affected Components	Recommended or Should Consider	Checklist Status
1	Perform a risk assessment to understand RFID threats, the likelihood that those threats will be realized, and the potential impact of realized threats on the value of the organization's assets.[a]	All risks should be considered, including the risk of RFID systems to other enterprise information systems and the risk that the existence of RFID will enable adversaries to collect information about an organization's activities that could adversely impact its ability to perform its mission. For supply chain applications, the risk assessment should consider threats that occur when the RFID tags are located outside the organization's control, such as when tagged items are in transit. The risk assessment is an important input to the development of the RFID usage policy because it identifies which RFID activities pose an acceptable risk to the organization's information resources and which do not. In particular, it can help determine which type of RFID technology may be appropriate for the desired application (e.g., active versus passive tags). The risks assessment should also determine whether the RFID system will collect, store, process or share PII, or enable PII to be inferred through direct or indirect means. A complete privacy impact assessment should be conducted for any RFID systems involving PII.	All	Recommended	
2	Establish an RFID usage policy that specifies what assets should be tagged, who is authorized to use RFID technology, and for what business purposes this authorization applies.	An RFID usage policy is the foundation on which subsequent security controls are based. The policy should cover all components of the RFID system, including tags, readers, and support systems (e.g., middleware and analytic systems). The policy should distinguish between the levels of access provided to those that use the system, those that administer it, and those that need access to its data, including external business partners. For instance, logistics administrators may be granted the ability to modify a reader's configuration (duty cycle, power output, network settings, RF frequency settings, Transmission Control Protocol [TCP] ports, etc.) while operations personnel may only be able to scan tags. External parties should almost never get access to an organization's readers, but they might need read access to certain database elements. The policy should also address the collection and handling of sensor data that might be transmitted over the RFID system.	All	Recommended	

(continued)

TABLE A.1 (continued)

RFID Security Checklist: Initiation Phase

No.	Security Practice	Rationale/Discussion	Affected Components	Recommended or Should Consider	Checklist Status
		The RFID usage policy should also integrate privacy policies and practices. All statements made in privacy compliance documentation should be reflected in and supported by the RFID usage policy.			
3	Establish an RFID privacy policy.	Federal government agencies are required to create a privacy impact assessment (PIA) if the RFID system will store or manage personal information. While privacy policy is not within the scope of this publication, the technical security controls that result from the policy are within the scope of the publication. For example, implementation of the privacy policy might require the use of the *kill* command or an alternative means to disable tags. Requirements related to data sharing limitations may need to be supported by certain authentication and access control methods. A privacy policy should be in place before RFID system architects determine the appropriate security controls.	All	Recommended	
4	Establish HERF/HERO/HERP policies if applicable.	If the risk assessment identifies risks related to human health, fuel, ordnance, or other sensitive materials (e.g., pharmaceuticals) that are not fully mitigated by the RFID usage policy, then the organization should require additional controls to prevent the associated hazard from being realized. A separate policy is needed for each hazard type (HERF/HERO/HERP/other sensitive materials) because each one has distinct issues. Organizations facing these hazards should also consult safety and regulatory experts in this area to ensure their approaches are valid and comply with legally-mandated FCC exposure limits.[b]	RF subsystem	Recommended	

5	Enhance the organization's information security policy to account for the presence of RFID systems.	The introduction of RFID technology represents a new challenge to the security of the enterprise network that should be mitigated by policy and associated technical, operational, and management controls. Elements of the network security policy that might require revision include (a) perimeter security (i.e., firewalls and extranets), (b) database security, (c) application security, and (d) wireless connections (i.e., between readers and the enterprise network). Typically a firewall separates readers from the enterprise network that hosts RFID database and application servers. Policies related to database and application security should cover authentication, access control, and development practices to reduce the likelihood of malicious code insertion, exploitation of buffer overflow vulnerabilities, and other attacks. In addition, if readers are connected to the enterprise infrastructure via a wireless link, then the policy should require mutual authentication between the reader and its network access point. It should also provide for data confidentiality and integrity services for wireless traffic, if needed.	All	Recommended
6	Establish an RFID security and privacy training program for operators of the RFID system.	Many RFID risks are best mitigated when the personnel operating the system are aware of the risks and the associated countermeasures. The training program should cover the RFID usage policy and teach administrators and operators how to identify and report violations of the policy. If the system involves PII, operator training should explain how individuals and PII should be handled to sustain privacy protections. NIST Special Publication 800–50, *Building an Information Technology Security Awareness and Training Program*, contains detailed guidelines on designing, developing, implementing, and monitoring an IT security awareness and training program.[c]	All	Should consider

[a] For more information on performing risk assessments, see G. Stoneburner, A. Goguen, and A. Feringa, *Risk Management Guide for Information Technology Systems*. NIST Special Publication 800–30, July 2002.

[b] R. Cleveland Jr. and J. Ulcek, *Questions and Answers about Biological Effects and Potential Hazards of Radiofrequency Electromagnetic Fields*. Federal Communications Commission Office of Engineering and Technology (OET), Washington, D.C., OET Bulletin 56, Fourth Edition, August 1999, pp. 11–16.

[c] M. Wilson and J. Hash, *Building an Information Technology Security Awareness and Training Program*. NIST Special Publication 800–50, October 2003.

TABLE A.2

RFID Security Checklist: Planning and Design Phase

Planning and Design Phase					
No.	**Security Practice**	**Rationale/Discussion**	**Affected Components**	**Recommended or Should Consider**	**Checklist Status**
7	Identify the RFID standards with which the RFID system will comply.	The selected RFID standards in effect determine the types of tags that will be deployed and the operating frequencies on which RF subsystem communication will occur. The standards also specify the available technical security mechanisms. For instance, some tags support passwords while others do not. An organization may also choose a standard to support a particular operating frequency to avoid unwanted RF interference, improve performance, and reduce technical problems. The choice of operating frequency is often closely associated with relevant regulations and the application area (e.g., healthcare, supply chain, security access control, and animal tracking).	RF subsystem	Recommended	
8	Include security and privacy considerations in RFID system investment and budget requests.	Including security and privacy planning in funding plans ensures that adequate resources are available for implementation of appropriate controls. Including these considerations in budget planning and analysis also increases the likelihood that cost-effective approaches will be selected to mitigate risk. Budget requests should also demonstrate that plans for the RFID system are consistent with the information technology architecture of the implementing organization.	All	Recommended	
9	Conduct a site survey to determine the proper location of readers and other devices given a desired coverage area.	The estimated usable range of readers and tags should not extend beyond the physical boundaries of the facility whenever possible. The survey should note the location of metal or reflective objects and RF absorbing materials such as water that have the potential to adversely affect the operation of the RFID system. The site survey should also identify potential radio interference between the RFID system and other RF sources at the site or in neighboring facilities.	RF subsystem	Recommended	
10	Determine approach to RF emissions control.	The approach should be based on the risk assessment and site survey. In many cases, physical security may offer the best mechanism to protect against unauthorized use of RFID technology, including attacks involving reader spoofing and jamming, modification of tag data, and eavesdropping. When this is not possible, countermeasures such as shielding and adjusting the power level of the reader may be employed. The selected approach might involve the location of readers and tagged assets, the placement of blocker devices, the power levels at which RF components operate, and the potential need for additional perimeter security (e.g., fences around warehouses).	RF subsystem	Recommended	
11	Identify an approach to securing network management traffic, using dedicated	If network management traffic is left unprotected, adversaries might be able to breach the RFID system, enabling a number of subsequent attacks, including those that could disable the system or compromise confidential data. The approach to securing network management traffic depends largely on the technical	Enterprise subsystem and readers	Recommended	

	networks and encryption when feasible.	architecture. If network management occurs over Web interfaces, then Secure Sockets Layer (SSL) or Transport Layer Security (TLS) should be employed. In some cases, devices such as readers will be managed using SNMP. In these cases, SNMP version 3 is the preferred protocol, and community strings should be changed from defaults to complex character strings (i.e., mix of upper and lower case, both alphabetic and numeric characters).		
12	Design a network firewall between the RF subsystem and the enterprise network.[a]	A firewall can enforce a security policy on the information flow between the RF subsystem and any attached network, allowing only authorized protocols and services to traverse this boundary, such as those needed for readers to communicate with middleware servers and for management consoles to monitor and configure readers. This configuration limits the ability of an adversary that compromises RFID equipment to exploit vulnerabilities on non-RFID systems that also reside on the network. Appropriate firewall placement depends on the network architecture. For example, if middleware is integrated into the switches to which the readers connect, the firewall may be included in the switch or may reside between the middleware and the enterprise network. On the other hand, if middleware servers are located inside an enterprise network (e.g., at a remote data center), then the firewall may reside between the readers and the middleware.	Enterprise subsystem	Should consider
13	Develop RFID audit processes and procedures that identify the types of security relevant events that should be captured, and determine how audit records will be securely stored for subsequent analysis.	Audit records are necessary for forensic analysis of security and privacy incidents and also support real-time intrusion detection capabilities in many cases. The audit procedures should be reviewed for privacy protection considerations to determine if audit records contain or could be used to create PII. Ideally, audit data should be forwarded to a dedicated audit server that can preserve the integrity of event logs even when other RFID system components have been compromised. To facilitate implementation and compliance, existing audit processes and procedures for other enterprise information systems should be leveraged whenever appropriate. Events to be captured should include, at a minimum, unsuccessful authentication attempts.	Enterprise subsystem and readers	Recommended
14	Develop a password management system for tags that support password-protected features.	The password management system should specify how passwords are generated, assigned, stored, shared, and discarded. Passwords should be randomly generated. When passwords are written to tags using over-the-air mechanisms, additional care should be taken to avoid eavesdropping. When passwords are stored in enterprise databases, the databases have authentication and access control mechanisms to prevent unauthorized reading of the passwords. MOUs and MOAs with external organizations should cover roles and responsibilities related to the handling of passwords.	Tags	Recommended
15	Determine approach to tag memory protection, if applicable.	Important considerations include what data elements require read or write protection and to whether write protection for certain elements must be permanent. In some applications, the tag identifier may be modifiable while in others it must be permanently fixed.	Tags	Recommended

[a] For more information on network firewalls, see J. Wack, K. Cutler, and J. Pole, *Guidelines on Firewalls and Firewall Policy*. NIST Special Publication 800–41, January 2002.

TABLE A.3

RFID Security Checklist: Procurement Phase

No.	Security Practice	Rationale/Discussion	Affected Components	Recommended or Should Consider	Checklist Status
16	Procure products that use FIPS-validated cryptographic modules.[a]	Federal agencies are required to use FIPS-validated cryptographic modules. Cryptographic modules that are not FIPS-validated cannot be assured of providing the level of cryptographic protection intended. Identify all expected uses of cryptography, including those that will be used to secure data traffic in the enterprise subsystem. Significant resource constraints on tags preclude the use of cryptography for many applications, but if an organization decides that the additional expense of cryptography is required to protect sensitive information, then the corresponding cryptographic modules must be FIPS-validated.	All	Recommended	
17	Procure products that are functionally capable of supporting the organization's security and privacy policy.	If a product that does not support the security and privacy policy is deployed, noncompliance is guaranteed. For example, if the RFID usage policy requires data confidentiality between the reader and the enterprise subsystem, then the readers need to support appropriate cryptographic services on their enterprise interface. In general, tags do not have cryptographic data functionality, but data encrypted elsewhere can be stored on a tag if it has sufficient capacity, which typically is the case for active tags only. If a requirement exists to read or write protect certain data elements on a tag, then the organization should procure tags that support the desired memory access protections.	All	Recommended	

18	Procure readers, middleware, and analytic systems that log security relevant events and forward them to a remote audit server.	Audit technology helps ensure that the organization can detect unauthorized behavior and take actions to prevent or limit the extent of a security breach. If software components do not support audit event forwarding, then the organization should ensure that the supporting operating systems do so. At a minimum, the events should contain the tag ID, reader ID, and the reader timestamp for security relevant events.	Readers and enterprise subsystem	Recommended
19	Procure readers and server platforms that support the selected approach to securing network management traffic.	The network management architecture only can be implemented if the selected products support it. Potential protocols include SNMP version 3 or the encapsulation of management traffic within SSL/TLS or Internet Protocol Security (IPsec) tunnels.	Readers and enterprise subsystem	Recommended
20	Procure readers and server platforms that support Network Time Protocol (NTP).	NTP allows distributed devices to synchronize timestamps, which is critical to effective log analysis because it allows audit personnel to establish accurate event sequences across multiple devices. Many applications also need to obtain very accurate measurements of the time elapsed between transactions.	Readers and enterprise subsystem	Recommended
21	Procure an auditing tool to automate the review of RFID audit data.	Audit tools often are more effective than humans at distilling relevant information from multiple sources. In large enterprise RFID deployments, reviewing the amount of data generated could overwhelm technical support staff if they do not have appropriate tools to assist them with this task.	Enterprise subsystem	Should consider
22	Procure readers that can be upgraded easily in software or firmware.	This capability enables the readers to receive security patches and enhancements released after product shipment.	Readers	Recommended

[a] The following reference provides a list of FIPS-validated cryptographic modules: National Institute of Standards and Technology, *Cryptographic Standards and Validation Programs at NIST*, December 19, 2006. Available at http://csrc.nist.gov/cryptval/.

TABLE A.4
RFID Security Checklist: Implementation Phase

Implementation Phase					
No.	**Security Practice**	**Rationale/Discussion**	**Affected Components**	**Recommended or Should Consider**	**Checklist Status**
23	Harden all platforms supporting RFID components (e.g., middleware, analytic systems, and database servers).	Organizations should apply secure operating system and database configurations to all relevant hosts. See other NIST guidelines for recommended configuration information.[a]	Enterprise subsystem	Recommended	
24	Ensure that readers that support user authentication have strong, unique administrative passwords.	To protect against dictionary attacks, administrator passwords on readers should not be easy to guess.	Readers	Recommended	
25	Secure wireless interfaces on readers.	If the reader is mobile, it likely will have a second wireless interface to connect to the enterprise subsystem. In this case, the second interface should have a secure configuration.[b]	Readers	Recommended	
26	Assign unique passwords to tags.	When tags support passwords, organizations should not use a common password for multiple tags. Otherwise, a compromised password on one tag could have much wider consequences. Managing unique passwords requires the implementing organization to maintain a password database and support remote queries of the database, which might not be feasible in all environments.	Tags	Should consider	
27	Lock tag memory.	The organization should lock tag memory to meet business and security requirements as determined in the planning and design phase.	Tags	Recommended	

28	Disable all insecure and unused management protocols on readers and enterprise subsystem components. Configure remaining management protocols for least privilege.	Disabling all insecure and nonessential management protocols eliminates potential methods that an adversary can use when attempting to compromise a host. Examples of insecure management protocols include SNMP version 1 and SNMP version 2. If SNMP version 3 is used, configure it for least privilege (i.e., read only) unless write access is required (e.g., to change configuration settings as part of an automated incident response procedure).	All	Recommended
29	Activate logging and direct log entries to a remote audit server.	Logs enable security and support staff to identify potential security issues and respond accordingly. Using a remote central logging server facilitates reviews of logs across the enterprise and ensures the integrity of log data when RFID components are compromised.	Readers and enterprise subsystem	Should consider
30	If applicable, initiate a HERF/HERO/HERP compliance program to include operator training, posting of notices, and application of labels to sensitive materials.	If personnel are reminded of risks to their safety, they are more likely to engage in behavior that will prevent the realization of those risks. The compliance program should comply with Occupational Health and Safety Administration (OSHA) regulations regarding workplace safety.[c] Notices should appear in the same or comparable locations as other OSHA notices.	RF subsystem	Recommended

[a] The NIST Security Configuration Checklists Program for IT Products contains a repository of checklists for securing various operating systems and applications. Additional information may be obtained at http://checklists.nist.gov/.

[b] For more information on how to secure common wireless protocols, see T. Karygiannis and L. Owens, *Wireless Network Security: 802.11, Bluetooth and Handheld Devices*. NIST Special Publication 800–48, November 2002 and S. Frankel, B. Eydt, L. Owens, and K. Scarfone, *Establishing Wireless Robust Security Networks: A Guide to IEEE 802.11I*. NIST Special Publication 800–97, February 2007.

[c] 29 CFR § 1910.97. Nonionizing radiation.

TABLE A.5

RFID Security Checklist: Operations/Maintenance Phase

No.	Security Practice	Rationale/Discussion	Affected Components	Recommended or Should Consider	Checklist Status
31	Test and deploy software patches and upgrades on a regular basis.[a]	Newly discovered security vulnerabilities of vendor products should be patched to prevent inadvertent and malicious exploits. Patches should also be tested before implementation to ensure that they work properly.	All	Recommended	
32	Review audit logs frequently.	Frequent reviews of audit logs allow security and support personnel to identify security issues and take corrective or preventative measures quickly. All components of the RFID system should generate event logs. Automated logging tools can assist with log review and send real-time alerts in response to critical events, such as repeated failed authentication attempts during a short time period. RFID middleware products often provide advanced audit capabilities, including "dashboards" that allow administrators to monitor the activities of readers in real time.[b]	All	Recommended	
33	Perform comprehensive RFID security assessments at regular and/or random intervals.	Security assessments, or audits, are an essential tool for checking the security posture of an RFID system and identifying corrective actions necessary to maintain acceptable levels of security. The assessments should include monitoring of the RF spectrum to determine potential sources of RF interference and to identify ongoing surveillance or attacks. The assessment should also verify configuration settings on all RFID components.	All	Recommended	
34	Designate an individual or group to track RFID product vulnerabilities and wireless security trends.	Assigning responsibility to an individual for tracking wireless security issues helps ensure continued secure implementation of the organization's RFID systems.	All	Should consider	

[a] For more information on patching, see P. Mell, T. Bergeron, and D. Henning, *Creating a Patch and Vulnerability Management Program.* NIST Special Publication 800–40, Version 2.0, November 2005.

[b] For additional information on log management, see K. Kent and M. Souppaya, *Guide to Computer Security Log Management.* NIST Special Publication 800–92, September 2006.

TABLE A.6

RFID Security Checklist: Disposition Phase

No.	Security Practice	Rationale/Discussion	Affected Components	Recommended or Should Consider	Checklist Status
35	When disposing of tags, disable or destroy them.	The appropriate disposal or destruction mechanism depends on the type of tag, the level of assurance required, and the cost of the destruction. When tags contain memory, this memory should be rendered inaccessible. Options include the *kill* command and physical destruction. Many tags can be rendered inoperable by cutting them with a box knife, scissors, or other sharp object. The antenna on some tags can be separated from their transmitters by tearing them by hand, although accessing the tag data is still possible through physical analysis. Even if a tag contains nothing but an identifier, destruction may be advisable if there is the potential for an adversary with knowledge of the tag encoding protocol to correlate the identifier with other information, such as tag ownership. This attack is particularly salient for EPCglobal tags, because of the potential to discern the identity of the EPC Manager from the pointers returned by the Root ONS. In many cases, the tag identifier also reveals the serial number of the asset. On the other hand, many organizations may determine that that this risk is acceptable, especially if database records corresponding to a particular identifier are erased or disabled when the tag is no longer needed.	Tags	Should consider	
36	When disposing of an RFID component, ensure that its audit records are retained or destroyed as needed to meet legal or other requirements.	Information contained in the audit records may be needed even after an RFID component is discarded (e.g., for an investigation of a subsequently discovered security breach). Organizations should identify the legal requirements for data retention that apply to their operations.[a] When log events are forwarded to a central audit server, regular backup of the server facilitates the retention of records. When a log server does not exist, the disposal process may include capturing the existing log data and storing it on alternative media, such as CD-ROM or tape. On the other hand, retention of audit records may raise a privacy concern in some applications. For example, records may reveal sensitive personal information or associate a person with particular items or transactions in a manner that violates privacy laws or policy. In these cases, the requirement may be to destroy the records after a certain period of time or after they are no longer needed.	All	Recommended	
37	Recycle retired tags.	In some cases, recycling may involve putting the tags back into service. This type of recycling is not recommended when tag memory contains confidential data, but may be cost effective otherwise. Recycling may also involve using discarded tag material for other purposes in a similar manner to recycling programs for household plastics and metals. Both forms of recycling address a concern about the environmental impact of large scale consumer and industrial uses of tags.	Tags	Should consider	

[a] For an example of a requirements document, see National Archives and Records Administration, General Records Schedule 24, Information Technology Operations and Management Records, April 2003. Available at http://www.archives.gov/records-mgmt/ardor/grs24.html.

31

Lightweight Cryptography for Low Cost RFID: A New Direction in Cryptography

Damith C. Ranasinghe, Raja Ghosal, Alfio Grasso, and Peter H. Cole

CONTENTS

31.1 Introduction

The primary functionality of a modern radio frequency identification (RFID) system is that of identifying objects through a uniquely formatted number kept on each label (also called a tag) as a data field identifier with the associated data being stored in a backend database system. The unique object identifier must have a global scope that is capable of identifying all objects uniquely and acting as a pointer to information stored about the object somewhere over a network.

Here we define the term "networked low cost RFID systems" as RFID systems based on passive RFID technology using microelectronic transponders. In such a system the readers (also called interrogators) are connected to a backend network. Conceptually, the backend network can be described as an intelligent ubiquitous infrastructure that automatically and seamlessly links physical objects to the global Internet. The network of physical objects is achieved by integrating a tag, or an RFID label, into each object. The system network objects seamlessly by communicating with these labels using a network of readers (Ranasinghe et al., 2005).

The safe delivery of pharmaceuticals and health care as well as homeland security are all areas that can significantly benefit from adopting RFID technology. Although the impact of RFID technology on these applications is immense, it is not possible to engage the technology because of its many shortcomings, especially the lack of methods and infrastructure for providing electronic information security (e-security) to networked low cost RFID technology. Although shortcomings of low cost RFID systems have been detailed in publications elsewhere (Juels, 2006; Ranasinghe and Cole, 2006a,b), solutions have not been forthcoming.

Networked RFID systems are now a pervasive form of computing. In the context of security and privacy, the most threatening (to privacy) and vulnerable (to insecurity) are the low cost RFID systems. Modern cryptographic mechanisms are unsuitable for use in securing RFID systems because of the lack of resources available on RFID microelectronic transponders, and a novel avenue in cryptography with new thinking is required.

A new view on the development of cryptography to address the security needs of networked resource constrained devices, relevant developments in the area of cryptography, and future avenues of promising research are presented in this chapter.

31.2 Contemporary Ideas of Security

The most commonly used encryption techniques in large information technology system deployments are cryptographic methods based on the following problems considered to be in a class of mathematically hard problems. A mathematical problem is said to be hard if the time it takes to solve the problem becomes immense compared with the increase in the size of the inputs to the problem. There are many other classes of problems, as discussed later in this Chapter, some of which are not hard, and hence not of interest to cryptography.

- Discrete-Logarithm Problem (examples of cryptographic systems are Diffie-Hellman, El-Gamal, Massey-Omura systems, and relatively new Elliptic Curve Cryptosystems)
- Integer Factorization Problem (the most popular example of a cryptosystem based on the latter problem is the RSA cryptosystem)

We can define the term well functioned secure communication system as one in which there are, at each end of the communication link, significant computing resources, such as provided by a microprocessor, with power supplied for operation for an indefinitely long period. In all communication systems a great variety of attacks on security are possible. Security against such attacks is normally provided by exploiting the computing resources of a well functioned secure communication system. In RFID labels it is impractical to provide such computing resources.

Cryptosystems such as RSA, Diffie-Hellman, and ElGamal have withstood tests of time, and proven to be secure, using sufficiently large keys, typically at least 1024 bits (Schneier, 1994), preferably 2048, 4096, or 8192 bits. The larger the key the higher is the computation load on the device, due to the correspondingly larger size of the prime p, used in typical arithmetic operations such as $m^p \bmod n$. However, affordability of computing power has evolved to support these cryptosystems in modern communication systems.

31.2.1 Quantifying Security

Shannon, in his information-theoretic approach, asked the following questions:

- Are there systems that withstand any attack (even those with unlimited computing power)?
- How much ciphertext does the opponent need before they can uniquely solve for the plaintext?

The answers to these questions have formulated modern ideas of defining the level of security provided by a cryptographic mechanism.

Perfectly secure (unconditional security): Shannon observed, that if plaintext X, and ciphertext Y, are independent by information-theoretic measurements I, and uncertainty H, then a perfectly secure or unconditionally secure system is one where $I(X; Y) = 0$ (Rueppel, 1989).

Perfect security is the nonplus-ultra security model. It assumes unrestricted computational power of the adversary. Therefore, for a cryptographic primitive to fall into this category there must not be an algorithm for breaking it, irrespective of the computational power available. An example of a simple primitive offering unconditional security is the one-time pad (Menezes et al., 1997). To generate the ciphertext a plaintext is XORed with a unique secret key of the same length as the plaintext. Because of the possible large key sizes required in modern communication systems for the transmitting large quantities of data, one-time pads are impractical for conventional message encryption.

Ideally secure: Shannon further observed, when $I(X; Y) < 0$ there is some residual uncertainty. Such systems are ideally secure. Cryptanalysts can sharpen probability distributions, but cannot find a unique X (plaintext) such that $I(X; Y) = H(X)$ (Ruppel, 1989).

Complexity theoretic approach refers to quantitative analysis of the difficulty of algorithms. In this classification scheme we consider a hierarchy of problem classes with varying degrees of difficulty. These classes and their potential applications to RFID are discussed in Section 31.4.2.1. The following is an application of complexity theory, in modern cryptography, to quantifying security.

Computationally secure (*practically secure*): A system is computationally secure if the best known algorithm for defeating the system requires at least unrealistic number of operations or unreasonably large memory size. The number of operations in the definition is chosen to be sufficiently high. Modern cryptographic primitives typically offer practical security. However, as opposed to unconditional security, there are relative bounds on the opponent. These bounds are described in terms of computing power and available memory for some hypothetical but realistic opponent.

Provably secure: A cryptosystem can be proved to be secure if it is possible to show that the complexity of a successful attack on a cryptosystem is equivalent to solving a well known, supposedly hard, mathematical problem such as the integer factorization problem. Typical cryptographic primitives based on public keys fall into this category (Menezes et al., 1997).

Since the adoption of probably secure or ideally secure cryptosystems inevitably requires computing resources beyond that of low cost RFID tags the prudent approach in RFID is to achieve practical security with a realistic adversary envisioned, with the capabilities and the potential application of the technology in mind.

31.3 New Direction in Cryptography

With the growth of inexpensive computing resources, modern cryptography has developed in a direction involving computationally intensive calculations. Most modern cryptosystems are based on some mathematically hard problem and the level of security provided by the system depends on the difficulty of the mathematical problem.

In the provision of security in RFID systems, the resources within the tags themselves are particularly constrained. The set of currently available security primitives is unsuitable for RFID microelectronic transponders for the reasons given below.

- Costly to be implemented on a cost constrained RFID Integrated Circuit (IC) with a price tag, in large volumes, of around 8 US cents (Ranasinghe and Cole, 2006a).
- Relative power consumption by the cryptographic hardware modules is too high for RFID labels that are passive (not self-powered).
- Unsuitable for RFID transponders with limited logic functionality (limited to a simple finite state machine) and limited memory (limited to a few kilobits).

Large key sizes and the resulting ciphertext sizes are generally unsuitable for narrow band communication systems where transmission of significant amount of data directly affects the performance of the system.

Most modern methods cannot be applied directly to lightweight cryptography because modern security concepts have evolved in a direction of utilizing increasing key sizes (typically 1024 bits in RSA) supported by increasing computing power to achieve adequate levels of computational security. The algorithm adopted for low cost RFID systems must match the processing capabilities of the devices. Thus there is scope for the design of lightweight algorithms. In the lightweight context, the designer has to analyze the computational complexity of the algorithm, with respect to the demands on the hardware and other limitations of the device. There is both a direction, and a constraining challenge in these limitations that guide the development of cryptography.

The emerging field of *lightweight cryptography* is a fusion of separate disciplines in cryptography, information technology, radio frequency engineering, and microelectronics. It can be considered as a novel branch of cryptography that aims to develop fast and efficient security mechanisms for resource constrained environments. This branch of cryptography is becoming the most promising avenue for generating secure cryptographic solutions for low cost networked RFID systems.

The following sections describe the development of the novel paradigm of lightweight cryptography for resource limited environments. Research outcomes from lightweight cryptography can be utilized to provide security services such as authentication, confidentiality and anonymity for resource limited microelectronic tags.

31.4 Foundations of Lightweight Concepts

Unlike more traditional forms of thinking, focusing on mathematically hard problems or provable security, lightweight cryptography tends to use all aspects of RFID technology from both a mathematical as well as an electrical and radio frequency engineering perspective to generate research and solutions. The following sections describe the fundamental concepts underpinning developments in lightweight cryptography.

31.4.1 Desired Level of Security

The flow of information in a traditional communication model using archetypical characters can be likened to an RFID system. Figure 31.1 illustrates a reader transmitting information and power over a communication channel that can be monitored and altered by an eavesdropper Eve and an active attacker Mallory. The tag is the recipient of power and commands from a reader. A tag responds to reader commands by transmitting replies back to the reader over the same insecure channel. While a reader's connection to backend resources may be either wired or wireless, this connection can be assumed to be secure. Taking into account modern network security measures based on established cryptographic primitives this assumption ought to be legitimate. Therefore considerations of securing networked low cost RFID systems solely address security and privacy issues of the RFID air-interface and tags, while the reader and the backend systems are treated as a single entity.

We will use the term *computationally feasible* to describe a cryptographic system that is capable of being implemented on a networked low cost RFID system for providing any or all of the required security services—authentication, integrity, confidentiality and privacy (anonymity and untraceability)—while meeting all of the end-user performance requirements (Ranasinghe and Cole, 2006a). For a computationally feasible cryptographic system to be considered secure it must be practically difficult for a cryptanalyst to recover the plaintext, without prior knowledge of the keys, under a practical adversarial model that considers the mobility of tags as well as the unique characteristics of low cost RFID systems outlined in (Ranasinghe and Cole, 2006b).

31.4.2 Computational Model

In contrast to conventional security systems we need cryptographic systems with computational resources that are comparable with the complexity of low cost RFID hardware.

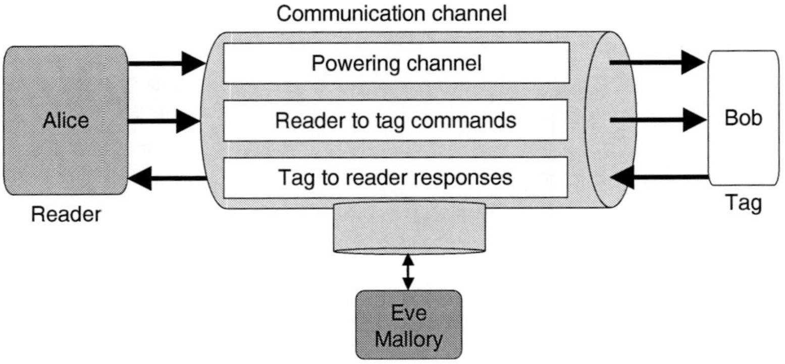

FIGURE 31.1
Communication channel model capturing the interface between an RFID reader and a tag.

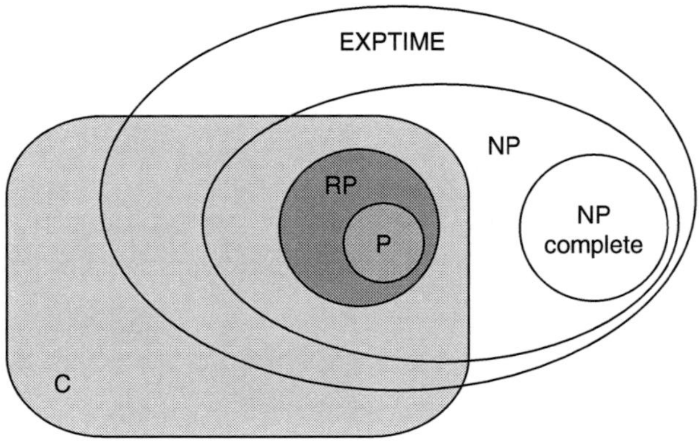

FIGURE 31.2
Complexity classes.

Complexity theory provides a theoretical and a procedural framework to any algorithmic design. It helps in quantifying problems in terms of computation resources and running time required for their solution as the size of inputs to the problem increases.

Complexity classes, as shown in Figure 31.2, provide ahead of any design or implementation a classification of problems based on the type of computing machine and the computing resources required to solve the problems or execute the algorithms. The simplest set P, are a set of problems that can be solved by an algorithm in polynomial time using a deterministic Turing machine. NP class of problems can be solved by an algorithm using a nondeterministic Turing machine in polynomial time, however, for NP-complete problems there is no known polynomial time solution.

Class C (circuits with polynomial order of the number of logic gates) has complete overlap with P, and RP classes. Class P and class RP (randomly polynomial) are proper subsets of C and class C has some overlap with NP but no overlap with the set NP-complete (Welsh, 1990).

31.4.2.1 Usefulness of Complexity Theory

There are several ways of applying complexity theory to the development of lightweight cryptographic primitives for RFID. One such approach is to first assess the complexity of the RFID circuits, and to locate the class of algorithms that can yield an acceptable circuit complexity. The primary constraining factor in resource constraint devices is the limited support for the complexity required for algorithms pertaining to hard problems. The number of logic gates in RFID tags and, in general, in lightweight devices such as mobile phones can be approximately modeled to lie in the C class of circuits. RFID tags belonging to the C class can perform tasks that belong to the RP class of algorithms. These tasks include processing of data, transmitting data, and other normal operational procedures. By complexity theory, all RP (randomly polynomial) class of algorithms can be implemented on the C class of circuits (Welsh, 1990).

P class is the simplest, but systems based on problems in this class can be easily broken into by an intruder (cryptanalyst) in polynomial time, for instance by a simple exhaustive search of the solutions space. However, NP class is more difficult and is currently thought to include the integer factorization problem and discrete logarithm problem. Discovering a solution by an exhaustive search for an NP class of problem will require an unreasonable amount of time. Cryptographic primitives based on the NP class of problems are very secure. The complexity of the algorithm presents a difficulty in the implementation of cryptographic primitives based on NP class of problems in RFID tags, in spite of the reduction in computation required due to the shorter keys, as in elliptic curve cryptography (ECC).

Therefore computational complexity (theoretical) results help in limiting the focus of the security algorithms those that fall between the P class and the NP class. This suggests that the focus of computationally feasible mechanisms should be on algorithms falling into the RP class. Clearly, complexity theory also ensures that we do not look at problems that are mathematically more hard and complex than RP class of problems, such as the integer factorization problem expounded in RSA. Analysis of existing security mechanism for networked low cost RFID systems, such as HB, HB +, and noisy tags protocols, discussed later in this Chapter, show that they can be anticipated to fall into the RP (randomly polynomial) class of algorithms.

31.4.3 Principles of Design

Fundamental thinking and design paradigms envisioned to create cryptographic solutions are guided by the limitations and computation boundaries of the underlying technology. The following is a set of valuable design principles.

- Remove IC complexity by transferring resource intensive tasks to networked backend systems or proxy servers and reduce the computation burden on resource constrained tags (Ranasinghe and Cole, 2006a).
- Use Shannon's concepts of confusion and diffusion (Schneier, 1994) instead of complexity to achieve a high degree of information entropy. Confusion is a core part of the process of a cryptographic primitive where the plaintext is transformed into a ciphertext. Examples of confusion can be seen from simple schemes such as the substitution ciphers to more modern techniques such as modular exponentiation used in RSA. Diffusion is a mechanism of using redundant information in the ciphertext to make it difficult for the intruder to discover the actual ciphertext in an encrypted communication. The discussion on the use of noisy tag protocols in Section 31.6.1.2 is an example of using diffusion.
- Use mathematical operations and problems which translate to simple hardware implementations to reduce the cost of implementation in silicon. This involves shifting the focus on to simple operations such as bit shift, XOR and other general $GF(2^n)$ operations, rather than complex arithmetic operations such as exponentiation operations.
- The security of the cryptosystem should depend only on the key(s) and not on the secrecy of the algorithm and its implementation (Kerckhoffs, 1883). Schneier (1994) points out that keeping the algorithm secret has the risk that once the algorithm is broken into, the entire organization has to redesign a new algorithm. On the other hand if only, the keys are broken, then a redistribution of keys does not have the same impact.
- The security systems should, ideally, have a level of security that is provably secure, if not they should at least be practically secure as discussed in Section 3.2.1.
- Design systems where the keys utilized can be changed with ease so that the long term storage of many keys on tags is unnecessary. This is because tags are constantly in an untrusted environment and cannot be trusted to store long term secrets.
- Proportioning computation burden and participation to resource availability, without attempting to increase resources to support complexity. This will ensure that tag cost is kept to a minimum while ensuring the participation of the tag to complete the execution of the security mechanism where the tag participation achieves the set security objectives.

Lightweight cryptography needs to encompass three separate but related areas of lightweight: hardware (low power, small IC footprint, and low cost), cryptographic protocols (efficient with low overheard), and primitives (distributed functions with low cost implementation on silicon) to provide a complete solution. The following sections discuss these topics in more detail.

31.5 Lightweight Primitives

Cryptographic primitives form the building blocks of cryptographic systems (for instance the RSA cryptographic primitive is used for building public key ciphers). Primitives for lightweight cryptography must be generated using mathematical operations requiring a simple translation into digital circuit designs. The functions used should be such that complexity can be transferred to backend systems where bulk of the operations can take place, while the hardware on an RFID label is only required to perform a minimal level of computation.

31.5.1 Developments in the Area of Lightweight Primitives

31.5.1.1 Ultra Wide Band Modulation

This simple prudent method is based on time division of transmission to 65,536 slots in which the data of an RFID tag can be transmitted. The security lies in the difficulty of an intruder guessing in which time slot the sender has sent the desired data. This uses a pulse position modulator (PPM). A CS PRNG (Cryptographically Secure Pseudo Random Number Generator) determines the time hopping codes (Avoine, 2007).

31.5.1.2 Physically Uncloneable Functions

Physically uncloneable functions is a lightweight primitive for implementing a challenge–response authentication protocol on RFID tags without the need for costly cryptographic hardware. A more recent implementation and application to low cost RFID can be found in Ranasinghe et al. (2007). These schemes outlined are based on a hardware based random function (PUF) being integrated on to a low cost RFID tag IC. The functions allow for calculating a unique responses vector $\boldsymbol{r}$, to a challenge vector $\boldsymbol{c}$, based on a key k. The unique feature of this primitive is the exploitation of statistical variations in the silicon fabrication process to build a hardware based key k (Lee et al., 2004), onto the each individual RFID IC. A formulation of the function is given in (1) and (2). The silicon implementation of the hardware costs only hundreds of logical gates (Ranasinghe et al., 2007).

$$\{c_1, c_2, c_3, \ldots, c_m\} \rightarrow f\ \{(c_1, c_2, c_3, \ldots, c_m), k\} \rightarrow \{r\} \tag{31.1}$$

$$\begin{aligned} c_i &= (c_1, c_2, c_3, \ldots, c_n) \in \{0,1\} \quad \text{for } i = 1, \ldots, m \\ r &= (r_1, r_2, r_3, \ldots, r_m) \in \{0,1\} \end{aligned} \tag{31.2}$$

31.5.1.3 Minimalist Cryptography

It will be challenging to provide security to RFID tags whose prices are likely to drop to 5 US cents over the next several years. Low cost RFID tags at present do not have enough computational power to perform even the most rudimentary cryptographic algorithms. RFID tags are fast becoming the alternate to barcodes technology in supply chain applications.

In Juels (2004) there is a formal framework for a minimalist cryptography system for such extremely resource scarce devices, based on one-time codes. A major idea in Juels (2004) is the application of pseudonyms to help enforce privacy in RFID tags. A tag may carry multiple, random-looking names. On each occasion a tag is queried, the tag releases a different name. In principle, only an authorized verifier can tell when two different names belong to the same tag. It is possible for an adversary to query a tag multiple times to collate all the names so as to weaken the security of the scheme. The approach described in Juels (2004) involves enhancements to overcome the latter weakness. First, tags release their names only at a certain (suitably slow) prescribed rate. Second, pseudonyms can be refreshed by authorized readers.

31.6 Lightweight Cryptographic Protocols

Establishing robust and secure protocols is essential to describe the execution of various cryptographic primitives. Lightweight protocols need to establish cryptographic protocols with extremely low computational overhead and minimal digital logic implementation costs. A promising avenue of research for developing lightweight protocols is based on the conjectured hardness of the Learning Parity with Noise (LPN) problem (Piramuthu, 2006).

Lightweight protocols should be able to meet the end-user performance goals (such as read rates) by delaying communications rounds with tags and their implementations should not consume excessive computing resources. Such fresh approaches are based on use of shared secrets between a tag and a database, with one or more of the features: (1) an initial supply of random one-time codes for use through the life of the labeled product, (2) update of the shared secrets in an environment secure against eavesdropping, and (3) use of self-updating architectures using high entropy encrypted signaling generated by computationally feasible mechanisms.

31.6.1 Developments in the Area of Lightweight Protocols

31.6.1.1 Hopper and Blum Protocols for RFID

Hopper and Blum have suggested two shared-key authentication protocols, HB and HB+ protocols (Hopper and Blum, 2001). Adaptation of HB and HB+protocols to RFID were presented in (Juels and Weis, 2005). There have been further improvements, by various authors as discussed in Piramuthu (2006). He has discussed modified HB and its variants, the HB +, and the most recent variant HB++ protocols. The HB protocol uses simple operators such as the XOR operator. The computational basis of the HB protocol is based on the hardness of the LPN. Piramuthu (2006) also indicates the classical symmetric key cryptography solutions by other researchers make the protocol vulnerable to active attacks. HB protocol is based on k-bit data binary vectors, k-bit secret key vectors, and some noise bits, 1 or 0, based on a probability function. It is possible for an intruder to correctly surmise the probability function from the pattern of data using standard stochastic estimation theory.

31.6.1.2 Noisy Tag Protocols

Castelluccia and Avoine (2006) present a key exchange protocol for RFID based on noisy tags. These tags belong to the system and are in the presence of the interrogation zone of a trusted reader. The underlying concept is to generate noise during tag responses. The pattern of the noise is known to the trusted reader and therefore can be subtracted in order to recover tag responses. Unlike the trusted reader an eavesdropper cannot understand tag responses because it cannot differentiate between noise and meaningful data.

31.6.1.3 One-Time Codes

Use of one-time codes coupled with the simple operators such as the XOR function has provided three different avenues of developing lightweight protocols. These avenues are based on use of shared secret strings between a tag and a database, with one or more of the features outlined below.

1. Tags are given an initial supply of random one-time codes for use through the life of the labeled product during a set-up phase.
2. Update of the tag one-time pad (the shared secrets strings) in an environment secure against eavesdropping.
3. Alternatively, use self-updating architectures with high entropy encrypted signaling generated by computationally feasible mechanisms.

A lightweight mutual authentication scheme for networked low cost RFID systems based on random tag identifiers and XOR-padding is presented in Ghosal et al. (2006). This scheme provides privacy in terms of preventing tags from being traced and security in terms of mutual authentication between RFID readers and tags (Ghosal et al., 2006). The main objective of the scheme described therein is to provide mutual authentication between RFID readers and tags.

As mentioned in Section 31.4.1, the backend database and the reader are treated as a single entity, which communicates with a tag via the RFID air-interface. The scheme uses three different encryption strings (T, R, and A), which are used in conjunction with an XOR operation to disguise transmissions of identifiers. These encryption strings are random binary sequences that are shared between each individual tag and the backend database (Ghosal et al., 2006).

31.7 Lightweight Hardware

Lightweight hardware focuses on research to develop efficient low cost hardware implementations of cryptographic primitives and protocols. The process of developing or optimizing existing hardware for security mechanisms has been an active area of research for smart card processors (Aigner, 2006). However, research into developing cryptographic hardware modules for low cost RFID ICs limited by available power, bandwidth, cost and timing constraints is a challenging area.

Ideally, for security purposes we have to look at mathematically hard problems, typically one-way functions whose forward (encryption) path is easy, normally in the P class, and whose reverse or inverse path (decryption) is hard, normally in the NP class. The challenge often in the development of lightweight hardware is to implement a hard problem, such as the integer factorization or the discrete logarithm problem, by way of simple methods such as simple Boolean logic or shift logic circuits.

31.7.1 Developments in the Area of Lightweight Hardware

31.7.1.1 Elliptic Curve Cryptography

Elliptic Curve Cryptography (ECC) is a newer approach than RSA with the advantage of smaller key sizes than RSA to achieve the same level of security as the more popular RSA systems. ECC has the hardness of exponential time challenge for cryptanalysts attempting to obtain plaintext from ciphertext with no knowledge of the keys used. As a comparison, in ECC, a 160 bits key provides the same security as a 1024 bits key in RSA, 224 key in ECC provides the same level of security as 2048 bits key in RSA. RSA systems require larger key sizes to achieve a level of computational security considered as being adequate given availability of relatively cheap computing power. ECC implementations require lower memory due to shorter keys and correspondingly lower computations. However, the nature of the ECC algorithm is such that their execution places greater demands on power consumption (Menezes, 1993). There are also no known sub-exponential time algorithms for successfully attacking elliptic curve cryptosystems (Basubramaniun, 2003).

Shorter key sizes make ECC a suitable candidate for computing power and memory limited devices. Sun Microsystems's Sizzle, a coin sized wireless secure web browser, uses ECC (Gupta et al., 2005; Morgan, 2005). However, the elliptic curve algorithm is fairly complex, and hence thus far it is not directly applicable to RFID tags. A design of an ECC processing unit on an RFID tag is presented in (Wolkerstorfer, 2005). Batina et al. (2006) have reported further progress on the design of an ECC unit suitable for RFID tags.

(Wolkerstorfer, 2003) presented a hardware implementation of an elliptic curve cryptography processor and examined the feasibility of implementing it into low cost RFID tags (Wolkerstorfer, 2005). He has implemented an ECC unit in RFID. He reported that 0.35 μm CMOS requires a silicon area of 1.31 mm^2 for an implementation of an ECC hardware unit. This silicon area is somewhat large for an RFID tag IC. Even if the proposed ECC processor is incorporated into an RFID tag, there are concerns about the power consumption, execution time, and potential doubling of the cost of the RFID tag IC.

31.7.1.2 Block Ciphers

A lightweight block-cipher hardware implementation, mCrypton , was presented by (Lim and Korkishko, 2005). mCrypton was especially designed for RFID and sensor applications and shows a hardware complexity of 3500 to 4100 gates depending on the key size (64–128 bits, respectively).

31.7.1.3 Symmetric Key Ciphers

Feldhofer et al. (2004) presented a hardware implementation of the AES algorithm for RFID tags for strong tag to reader authentication. The estimated hardware complexity of the implementation is 3600. However, a drawback of the implementation is the relatively large number of clock cycles (1000) required to encrypt a block of 128 bits. Feldhofer et al. (2004) additionally proposed an enhancement for the ISO/IEC 18000 standard in order to integrate the authentication mechanism and to deal with the long encryption time in the tag.

31.8 Future Directions

31.8.1 Multivariate Polynomials and Hidden Field Equations for Lightweight Systems

Wolf and Leuven (2003) methods are along the lines of some 1980s asymmetric cryptographic systems using multivariate polynomials. These methods can be used for both signing and encryption. They use polynomials over finite fields of different sizes based on the intractability (hardness) of the "MQ" problem, which denotes that, multivariate quadratic (MQ) equations over finite fields are difficult to solve. While some of the systems based on multivariate polynomials have been broken, they offer alternate interesting challenges using algorithms based on low computation costs.

Barkan et al. (2003) showed the use of the Hidden Field Equations (HFE) and multivariate quadratic residue quotient systems (MCQ) in the analysis of cryptographic systems. Use of HFE and MCQ was demonstrated in the analysis of the A5 algorithms used in GSM mobile communication standard (Barkan et al., 2003). The methods used have a direct application to the analysis of bit oriented stream generators based on shift registers with no formal proofs of security but with the ability to represent the encryption hardware using polynomial equations (Barkan et al., 2003).

The earlier-mentioned methods may be used to formulate attacks and analyze the feasibility of attacks against lightweight cryptosystems based on bit shift and XOR operations. Unlike provably secure systems based on mathematically hard problems, most lightweight systems do not have such rigorous formal quantitative analysis. The earlier-mentioned methods provide directions to analyze lightweight cryptosystems.

31.8.2 Can We Look to Quantum and Quasar Cryptography?

Quantum Cryptography is one possible area of research in one-time codes. This is based on the quantum states of spin and other variables of subatomic particles like photons or electrons (Hershey, 2003). This area is an active area of research where its current applications include the secure transmission of cryptographic keys (Stix, 2005). It is not yet feasible for full data encryption and its relevance to RFID systems can be questioned.

In its potential application to RFID, one may note the need for both a quantum transmission channel and a classical channel, and the emergence of only partial keys and then only after classical channel communication between sender and receiver of many quantum channel communications.

Quasar Cryptography based on random noise in space, galaxies, via radio astronomy (Hershey, 2003; Schneier, 2006), is claimed to be another promising area of one-time codes generation. Any intruder must know exactly which of the many radio telescopes pointing to a specific galaxy is being used for the random noise sequence (Schneier, 2006). Herschey (2003) also pointed out the importance of channel based cryptography where the channel is the cryptographic-variable. A short burst of random noise injected into the communication channel affects the receiver and the transmitter in the same manner. Quantum computing and Quasar Computing are both applications of channel based cryptography (Hershey, 2003). At the moment these concepts are infeasible alternatives in lightweight devices.

31.8.3 Learning from the Smart Card Industry

Aigner (2006) presents an analysis and an overview of crypto implementations for RFID tags by reviewing the developments in the smart card industry. Oren and Shamir (2006) have highlighted the current practical limitations of hardware implementations which are theoretically proven secure. They have, on an optimistic note, pointed out the experiences and evolution of smart cards, especially the outcomes of making them immune to power and side channel attacks. These outcomes could eventually be applied to RFID tags without reinventing the wheel.

31.8.4 Feed Back Shift Registers and XOR Cryptography

Developments in pseudo random bit generators (PRBG), based on LFSR (Linear Feedback Shift Registers), have been used widely because of the information-theoretic security provided by one-time pads. The barrier preventing the use of one-time pads is the need to keep or transmit a key of the same length as the plaintext. Stream ciphers overcome this impracticality by allowing the fast generation of pseudo random numbers at the transmitter, for encrypting, and at the receiver, for decrypting the received data. The encryption operation is a simple XOR operation of the plaintext with the key stream, while the decryption operation is an XOR operation of the key stream with the plaintext. In an RFID context, they also provide the most promising avenue for achieving mutual authentication between a tag and a reader (Ranasinghe, 2007). However, bit stream generators do not provide the same perfect secrecy as one-time pads as the key sequence is only pseudo random.

There are several measures of the level of security provided by a stream cipher. These are based on metrics such as the linear complexity of the key stream, and randomness and correlation-immunity of the output sequence. However, these metrics are not sufficient to guarantee the security of stream ciphers (Menezes et al., 1997). Most of the vulnerabilities of stream ciphers result from the linearity of the feedback scheme. A simple solution is to input the feedback values in LFSRs through a nonlinear function before they are fed back into the shift registers. The result is to increase the level of confusion in the output of the generator. An example of a nonlinear combination generator is the Geffe generator consisting of three maximum length LFSRs of pairwise relatively prime period lengths that are combined applying a nonlinear Boolean function (Menezes et al., 1997).

The shrinking generator (Coppersmith et al., 1994) achieves such nonlinearity by utilizing two LFSRs, wherein the output of one LFSR is used to decide where to use or discard the output of the second LFSR. By using maximum length LFSRs of length $L1$ and $L2$ where $gcd(L1,L2)=1$, the shrinking generator has a linear complexity of $2^{L2}.L1$. With a prudent choice of initial seeds and the tap polynomials for the 2 LFSRs, the shrinking generator is very hard to defeat (Menezes et al., 1997). The order of complexity of known attacks has exponential time complexity as they are a function of the length of both LFSRs. A shrinking generator implementation offers the most promising avenue for achieving confidentiality and mutual authentication between a tag and a reader given its many practical advantages related to efficiency, simplicity in implementation requiring only XOR cryptography, shift registers and Boolean logic, and requiring very little memory (the only storage requirement being the initial seeds and/or the connection polynomial weights).

31.8.5 How Can We Advance the Study of Lightweight Protocols?

We need to examine the properties of all existing cryptographic protocols to ascertain what can be considered as lightweight cryptographic protocols. Such an exploration will allow us to assess whether they are applicable to low cost RFID, and also to examine whether they are indicative of directions in which to search for cryptographic protocols useful in low cost RFID.

31.9 Conclusion

It is evident that RFID privacy and security are challenging areas of research. Limitations and constraints of the enabling technology prevent the use of contemporary cryptographic solutions. We have expounded upon a new direction in cryptography needed to address the security and privacy needs of networked low cost RFID systems as well as to enable many applications that can benefit from security services provided to low cost RFID. Lightweight cryptographic design ideas elaborated have recognized the resource limitation of low cost labels that require simplicity at the tag silicon level.

It is evident that RFID privacy and security are challenging areas for both implementation and research. There are three specific areas of research (lightweight primitives, lightweight hardware and lightweight protocols) which will greatly benefit low cost RFID security and privacy, and successful results from research on these topics will lead to the widespread adoption of this technology. Lightweight hardware based on pseudo random number generators such as the shrinking generator, lightweight primitives such as the PUF, and lightweight protocols based on one-time codes (Ghosal et al., 2006), which involve one or more small shared secrets between a label and an interrogator, provides the best avenues for generating practicable solutions.

It is important to note that the level of security and privacy will depend on the application. Clearly there is no universal solution but a collection of solutions suited to different applications based on compromises and on security services required.

References

Aigner, M. 2006, Crypto implementations for RFID tags: Learning from the smart card industry. PROACT, Crypto for RFID Series 2006, Graz.

Avoine, G. 2007, *Bibliography on Security and Privacy in RFID Systems*, MIT, Cambridge, MA. Available from http://lasecwww.epfl.ch/~gavoine/rfid/.

Barkan, E., Biham, E., and Keller, N. 2003, Instant ciphertext-only cryptoanalysis of GSM encrypted communications, *Crypto 2003*, Santa Barbara, CA.

Basubramanian, R. 2003, *Elliptic Curres, Modular Forms and Cryptography*, Hindustan Book Agency, New Delhi, pp. 325–345.

Batina, L., Guajardo, J., Kerins, T., Mentens, N., Tuyls, P., and Verbauwhede, I. 2006. An Elliptic Curve Processor suitable for RFID-tags, Cryptology ePrint Archive: Report 2006/227 (4th July 2006).

Castelluccia, C. and Avoine, G. 2006. Noisy tags: A pretty good key exchange protocol for RFID TAGS, *Proceedings of the International Conference on Smart Card Research and Advanced Applications*, Springer Lecture Notes in Computer Science, Vol. 3928, 2006, pp. 289–299.

Coppersmith, D., Krawczyk, H., and Mansour, Y. 1994, The shrinking generator, *Proceedings of Crypto 93*, Springer-Verlag, 1994, pp. 22–39.

Feldhofer, M., Dominikus, S., and Wolkerstorfer, J. 2004, Strong authentication for RFID systems using the AES Algorithm, *Proceedings of the 6th International Workshop on Cryptographic Hardware and Embedded Systems*, Springer Lecture Notes in Computer Science, Vol. 3156, pp. 357–370.

Ghosal, R., Jantscher, M., Grasso, A., and Cole, P.H. 2006, One Time Codes, White Paper Series, Auto-ID Labs, WP030, Adelaide.

Gupta, V., Wurm, M., Zhu, Y., Millard, M., Fung, S., Gura, N., Eberle, H., and Sheuel Chang Shantz, S. 2005, Sizzle: A standards-based end-to-end security architecture for the embedded internet, *Pervasive and Mobile Computing*, 1, 2005, 425–445.

Hershey, J.E. 2003, *Cryptography Demystified*, McGraw-Hill, New York, NY. 2003.

Hopper, N.J. and Blum, M. 2001, Secure human identification, protocols, LNCS Vol. 2248, 2001, pp. 52.

Juels, A. 2004, Minimalist cryptography for low-cost RFID tags, in *Proceedings of the 4th International Conference on Security in Communication Networks*, C. Blando (ed), Springer Lecture Notes in Computer Science, Vol. 3352, 2004, pp. 149–164.

Juels, A. 2006, RFID, Security and privacy: A research survey. *Journal of Selected Areas in Communication*, 24(2), 2006, 381–395.

Juels, A. and Weis. S. 2005, Authenticating pervasive devices with human protocols, in *Advances in Cryptology—Crypto 2005*, LNCS, Vol. 3621, Springer-Verlag, pp. 293–308.

Kerckhoffs, A. 1883, La cryptographie militaire, *Journal des Sciences Militaires*, 9, Jan. 1883, 5–83; Feb. 1883, 161–191.

Lee, J., Lim, D., Gassend, B., Suh, G.E., Dijk, M., and Devadas, S. 2004, A Technique to Build a Secret Key in Integrated Circuits for Identification and Authentication Applications. Symposium on VLSI circuits, 2004.

Lim, C. and Korkishko, T. 2005, mCrypton—A Lightweight Block Cipher for Security of Low-Cost RFID Tags and Sensors, in Proceedings of the 6th International Workshop on Information Security Applications, Springer Lecture Notes in Computer Science, Vol. 3786, 2005, pp. 243–258.

Menezes, A.J. 1993, *Elliptic Curve Public Key Cryptosystems*, Kluwer Academic Publishers, Boston, MA.

Menezes, A., van Oorschot, P. and Vanstone, S. 1997, *Handbook of Applied Cryptography*, CRC Press, Boca Raton, FL.

Morgan, T.P. 2005, Sun Creates World's Smallest SSL Web Server, http://www.computerwire.com/industries/research/?pid=C55355B9-B6CD-42EC-80BC-ACFDA6F2CDD3.

Oren, Y. and Shamir, A. 2006, Power Analysis of RFID Tags. Available at http://www.wisdom.weizmann.ac.il/~yossio/rfid/.

Piramuthu, S. 2006, HB and related lightweight authentication protocols for secure RFID tag/reader authentication, CollECTeR Europe Conference, Basel, Switzerland, 2006.

Ranasinghe, D.C., Leong, K.S., Ng, M.L., Engels, D.W., and Cole, P.H. 2005, A distributed architecture for a ubiquitous item identification network, Seventh International Conference on Ubiquitous Computing, Tokyo, Japan, Sept. 2005.

Ranasinghe, D. and Cole, P.H. 2006a, Confronting security and privacy threats in modern RFID systems, Asilomar Conference on Signals, Systems, and Computers, California, October 2006.

Ranasinghe, D. and Cole, P.H. 2006b, Security in low cost RFID, Auto-ID Labs, White Paper Series on Anti-Counterfeiting and Secure Supply Chain, 2006.

Ranasinghe, D.C. 2007, New directions in advanced RFID system, PhD dissertation, School of Electrical and Electronic Engineering, The University of Adelaide, Australia.

Ranasinghe, D.C., Devadas, S., and Cole, P.H. 2007, A Low Cost Solution to Authentication Based on a Lightweight Primitive, Auto-ID labs white paper series on anti-counterfeiting, 2007.

Rueppel, R.A. 1989, Good stream ciphers are hard to design, in *Proceedings of the International Carnahan Conference on Security Technology*, pp. 163–174.

Schneier, B. 1994, *Applied Cryptography: Protocols, Algorithms and Source Code in C*, Wiley, New York, NY, 1994.

Schneier, B. 2006, Quasar Encryption, in Schneier on Security, A Weblog Covering Security, and Security Technology, 27th March 2006.

Stix, G. 2005, Best-kept secrets, quantum cryptography has marched from theory to laboratory to real products, Scientific American Magazine, January 2005.

Welsh, D. 1990, *Codes and Cryptography*, Oxford University Press, Oxford, 1990.

Wolf, C. and Leuven K.U. 2003, Efficient Public Key Generation for Multivariate Cryptosystems, 2003.

Wolkerstorfer, J. 2003, Hardware Aspects of Elliptic Curve Cryptography, Abstract of PhD Thesis, IAIK, Graz, Austria.

Wolkerstorfer, J. 2005, Is elliptic-curve cryptography suitable to secure RFID tags? Workshop on RFID and Lightweight Crypto, July 14–15, 2005, IAIK, Graz University of Technology, Graz, Austria.

32

*Low Overhead RFID Security**

Shlomi Dolev, Marina Kopeetsky, Thomas Clouser, and Mikhail Nesterenko

CONTENTS

32.1 Introduction

A radio frequency identification device (RFID) technology is poised to revolutionize supply-chain management and the retail industry [29]. An RFID system consists of a tag, a reader, and a database. An RFID *tag* is a miniature electronic circuit that is capable of elementary information storage, processing, and radio communication. An RFID *reader* is a device that is designed to communicate with the tag. A reader can extract the information from the tag that identifies the tagged item. The reader is connected to a *database* that contains additional information about the tag and the item. A tag can be self-powered or it can use the power of the reader to do its processing and communication through a mechanism called backscatter. Depending on the design, the range of this communication varies from a few centimeters to hundreds of meters.

* The preliminary version of some of the material of this chapter appeared in [6, 17].

As line-of-sight between the tag and the reader is not required for communication, RFID systems reduce the time and cost of processing tagged items compared to optical bar-codes. For example, purchased goods can be processed right inside a shopping cart as the customer walks through an automated checkout gate. Potential applications of RFID systems range from inventory-control to smart credit cards, automated toll collection, and counterfeit protection. To be a viable alternative to optical bar-codes, the price of an individual tag should be under ten cents [29].

Since the reader can potentially communicate with multiple tags, the problem of *singulation* arises: the reader should be able to identify multiple tags or be able to communicate with each tag individually. Juels et al. [15] propose *treewalking singulation*. The tag identifiers are arranged as leaves of a binary tree. The reader poses queries to individual bits of each tag and descends this binary tree depth-first to identify individual tags.

One of the main hurdles for the widespread adoption of RFID systems is privacy concerns. The concerns become particularly salient as the retail industry contemplates moving from pallet and crate tagging to individual item tagging [29]. RFID use substantially differs from that of other systems. The tag has a close association with the item it identifies. Moreover, the sensitive information usually does not pertain to the tag itself but to the item. This close association between the tag and the item that it identifies gives rise to novel threats such as *tracking* [22] that are not usually addressed in conventional security systems. For example, a conventional system is considered secure if the principal is capable of recognizing the intruder and aborting the communication session before the intruder is able to learn any sensitive information. However, even if the sessions are aborted, the intruder may be able to match the tag across several communication sessions. This gives the intruder the information about the location of the item or the person who carries it. The tag is often used in the environment where the intruder can easily approach it and either eavesdrop on the communication or interfere with it without the knowledge of the communicating parties [10]. For example, an RFID-enabled credit card can be read through the envelope as it is en route from the bank to its owner [11].

To be economically viable for most applications, the tag is not allowed to possess sophisticated data processing capabilities. Thus, the design of security protection for RFID systems is challenging. For example, extensive cryptosystems such as AES, DES, ECC [18], or high-quality random number generators may not be available on the tag. Hence, a substantial amount of recent research effort has been dedicated to design security techniques with sufficiently low overhead to be feasible on RFID systems.

The remainder of the chapter is organized as follows. We survey various approaches to RFID security in Section 32.2. We then present two examples of low overhead authentication algorithms. In Section 32.3, we present an algorithm for mutual tag and reader authentication—M2M that is specialized for the rental agency setting. In Section 32.4, we describe a reader-only authentication algorithm—PISP. In Section 5, we describe how low overhead algorithms such as M2M and PISP can be concurrently applied to multiple tags. We conclude the chapter in Section 32.6 with the discussion of the potential of low overhead cryptography in RFID applications.

32.2 Approaches to RFID Security

32.2.1 Using Traditional Cryptography

There is a large number of recent studies that consider RFID security. Avoine [1] maintains an extensive bibliography on the subject. Juels [13] provides a comprehensive survey of the general area. Most articles focus on RFID security using a variation of traditional symmetric or asymmetric cryptosystems [18].

However, these mechanisms may be unavailable. The manufacturing costs limit the functionality of the tags that are expected to replace optical bar-codes in retail industry. Such tags contain between 500 and 5000 gates [8] most of which are dedicated to basic operations. Only on the order of a few hundred gates are available for cryptographic operations. For example, Feldhofer et al. propose to use AES [7]. Their solution requires 3595 equivalent gates for the security component of an RFID tag. Poschmann et al. [27] present a lightweight version of DES and apply it to RFID. Their implementation requires 1848 gates. Ohkubo et al. [23] present a solution that utilizes two cryptographic hashes. Their estimate is that their solution requires from 6000 to 13,000 gates. Several other keyed hash-based approaches to secure tag authentication are proposed [4,33]. Batina et al. [3] evaluate the feasibility of using elliptic curve cryptography for RFID. They design a processor dedicated to computing ECC on a tag. Their estimate is that such processor will require either over 12,000 equivalent gates or over 8000 gates of dedicated circuits. Skiyama et al. [30] present a specialized processor for ECC cryptography which, they claim, needs only 2171 gates. Molnar and Wagner [22] analyze the security threats of library RFID tags and propose a solution that requires an RFID tag to use random numbers. They do not provide an estimate of the implementation of their algorithm. However, it is known [18, Chapter 11] that obtaining high-quality pseudorandom numbers is equivalent to producing cryptographic hashes or digital signatures. Avoine and Oechslin [2] underline the risks of inadequate pseudorandom number generation in RFID tags. Juels [12] describes a *one-time random pad* security scheme where the communication between the tag and the reader is padded by a random sequence of bits. Both the tag and the reader store this sequence. The communication is secure as long as the sequence is used only once. Because of the limited tag storage, this method restricts the frequency of communication. Thus, an alternative to conventional cryptography may be required for RFID security.

32.2.2 Using No Cryptography

The problems with consumer privacy prompted the suggestions to disable the tags after their services are no longer required. Juels et al. [15] propose to erase all information from the tag after it has been read. Certainly, this diminishes the usefulness of RFID. However, the main difficulty in this approach is to reliably verify if the tag is indeed disabled. Karjoth and Moskowitz [16] propose to physically separate the antenna from the tag. This, however, requires physical contact with the device, which diminishes the applicability of this technology.

Several researchers [9,14,15,28] propose a *blocker tag* or a guardian—a device that monitors communication to protected tags and, if necessary, blocks unauthorized inquiries. For example, Juels et al. [15] propose a blocker tag that does not allow the reader to descend past a certain depth in the singulation tree without authorization. This device answers to the reader as if the identifiers are present in every leaf of the tree thus foiling identification. However, a blocker tag requires the user to inform it which authentication requests are legitimate and which tags need to be blocked. Thus, this approach may not be appropriate for some applications.

32.2.3 Using Low Overhead Cryptography

The advent of RFID technology and the need for lightweight security renewed interest in the solutions that do not require extensive resources from one of the principals: these solutions use limited memory and relatively simple operations such as *XOR*, addition, and possibly multiplication. Such mechanisms might work in RFID security because of the specifics of the application. For example, classic cryptographic protocols are designed to be secure against open or chosen plaintext attacks [18] where the intruder needs to learn the keys and either has access to the unencrypted text or can force one of the legitimate principals to encrypt the text of its choosing. These attacks do not not seem to be applicable to RFID systems.

A number of low overhead RFID security algorithms are proposed [24–26,34]. Peris-Lopez et al. present a series of simple RFID authentication algorithms: LMAP [24], M2AP [25], and EMAP [26]. The proposed algorithms use bitwise *XOR*, *AND*, and other simple operations and require from 100 to 500 gates. However, recent publications [19,20] demonstrate that all three protocols are vulnerable to *desynchronization attack*. To maintain the freshness of its keys, the tag relies on the reader to provide them in each communication session. If the intruder sends arbitrary information to the tag, the tag may accept it and refresh its keys. This leaves the tag and the legitimate reader unable to communicate. It is also shown that the intruder may determine the keys by observing a sufficient number of communication sessions. Vajda and Buttyán [34] propose another low overhead authentication protocol. To encrypt the communication, in each session the reader provides the tag with a one-time pad. However, Defend et al. [5] show that due to random pad reuse, the protocol can be compromised by a passive intruder within 70 sessions. Even quicker cryptanalysis is possible if an active intruder is considered.

32.2.4 Application of Low Overhead Cryptography through Domain Restriction

The apparent vulnerability of low overhead cryptography appears to limit its usability for RFID security. However, we believe that such approaches are still applicable if the domain of RFID use is clearly defined. For example, libraries or other rental agencies have a rather particular routine for tag use. As the tagged items are in storage, the tags may be read for inventory purposes but the security of tag reading is not of primary concern. The tags are read at checkout. Upon return, the tagged items may have to be inspected for damage or otherwise manually checked in. Thus, the tags are exposed to the intruder only between the check-out and check-in time. At the return, there is a possibility to use a back-channel to reinitialize the tags. In Section 32.3, we present a mutual authentication algorithm M2M that is specialized for such environments. Alternatively, there may be some applications where only the reader authentication is required. In Section 32.4, we describe PISP, which is such an algorithm.

32.3 Mutual Authentication with M2M*

32.3.1 Algorithm Description

Each tag stores two square $p \times p$ matrices: M_2 and M_2^{-1}. The reader maintains another two matrices: M_2 and M_1^{-1} of the same size. The matrices M_1^{-1} and M_2^{-1} are the inverses of M_1 and M_2 respectively. The tag and the reader also share a key K which is a vector of size q, where $q = rp$. Factor r is an integer. The matrices and the key are randomly chosen per each tag.

* This algorithm was presented by Karthikeyan and Nesterenko [17].

$$A = (a_1, \ldots, a_{pr}),\ B = (b_1, \ldots, b_{pr}),\ M = (m_{p \times p})$$

$$(a_{p(i-1)+1}, \ldots, a_{pi}) = (m_{p \times p}) \begin{pmatrix} b_{p(i-1)+1} \\ \vdots \\ b_{pi} \end{pmatrix}, \text{ where } 1 \leq i \leq r$$

FIGURE 32.1
Explanation of $A = MB$ notation. (From Karthikeyan, S. and Nesterenko, M., *Proceedings of the ACM Workshop on Security of Ad Hoc and Sensor Networks*, November 2005. With permission.)

As a slight abuse of notation, we denote $A = MB$, where M is a $p \times p$ matrix and B is a vector of size q, a component-wise multiplication of M and B (see Figure 32.1). That is, each p-element component A_i of vector A, where $1 \leq i \leq r$, is obtained by multiplying M and the following elements of B: $b_{p(i-1)+1}, \ldots, b_{pi}$. Also, we assume that in our calculations the vector is always properly transposed so as to be compatible with the matrix.

Key K is selected such that product $X = M_1K$ is unique for each tag in the system. The tag information stored in the reader's database is indexed by X. A fresh key is used for every authentication session.

The authentication session has two parts: the tag identification and reader authentication. A complete session of the algorithm is shown in Figure 32.2. At first, the tag is identified by the reader. The reader initiates the session by contacting the tag. The tag replies with $X = KM_1$. After replying, the tag starts a timer. Product X uniquely identifies the tag. Thus, when the reader receives X, the reader can obtain the rest of the information about the tag and the tagged item from its database.

In the second phase, the reader authenticates itself to the tag and supplies it with a new key. For authentication, the reader proves to the tag that it is in possession of the key. To save tag resources, rather than sending the whole key back to the tag, the reader uses exclusive OR bitwise on the p-size components of K and multiplies the result by M_2. To calculate a fresh key, the reader selects unique X_{new} and obtains the key as $K_{new} \leftarrow X_{new}M_1^{-1}$. The reader sends both vectors to the tag. The tag verifies the reader's credentials and accepts the new key. In case the reader authentication fails or the reader fails to respond before the timeout expires, the tag stops further communication until reset. The tag is allowed to participate in only one authentication session at a time.

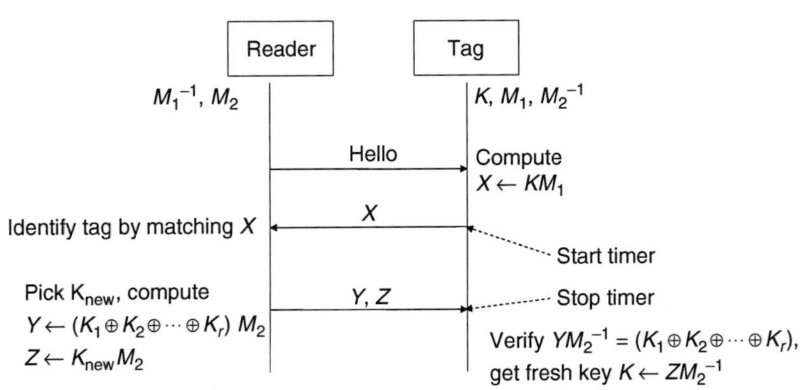

FIGURE 32.2
M2M authentication scenario. (From Karthikeyan, S. and Nesterenko, M., *Proceedings of the ACM Workshop on Security of Ad Hoc and Sensor Networks*, November 2005. With permission.)

32.3.2 Security Discussion

The security of our algorithm is based on the difficulty of recovering the multiplicand or multiplier from the product of matrix multiplication [31, Chapter 2]. Hence, the intruder cannot discover the key or the matrix used by the tag and the reader. This prevents the intruder from identifying the tag. Observe that the algorithm is only secure against known-ciphertext attacks. However, we assume that such guarantee is sufficient for RFID systems.

Let us consider the security of our algorithm against tracking. The tag does not participate in multiple concurrent authentication sessions, neither does it respond to identification requests after an unsuccessful session. Thus, there may be at most one aborted session per tag. Observe that during each session, including the single aborted session, the tag and the reader send data based on a fresh key. Since the intruder cannot decode the transmission, he cannot match the tag across multiple sessions. Hence, the intruder may not be able to track the tag.

Note we assume that the intruder is not capable of matching multiple authentication sessions of the same tags through nonradio means (e.g., by observing the tagged objects).

32.4 Reader Authentication with PISP*

32.4.1 Security Model

The proactive informational secure protocol (PISP) is based on the limited intruder capabilities. The underlining assumption of this protocol is that the intruder is not eavesdropping in at least one of each n successive interactions between the tag and the reader. The underlying assumption of PISP is that each communication session is atomic. We mean that the intruder cannot modify part of the communication in a session. The intruder may either listen to the communication during a session, or try to communicate (on behalf of the RFID tag) during an entire session. The intruder may not impersonate the tag.

32.4.2 PISP Description

The pseudocode for PISP is shown in Figure 32.3. The tag and the reader share a square $n \times n$ matrix B. Each element of this matrix contains two parts: a_{ij} and b_{ij}. The first part a_{ij} is a key assigned to the tag and the reader at initialization. The second part b_{ij} is a random number obtained by the tag and the reader during their communication.

During initialization, both the tag and the reader get a square matrix $B = (a_{ij})$ such that dim $(B) = n$ (see Figure 32.3, protocols for the tag and the reader, lines 1–5). To authenticate the reader, the tag initiates the session by sending the reader a message $s_1 = (X_1, b_1)$ (Figure 32.3, protocol for RFID Tag, lines 6–9). The message contains the following two elements: XOR of the nth column $X_1 = a_{1n} \oplus a_{2n} \oplus \cdots \oplus a_{nn}$ and a randomly generated n-dimensional vector $b_1 = (b_{11}, b_{12}, \ldots, b_{1n})$. After message transmission both the tag and the reader shift B's rows down so that $b_1 = (b_{11}, b_{12}, \ldots, b_{1n})$ is treated as the first B's row and the last row is deleted (Figure 32.3, protocol for RFID Tag, lines 10 and 11). This procedure repeats in the next authentication session. That is, the tag generates a new random n-dimensional vector $b_2 = (b_{21}, b_{22}, \ldots, b_{2n})$, calculates XOR of $(n-1)$ B's column

* This algorithm was presented by Dolev and Kopeetsky [6].

Protocol for RFID Tag

```
1:   Initialization:
2:   Define Data Structure E of
3:   int array [1..n], int XOR, E · next
4:   Create Linked List L of n
         elements of type E
5:   int i := 1;
     column = n – (i – 1)mod n

6:   Upon user request
7:       Call function
         XOR X of column i
8:       Create new random array b
9:       Send = (X,b) to Reader
10:      Call Updating procedure
11:  End user request

u1:  Updating procedure
u2:      Add b into the head of List
u3:      Remove last element of List
u4:      i := i + 1

c1:  Function
     XOR X of index column
c2:      X := 0
c3:      current := head
c4:      while current.next not equal to
         NULL do
c5:          X := X ⊕
             current.array[column]
c6:          current := current.next
c7:      end while
c8:      Return X
```

Protocol for RFID Reader

```
1:   Initialization:
2:   Define Data Structure E of
3:   int array [1..n], int XOR, E · next
4:   Create Linked List L of n
         elements of type E
5:   int i = 1;
     column = n – (i – 1)mod n

6:   Upon reception of key message
7:       Call function
         XOR Y of column i
8:       if X = Y
9:           Send "OPEN" to Tag and
             Call Updating procedure
10:      else Send "DoNotOpen"
         to Tag
11:  End of key message reception
```

FIGURE 32.3
PISP pseudocode.

elements $X_2 = b_{1n-1} \oplus a_{1n-1} \oplus \cdots \oplus a_{n-1n-1}$ and sends the newly generated message $s_2 = (X_2, b_2)$ to the reader. The reader generates the response message r_{i+1} as described above. That is, the ith authentication procedure consists of scanning the matrix columns and shifts the B's rows down so that the last matrix' B row is deleted and the vector b_i occupies the first row of B. Note that b_i is randomly generated by the tag and sent to the reader in the message s_{j-1}. See Figure 32.3 for the matrix update and *XOR* calculation procedure (protocol for RFID Tag, lines u1–u4 and c1–c8).

To verify the authentication the reader executes the following authentication procedure: after receiving message $s_i = (X_i, b_i)$, the reader verifies that X_i is the correct *XOR* of the appropriate $(n - (i - 1)(\text{mod}(n)))$th column. If so, the reader confirms the authentication by sending message $r_i = Open$ to the tag. The reader also updates B (Figure 32.3, protocol for RFID Reader, lines 6–11). Otherwise, the reader sends message $r_i = DoNotOpen$ to the tag and does not update B.

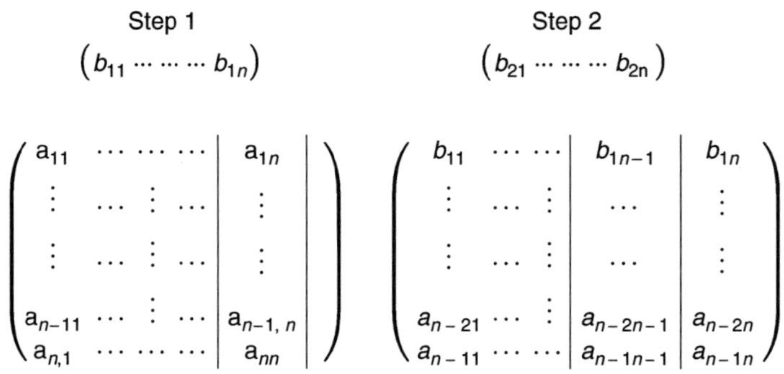

FIGURE 32.4
Operation of PISP.

Suppose that for any sequence of authentication messages of length n, at least one message is not received by the intruder. In order to compromise the authentication algorithm the intruder has to perform authentication procedure similar to the tag. To do so the intruder has to forge the key message s_{j_i} in any authentication session S_j^{th}. For that the intruder has to correctly guess the *XOR* of the corresponding $(n - (j_i - 1) \pmod{n}))$th column elements of the matrix B.

Recall that dim $(B) = n$. Let the intruder be unfamiliar with the authentication message $s_1 = (X_1, b_1)$ sent by the tag to the reader during the first authentication session (Figure 32.4, Step 1). That is, the intruder does not know the nth column of B—$(a_{1n}, a_{2n}, \ldots, a_{nn})$ and the appropriate row vector is $b_1 = (b_{11}, b_{12}, \ldots, b_{1n})$.

After the tag transmits the authentication message $s_1 = (X_1, b_1)$, $X_1 = (a_{1n} \oplus a_{2n} \oplus \cdots \oplus a_{nn})$, $b_1 = (b_{11}, b_{12}, \ldots, b_{1n})$, both the tag and the reader shift the rows of B down as described earlier. Note that in the next session the tag will send to the reader the *XOR* of the updated $(n - 1)$th B's column $X_2 = (b_{1n-1} \oplus a_{1n-1} \oplus \cdots \oplus a_{n-1n-1})$ and a new randomly generated vector $b_2 = (b_{2n}, b_{2n-1}, \ldots, b_{21})$ (Figure 32.4, Step 2). Now matrix B differs from the previous one by the newly inserted first row and the appropriate deletion of the last row.

The PISP is secure from the information-theoretic standpoint, or unconditionally secure as was defined in [32, Chapter 2]. That is, the probability that the intruder will forge the key message and successfully perform session on behalf of the reader is negligible for long enough l, where l is the number of bits of the element of B. Note also that the matrix size n factors in the security of PISP. If n is large the probability that the intruder does not overhear at least one authentication session is greater. However, large n and l require greater storage resources on the tag.

Note that if the intruder is allowed to eavesdrop n consecutive sessions, it can desynchronize the tag and the reader by forcing the reader to replace a row in B which is unknown to the tag.

32.4.3 Proactive Computationally Secure Protocol

We now describe the proactive computationally secure protocol (PCSP). It is the extension of PISP that allows us to lift the assumption on limited intruder eavesdropping. The pseudocode for the protocol is shown in Figure 32.5. As in PISP, the reader and the tag

```
Protocol for RFID Tag

1:   Initialization:
2:   Define Data Structure E of
3:   int array [1..n], int XOR, E · next
4:   Create Linked List L of n
         elements of type E
5:   int j := 1,  seed = 0
         column = n – (j – 1)mod n
6:   int keyword[k]

7:   Upon user request
8:       Call function XOR X [column]
9:       Create new random array b
10:      Create pseudo-random sequence
         (c[column]) of length m
11:      from seed = X[column] ⊕ seed
12:      Y = (b||keyword[k]) ⊕ (c)
13:      Send s = (Y) to Receiver
14:      Call Updating procedure
15:  End user request

u1:      Updating procedure
u2:      Add b into the head of List
u3:      Remove last element of List
u4:      j := j + 1

c1:  Function XOR X [column]
         of index column
c2:      X [column] := 0
c3:      current := head
c4:      while current.next not equal to
         NULL do
c5:          X [column] := X [column]⊕
             current.array [column]
c6:          current := current.next
c7:      end while
c8:      Return X
```

```
Protocol for RFID Reader

1:   Initialization:
2:   Define Data Structure E of
3:   int array [1..n], int XOR, E·next
4:   Create Linked List L of n
     elements of type E
5:   int j := 1, seed = 0
     column = n – (j – 1)mod n
6:   int keyword [k]

7:   Upon key message reception
8:       Call function XOR X [column]
9:       Create pseudo-random sequence
             (c[column]) of length m

10:      from seed = X[column] ⊕ seed
11:          Z = Y ⊕ c[column]

12:      if Z [(n+1)..m] = keyword[k]
13:          send "OPEN" to Tag and
             call Updating procedure
14:      else
15:          send "DoNotOpen" to Tag
16:  End of key message reception
```

FIGURE 32.5
Pseudocode for PCSP.

share a square $n \times n$ matrix B (Figure 32.5, protocols for the tag and the reader, lines 1–6). In addition, the tag and the reader also share a string keyword $[k]$.

During the first authentication session the tag acts as follows. As in PISP, the tag calculates the *XOR* of the nth column of B and $X_1 = a_{1n} \oplus a_{2n} \oplus \cdots \oplus a_{nn}$. Similar to PISP, a new row $b_1 = (b_{11}, \ldots, b_{1n})$ is also created as in the proactive information secure protocol case. Tag uses X_1 to initialize its pseudorandom number generator [21, Chapter 12].

The tag creates a new vector row Y_1 that it sends to the reader in the first authentication message. Y_1 is the *XOR* of the previously generated pseudorandom sequence $(c_{11}, \ldots, c_{1m})$

and b_1 concatenated with the keyword: $Y_1 = (c_{11}, \ldots, c_{1m}) \oplus (b_1 \| keyword[k])$ (Figure 32.5). Eventually, the secure information encapsulation is provided. The first key message sent from the tag to the reader during the first communication session is $s_1 = (Y_1)$ (Figure 32.5, protocol for RFID Tag, lines 7–13).

Upon receiving this message $s_1 = Y_1$, the reader decrypts it by calculating $Y_1 \oplus (c_{11}, \ldots, c_{1m})$. If the decrypted suffix of the string is equal to the predefined string *keyword*[*k*], then the reader authenticates the tag and returns the message $r_1 = Open$ to the tag. The matrix *B* updating is provided by the prefix of the decrypted string as in the basic information secure protocol. Otherwise, the reader sends message $r_1 = DoNotOpen$ (Figure 32.5, Protocol for RFID Reader, lines 7–16). The update procedure and calculation of *XOR* for the corresponding column elements of *B* is described in Figure 32.5 (protocol for RFID Tag, lines u1–u4 and c1–c8 respectively).

During any *j*th authentication session $S_j, j = 1, 2, \ldots,$ the tag sends authentication message $s_j = Y_j = (c_{j1}, \ldots, c_{jm}) \oplus (b_j \| keyword[k])$, where $c_j = (c_{j1}, \ldots, c_{jm})$ is the pseudorandom sequence generated by the seed $= X_j \oplus seed$, where the initial value of *seed* is zero (Figure 32.5, protocols for RFID Tag and for RFID Reader, line 5). X_j is the *XOR* of $(n - (j-1)(\text{mod}(n)))$th column elements, and b_j is a newly generated random vector that updates matrix *B*.

Note that the keyword and the pseudorandom number generation function can be known to the intruder. The random seed ensures the security of PCSP. The recursive reuse of the seed used in the previous communication session enhances the security of PCSP.

32.5 Tag Singulation*

Observe that the tag identification algorithms assume that the reader and the tag use the radio channel exclusively. In practice, multiple tags may potentially share the channel. However, the tags do not have sophisticated channel arbitration capabilities. In this section we discuss the scheme that augments our tag identification algorithms to enable the reader to communicate with multiple tags. Notice that the singulation proceeds concurrently with authentication. Thus, multiple tags can be authenticated concurrently.

The main change in the algorithms is in the identification phase. Recall that in this phase the reader obtains the key from the tag. In the multiple-tag version, the reader learns the keys of all the tags present. Moreover, each tag learns its key's position in the order (e.g., ascending) of the keys of the tags participating in the identification session. Once the tag knows its position, the second phase of the identification algorithm can proceed sequentially. The reader broadcasts the messages for the tags in the order of their keys. Each tag receives the message sent specifically to it and ignores the rest.

We assume that each tag is capable of broadcasting its key bit-by-bit. If multiple tags broadcast the same bit—0 or 1 simultaneously—the reader is able to receive the bit successfully. If some tags broadcast 0 and others 1, then all tags and the reader sense a message collision [15]. In case the tags are incapable of sensing the collision on their own, the reader has to notify the tags if the collision has occurred.

Reader-side singulation. Our scheme is based on breadth-first descent of the binary tree of the key-space. See Figure 32.6 for the illustration. Note that for the reader, learning the

* This algorithm was presented by Karthikeyan and Nesterenko [17].

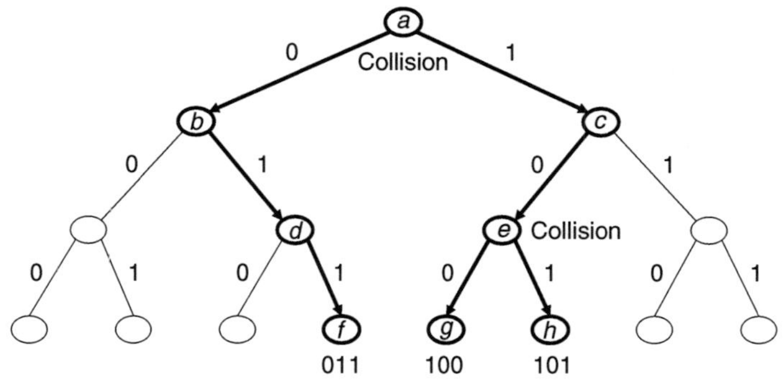

FIGURE 32.6
Example tag singulation.

tag's key is equivalent to establishing the path from the root of the tree to the particular leaf. The reader discovers this path as it descends the tree. The part of the path already learned by the reader terminates in a *growth point*. The reader iterates through growth points in a sequence of *trials*. Observe that all paths share prefixes of various lengths. The objective of the trial is to let the reader know what the next bit on the path after the growth point is and whether the paths split.

In each trial the reader requests that every tag whose key contains the path from the root to the the particular growth point sends its next bit. The reader appends the received bit to the growth point. If there is a collision, the path splits producing two growth points.

We illustrate the principle of multiple tag singulation scheme with the example shown in Figure 32.6. Assume that the key length is three bits. The tags participating in the identification session have keys: (011), (100), and (101). The reader starts from the growth point *a* which is the root of the tree. The first trial results in collision. This produces two growth points—*b* and *c*. The reader examines *b* first. The trial produces the next bit without collision, the reader moves the growth point to *d*. Then the reader examines *c* and moves it to *e*. In the next two trials the complete keys of the tags are discovered.

Tag-side singulation. The pseudocode for the algorithm executed by the tag is shown in Figure 32.7. The tag has to participate in trials as well as determine its position in the sequence of keys. To be able to do that, the tag maintains the number of growth points in front and behind the growth point that leads to its own key. The tag keeps track as to which growth point is being examined at the current trial. If there is a collision the appropriate number of growth points is incremented. After the entire tree is descended the growth points terminate in the concrete keys and the tag learns its position in the key sequence.

32.6 Conclusion

The need to adequately address RFID security and privacy is important for the technology to fully realize its potential. Certainly, this need will be addressed in part by the conventional cryptography algorithms that are adapted to use for RFID. However, for the systems

```
Initialization
constants
q: integer {key size}
k[1..q]: integer {key}

variables
collide : boolean {trial outcome}
cfront, pfront: integer, initially 0
    {currently and previously number of growth points in front}
cback, pback: integer, initially 0,
    {currently and previously number of growth points behind}

Operation

for i := 1 to q do

for j := 1 to pfront do
    collide := trial()
    cfront := cfront + 1
    if collide = true, then cfront := cfront + 1

collide := trial()
if collide = true, then
    if key[i] = 0 then
        cback := cback + 1
    else
        cfront := cfront + 1

for j := 1 to pback do
    collide := trial()
    cback := cback + 1
    if collide = true, then cback := cback + 1

pback := cback, cback := 0, pfront := cfront, cfront := 0
```

FIGURE 32.7
Tag-side singulation algorithm.

such as retail item-tagging where such heavyweight solutions are prohibitively expensive, low overhead cryptography may provide a convenient alternative.

Acknowledgments

This work was partially supported by Rita Altura Trust Chair in Computer Sciences, Lynne and William Frankel Center for Computer Sciences and Deutsche Telekom, DARPA contract OSU-RF #F33615-01-C-1901, NSF CAREER Award 0347485, and a grant from the Wright Center for Sensor System Engineering.

References

1. G. Avoine. Bibliography on security and privacy in RFID systems. Available Online, 2006.
2. G. Avoine and P. Oechslin. RFID traceability: A multilayer problem. In *Financial Cryptography (FC)*, Roseau, The Commonwealth Of Dominica, February–March 2005. *Lecture Notes in Computer Science*, vol. 3570, pp. 125–140. IFCA, Springer-Verlag, 2005.
3. L. Batina, J. Guajardo, T. Kerins, N. Mentens, P. Tuyls, and I. Verbauwhede. An elliptic curve processor suitable for RFID-tags. Cryptology ePrint Archive, Report 2006/227, 2006.
4. M. Burmester, T. van Le, and B. de Medeiros. Provably secure ubiquitous systems: Universally composable RFID authentication protocols. In *Conference on Security and Privacy for Emerging Areas in Communication Networks (SecureComm)*, Baltimore, Maryland, USA, August–September 2006. IEEE, 2006.
5. B. Defend, K. Fu, and A. Juels. Cryptanalysis of two lightweight RFID authentication schemes. In *International Workshop on Pervasive Computing and Communication Security (PerSec) 2007*, New York, USA, March 2007. IEEE Computer Society Press.
6. S. Dolev and M. Kopeetsky. Secure communication for RFIDs proactive information security within computational security. In *Stabilization, Safety, and Security of Distributed Systems (SSS), Lecture Notes in Computer Science*, vol. 4280, pp. 290–303. Springer, 2006.
7. M. Feldhofer, S. Dominikus, and J. Wolkerstorfer. Strong authentication for RFID systems using the AES algorithm. In *Workshop on Cryptographic Hardware and Embedded Systems (CHES)*, Boston, Massachusetts, USA, August 2004. *Lecture Notes in Computer Science*, vol. 3156, pp. 357–370. IACR, Springer-Verlag, 2004.
8. K. Finkenzeller. *RFID Handbook: Fundamentals and Applications in Contactless Smart Cards and Identification*. John Wiley, New York, NY, USA, 2003.
9. C. Floerkemeier, R. Schneider, and M. Langheinrich. Scanning with a purpose—supporting the fair information principles in RFID protocols. In *International Symposium on Ubiquitous Computing Systems (UCS)*, Tokyo, Japan, November 2004. *Lecture Notes in Computer Science*, vol. 3598, pp. 214–231. Springer-Verlag, 2004.
10. G. Hancke. Practical attacks on proximity identification systems (short paper). In *IEEE Symposium on Security and Privacy*, Oakland, California, USA, May 2006, pp. 328–333. IEEE Computer Society Press, 2006.
11. T.S. Heydt-Benjamin, D.V. Bailey, K. Fu, A. Juels, and T. O'Hare. Vulnerabilities in first-generation RFID-enabled credit cards. Manuscript, October 2006.
12. A. Juels. Minimalist cryptography for low-cost RFID tags. In *International Conference on Security in Communication Networks (SCN)*, Amalfi, Italia, September 2004. *Lecture Notes in Computer Science*, vol. 3352, pp. 149–164. Springer-Verlag, 2004.
13. A. Juels. RFID security and privacy: A research survey. Manuscript, September 2005.
14. A. Juels and J. Brainard. Soft blocking: Flexible blocker tags on the cheap. In *Workshop on Privacy in the Electronic Society (WPES)*, Washington, DC, USA, October 2004, pp. 1–7. ACM Press, 2004.
15. A. Juels, R. Rivest, and M. Szydlo. The blocker tag: Selective blocking of RFID tags for consumer privacy. In *Conference on Computer and Communications Security (CCS)*, Washington, DC, USA, October 2003, pp. 103–111. ACM Press, 2003.
16. G. Karjoth and R. Moskowitz. Disabling RFID tags with visible confirmation: Clipped tags are silenced. In *Workshop on Privacy in the Electronic Society (WPES)*, Alexandria, Virginia, USA, November 2005. ACM Press, 2005.
17. S. Karthikeyan and M. Nesterenko. RFID security without extensive cryptography. In *Proceedings of the 3rd ACM workshop on security of ad hoc and sensor networks (SASN)*, New York, NY, USA, 2005, pp. 63–67. ACM Press, 2005.
18. C. Kaufman, R. Perlman, and M. Speciner. *Network Security: Private Communication in a Public World*. Prentice-Hall, 1995.
19. T. Li and R.H. Deng. Vulnerability analysis of EMAP—an efficient RFID mutual authentication protocol. In *Second International Conference on Availability, Reliability and Security (AReS)*, Vienna, Austria, April 2007.

20. T. Li and G. Wang. Security analysis of two ultra-lightweight RFID authentication protocols. In *International Information Security Conference (IFIP SEC)*, Sandton, Gauteng, South Africa, May 2007. IFIP, 2007.
21. A.J. Menezes, S.A. Vanstone, and P.C. Van Oorschot. *Handbook of Applied Cryptography*. CRC Press, Boca Raton, FL, USA, 1996.
22. D. Molnar and D. Wagner. Privacy and security in library RFID: Issues, practices, and architectures. In *ACM Conference on Computer and Communications Security (CCS)*, Washington, DC, USA, October 2004, pp. 210–219, ACM Press, 2004.
23. M. Ohkubo, K. Suzuki, and S. Kinoshita. Cryptographic approach to "privacy-friendly" tags. In *RFID Privacy Workshop*, MIT, MA, USA, November 2003.
24. P. Peris-Lopez, J.C. Hernandez-Castro, J. Estevez-Tapiador, and A. Ribagorda. LMAP: A real lightweight mutual authentication protocol for low-cost RFID tags. In *Workshop on RFID Security (RFIDSec)*, Graz, Austria, July 2006. Ecrypt.
25. P. Peris-Lopez, J.C. Hernandez-Castro, J. Estevez-Tapiador, and A. Ribagorda. M2AP: A minimalist mutual-authentication protocol for low-cost RFID tags. In *International Conference on Ubiquitous Intelligence and Computing (UIC), Lecture Notes in Computer Science*, vol. 4159, pp. 912–923. Springer-Verlag, 2006.
26. P. Peris-Lopez, J.C. Hernandez-Castro, J.M. Estevez-Tapiador, and A. Ribagorda. EMAP: An efficient mutual authentication protocol for low-cost RFID tags. In *OTM Federated Conferences and Workshop: IS Workshop, Lecture Notes in Computer Science*, vol. 4277, pp. 352–361. Springer-Verlag, 2006.
27. A. Poschmann, G. Leander, K. Schramm, and C. Paar. A family of light-weight block ciphers based on DES suited for RFID applications, July 2006.
28. M. Rieback, G. Gaydadjiev, B. Crispo, R. Hofman, and A. Tanenbaum. A platform for RFID security and privacy administration. In *USENIX/SAGE Large Installation System Administration conference (LISA)*, Washington DC, USA, December 2006.
29. G. Roussos. Enabling RFID in retail. *IEEE Computer*, 39(3):25–30, 2006.
30. K. Sakiyama, L. Batina, N. Mentens, B. Preneel, and I. Verbauwhede. Small-footprint ALU for public-key processors for pervasive security. In *Workshop on RFID Security (RFIDSec)*, Graz, Austria, July 2006. Ecrypt.
31. W. Stallings. *Cryptography and Network Security: Principles and Practices*, 2nd end. Prentice-Hall, Upper Saddle River, NJ, USA, 1999.
32. D. Stinson. *Cryptography: Theory and Practice*, 3rd end CRC/C&H, 2006.
33. G. Tsudik. YA-TRAP: Yet another trivial RFID authentication protocol. In *International Conference on Pervasive Computing and Communications (PerCom)*, Pisa, Italy, March 2006. IEEE Computer Society Press, 2006.
34. I. Vajda and L. Buttyán. Lightweight authentication protocols for low-cost RFID tags. In *Second Workshop on Security in Ubiquitous Computing (Ubicomp)*, Seattle, WA, USA, October 2003.

33

Layers of Security for Active RFID Tags

Shenchih Tung, Swapna Dontharaju, Leonid Mats, Peter J. Hawrylak, James T. Cain, Marlin H. Mickle, and Alex K. Jones

CONTENTS

33.1 Introduction

Applications for RFID continue to expand into domains such as electronic passports, electronic payment systems, and electronic container seals. These applications have a risk of unauthorized access to sensitive biometric or financial information through the RFID tag or tag communication.

However, as RFID devices are intended to be small and relatively simple devices, security protocols and techniques can significantly lag behind the other details such as correctness, read rate, power consumption, etc. As such, the state of security in RFID systems is generally weak compared with other mature computational technologies such as Internet servers, shared computing workstations, and even smartcards.

RFID systems require security features to be implemented using techniques that provide a high strength of protection while not significantly increasing the complexity of the system as complexity can increase power consumption and implementation cost. Thus, it is important to develop new security techniques that take advantage of the fundamental properties of RFID communication such as physical layer protocols, low-power communication extensions, sleep modes, physical implementation variations, etc.

This chapter provides a survey of existing security techniques employed in RFID systems, including authentication and encryption as well as commonly employed types of attacks, and security features of existing RFID standards. This discussion is contained in Section 33.2. Subsequently, we present a security scheme called *layers of security* for an RFID tag that combines features from passive and active tags. Our security system takes advantage of the fundamental properties of the tag to create multiple levels of low-complexity security features, that when combined provide a strong protection. This is described in Section 33.3. Some concluding remarks are related in Section 33.4.

33.2 Security in RFID Systems

RFID systems in general and RFID tags in particular have the requirement of a power versus area tradeoff. The power consumed by the tags impacts the range of the device (passive) or lifetime of the tag (active) where the area dictates the cost of the device. In general, the industry has focused more on the area constraint than the power constraint. The 5¢ RFID tag is a famous example of an area optimized device for reduction in cost. The 5¢ tag is composed of less than 10,000 gates.

Security concerns have often been a secondary concern for vendors primarily because strong authentication and encryption algorithms are complex and would significantly increase the cost and power budget of the tag. Sarma et al. [1] discusses the security risks and challenges of low-cost RFID tags. For example, in the 5¢ tag, $<$1000 gates are dedicated to security. In contrast, commercial implementations of the Advanced Encryption Standard (AES) require $\sim$20,000–30,000 gates.

As a result, security techniques for RFID tags must leverage specific details of RFID communication to create low-overhead secure transmissions. It is important to ensure that neither the tag nor the reader is malicious with the intent to access restricted data or destroy data. To prevent this, authentication techniques for RFID are described in Section 33.2.2. Additionally, once a reader and tag have entered into a trusted communication, it is important to prevent a malicious device from overhearing the communication and stealing data. To prevent this, encryption techniques for RFID are described in Section 33.2.3.

33.2.1 Types of Attacks for RFID Systems

An intrusion or an attack to a computer on a traditional network has a physical limitation due to physical network links. An intruder attempting to perform an attack requires either some physical connection to the network through tapping a network cable or access to some terminal on the victim's network. Those attacks may be traceable by detecting traffic over particular network links or tracing with IP addresses, etc. However, a malicious access to computing nodes in an open RF environment, such as RFID network, wireless sensor network, or other wireless or RF network is not easy to be detected and much more difficult to track. For example, data travels over the air, which is easily detectable by many nontrusted devices. As a result, there is a high demand to increase the security and privacy for RFID networks. To understand potential attacks to RFID networks we have summarized several common attacks in the following subsections.

33.2.1.1 Eavesdropping

An unauthorized observer attempts to capture the exchanged information between the reader and tags without permission. This is a passive style attack. RFID communication occurs over the air and is thus an easy target for an eavesdropper to monitor data transmitted without being detected. As a result, sensitive corporate or personal information may be captured, recorded, and analyzed by a malicious attacker. Complex data encryption and decryption algorithms would be suggested as good approaches to prevent this type of attack. Unfortunately, complex data encryption such as AES encryption is often too expensive for low-cost or passive RFID tags (or both).

33.2.1.2 Spoofing

In the spoofing attack, a malicious device forges an existing tag with a cloned tag that can interact in a trusted manner by a reader. For example, after employing the eavesdropping attack, an unauthorized third party may be able to analyze and reverse engineer the captured encrypted information. Often, this information is easy to crack as the encryption is typically simple because of the limited design complexity of these tags. After the attacker breaks the architecture or the secret key of the encryption algorithm, they are capable of cloning the tag to deceive a legitimate reader.

This type of attack is able to defeat access control [2], allow unauthorized access to personal property and systems such as e-payment [3]. Mutual authentication and complex data encryption are suggested for preventing this type of attack. As before complex data encryption is often too expensive for many RFID systems. When adding mutual authentication as a requirement, the cost is increasingly prohibitive.

33.2.1.3 Denial of Service

In computer networks, a denial-of-service (DoS) attack attempts to paralyze computers so that they are not available for a legitimate access. One DoS attack is to flood a network in order to disrupt services to legitimate users or occupy the most of the network bandwidth. SYN flood and PING flood are two well-known DoS attacks in computer networks.

From the RFID system perspective, a DoS attack may cause either RFID readers or RFID tags to malfunction [4]. In consequence, it will cause a tag temporary or permanent paralysis where the tag becomes unreadable or untraceable. This type of attack may cause problems to applications such as automated inventory processing (e.g., shopping) or tracking military shipments.

The DoS attack can be easily performed on active RFID tags where tags utilize batteries. The energy of battery will be exhausted at a drastic rate when an attacker surreptitiously transmits a large number of RFID commands.

33.2.1.4 Brute Force

A brute force attack is an approach to guess a secret key of an encryption algorithm or to gain an access of a computer resource by exhaustively trying all possible keys. For example, to search the 56-bit secret key of a DES encryption system, a hacker needs to try 256 possible keys, meaning 255 trials (36 × 1015 trials), to find the match. To exhaustively compute this is easily accomplished within ~7 days as illustrated by the study of Kumar et al. [5].

Researchers at the Johns Hopkins University Information Security Institute and RSA Laboratories (JHU-RSA) successfully demonstrated the vulnerability of a 40-bit RFID-enabled cryptographic tag [3]. The Digital Signature Transponder (DST) by Texas Instruments utilizes a 20-round unbalanced Feistel cipher encryption algorithm with 40-bit secret key [3]. This is the RFID technology behind ExxonMobil speedpass, a wireless payment system tied into a credit card. The research group at JHU-RSA built a parallel Field Programmable Gate Array (FPGA) platform to perform a brute force attack to crack the 40-bit secret key. By utilizing parallel computation of 16 FPGA devices, the 40-bit secret key can be recovered on average in <1 h. This group suggests that a 40-bit cipher key is vulnerable for brute-force attack and a 128-bit cryptographic algorithm is recommended for commercial and homeland security applications.

33.2.1.5 Relay

The relay attack, also called the "man in the middle" attack, is an example of a spoofing attack where a third party deceives the first legitimate party thinking he is the second legitimate party and deceives the second party into thinking he is the first party. In effect the third party relays the message from one party to the second and vice versa. Thus, the two authorized parties intending to communicate with each other are tricked into a three party conversation without being aware of the third party. Two relay-attack examples are shown in Refs. [6,7].

The longest distance to make this type of attack is ~50 m [7]. However, the relay attack is limited by the distance between the reader and the tag. For example, as the distance between the tag and reader increases, the time of relaying signals by the third party between the reader and the tag increases as well. According to the ISO 18000 Part 6C standard, the reader waits for the tag response within 77 μs after transmitting a Query to the tag. Otherwise, the reader will terminate this communication attempt. Therefore, a possible prevention for this type of attack is to set a more strict time requirement to prevent this attack even at reasonably close distances.

33.2.1.6 Side Channel

Unlike previous attack types, in a side channel attack a cryptanalyst exploits some information measured externally from the device [8], such as timing [9], power [10], or electromagnetic [11,12] details. Side channel attacks based on power analysis have seen particular interest for RFID and similar systems.

Power analysis techniques measure or monitor power consumption of a cryptosystem by physically probing a ground pin [10]. These analysis techniques have been classified into two sets: simple power analysis (SPA) and differential power analysis (DPA) [10]. The accuracy of performing SPA can be easily affected by the signal-noise-ratio (SNR) on the probed pin. However, the DPA technique not only measures power consumption of a cryptosystem, but also requires a statistical analysis based on the collected power consumption information. Thus, it can minimize the effect of SNR because it differentiates output power traces.

The statistical technique of DPA works using a *correlation function* that indicates the correlation between using a guessed key and the unknown secret key of the cryptosystem. The key is subdivided into sub keys and each permutation of the sub key is tested while the remainder of the key is held constant. If the correct guessed sub key is used, the correlation function will indicate high correlation (e.g., spikes of differential power traces) between the correct guessed sub key and the actual secret sub key. At the end the entire correct key is composed of the high correlation sub keys. By using this technique the number of keys to try is reduced by orders of magnitude over a brute force attack.

In 1999, Kocher et al. [10] announced a DPA attack against a Data Encryption Standard (DES) coprocessor. This work describes DPA traces with correct and incorrect guessed sub keys against a secret key of the DES coprocessor and shows the actual power spikes when the correct sub key is guessed. DPA attacks have been proven effective on various cryptosystems such as an AES ASIC coprocessor [13], AES hardware-based implementations [14,15], and a DES FPGA implementation [16]. To increase resistances of DPA, many protection countermeasures are proposed [14,17,18].

For RFID systems, Rakers et al. [19] describe a DPA attack on an ISO 14443 RFID enabled "contactless" smartcard. In 2002, Messerges et al. [20] illustrate an implementation of a similar DPA attack on smartcards. Rakers also describes an approach to protect the contactless smart card from DPA attacks by exploiting an isolation circuit. An isolation circuit is primarily used on an ASIC to prevent bit error rate degradations due to digital interference. It behaves like a current source that is independent of power consumption of the digital circuitry. Thus, the power signature of the digital circuitry is reduced by a factor of 2000 or 66 dB, which significantly increases the difficulty to perform DPA.

33.2.2 Authentication Techniques in RFID Systems

Authentication is a mechanism or process to either verify the identification of a party with which communication occurs or to verify the integrity of received messages without their having been modified. For example, symmetric encryption mechanism provides a form of authentication among those who share the same secret key. In addition, asymmetric encryption provides authentication and confidentiality among those who share public keys.

An alternative approach to authenticate a message is the use of a hash function. The hash function maps a variable-length message into a fixed-length hash value. It is easy to convert a message into a hash value, but it is difficult to recover the message from a hash value without appropriate information. An advanced hash function, such as Secure Hash Algorithm, combines hash function with secret keys.

33.2.2.1 Lightweight Authentication Protocols

Vajda and Buttyan [21] describe several lightweight authentication protocols for RFID tags, including encrypting a challenge message using XOR, subset, square, RSA, and KNAPSACK. For example, in the XOR authentication protocol, a reader and a tag initially share two different, independent random keys, k_R^0 and k_T^0, respectively. The RFID reader initiates the authentication protocol, shown in Equation 33.1, by transmitting an initial challenge message of x^0 XOR k_R^0 to the tag. The tag retrieves the challenge message by computing XOR with k_R^0 and responds with another challenge message computed by XOR with k_T^0 to the reader. The reader extracts x^0 and compares with the one it transmitted to the tag.

$$\begin{aligned} R \rightarrow T &: x^0 \oplus k_R^0, \\ T \rightarrow R &: x^0 \oplus k_T^0. \end{aligned} \tag{33.1}$$

For subsequent runs of the protocol, x^i and k_R^i are random numbers generated by the reader. The reader transmits not only the encrypted challenge message but also the k_R^i XOR k_R^{i-1}. Equation 33.2 details the protocol for XOR. The other four authentication protocols utilize a similar flow but apply additionally complex encryption methodologies to encrypt challenge messages.

$$\begin{aligned} R \rightarrow T &: x^i \oplus k_R^i \| k_R^i \oplus k_R^{i-1}, \\ T \rightarrow R &: x^i \oplus k_T^0. \end{aligned} \tag{33.2}$$

Peris-Lopez et al. [22] propose a new lightweight mutual authentication protocol (LMAP) technique targeting area optimized implementation for low-cost devices such as RFID tags. The LMAPs is decomposed into four steps: tag identification (TID), mutual authentication, index updating, and key updating. The implementation of their proposed protocol needs 86, 173, 346, 691, and 1037 gates for 8, 16, 32, 64, and 96-bit authentication messages, respectively.

Bernardi et al. propose a new architecture of a UHF RFID transponder compatible with the ISO 18000 Part 6C standard and extending the system with secure authentications. Additionally, they proposed a multilayer network protocol for ISO 18000 Part 6C. The architecture includes a microprocessor, output control unit, memory blocks, I/O encoder and decoder, and an encryption module. A simplified asymmetric RSA encryption algorithm is implemented in the encryption module, which encrypts 16-bit input plain text with a 1024-bit fixed secret key and provides 1024 bits of encrypted data. For a 90 nm HMOS process, a design running at 1.9 MHz required 1.4 mm^2 of area and consumed ~1.5 mW of power.

Bolotnyy and Robin [23] proposed a hardware-based authentication approach based on physically unclonable functions (PUFs). PUFs take advantages of variations in silicon processes and operating conditions to increase authentication strength. The PUF is based on a silicon random number generator [24–26]. The value generated from the silicon random number generator depends highly on the wire delay, temperature, thermal gradients, etc., of the physical device. The proposed RFID architecture combines the PUF with a universal hash function [27]. The proposed 64 PUF hash function circuit is composed of 64 single-bit units and 1 oscillating counter circuit. Each single-bit unit requires an estimated 8 gates and the oscillating counter circuit are estimated at 33 gates, which when extrapolated to a 64-bit PUF hash function requires 545 gates.

Leung et al. [28] proposed the use of a cryptographic nonce as the tag identifier designed to avoid the tracing and cloning style attacks. A cryptographic nonce is a one-time-use number. Because the identifiers are used only once, it is not possible to trace the tag with its identifier unless the sequence of identifiers is known to the attacker. Similarly, it is not possible to clone the device. However, it is relatively easy for the tag and reader to be desynchronized through a denial of service attack.

33.2.2.2 Symmetric Authentication Protocols

A symmetric authentication protocol is an authentication mechanism for those who share the same secret key to verify authenticated messages. In describing the tradeoffs of security versus cost of RFID tags, Weis et al. [29,30] proposed hash-lock and randomized hash-lock authentication protocols. As hash-lock authentication uses hash-based access control using one-way hash functions. By comparing an internal "metaID" with the value of a hash function of a received key from the reader, the tag unlocks itself to the reader upon a match.

The randomized hash-lock authentication protocol [29,30] extends the hash-lock technique to include randomly generated numbers as identifiers in the exchange. For example, a tag is equipped with a random number generator. When the tag is queried it responds with a pair $(r, h(ID_k \| r))$ where r is a random number, h is a hash function, and $ID_k \| r$ is the kth ID concatenated with the random number. The reader performs a hash function of all ID in the database concatenated with r until it finds a match. To unlock the tag, the reader transfers ID_k back to the tag.

33.2.2.3 Asymmetric Authentication Protocols

An asymmetric authentication protocol utilizes a public and private key to perform the authentication process. For example, the reader transmits an authentication message to the tag encrypted by the tag's public key. The tag decrypts the message with its private key. The tag responds with the same authentication message back to the reader encrypted by the reader's public key. The reader then decrypts it with its private key and checks it against the message it initially sent. General asymmetric encryption implementations such as Elliptic Curve Cryptography (ECC) are exponential computations inefficient for pervasive computing environments [31].

Niederreiter asymmetric encryption is an existing asymmetric encryption algorithm that does not require exponential complexity but requires matrix operations [32]. Cui et al. [31] introduced a lightweight asymmetric authentication protocol designed for RFID devices. This technique employs the Niederreiter asymmetric encryption technique to perform a challenge response authentication protocol by using public and private keys. The public and private keys are built in Niederreiter encryption algorithm by generating several matrices and an (n,k)-linear code.

33.2.2.4 Strong Authentication Protocols

Feldhofer proposed a strong authentication technique for RFID based on AES [33]. First the reader generates a random number N_{reader} and sends it to the tag. The tag returns a random number N_{tag} and both the N_{tag} and N_{reader} encrypted with the AES secret key. The reader decrypts and verifies the encrypted N_{reader} is the same as originally sent. It then reverses the order of N_{reader} and N_{tag} and returns this, encrypted, back to the tag. If the reader decrypts the correct N_{tag}, then the reader and tag continue with their transaction.

For this implementation the system for AES encryption is an 8-bit coprocessor for and thus requires only 3600 gates. However, it requires ~1000 clock cycles to accomplish each encryption operation thus requiring ~10 ms to computation time. From a power consumption perspective, this implementation of the design utilizes 0.35 μm technology and requires 8.15 μA at 100 kHz.

33.2.3 Encryption Techniques in RFID Systems

Authentication provides a procedure to verify the identity of the parties to communication. In the case of RFID, this includes tags and readers. However, after the authentication has been completed, the readers and tags transmit insecure information. As the communication is wireless, it is very easy for a snooping device to pick up and steal this information. In many cases this information may be of a sensitive nature, such as biometric data, identification numbers such as credit card or social security numbers, corporate infrastructure details, etc.

To secure the information exchanged between readers and tags, data encryption is needed to protect these systems. However, current commercial data encryption modules may be too expensive in terms of area (cost) and power (range or lifetime) to be feasible for an RFID system. Therefore, several efforts have studied how to find "low-cost" encryption techniques suitable for RFID systems.

33.2.3.1 Symmetric Key Encryption

A symmetric key encryption mechanism utilizes a single private key shared among readers and tags to encrypt or decrypt a message or data. The length of the key controls the level of security for this encryption quantified by the time taken to break the encryption with a brute-force attack. The length of key also correlates with the implementation complexity of the security technique. For example, a modern 128-bit symmetric-key encryption core costs more than 100,000 gates in area [34] while performing high throughput of encrypted data with the strength of 2^{128} possible key values. However, for low-cost RFID systems, an RFID tag has much stricter area constraints, making this algorithm infeasible. Thus, implementations for RFID systems must trade off encryption strength against implementation complexity.

The Tiny Encryption Algorithm (TEA) [35] has been proposed for low-cost RFID data encryption [36]. A 32-bit TEA algorithm has been implemented and synthesized for a 0.35 μm CMOS technology. The implementation required 0.21 mm^2 of area and could achieve a maximum clock frequency of 50 MHz. The dynamic power dissipation is estimated at 7.37 μW with a 25.6 kHz clock.

The TinyAES architecture is based on AES-128 for RFID systems [33]. The data path of the hardware-based architecture is 8-bits wide. The silicon implementation result shows that the TinyAES module draws a current of 3.0 μA and consumed the power of 4.5 μW when operated at 100 kHz and 1.5 V. The core needs an area of 0.25 mm^2 on a 0.35 μm CMOS technology, which compares roughly to 4400 gates.

Another approach describes a tradeoff between area and timing of an AES encryption implementation [37]. This implementation was introduced to reduce timing requirement of AES encryption by optimizing the data processing of the algorithm. The 8-bit hardware-based architecture implemented in 0.25 μm CMOS technology has a gate count of 3868 and requires 870 clock cycles at 10 MHz. However, the power dissipation information for this implementation is not provided.

33.2.3.2 Asymmetric Key Encryption

An asymmetric encryption algorithm relies on one key, called a public key, for encryption and a different, but related key, called a private key for decryption [38]. Thus, to send an encrypted message, first the receiver provides the sender with its public key. The sender encrypts the message with the public key and the receiver decrypts the message with its private key. It is important not to be able to determine the private key from the public key. To increase security by adding an authentication component, to verify that the message came from a trusted sender, it would be required to decrypt the message with the receiver's private key and sender's public key.

Asymmetric key encryption requires more complex algorithms compared with symmetric key encryption for comparable encryption strengths, but also provides many advantages such as not having to divulge and share a secret key and the capability to manage the system for individual devices and groups of devices. For RFID systems, implementing an asymmetric encryption scheme is a significant challenge.

ECC is a type of public-key cryptography proposed for low-cost RFID systems [39] and in particular for ISO 18000 Part 6C [40]. The ECC processor is composed of a control unit, arithmetic unit (ALU), and memory (RAM and ROM). The ECC synthesized in 0.25 μm CMOS technology shows that the ALU requires 6300–7800 gates, depending on the size of the Galois Field adder GF(2^m), where m is the length of the key. It can also be much larger depending on the size of the memory block.

The study of Leung et al. [28] describes an implementation of a 173-bit ECC crypto-processor based on an Optimal Normal Basis (ONB) methodology. This implementation is designed for a low-power inductive RFID application such as the ISO 14443 contactless smartcard standard, which operates at 13.56 MHz. The 173-bit ECC crypto-processor is capable of operating at 18 MHz and executing an ECC encryption process within 7.56 ms and requires 95 mW with a 3.3 V power supply.

33.2.4 Security in RFID Standards

33.2.4.1 ISO 18000 Part 7

The ISO 18000 Part 7 standard [41] is the most popular protocol for active ultra high frequency (UHF) systems. To prevent malicious access to the tags, a password style authentication mechanism is provided in the standard by the `set password`, `set password protect`, and `unlock` commands. The `set password` command sets an internal password required for further tag accesses. The `set password protect` command enables or disables password authentication. The `unlock` command allows unprotected access to the tag. Unfortunately, password protection is not a strong authentication technique, and passwords can easily be stolen and spoofed.

33.2.4.2 ISO 18000 Part 6C

The ISO 18000 Part 6C standard [42] is a recent adoption by ISO of the Class 1 Generation 2 "Gen 2" RFID specification [43] for passive UHF tags from electronic product code (EPC) global, Inc. Even prior to its standardization by ISO, Gen 2 tags have become very popular and widely used for applications requiring passive RFID.

Neither the ISO standard nor the Gen 2 specification contain any particular requirements for security in compliant tag or reader implementations and typical implementations employ minimal security features.

The interrogator or reader can lock or unlock each individual area of memory. This includes access to the access or kill passwords, the electronic product code (EPC) memory bank, or TID memory bank. When the tag is locked, the passwords cannot be read or modified and all other memory banks are write protected. Additionally, the reader can *permalock* the lock status for a password or memory bank so that it is not changeable. As the name implies, one permalock is asserted to a particular memory block, and it cannot be changed. Finally, the tag can be permanently deactivated if the tag receives the `kill` command with the correct password. After the tag is killed it no longer responds to interrogator commands. Interestingly, killed tags are actually alive and can still receive passive power and actually buffer incoming packets, but never respond.

Juels proposes an authentication protocol to combat tag cloning even in environments with untrusted readers [44]. This protocol assumes an authenticated tag that contains its own unique EPC identifier and its own 32-bit secret key. The readers authenticated to communicate with the tag contain a database of secret keys associated with the authenticated tags. During a transaction, a tag identifies itself and the reader verifies the identifier with its database. The reader responds with the appropriate secret key. Finally, the tag verifies the secret key from the authenticated reader. Unfortunately, if any adversary eavesdrops during this authentication process the tag identifier and secret key can be captured. Data encryption is required to protect this technique for this information leakage.

33.2.4.3 ISO 14443

The ISO 14443 standard [45] is the most popular standard for high frequency proximity cards or what are also called contactless RFID cards or contactless smartcards about to the size of a credit card. ISO 14443 compliant devices typically have a range of $\sim$3 inches and cost $<$\$5. The devices are passively powered. The ISO 14443 standard does not specify any specific cryptographic algorithm or authentication protocol. However, security features, such as authentication and encryption, available in the smart card standard ISO 7816 [46] are also compatible for ISO 14443 devices [47]. The ISO 7816 smartcards, which are in contact with the readers when they are accessed, can be directly powered. Thus, the passively powered ISO 14443 devices have a significantly lower power budget, making many of these cryptographic algorithms difficult to implement.

Mandel et al. [2] describe how to use a \$30 device to perform a relay attack on proximity cards that adhere to the ISO 14443 standard. The technique takes advantage of the passive authentication feature employed by these devices rather than more secure alternatives such as active authentication and a challenge–response protocol. The broadcast signal from the proximity card is captured and then relayed using a \$30 AM transmitter to a gate that subsequently opens. In addition to being able to recreate the signal, after snooping several proximity ID and broadcast signals, the passive authentication structure in the card was easily reverse engineered.

Kfir and Wool [6] analyzed the man-in-the-middle attack approach to perform relay attacking on the RFID-enabled contactless smartcard. In their approach they use a theoretical model to describe the behavior of the device and show that it is possible to perform relay attacks when the authorized tag and reader are in relatively close proximity to each other.

Hancke implemented prototypes called a "Mole" and "Proxy" system to perform a practical relay attack against an ISO 14443 type-A contactless smartcard [7]. In this experiment, the distance of the relay attack is shown to be as far as 50 m.

The MiFare microprocessor [48] from Philips, a leading manufacturer of ISO 14443 chips, contains a crypto-coprocessor capable of processing 3DES algorithms in hardware. 3DES is a variant of the DES encryption algorithm originally proposed by IBM [49], which avoids brute force attack vulnerabilities of DES and meet-in-the-middle vulnerabilities of double DES.

33.2.4.4 ICAO Extension to ISO 14443

Personal information contained within a travel document, credit card, driver's license, or other identification is often extremely sensitive and results in concerns about information privacy. One of the applications that employs the ISO 14443 standard is electronic passports (E-passports). As this application raises a significant privacy concern, efforts have been made to add additional security to ISO 14443 specifically for this application.

Juels et al. [50] employ basic access control to prevent the RFID chip from being accessed by an unfamiliar reader unless the reader can prove it is authorized. To accomplish this, they employ a digital signature cryptographic technique based on public key cryptographic algorithms [38]. However, as these devices are passively powered, the power budget for operating the tag is limited, which also limits the strength of authentication that can be employed. In this implementation the key length is very small, making it relatively easy to defeat. Additionally, the key is hard wired into the device, making it quickly obsolete.

E-passport cryptography is further addressed in the Machine Readable Travel Documents (MRTDs) specification published by the International Civil Aviation Organization [51]. The ICAO requires a baseline security technique of passive authentication based on RSA, DSA, or a similar variant. It also goes further to describe techniques for active authentication, basic access control, extended access control, and data encryption.

The passive and active authentication measures ensure that the content of the RFID chip in the E-passport has not been modified or that the RFID tag has been improperly replaced or forged. Extended access control is a protection protocol to prevent unauthorized access to biometric information. Data encryption prevents against a third party listening to the transmitted biometric information.

33.2.4.5 ISO 18185

The ISO 18185 standard [52] describes RFID tags that can be used for electronic container seals. Electronic container seals allow the tracking of a container all the way through a supply chain to verify that the items have not been opened, or tampered with. When the container is initially filled with cargo and sealed the ISO 18185 compliant tag controls a locking bolt that is put in place. After the container has been sealed it may not be resealed. Once the container reaches its destination it is opened by breaking the bolt and the RFID tag moves to the "opened" state, where it remains and cannot be changed. After use the tag is disposed of.

The container seal tags contain no significant security features; data, such as cargo contents, is transmitted in clear text. The only form of authentication is the use of a unique tag identifier to avoid replication. The 18185 tag is susceptible to a snooping and spoofing attack. For example, an attacker can listen to a tag access to get the unique information.

Then they can break the seal and replace it with a spoofed tag that uses the same unique ID as the tag placed prior to tampering.

33.2.4.6 ANSI NCITS 256

The American National Standards Institute (ANSI) published a standard for active RFID systems (ANSI NCITS 256) [53]. It defines an equivalent air interface characteristics and specifications to the ISO 18000 Part 7 standard, including the same preamble signal and data format. The differences from ISO 18000 Part 7 deal with the types of RFID commands. Except the collection command, there are 13 data commands in ANSI 256 standard. However, none of them is used for access control, authentication, or data protection. ANSI NCITS 256 is vulnerable to eavesdropping, tag cloning, data disclosure, and other standard attacks.

33.2.5 Conclusions about Existing Techniques

Based on the study of existing standards and security techniques, we can conclude that the state of security in RFID systems has several key problems that must be addressed. Security in existing standards is fairly minimal, when it exists at all. Often, it is left to designers of standard compliant systems to integrate their own security methods as they see fit. Additionally, the cost and power requirements of RFID tags often make it difficult to incorporate strong security in the tags. As a result, many techniques propose lightweight authentication or encryption, which may be broken relatively easily. Stronger authentication and encryption are suggested, but often have high power budgets or area requirements (increasing cost) that exceed what is possible for RFID.

In the next section, we describe a security alternative called *layers of security* that addresses the strength of security and the power budget issue for a passive active hybrid RFID tag system. This tag does tradeoff an increase in area for savings in power and improvements in performance. However, in comparison of active tags, which typically are significantly more expensive, this increase in cost is nominal. This approach differs with the existing approaches described here as it leverages details of how the RFID tag communicates to find low power consumption security extensions that protect the data and tag access.

33.3 Layers of Security

Our approach for securing RFID transactions is to employ security at multiple levels during the RFID transaction. These levels are applied to different layers in the communication scheme similar to the layers in the Open System Interconnection (OSI) Model. As such, our system assumes as the baseline architecture the passive active RFID tag (PART) [54] described in Chapter 11 and the use of a design automation technique to program the architecture [55–57] described in Chapter 3. Thus, our approach provides layers of security in the final RFID system, including (1) the passive activation layer (burst switch), (2) the active communication encoding (physical layer), (3) the use of encrypted data in the communication primitives (specified with the RFID design automation), and (4) physical security protection.

33.3.1 Passive Activation Layer Security

The burst switch alone is not sufficient to replace the active receiver for RFID tags. The simple presence of RF energy such as noise or a communication with other wireless

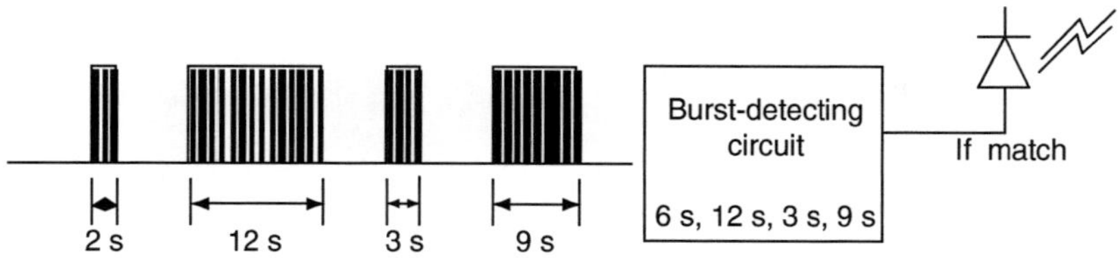

FIGURE 33.1
Example encoding for the burst switch. The wake-up signal contains four bursts of 2, 12, 3, and 9 time units, respectively.

systems or even a malicious reader can cause the tag to wake up the active transceiver. To solve this problem we have developed a signal encoding methodology using both hardware and software prototypes. The RFID reader generates bursts of energy of different durations like those shown in Figure 33.1. In the example from Figure 33.1, the reader generates four pulses with lengths of 2, 12, 3, and 9 time units. The tag must detect a unique code from these bursts so as to activate the remainder of the tag. The software-based system is implemented with a PIC microprocessor [58]. The hardware-based system is designed for implementation in an ASIC or SoC. The strength of the encoding is related to two components: the number of bursts in the sequence n and the unique number of different burst lengths detectable by the receiver b. Thus, the resulting number of unique codes is nb. The main components of the detection circuit, shown in Figure 33.2, are two counters and a comparator. The first counter detects the value of the burst determined by the burst length requiring [lg b] bits and the second to track which burst is being checked in the sequence requiring [lg n] bits.

The clock speed of the circuit depends on the detection precision of the burst switch. For example, if we consider the detection of a 6 μs burst with a granularity of 1 μs we clock our circuit at 10 MHz. However, as shown in Figure 33.3 with the solid line, a 6 μs burst was detected with nonzero probability for bursts ranging from 5.1 to 6.9 μs. This can be

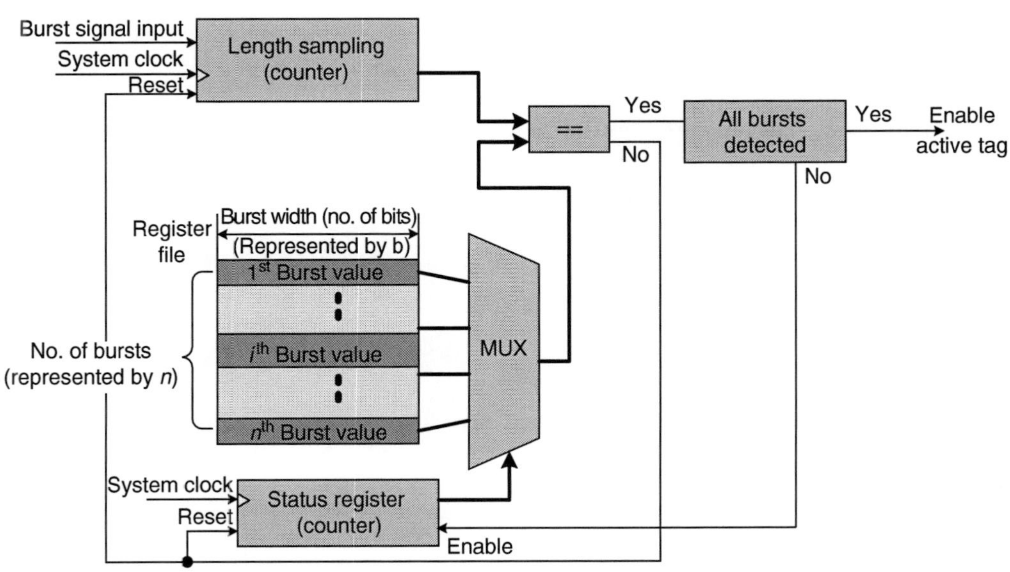

FIGURE 33.2
Architecture of the burst sequence detection circuit.

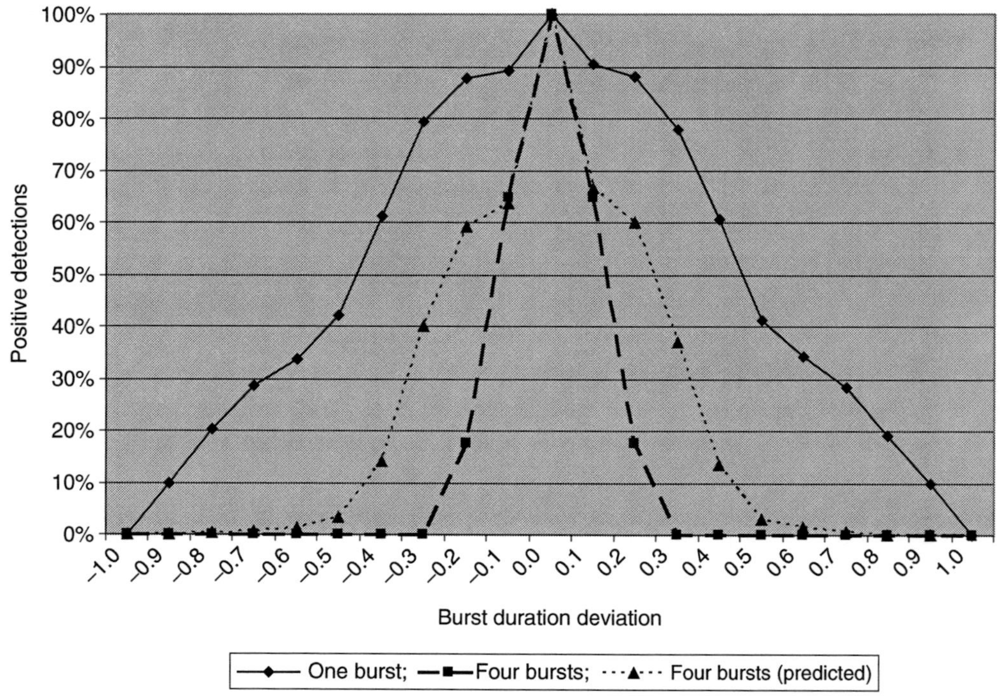

FIGURE 33.3
Percentage of false positives for one burst and four bursts detection between ±1 μs and granularity of 0.1 μs.

corrected by oversampling of 10× (e.g., clocking at 100 MHz instead of 10). Unfortunately, increasing the clock speed by 10× also increases the power consumed dramatically. However, only at 6.0 μs does detection occur with 100% probability.

Thus, rather than oversample, we consider the impact of increasing the number of pulses in the sequence. The heavy dashed line from Figure 33.3 shows the impact of placing four bursts together with the same deviations. Interestingly, the false positives drop off much more quickly than predicted (dotted line), which still predicts as high as 40% false positives after the actual number of false positives have already dropped to zero. By including a small amount of oversampling (2–3×) and breaking the burst detection into multiple bursts, it is possible to see negligible false positives and keep our power budget at a minimum.

Once the proper burst switch encoding has been received, the active transceiver is wakened to begin the active stage of communication. The encoded burst switch prevents a malicious reader from executing a denial of service attack on the passive–active tag to dramatically drain the battery.

33.3.1.1 Results

To determine the minimum time increment for differentiating a different length pulse, we prototyped the digital portion of the hardware with Spartan 3 FPGAs and connected the generator and detector with a wire. Figure 33.3 shows the percent of reads that were detected as 6 μs for values ranging from 5 to 7 μs. In this experiment the deviation was ~1 μs. By considering four pulses, this deviation drops to 0% for 0.3 μs. When testing with the Lynx transmitter and receiver we found that the clock speed should be reduced significantly below 1 MHz as the resolution of the transceiver is at least an order of magnitude (100 kHz) slower [59].

TABLE 33.1

Hardware Burst Switch Detection Circuit Implemented in 0.16 μm Technology at 1.8 V

	2 bit	4 bit	8 bit	16 bit	32 bit
Area (no. of cells)	42	52	78	120	205
Area (μm^2)	124	149	235	327	566
10 MHz Clock Power (μW)	14.95	17.82	24.33	33.96	53.45
1 MHz Clock Power (μW)	1.78	2.21	2.85	3.84	5.86
100 kHz Clock Power (μW)	0.17	0.22	0.28	0.38	0.58
Gated Clock Power (μW)	0.11	0.17	0.22	0.32	0.55

The burst switch detector was implemented in ASIC hardware using Synopsys Design Compiler targeting 0.16 μm Oki cells and examined for power consumption using Synopsys PrimePower at 10 MHz, 1 MHz, and 100 kHz system clock rates. The number of unique representations for each burst was represented using 2, 4, 8, 16, and 32 bits and for each experiment the number of bursts was held constant at four. The area and power results are shown in Table 33.1. For comparison, the same functionality was implemented in a Microchip PIC12F635 8 bit ultra low power microprocessor requiring 200 μW of power [58]. Even operating at the highest clock speed of 10 MHz, the power consumption for the largest design is less than 30% of the PIC. However, a 100 kHz clock speed design, closer to matching the capability of the transceiver, requires 300× less power than the PIC when processing. Interestingly, at such low clock speeds, the clock gated circuit provides little power advantage.

33.3.2 Physical Layer Security

Active RFID tags generally communicate using some form of Manchester encoding. Manchester encoding combines data communication with a synchronization clock. Each bit is contained within a *window* in the signal that contains a transition in the middle. If the transition goes from high to low (falling edge), the bit is a "0." Similarly, if the transition goes from low to high (rising edge), the bit is a "1." Differential Manchester encoding considers the same window, while again, there is always a transition in the middle. However, the bit value is determined if there is a transition between the end of one window and the beginning of the next. If there is a transition, the bit is a "0," and if there is no transition the bit is a "1."

Figure 33.4 provides an example of Manchester and Differential Manchester encoding of the bit vector "011001." The interesting thing about these two different encodings is that they appear superficially the same; however, unless you know which code is being used, it is impossible to determine the encoding just by examining the data itself. For example, in Figure 33.4, if a Differential Manchester encoding of "011001" is read as Manchester code, it appears to be "101110," which is entirely unrelated to the original value.

To further protect data in our system, we mix the use of Manchester and Differential Manchester encoding in the same data sequence. A key will be used to specify the encoding employed, for example, a "0" would imply Manchester encoding while a "1" would suggest Differential Manchester encoding.

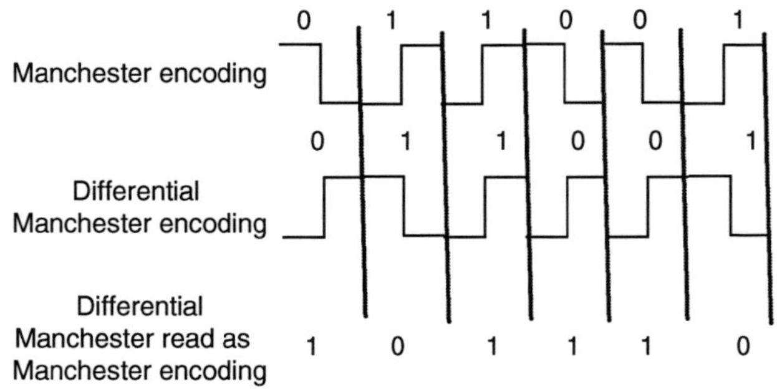

FIGURE 33.4
Example of the bit-vector "011001" encoded as Manchester encoding, Differential Manchester encoding, and Differential Manchester encoding read as Manchester encoding.

To convert Manchester or Differential Manchester code into bit-vectors, the signal is sampled so as to detect the transition in the middle or beginning of the window, respectively. To extract the values from a signal that combines Manchester and Differential Manchester code in the same signal it is necessary to sample for transitions at both the beginning of the window and the middle of the window, simultaneously. This requires doubling the sampling rate over either encoding alone. Additionally, it will be necessary to create logic to detect the existence or absence of a transition at the beginning of the window and the direction of the transition in the middle of window to determine the value. However, the key can be used as an enable signal to these blocks, so the inactive block does not consume power.

The architecture of the decoder is described in Figure 33.5. The input signal is sampled to detect the Manchester value and Differential Manchester value for the same window. The Manchester circuit keeps the decoder in synchronization as it detects the transition that occurs in the middle of the window with the direction of this transition dictating the "Manchester" value. It also sends a synchronization signal to the Differential Manchester

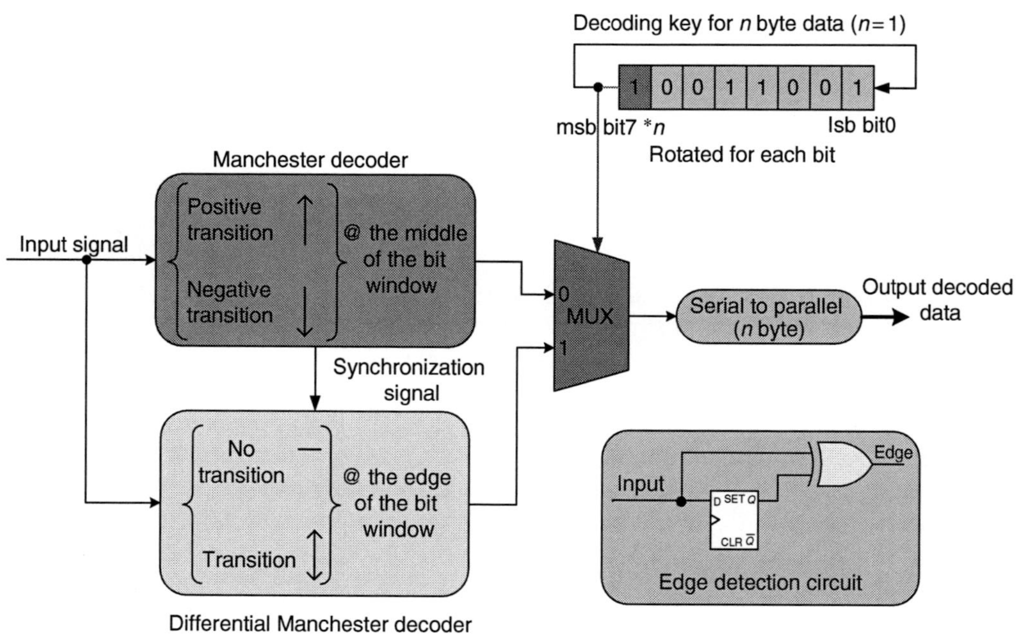

FIGURE 33.5
Encoder and decoder block for combining Manchester and Differential Manchester encodings using a key.

decoder to tell it when to sample for the beginning of the next window. This is necessary as there is no guarantee for a transition between the windows. The existence or lack of a transition dictates the "Differential Manchester" value. The transition detection is completed by some variation of the *edge detection circuit*. This circuit compares the current and previously sampled value and generates a "1" if there is an edge or a "0" if there is not an edge. The direction of the edge can be detected by checking the current or previously sampled value.

The final decoded value is selected by the key, where a "0" means select the Manchester value and a "1" means select the Differential Manchester value, indicated with the multiplexer in the figure. The bits arrive serially and are grouped into bytes for processing in the tag controller. Keys may be updated by adding new primitives into the RFID system using the design automation approach described in Chapter 3.

33.3.2.1 Results

We built a Manchester decoder in 0.16 μm Oki cell-based ASIC hardware synthesized with Synopsys Design Compiler and profiled for power with Synopsys PrimePower. The design was analyzed by simulating the design at 500 kHz. The selected appropriate sampling rate for a 27.7 kHz signal is 500 kHz, which is the standard data rate for ISO 18000 Part 7 [41]. This design is based on the Manchester decoder developed for a power saving smart buffer for active tags described in Ref. [55]. The design was extended as shown in Figure 33.5 to include concurrent Differential Manchester decoding and selection based on a key. The results of these two designs are summarized in Table 33.2. The overhead of adding Differential Manchester decoding increases the design area by ~42% and the power consumed by ~38%.

33.3.3 Encrypted Data Transmission with AES

Once the correct encoding has been received by the burst switch and the physical layer has been successfully decoded creating a packet comprised of several bytes of data, the data must be decrypted. The AES algorithm from the National Institute of Standards and Technology (NIST) was selected for our third layer of security. Our implementation goal for AES was to create an implementation that was low-energy to allow the longest battery life possible for the tag. As a result, we employed the SuperCISC compiler [60] to assist with the hardware generation for the encryption and decryption completed in the tag.

The SuperCISC compiler generates extremely energy-efficient hardware descriptions from C applications. The primary concern for active tags is battery life. It is often inconvenient or impossible to change a battery. Additionally, active tags are typically much more expensive than passive tags, costing on the order of $100, compared with several cents for passive tags. Silicon area in the tag controller design impacts the per part cost of

TABLE 33.2

Overhead for Adding Differential Manchester Decoding Using a Key to a Manchester Decoder

	Area (No. of cells)	Area (μm^2)	Power (μW)
Manchester Decoder alone	466	3780	2.886
With Differential Manchester	664	5386	3.996

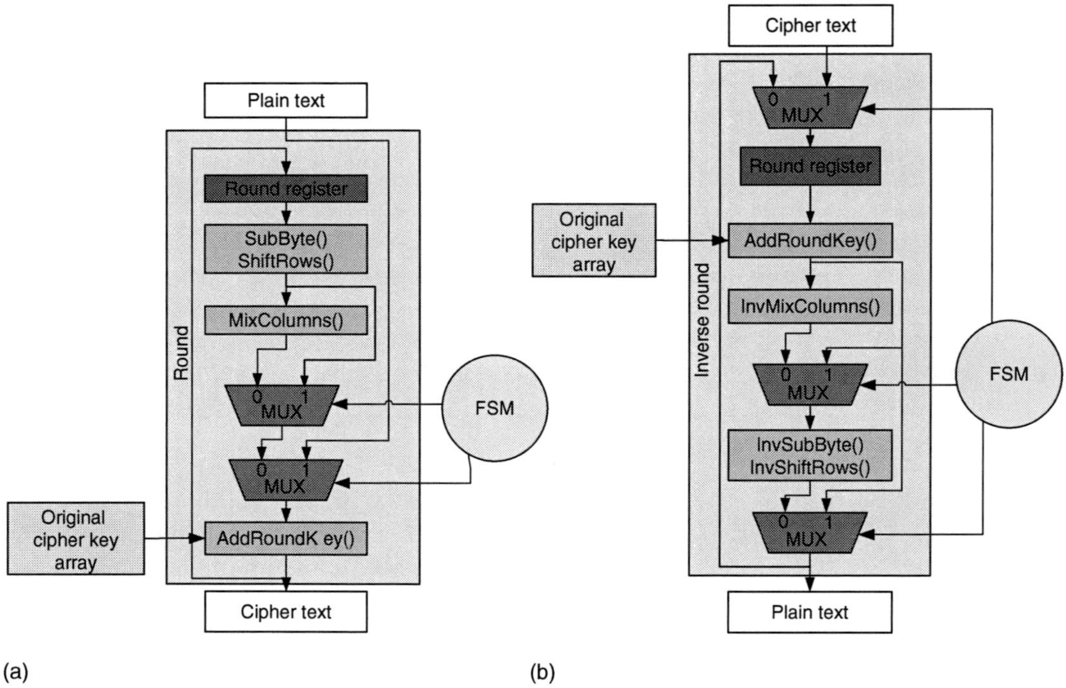

FIGURE 33.6
AES architectural overview: (a) encryption and (b) decryption.

the chip that goes into the tag. The SuperCISC approach often trades area increases for energy savings. However, with the cost of active tags being so high, a slight increase in tag cost versus a longer battery life is easily acceptable.

An overview of the AES implementation approach is shown in Figure 33.6. A total of five blocks for the C implementation of AES were synthesized using SuperCISC. For encryption, AES requires 12 rounds of computation with four functions occurring back to back: `SubByte ()`, `ShiftRows ()`, `MixColumns ()`, and `AddRoundKey ()`, with `AddRoundKey ()` being executing prior to these rounds and all functions with the exception of `MixColumns ()` being executed once after the rounds. Thus for encryption, three blocks were synthesized: `AddRoundKey ()`, which is primarily a Galois Field Adder, the combination of `SubByte () and ShiftRows ()`, which simplify to table lookups and routing, and `MixColumns ()`. Using a simple finite state machine (FSM), the resulting encryption block is shown in Figure 33.6a. Decryption requires complementary functions `InvMixColumns ()` and the combination of `InvSubByte ()` and `InvShiftRows ()` in addition to the `AddRoundKeys ()` block from encryption. Decryption is implemented in a similar manner to encryption, as shown in Figure 33.6b.

Figure 33.7 shows core synthesized hardware for the `MixColumns ()` block. For ease of visualization, only a single iteration is shown here. `MixColumns` and its decryption complement `InvMixColumns ()` are the most complex blocks in the AES algorithm. Even this function is built from relatively simple computations, constant modulus, constant increment, compare against zero and select, table lookups, and exclusive OR. The SuperCISC compiler provides a low latency combinational implementation of all these blocks to avoid the power overheads of a highly pipelined implementation with a possible increase in area. However, the performance of these blocks far exceeds the relatively slow communication speeds of ~30 kHz employed by active tags.

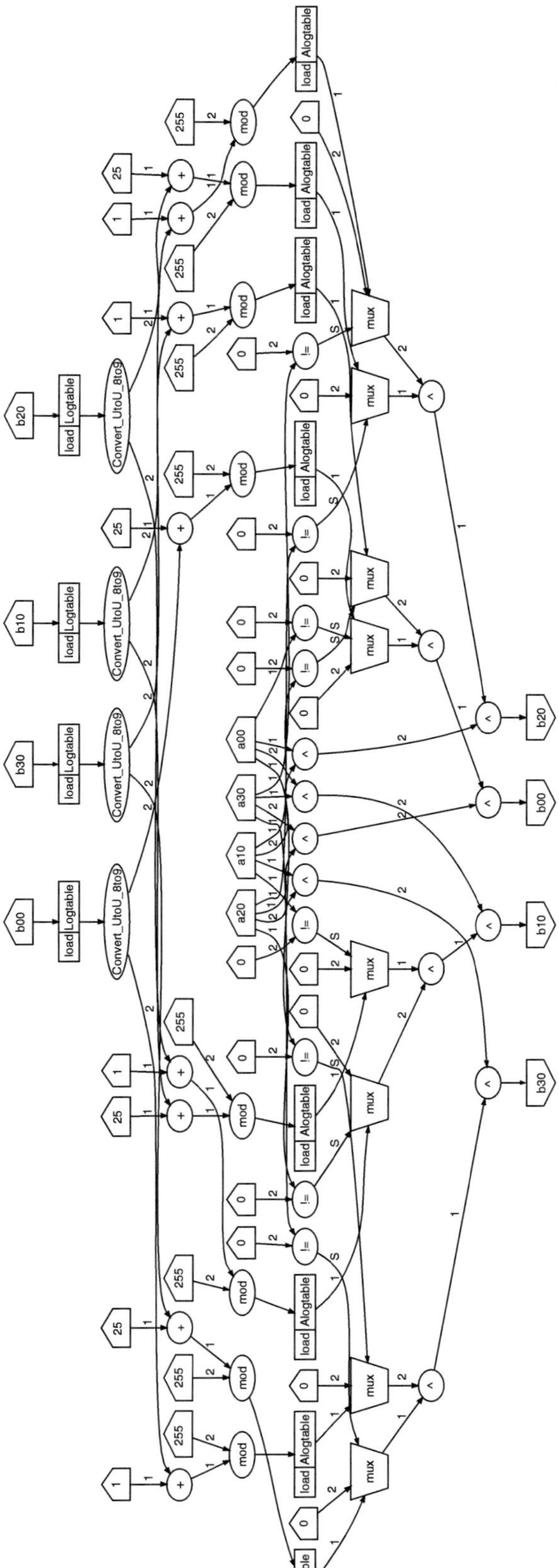

FIGURE 33.7
One loop iteration of the `MixColumns ()` function from AES encrypt.

33.3.4 Physical Security

RFID tags are often deployed in environments that do not provide physical security for the device. This opens the RFID tag to a class of attacks that involve analysis of the tag power consumption to determine the secret key stored in the device. These types of power attacks have been effective for breaking the encryption of smart cards [20]. The first class of attack is SPA, which involves correlating the power consumption of the device to the input provided [9]. The second class of attack is DPA, which uses statistical analysis to help discover the secret key [10]. While both SPA and DPA require knowledge of the encryption algorithm's behavior, SPA is only effective when the algorithm has a control flow that depends on the secret key, where DPA is effective (independent of whether the control flow depends on the secret key [17]), making it a much more powerful attack. As a result, we consider the effectiveness of DPA attacks on the RFID tag and discuss methods to protect against these attacks.

DPA attacks attempt to discover the secret key by subdividing the key into subgroups of a limited number of bits and examining power traces for all the combinations of each of these subgroups. For example, in 128-bit AES, the 128-bit key is subdivided into 32 groups of 4 bits each. During DPA, a subgroup is examined between "0000" and "1111," with all the remaining groups set to a fixed value. The power analysis compares a known implementation, that is, using the guessed key value with the system to be broken where the internal value of the key is unknown. DPA takes the difference between these two implementations (averaged over the course of many input text samples) for each guessed key value where on average the power differences cancel out; however, there are spikes when key values match between the two systems.

A more rigorous explanation is based on the descriptions in Refs. [10,20]. The AES algorithm is executed with N values of plain-text input for each guessed key. A discretized power output signal $S_i[j]$ is recorded for each of the guessed keys, where i is a particular plain-text input and j is a particular key. For AES, $0 \leq i < N$, with N typically being 512, and $0 \leq j < 32$, where the upper bound is the product of the number of discrete keys per group and the number of groups. Using the partitioning function $D(\cdot)$, the power traces are subdivided into sets where the guessed key contains a "0" or "1" as follows:

$$S_0 = \{S_i[j] | D(\cdot) = 0\},$$
$$S_1 = \{S_i[j] | D(\cdot) = 1\}. \quad (33.3)$$

The average power for each guessed "0" and "1" is computed as $A_i[j]$ as follows:

$$A_0[j] = \frac{1}{|S_0|} \sum_{S_i[j] \in S_0} S_i[j],$$
$$A_1[j] = \frac{1}{|S_1|} \sum_{S_i[j] \in S_1} S_i[j]. \quad (33.4)$$

The DPA bias signal, $T[j]$, is the difference of the average power:

$$T[j] = A_0[j] - A_1[j]. \quad (33.5)$$

$T[j]$ is the value that shows spikes when the correct secret key is guessed.

TABLE 33.3

Power, Energy, Throughput, and Area of the AES Hardware Block Implemented in Oki 0.16 μm Cell-Based ASIC Design

AES Mode	Clock Speed	Power	Energy (nJ)	Rate
Encryption	Area, 0.46 mm^2			
	10 MHz	4.62 mW	5.08	1 Gb/s
	500 kHz	0.19 mW	4.18	5 Mb/s
	100 kHz	0.047 mW	5.17	1 Mb/s
	10 kHz	4.056 μW	4.46	106 Kb/s
Decryption	Area, 0.72 mm^2			
	10 MHz	7.05 mW	7.76	1 Gb/s
	500 kHz	0.28 mW	6.16	5 Mb/s
	100 kHz	0.069 mW	7.59	1 Mb/s
	10 kHz	7.031 μW	7.73	106 Kb/s

33.3.4.1 Results

The AES encryption and decryption components were implemented in 0.16 μm Oki cell-based hardware synthesized using Synopsys Design Compiler and simulated at different speeds with Mentor Graphics Modelsim. The power consumption was analyzed using Synopsys PrimePower. The results are reported in Table 33.3. The results show that while there is a large variation in power consumption dependent on the speed, the energy consumption for the entire AES operation is relatively similar varying between 4.18 and 5.17 nJ for encryption and 6.16 and 7.76 nJ for decryption. If the minimal energy solution was selected, it would be the 500 kHz speed allowing a throughput of 5 Mb/s, which matches the clock speed for the Manchester decoder. However, for line speeds of 27.7 kHz as is defined by the ISO 18000 Part 7 standard for active tags, the 10 kHz clock yielding a throughput of 106 Kb/s is more than sufficient.

This design may also be possible for passive tags, which operate on a very strict power budget. Typically, passive tags consume a few microwatts of power. The AES design at 10 kHz is already in the microwatt range, and if the speed is reduced to ~30 kHz, the power may be reduced by as much as 3.5–4 times to 1–2 μW.

33.3.4.2 Resistance to the DPA Attack

It has been shown that the predominant power consuming component of the AES algorithm is the Galois Field Adder, $GF(2^n)$, which is typically implemented as an XOR function. The $GF(2^n)$ adder is from the `AddRoundKey ()` function, and the SDFG for this function is shown in Figure 33.8 as parallel XOR gates. To ensure that our RFID tag is not susceptible to DPA, it is necessary to mask the power signature of this component so that it does not change in the predicted manner dependent on the key.

The standard technique to accomplish the power signature obfuscation is to calculate each bit of the $GF(2^n)$ (XOR) calculation as if the bit is a "0" and a "1" and then use the actual key value to select which result is actually used in the computation. While this method obfuscates the power consumption, it does require twice as much work increasing the area and power consumption of the device. Our goal is to protect the implementation from DPA while minimizing the area and power increase.

The DPA technique presented by Kocher [9,10] assumes that the encryption algorithm is run on a microprocessor and that based on the output testing, the test can determine when the application is processing the portion of the key containing the 4-bit window currently being tested. By using the SuperCISC technique for our implementation, the entire key

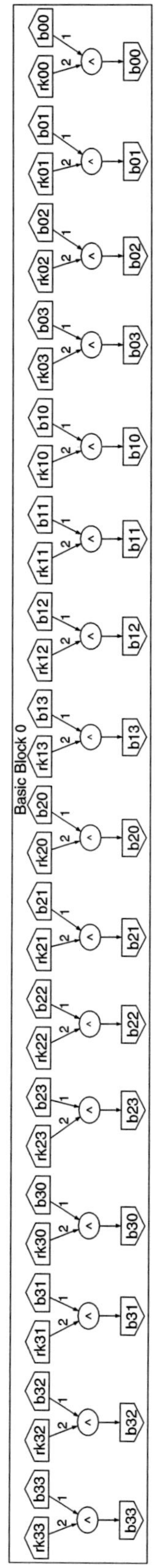

FIGURE 33.8
`AddRoundKey()` SDFG, which is basically a 2^n Galois Field Adder.

is operated on in parallel, making it impossible to distinguish the operations on the 4-bit window from the other 124 bits of the key.

The DPA attack was applied to our AES design holding the secret key 0x2B7E151628AED2A6ABF7158809CF4F3C. Table 33.4 shows the DPA analysis (power difference) between our design and a reference design. In the table the rows represent the differences for a particular 4 bit window of the key. The bold values represent the maximum power difference with a particular key and the asterisk number is the actual key. Based on the results, the predicted key would be 0x4B33D3CD2112179A628B3A11672FB053, which is different from the actual key.

To prove that there is no relationship between the actual key and the predicted key we performed a *t*-test between the maximum difference magnitude and the difference magnitude when the key matched the secret key. For our test, we used an α value of 0.05 and the resulting *p*-value was <0.01, which shows that there is sufficient resolution for the maximum difference and that the maximum difference is unrelated to the difference from guessing the correct key. We also calculated the number of bits that differ between the key with the maximum difference and the correct key. The average number of bits that differ is

TABLE 33.4

Differences for Each 4 Bit Window of the 128 Bit Key

	Difference in μW for the Guessed Key															
Window	**0**	**1**	**2**	**3**	**4**	**5**	**6**	**7**	**8**	**9**	**A**	**B**	**C**	**D**	**E**	**F**
0	−1	6	5*	1	**12**	5	1	6	1	0	−7	2	−1	−4	−1	−3
1	5	4	6	1	−4	0	−12	−100	3	−1	5	**−108***	1	3	6	−1
2	7	8	5	**−9**	−2	−1	0	−4	2	3	4	0	−4	3	1*	−4
3	7	8*	5	**−9**	−2	−1	0	−4	2	3	4	0	−4	3	1	−4
4	1	3*	−1	−4	−2	−1	0	−1	−4	3	4	−6	−4	**9**	2	−1
5	7	8	5	**−9**	−2	−1*	0	−4	2	3	4	0	−4	3	1	−4
6	7	8*	6	−4	−2	−1	−1	0	2	3	4	−8	**9**	6	1	−4
7	−1	4	9	−5	−4	−1	2*	−4	2	3	3	7	−1	**−8**	−7	−5
8	−1	−1	**6***	−3	−2	−1	5	−4	−4	−3	0	3	−4	−5	1	−4
9	7	**8**	5	−1	−2	−1	0	−4	2*	3	4	0	−4	3	1	−4
10	2	**−9**	1	0	4	7	0	3	8	−7	3*	4	−6	4	−1	−2
11	−6	−8	**−9**	−8	−4	−5	7	5	5	7	4	−3	−4	**−9**	−6*	−7
12	7	**8**	**8**	−3	−2	−3	0	−4	2	3	4	0	−4	3*	1	−4
13	7	8	5*	−4	−2	−1	−2	**−9**	−1	2	4	0	3	1	3	−4
14	−2	2	−1	−5	8	−8	−5	−2	0	**−9**	−5*	4	−2	2	0	−5
15	−1	−1	1	0	3	2	6*	−2	0	−6	**−9**	6	7	5	−2	3
16	6	4	2	−1	−2	3	**9**	2	1	−3	−7*	−7	−3	0	2	−1
17	−2	−8	**10**	4	−2	0	5	1	−1	−2	−1	−6*	−1	7	1	−6
18	−1	−1	5	−4	−3	1	0	5	**9**	3	−1	1	−7	−3	−1	1*
19	−2	−4	10	−1	−5	7	2	−9*	2	4	−1	**12**	−4	1	−5	−7
20	−9	5*	6	**11**	0	−2	7	−2	−4	8	−1	5	−1	−6	3	−2
21	−2	−8	−1	−2	1	−1*	−1	6	2	−5	**10**	6	6	1	**10**	1
22	3	**−11**	9	−4	−2	−6	4	0	2*	1	−8	0	−6	−4	−3	−3
23	3	**8**	−4	−5	0	−2	−1	−4	−3*	0	−3	4	1	2	1	**−8**
24	3*	4	−5	−6	5	3	**11**	3	4	−9	6	8	−8	0	−3	−1
25	−6	2	7	−5	4	−5	−1	**−11**	0	−5*	−5	2	1	0	2	−3
26	1	0	**−7**	−6	4	2	4	−2	−1	1	6	−6	0*	−3	−2	4
27	−1	0	−6	0	3	5	−3	**7**	−4	−1	3	1	−6	2	2	**7***
28	2	3	−4	−2	−5*	1	−3	5	1	1	2	**6**	3	3	2	−2
29	**9**	−4	2	7	−1	0	1	0	4	−5	1	2	−1	7	3	−1*
30	5	2	2	1*	−7	**−8**	0	−1	−2	0	**8**	6	3	−1	1	5
31	−2	−5	−6	**9**	2	−2	1	0	−7	2	−2	5	2*	1	1	4

Note: Maximum difference is in bold and correct key is denoted with a *.

2.00, with a standard deviation of 0.95. The median is 2. This shows that even considering the deviation the result is a positive number of bits typically ranging from 1 to 3 bits that differ between the correct and maximum difference guess. This confirms that the guess from DPA of our AES design is truly unrelated to the actual correct key.

33.4 Conclusions

In this chapter we have described an overview of the state of the art of security in RFID systems. In particular we have outlined the classes of security applied to RFID, including various classes of authentication and encryption. We have also described the types of attacks employed on RFID systems and the facilities and extensions for securing RFID standards.

In this light, we have presented a secure transmission methodology for PART, where our primary concern is functionality and power (battery lifetime). Our technique provides layers of security as multiple levels of defense against attack. The passive activation layer is encoded to prevent a malicious reader from continuously activating the tag to drain its battery or to attempt to issue malicious commands. The physical layer is encrypted using a mixture of Manchester and Differential Manchester encoding. Finally, the data itself is encrypted using the AES implemented to avoid the key from being broken using DPA.

These security modifications are implemented with a minimal power overhead from the actual RFID transaction power required. The burst switch detection circuit requires <1 μW when operating at 100 kHz even for 32-bit values (0.58 μW). The modified Manchester, Differential Manchester decoder requires ~1 μW of additional power (1.11 μW). Finally, the AES operation requires a maximum of ~7 μW at 10 kHz (7.031 μW), which is more than sufficient for the data transmission speed. Based on the active tag primitive logic controller implemented in the same hardware, depending on the number of primitives included (up to 40), the power consumed is 63–65 μW [57]. Thus, the overhead for the security is just under 9 μW (8.721 μW), which is 13.4% increase over the active tag logic alone.

References

1. S.E. Sarma, S.A. Weis, and D. Engels, "Radio-frequency-identification security risks and challenges," *CryptoBytes*, 6(1): pp. 2–9, 2003.
2. J. Mandel, A. Roach, and K. Winstein, "MIT proximity card vulnerabilities," Tech. Rep., Massachusetts Institute of Technology, March 2004.
3. S. Bono, M. Green, A. Stubblefield, A. Juels, A. Rubin, and M. Szydlo, "Security analysis of a cryptographically-enabled RFID device," *Proceedings of the USENIX Security Symposium*, pp. 1–16, 2005.
4. M. Burmester and B. de Medeiros, "RFID security: Attacks, counter-measures and challenges," *Proceedings of the RFID Academic Convocation, The RFID Journal Conference*, 2007.
5. S. Kumar, C. Paar, J. Pelzl, G. Pfeiffer, and M. Schimmler, "Breaking ciphers with COPACO BANA—A cost-optimized parallel code breaker," *CHES*, pp. 101–118, 2006.
6. K. Kfir and A. Wool, "Picking virtual pockets using relay attacks on contactless smartcard systems," *Proceedings of SecureComm*, 2005.
7. G. Hancke, "A practical relay attack on ISO 14443 proximity cards," Tech. Rep., University of Cambridge, 2005.

8. J. Coron, D. Naccache, and P. Kocher, "Statistics and secret leakage," *ACM Transactions on Embedded Computing Systems*, 3: 492–508, August 2004.
9. P. Kocher, "Timing attacks on implementations of Diffle-Hellman, RSA, DSS and other systems," *Proceedings of the International Advances in Cryptology Conference (CRYPTO)*, pp. 104–113, 1996.
10. P. Kocher, J. Jaffe, and B. Jun, "Differential power analysis," *Proceedings of the International Advances in Cryptology Conference (CRYPTO), Lecture Notes in Computer Science*, Vol. 1666, pp. 388–397, Springer-Verlag, 1999.
11. E. De Mulder, P. Buysschaert, S.B. Ors, P. Delmotte, B. Preneel, G. Vandenbosch, and I. Verbauwhede, "Electromagnetic analysis attack on an FPGA implementation of an Elliptic Curve Cryptosystem," *The International Conference on Computer as a Tool, 2005, EUROCON 2005*, Vol. 2, pp. 1879–1882, 2005.
12. K. Gandolfi, C. Mourtel, and F. Olivier, "Electromagnetic analysis: Concrete results," *CHES '01: Proceedings of the Third International Workshop on Cryptographic Hardware and Embedded Systems* (London, UK), pp. 251–261, Springer-Verlag, 2001.
13. S. Mangard, N. Pramstaller, and E. Oswald, "Successfully attacking masked AES hardware implementations," *Proceedings of Cryptographic Hardware and Embedded Systems (CHES)* (J.R. Rao and B. Sunar, eds.), *Lecture Notes in Computer Science*, Vol. 3659, pp. 157–171, Springer, 2005.
14. D.D. Hwang, K. Tiri, A. Hodjat, B.-C. Lai, S. Yang, P. Schaumont, and I. Verbauwhede, "AES-based security coprocessor IC in 0.18-μm CMOS with resistance to differential power analysis side-channel attacks," *IEEE Journal of Solid-State Circuits*, 41: 781–792, April 2006.
15. S.B. Örs, E. Oswald, and B. Preneel, "Power-analysis attacks on FP-GAs—First experimental results," *Proceedings of Cryptographic Hardware and Embedded Systems (CHES)* (C.D. Walter, Çetin Kaya Koç, and C. Paar, eds.), *Lecture Notes in Computer Science*, Vol. 2779, pp. 35–50, Springer, 2003.
16. F.-X. Standaert, S.B. Örs, J.-J. Quisquater, and B. Preneel, "Power analysis attacks against FPGA implementations of the DES," *Proceedings of Field-Programmable Logic and its Applications, Lecture Notes in Computer Science*, Vol. 3203, pp. 84–94, 2004.
17. L. Benini, A. Macii, E. Macii, E. Omerbegovic, F. Pro, and M. Poncino, "Energy-aware design techniques for differential power analysis protection," *Proceedings of the Design Automation Conference (DAC)*, pp. 36–41, 2003.
18. C. Gebotys, "A table masking countermeasure for low-energy secure embedded systems," *IEEE Transactions on Very Large Scale Integration (VLSI) Systems*, 14: 740–753, July 2006.
19. P. Rakers, L. Connell, T. Collins, and D. Russell, "Secure contactless smart-card ASIC with DPA protection," *IEEE Journal of Solid-State Circuits*, 36: 59–565, March 2001.
20. T.S. Messerges, E.A. Dabbish, and R.H. Sloan, "Examining smart-card security under the threat of power analysis attacks," *IEEE Transaction on Computation*, 51(5): 541–552, 2002.
21. I. Vajda and L. Buttyan, "Lightweight authentication protocols for low-cost RFID tags," *Proceedings of Ubiquitious Computing (UBICOMP)*, 2003.
22. P. Peris-Lopez, J.C. Hernandez-Castro, J. Estevez-Tapiador, and A. Ribagorda, "LMAP: A real lightweight mutual authentication protocol for low-cost RFID tags," *Workshop on RFID Security (RFIDSec)*, 2006.
23. L. Bolotnyy and G. Robins, "Physically unclonable function-based security and privacy in RFID systems," *Fifth Annual IEEE International Conference on Pervasive Computing and Communications, 2007. PerCom '07*, pp. 211–220, March 2007.
24. B. Gassend, D. Clarke, M. van Dijk, and S. Devadas, "Silicon physical random functions," *Proceedings of the ACM Conference on Computer and Communications Security*, 2002.
25. B. Gassend, D. Clarke, M. van Dijk, and S. Devadas, "Controlled physical random functions," *Proceedings of the Computer Security Applications Conference*, pp. 149–160, 2002.
26. D. Lim, J. Lee, B. Gassend, G. Suh, M. van Dijk, and S. Devadas, "Extracting secret keys from integrated circuits," *IEEE Transactions on Very Large Scale Integration (VLSI) Systems*, 13: 1200–1205, October 2005.
27. K. Yuksel, "Universal hashing for ultra-low-power cryptographic hardware applications," Tech. Rep., Worcester Polytechnic Institute, 2004.
28. P.-K. Leung, C.-S. Choy, C.-F. Chan, and K.-P. Pun, "An optimal normal basis elliptic curve cryptoprocessor for inductive RFID application," *Proceedings of the IEEE International Symposium on Circuits and Systems (ISCAS)*, pp. 309–312, 2006.

29. S.A. Weis, S.E. Sarma, R.L. Rivest, and D.W. Engels, "Security and privacy aspects of low-cost radio frequency identification systems," *Security in Pervasive Computing, Lecture Notes in Computer Science*, Vol. 2802 pp. 201–212, 2004.
30. S.A. Weis, "Security and privacy in radio-frequency identification devices," Master's thesis, Massachusetts Institute of Technology, 2003.
31. Y. Cui, K. Kobara, K. Matsuura, and H. Imai, "Lightweight asymmetric privacy-preserving authentication protocols secure against active attack," *Proceedings of the Pervasive Computing and Communications Workshop*, pp. 223–228, 2007.
32. H. Niederreiter, "Knapsack-type cryptosystems and algebraic coding theory," *Problems of Control Information Theory*, 15(2): 159–166, 1986.
33. M. Feldhofer, S. Mominikus, and J. Wolkerstorfer, "Strong authentication for RFID systems using the AES algorithm," *Proceedings of CHES 2004, Lecture Notes on Computer Science*, Vol. 3156, pp. 357–370, 2004.
34. I. Verbauwhede, P. Schaumont, and H. Kuo, "Design and performance testing of a 2.29Gb/s rijndael processor," *IEEE Journal of Solid-State Circuits*, 38: 569–572, March 2003.
35. D.J. Wheeler and R.M. Needham, "TEA, a tiny encryption algorithm," *Proceedings of the Workshop Fast Software Encryption, Lecture Notes in Computer Science*, Vol. 1008, pp. 363–366, 1994.
36. P. Israsena, "Securing ubiquitous and low-cost RFID using tiny encryption algorithm," *International Symposium on Wireless Pervasive Computing*, 2006.
37. M. Kim, J. Ryou, Y. Choi, and S. Jun, "Low-cost cryptographic circuits for authentication in radio frequency identification systems," *IEEE International Symposium on Consumer Electronics (ISCE)*, pp. 1–5, 2006.
38. W. Stallings, *Cryptography and Network Security*. Pearson Prentice Hall, 2006.
39. L. Batina, J. Guajardo, T. Kerins, N. Mentens, P. Tuyls, and I. Verbauwhede, "An elliptic curve processor suitable for RFID-tags." Cryptology ePrint Archive, Report 2006/227, 2006.
40. L. Batina, J. Guajardo, T. Kerins, N. Mentens, P. Tuyls, and I. Verbauwhede, "Public-Key Cryptography for RFID-Tags," *Proceedings of the IEEE International Conference on Pervasive Computing and Communications Workshops (PerComW)*, pp. 217–222, 2007.
41. International Standards Organization, "ISO/IEC FDIS 18000–7:2004(E)." Standard Specification, 2004.
42. International Standards Organization, "ISO/IEC FDIS 18000–6:2004/Amd 1:2006(E)." Standard Specification, 2006.
43. EPCglobal, Inc., *EPC Radio-Frequency Identity Protocols: Class-1 Generation-2 UHF RFID Protocol for Communications at 860 MHz – 960 MHz*, version 1.0.9 ed., January 2005.
44. A. Juels, "Strengthening EPC tags against cloning," *WiSe '05: Proceedings of the 4th ACM Workshop on Wireless Security* (New York, NY, USA), pp. 67–76, ACM Press, 2005.
45. International Standards Organization, "ISO/IEC 14443–4:2001." Standard Specification, 2001.
46. International Standards Organization, "ISO/IEC 7816–3:2006." Standard Specification, 2006.
47. K. Finkenzeller, *RFID Handbook: Radio-Frequency Identification Fundamentals and Applications*. Wiley, 1999.
48. NXP/Philips, *MiFare DESfire: Contactless Multi-Application IC with DES and 3DES Security MF3 IC D40*, 3 ed., 2002.
49. D. Coppersmith, "The data encryption standard (DES) and its strength against attacks," *IBM Journal of Research and Development*, 30(3): 243–250, 1994.
50. A. Juels, D. Molnar, and D. Wagner, "Security and privacy issues in E-passports," *Proceedings of Security and Privacy for Emerging Areas in Communications Networks*, pp. 74–88, 2005.
51. International Civil Aviation Organization, *PIK for Machine Readable Travel Documents offering ICC Read-Only Access*, 1.1 ed., October 2004.
52. International Standards Organization. "ISO/IEC FDIS 18185–1,4:2006." Standard Specification, 2006.
53. American National Standards Institute, "ANSI NCITS 236:2001." Standard Specification, 2002.
54. A.K. Jones, S. Dontharaju, S. Tung, P.J. Hawrylak, L. Mats, R. Hoare, J.T. Cain, and M.H. Mickle, "Passive active radio frequency identification tags (PART)," *International Journal of Radio Frequency Identification Technology and Application (IJRFITA)*, 1(1): 52–73, 2006.

55. A.K. Jones, R. Hoare, S. Dontharaju, S. Tung, R. Sprang, J. Fazekas, J.T. Cain, and M.H. Mickle, "An automated, FPGA-based reconfigurable, low-power RFID tag," *Journal of Microprocessors and Microsystems*, 31(2): 116–134, March 2007.
56. A.K. Jones, R.R. Hoare, S.R. Dontharaju, S. Tung, R. Sprang, J. Fazekas, J.T. Cain, and M.H. Mickle, "A field programmable RFID tag and associated design flow," *Proceedings of the 14th Annual IEEE Symposium on Field Programmable Custom Computing Machines*, pp. 165–174, 2006.
57. A.K. Jones, R. Hoare, S. Dontharaju, S. Tung, R. Sprang, J. Fazekas, J.T. Cain, and M.H. Mickle, "An automated, FPGA-based reconfigurable, low-power RFID tag," *Proceedings of the 43rd Design Automation Conference (DAC)*, pp. 131–136, ACM, July 2006.
58. P.J. Hawrylak, L. Mats, J.T. Cain, A.K. Jones, S. Tung, and M.H. Mickle, "Ultra low-power computing systems for wireless devices," *International Review on Computers and Software (IRECOS)*, 1(1): 1–10, 2006.
59. A.K. Jones, S. Dontharaju, S. Tung, L. Mats, P. Hawrylak, R. Hoare, J.T. Cain, and M.H. Mickle, "Radio frequency identification prototyping," *ACM Transactions on Design Automation for Electronic Systems (TODAES)* (in press).
60. A.K. Jones, R. Hoare, D. Kusic, G. Mehta, J. Fazekas, and J. Foster, "Reducing power while increasing performance with SuperCISC," *ACM Transactions on Embedded Computing Systems (TECS)*, 5(3): 1–29, August 2006.

34

Cryptographic Approaches to RFID Security and Privacy

Koutarou Suzuki, Miyako Ohkubo, and Shingo Kinoshita

CONTENTS

34.1 Introduction

Radio frequency identification (RFID) is an auto-identification technology that utilizes wireless communications, and is considered to be the next-generation bar code. The RFID system consists of wireless tags and wireless readers.

A tag consists of an IC chip and antenna, and sends information to the reader via a wireless channel. A reader receives the information sent by a tag via a wireless channel, and determines the ID of the tag from the received information.

Wireless communications between the tag and the reader usually employ the LF (Low Frequency) band (124–135 KHz), HF (High Frequency) band (13.56 MHz), or UHF (Ultra High Frequency) band (860–960 MHz), depending on the tag and reader specifications.

EPCglobal [1] establishes the specifications for the ID code called the EPC (Electronic Product Code) and the total RFID system called the EPCglobal network, which includes the tags, readers, ONS (Object Name Service), and EPCIS (EPC Information Service). EPCglobal also works on standardization, promotion, and administration of the EPC and the EPCglobal network.

RFID has the potential to be used widely in supply-chain management instead of bar codes. There are, however, several barriers to the widespread use of RFID tags. One is the privacy problem and another is the high cost of current units.

The privacy problem occurs due to the basic properties of RFID tags:

- Each tag can be identified freely using a wireless probe.
- Each tag has a unique ID enabling tag-by-tag identification.

Because of these properties, a specific consumer or object can be identified and tracked over wide areas. Because this issue is becoming a wide public concern [2–4], addressing the privacy problem is essential to the success of RFID tags. Hence, many investigations have been conducted in recent years concerning RFID privacy protection. In this chapter, we survey the studies pertaining to the security and privacy of RFID tags, especially in the context of cryptography. The cost limitation of a tag unit yields another basic property of RFID tags: each tag has weak computational power.

Because RFID tags have only weak computational power, commonly used public key cryptographic primitives cannot be used in RFID tags. Therefore, the main and specific concern of these studies is how to address the security and privacy problems in the light of these weak computational tags. The security and privacy problems of RFID tags appear in various scenarios. These are some examples.

Leakage of information: People have items that are quite personal and do not want anyone to have information related to these products, e.g., books, money, and medicine. If such items are tagged, various types of personal information can be obtained through illegal scanning of the wireless tags.

Tracing a user: People have objects that they always carry with them, e.g., a watch, glasses, and shoes. If such things are tagged, personal movements can be tracked through illegal tracing of the wireless tags.

We discuss these security and privacy issues in detail in Section 34.2, and present definitions for RFID security and privacy. Various countermeasures are overviewed in Sections 34.3 through 34.6, as well as the existing studies concerning physical approaches and cryptographic approaches for privacy, authenticity, and implementation of cryptographic primitives.

There is another survey of studies on RFID security and privacy in [5], and a bibliographic list of studies in [6].

In Section 34.7, we present our conclusions.

34.2 Security and Privacy of RFID

In this section, we discuss security and privacy requirements and issues pertaining to the RFID system from various viewpoints.

34.2.1 Authenticity and Privacy

To identify a tag, which represents the most basic function of the tag, the RFID tag scheme should have authenticity. On the other hand, to protect user privacy, the RFID tag scheme should have indistinguishability. The harmonization of authenticity and indistinguishability is an important topic in cryptographic research pertaining to RFID security and privacy.

In this section, we give the definition of authenticity and indistinguishability for RFID tags. To identify and check the validity of the tags, the tag scheme should incorporate authenticity, i.e., the reader accepts the output only from a valid tag. To avoid ID leakage and ID tracing of the RFID tag, the tag scheme should incorporate indistinguishability, i.e., one cannot distinguish two tags by observing the output of the tags.

Avoine [7] proposes indistinguishability for RFID tags, Juels and Weis [8] propose indistinguishability for a case involving correlated secrets, and Damgård and Østergaard [9] propose authenticity. Intuitive definitions for these properties are given hereafter. Further information and detailed definitions are given in the respective papers.

The security definition consists of an attack environment and attack goal, i.e., an adversary tries to achieve the attack goal utilizing the attack environment. We call this the attack game. First, we describe the attack environment for both the authenticity and indistinguishability games. There are n tags $T_1, \ldots, T_n$ and a reader R. Each tag T_i contains secret s_i, and communicates with reader R. Reader R contains n secrets $s_1, \ldots, s_n$, communicates with the ith tag T_i, and outputs "reject" or "accept the ith tag." In the attack environment, the adversary can (1) communicate with reader R, (2) communicate with tag T_i, and (3) corrupt tag T_i to obtain secret s_i, multiple times and in any order.

The attack goal of the authenticity game is for the adversary to communicate with reader R and make reader R output "accept the ith tag," providing tag T_i is not corrupted and tag T_i is not involved in the session, i.e., a man-in-the-middle attack is not allowed.

The attack goal of the indistinguishability game is for the adversary to communicate with tag T_i and T_j without seeing indices i and j and to distinguish these two tags, provided tag T_i and T_j are not corrupted.

We say an RFID scheme has authenticity/indistinguishability if and only if achieving the attack goal of the authenticity/indistinguishability game in the attack environment is computationally infeasible, respectively.

The secure RFID tag scheme must have at least authenticity to avoid impersonation, and should have both authenticity and indistinguishability to protect user privacy.

34.2.2 Forward Security

After a tagged item is thrown away, it is easy to retrieve the tag from the trash and tamper with the tag to obtain the secret information stored in the tag. Moreover, because of the cost limitation, a tamper resistant implementation cannot be expected for RFID tags. Therefore, we need to be mindful of the leakage of secret information stored in the tag, and its impact on the security and privacy of RFID tags.

Ohkubo et al. [10] point out that if the secret key in a tag is leaked due to tampering, the history of the tag can be traced using the secret key and the output records of the tag. They propose the concept of forward security for the privacy of RFID tags, i.e., even if the secret key in a tag is leaked due to tampering, the history of the tag remains untraceable. This corresponds to allowing the adversary to corrupt tags T_i and T_j, only after communicating with these two tags, in the indistinguishable game described in the previous subsection. Moreover, Ohkubo et al. propose a scheme that satisfies forward security using the hash chain technique.

34.2.3 Restriction and Delegation of Traceability

If we reuse a tag, we encounter another issue, i.e., after a tag is transferred to a new owner, the previous owner should not be able to trace the tag.

Molnar et al. [11] point out the above issue of ownership transfer, and discuss the requirements for restricting the ability to trace a tag within a certain period, to prevent the previous owner from tracing the tag after the tag is transferred to the new owner. They propose a scheme where the ability to trace the tag can be restricted to a period of time and delegated to readers.

34.2.4 Proof of Existence

The main application of RFID tags is supply-chain management, i.e., each product is tagged and traced via readers along a distribution route. Hence, another important security requirement is to guarantee the existence of a specific tag in a specific location, at a specific time, and with other specific tags.

Juels [12] proposes a protocol called the yoking-proof that provides proof that two tags are scanned by a reader simultaneously. This can be applied, e.g., to pharmaceutical distribution where medicine and the description of that medicine should be distributed together.

34.2.5 Low-Level Layer Security

In the previous sections, we focused on the security and privacy of only the high-level layer of communications between the RFID tag and reader. However, we must consider low-level layer security and privacy in a practical application.

Avoine and Oechslin [13] points out that communication between an RFID tag and a reader has multiple layers, i.e., application, communication, and physical layers, and that we need to consider all these communication layers to avoid tracing. Let us consider a situation in which a person carries 10 tags where 2 tags respond in the LF (Low Frequency) band, 3 tags respond in the HF (High Frequency) band, and 5 tags respond in the UHF (Ultra High Frequency) band. These kinds of characteristics of wireless signals from a person can be used to identify and trace the person, even if the privacy of the ID of the tags is protected in the high-level layer.

Oren and Shamir [14] propose the power analysis of RFID tags in which secret information stored in a tag is obtained by observing the field strength of the wireless signals from the tag. The attack can be performed without physical contact with the tag, so the attack can be implemented easily and detection of the attack is difficult. We need to implement carefully a security mechanism inside the tag considering this kind of side-channel analysis.

34.2.6 Security of Total System

In the previous sections, we focused on the security and privacy of only the RFID tag and reader. However, in practice, we need to consider the security and privacy of the total RFID system.

Fabian et al. [15] point out the security and privacy issues concerning ONS (Object Name Service) of the EPC-global network [EPC]. When the reader requests the ONS with the ID of the tag, the ONS responds with the location of a database that contains information pertaining to the tag. The ONS is constructed based on DNS (Domain Name Service) technology. Fabian et al. point out issues pertaining to confidentiality and integrity of the communications between the reader and ONS and the availability of the ONS service. For instance, the query from the reader to the ONS can be tapped and/or modified to obtain the ID and information of the tag. They also propose solutions for these ONS issues.

Rieback et al. [16] demonstrate the possibility of transmitting a virus via an RFID tag. They construct a virus code that can be stored in the memory of the tag. The virus code is read and executed by a server using a security flaw in the server system. The virus writes copies of itself into other tags to spread the infection. This shows that tags can be a new route for viruses.

34.3 Physical and Low-Level Layer Approaches

In this section, we discuss the physical and low-level layer approaches to security and privacy for the RFID system. By covering a tag with metal film, which is called a Faraday cage, wireless communications between the tag and a reader can be blocked. With the Faraday cage, we can maintain privacy by preventing the illegal scanning of a wireless tag.

However, it is difficult to cover all tags that a person may carry, as that person may carry many tags, regarding some of which the person may not be aware. A jamming signal can also be used to block wireless communications between a tag and a reader to maintain privacy. However, this is not practical, since a jamming signal can affect other wireless devices.

An EPCglobal tag [1] is equipped with a kill command to protect user privacy. A kill command can be sent to a tag to deactivate the tag forever. To avoid the illegal use of this function, a kill password should be sent to the tag with the kill command and the kill command should be executed only if the kill password is correct. Since the tag never responds to a reader after the kill command is executed, privacy is perfectly protected thereafter. However, the permanent deactivation removes the possibility of using the tag thereafter.

Juels et al. [17] propose a blocker tag that is a special tag that blocks the access to readers in the vicinity of the tag. The blocker tag interferes with the anticollision protocol of the RFID system that is established between tags and the reader to block the access of the reader. This is accomplished in the following manner. To avoid collision of communications with many tags, the reader performs the following anticollision protocol. First, the reader requests all the tags in the vicinity to transmit the first bit of the ID. If the reader receives 0 as the answer, the reader knows that there are tags in the vicinity that have an ID with 0 as the first bit. Then, the reader requests all the tags that have an ID with 0 as the first bit in the vicinity, to transmit the second bit of the ID. If the reader receives 1 as the answer, the reader knows that there are tags in the vicinity that have an ID with 0 as the first bit and 1 as the second bit. By iterating this process the reader can identify the IDs of all the tags in the vicinity. To interfere with the anticollision protocol, the blocker tag always responds with 0 and 1 to reader requests. From the viewpoint of the reader, there exists an exponentially large number of tags, e.g., 2^{96} tags in the case of a 96 bit ID. This causes the reader to make an exponentially large number of inquiries, so the reader cannot complete the anticollision protocol, and the access of the reader to the tags is blocked, i.e., the blocker tag performs a kind of DoS (Denial of Service) attack on the reader.

Weis et al. [18] propose a secure communication channel between the tag and reader that utilizes the difference in distance that the signals of the tag and reader can reach. A passive tag is not equipped with a power supply, so the distance from the tag that the signal can reach is short, e.g., approximately 50 cm. On the other hand, the distance from the reader that the signal can reach is much longer than that for the tag, e.g., approximately 50 m. Hence, an eavesdropper who cannot approach the tag and reader, e.g., within 1 m, cannot tap the communications from the tag to the reader, even though the eavesdropper can tap the communications from the reader to the tag. By using this fact, a secure communication channel from the reader to the tag is achieved as follows. First, the tag sends random r to the reader, then the reader sends encryption $r \oplus m$ of message m with the random r to the tag. Since the eavesdropper cannot tap random r, he cannot know message m even though he can tap encryption $r \oplus m$.

34.4 Cryptographic Approaches to Privacy

In this section, we discuss the cryptographic approaches to privacy for the RFID system.

34.4.1 Approaches Using Hash Function

In this subsection, we discuss RFID privacy protection schemes that use a hash function, which is a light weight cryptographic function, and can be computed inside the RFID tag. In these schemes, a tag computes the hash function inside it to protect privacy.

Sarma et al. [19] mention RFID security benefits and threats in the very early period of research of RFID security and privacy, and propose a privacy protection scheme at a low cost as all it requires is a hash function. In the scheme, the validity of the tag reader is checked using the following procedure to avoid illegal ID scanning and ID leakage. The reader has key k for each tag, and each tag holds hash value $h = H(k)$, called a meta-ID, where H is the hash function. A tag receives a request for ID access and then sends meta-ID h in response. The reader sends key k that is related to the meta-ID h received from the tag. The tag then calculates the hash function from the received key k and checks the relation of $h = H(k)$ with meta-ID h held in the tag. The tag responds with its own ID to the reader only if the relation holds.

However, the scheme is still susceptible to tracing, since the meta-ID is fixed and the adversary can trace the tag via the meta-ID, while the scheme offers protection against Id leakage at low cost. To avoid this, the meta-ID should be changed repeatedly. This requirement may become a problem in practical use.

Weis et al. [18] propose a randomized hash scheme that is an extension of the hash lock scheme. In this scheme, the tag is required to have key k, a hash function circuit, and a random generator. Each tag calculates hash function H with the inputs of key k and random r, i.e., $c = H(k|r)$. The tag then sends c and r to the reader. The reader maintains the database for key k and the tag ID. The reader checks relation $c = H(k|r)$ of received key c and r for all keys k, and finds the ID of the tag from the database. The tag output changes with each access, so this scheme deters tracing.

However, this scheme allows the location history of the RFID tag to be traced if key k in the tag is exposed, i.e., this scheme cannot satisfy the concept of forward security. Moreover, the cost imposed by the random generator is significant.

Ohkubo et al. [10] describe the new security concept called forward security, and propose a forward secure scheme using a hash chain. Forward security means that even if the adversary acquires the secret data stored in the tag, the adversary cannot trace the tag back to past events in which the tag was involved. To achieve forward security, Ohkubo et al. use the hash chain technique to renew the secret stored in the tag. More precisely, in the ith transaction, the tag (1) sends answer $a_i = G(s_i)$ to the reader, (2) renews secret

$s_{i+1} = H(s_i)$ as determined from previous secret s_i, and (3) erases previous secret s_i where H and G are hash functions. The reader maintains the database for initial secret s_0 and the tag ID. The reader then checks relation $a_i = G(H^i(s_0))$ of the hash chain for all initial secrets s_0, and finds the ID for the tag from the database.

However, the reader must search for all indices i and all tags that are relatively heavy, although computation of the tag is sufficiently efficient since the tag performs only hash operations that require a small gate. Thus, the scheme is practical, while still ensuring privacy even in the face of tampering.

Molnar et al. [11] propose a privacy protection scheme employing tree structured keys that are efficient and can delegate the ability to identify tags. In this scheme, keys are generated using a hash tree, and the root value of the hash tree is stored in the tag. At each session with a reader, the tag generates its output by computing the hash tree from the root value. This can be done efficiently by virtue of the tree structure. The ability to identify a tag restricted within a period of sessions can be delegated to another person by providing the root value of the subtree that corresponds to the period. By using this delegation function, the ownership of the tag can be transferred to another person without violation of privacy. The new owner makes the tag update its output a sufficient number of times when ownership of the tag is transferred, so that the previous owner can no longer identify the tag since the identification ability of the previous owner is restricted within a certain number of times.

Juels [20] proposes a significantly light scheme that requires only XOR calculations, and is therefore of very low cost. In this scheme, the reader and tag share a common list of random keys, and in some interactions they confirm that the partner has the common list. If it passes the check, the tag sends its ID. This scheme does not require any heavy calculations and the tag need only perform the XOR operation. This scheme requires several interactions between the tag and reader, and the common list should be overwritten completely as needed to ensure security.

34.4.2 Approaches Employing Rewriting

In this section, we discuss an RFID privacy protection scheme that rewrites the label stored in the tag to protect privacy. In these schemes, a tag is required to have only a rewritable ROM. The reader receives the label from a tag, updates it, and rewrites the updated label in the tag, to protect privacy.

Inoue and Yasuura [21] propose an ID rewriting scheme (that does not use cryptography) to maintain user privacy. When a user purchases a tagged item, the user writes a random label in the tag. The tag responds with the random label thereafter, so no information of the tagged item is leaked. However, the user must maintain a table of the random labels and the tagged items.

Juels and Pappu [22] propose an encrypted ID scheme employing the ElGamal encryption and re-encryption technique. ElGamal encryption is a public key encryption with secret key x, public key $(g, y = g^x)$, and ciphertext $E(m,r) = (g^r, my^r)$, where m is the message and r is a random value. It can be randomized by multiplying $(g^r \times g^s, my^r \times y^s)$ using only a public key (g, y); this is called re-encryption. In the scheme, the ElGamal encryption of ID $E(ID, r)$, is stored in the tag. The reader can update the ElGamal encryption of ID $E(ID, r)$ by randomizing it without decryption and secret key x. Thus, readers are not required to have a secret key.

Ishikawa et al. [23] propose an encrypted ID scheme with any encryption scheme. In the scheme, the encrypted ID $E(ID, r)$ is stored in the tag, where E is a probabilistic encryption and r is a random value. The reader receives the encrypted ID $E(ID, r)$, and decrypts it to obtain the ID of the tag. After this, the reader again encrypts ID $E(ID, r')$ with a different random r', and rewrites the new encrypted ID $E(ID, r')$ in the tag. Since the new encrypted ID $E(ID, r')$ looks totally different, the encrypted ID $E(ID, r)$ cannot be linked to the new encrypted ID $E(ID, r')$, and so tracing the tag is prevented. A scheme that utilizes the ElGamal re-encryption technique is also proposed in this paper.

However, we need to use a single key for a group within which tags are indistinguishable, as we cannot determine which key corresponds to a ciphertext if we use multiple keys.

Golle et al. [24] propose an encrypted ID scheme with universal ElGamal re-encryption. In this scheme, the public key (g, y) and ElGamal encryption of ID $E(ID, r)$ are stored in the tag. The reader can update them by randomizing both the public key and encryption. Thus, multiple public keys can be employed and readers do not need to maintain public keys, since the corresponding public key is stored with the ciphertext in the tag.

However, the ciphertext stored in a tag can be subverted if the adversary replaces the ciphertext of the message with his own public key. Later, the adversary can decrypt the output of the tag using his own secret key to trace the tag.

Ateniese et al. [25] propose an encrypted ID scheme with insubvertible encryption. In this scheme, the certificate from a CA (Certificate Authority), public key, and encryption of ID are stored in the tag. The reader can check the validity of the certificate and update it by randomizing the certificate, public key, and encryption of ID. Thus, the stored ciphertext cannot be subverted since the adversary cannot forge the certificate from the CA.

34.5 Cryptographic Approaches to Authenticity

In this section, we discuss cryptographic approaches to the authenticity of the RFID system.

Hopper and Blum [26] present an authentication scheme called the HB protocol based on the LPN (Learning Parity with Noise) problem. The scheme uses only XOR and requires only two moves. The details of the scheme are given hereafter. In the initialization phase, reader R holds secret x, tag T holds secret (x, η), and $v \in \{0, 1 | Prob[v = 1] = \eta\}$ is a biased random bit, where $0 < \eta < ½$. Then, tag T and reader R iterate the following steps: (1) R selects random $a \in \{0, 1\}^k$ and sends a to T as a challenge, (2) T calculates $z = (a \cdot x) \oplus v$ and sends z to R as a response, and (3) R accepts, if $a \cdot x = z$. Here, we denote the inner-product of the vectors over the finite field by $\cdot$.

However, the HB protocol is only secure against passive eavesdroppers, and it is not secure against an active adversary with the ability to query tags. If an active adversary repeats the same challenge a multiple times, he can learn the error-free value of $a \cdot x$ with very high probability.

Juels and Weis [27] present an extended authentication scheme to the HB protocol, called the HB+ protocol, that is secure against an active adversary. The scheme uses only the XOR calculation and requires three moves. The details of the scheme are given hereafter. In the initialization phase, reader R holds secret (x, y), tag T holds secret (x, y, η), and $v \in \{0, 1 | Prob[v = 1] = \eta\}$ is a biased random bit. Then, tag T and reader R iterate the following steps: 0) T selects random $b \in \{0, 1\}^k$ and sends b to R as a blinding factor, (1) R selects random $a \in \{0, 1\}^k$ and sends a to T as a challenge, (2) T calculates $z = (a \cdot x) \oplus (b \cdot y) \oplus v$ and sends z to R as a response, and (3) R accepts, if $(a \cdot x) \oplus (b \cdot y) = z$. By selecting its own random blinding factor b, tag T effectively prevents an active adversary who tries to extract x or y with a nonrandom challenge.

However, more precise analyses against HB+ protocol have been presented recently.

Gilbert et al. [28] present a man-in-the-middle attack against the HB+ protocol. The attack is performed as follows to obtain secret x. The adversary selects a constant $\delta \in \{0, 1\}^k$ and perturbs challenge a sent by reader R to tag T to $a \oplus \delta$ in all rounds of the HB+ protocol. If the authentication process is successful, then it means that $\delta \cdot x = 0$ with overwhelming probability. Hence, the adversary can obtain partial information concerning secret x, and by repeating this with different δ, he can obtain secret x. Once x is derived, the adversary can immediately impersonate the tag with blinding factor $b = 0$. Another side effect of the disclosure of x is that the privacy of the tag ID is compromised. To avoid this attack, we can set an alarm to go off when a threshold for the number of failed authentications is exceeded.

Katz and Shin [29] analyze the HB+ and HB protocol, and show that the HB+ protocol is secure under concurrent and parallel composition if $0 < \eta < ¼$. Moreover, Katz and Smith [30] provide analysis in the case $¼ \leq \eta < ½$.

34.6 Implementation

In this section, we discuss light implementations of the cryptographic primitives for RFID tags.

Tuyls and Batina [31] implement an elliptic curve computation that has the possibility to achieve a public-key based protocol for RFID tags. In their paper, as an example, the Schnorr identification scheme is implemented based on an elliptic curve for the RFID tag authentication. Batina et al. [32] implement the Okamoto identification scheme based on an elliptic curve.

McLoone and Robshaw [33] implement a public-key based authentication protocol for the RFID tag authentication. In the protocol, by utilizing precomputation, the required calculation load in the on-line phase is sufficiently light for a tag to calculate. The protocol is based on an elliptic curve version of the GPS identification scheme, which is a public-key three move Schnorr-like identification scheme proposed by Girault, Poupard, and Stern. In the precomputation phase, the first message of the three move proof is computed, and the result, called a coupon, is stored in the tag. The on-line phase is efficient and the tag requires no elliptic scalar multiplication. However, there is a storage cost for coupons for each tag.

Feldhofer et al. [34] implement AES so that AES can be embedded in small devices, e.g., in RFID tags. The AES implementation requires 3595 gates and 8.15 A consumption at a frequency of 100 KHz. The number of clock cycles to encrypt 128 bits is 1000. They implement the AES algorithm as an 8-bit architecture, which is different from the usual implementation as a 32-bit architecture. By implementing the AES algorithm as an 8-bit architecture, the number of S-boxes is reduced from 4 to 1 to conserve silicon resources, and the consumed power is lower than that required by the 32-bit operations. On the other hand, the number of clock cycles for encryption is increased by more than 10 fold.

Poschmann et al. [35] present a new light weight block cipher DESL based on DES, which is compact and requires only light calculations. In the DESL, the eight original DES S-boxes are replaced by a single S-box, which is repeated eight times. The construction implies a lightweight implementation, which requires 50% smaller chips, 85% fewer clock cycles, and 90% less energy than the best AES implementations with regard to RFID applications [34]. As a detailed example, DESL requires 144 clock cycles to encrypt one 64-bit block of plaintext. For one encryption at 100 KHz the average power consumption is 0.89 μA, and in the case of 500 KHz it is 4.45 μA. The throughput reaches 5.55 KB/s at 100 KHz and 27.78 KB/s at 500 KHz.

Vaudenay [36] presents a new lightweight public-key encryption scheme for the RFID tag authentication, which requires only a small computation load that a tag can calculate. Moreover, he describes new security definitions and their relationships. The new public-key encryption scheme is based on the problem of finding a sparse polynomial that is a multiple of the public key. To encrypt a block of messages, only a computation of LFSR (Linear Feedback Shift Register) and the generation of a biased random string are required. So, the new public-key encryption scheme can be implemented with a small gate size making it suitable for RFID tags.

34.7 Conclusion

In this chapter, we have surveyed studies pertaining to the security and privacy for RFID tags, especially from the context of cryptography. We have presented definitions for RFID security and privacy from various viewpoints. We have also overviewed the existing studies concerning physical approaches, cryptographic approaches for privacy, authenticity, and implementation of cryptographic primitives.

References

1. EPCglobal. EPCglobal web site. http://www.epcglobalinc.org.
2. CASPIAN (Consumers Against Supermarket Privacy Invasion and Numbering). http://www.nocards.org, 2002.
3. Associated Press. Benetton undecided on use of 'smart tags,' 8 April 2003.
4. CNET. Wal-mart cancels 'smart shelf' trial. http://www.cnet.com, 9 July 2003.
5. Ari Juels. RFID security and privacy: A research survey. *IEEE Journal on Selected Areas in Communications*, 24(2), pages 381–394. IEEE Computer Society Press, 2006.
6. Gildas Avoine. Bibliography on security and privacy in RFID systems. Available online, 2006.
7. Gildas Avoine. Adversary model for radio frequency identification. Technical Report LASEC-REPORT-2005-001, Swiss Federal Institute of Technology (EPFL), Security and Cryptography Laboratory (LASEC), 2005.
8. Ari Juels and Stephen Weis. Defining strong privacy for RFID. *IACR Cryptology ePrint Archive*, 2006/137, 2006.
9. Ivan Damgård and Michael Østergaard. RFID security: Tradeoffs between security and efficiency. *IACR Cryptology ePrint Archive*, 2006/234, 2006.
10. Miyako Ohkubo, Koutarou Suzuki, and Shingo Kinoshita. Cryptographic approach to "privacy-friendly" tags. In *RFID Privacy Workshop*, 2003.
11. David Molnar, Andrea Soppera, and David Wagner. A scalable, delegatable pseudonym protocol enabling ownership transfer of RFID tags. In *Selected Areas in Cryptography—SAC 2005*, Lecture Notes in Computer Science, Vol. 3897, pages 276–290. Springer-Verlag, 2005.
12. Ari Juels. "yoking-proofs" for RFID tags. In *International Workshop on Pervasive Computing and Communication Security—PerSec 2004*, pages 138–143. IEEE Computer Society, 2004.
13. Gildas Avoine and Philippe Oechslin. RFID traceability: A multilayer problem. In *Financial Cryptography—FC 2005*, Lecture Notes in Computer Science, Vol. 3570. pages 125–140. Springer-Verlag, 2005.
14. Yossi Oren and Adi Shamir. Power analysis of RFID tags. Panel discussion in RSA Conference 2006, 2006.

15. Benjamin Fabian, Oliver Günther, and Sarah Spiekermann. Security analysis of the object name service for RFID. In *International Workshop on Security, Privacy and Trust in Pervasive and Ubiquitous Computing—SecPerU 2005*. IEEE Computer Society Press, 2005.
16. Melanie Rieback, Bruno Crispo, and Andrew Tanenbaum. Is your cat infected with a computer virus? In *Pervasive Computing and Communications—PerComm 2006*, pages 169–179. IEEE Computer Society Press, 2006.
17. Ari Juels, Ronald Rivest, and Michael Szydlo. The blocker tag: Selective blocking of RFID tags for consumer privacy. In *Conference on Computer and Communications Security—ACM CCS 2003*, pages 103–111. ACM Press, 2003.
18. Stephen Weis, Sanjay Sarma, Ronald Rivest, and Daniel Engels. Security and privacy aspects of low-cost radio frequency identification systems. In *International Conference on Security in Pervasive Computing—SPC 2003*, Lecture Notes in Computer Science, Vol. 2802, pages 454–469. Springer-Verlag, 2003.
19. Sanjay Sarma, Stephen Weis, and Daniel Engels. RFID systems and security and privacy implications. In *Cryptographic Hardware and Embedded Systems—CHES 2002*, Lecture Notes in Computer Science, Vol. 2523, pages 454–469. Springer-Verlag, 2002.
20. Ari Juels. Minimalist cryptography for low-cost RFID tags. In *International Conference on Security in Communication Networks—SCN 2004*, Lecture Notes in Computer Science, Vol. 3352, pages 149–164. Springer-Verlag, 2004.
21. Sozo Inoue and Hiroto Yasuura. RFID privacy using user-controllable uniqueness. In *RFID Privacy Workshop*, 2003.
22. Ari Juels and Ravikanth Pappu. Squealing euros: Privacy protection in RFID-enabled banknotes. In *Financial Cryptography—FC 2003*, Lecture Notes in Computer Science, Vol. 2742, pages 103–121. Springer-Verlag, 2003.
23. Toshiharu Ishikawa, Yukiko Yumoto, Michio Kurata, Makoto Endo, Shingo Kinoshita, Fumitaka Hoshino, Satoshi Yagi, and Masatoshi Nomachi. Applying auto-id to the japanese publication business. White Paper KEIAUTOID-WH-004, Auto-ID Center, 2003.
24. Philippe Golle, Markus Jakobsson, Ari Juels, and Paul Syverson. Universal re-encryption for mixnets. In *The Cryptographers' Track at the RSA Conference—CT-RSA 2004*, Lecture Notes in Computer Science, Vol. 2964, pages 163–178. Springer-Verlag, 2004.
25. Giuseppe Ateniese, Jan Camenisch, and Breno de Medeiros. Untraceable RFID tags via insubvertible encryption. In *Conference on Computer and Communications Security—ACM CCS 2005*, pages 92–101. ACM Press, 2005.
26. Nicholas J. Hopper and Manuel Blum. Secure human identification protocols. In *Advances in Cryptology—ASIACRYPT 2001*, Lecture Notes in Computer Science, Vol. 2248, pages 52–66. Springer-Verlag, 2001.
27. Ari Juels and Stephen Weis. Authenticating pervasive devices with human protocols. In *Advances in Cryptology—CRYPTO 2005*, Lecture Notes in Computer Science, Vol. 3126, pages 293–308. Springer-Verlag, 2005.
28. Henri Gilbert, Matthew Robshaw, and Hervé Sibert. An active attack against HB^+—a provably secure lightweight authentication protocol. In *IEE Electronic Letters* 41(21), pages 1169–1170, 2005.
29. Jonathan Katz and Ji Sun Shin. Parallel and concurrent security of the HB and HB^+ protocols. In *Advances in Cryptology—EUROCRYPT 2006*, Lecture Notes in Computer Science, Vol. 4004, pages 73–87. Springer-Verlag, 2006.
30. Jonathan Katz and Adam Smith. Analyzing the HB and HB+ protocols in the "large error" case. *IACR Cryptology ePrint Archive*, 2006/326, 2006.
31. Pim Tuyls and Lejla Batina. RFID-tags for anti-counterfeiting. In *The Cryptographers' Track at the RSA Conference—CT-RSA 2006*, Lecture Notes in Computer Science, Vol. 3860, pages 115–131. Springer-Verlag, 2006.
32. Lejla Batina, Jorge Guajardo, Tim Kerins, Nele Mentens, Pim Tuyls, and Ingrid Verbauwhede. Public-key cryptography for RFID-tags. In *Pervasive Computing and Communications Workshops—PerComm Workshops 2007*, pages 217–222. IEEE Computer Society Press, 2007.
33. Máire McLoone and Matthew J.B. Robshaw. Public key cryptography and RFID tags. In *The Cryptographers' Track at the RSA Conference—CT-RSA 2007*, Lecture Notes in Computer Science, Vol. 4377, pages 372–384. Springer-Verlag, 2007, available at http://www.springerlink.com/content/554p55777q2181w7/

34. Martin Feldhofer, Sandra Dominikus, and Johannes Wolkerstorfer. Strong authentication for RFID systems using the AES algorithm. In *Workshop on Cryptographic Hardware and Embedded Systems—CHES 2004*, Lecture Notes in Computer Science, Vol. 3156, pages 357–370. Springer-Verlag, 2004.
35. Axel Poschmann, Gregor Leander, Kai Schramm, and Christof Paar. New light-weight crypto algorithms for RFID. *Circuits and Systems—ISCAS 2007*, pages 1843–1846. IEEE Computer Society Press, 2006.
36. Serge Vaudenay. RFID privacy based on public-key cryptography (abstract). In *Information Security and Cryptology—ICISC 2006*, Lecture Notes in Computer Science, Vol. 4296, pages 1–6. Springer-Verlag, 2006.

35

RFID Authentication: Reconciling Anonymity and Availability

Breno de Medeiros and Mike Burmester

CONTENTS

35.1 Introduction

This chapter explores the issues attending to the provision of two security services, namely privacy and availability, in context of RFID applications.

Since the word privacy is used in different settings with distinct meanings, it is necessary to provide its scope within the RFID context. Automatic identification is the primary functionality provided by RFID, enabling recognition of tagged objects and supporting applications such as resource tracking. Consequently, RFID-tagged objects or persons can be easily recognized and located, and their movements and whereabouts can be monitored and recorded using relatively inexpensive readers and without necessity of human intervention. This information being available in digital form, inference and correlations can be drawn using information retrieval and understanding technologies. Such ease of acquisition of location data from RFID-tagged objects, and the potential for abuse of the derived information motivate the provision of *location privacy* services in RFID.

Availability concerns in RFID are related to the ability of the system to function correctly through its projected lifetime. It is compromised when parties can perform unauthorized actions that render components of the RFID system either temporarily or permanently incapable of exercising its proper role. Disabling attacks can involve jamming the communication medium, targeting battery consumption rates (against active or semi-active tags), or corrupting tag-stored data or other manipulation of tag state leading to it becoming unusable.

35.2 Characteristics of RFID Systems

A typical deployment of an RFID system involves three types of legitimate entities: *tags*, *readers*, and a *back-end server* (Sharma et al., 2001). When referring to an object or person that is identified by carrying one or more RFID tags, we may use the term *RFID host*, or simply *host*.

The tags, also called *transponders*, are attached to, or embedded in, objects to be identified (hosts). They consist of a microchip and an RF coupling element and antenna that enables communication via radio waves. The microchip component of a tag can include features such as a CMOS integrated circuit, ROM, RAM, and nonvolatile EEPROM. Tags may also have a source of autonomous power, typically a battery. In case the battery is absent, the coupling element is configured so as to capture power and clock pulses from the electromagnetic field of the radio waves received at the antenna. The type of power source in the tag has great impact in defining constraints on the availability of overall resources, and in particular of those that can be used to implement security features. The possibilities with regard to power source are:

- *Passive tags*: Power captured from reader's radio waves through induction at the antenna.
- *Active tags*: Power supplied autonomously (usually a battery).
- *Semi-passive tags*: Autonomous power provides for an onboard clock and/or powers the tag circuitry, but does not supply the radio interface. Communication is powered by induction as in the passive case.

The readers (also called *transceivers*) include a radiofrequency module, a control unit, and a coupling element to scan for, and communicate with tags. Some readers are strictly stand-alone, units—for instance, handheld readers may be carried to facilitate the work of human monitors in verifying the arrival of shipped goods. Other readers have an interface for communication with a back-end server.

The back-end server is typically a PC-type device, hosting business applications, e.g., a database server augmented with RFID-specific middleware for communication with the readers. Custom applications are also typically deployed that interface with the middleware or the database to implement specific business logic—including support of security functions. There are, however, alternative scenarios. As an example, RFID readers have been integrated with cell-phone devices, and in this case the back-end functionality is properly described as being a distributed computing application that includes custom software in the cell phone, and other components in remote databases accessed through the cell network (Pering et al., 2005; Want, 2006). This variant scenario could become a commonplace if consumer-oriented RFID deployments are embraced by the public.

35.3 RFID Communication Model

Communication in RFID is modeled in three distinct layers, which are the lowest, or *physical layer*, involving the RF coupling element; the *data link* or *communication layer*, including the collision-avoidance protocols, and the *protocol* or *application layer*, wherein higher-level mechanisms such as authentication protocols can be implemented (Sharma et al., 2001).

At every layer of this communication model there are issues that affect RFID security. At the physical layer, powerful attacks against some RFID architectures have been described that directly recover the internal tag state (Oren and Shamir, 2006). These attacks analyze the reflected electromagnetic emanations from passive RFIDs. Since passive RFIDs must draw their power from the antenna, they tend to dampen the reflected emanations when performing a computation that requires more power, and this can be used to derive the values of bits under manipulation in the RFID internal circuit. From this data, information about cryptographic secrets can be gleaned. New RFID designs—perhaps including *Faraday cage** coatings or larger capacitors—will be needed to eliminate the possibility of such physical-observation attacks. US e-passports incorporate Faraday shielding pages.

More broadly, it is necessary to consider physical attacks that go beyond observation, and manipulate the amount of energy available to passive tags for computation. Smart cards, which are power-assisted, have been successfully attacked through the exploitation of subtle faults that can be triggered by manipulation of the power source feed. After a fault is induced—for instance, a state reset that forces the card to reuse randomness or other one-time parameters—the attacker can compare the outputs produced in each case and perform differential-type attacks. It is likely that some RFID cryptographic implementations will be vulnerable to similar attacks, unless the design process takes this type of threat into consideration and develops specific approaches to counter it.

At the communication layer, it is also possible to achieve security violations (and in particular location privacy exploits) by abuse of some mechanisms, such as the *singulation* protocol in the EPC Gen2 standard (EPCglobal, 2005). Since multiple tags may be simultaneously within reach of a reader, singulation protocols are used to allow the reader to identify (isolate or singulate) specific tags, establishing listening channels for each (Sharma et al., 2001). By keeping the singulation protocol in open stage, it is possible for a reader to continuously track a tag that remains within reach, even if the tag were to implement privacy-preserving protocols at higher layers.

To achieve security (or any particular security service, such as location privacy) in RFID, therefore, it is necessary to consider this issue at each layer, and to properly compose the results. In this chapter, however, we focus on the protocol layer exclusively.

* Enclosures made of electrically conducting material, named after physicist M. Faraday. They block external electrostatic fields, and if sufficiently thick can also dampen magnetic fields.

35.3.1 RFID Security at the Protocol Layer

From an engineering perspective, the definition of RFID technology follows from what is commonly agreed to constitute their physical layer (principally) and the communication layers—increasingly defined by common standards, such as the EPC Gen2 standard by EPC Global (EPCglobal, 2005)—and this approach makes the definition of the RFID domain incontrovertible. However, taking a protocol-layer only view introduces conceptualization issues and obscures the boundaries of the RFID domain.

From the viewpoint of security protocols, what are the differences between RFID systems and other constrained-resource or ubiquitous architectures, such as sensor networks, mobile ad hoc networks (MANETs), or vehicular ad hoc networks? It would be difficult to claim a well-defined RFID security research area if the resources provided by RFID technology (at the protocol layer) were to substantially overlap with those characterizing other constrained devices.

In practice, it has become common among researchers that focus on security at the protocol layer to restrict their attention to the passive RFID tags. This empirical rule-of-thumb provides a useful distinction, as passive RFIDs are strongly restricted in their maximum circuit area by the amount of voltage provided by electromagnetic energy captured by the antenna. This restriction provides an envelope of a few thousand gates of circuitry available for the implementation of all protocol-layer services. Additionally, passive RFIDs are not vulnerable to battery-depletion denial-of-service attacks that affect other constrained settings, such as sensor networks.

Semipassive tags also represent a unique security domain, though one that has not been widely studied. Semipassive tags do not present concerns about minimization of the communication cost (as their radio interface is powered by the readers) nor such strong restrictions on circuit area (as the battery can power comparatively larger circuits). On the other hand, they are susceptible to battery-depletion denial-of-service attacks of a special type (computational-intensive approach). Although battery-depletion attacks have been studied in the context of other constrained devices, the situation with semipassive RFID is markedly different as they are vulnerable only through computation cost, not (the typically larger) communication cost. Therefore, the types of strategies useful to counter battery-depleting DoS attacks against semipassive RFID tags should differ from those applicable to the protection of other constrained devices, such as sensors.

From the perspective of security at the protocol layer, it may not be necessary to consider active tags separately from sensor networks and related technologies. Only when discussing security at the physical and communication layers do such distinctions become significant.

35.4 The Threat Model for RFID

Providing strong security assurances for RFID is challenging due to the constraints of the technology. Accordingly, it is rarely suitable to employ standard security protocols for other domains in this setting. Yet, the robustness and security requirements for RFID applications can be quite significant. Ultimately, security solutions for RFID applications must take as rigorous a view of security as other types of applications. In particular, whenever departure from standard practice is adopted, a rationale for any simplifications should be given, grounding all arguments on recognizable particularities of the RFID domain.

For instance, some solutions assume a bound on the communication complexity available to attackers, limiting the number of messages exchanged between successive interactions of a tag with legitimate readers (Juels, 2004a). Such an assumption should be validated against the projected threats in each specific deployment scenario.

35.4.1 Byzantine Adversaries

Since RFID is a wireless networking technology, it is natural to consider traditional threat models for the communication, and to assume a Byzantine adversary (Dolev and Yao, 1983). In this approach, all legitimate entities (tags, readers, the back-end server), as well as the adversary, have polynomial-bounded resources. The adversary controls the delivery schedule of the communication channels, and may eavesdrop into, or modify, their contents. The adversary may instantiate new channels and directly interacts with honest parties. Only recently such strong adversarial models have been advocated as feasible for the RFID setting (Burmester et al., 2006; Juels and Weiss, 2006).

35.4.2 Handling Multiple Security Requirements

It is a common practice to except the channels between the readers and the back-end servers from the security model. This is justifiable, since that communication can use standard protocols for establishment of secure channels. One result is that RFID security models capture only partial aspects of the entire system. Taking also into account that RFID components are designed for concurrent communication,* RFID security models have been proposed that provide security guarantees in the presence of concurrent executions and modular composition of protocols (Burmester et al., 2006; van Le et al., 2007).

To achieve harmonization of the multiple security requirements of RFID authentication protocols, and in particular to capture the tension between the privacy and availability requirements, it is useful to have a unified security model. Recently, a model has been formulated that simultaneously captures (forward-) secure authentication and anonymity, for both RFID entity authentication and key exchange, but that model does not address availability requirements (Juels and Weiss, 2006). An independent, and concurrent formalization captures authenticity, anonymity, and additionally supports availability for RFID entity authentication (Burmester et al., 2006). Later it was extended to include forward-security considerations attending both entity authentication and key exchange (van Le et al., 2007).

35.5 Aspects of RFID Security

Some types of attacks and threats against RFID are classical strategies such as replay attacks (authenticating values are captured and replayed later for impersonation, compromising authenticity), and interleaving attacks (authentication flows from different transactions are combined to create new, valid authentication transcripts). Others exploit unique features of the RFID setting, such as mobility and *promiscuity*, i.e., the common circumstance of tags being always available to respond to queries. This feature, problematic in view of privacy concerns, can be attenuated by introducing a temporary deactivation mechanism—which may introduce its own set of vulnerabilities, e.g., threatening availability—or via physical mechanisms such as a switch. The latter strategy is only reasonable for tags carried by persons.

* Standard-compliant RFID readers are capable of simultaneous communication with large numbers of tags (EPCglobal, 2005).

Among the various security interests in RFID technology, authenticity of identification and data, (location) privacy, and availability are arguably the most critical.

35.5.1 Authentication and Integrity

RFID-facilitated identification is being used to improve the performance of secure identity verification. For instance, RFIDs have been added to the new international passport standard (ICAO, 2004), to enhance the security and efficiency of database-driven checks at ports of entry. It is useful to consider this example in some detail to appreciate the real-world consequences of the adoption of RFID as a security mechanism. In this case, while RFIDs are not supposed to replace manual verification by the agent/officer, documents guiding its adoption in the United States indicate an intention to reduce the time taken to process passports, including eliminating the need to use optical scanning to verify them (Juels et al., 2005; U.S. Department of State, 2005). Focus on processing time may lead to RFID-based identification becoming the primary mechanism providing passport authenticity guarantees, such as protection from forgery.

It is claimed that the additional difficulty of forging an RFID tag will make the passport and ID systems more secure in the face of terrorism threats. However, research has shown that the RFID tags used in the e-passport system may not be difficult to forge, in particular for countries that do not adopt active authentication measures (Juels et al., 2005).

Secure facilities have adopted RFID for controlled access to restricted areas, and RFID is being deployed to increase the efficiency of container shipping tracking, including supporting international initiatives on counter-terrorism, and on combating drug and weapons smuggling (Willis, 2005). In this case, containers are inspected, tagged, and sealed at the point of origination. Upon arrival at the local destination (or perhaps a transshipment site), secondary inspections and opening of the container by authorities may be dispensed with if the RFID reader can verify the tag's identity.

In each of these examples, the RFID system substitutes for other security systems, providing more convenience and potentially more security by increasing the ability to consistently apply security checks (for instance, shipping containers are currently only inspected by sampling). However, integrity of these RFID-based verification systems is entirely dependent on the (frequently false) assumption that tag cloning is not easily accomplished.

The use of strong authentication mechanisms in RFID tags is one element of making such systems more resilient against cloning attacks. At the physical layer, protection against extraction of authentication keys through observation of the reflected electromagnetic field must also be addressed.

Authentication and integrity are also important in connection with privacy concerns, to be considered next. For instance, tags that do not support authentication of readers and allow for arbitrary writes into their data fields are vulnerable to attacks that introduce markers in the tags.

35.5.2 Privacy

RFID tags have low intrinsic value, and even any data they carry may itself be of low relevance. However, such tags are attached to resources that are worth monitoring. As standard RFID technology dictates that tags must satisfy interrogation requests by arbitrary readers, it is possible for covert and unauthorized readers to be deployed (Sharma et al., 2003). An adversary could use a network of such readers to automatically acquire information from circulating tags, accumulating transit information (location/times) of RFID hosts.

If an individual carries an RFID or RFID host object, this person's location information is at risk from being compromised by covert means. If, on the other hand, the RFID tags are used in their typical functions of automated shipment verification and inventory control, an attacker's ability to monitor RFIDs using unauthorized readers provides a threat to the confidentiality of the enterprise's logistic operation—facilitating industrial or military espionage of a tactical asset.

Note that simply hiding the identity of the tag, for instance, via pseudonyms, is not sufficient, because the identifier embedded in the tag may be of little value to an observer. Instead, anonymity of the identifier must provide for unlinkability, i.e., must prevent an observer from correlating two instances of interaction/communication *by the same tag*. In this chapter, whenever we refer to *location privacy*, or to *anonymity*, we refer to unlinkable anonymity.

RFID tags may contain auxiliary data as needed by the particular application (EPCglobal, 2005). This transport of related data provides a secondary channel that can be exploited by information gathering efforts. If such auxiliary data is stored in plaintext in the tag, the target of an attack may be simply reading it to use in combination with location information for data-mining purposes. Alternatively, even if the information is encrypted with a key unbeknownst to the attacker, if the enciphered value is constant between successive readings, it can serve as a pseudo-identifier that enables the attacker to violate the location privacy of the RFID host. This threat to privacy through exploitation of *hidden channels* may also utilize unauthorized readings, for instance, of counters or other data structures that change in predictable ways between instances of the communication protocol involving a particular tag.

In addition to considering the threat of privacy compromises by outsiders, one must also evaluate the potential for abuse by system operators. Such privacy concerns are referred to as *Big Brother* privacy concerns. They involve considerations of a different nature, including legal aspects, individual privacy expectations, and other public policy matters. For instance, in the United Kingdom, RFID tags are being embedded in license plates (e-Plates). Police officers furnished with readers may be able to ascertain such facts as whether a license plate is legitimate or forged, and if it is attached to the vehicle for which it was issued. In the absence of adequate safeguarding measures, this infrastructure is exploitable by outsiders (e.g., private eyes). It may also give unscrupulous parties access to a wealth of information about activities of members of the public. A good discussion on Big Brother issues in RFID is given in (Sharma et al., 2003).

At a finer granularity, one may exempt the back-end server from suspicion, and consider insider threat risks coming only from corrupt readers. Such threats are certainly worthy of evaluation, since the frequently mobile readers may be as vulnerable as the tags themselves.

35.5.3 Availability

Availability refers to the security guarantee that system resources will be accessible when needed. Clearly, availability implies some preexisting level of expectation for performance parameters, and it is both a security as well as a reliability and performance concern. Concentrating on security aspects, it is possible to enumerate a number of different attacks against availability and/or available counter-measures:

- *Killing attacks*: Some RFID tags support a *kill* functionality or *kill key* (EPCglobal, 2005). If a particular value is broadcast to a tag, it will be de-activated, either temporarily (until an enabling value is received) or permanently. For instance, the kill-key feature is available in RFID tags used to prevent shoplifting; these are disabled at the point-of-sale to allow for handling and reuse without raising false alarms. If the kill functionality is available, the system may be vulnerable to unauthorized disabling attacks that defeat its security functions. To prevent against such attacks, the kill functionality should require the disclosure of a secret by the back-end server to the tag. The secret should be tag-specific and unpredictable to an eavesdropper that has witnessed previous kill sessions with different tags, or perhaps even with the same tag, if the system allows for temporary tag de-activation.

- *Disabling attacks*: This generic class of attacks exploits state synchronization requirements of authentication mechanisms for tags. Some existing and proposed protocols for secure RFID authentication require tags to maintain state information that should match with other information available to the readers or (more commonly) to the back-end server. Disabling attacks interfere with the communication between tag and authorized reader to cause divergence between the state information among the parties, preventing further use of the tag. To prevent against such attacks, either mechanisms must be provided for recovery of a convergent state, or mutual authentication must be used to ensure integrity of exchanged messages before a state update is performed.
- *Jamming attacks*: The communication frequencies can be filled with noise by a reader (or other broadcasting device) that does not comply with accepted standards. If the level of introduced noise is high, it may be difficult or impossible to prevent against such attacks. Available techniques to tolerate (some level of) jamming attacks have been described in (Engels, 2001; Sharma et al., 2001; EPCglobal, 2005). Naturally, these solutions involve mechanisms at the communication layer.

35.6 Conflicts between Anonymity and Availability Requirements

In RFID technologies, anonymity and availability appear often to conflict. For instance, a privacy preserving technique can be created from an availability threat (jamming), as exemplified by the blocker tag proposed in a seminal paper on RFID privacy (Juels et al., 2003). In the following, we discuss how privacy techniques lead to availability risks.

Of the earlier-mentioned classes of attacks against availability, disabling attacks represent a particularly difficult challenge to address, particular in conjunction with location privacy requirements. Privacy implies that tags must change the values they use to authenticate or identify themselves to the system, to prevent from recognition and tracking by unauthorized parties. This implies that some form of changeable shared state must be maintained between tags and server.

Consider the case of single-side authentication, i.e., the tag authenticates itself to the reader but not conversely. In some protocols, this is achieved through the use of a shared-state, such as the seed for generating an unpredictable sequence. One of the first protocols in this class was introduced in Ohkubo et al. (2003). Disabling attacks against such protocols involves impersonating a reader to lead the tag in stepping the sequence, reaching the next state. Typically, as tag authentication protocols are designed to take a few hundred milliseconds, the fake reader has ample opportunity to cause the tag to become significantly desynchronized from the back-end server, by repeated application of this attack. Since there is no obvious bound on the number of state updates that an adversary could force on a tag, there is accordingly no guarantee that the back-end server will recognize the tag after speculatively stepping the state for every tag in its database a fixed number of times.

In general, protecting the exchanges between tags and the other system components to prevent faulty state changes is difficult. Passive tags do not maintain a clock and cannot use (even loosely synchronized) timing information as a security mechanism; they allow for implementation of only a limited set of cryptographic operations and have limited transient and permanent storage, therefore being less capable of detecting attacks. We now consider some strategies that are available to prevent against such desynchronization attacks.

35.6.1 Tag Identification and Server Authentication

It may be possible to use back-end server (or reader) authentication to reconcile anonymity and availability, by allowing the tag to detect malicious queries. Intuitively, one would have the tag generate a fresh (pseudo-) random challenge. After receiving a recognizable response authenticating the server, it could update its state and perform authentication without fear that the state might become desynchronized with the server.

However, there are difficulties with the mutual authentication approach as described. If, as we assume, the tag is incapable of performing public-key cryptographic operations, it can only authenticate the server based on shared, unique secrets. As a result, prior to identifying the tag, the back-end server has no ability to authenticate because it does not know which secret to use in computing the response to the tag's challenge. Therefore, server authentication must follow tag identification. Since identification is a weaker requirement* than authentication, if the former were amenable to a solution reconciling availability and privacy, then by following an identification step with mutual authentication, it may be possible to eliminate the threat of disabling attacks altogether. In the following, several strategies to tag identification are discussed.

35.6.2 Resynchronization

This is used by optimistic protocols (Burmester et al., 2006; van Le et al., 2007). When an attacker performs a desynchronization attack, these protocols revert to exhaustive search among the valid keys. Once the key is found, the server automatically resynchronizes with the tag. The advantage of this approach is that it requires no additional circuitry or computational capabilities in the tag, with the entire recovery process transparently handled by the server.

However, this strategy suffers from scalability issues. More precisely, to desynchronize n tags, the attacker needs to perform a number of computations proportional to n, whereas the work required by the server to recover is proportional to the product mn, where m is the number of tag keys currently valid (recognizable by the server). At the limit when the capacity of the system is tested, timing information (in the form of server delays) may allow for (partial) identification of tags, threatening privacy as well.

35.6.3 Exhaustive Key Search

A different approach to tag identification is to exploit the fact that the key-space to be exhausted by an adversary is much higher than the set of valid keys. In particular the latter may be easily exhaustable by the server in a full database search.

To facilitate the work of the back-end server during exhaustive key search, it is possible to use two or more keys in a hierarchical fashion. If t keys are used, then the search space

* Identification does not guarantee *freshness:* In an identification mechanism, it may be possible to impersonate the tag by capturing the contents of its transmission and replaying them later.

for the server is reduced to $m^{1/t}n$ to identify n tags in a database containing m entries. This approach was put forward by Molnar et al. (2005), and has the benefit of making the attack more scalable, as the ratio between the back-end server identification workload and the attacker desynchronization workload, $m^{1/t}$, can be reduced to a constant by choosing t of order log(m). However, even if relatively smaller values for t are used—say, $t=2$ or 3—this implies a significant increase in the complexity of the tag computations, for instance, the tag will require logic to rekey its cryptographic function several times during one transaction. Another difficulty of this method is that the higher levels of the hierarch include keys that are shared among many tags, leading to privacy risks, in particular if key compromise (known-key attacks) are considered as part of the threat model.

A different approach involves using time-memory trade-offs to speed up key search at the back-end server (Avoine and Oeschlin, 2005). The approach reduces the back-end server workload to identify a single tag from $O(m)$ to $O(m^{2/3})$. Although a significant improvement, this method requires expensive precomputation steps.

A different strategy to private identification that does not require cryptographic operations, and instead uses only rotations (bit shifts and selections) was described in (Juels, 2004a). However, that minimalist solution assumes a bound on the number of interactions between the tag and reader. Under such assumption, the resynchronization approach as described earlier is scalable and provides information-theoretic security if the tags can afford to compute pseudo-random functions.

Recently, a process has been described that strengthens the minimalist approach, to include also bit complementation (Castelluccia and Soos, 2007). To discover the key after a number of (rotated and potentially complemented) bit disclosures, the back-end server (or an adversary) must first solve a set of linear constraints, which can be shown to be equivalent to solving a SAT instance. The security therefore depends on the difference between the hardness of solving SAT instances with prior knowledge of a relatively small set of possible solutions—i.e., the set of valid keys (known to the back-end server)—versus solving the same instances without such special knowledge, as in the case of a would-be attacker. This approach, however, does not scale as it requires search through all valid keys.

35.6.4 Future Directions

The earlier-mentioned solutions represent different compromises. They suffer from various scalability issues, may present secondary privacy threats, or may have drawbacks such as requiring more complex tags or extensive precomputation. Indeed, at the time of the writing of this chapter, the authors are not aware of scalable solutions that fully reconcile anonymity and availability requirements, while using only efficient and simple symmetric-key constructions. However, at the pace with which new ideas and strategies are being introduced, further research is likely to provide practical and elegant solutions to this issue in the not distant future.

35.7 Resources Available for RFID Security Mechanisms

Having discussed the remaining challenges attending the simultaneous provision of anonymity and availability in RFID authentication protocols, we now describe the set of tools that are available to their solution. This section describes capabilities available to RFID tags for security measures, considering primarily the requirements of passive tags.

The minimal circuit area used by each feature, in terms of NAND gate-equivalents (GE), is a crucial measure of their feasibility for RFID implementations, with the lowest gate-count being preferred. This is because current technology can only provide a few thousand GEs for security in the higher-end tags. A common rule-of-thumb is that developers will count with no more than 1/3 of the circuitry to be devoted to security operations.

An even more significant concern than the circuit area is the power demand of the RFID tag's circuit pipeline. Passive RFIDs are powered by inductance, and the lower the wattage that their operation requires, the longer the distances from which they can be scanned. Since reader reach is a crucial measurement of usability for RFIDs, both the per-cycle maximum power and the per-cycle average power required by RFID circuits are restricted (Feldhofer and Rechberger, 2006).

35.7.1 Transient Storage

Transient storage is implemented using latches, and, therefore, is nothing less than additional circuitry. The cost of one bit of transient storage is estimated as approximately 8 gate equivalents (Feldhofer and Rechberger, 2006).

35.7.2 EPROM and EEPROM

Both require higher resources and higher power utilization than transient storage. Therefore, they should be used sparingly for security features. Among the types of measures that become available through the use of EPROM and EEPROM are key updates to provide forward-security, and key changes to deal with revocation of tags without need for issuance of new ones.

35.7.3 Time-Out Mechanisms

RFID passive tags do not maintain clocks or keep time. However, the activity time span of a tag during a single session can be limited using techniques such as measuring the discharge rate of capacitors (Juels, 2004b).

35.7.4 Asymmetric Cryptographic Primitives

The use of special architectural constructions (digit-serial multipliers) may make it possible to achieve elliptic curve cryptography (ECC) implementations in as few as 6300 GEs (Batina et al., 2007), though the numbers get worse if such constructions are not available (approx. 11,446 GE). These numbers indicate at least the possibility that ECC may eventually be available in high-end, passive tags. Still, taking into consideration pricing pressures and the current state of the art in passive tag technology, it appears unlikely that RFID deployments based on public-key cryptography will be common in the next few years.

35.7.5 Symmetric Cryptographic Primitives

Among the symmetric key primitives, not all are equally amenable to implementation.

- *Block ciphers*—Block ciphers suitable for RFID implementations are those that have been designed to achieve highly efficient hardware optimizations under constrained memory conditions. In particular, substantial work has been done that validates the suitability of AES for RFID tags (Feldhofer and Rechberger, 2006), with full implementations requiring as few as 3300–3400 GE.

- *Pseudo-random number generators*—PRNGs are a flexible primitive, and can be build from block ciphers in counter mode, from stream ciphers (the key stream is pseudo-random), and independently using other technologies, such as LFSR-based generators (Coppersmith et al., 1994), providing flexible trade-off opportunities between security and efficiency needs.
- *Stream ciphers*—Choices of stream ciphers for RFID are made more complex due to the lack of standardized choices. This situation could improve soon with the conclusion (projected for May 2008) of the stream cipher selection process (project eSTREAM), which is part of the European Network of Excellence for Cryptology (ECRYPT) research initiative. Several candidate stream ciphers have been selected for the final phase of the evaluation process on the basis of their performance characteristics in hardware implementations. An evaluation of the candidate ciphers shows a count of fewer than 1300 gates for Grain80 up to approximately 4760 gates for F-FCSR-H among ciphers claiming 80-bit security, and a low of fewer than 1860 gates for Grain128 up to approximately 5040 gates for Mickey128 for 128-bit security ciphers. Considering the number of finalists which have so far resisted cryptanalysis and the good miniaturization characteristics of several of the project finalists, it looks like the promise of a standardized, RFID-suitable stream cipher will be fulfilled in the near future (Good and Benaissa, 2007).
- *Pseudo-random functions*—One advantage in designing PRF-based security for RFID is the amount of choice available in the construction of PRFs. They can be composed using the strategy of cascading a PRG, which while being a relatively slow method, results in little gate-count and per-cycle power overhead over the underlying PRG (Goldreich et al., 1986). Block ciphers also provide ready implementations of the PRF primitive.
- *Hash functions*—Counter to intuition, hash functions seem at the moment to be less suitable for RFID implementation than the most efficient block ciphers. The reason is that the design of many collision-resistant hash functions uses an underlying block cipher with a large block length—which optimizes the software performance of the hash function by allowing it to process large chunks of data at a time, but makes it inefficient to implement under constrained-memory settings. Their use is discouraged in favor of other primitives whenever collision-resistance is not a requirement (Feldhofer and Rechberger, 2006).
- *Message authentication codes*—Implementations based on block ciphers, e.g., CBC-MAC, or on universal hashing, are preferred to those based on hashes, such as HMAC, due to the earlier-mentioned disadvantages of implementing hashes in RFID tags.

35.7.6 Other Primitives

Some RFID protocols, in particular those devoted to identification—as opposed to authentication—employ noncryptographic primitives, such as XOR operations, fixed-precision arithmetic operations, bit shifts, or linear-feedback shift registers. The entire reason for proposing such primitives is that they can be implemented with a low number of gates and within the per-cycle power envelope available for typical RFID architectures. Accordingly, their suitability is dependent on whether they possess the characteristics demanded by their role in supporting security, and must be assessed in each protocol specification.

35.8 Conclusion

RFID is a recent technology, and its attending security concerns have only recently been recognized, starting with the seminal work of the Auto-ID Labs at MIT (Sharma et al., 2001). Since then, it has attracted the attention of several imminent researchers who proposed innovative and creative security approaches. Important security issues are still at the forefront of research, and as RFID security applications increase, this research area is guaranteed to remain a fertile and relevant field of study.

References

Avoine, G., and Oechslin, P. 2005. A scalable and provably secure hash-based RFID protocol. In *IEEE International Conference on Pervasive Computing and Communications (PerCom 2005)*, Kauai, Hawaii, March 2005, IEEE Press, pp. 110–114.

Batina, L., Guajardo, J., Kerins, T., Mentens, N., Tuyls, P., and Verbauwhede, I. 2007. Public-key cryptography for RFID-Tags. *IEEE International Conference on Pervasive Computing and Communications Workshops*, New York, March 2007, pp. 217–222.

Burmester, M., van Le, T., and de Medeiros, B. 2006. Provably secure ubiquitous systems: Universally composable RFID authentication protocols. *2nd IEEE/CreateNet International Conference on Security and Privacy in Communication Networks (SECURECOMM 2006)*, Baltimore, Maryland, August 2006.

Castellucia, C. and Soos, M. 2007. Secret shuffling: A novel approach to RFID private identification. *International Conference on RFID Security*, Malaga, Spain, July 2007.

Coppersmith, D., Krawczyk, H., and Mansour, Y. 1994. The shrinking generator, *Advances in Cryptology—CRYPTO'93*, LNCS, Springer-Verlag, pp. 22–39.

Dolev, D. and Yao, A. 1983. On the security of public key protocols. *IEEE Transactions on Information Theory*, 29(2): 198–208.

Engels, W.D. 2001. The reader collision problem. Whitepaper. #MIT-AUTOID-WH-007, Auto-ID Center, Massachusetts Institute of Technology. http://autoid.mit.edu/whitepapers/MIT-AUTOID-WH-007.PDF.

EPCglobal. 2005. Class 1 generation 2 UHF air interface protocol standard. http://www.epcglobalinc.org/standards/uhfc1g2/.

Feldhofer, M. and Rechberger, C. 2006. A case against currently used hash functions in RFID protocols. On the move to meaningful internet systems: OTM 2006 workshops. *Lecture Notes in Computer Science*, 4277: 372–381.

Goldreich, O., Goldwasser, S., and Micali, S. 1986. How to construct random functions. *Journal of the ACM*, 33(4): 792–807.

Good, T. and Benaissa, M. 2007. Hardware results for selected stream cipher candidates. eSTREAM Phase 3, ECRYPT Network of Excellence within the Information Societies Technology (IST) Programme of the European Commission. http://www.ecrypt.eu.org/stream/papersdir/2007/023.pdf.

ICAO. October 2004. Document 9303, Machine readable travel documents.

Juels, A. 2004a. Minimalist cryptography for low-cost RFID tags, in Blundo, C. and Climato, S. (eds.), *International Conference on Security in Communication Networks*, Amalfi, Italy, September 2004, *Lecture Notes in Computer Science*, Vol. 3352, pp. 149–164.

Juels, A. 2004b. "Yoking-proofs" for RFID tags. *IEEE Conference on Pervasive Computing and Communications Workshops*, Orlando, March 2004, pp. 138–143.

Juels, A., Molnar, D., and Wagner, D. 2005. Security and privacy issues in e-passports. *First International Conference on Security and Privacy for Emerging Areas in Communications Networks (SecureComm)*, Athens, Greece, September 5–9, pp. 74–88.

Juels, A., Rivest, R.L., and Szydlo, M. 2003. The blocker tag: selective blocking of RFID tags for consumer privacy. *ACM Conference on Computer and Communication Security*, pp. 103–111.

Juels, A. and Weiss, S. 2006. Defining strong privacy for RFID. International Association for Cryptologic Research, Technical report # 2006/137. http://eprint.iacr.org/2006/137.

van Le, T., Burmester, M., and de Medeiros, B. 2007. Universally composable and forward-secure RFID authentication and authenticated key exchange. *ACM Symposium on Information, Computer, and Communications Security (ASIACCS 2007)*, Singapore, March 2007, ACM Press, pp. 242–252.

Molnar, D., Soppera, A., and Wagner, D. 2005. A scalable, delegatable pseudonym protocol enabling ownership transfer of RFID tags, Workshop on selected areas in cryptography, *Lecture Notes in Computer Science*, Vol. 3897, pp. 276–290.

Ohkubo, M., Suzuki, K., and Kinoshita, S. 2003. Cryptographic approach to "privacy-friendly" tags, *RFID Privacy Workshop*, Massachusetts Institute of Technology, Cambridge, November 2003.

Oren, Y. and Shamir, A. 2006. Power analysis of RFID tags. Advances in Cryptology, CRYPTO Rump Session.

Pering, T., Ballagas, R., and Want, R., 2005. Spontaneous marriage of mobile devices and interactive spaces. *Communications of the ACM*, 48(9): 53–59.

Sharma, S.E., Weiss, S.A., and Engels, W.D. 2001. RFID systems, security and privacy implications. Whitepaper #MIT-AUTOID-WH-014, Auto-ID Center, Massachusetts Institute of Technology, http://autoid.mit.edu/whitepapers/MIT-AUTOID-WH-014.PDF.

Sharma, S.E., Wang, S.A., and Engels, D.W. 2003. RFID systems and security and privacy implications. *Proceedings of the Workshop on Cryptographic Hardware and Embedded Systems* (*CHES 20002*). *LNCS*, Vol. 2523, pp. 454–469.

U.S. Department of State. 2005. 22 CFR Part 51, Public Notice 4993, RIN 1400-AB93, Electronic Passport. Federal Register, 70(33), Proposed Rule (18 February 2005). http://a257.g.akamaitech.net/7/257/2422/01jan20051800/edocket.access.gpo.gov/2005/05-3080.htm.

Want, R. 2006. An introduction to RFID technology. *IEEE Pervasive Computing*, 5(1): 25–33.

Willis, H.H. 2005. Evaluating the security of the global containerized supply chain, TR-214-RC, RAND Corporation.

36

Security and Privacy of RFID for Biomedical Applications: A Survey

Ellen Stuart, Melody Moh, and Teng-Sheng Moh

CONTENTS

36.1 Security and Privacy of RFID for Biomedical Applications

Implantable RFID tags are a commercially available technology, used for implantation in animals since the 1980s and more recently approved by the FDA for use in humans. Troyk describes the development of this class of RFID tags, citing a patent application in 1979 for an RFID type device intended for subdermal implantation designed to be used to locate and track pets, livestock, and migratory and endangered animals (Troyk, 1999). Garfinkel also describes the evolution of these devices, originating with a series of patent applications in 1986 for a syringe-implantable identification transponder to be used for identifying horses (Garfinkel, 2002). This section will describe possible applications of RFID technology within the field of biomedical engineering and discuss the applicability of various security-enabled protocols to these scenarios.

36.1.1 Biomedical Applications of RFID Systems—Subdermal Tags

Literature on implementations of RFID technology within the medical industry focuses on identification and tracking use cases that are common in the supply chain management arena. RFID shows promise in reducing not only the losses due to lost pharmaceuticals and medical equipment, but also the mistakes due to forgotten equipment and treatment miscommunications (www.drugresearcher.com, 2004a,b). The biomedical applications detailed here will move away from the use of typical RFID tags and protocols into implementations that utilize RFID tags to transcutaneously return sensory information or receive commands that initiate action after receiving power in a similar transcutaneous manner.

Both Troyk (Troyk, 1999) and Garfinkel acknowledge the first patented device, applied for in 1991 and issued in 1993 to Destron/Identification Devices Inc. and Hughes Aircraft. Over the past decade, this RFID tag footprint has been reduced to the size of a grain of rice, allowing for subdermal placement using a 12-gauge needle (Garfinkel, 2002). Researchers have also managed to increase the functionality and transmission capabilities of the tags, making marked improvements in both speed and distance. In 1999 Troyk described the RFID equipped injectable micromodule with a cylindrical hermetically sealed glass capsule about 2–3 mm in diameter, 1–1.5 cm in length, and containing a microcoil antennae, capacitor, and a chip (Figure 36.1).

Currently, Applied Digital Solutions (ADS) manufactures implantable tags, with intended functions such as subdermal GPS personal location devices (PLD), for identification and authentication. VeriChip, a subsidiary of Applied Digital, has the only patented implantable RFID tag that is cleared by the FDA for use with humans. The tag operates at a frequency of 125 kHz and measures 11.1×2.1 mm and can be implanted using local anesthesia and either 12-gauge needle or a very small incision.

The VeriChip tags are similar to other RFID tags, but are designed to be inserted just below the skin and have a very limited reading range because of the limited power and lower operational frequencies. Current research indicates that implantable RFID tag technology is completely limited to the passive tag systems that have been described in this work. The passive tag devices, because of their lack of a battery, lack the power capabilities of active tags, but have a superior life span because they are not limited by an internal battery life span.

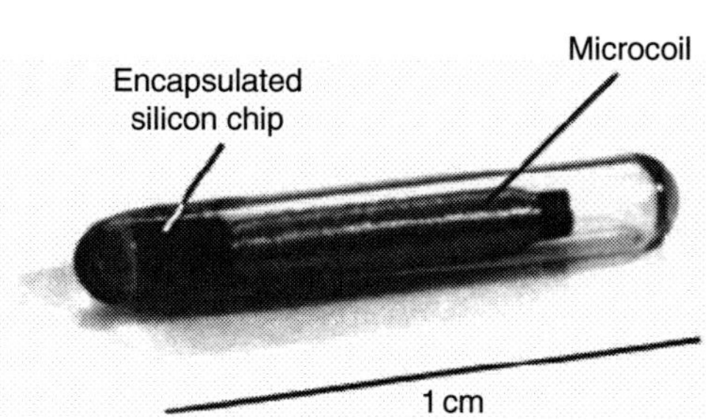

FIGURE 36.1
Example of an implantable RFID tag, also referred to as a micromodule. (From Troyk, P.R., *Ann. Rev. Biomed. Eng.*, 1, 177, 1999.)

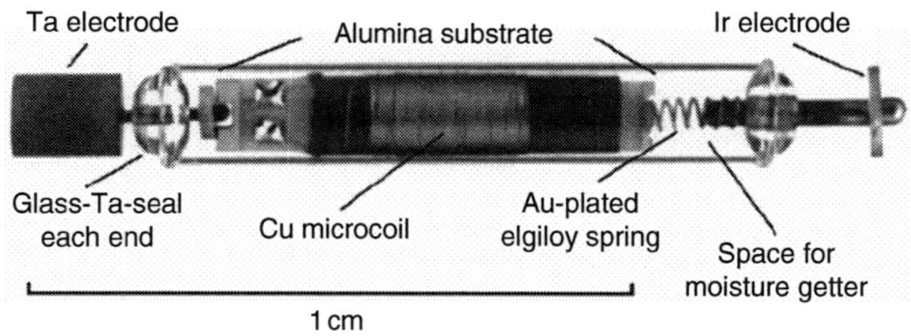

FIGURE 36.2
Example of a microstimulator. (From Troyk, P.R., *Ann. Rev. Biomed. Eng.*, 1, 177, 1999.)

Troyk also introduces an enhanced version of the implantable RFID micromodule, the microstimulator (Figure 36.2), which supplements the traditional RFID micromodule tag with electrodes that are used to stimulate muscle tissue when given instruction from the extracorporeal reader. Applications for this RFID system would include functional neuromuscular stimulation (FNS) (stimulation of paralyzed muscle), treatment of foot-drop in hemiplegic stroke patients with damage to the peroneal nerve, joint instability, post-operative rehabilitation, sleep apnea, and other neuromuscular conditions. Potential and partially realized applications include the following.

36.1.1.1 Chipping Humans

The practice of chipping, inserting an RFID tag below the surface of the skin, has a diverse set of applications, such as those marketed by VeriChip and Digital Angel Corporation. These applications include systems that would prevent elderly members of our society from being lost and babies from being accidentally switched or intentionally stolen by integrating RFID tags with automated door locks. The systems would prevent the tag, either worn by an elderly patient wandering around a care facility or held by a kidnapper carrying a tagged-baby, from leaving a building. RFID tag devices have already been implemented into some of the prison systems, used to identify and monitor inmates.

36.1.1.2 Cardiac Monitoring/Electrocardiography

A proposed RFID-enabled Holter monitor would utilize a personal RFID system reader to transmit event data to a store-and-forward application host device that would store irregular EKG events and forward the event data to the hospital at regular intervals. This version of the Holter monitor would provide patients with a wireless wearable tag and reader system, allowing them to be more mobile while potentially providing them with enhanced sensory and reactive functions. More advanced RFID systems could include chips programmed to recognize the patterns corresponding to severe heart events and automatically trigger an emergency request for paramedics (Weiss, 1998).

36.1.1.3 Pacemaker Augmentation

Digital Angel Corporation recently introduced BioThermo, a new temperature sensing implantable RFID tag device (www.adsx.com). These advances in sensory enabled tagging show promise for sensory enabled applications of RFID systems. Sensory enabled RFID microstimulators could potentially replace the wired architecture of current systems.

Implanted passive tags would have a longer life span and readers either could be worn externally or could continue to be placed subdermally. A wireless transdermal system could provide longer, more efficient recording of heart activity and the absence of wiring would result in a less invasive pacemaker implantation procedure.

36.1.1.4 Seizure Disorder Monitoring and Treatment

Similar to the pacemaker system described earlier, sensory enabled RFID tags have the potential for integration into devices to monitor other electrical abnormalities in the neurosystem. Utilizing Electroencephalography (EEG) monitoring, abnormal electrical activity can be localized and can guide treatment efforts for seizure causing disruptions.

Cyberonics, Inc. developed a product called the NeuroCybernetic Prosthesis (NCP), a battery-powered transmitter, which treats depression and seizures by sending timed electrical impulses to the brain (CNN, 1999; Koren and Holmes, 2006). This pacemaker for the brain has been available since 1997. Potential implanted sensory microstimulators could report data and respond to external sources with increased processing capacities. Using RFID, NCP could reach a longer life span and would again eliminate the wired solutions currently used. In his paper *Implications of Silicon Monolithic RFICs for Medical Instrumentation and Telemetry*, Weiss proposes a device that can detect and treat seizures. Despite Weiss' commentary regarding the speculative state of the proposed implementation, concurrent developments by Cyberonics provided the proof of concept.

36.1.1.5 Infant and Pediatric Thermometry, Data Reporting

The recent developments by Digital Angel Corporation also allow for growth in the field of thermometry devices. Doctors might use these systems in cases such as those involving hypothermic exposure, where changes in temperature require immediate diagnosis and treatment.

36.1.1.6 Continuous Blood Pressure and Glucose Monitoring

Currently, blood pressure is measured and monitored using a computer to analyze measurements from an arterial line fixed with a transducer, a hollow bore needle that gauges pressure changes. With a wireless RFID implementation, doctors and patients would not be a hindered by connections when attempting to relocate or position themselves around patients. Similarly, diabetic patients require continual monitoring of blood glucose levels, especially for a new onset diabetic. Weiss describes RFID sensor technology (Figure 36.3) as ideally suited for utilization with chemical analysis applications for the human body. Weiss presents an implantable monitoring system, placed directly onto the wall of an artery. The system would alleviate the current need to withdraw blood to complete the test. The implanted sensor would detect changes in blood glucose levels by measuring the change in oscillation of the acoustic wave generated by a surface acoustic wave resonator (SAW).

36.1.1.7 Infant and Pediatric Sleep Apnea Monitoring

Weiss cites the monitoring of infants with sleep apnea as another possible implementation of RFID technology. Independent of actual causes of sudden infant death syndrome (SIDS), it seems reasonable to believe that some type of portable monitoring system, not necessarily invasive, could be developed as an enhanced baby monitor (www.sleepfoundation.org).

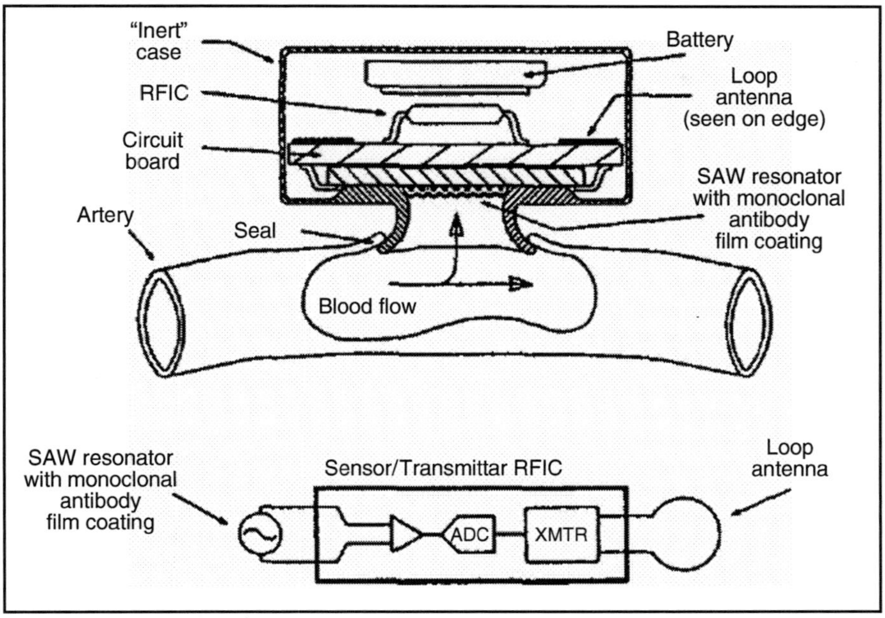

FIGURE 36.3
A blood glucose RFID diagram. (From Weiss, F.G., *Topical Meeting on Silicon Monolithic Integrated Circuits in RF Systems, Digest of Papers*, Ann Arbor, Michigan, September 17–18, 1998, pp. 195–204.)

36.1.1.8 Prosthetic Control—Implantable Myoelectric Sensors (IMES)

Weir, Troyk, DeMichele, and Kuiken describe their preliminary work in the creation of an upper limb prosthetic control system that offers up to eight degrees-of-freedom, integrating myoelectric sensors and RFID technology to provide a more seamless sequential control, eliminating intermediate steps and greatly increasing control of more refined movements (Weir et al., 2003). The reader relays data to a controller that processes and interprets the information into commands for desired movement (Weir et al., 2003). The resulting system would provide mobility in range of movement and in the lack of connected wires. Although Weir and his colleagues had only proposed and conducted preliminary tests for an RFID-enabled prosthetic arm and hand system, their results indicate such a system is indeed possible. Recent work with monkeys and humans has shown promise in the further advancement of prosthetic limb control (//news.bbc.co.uk/1/hi/health/4275245.stm).

36.1.1.9 RFID with Muscle Paralysis

One of Troyk's earlier papers (Troyk, 1999) details the use of RFID in creating an FNS system for patients with permanent muscle paralysis (Figure 36.4). Troyk proposes using

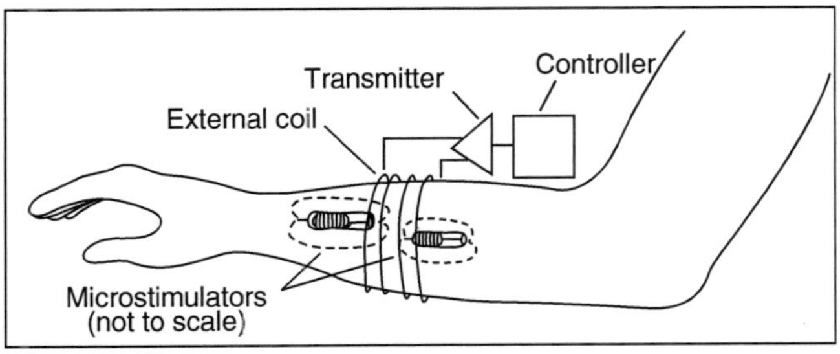

FIGURE 36.4
General concept proposed by Troyk to address muscle paralysis. (From Troyk, P.R., *Ann. Rev. Biomed. Eng.*, 1, 177, 1999.)

implanted microstimulators to respond to control signals emitted from an external transmitter by emitting a charge from the electrodes and consequentially stimulating a paralyzed muscle. In his description of the FNS system, the communication from the reader signals action by the transponder, but the RFID microstimulator has no mandated reverse telemetry exchange, contrary to mainstream RFID applications that focus on data sent from tag to reader. When an activation signal is received by the microstimulator, the stimulating electrode is activated, emulating a neurological signal sent to the muscle, and causes the muscle to contract. Systems using wired transdermal connections are already in place, but because of the intent of the system, to create movement, wireless solutions provide an obvious advantage.

36.1.2 Security Issues in Biomedical RFID Systems

RFID enhancement of existing biomedical systems offers many benefits to patients and to the medical industry. System life span can be extended, maintenance reduced, and monitoring systems more automated. With this come certain drawbacks, many of them analogous to the vulnerabilities created when moving from a wired network solution to wireless access. This section will examine considerations, risk factors, and security and privacy issues associated more specifically with biomedical applications.

36.1.2.1 Considerations and System Vulnerabilities

36.1.2.1.1 Health Risks of Implant

Prior to clearing the VeriChip tag for subcutaneous applications in humans, the FDA completed a risk assessment in which it identified the potential health risks of the implants. The FDA cites adverse tissue reaction, transponder migration, transponder failure, electromagnetic interference, electrical hazards, MRI interaction, and needle stick/trauma (Nsanze, 2005). In a study of microstimulator implants, Walter et al. analyze the short-term effects of implantation into cats (Walter et al., 1997). In this study, inactive implants were attached to the bladder wall lining for eight weeks, during which all of the cats appeared healthy. The bladder tissue surrounding the implants was removed and the implants were found to be attached, encapsulated by a thin layer of connective tissue without any apparent defects, physical damage, migration, or inflammation beyond what was considered to be in range of the normal reaction to implantation of a foreign body.

Further studies by Cameron et al. noted the variation of inflammatory responses depending on the placement of microstimulator devices implanted for up to 3 months (Cameron, 1998). She compares the reaction to other types of implanted foreign bodies, noting the progression of healing, tissue development, and the effectiveness of stimulation in producing a stable desired muscle contraction, noting that damage to muscles tissue caused by the improper or overuse of a stimulator was present, but also present with vigorous exercise.

Lamberg studied the effects of MRI on implanted medical devices containing metal, focusing on the VeriChip RFID tag device cleared by the FDA and organizing his study into four test categories: device movement, device heating, image distortion, and device operation (Lamberg, 2004). He determines the likelihood of device migration by measuring the worst-case scenario of forces exerted on the device against the force needed to tear the VeriChip device away from connective tissues. It resulted in a slight "tugging" sensation, not considerable enough to cause pain, but enough to require caution and further testing, before placing them near large vessels or organs, given the severity of possible side effects. The device showed no noticeable heating, but resulted in a loss of signal around the antenna area and a distortion of more than 10% around the barrier of the device within the image testing. However, these effects were not enough to obscure the reading of the image and, depending on the implant location, would not likely cover up relevant information. After completing the three previous tests, Lamberg retested the VeriChip devices and found that one of the two devices did not produce a signal.

36.1.2.1.2 Weak Encryption

In many of the more common applications, RFID system vulnerabilities have serious risks to personal privacy. For biomedical applications of RFID technology; however, compromises to an implanted tag can be life threatening, causing serious bodily harm or possibly death. System weaknesses described in the typical large-scale supply chain and tracking systems are due in part to the limitations of RFID Cryptosystems, a result of the inadequate length of the key, reusing of key between different owner entities, and the limited resources that constrain algorithm improvement.

As with other cryptosystems, it is important to implement proven, industry standards using sufficiently long keys. While this does not seem feasible in systems that are strictly regulated by the difference in tag cost of a penny, the relatively small, closed systems that characterize the biomedical applications lack these severe restrictions on tag cost. Moore's Law seems to hold weight in systems where the reduction of the tag footprint and the advancement of integrated functionality far out-weigh any cost restriction. Proven industry cryptographic standards and long key lengths do not appear to be out of reach given this pace of development.

Possible compromise to one of these systems is completely unacceptable and should deter implementation. The trade-off made in replacing the use of a proven cryptographic that might reveal traceable signals with the use of a protocol that sacrifices system security in order to create obscurity in terms of tracking, does not seem reasonable.

In the examples provided in the previous section, the RFID system is limited to a set membership of tags and readers. Except in the subdermal information systems example, the tag is owned by a single or set group of readers. The readers in turn have a set group of tags. Security protocols determined to be unrealistic in large-scale systems that require brute force processing might be more plausible in these limited systems.

The subcutaneous placement of tags in these systems prevents the tags from being available for tampering. Because tags in these systems are in reasonably fixed locations relative to their readers, transmission range of the tags can be specified to allow for increased security. Conversely, tags may still receive transmissions from foreign readers and faraday cage solutions are not a feasible solution.

36.1.2.2 Personal Privacy

Monitoring and tracking of implanted devices outside of the designed purpose is a critical issue, as it not only describes ownership of a marked item but the actual person. A trackable tag communication system can result in the direct identification and monitoring of an individual with an implanted medical device. This might be especially compromising with the identification of tag communication associated with a given medical device. This issue seems is addressed using the combination of privacy ensuring protocols and strictly tuned transmission ranges for all system devices.

36.1.2.3 Interference

Any devices operating at the same frequency can interfere with each other. Depending on the specific implementation, RFID systems might experience interference from other RFID devices, cell phones, automatic door openers, microwaves, and other consumer devices that operating mainly in the 900 MHz band (commercial RFID operates mainly in the 900 and 13.56 MHz frequency bands). Also, as expected, devices operating within the same system often interfere with each other. This is especially true in biomedical applications where multiple tags are implanted in close proximity under the control of a single reader. Solutions for interference among RFID tags and other devices must be addressed in anti-collision or singulation layer 2 MAC protocols.

36.1.2.3.1 RF Exposure

Although there is no proven association between prolonged exposure to radio frequency electromagnetic radiation (RFR) sources and cancer, the proliferation of cell phones in our society has generated considerable debate regarding the health effects of exposure to RFR and discussion of various transmission levels (RFR covers the frequency range of 3 kHz–300 GHz). One of the systems described by Troyk utilized the 2 MHz band with tag stimulus amplitude fixed at a constant current of 10 mA. Relative to Table 36.1, the various microstimulators used in Troyk's studies appear to be well within the safety levels defined by IEEE (IEEE, 1999).

TABLE 36.1

Effectiveness of Referenced Security Protocols

Method	Mechanism	Suitable for Biomedical Applications
Rieback, Crispo, & Tanenbaum	Proxy—RFID Guardian	Yes, implemented at the reader
Juels, Rivest, & Szydlo	Blocker tag	No, detrimental to surrounding systems
Feldhofer, Dominikus, & Wolkerstorfer	AES variation	Yes, must be coupled with other mechanism to protect privacy
Ayoade, Takizawa, & Nakao	Centralized authority	Inefficient
Sarma	Centralized authority	Inefficient
Sarma	Update algorithm using meta-id	Not secure
Dimitriou	Update algorithm, shared secret	No, highly susceptible to desynchronization
Dimitriou	Update algorithm, shared secret, mutual authentication	No, susceptible to desynchronization (although more challenging, would render system useless)
Molnar & Wagner (library check out)	Challenge–response using updated meta-id	No, not secure
Molnar & Wagner (enhanced version)	Mutual authentication challenge–response using PSF generated nonce	Yes, implanted tags are not susceptible to tampering without notice
Ohkubo, Suzuki, & Kinoshita	Hash chain protocol	Precomputation and rebuilding of the hash chain is a deterrent in time sensitive systems
Avoine, Dysli, & Oechslin	Mutual authentication using hash chains	Precomputation and rebuilding of the hash chain is a deterrent in time sensitive systems
Molnar, Soppera, & Wagner	Centralized authority	No, temporary access does not provide additional security and requires increased processing
Hopper & Blum	Challenge–response rounds	No, not secure against active attack
Juels & Weis	Challenge–response rounds using random binding factors	No, not secure against active attack

36.1.3 Biomedical Security Solutions

This section will assume a biomedical system model: a functional electrical stimulation system, such as the one described by Troyk containing between 10–20 tags placed into muscles tissue and a single external reader (Troyk, 1999). This model is designed to represent the general requirements and constraints of these and other biomedical RFID systems in the discussion of the applicability of proposed RFID protocols and security mechanisms to this application area.

While RFID systems include substantial communication between the reader and the application host, we will focus our study in this chapter on the protocols involving tag-reader communications utilized with passive devices. The security and privacy solutions presented will include proposals of that implement one or more of the following mechanism to achieve a security or privacy solution:

- Layer 2 privacy control
- Proxy—the use of an intermediary device to prevent direct interaction between a reader and the tags it is assigned to
- Encryption—depending on tag, may or may not be feasible
- Centralized authority—grants and restricts access to time sensitive key; allows for items to pass between different owners while preventing residual ownership
- Challenge–response authentication protocols
- Update algorithm for secret keys—perhaps based on shared secret; related to use of centralized authority

36.1.3.1 Layer 2 Privacy Control

As Molnar and Wagner cite in their study of library-based RFID systems (Figure 36.5), much of the current RFID hardware is hard coded to the degree that tracking of the individual items and their associated owners could not be prevented (2004). This lack of privacy stems from the implementation of layer 2 anticollision MAC protocols that use a unique identifier to determine the computed wait time. While systems that utilize these practices need to be reevaluated, a possible patch to this weakness would be the use of either a Faraday cage type of shield or the implementation of some other type of proxy system that shields tag transmissions from outside readers.

36.1.3.2 Proxy

Rieback, Crispo, and Tanenbaum present a solution in the form of an intermediary personal device that shields interaction between member tags and inquiries from a foreign environment (Rieback et al., 2005). The platform they suggest, the RFID Guardian, is a portable, battery-powered device that would provide a centralized entity, with a higher

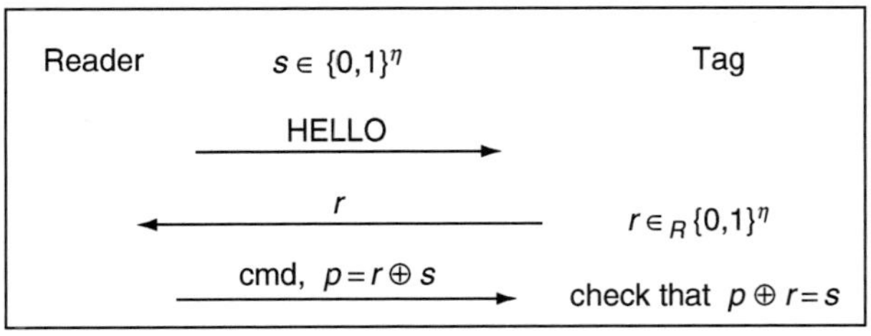

FIGURE 36.5
Protocol for Library System Checkout Scenario. (From Molnar, D. and Wagner, D., in *Computer and Communications Security*, Pfitzmann, B. and McDaniel, P., eds., ACM Press, 2004.)

capacity for processing and communication between tag and reader. It would be a secure, singly authorized proxy for tags within the head to toe range of an individual that would be portable and owner-activated. The Guardian would not only be used to monitor and regulate all tag communications for all tags within range that it has access to, but also to perform direct communication in response to readers' queries.

This solution would utilize RFID functionalities within a secure location (e.g., a person's home) but prevent use once outside of the secure location. The development of this technology must coexist with development of a mechanism for extended control over tag devices, specifically, the ability for owners to interact with tags, authenticating readers and providing information to items such as laundry, doors, refrigerators, and such, while restricting unauthorized readers. The idea of a personal shield around RFID communications seems like a feasible concept for consumer and other user-based applications, with promising benefits to systems that interact with multiple readers and have a pressing need for authentication resources.

A predecessor to the Guardian, Juels et al. (2003) describe the use of another physical mechanism that acts as a shield rather than a proxy. This blocker tag would be used to prevent a reader from effective querying and identifying tags within a given range. While these types of devices might provide some security, Juels calls attention to the potential for this type of instrument to be misused to block legitimate RFID systems in denial-of-service (DoS) attacks (unintentional and intentional) and to circumvent other systems.

In the model system, Rieback's RFID Guardian would function as an intermediary personal device, shielding interaction between subdermal member tags and inquiries from the surrounding environment. This Guardian could be implemented with the reader associated with the system, and enforced using MAC protocols and authentication schemes. The reader would be the logical centralized authority as it is consistently located within range of the tags.

In the model system, however, the majority of information should be blocked from reaching the tags, as there should be no access by other devices, exchanges should be limited to the closed reader and tag system. The application of a proxy device in the model system either would most likely resemble the blocker tag idea (Juels et al., 2003) or, if implemented at layer 2, would result in a less disruptive reader response to outside queries of silence. Since the blocker tag might create denial-of-service attack in surrounding systems and the reader cannot be reasonably shut-off for intervals when the patient enters an RFID-rich environment, a separate MAC layer solution would be necessary to implement the silencer. This solution most likely would not implement an owner controlled deactivation switch except to disable the reader to enable testing, tuning, or updates by another authority.

36.1.3.3 Encryption

With current supply chain systems, emphasis is placed on reducing implementation cost of the new, often mandated systems. The lack of adequate security available to the lower cost tags does not seem to deter the push to minimize cost. In medical devices such as the model system, the limited number of tags, importance of their function, expected lifetime, and effect of their failure leads to an emphasis on quality and functional capability. The limitation in terms of functionality and security is generated by the restricted size of implantable tag devices.

Following Moore's Law at our rate of development, the complexity of the integrated circuit embedded in the RFID tag, with respect to minimum component cost, will double approximately every 18 months. With supply chain systems, the resulting decrease in cost is not necessarily an incentive to upgrade functionality, but more promise of reduced implementation costs, making widespread acceptance more feasible. In medical systems, this does not translate to reduced cost, but increased functionality and an increased allotment of resources for security enhancement.

The challenge–response based authentication protocol developed by Feldhofer et al., which implements a variation of the Advanced Encryption Standard (AES), is achievable for RFID tags meeting the hardware requirement of 3595 gates (Feldhofer, 2004). This would be plausible for the less cost sensitive medical system model being considered.

36.1.3.4 Centralized Authority

The use of a centralized authority has been suggested in various RFID security schemes. Although these systems might prove impractical in a broad supply chain implementation, the approach might prove relevant for the closed model system.

Ayoade et al. (2005) present a framework that uses a centralized authority to provide decryption keys to registered readers. This authority contains decryption keys for registered tags. When a registered reader receives an encrypted response from a tag, it queries the authority for the decryption key, and the central authority determines the access rights for the reader. While this protects the information contained within the tag, Ayoade's system does not protect the privacy of the owner of the tag.

In the framework that Ayoade presents, the reader in our model system would have access to the decryption keys and because of the limited amount of keys, could easily store specific information for all of the tags. Because the system involves a single, seldom changing reader, a protocol that would require the reader to continually refresh authority over the tags, is inefficient.

Sarma proposes a minimalist tracking system involving the EPC code, an Object Name Service (ONS), the Savant, and tags that implement password protected self-destruct commands (Sarma et al., 2002). The system uses the ONS (similar idea to domain name system (DNS)) to map the EPC to an IP address where tag information can be accessed. The Savant is the management system used to process reader captured data, retrieving more detailed information regarding the tag. The system was designed to reduce the processing, power, and memory requirements on the tags and create more stability and scalability. Using the IP mapping might instead limit the scalability by creating the need for a centralized ownership of the tags, an undesirable quality in supply chain systems, but possibly favorable for other types of centralized wholly owned systems. Since the ownership in the model system remains constant, the added communication involved in accessing tag information would not produce benefits while also presenting vulnerability to attack. The concept of a centralized authority, while relevant to the model system, is most likely to be integrated into the single reader device.

36.1.3.5 Update Algorithms

Sarma notes that, in addition to the goals listed in the beginning of this section, designers of RFID protocols need to be especially mindful that the RFID system be tolerant of lapses in power. He suggests a protocol that uses a one-way hash function to authenticate the reader to the tag, effectively unlocking and locking the tag. The system uses a periodically changing meta-id, supplied by a hash value created at random by the owner (authorized reader). This meta-id is used to respond to inquiries while the tag is in its locked state. The tag may only be unlocked by the key that was used to create the hash value, created by the authorized reader. The tag defaults to a locked state in the event of power loss, and has a physical self-destruct mechanism to eliminate tracking past a signaled event. This would prevent some level of tracking over the life span of the tag, but similar to Dimitriou's enhanced protocol (Dimitriou, 2005), tracking between authorized interactions would be unprotected. This vulnerability could be addressed by the implementation of a periodic refresh on the meta-id that is transmitted by the tag while in a locked state, but might be challenging given possible limitations on the resources of the tag.

The implementation of a scheme such as Sarma's, preventing the tracking of a single tag or a group of tags, would be beneficial to the model system. By obscuring the tag signature and limiting both forward and reverse telemetry channels, tracking by foreign readers is limited to periods between the authorized reader updates. Depending on the frequency of authorized reader interaction with the tag, this could allow extended tracking with readers within range.

36.1.3.6 More Challenge–Response Protocols

Dimitriou further develops this practice of using one-way hash functions to secure RFID communications (Dimitriou, 2005). He presents two versions of a challenge–response protocol that utilizes hashing algorithms to provide for authentication based on shared secret. The first version is a simplified two-step algorithm that provides for authentication of the tag, after which the tag and the reader update their shared secret to deter tracking from transaction to transaction. Dimitriou points out that this protocol is very susceptible to simpler unsophisticated attacks, such as the possible unintentional event where a single unauthorized reader interrogates a tag. The tag would be rendered completely useless once it increments its secret, losing its synchronization with the authorized reader and associated data system.

Since the first of Dimitriou's two challenge–response protocols that utilizes shared secrets and hashing algorithms to provide for authentication is very susceptible to simple unsophisticated attacks, it is clearly unsuitable for any practical implementation. Implanted systems would quickly lose synchronicity once any outside query was received.

Dimitriou's algorithm is similar to Sarma's in that queries between legitimate reader attempts will be constant and traceable. Again, because of the limited range of tag device transmissions, this may or may not be feasible. The issue of synchronization is the primary concern for this system. If the last transmission from reader to tag is somehow disrupted, the tag is effectively disabled until, if it is possible, the reader and tag can be reinitialized.

In the revised algorithm, the tag authenticates the reader before updating the secret, avoiding the desynchronization resulting from the previous version. Between these legitimate inquiries, the tags will however exhibit traceable signals. Although this protocol does protect the tag from being traced over the life span of the tag, it is susceptible to privacy attacks within the timeframe between legitimate reads.

Dimitriou's algorithm prevents unsophisticated attempts to desynchronize the tag and database, but is still open to more involved, complex attacks that would block communication at critical points in order to desynchronize the system. Dimitriou discredits this category of attack, citing the physical weakness of electromagnetic tagging system as more easily broken in such a situation, maintaining the relative security of this enhanced protocol. Compared to more common, extracorporeal systems where tags are vulnerable, the tags in the model system are secure against tampering without known disturbance to the device. Because the owner would most likely be aware of any compromise to the tag, there would be a clear signal of system vulnerability.

Additionally, because the model system is not limited by the resources of the low cost tags used in supply chains and other large-scale tracking systems, the brute force attack would not be feasible with adequate key length and hash function. The man-in-the-middle style of attack described by Dimitriou (resulting in desynchronization) and common in many wireless systems would not be an issue in an environment such as the model system where an attacker would have to block communications between the legitimate reader and tag, while still receiving and transmitting to both.

Molnar and Wagner present a simple library check-out/in procedure that includes a simple three-step challenge–response protocol that is intended to prevent the most passive form of attack (2004). This simple three-step approach does not support the security requirements for privacy or security for biological systems. During the check-out phase, the authorized reader generates a unique random number r, reads the tag data D (not explained how), stores the record (r, D) in a library database, erases D from the tag, and writes r to the tag. During the check-in phase, the library reader sends the tag a hello, receives r from the tag, looks up R in the library database to retrieve the corresponding D, and writes D back to the tag. Figure 36.5 depicts the check-in identification procedure.

Molnar and Wagner attempt to exploit the relative difficulty of eavesdropping on tag to reader by making the tag to reader communication contain the cipher information (r). While this provides some measure of increased security, an unauthorized reading device could retrieve information between check-out and check-in by making an inquiry. Given the opportunity that checked out book is within range of an unauthorized device, the algorithm presents a solution that is one computation away from a system that sends its identification in the clear.

Molnar and Wagner then present an enhanced algorithm that provides for mutual authentication of tag and reader, as well as privacy (shown below).

This algorithm uses nonce (r) as input to a pseudo-random function that generates a single use value to create the cipher. Molnar and Wagner accompany this communication protocol with a tree-based reader authentication scheme for reduced time complexity in retrieving the (s, ID) record where $ID = \sigma \oplus f_s(0, r1, r2)$, from $O(n)$ to $O(\log n)$. Each tag would require the capacity to store its secret information, the path from the base of the tree to the location where the tag information is stored.

Molnar and Wagner's enhanced algorithm provides for mutual authentication of tag and reader, as well as privacy. Avoine et al. (2005) identify the weaknesses in Molnar and Wagner's Pseudorandom Function-Based Private Authentication Protocol (Figure 36.6), describing the vulnerability of their authentication scheme to tracing when a tag is susceptible to tampering and the increased weakness when additional devices are compromised. As was previously identified, the implanted tags in the model system are reasonably protected against tampering making this weakness irrelevant.

Avoine et al. (2005) also analyze hash chain protocol of Ohkubo et al. (2003) (Figure 36.7) and present an enhanced version of the protocol (see later) (Figure 36.8).

The original protocol uses the hash chain mechanism to prevent tracking. Upon each query request, the tag updates its identifier to the hashed value of the previous identifier, until reaching a maximum threshold and recomputing the hash tables. Per query, the

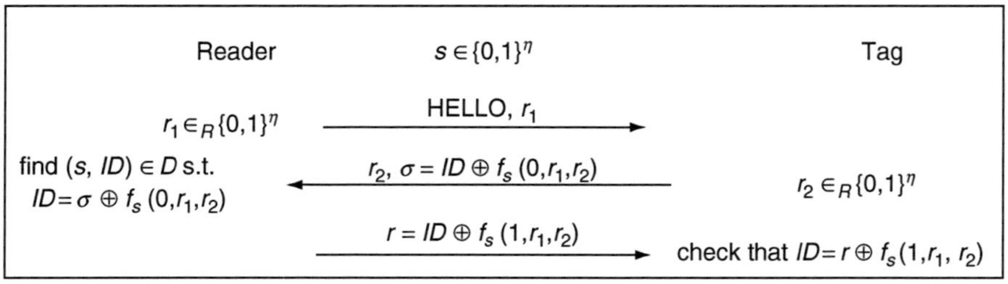

FIGURE 36.6
PRF-Based Private Authentication Protocol. (From Molnar, D. and Wagner, D., in *Computer and Communications Security*, Pfitzmann, B. and McDaniel, P., eds., ACM Press, 2004.)

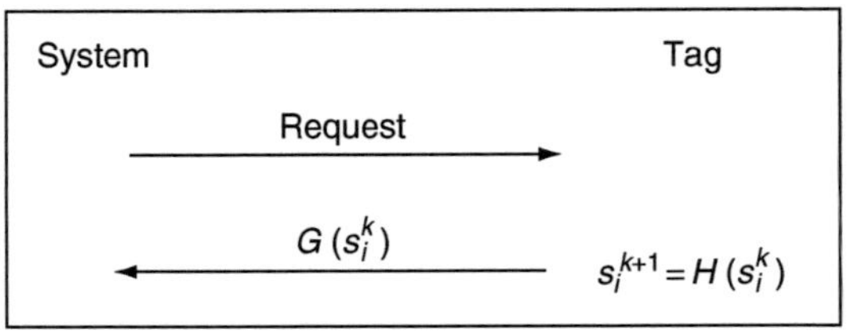

FIGURE 36.7
Ohkubo, Suzuki, and Kinoshita's Hash Chain Protocol. (From Avoine, G., Adversarial model for Radio frequency identification; Cryptology ePrint archive, Report 2005, http://eprint.iacr.org.)

system would be required to build hash chains for each of the *n* initial values until it finds the returned value of the tag or the reaches the threshold for maximum chain length. Ohkubo, Suzuki, and Kinoshita's protocol requires a significant amount of precomputation, yielding an equivalent order of time complexity to that of Molnar and Wagner's, while additionally enforcing forward privacy. However, given the limited scale of the RFID model system, this time complexity would be within reason.

Avoine, Dysli, and Oechslin suggest the addition of a third message in Ohkubo's protocol (see later) from system to tag containing the hash of the new id and public identifier.

This last step would require the reader to send a hashed value of the updated id value combined with a public identifier, resulting in the authentication of the reader to the tag. Avoine, Dysli, and Oechslin present another modification to the Ohkubo, Suzuki, and Kinoshita protocol, intended to prevent replay attacks by altering the generic reader-to-tag request to a random identifier to be used in the hash calculation. Avoine's resulting protocol necessitates an increase of required memory for the precomputation of all of the tables for the hash chains, taking into account all of the random identifiers.

While Avoine details how to achieve the time complexity of Molnar's results with precomputation, he emphasizes the time memory trade-off that exists and acknowledges the limiting factors on his algorithm, ending by citing the advantage of Molnar and Wagner's algorithm in the lack of required precomputation and suggesting the use of larger branching values within their tree structure to limit the effect of a compromised tag.

Avoine's modified version of the Ohkubo, Suzuki, and Kinoshita protocol, necessitates an increase of required reader memory for the precomputation of all of the tables for the hash chains, taking into account all of the random identifiers. As in the original version, because the system size is limited, this requirement would most likely be within reader capabilities. The time required in precomputation is not as significant as the time required for tag-reader communication cycles, but might be unfeasible because both the Avoine and Ohkubo algorithms require that the hash chains and the hash chain tables be periodically refreshed, rendering the model system temporarily inactive and delaying delivery of tag instruction.

Molnar, Soppera, and Wagner present two protocols based on the use of a centralized authority. They describe their RFID tag pseudonym protocol as an improvement to the Avoine, Dysli, and Oechslin protocol and the Ohkubo, Suzuki, and Kinoshita protocol in time complexity, designed to protect privacy against attackers and through the change of

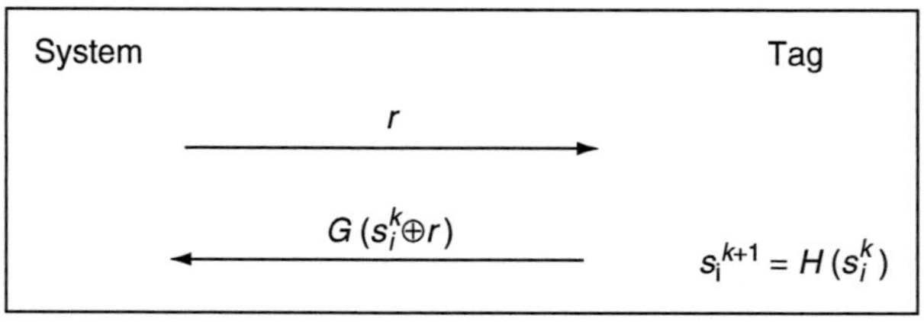

FIGURE 36.8
Modified Protocol. (From Avoine, G., Adversarial model for Radio frequency identification; Cryptology ePrint archive, Report 2005, http://eprint.iacr.org.)

ownership (Molnar et al., August 2005). Utilizing a trusted center, access to a time sensitive pseudonym is granted and restricted, allowing items to pass between different owners while preventing residual ownership.

Molnar's system implements an updating pseudonym system where the centralized authority may be queried by the reader, and may authorize the reader with updated tag information. The reader maintains a list of possible updated values and queries the central authority for extended access when this list is surpassed. While their protocol once again uses a tree structure to store tag information (that again may be susceptible to privacy attack once a tag is compromised, improved with higher branching factor), this information is controlled by an additional entity in the RFID system.

Although Molnar, Soppera, and Wagner describe their pseudonym protocol as providing for the transfer of ownership, their claim is qualified by the assumption of a central authoritative entity. This Trusted Center, while appropriate in systems involving various access rights to tag information within an internal organization (such as that of a hospital system that differentiates between departments), might not provide the flexibility required to be effective across supply chain systems that span various independent organizations. Because reader authorization in the model system is comprised of a more limited set of tag devices that are constant, the granting and expiring of access to tags would be unnecessary.

Juels and Weis introduce another symmetric key-based challenge authentication protocol (Juels and Weis, 2005) based on the ideas introduced by Hopper and Blum in their human identification protocols (Hopper and Blum, 2001) (Figure 36.9). Juels draws a comparison between the human-to-computer authentication scenario introduced by Hopper and Blum and the problem of tag-reader authentication.

Hopper and Blum supply definitions of identification protocols and secure identification protocols and present two round-based protocols that attempt to provide secure authentication appropriate for human executable computation. Their first solution is a repeated challenge–response exchange based on a shared secret with noise added to prevent Gaussian elimination. It is acknowledged secure against passive attacks, but insecure in the situation where an attacker is able to disturb the communication between tag and reader (susceptible to Gaussian elimination). They advocate their solution as a practical solution (within or not too far out of the range of computation for humans) for achieving reasonable security and as an invitation for improvement.

Juels and Weiss cite earlier works that propose human authentication protocols, but focus on the algorithm described by Hopper and Blum. They advocate Hopper and Blum's solution as a practical solution for low cost devices, safe from passive eavesdropping, but

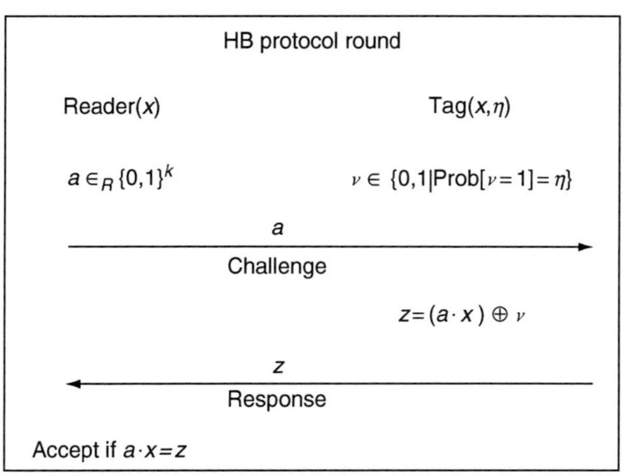

FIGURE 36.9
One round of the Hopper and Blum Human Identification Protocol. (From Juels, A. and Weis, S., in *Advances in Cryptology, Lecture Notes in Computer Science*, Santa Barbara, CA, USA, 2005.)

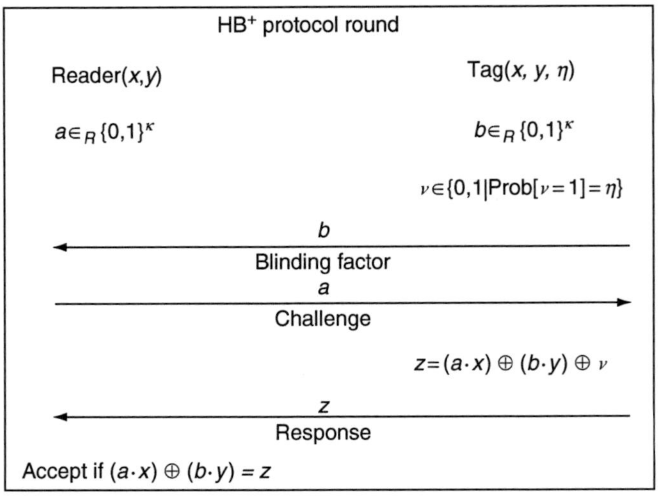

FIGURE 36.10
One round of the HB + Protocol. (From Juels, A. and Weis, S., in *Advances in Cryptology, Lecture Notes in Computer Science*, Santa Barbara, CA, USA, 2005.)

also identify its vulnerability to active attacks. Juels and Weiss define an enhanced version of this protocol, HB + (Figure 36.10), which they claim is secure against active attacks such as skimming, swapping, and DoS attacks. They offer an interesting comparison between humans and inexpensive tagging devices, describing both as having a lack of complex computational abilities, and suggest the adoption of human authentication protocols for tags.

Following Hopper and Blum's suggestion, Juels and Weis extend the application of the HB protocol beyond the arena of human protocols and define the HB + protocol using a challenge–response mechanism that is enhanced using an additional independent secret and a binding factor. This is implemented similarly to the HB protocol using multiple rounds of the challenge–response cycle. The added security is in the randomness that the tag guarantees by sending the random binding factor and calculating its response to the reader using the XOR-ed value of the secret, binding factor, challenge and weight vector. This added layer of security prevents an attacker from finding the value of the original shared secret.

Hopper and Blum's symmetric key authentication protocol, only secure against passive attacks, would not be appropriate for the model implanted RFID system. Although Juels and Weiss claim that their enhanced version of Hopper and Blum's symmetric key-based challenge authentication protocol, HB +, is secure against active attacks such as skimming, swapping, and DoS attacks (Juels and Weis, 2005), Gilbert, Robshaw, and Sibert describe a simple and feasible linear time attack on the HB + RFID protocol (Gilbert et al., 2005). This modified MiM attack uses a false reader and tag to decipher the shared secret. In each round, the reader might work with the false tag to intercept both the binding vector from the tag and the challenge from the reader. Using a constant vector to alter the challenges and responses, the attacker is able to learn the secret, bit by bit from each round. Once the tag is compromised, it can be cloned and impersonated, compromising the system. As noted previously, because the model system has a set number of tags that are bound within close proximity to the reader, the feasibility of a man-in-the-middle style of attack is questionable.

36.1.4 Summary

As sensor technology continues to improve, the applications of RFID seem limitless. The security and privacy issues at hand make it imperative that the development of communication protocols continue to be considered for these systems as they are being spearheaded in other arenas. A summary of security schemes surveyed, along with their suitability, is presented in Table 36.1. With the growing presence of RF communications, it is vital that the effects of ubiquitous RFID are studied when designing communication schemes for these sensitive systems, as well as the large-scale tracking systems that will surround them.

References

Avoine, G. Adversarial model for Radio frequency identification; Cryptology ePrint archive, Report 2005, http://eprint.iacr.org.

Avoine, G., Dysli, E., and Oechslin, P. Reducing time complexity in RFID Systems. In Preneel, Bart and Tavares, Stafford, editors, *Selected Areas in Cryptography—SAC 2005, Lecture Notes in Computer Science*; Vol. 3897; pp. 291–306; August 2005; Kingston, Canada; Springer-Verlag.

Ayoade, J., Takizawa, O., and Nakao, K. A prototype system of the RFID authentication processing framework, *Proceedings of the 3rd International Workshop in Wireless Security Technologies*; London, U.K.; April 2005.

Cameron, T., Liinamaa, T.L., Loeb, G.E., and Richmond, F.J.R. Long-term biocompatibility of a miniature stimulator implanted in feline hind limb muscles. *IEEE Trans. Biomed. Engng.*, 45:1024–1035, 1998.

CNN News website (1999). Retrieved March 2005 from http://www.cnn.com/HEALTH/9908/25/brain.pacemaker/.

Dimitriou, T. A lightweight RFID protocol to protect against traceability and cloning attacks, *Conference on Security and Privacy for Emerging Areas in Communication Networks—SecureComm*; September 2005; Athens, Greece; IEEE.

FDA clears RFID chip for humans (October 18, 2004a). Retrieved February 2005 from http://www.drugresearcher.com/news/news-NG.asp?n=55448-fda-clears-rfid http://montekids.org/healthlibrary/peds/neuro/glossary.htm.

Feldhofer, M. Strong authentication for RFID systems using the AES algorithm, *Cryptographic hardware and embedded systems—CHES 2004; Proceedings of the 6th International Workshop*; August 2004; Cambridge, MA; pp. 357–370.

Garfinkel, S. Adopting fair information practices to low cost RFID Systems. Presented at *Ubiquitous Computing 2002 Privacy Workshop*; Gotenborg, Sweden; September 29, 2002.

Gilbert, H., Robshaw, M., and Sibert, H. An active attack against HB^{+}—A provably secure lightweight authentication protocol; Manuscript; July 2005; France Telecom.

Hopper, N.J. and Blum, M. Secure human identification protocols, In Boyd, C., editor, *Advances in Cryptology—Asiacrypt '01, Lecture Notes in Computer Science*; Vol. 2248; pp. 52–66; 2001; Springer-Verlag, Berlin/Heidelberg.

IEEE, *IEEE Standard Safety Levels with Respect to Human Exposure to Radio Frequency Electromagnetic Fields, 3 KHz to 300 GHz*. Institute of Electrical and Electronics Engineers, 16-Apr-1999, 76 pages.

Implantable RFID chip decision draws criticism (October 20, 2004b). Retrieved February 2005 from http://www.drugresearcher.com/news/news-ng.asp?n=55533-implantable-rfid-chip.

Juels, A., Rivest, R., and Szydlo, M. The blocker tag: Selective blocking of RFID tags for consumer privacy; 2003; http://www.rsasecurity.com/rsalabs/staff/bios/ajuels/ publications/blocker/blocker.pdf.

Juels, A. and Weis, S. Authenticating pervasive devices with human protocols. In *Advances in Cryptology, Lecture Notes in Computer Science*; Vol. 3126; pp. 293–308; August 2005; Santa Barbara, California, USA; IACR.

Koren, M. and Holmes, M. (2006). Vagus nerve stimulation does not lead to significant changes in body weight in patients with epilepsy; *Epilepsy Behaviour*; 8(1):246–249. Epub 2005 December 15.

Lamberg, J. Magnetic resonance imaging & verchip^{TM} (RFID human implant at 1.5 tesla. MRI and RFID human implants; December 2004. Retrieved February 2005, from http://www.rfidjournal.com/whitepapers/download/23.

Molnar, D., Soppera, A., and Wagner, D. A Scalable, Delegatable pseudonym protocol enabling ownership transfer of RFID tags; Selected areas in cryptography—SAC 2005, *Lecture Notes in Computer Science*; Volume 3897/2006; pp. 276–290; August 2005; Kingston, Canada; Springer-Verlag.

Molnar, D. and Wagner, D. Privacy and security in library RFID: Issues, practices, and architectures, In Pfitzmann, B. and McDaniel, P., editors, *Computer and Communications Security*; ACM Press, New York; pp. 210–219; 2004.

Nsanze, F. ICT implants in the human body—A review, *Ethical Aspects of ICT Implants in the Human Body*, Opinion No 20, European Group on Ethics in Science and New Technologies to the European Commission, 2005.

Ohkubo, M., Suzuki, K., and Kinoshita, S. Cryptographic approach to "privacy-friendly" tags. In *RFID Privacy Workshop*; November 2003; MIT, MA, USA.

Rieback, M., Crispo, B., and Tanenbaum, A. RFID guardian: A battery-powered mobile device for RFID privacy management, *Australasian Conference on Information Security and Privacy—ACISP'05, Lecture Notes in Computer Science*; Vol. 3574; pp. 184–194; July 2005; Brisbane, Australia; Springer-Verlag.

Sarma, S.E., Weis, S.A., and Engels, D.W. RFID Systems and Security and Privacy Implications, *Lecture Notes in Computer Science*; Vol. 2523; Revised Papers from the *4th International Workshop on Cryptographic Hardware and Embedded Systems*; pp. 454–469; 2002.

Troyk, P.R. (1999). Injectable electronic identification, monitoring, and stimulation systems; *Annual Reviews Biomedical Engineering*; 1:177–209.

Walter, J.S., Reidy, L., King, W., Dunn, R., Wheeler, J.S., Najafi, K., and Dokmeci, M. Histological response to implantation of microstimulators on the bladder wall: Short term results, *Proceedings of the 19th International Conference of the IEEE Engineering in Medicine and Biology Society*, Chicago, Illinois, October 1997, pp. 1796–1798.

Weir, R.F., Troyk, P.R., DeMichele, G., and Kuiken, T. Implantable myoelectric sensors (IMES) for upper-extremity prosthesis control—Preliminary work, *Proceedings of the 25th Annual Conference of the IEEE EMBS*; Cancun, Mexico; September 2003; pp. 1562–1565.

Weiss, F.G. (1998). Implications of silicon monolithic RFICs for medical instrumentation and telemetry, *Topical Meeting on Silicon Monolithic Integrated Circuits in RF Systems, 1998. Digest of Papers*. September 17–18 1998; Ann Arbor, Michigan; pp. 195–204.

Index

C

D

E

N

O

P